Protecting Groundwater for Health

Protecting Groundwater for Health

Managing the Quality of Drinking-water Sources

Edited by Oliver Schmoll, Guy Howard,
John Chilton and Ingrid Chorus

World Health Organization

IWA Publishing

LONDON • SEATTLE

Published on behalf of the World Health Organization by
IWA Publishing, Alliance House, 12 Caxton Street, London SW1H 0QS, UK
Telephone: +44 (0) 20 7654 5500; Fax: +44 (0) 20 7654 5555; Email: publications@iwap.co.uk
www.iwapublishing.com

First published 2006

Printed by TJ International (Ltd), Padstow, Cornwall, UK

Disclaimer

The opinions expressed in this publication are those of the authors and do not necessarily reflect the views or policies of the International Water Association or the World Health Organization. IWA, WHO and the editors will not accept responsibility for any loss or damage suffered by any person acting or refraining from acting upon any material contained in this publication.

In addition, the mention of specific manufacturers' products does not imply that they are endorsed or recommended in preference to others of a similar nature that are not mentioned. Errors and omissions excepted, the names of proprietary products are distinguished by initial capital letters.

The designations employed and the presentation of the material in this publication do not imply the expression of any opinion whatsoever on the part of the World Health Organization concerning the legal status of any country, territory, city or area or of its authorities, or concerning the delimitation of its frontiers or boundaries. Dotted lines on maps represent approximate border lines for which there may not yet be full agreement.

The views expressed in Chapter 1 are those of the authors and do not necessarily reflect the views or policies of the US EPA.

British Library Cataloguing-in-Publication Data
A CIP catalogue record for this book is available from the British Library

WHO Library Cataloguing-in-Publication Data
Protecting ground water for health.
1.Potable water 2 Water pollution - prevention and control 3.Water supply 4.Risk management - methods I.World Health Organization.
ISBN 92 4 154668 9 (NLM classification: WA 675)

ISBN13: 9781843390794 (IWA Publishing)

Contents

SECTION III: SITUATION ANALYSIS

SECTION IV: APPROACHES TO DRINKING-WATER SOURCE PROTECTION MANAGEMENT

Structure of this book

This book is a tool for developing strategies to protect groundwater for health by managing the quality of drinking-water sources. For this purpose it provides different points of entry. As illustrated in the Figure below, the book consists of five sections.

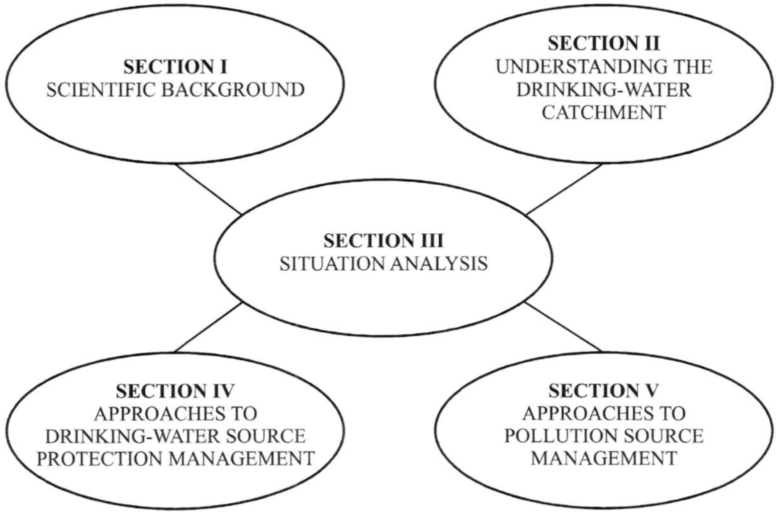

Structure of Protecting Groundwater for Health

Section I covers the scientific background needed to understand which pathogens and chemicals are relevant to human health, how they are transported in the underground and how they may be reduced, removed or retarded (Chapters 3 and 4). The criteria for inclusion of agents in this overview are their relevance to human health and their relevance in groundwater. Further the concept of groundwater recharge areas is introduced in Chapter 2, and basic hydrological and hydrogeological background information is provided. The section is concluded by Chapter 5 which introduces socio-economic and institutional considerations relevant to developing the protection of groundwater resources.

Section II provides background information for characterizing and understanding the drinking-water catchment. The chapters in this section explain how conditions and human activities in the catchment may lead to the occurrence of pathogens or hazardous substances in groundwater. The section begins with general guidance on collecting information (Chapter 6). Chapter 7 discusses assessing the socio-economic and institutional setting as a necessary basis for choosing and implementing feasible management actions. Chapter 8 outlines the background and information required for understanding the hydrogeological conditions determining the likelihood of pollutants to reach aquifers. Chapters 9-13 address the range of human activities potentially releasing pollutants to the underground, i.e. agriculture, sanitation practices, industry, mining, military sites, waste disposal and traffic. These chapters end with checklists highlighting the type of information needed about the setting and the human activities in it for assessing health hazards potentially affecting groundwater.

Section III provides conceptual guidance on prioritizing both hazards and management responses. Chapter 14 describes how information on the hydrogeological conditions, particularly on aquifer vulnerability, can be related to human activities in the drinking-water catchment area in order to assess the potential for pollutants emitted from these activities to reach the aquifer. Chapter 15 discusses how to prioritize pollutants according to their public health burden as well as to their likelihood of long-term accumulation in the aquifer. It also addresses the need to consider the socio-economic context in choosing feasible options from the range of technically appropriate management responses for protection, control or remediation.

Section IV provides an overview of the potential management actions that may be taken to protect drinking-water sources. These begin with their integration into a comprehensive Water Safety Plan that covers all supply steps from catchment to consumer (Chapter 16). Two chapters specifically cover protection of the drinking-water source: Chapter 17 at the scale of designating and managing groundwater protection zones in the catchment and Chapter 18 at the scale of protecting wellheads. Lastly, Chapter 19 addresses the management of groundwater abstraction in order to avoid impacts upon quality and quantity and thus on human health.

Section V provides an overview of control measures to prevent pollution from human activities in the catchment, beginning with the overarching issues of policy, land-use planning and implementation of management options for protecting groundwater (Chapter 20). Chapters 21-25 follow with overviews of the specific management approaches that help avoid groundwater pollution from the range of human activities in the catchment, i.e. agriculture, sanitation practices, industry, mining, military sites, waste disposal and traffic.

Acknowledgements

The World Health Organization wishes to express its appreciation to all those whose efforts have made the production of this book possible. Special thanks are due to Ministry of Health (Ministerium für Gesundheit) and the Federal Environmental Agency (Umweltbundesamt), Germany, which provided financial and organizational support for the development of this book.

An international group of experts provided the material for the book and undertook a process of mutual review. While authorship of individual chapters is noted below, the quality of the volume as a whole is due in large part to the review and comments provided by many individuals. Drafts of the text were discussed and reviewed at editorial meetings at Bad Elster (2001 and 2002). Intellectual input and review by the following individuals is gratefully acknowledged: Robert Bannerman, Gudrun Abbt-Braun, Houssain Abouzaid, Richard Carter, Claudia Castell-Exner, Joseph Cotruvo, Annette Davison, Daniel Deere, Braam de Villiers, John Fawell, Norbert Feldwisch, Richard Franceys, Walter Giger, Alan Godfree, Andreas Grohmann, Helmut Höring, Terence Lee, Lucy Lytton, Scott Miller, Christian Nels, Ute Ringelband, Ruprecht Schleyer, Sally Sutton, Michael Taylor and Peter Waggitt.

Special thanks are due to Kathy Pond and Edie Campbell of the Robens Centre for Public and Environmental Health for editing the text; and to Monika Jung, Gertrud Schlag and Penny Ward for technical and administrative support.

The following authors are acknowledged for their major contributions to the chapters of this book:

Alistair Allen, Department of Geology, University College Cork, Cork, Ireland
William Alley, Office of Ground Water, United States Geological Survey, San Diego, USA
Steve Appleyard, Department of Environment, Western Australia, Perth, Australia
Patricia Hotchkiss Bakir, Amman, Jordan
Mike Barrett, Department for Environment Food and Rural Affairs, London, United Kingdom (formerly at the Robens Centre for Public and Environmental Health, University of Surrey, Guildford, United Kingdom)
Jamie Bartram, Water, Sanitation and Health Programme, World Health Organization, Geneva, Switzerland
Philip Berger, Office of Ground Water and Drinking Water, United States Environmental Protection Agency, Washington DC, USA
Theechat Boonyakarnkul, Health Department, Ministry of Public Health, Nonthaburi, Thailand
Ellen Braun-Howland, New York State Department of Health, New York, USA
Peter Chave, Credition, United Kingdom
John Chilton, British Geological Survey, Wallingford, United Kingdom
Ingrid Chorus, Federal Environmental Agency, Berlin, Germany
Hermann H. Dieter, Federal Environmental Agency, Berlin, Germany
Jörg Drewes, Environmental Science & Engineering Division, Colorado School of Mines, Colorado, USA
Jutta Fastner, Federal Environmental Agency, Berlin, Germany
Frank Fladerer, Bremen Overseas Research and Development Association (BORDA), Yogyakarta, Indonesia (formerly at GTZ, Indonesian-German Government Cooperation on Drinking Water Quality Surveillance)
Fritz Frimmel, Engler-Bunte-Institute, Department of Water Chemistry, University of Karlsruhe, Karlsruhe, Germany
Sam Godfrey, United Nations Children's Fund, Bhopal, India (formerly at the Water, Engineering and Development Centre, Loughborough University, Loughborough, United Kingdom)
Arthur Golwer, Wiesbaden, Germany
Bilqis Amin Hoque, Environment and Population Research Centre, Dhaka, Bangladesh
Guy Howard, Department for International Development, Glasgow, United Kingdom (formerly at the Water, Engineering and Development Centre, Loughborough University, Loughborough, United Kingdom)
Jutta Jahnel, Engler-Bunte-Institute, Department of Water Chemistry, University of Karlsruhe, Karlsruhe, Germany
Michael Kuperberg, Institute for International Cooperative Environmental Research, Florida State University, Tallahassee, Florida, USA
Dennis McChesney, Groundwater Compliance Section, United States Environmental Protection Agency, New York, USA
Broder Merkel, Geology Department, Technical University of Freiberg, Freiberg, Germany
Hans-Martin Mulisch, Umweltbüro Mulisch GmbH, Potsdam, Germany

Steve Pedley, Robens Centre for Public and Environmental Health, University of Surrey, Guildford, United Kingdom

Brian Reed, Water, Engineering and Development Centre, Loughborough University, Loughborough, United Kingdom

Mike Rivett, School of Geography, Earth & Environmental Sciences, University of Birmingham, Birmingham, United Kingdom

Robert Sage, Veolia Water Partnership, Bushey, United Kingdom

Jack Schijven, Microbiological Laboratory for Health Protection, National Institute of Public Health and the Environment, Bilthoven, The Netherlands

Wilfried Schimon, Federal Ministry of Agriculture, Forestry, Environment and Water Management, Vienna, Austria

Oliver Schmoll, Federal Environmental Agency, Berlin, Germany

Klaus-Peter Seiler, GSF – National Research Centre for Environment and Health, Neuherberg, Germany

Richard Taylor, Department of Geography, University College London, London, United Kingdom

Christopher Teaf, Center for Biomedical and Toxicological Research and Waste Management, Florida State University, Tallahassee, Florida, USA

Don Wauchope, United States Department of Agriculture, Agricultural Research Service, Tifton, Georgia, USA

Eleonora Wcislo, Environmental Risk Analysis Department, Institute for the Ecology of Industrial Areas, Katowice, Poland

Julie West, British Geological Survey, Nottingham, United Kingdom

Marylynn Yates, Department of Environmental Sciences, University of California, Riverside, California, USA

Special thanks are also due to the following experts who provided case studies and text boxes:

Steven Bailey (Table 9.6), Royal Society for the Protection of Birds, Bedfordshire, United Kingdom (formerly at ADAS, Wolverhampton, United Kingdom)

Hartmut Bartel (Box 6.1), Federal Environmental Agency, Berlin, Germany

Rainer Hallebach (Box 5.1), Federal Environmental Agency, Bad Elster, Germany

Manfred Hobler (case study in Chapter 8.1.5), Federal Institute for Geosciences and Natural Resources, Hannover, Germany

Helmut Kerndorff (Box 12.3), Federal Environmental Agency, Berlin, Germany

Klaus Kümmerer (Boxes 12.1 and 24.2), Institute of Environmental Medicine and Hospital Epidemiology, University Hospital of Freiburg, Freiburg, Germany

Benjamin Mafa (Box 10.2), Department of Water Affairs, Gaborone, Botswana

Armin Margane (case study in Chapter 8.1.5), Federal Institute for Geosciences and Natural Resources, Hannover, Germany

Nicola Martin (Box 10.2), Federal Institute for Geosciences and Natural Resources, Hannover, Germany

Lars Matthes (Box 11.4), Institute of Applied Geosciences, Technical University Berlin, Berlin, Germany

Paito Obote (Box 22.1), WaterAid, Uganda

Asaf Pekdeger (Box 17.1), Institute for Geological Sciences, Free University Berlin, Berlin, Germany

Manfred Scheu (Box 24.3), GTZ, Eschborn, Germany
Gerhard Schmidt (Box 19.1), Federal Institute for Geosciences and Natural Resources, Hannover, Germany
Bernd Söfner (Box 19.1), Federal Institute for Geosciences and Natural Resources, Hannover, Germany
Ali Subah (case study in Chapter 8.1.5), Ministry of Water and Irrigation, Amman, Jordan
Horst Vogel (Boxes 10.2 and 11.4), Federal Institute for Geosciences and Natural Resources, Hannover, Germany
Michael von Hoyer (Box 24.1), Federal Institute for Geosciences and Natural Resources, Hannover, Germany
Hans-Jürg Weber (Box 16.1), SWL Energie AG, Lenzburg, Switzerland
Gerhard Winkelmann-Oei (Box 23.2), Federal Environmental Agency, Berlin, Germany

Acronyms and abbreviations

ARMCANZ	Agriculture and Resource Management Council of Australia and New Zealand
ASS	acid sulphate soils
BAT	best available technology
BC MAFF	British Columbia Ministry for Agriculture, Fisheries & Forestries, Canada
BGS	British Geological Survey
BMP(s)	best management practice(s)
BOD	biochemical oxygen demand
BTEX	benzene, toluene, ethylbenzene, xylene
C&D	construction and demolition
CCTV	close circuit television
cDCE	cis-dichloroethene
cf	contamination factor
CJD	Creutzfeldt-Jakob disease
COD	chemical oxygen demand
CSOs	combined sewer overflows
CTC	carbon tetrachloride/tetrachloromethane
CVM	contingent valuation methodologies
CW	chemical warfare
2,4-D	(2,4-dichlorophenoxy)acetic acid
DALY	Disability Affected Life Years
2,4-DB	(2,4-dichlorophenoxy)butyric acid
1,2-DCA	1,2-dichloroethane

1,2-DCB	1,2-dichlorobenzene
1,4-DCB	1,4-dichlorobenzene
1,1-DCE	1,1-dichloroethene
DCM	dichloromethane
DDT	dichlorodiphenyltrichloroethane
DFID	Department for International Development, UK
DNAPL	dense non-aqueous phase liquid
DNB	dinitrobenzene
DNT	dinitrotoluene
DOC	dissolved organic carbon
DOE	Department of the Environment
DWI	Drinking Water Inspectorate
EA	Environment Agency
EDCs	endocrine disrupting chemicals/compounds
EDTA	ethylendiamine tetraacetic acid
EED	Environmental Engineering Division
EIA(S)	Environmental Impact Assessment (Study)
EU	European Union
FS	faecal streptococci
FAO	Food and Agriculture Organization
GIS	Geographical Information System
GDWQ	*Guidelines for Drinking-water Quality*, WHO
HACCP	Hazard Analysis and Critical Control Points
GV	guideline value
Hb	haemoglobin
HCB	hexachlorobenze
HD	mustard gas
HIV	human immunodeficiency virus
HMX	High Melting Explosive cyclotetramethylenetetranitramine
IARC	International Agency for Research on Cancer
ICPE	International Commissions for the Protection of the Elbe
ICPR	International Commissions for the Protection of the Rhine
IDWSSD	International Drinking Water Supply and Sanitation Decade
ISL	in situ leaching
IUPAC	International Union of Pure Applied Chemistry
LNAPLs	light non-aqueous phase liquid
LWS	Lenzburg water supply
MCPA	(4-chloro-2-methylphenoxy)acetic acid
MCPP	2-(4-chloro-2-methylphenoxy)propanoic acid (mecoprop)
MDG	Millennium Development Goal
metHb	methaemoglobin
MNA	monitored natural attenuation
MSW	municipal solid waste
MTBE	methyl tertiary-butyl ether
NA	natural attenuation
NAPL	non aqueous phase liquid
NCRP	National Council on Radiation Protection Measurements, USA
NGOs	non-governmental organizations
NRC	National Research Council, USA

NSW	New South Wales, Australia
PAH	polynuclear aromatic hydrocarbon
PCB	polychlorinated biphenyl
PCE	perchloroethylene/tetrachloroethene
PCP	pentachlorophenol
PCPs	personal care products
PCR	polymerase chain reaction
PHAST	participating hygiene and sanitation transformation
POPs	persistent organic pollutants
PPP	purchasing power parity
RDX	Royal Dutch Explosive cyclotrimethylenetetranitramine
REC	Regional Environmental Council
RNA	ribonucleic acid
SHI	Sanitary Hazard Index
SPA	Source Protection Areas
2,4,5-T	(2,4,5-tricholorophenoxy)acetic acid
TCA	trichloroethane
TCE	trichloroethene
TCM	trichloromethane
tDCE	trans-dichloroethene
TDS	total dissolved solids
TeCE	tetrachloroethene
TNT	trinitrotoluene
TON	total organic nitrogen
2,4,5-TP	(2,4,5-tricholorophenoxy)propanoic acid (fenoprop)
TTC	thermotolerant coliforms
UNDP	United Nations Development Programme
UNECE	United Nations Economic Commission for Europe
UNEP	United Nations Environment Programme
UNICEF	United Nations International Children's Emergency Fund
UNESCO	United Nations Educational, Scientific and Cultural Organization
UNSCEAR	United Nations Scientific Committee of Effects of Atomic Radiation
US EPA	United States Environmental Protection Agency
USGS	United States Geological Survey
VBNC	viable but non-culturable
VC	vinyl chloride
VFAs	volatile fatty acids
VOC	volatile organic compounds
WEDC	Water Engineering and Development Centre, University of Loughborough, United Kingdom
WHO	World Health Organization
WMO	World Meteorological Organization
WSP	Water Safety Plan
γ-HCH	1α,2α,3β,4α,5α,6β-hexachlorocyclohexane (lindane)

Section I

Scientific background

1

Groundwater and public health

G. Howard, J. Bartram, S. Pedley, O. Schmoll, I. Chorus and P. Berger

Water-related disease remains one of the major health concerns in the world. Diarrhoeal diseases, which are largely derived from poor water and sanitation, accounted for 1.8 million deaths in 2002 and contributed around 62 million Disability Adjusted Life Years per annum (WHO, 2004a). On a global scale, this places diarrhoeal disease as the sixth highest cause of mortality and third in the list of morbidity and it is estimated that 3.7 per cent of the global disease burden is derived from poor water, sanitation and hygiene (Prüss-Üstün *et al.*, 2004). This health burden is primarily borne by the populations in developing countries and by children.

At 2002 estimates, roughly one-sixth of humanity (1.1 billion people) lack access to any form of improved water supply within 1 kilometre of their home, and approximately 40 per cent of humanity (2.6 billion people) lack access to some form of improved excreta disposal (WHO and UNICEF, 2004). These figures relate to the clear definitions provided in the updated Global Water Supply and Sanitation Assessment Report and are shown in Table 1.1 below.

If the quality of water or sanitation were taken into account, these numbers of people without access to water supplies and sanitation would increase even further.

Endemic and epidemic disease derived from poor water supply affects all nations. Outbreaks of waterborne disease continue to occur in both developed and developing countries, leading to loss of life, avoidable disease and economic costs to individuals and communities. The improvement of water quality control strategies, in conjunction with improvements in excreta disposal and personal hygiene can be expected to deliver substantial health gains in the population.

Table 1.1. Definition of improved and unimproved water supply and sanitation facilities (WHO and UNICEF, 2000)

Water supply		Sanitation	
Improved	Unimproved	Improved	Unimproved
Household connection	Unprotected well	Connection to a public sewer	Service or bucket latrines (excreta removed manually)
Public standpipe	Unprotected spring	Connection to a septic system	
Boreholes	Vendor-provided water	Pour-flush latrine	Public latrines
Protected dug well	Bottled water	Simple pit latrine	Latrines with an open pit
Protected spring	Tanker-truck provided water	Ventilated improved pit latrine	
Rainwater collection			

This monograph provides information on strategies for the protection of groundwater sources used for drinking-water as a component of an integrated approach to drinking-water safety management (WHO, 2004b). The importance of source protection as the first stage of managing water quality has been an important component in both national and international efforts. In their *Guidelines for Drinking-water Quality*, WHO (2004b and previous editions) emphasize the need for effective source protection.

The focus of this monograph is the public health aspects of groundwater protection. It does not address environmental concerns, such as ecological protection. The control of some pollutants, whilst of little importance for health, may be very important to environmental protection. For guidance on these areas, readers should consult appropriate texts such as Chapman (1996).

1.1 GROUNDWATER AS A SOURCE OF DRINKING-WATER

Groundwater is the water contained beneath the surface in rocks and soil, and is the water that accumulates underground in aquifers. Groundwater constitutes 97 per cent of global freshwater and is an important source of drinking-water in many regions of the world. In many parts of the world groundwater sources are the single most important supply for the production of drinking-water, particularly in areas with limited or polluted surface water sources. For many communities it may be the only economically viable option. This is in part because groundwater is typically of more stable quality and better microbial quality than surface waters. Groundwaters often require little or no treatment to be suitable for drinking whereas surface waters generally need to be treated, often extensively. There are many examples of groundwater being distributed without treatment. It is vital therefore that the quality of groundwater is protected if public health is not to be compromised.

National statistics for the use of groundwater as a source of drinking-water are sparse, but the importance of this resource is highlighted by figures published in Europe and the USA. The proportion of groundwater in drinking-water supplies in some European countries is illustrated in Table 1.2 and for the USA in Table 1.3. The data show that reliance upon groundwater varies considerably between countries; for example, Norway takes only 13 per cent of its drinking-water from groundwater sources, whereas Austria and Denmark use groundwater resources almost exclusively for drinking-water supply. A global estimate of one-third of the world's population depending on groundwater supply is given by Falkenmark (2005).

Table 1.2. Proportion of groundwater in drinking-water supplies in selected European countries (EEA, 1999; UNECE, 1999)

Country	Proportion	Country	Proportion
Austria	99%	Bulgaria	60%
Denmark	98%	Finland	57%
Hungary	95%	France	56%
Switzerland	83%	Greece	50%
Portugal	80%	Sweden	49%
Slovak Republic	80%	Czech Republic	43%
Italy	80%	United Kingdom	28%
Germany	72%	Spain	21%
Netherlands	68%	Norway	13%

The data from the USA demonstrates the importance of groundwater particularly for smaller supplies, reflecting the generally limited treatment requirements. However, this has implications for control of public health risks as the management and maintenance of smaller supplies is often weaker than for larger, utility operated supplies (Bartram, 1999).

Table 1.3. Proportion of groundwater in drinking-water supplies in the USA by size of supply (US EPA, 2004)

Population served	Proportion groundwater	Proportion surface water
<500	89%	11%
500-1000	78%	22%
1001-3300	70%	30%
3301-10 000	57%	43%
10 000-50 000	43%	57%
>50 000	26%	74%

Within countries the usage of groundwater may also vary substantially, depending on the terrain and access to alternative water sources. For instance, in the USA it ranges from 25 per cent or less in Colorado and Kentucky to more than 95 per cent in Hawaii and Idaho. In rural areas of the USA, 96 per cent of domestic water comes from groundwater. In the United Kingdom, although the national average for groundwater usage is 28 per cent, the southern counties of England depend more heavily on groundwater than the northern counties, Wales and Scotland.

In Latin America, many of the continent's largest cities – Mexico City, Mexico, Lima, Peru, Buenos Aires, Argentina and Santiago de Chile, Chile – obtain a significant proportion of their municipal water supply from groundwater. In India, China, Bangladesh, Thailand, Indonesia and Viet Nam more than 50 per cent of potable supplies are provided from groundwater. In Africa and Asia, most of the largest cities use surface water, but many millions of people in rural areas and low-income peri-urban communities are dependent on groundwater. These populations are most vulnerable to waterborne disease. Pedley and Howard (1997) estimate that as much as 80 per cent of the drinking-water used by these communities is abstracted from groundwater sources.

Where it is available, groundwater frequently has important advantages over surface water. It may be conveniently available close to where it is required, can be developed at comparatively low cost and in stages to keep pace with rising demand. Although small, simple surface water supplies can be achieved relatively cheaply and pumping groundwater from deep aquifers may create significant operating costs, overall the capital costs associated with groundwater development are usually lower than with large-scale surface water supplies. For the latter, large, short-term capital investments in storage reservoirs often produce large, step-wise increments in water availability and temporary excess capacity that is gradually overtaken by the continuing rising demand for water. An additional disadvantage in some circumstances is that surface water reservoirs may have multiple, sometimes conflicting functions – water supply, flood control, irrigation, hydroelectric power and recreation – and cannot always be operated for the optimum benefit of water supply.

Furthermore, aquifers are often well protected by layers of soil and sediment, which effectively filter rainwater as it percolates though them, thus removing particles, pathogenic microorganisms and many chemical constituents. Therefore it is generally assumed to be a relatively safe drinking-water source.

However, groundwater has been termed the 'hidden sea' – *sea* because of the large amount of it, and *hidden* because it is not visible, thus pollution pathways and processes are not readily perceived (Chapelle, 1997). This highlights a key issue in the use of aquifers as drinking-water source, showing that particular attention is needed to ascertain whether the general assumption of groundwater being safe to drink is valid in individual settings. As discussed below, understanding the source-pathway-receptor relationship in any particular setting is critical to determine whether pollution will occur.

Whilst there is a large volume of groundwater in this 'hidden sea', its replenishment occurs slowly – at rates varying between locations. Over-exploitation therefore readily occurs, bringing with it additional quality concerns.

1.2 THE PUBLIC HEALTH AND SOCIOECONOMIC CONTEXT OF GROUNDWATER PROTECTION

The use of groundwater as a source of drinking-water is often preferred because of its generally good microbial quality in its natural state. Nevertheless, it is readily contaminated and outbreaks of disease from contaminated groundwater sources are reported from countries at all levels of economic development. Some groundwaters naturally contain constituents of health concern: fluoride and arsenic in particular. However, understanding the impact of groundwater on public health is often difficult and the interpretation of health data complex. This is made more difficult as many water supplies that use groundwater are small and outbreaks or background levels of disease are unlikely to be detected, especially in countries with limited health surveillance. Furthermore, in outbreaks of infectious disease, it is often not possible to identify the cause of the outbreak and many risk factors are typically involved.

Throughout the world, there is evidence of contaminated groundwater leading to outbreaks of disease and contributing to background endemic disease in situations where groundwater sources used for drinking have become contaminated. However, diarrhoeal disease transmission is also commonly due to poor excreta disposal practices and the improvement of sanitation is a key intervention to reduce disease transmission (Esrey *et al.,* 1991; Curtis *et al*., 2000). Furthermore, water that is of good quality at its source may be re-contaminated during withdrawal, transport and household storage. This may then require subsequent treatment and safe storage of water in the home (Sobsey, 2002).

Ensuring that water sources are microbially safe is important to reduce health burdens. However, a balance in investment must be maintained to ensure that other interventions, also important in reducing disease, are implemented. Diverting resources away from excreta disposal, improved hygiene practices in order to achieve very good quality water in sources may be counter-productive (Esrey, 1996). Balancing investment decisions for public health gain from water supply and sanitation investment is complex and does not simply reflect current knowledge (or lack of) regarding health benefits, but also the demands and priorities of the population (Briscoe, 1996).

Groundwater is generally of good microbial quality, but may become rapidly contaminated if protective measures at the point of abstraction are not implemented and well maintained. Further problems are caused by the creation of pathways that short-circuit the protective measures and natural layers offering greatest attenuation, for instance abandoned wells and leaking sewers. Pollution may also occur in areas of recharge, with persistent and mobile pollutants representing the principal risks.

The control of the microbial quality of drinking-water should be the first priority in all countries, given the immediate and potentially devastating consequences of waterborne infectious disease (WHO, 2004b). However, in some settings the control of chemical quality of groundwater may also be a priority, particularly in response to locally important natural constituents such as fluoride and arsenic. Furthermore, hazardous industrial chemicals and pesticides which can accumulate over time may potentially render a source unusable. The scale, impact and the often lack of feasible clean-up technologies for some chemical contamination in groundwater means that they should receive require priority for preventative and remedial strategies.

Groundwater also has a socioeconomic value. It is often a lower cost option than surface water as the treatment requirements are typically much lower. In many countries, groundwater is also more widely available for use in drinking-water supply. This may provide significant advantages to communities in obtaining affordable water supplies, which may have benefits in terms of promoting greater volumes of water used for hygiene and other purposes. The natural quality of groundwater also makes its use valued in industry, and it may provide environmental benefits through recharge of streams and rivers or for the growth of vegetation. These other benefits reinforce the need for its protection.

The actions taken to protect and conserve groundwater will also create costs to society, through lost opportunity costs for productive uses of land and increased production costs caused by pollution containment and treatment requirements. When developing protection plans and strategies, the cost of implementing such measures should be taken into consideration, as well as the cost of not protecting groundwater, in order for balanced decisions to be made.

1.3 GROUNDWATER QUANTITY

The interrelated issues of groundwater quality and quantity can best be addressed by management approaches encompassing entire groundwater recharge areas or groundwater catchments. These units are appropriate both for assessing pollution potential and for developing management approaches for protection and remediation.
Excessive groundwater abstraction in relation to recharge will lead to depletion of the resource and competition between uses, e.g. between irrigation and drinking-water supply. Strong hydraulic gradients ensuing from abstraction can induce the formation of preferential flow paths, reducing the efficacy of attenuation processes, and thus lead to elevated concentrations of contaminants in groundwater. Furthermore, changes in groundwater levels induced by abstraction may change conditions in the subsurface environment substantially, e.g. redox conditions, and thus induce mobilization of natural or anthropogenic contaminants.

Groundwater quantity issues may have substantial impacts on human health. Lack of a safe water supply affects disease incidence for instance by restricting options for personal and household hygiene. Competing demands for groundwater, often for irrigation and sometimes for industry, may lead to shortage of groundwater for domestic use. In such situations it is important to ensure allocation of sufficient groundwater reserves for potable and domestic use and health authorities often play an important role in this. This monograph largely focuses on water quality issues, as these are of direct relevance to the provision of safe drinking-water. Quantity issues are therefore addressed in the context of their impact on groundwater quality.

This text is concerned with groundwater as a source of drinking-water supply. However, in many locations other uses, for example irrigation, account for the largest fraction of groundwater abstraction, and inter-sectoral collaboration may be needed to develop effective groundwater allocation schemes.

1.4 DISEASE DERIVED FROM GROUNDWATER USE

Groundwater contributes to local and global disease burdens through the transmission of infectious disease and from chemical hazards.

1.4.1 Infectious disease transmission through groundwater

The global incidence of waterborne disease is significant, though it can only be estimated since reliable data are not sufficiently available for direct assessment of disease cases (Prüss-Üstün *et al.*, 2004). The contribution of groundwater to the global incidence of waterborne disease cannot be assessed easily, as there are many competing transmission routes; confounding from socioeconomic and behavioural factors is typically high; definitions of outcome vary; and, exposure-risk relationships are often unclear (Esrey *et al.*, 1991; Payment and Hunter, 2001; Prüss and Havelaar, 2001). Many waterborne disease outbreaks could have been prevented by good understanding and management of groundwaters for health. Pathogen contamination has often been associated with simple deficiencies in sanitation but also with inadequate understanding of the processes of attenuation of disease agents in the subsurface.

The most comprehensive reports of waterborne disease outbreaks come from two countries, the USA and the United Kingdom, and some indications of the role of groundwater in the infectious diarrhoeal disease burden can be estimated in these countries (Craun, 1992; Hunter, 1997; Payment and Hunter, 2001; Craun *et al.*, 2003; 2004).

Lee *et al.* (2002) identified that of 39 outbreaks of waterborne disease in the USA between 1999 and 2000, 17 were due to consumption of untreated groundwater, although approximately half of these outbreaks were reported from individual water supplies, which are not operated by a utility and served less than 15 connections or less than 25 persons. A further eight were reported in non-community supplies, which serve facilities such as schools, factories and restaurants.

A detailed analysis of the incidence of waterborne disease in the USA was published in the mid-1980s by Craun (1985), which is still relevant. In his summary of data from the period between 1971 and 1982, Craun reports that untreated or inadequately treated groundwater was responsible for 51 per cent of all waterborne disease outbreaks and 40 per cent of all waterborne illness. A recent analysis of public health data in the USA showed little change to the epidemiology of disease outbreaks (Craun *et al.*, 1997). Between 1971 and 1994, 58 per cent of all waterborne outbreaks were caused by contaminated groundwater systems, although this is in part is due to the higher number of water supplies using groundwater than those using surface water.

Craun *et al.* (2003) report that for the period 1991-1998, 68 per cent of the outbreaks in public systems were associated with groundwater, an increase from previous reports (Craun, 1985; Craun *et al.*, 1997). However, this apparent increase is likely to be due in part to the introduction of the USA Surface Water Treatment Rule in 1991, which requires 'conventional filtration' of most surface water supplies. In general it appears that waterborne outbreaks in the USA decreased after 1991, with the introduction of more stringent monitoring and treatment requirements.

Craun *et al.* (2004) provide a detailed discussion of waterborne outbreaks in relation to zoonotic organisms (organisms with an animal as well as human reservoir) between 1971 and 2000 in the USA. They note that 751 outbreaks were reported linked to drinking-water supplies during this period, the majority (648) being linked to community (year-round public service) water supplies. The aetiology was either known or suspected in 89 per cent of the outbreaks and zoonotic agents caused 118 outbreaks in community systems representing 38 per cent of outbreaks associated with these systems and 56 per cent of those where aetiology was identified. The data show that the majority of illnesses and deaths were caused by zoonotic agents in the reported waterborne outbreaks.

The zoonotic agents of greatest importance were *Giardia*, *Campylobacter*, *Cryptosporidium*, *Salmonella*, and *E. coli* in outbreaks caused by contaminated drinking-water. The majority of outbreaks caused by zoonotic bacteria (71 per cent) and *Cryptosporidium* (53 per cent) were reported in groundwater supplies. The use of contaminated, untreated or poorly treated groundwater was responsible for 49 per cent of outbreaks caused by *Campylobacter*, *Salmonella*, *E. coli*, and *Yersinia*. Groundwater that was contaminated, untreated or poorly treated contributed 18 per cent of all outbreaks caused by *Giardia* and *Cryptosporidium*.

Kukkula *et al.* (1997) describe an outbreak of waterborne viral gastroenteritis in the Finnish municipality of Noormarkku that affected some 1500-3000 people, i.e. between 25 and 50 per cent of the exposed population. Laboratory investigations confirmed that adenovirus, Norwalk-like virus and group A and C rotaviruses were the principal causative agents. The source of the outbreak was thought to be a groundwater well situated on the embankment of a river polluted by sewage discharges. In 1974 an outbreak of acute gastrointestinal illness at Richmond Heights in Florida, USA was traced to a supply well that was continuously contaminated with sewage from a nearby septic tank (Weissman *et al.*, 1976). The main aetiological agent was thought to be *Shigella sonnei*. During the outbreak approximately 1200 cases were recorded from a population of 6500.

Outbreaks of cryptosporidiosis have also been linked to groundwater sources, despite being usually regarded as a surface water problem. A large outbreak of cryptosporidiosis occurred in 1998 in Brush Creek, Texas, USA from the use of untreated groundwater drawn from the Edwards Plateau karst aquifer (Bergmire-Sweat *et al.*, 1999). There were 89 stool-confirmed cases and the estimated number of cases was between 1300 and 1500. This outbreak was associated with the consumption of water drawn from deep wells of over 30 m located more than 400 m from Brush Creek.

In 1997, epidemiological investigations traced an outbreak of cryptosporidiosis in the United Kingdom to water abstracted from a deep chalk borehole. Three hundred and forty five confirmed cases were recorded by the investigation team, who claimed this to be the largest outbreak linked to groundwater to have been reported (Willcocks *et al.*, 1998). This incident has particular significance because the water used in the supply was drawn from a deep borehole and was filtered before distribution.

In the outbreak of *E. coli* O157:H7 and *Campylobacter* in Walkerton, Ontario in Canada in 2000, the original source of pathogens appears to have derived from contaminated surface water entering into a surface water body directly linked to an abstraction borehole (Health Canada, 2000). Although the series of events leading to the outbreak indicate a failure in subsequent treatment and management of water quality, better protection of groundwater would have reduced the potential for such an outbreak. An outbreak of *E. coli* O157:H7 occurred among attendees at the Washington Country Fair, New York, USA and was shown to be caused by consuming water from a contaminated shallow well that had no chlorination (CDC, 1999). A total of 951 people reported having diarrhoea after attending the fair and stool cultures from 116 people yielded *E .coli* O157:H7. This outbreak resulted in hospitalization of 65 people, 11 children developed haemolytic syndrome and two people died.

In developing countries evidence of the role of groundwater in causing disease outbreaks is more limited, although there have been numerous studies into the impact of drinking-water, sanitation and hygiene on diarrhoeal disease. In part the limited data on groundwater related outbreaks reflects the often limited capacity of local health surveillance systems to identify causal factors and because it is common that several factors may be implicated in the spread of disease. However, the limited data on outbreaks specifically linked to groundwater may also reflect that improved groundwater sources are generally of relatively good quality. Diarrhoeal disease related directly to drinking-water is most likely to result from consumption of poorly protected or unimproved groundwater sources, untreated or poorly treated surface water, contamination of distribution systems and recontamination of water during transport.

Pokhrel and Viraraghavan (2004) in a review of diarrhoeal disease in Nepal in relation to water and sanitation, cite examples from South Asia where contamination of groundwater supplies has led to outbreaks of disease.

A study of local populations in Kanpur, India recorded an overall incidence rate of waterborne disease of 80.1 per 1000 population (Trivedi *et al.*, 1971). The communities in the study areas took water from shallow groundwater sources, analysis of which revealed that over 70 per cent of the wells were contaminated. Of the cases of waterborne disease investigated, the greatest proportion was of gastroenteritis, followed by dysentery.

In addition to outbreaks, there is some evidence of contaminated groundwater contributing to background levels of endemic diarrhoeal disease. For example, Nasinyama *et al.* (2000) showed that the use of protected springs in Kampala, Uganda which were in generally in poor condition was associated with higher rates of diarrhoea than the use of piped water supplies. Much of this disease burden is thought to occur in developing countries where the use of untreated water from shallow groundwater sources is common in both rural and peri-urban settlements (Pedley and Howard, 1997).

1.4.2 Chemical hazards

The risk to health from chemicals is typically lower than that from pathogens. The health effects of most, but not all, chemical hazards arise after prolonged exposure, and tend to be limited to specific geographical areas or particular water source types. Much remains to be understood about the epidemiology of diseases related to chemical hazards in water and the scale of disease burden remains uncertain. However, some data do exist. Craun *et al.* (2004) report that 11 per cent of waterborne outbreaks in the USA between 1971 and 2000 were associated with acute effects following ingestion of a chemical.

Ensuring that chemicals of health concern do not occur at significant concentrations in groundwaters implies understanding sources of pollution, aquifer vulnerability and specific attenuation processes as well as recognizing the importance of naturally-occurring chemicals of health concern. In groundwater, however, there are two contaminants in particular that represent particular hazards of concern: fluoride and arsenic.

Fluoride affects bone development and in excess leads to dental or, in extreme form, skeletal fluorosis. The latter is a painful debilitating disease that causes physical impairment. However, too little fluoride has also been associated with dental caries and other dental ill-health (WHO, 2004b). Drinking-water is the principal route of exposure to fluoride in most settings, although burning of high fluoride coal is a significant route of exposure in parts of China (Gu *et al.*, 1990).

Arsenic causes concern given the widespread occurrence in shallow groundwaters in Bangladesh, West Bengal, India and in groundwater in several other countries. The scale of arsenic contamination is most severe in the shallow groundwater of Bangladesh. At present, the total population exposed to elevated arsenic concentrations in drinking-water in Bangladesh remains uncertain, but is thought to be somewhere between 35 and 77 million and has been described as the largest recorded poisoning in history (Smith *et al.,* 2000; BGS and DPHE, 2001). Problems are also noted in countries as diverse as Mexico, Canada, Hungary and Ghana, although the source of arsenic and control strategies available vary. The true scale of the public health impact of arsenic in groundwater remains uncertain and the epidemiology is not fully understood.

In the case of Bangladesh, the lack of country-wide case-controlled studies makes estimating prevalence of arsenicosis difficult. In a recent evaluation of data collected by the DPHE-Unicef arsenic mitigation programmes, Rosenboom *et al.* (2004) found a prevalence rate of arsenicosis (keratosis, melanosis and de-pigmentation) of 0.78 per 1000 population exposed to elevated arsenic (above 50 μg/l) in 15 heavily affected Upazilas (an administrative unit in Bangladesh). These authors note, however, that the data were difficult to interpret and that exposure had been relatively short and therefore these numbers could increase. The lack of a national cancer prevalence study makes estimations outside small cross-section studies problematic.

Increasing numbers of countries in Asia are now identifying arsenic contamination of groundwater (including Cambodia, China, Laos, Myanmar, Nepal, Pakistan and Viet Nam). In India, increasing numbers of areas are being identified as arsenic affected beyond West Bengal (School of Environmental Studies, Javadpur University, 2004). This demonstrates that arsenic is an important contaminant for public health and concern is growing.

Other chemical contaminants of concern in groundwater may also lead to health problems. These include nitrate, uranium and selenium. Of these, nitrate is of concern as it is associated with an acute health effect (methaemoglobinaemia or infantile cyanosis). The scale of the health burden derived from nitrate remains uncertain although it has been suggested to cause significant health problems in some low-income countries where levels in groundwater reach extremely high values (Melian *et al.*, 1999). Nitrate is also of concern given that it is stable once in groundwater with reasonably high oxygen content, where it will not degrade. Thus it may accumulate to a long-term water resource problem that is expensive and difficult to remediate and whose effect may not be noticed until concentrations become critical.

1.5 GROUNDWATER IN THE CONTEXT OF INTERNATIONAL ACTIVITIES TO REDUCE WATER-RELATED DISEASE

The International Drinking Water Supply and Sanitation Decade (IDWSSD; 1980-1990) provided a sustained focus on the need for concerted efforts to accelerate activities to increase global access to safe water supply and to sanitation. The Rio Earth Summit (1992) placed water both as resource and as water supply on the priority agenda and the World Summit on Sustainable Development in 2002 also placed safe drinking-water as a key component of sustainable development. In September 2000, 189 UN Member States adopted the Millennium Development Goals (MDGs). Target 10 of the MDGs is to halve by 2015 the proportion of people without sustainable access to safe drinking-water and basic sanitation; the baseline for this target has been set as 1990. Other important initiatives have included a Protocol on Water and Health to the 1992 Convention on Use of Transboundary Watercourses and International Lakes (Box 1.1).

Box 1.1. The WHO-UNECE Protocol on Water and Health (UNECE and WHO, 1999)

The WHO-UNECE Protocol on Water and Health to the 1992 Convention on the Protection and Use of Transboundary Watercourses and International Lakes is an international legal instrument on the prevention, control and reduction of water-related diseases in Europe.

A major product of the Third European Ministerial Conference on Environment and Health (1999), the Protocol was signed at the Conference by 35 countries and represents the first major international legal approach for controlling water-related disease. It has become legally binding for the 16 countries that have ratified it in 2005. By adopting the Protocol, the signatories agreed to take all appropriate measures towards achieving:

- adequate supplies of wholesome drinking-water;
- adequate sanitation of a standard which sufficiently protects human health and the environment;
- effective protection of water resources used as sources of drinking-water and their related water ecosystems from pollution from other causes;
- adequate safeguards for human health against water-related diseases;
- effective systems for monitoring and responding to outbreaks or incidents of water-related diseases.

The Global Environmental Monitoring System Water programme, launched in 1977 by UNEP in collaboration with UNESCO, WHO and WMO, has the overall objective of observing and assessing global water quality issues in rivers, lakes and groundwater by collecting together and interpreting data from national monitoring networks. A first assessment of freshwater quality published in 1989 (Meybeck *et al.*, 1989) included discussion of links between water quality and health. The programme was reviewed and evaluated in 2001 with a view to enhancing its ability to contribute to inter-agency global programmes, including the Global International Waters Assessment and the UN-wide World Water Assessment Programme.

Recently, the World Bank has established a groundwater management advisory team (GW·MATE) to develop capacity and capability in groundwater resource management and quality protection through World Bank programmes and projects and through the activities of the Global Water Partnerships regional networks.

WHO's activities in support of safe drinking-water span across the range of its functions as a specialized agency of the UN system (Box 1.2).

Box 1.2. WHO activities related to safe drinking-water

Evidence and information: Burden of disease estimates (at global level and guidance on their conduct at other levels); and cost-effectiveness water interventions (generically at global level and guidance on their conduct at other levels).

Status and trends: Assessing coverage with access to improved sources of drinking-water, and to safe drinking-water (with UNICEF through the Joint Monitoring Programme).

Tools for good practice: Evidence-based guidance on effective (and ineffective) technologies, strategies and policies for health protection through water management.

Normative guidelines: Evidence-based and health-centred norms developed to assist development of effective national and regional regulations and standards.

Country cooperation: Intensive links to individual countries through its network of six regional offices, regional environment centres and country offices.

Research and testing: Encouraging and orienting research; developing and encouraging the application of protocols to increase harmonization, exchange and use of data. Publication with IWA of the Journal of Water and Health (http://www.iwapublishing.com/template.cfm?name=iwapwaterhealth).

Tools for disease reductions: Focussing especially on settings such as healthy cities, healthy villages, healthy schools.

1.6 GROUNDWATER IN THE WHO GUIDELINES FOR DRINKING-WATER QUALITY

Since 1958 WHO has published at about ten year intervals several editions of *International Standards for Drinking-water* and subsequently *Guidelines for Drinking-water Quality*. The third edition of the Guidelines, published in 2004, includes a substantial update of the approach towards the control of microbial hazards in particular based on a preventive management approach. In preparing the third edition of the Guidelines a series of state-of-the-art reviews was prepared on aspects of water quality management and human health (Box 1.3) of which *Protecting Groundwater for Health* is one.

In the overall context of the *Guidelines for Drinking-water Quality* (GDWQ), this monograph serves two purposes: it provides the background information on potential groundwater contamination as well as approaches to protection and remediation that were taken into account in developing the third edition of the GDWQ. Further, *Protecting Groundwater for Health* supplements it by providing comprehensive information on: assessing the potential for contamination of groundwater resources, prioritizing hazards and selecting management approaches appropriate to the specific socioeconomic and institutional conditions.

Box 1.3. State of the art reviews supporting the third edition of WHO *Guidelines for Drinking-water Quality* (selected titles)

Water Safety Plans: Managing Drinking-water Quality from Catchment to Consumer (Davison *et al.*, 2005)

Safe Piped Water: Managing Microbial Water Quality in Piped Distribution Systems (Ainsworth, 2004)

Water Treatment and Pathogen Control: Process Efficiency in Achieving Safe Drinking-water (LeChevallier and Au, 2004)

Assessing Microbial Safety of Drinking-water: Improving Approaches and Methods (Dufour *et al.*, 2003)

Quantifying Public Health Risks in the WHO Guidelines for Drinking-water Quality: A Burden of Disease Approach (Havelaar and Melse, 2003)

Rapid Assessment of Drinking-water Quality: A Handbook for Implementation (Howard *et al.*, 2003)

Domestic Water Quantity, Service Level and Health (Howard and Bartram, 2003)

Managing Water in the Home: Accelerated Health Gains from Improved Water Supply (Sobsey, 2002)

Water Quality: Guidelines, Standards and Health: Assessment of Risk and Risk Management for Water-related Infectious Disease (Fewtrell and Bartram, 2001)

Chemical Safety of Drinking-water: Assessing Priorities for Risk Management (in preparation)

A central approach of the third edition of the GDWQ is the development of a reliable preventive safety management approach: a Framework for Safe Drinking-water, the three key requirements of which are described in Box 1.4. This includes the introduction of Water Safety Plans (WSPs) as a management tool for avoidance and control of groundwater contamination. These are described in Chapter 16 of this book.

Box 1.4. The three key requirements of WHO's Framework for Safe Drinking-water (WHO, 2004b)

1. *Health-based targets* based on an evaluation of health concerns.
2. Development of a *Water Safety Plan* (WSP) that includes:
    - *System assessment* to determine whether the water supply (from source through treatment to the point of consumption) as a whole can deliver water of a quality that meets the health based targets.
    - *Operational monitoring* of the control measures in the drinking-water supply that are of particular importance in securing drinking-water safety.
    - *Management plans* documenting the system assessment and monitoring plans and describing actions to be taken in normal operating and incident conditions, including upgrading, documentation and communication.
3. A system of *independent surveillance* that verifies that the above are operating properly.

This approach meets the need for developing an understanding of the key steps in the supply chain at which pollution may be introduced or prevented, increased or reduced. Effective management implies identifying these, ideally through a quantitative system assessment. The framework also includes identification of the appropriate measures to ensure that processes are operating within the bounds necessary to ensure safety. For drinking-water supply from groundwaters these controls may extend into the recharge area but may also relate to more immediate source protection measures, such as well-head protection. Some of the measures to verify safe operation of processes relevant to groundwater safety may be amenable to sophisticated approaches such as on-line monitoring of levels (e.g. of landfill effluents). Others (such as the ongoing integrity of a well plinth) are best approached through periodic inspection regimes.

1.7 REFERENCES

Ainsworth, R. (ed.) (2004) *Safe Piped Water: Managing Microbial Water Quality in Piped Distribution Systems.* WHO Drinking-water Quality Series, IWA Publishing, London. http://www.who.int/water_sanitation_health/dwq/924156251X/en/index.html (accessed January 11, 2006).

Bartram, J.K. (1999) Effective monitoring of small drinking water supplies. In *Providing Safe Drinking Water in Small Systems: Technology, Operations and Economics*, (eds. J.A. Cotruvo, G.F. Craun and N. Hearne), pp. 353-365, Lewis Publishers, Washington DC.

Bergmire-Sweat, D., Wilson, K., Marengo, L., Lee, Y.M., MacKenzie, W.R., Morgan, J., Von Alt, K., Bennett, T., Tsang, V.C.W. and Furness, B. (1999) Cryptosporidiosis in Brush Creek:

Describing the epidemiology and causes of a large outbreak in Texas, 1998. In *Proc. Internat. Conf. on Emerging Infectious Diseases*, Milwaukee, WI, AWWA, Denver, Colorado.

BGS and DPHE (2001) *Arsenic Contamination of Groundwater in Bangladesh*. BGS Technical Report WC/00/19.

Briscoe, J. (1996) Financing water and sanitation services: the old and new challenges. *Water Supply*, **14**(3/4), 1-17.

CDC (1999) Outbreak of *Escherichia coli* O157:H7 and *Camplyobacter* among attendees of the Washington County Fair – New York 1999. *Morbidity and Mortality Weekly Report*, **48**(36), 803-804.

Chapelle, F.H. (1997) *The Hidden Sea: ground water, springs and wells*. Geoscience Press Inc., Arizona.

Chapman, D. (ed.) (1996) *Water Quality Assessments*. E&F Spon, London.

Craun, G.F. (1985) A summary of waterborne illness transmitted through contaminated groundwater. *J. Environ. Health*, **48**, 122-127.

Craun, G.F. (1992) Waterborne disease outbreaks in the United States of America: causes and prevention. *World Health Statistics Q.*, **45**(2-3), 192-199.

Craun, G.F., Berger, P.S. and Calderon, R.L. (1997) Coliform bacteria and waterborne disease outbreaks. *JAWWA*, **89**, 96-104.

Craun, G.F., Calderon, R.L. and Nwachuku, N. (2003) Causes of Waterborne Outbreaks Reported in the United States, 1991-1998. In *Drinking Water and Infectious Disease, Establishing the Links*, (eds. P. Hunter, M. Waite and E. Rochi), pp. 105- 117, CRC Press and IWA Publishing, Boca Raton, Florida.

Craun, G.F., Calderon R.L. and Craun, M.F. (2004) Waterborne Outbreaks Caused by Zoonotic Pathogens in the United States. In *Waterborne Zoonoses: Identification, causes and control*, (eds. J.A. Cotruvo, A. Dufour, G. Rees, J. Bartram, R. Carr, D.O. Cliver, G.F. Craun, R. Fayer and V.P.J. Gannon), pp. 120-135, WHO, IWA Publishing, London.

Curtis, V., Cairncross, S. and Yonli, R. (2000) Domestic hygiene and diarrhoea - pinpointing the problem. *Tropical Medicine and Internat. Health*, **5**(1), 22-32.

Davison, A., Howard, G., Stevens, M., Callan, P., Fewtrell, L., Deere, D. and Bartram, J. (2005) *Water Safety Plans: Managing Drinking-water Quality from Catchment to Consumer*. World Health Organization, Geneva. http://www.who.int/water_sanitation_health/dwq/wsp0506/en/index.html (accessed on January 11, 2006).

Dufour, A., Snozzi, M., Koster, W., Bartram, J., Ronchi, E. and Fewtrell, L. (eds.) (2003) *Assessing Microbial Safety of Drinking-water: Improving Approaches and Methods*. WHO Drinking-water Quality Series, IWA Publishing, London http://www.who.int/water_sanitation_health/dwq/9241546301/en/index.html (accessed on January 11, 2006).

EEA (1999) *Groundwater Quality and Quantity in Europe*. Environmental Assessment Report, No. 3.

Esrey, S.A., Potash, J.B., Roberts, L. and Shiff, C. (1991) Effects of improved water supply and sanitation on ascariasis, diarrhoea, dracunculiasis, hookworm infection, schistosomiasis, and trachoma. *Bulletin of the WHO*, **69**(5), 609-621.

Esrey, S.A. (1996) Water, wastes and well-being: a multi-country study. *Am. J. Epidemiology*, **143**(6), 608-623.

Gu, S.L., Ji, R.D. and Cao, S.R. (1990) The physical and chemical characteristics of particles in indoor air where high fluoride coal burning takes place. *Biomed. Environ. Sci.*, **3**(4), 384-390.

Falkenmark, M. (2005) Water usability degradation – economist wisdom or societal madness? *Water International*, **30**(2), 136-146.

Fewtrell, L. and Bartram, J. (eds.) (2001) *Water Quality: Guidelines, Standards and Health: Assessment of Risk and Risk Management for Water-related Infectious Disease*. IWA Publishing, London.

Havelaar, A.H. and Melse, J.M. (2003) *Quantifying Public Health Risks in the WHO Guidelines for Drinking-water Quality: A Burden of Disease Approach*. RIVM report 734301022/2003,

Bilthoven.
http://www.who.int/water_sanitation_health/dwq/quantifyinghealthrisks/en/index.html (accessed January 11, 2006).

Health Canada (2000) Waterborne outbreak of gastroenteritis associated with a contaminated municipal water supply, Walkerton, Ontario, May-June 2000. *Canada Communicable Disease Report*, **26**(20), 170-173.

Howard, G. and Bartram, J. (2003) *Domestic Water Quantity, Service Level and Health*. World Health Organization, Geneva.
http://www.who.int/water_sanitation_health/diseases/wsh0302/en/index.html (accessed January 11, 2006).

Howard, G., Ince, M. and Smith, M. (2003) *Rapid Assessment of Drinking-water Quality: A Handbook for Implementation*. Water, Engineering and Development Centre (WEDC), Loughborough.

Hunter, P.R. (1997) *Waterborne Disease: Epidemiology and Ecology*, John Wiley & Sons, Chichester, New York.

Kukkula, M., Arstila, P., Klossner, M.L., Maunula, L., Bonsdorff, C.H. and Jaatinen, P. (1997) Waterborne outbreak of viral gastroenteritis. *Scandinavian J. Infectious Disease*, **29**, 415-418.

LeChevallier, M.W. and Au, K.-K. (2004) *Water Treatment and Pathogen Control: Process Efficiency in Achieving Safe Drinking-water*. WHO Drinking-water Quality Series, IWA Publishing, London.
http://www.who.int/water_sanitation_health/dwq/9241562552/en/index.html (accessed January 11, 2006).

Lee, S.H., Levy, D.A., Craun, G.F., Beach, M.J. and Calderon, R.L. (2002) Surveillance for waterborne-disease outbreaks - United States, 1999-2000. *CDC Morbidity and Mortality Weekly Report*, **51**(SS-8), www.cdc.gov.mmwr/PDF/SS/SS5108.pdf (accessed April 29, 2005).

Melian, R., Myrlian, N., Gouriev, A., Moraru, C. and Radstake, F. (1999) Groundwater quality and rural drinking-water supplies in the Republic of Moldova. *Hydrogeology J.* **7**, 188-196.

Meybeck, M., Chapman, D. and Helmer, R. (1989) *Global Freshwater Quality: a first assessment*. Published for WHO/UNEP by Blackwell Scientific, Oxford.

Nasinyama, G.W., McEwen, S.A., Wilson, J.B., Waltner-Towers, D., Gyles, C.L. and Opuda-Asibo, J. (2000) Risk factors for acute diarrhoea among inhabitants of Kampala District, Uganda. *SA Medical J.* **90**(90), 891-898.

Payment, P. and Hunter, P.R. (2001) Endemic and epidemic infectious intestinal disease and its relationship to drinking water. In *Water Quality: guidelines, standards and health*, (eds. L. Fewtrell and J. Bartram), pp. 61-88, IWA Publishing, London.

Pedley, S. and Howard, G. (1997) The public health implications of microbiological contamination of groundwater. *Q. J. Eng. Geol.*, **30**, 179-188.

Pokhrel, D. and Viraraghavan, T. (2004) Diarrhoeal disease in Nepal vis-à-vis water supply and sanitation status. *J. Water and Health*, **2**(2), 71-81.

Prüss, A. and Havelaar, A. (2001) The global burden of disease study and applications in water, sanitation and hygiene. In *Water Quality: guidelines, standards and health*, (eds. L. Fewtrell and J. Bartram), pp. 43-59, IWA Publishing, London.

Prüss-Üstün, A., Kay, D., Fewtell, L. and Bartram, J. (2004) Unsafe water, sanitation and hygiene. In: *Comparative Quantification of Health Risks: Global and Regional Burden of Disease Attributable to Selected Major Risk Factors*, (eds. Ezzati, M., Lopez, A. D., Rodgers, A. and Murray, C. J. L), Volume 2, pp. 1321-1352, World Health Organization, Geneva..

Rosenboom, J.W., Ahmed, K.M., Pfaff, A. and Madajewicz, M. (2004) Arsenic in 15 Upazilas of Bangladesh: Water supplies, health and behaviour. *Report prepared for APSU, Unicef and DPHE*, Dhaka.

School of Environmental Studies (2004). Javadpur Univeristy, Kolkatta, West Bengal, www.soesju.org/arsenic/ (accessed April 29, 2005).

Smith, A.H., Lingas, E.O. and Rahman, M. (2000) Contamination of drinking-water by arsenic in Bangladesh: a public health emergency. *Bulletin of the WHO*, **78**(9), 1093-1103.
Sobsey, M.D. (2002) *Managing Water in the Home: accelerated health gains from improved water supply*. World Health Organization, Geneva. http://www.who.int/water_sanitation_health/dwq/wsh0207/en/index.html (accessed January 11, 2006).
Trivedi, B.K., Gandhi, H.S. and Shukla, M.K. (1971) Bacteriological water quality and the incidence of waterborne diseases in a rural population. *Indian J. Med. Sciences*, **25**, 795-801.
UNECE (1999) *Inventory of Transboundary Groundwaters*. UNECE Task Force on Monitoring and Assessment. Guidelines on monitoring of transboundary groundwaters, vol. 1, Supporting Technical Documents.
UNECE and WHO (1999) Protocol on Water and Health to the 1992 Convention on the Protection and use of Transboundary Watercourses and International Lakes. http://www.euro.who.int/Document/Peh-ehp/ProtocolWater.pdf (accessed January 11, 2006).
US EPA (2004) www.epa.gov/safewater/electronic/presentations/sdwq/pt4/sdwa57.html (accessed April 29, 2005).
Weissman, J.B., Craun, G.F., Lawrence, D.N., Pollard, R.A., Suslow, M.S. and Gangarosa, E.J. (1976) An epidemic of gastroenteritis traced to a contaminated public water supply. *Am. J. Epidemiology*, **103**, 319-398.
Willocks, L., Crampin, A., Milne, L., Seng, C., Susman, M., Gair, R., Moulsdale, M., Shafi, S., Wall, R., Wiggins, R. and Lightfoot, N. (1998) A large outbreak of cryptosporidiosis associated with a public water supply from a deep chalk borehole. Outbreak Investigation Team. *Communicable Disease and Public Health*, **1**, 239-243.
WHO (2004a) *World Health Report 2004*. WHO, Geneva.
WHO (2004b) *Guidelines for Drinking-Water Quality: Recommendations*. 3rd edn, vol 1, WHO, Geneva.
WHO and UNICEF (2000) *Global Water Supply and Sanitation Assessment 2000 Report*. WHO, Geneva and UNICEF, New York.

2

Groundwater occurrence and hydrogeological environments

J. Chilton and K.-P. Seiler

Many people are surprised to discover that groundwater is widely and heavily used throughout the world. During severe droughts in arid regions of the world, newspapers and television carry dramatic pictures of dry wells in rural communities and people walking long distances for small amounts of household water. However, groundwater usage is important in both humid and arid regions, and it can be a revelation that many cities are dependent on groundwater and use such large volumes of groundwater in their public water supplies.

One reason for this general lack of awareness is that groundwater is usually a hidden resource, out of sight and therefore out of mind. It is, nevertheless, as valuable an asset in water supply terms as rivers, lakes and reservoirs, and deserves to be equally protected. As a consequence of this lack of awareness, the main features of groundwater systems are poorly known or even misunderstood. To provide the necessary basic knowledge of hydrogeology for the reader to fully appreciate the rest of the monograph, this chapter aims to rectify the situation by placing place groundwater in its appropriate context within the wider water cycle. It then summarizes the ways in which groundwater occurs and moves, and how it is replenished. The characteristics of the main types of geological settings are described so that the reader is able to see how different hydrogeological environments vary in their response to the pressures of water abstraction and pollution.

This knowledge will be used to help guide the information requirements outlined in Chapter 8. In relation to the overall source-pathway-receptor approach to the assessment of pollution this chapter is mainly focussed on the pathway through groundwater systems to the receptor, and should be read in this context.

Providing an adequate technical basis would be difficult without defining at least some of the most important terms related to groundwater and pollution. The most important definitions are highlighted through the chapter and key concepts illustrated by figures. A short list of suitable standard texts which can provide further details for the interested reader is given at the end of the chapter, along with the references actually quoted.

2.1 GROUNDWATER IN THE HYDROLOGICAL SYSTEM

2.1.1 The hydrological cycle

The continuous movement of water between oceans, atmosphere and land is known as the hydrological cycle (Figure 2.1). Considering the freshwater component of the system, which is the part of greatest significance for this monograph, inflow is from precipitation in the form of rainfall and from melting snow and ice. Outflow occurs primarily as stream flow or runoff and as evapotranspiration, a combination of evaporation from water surfaces and the soil and transpiration from soil moisture by plants. Precipitation reaches streams and rivers both on the land surface as overland flow to tributary channels, and also by subsurface routes as interflow and baseflow following infiltration to the soil. Part of the precipitation that infiltrates deeply into the ground may accumulate above an impermeable bed and saturate the available pore spaces to form an underground body of water, called an aquifer. The water contained in aquifers contributes to the groundwater component of the cycle (Figure 2.1), from which natural discharge reaches streams and rivers, wetlands and the oceans.

Figure 2.1 simplifies the hydrological cycle, illustrating only its natural components. There are few areas of the world in which the cycle has not been interfered with and modified by human settlement and associated activities. Large urban areas alter the processes of infiltration and drainage (e.g. Lerner *et al.*, 1990; Lerner, 1997), as do big irrigation schemes. Negative and costly impacts of waterlogging and salinity are widely experienced where excess infiltration from irrigation with diverted surface water raises groundwater levels beneath the irrigated land. This is seen most dramatically in the lower Indus Valley in Pakistan. Estimates of the area affected vary, but of 16.1 million ha irrigated, some 4.6 million ha are affected to some extent by waterlogging and salinity, of which perhaps 2 million ha have suffered serious deterioration (Ghassemi *et al.*, 1995). Engineering works for flood control, irrigation, hydropower and navigation can all change the surface water component of the cycle locally but sometimes dramatically, and groundwater abstraction can intercept discharge to rivers, wetlands and the oceans. An example of modification of the hydrological cycle that clearly has the potential to cause negative health impacts is the uncontrolled discharge of untreated urban wastewater or industrial effluents to surface water or groundwater. Modifying the hydrological cycle by human intervention also implies changing pollutant pathways and transport mechanisms, and these changes must be understood in developing strategies for protecting the health of water users.

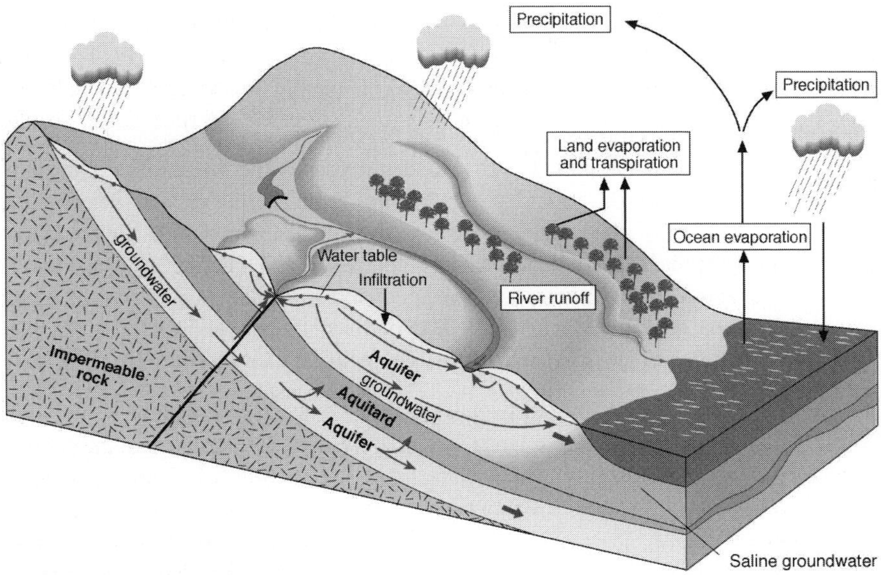

Figure 2.1. The natural hydrological cycle (modified from Morris *et al.*, 2003).

DEF ▶ *Aquifers are layers of rock or sediments which are sufficiently porous to store water and permeable enough to allow water to flow through them in economically viable quantities.*

The river basin or catchment is the geographical expression of the hydrological cycle, and the spatial unit within which water resource balances can be estimated and through which pollutants are transported within the cycle. Figure 2.1 demonstrates that both surface water and subsurface processes occur. This means that the river basin or sub-basin, and all of the activities within it, should be the unit or basis for the management of water resources, rather than political or administrative subdivisions. River Basin Management Plans are, therefore, an essential feature of the European Union (EU) Water Framework Directive (EC, 2000), which is intended to establish the overall approach to long-term management of water resources by EU Member States. It also follows that catchments can contain land from more than one or indeed several countries and transboundary or multi-national authorities, such as for the Rhine, Danube, Zambezi and others, have been established to oversee their management. Thus a key principle is that:

NOTE ▶ *The catchment boundaries of a river basin or sub-basin should define the management unit for water resources, rather than administrative or political boundaries.*

2.1.2 Groundwater in the hydrological cycle

While the definition of groundwater as the water contained beneath the surface in rocks and soil is conceptually simple and convenient, in practice the picture is a little more complex, and confusion can arise. The water beneath the ground surface includes that contained in the soil, that in the intermediate unsaturated zone below the soil, that comprising the capillary fringe and that below the water table (Figure 2.2). The soil is commonly understood to comprise the broken down and weathered rock and decaying plant debris at the ground surface. The region between the soil and the water table is commonly referred to as the unsaturated zone or sometimes the vadose zone.

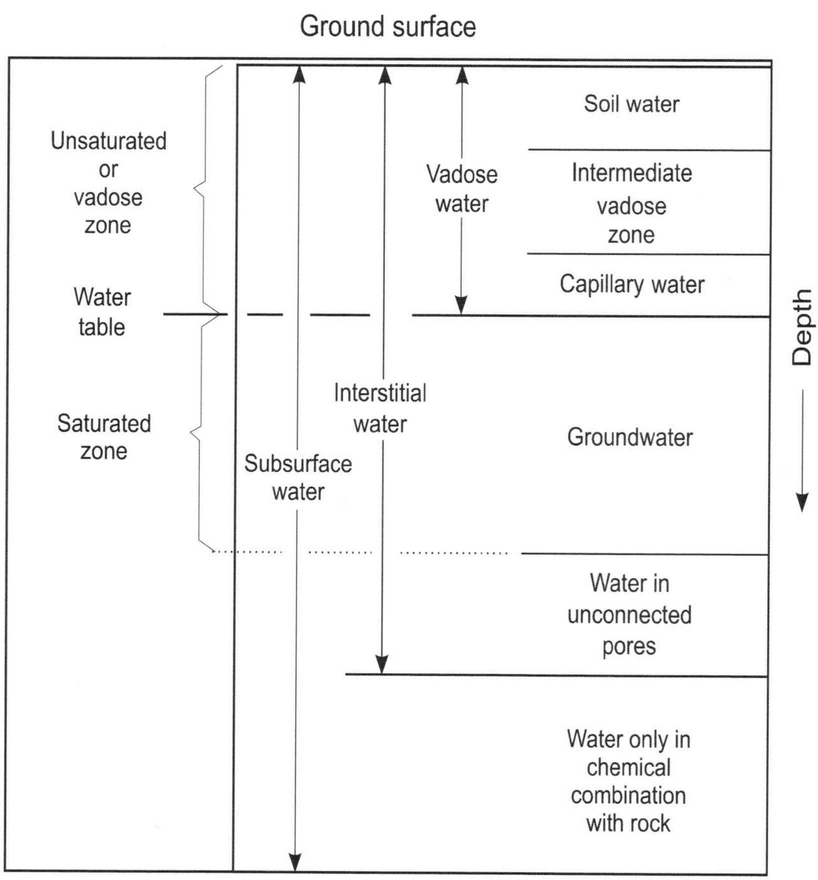

Figure 2.2. Classification of subsurface water (modified from Driscoll, 1986)

DEF ▶ *The* ***unsaturated zone*** *contains both air and water, while in the* ***saturated zone*** *all of the voids are full of water. The* ***water table*** *marks the boundary between the two, and is the surface at which fluid pressure is exactly equal to atmospheric pressure.*

Strictly speaking, therefore, groundwater refers only to water in the saturated zone beneath the water table, and the total water column beneath the earth's surface is usually called subsurface water (Figure 2.2). In practice, of course, the saturated and unsaturated zones are connected, and the position of the water table fluctuates seasonally, from year to year and with the effects of groundwater abstraction.

Appreciating this distinction is especially important in relation to protecting groundwater from pollutants originating from activities at the surface. Such pollutants can either be retained in the soil or they may be carried downwards by infiltrating water, depending on the physicochemical properties of both the soil material and of the pollutants. The soil, and the unsaturated zone beneath it, can be considered to serve as a reactive filter, delaying or even removing pollutants by the range of processes described in Chapters 3 and 4. The properties of the materials comprising the soil and the unsaturated zone are, therefore, critical factors in defining the vulnerability of groundwater to pollution, as described in Chapter 8.

In volume terms, groundwater is the most important component of the active terrestrial hydrological cycle, as shown in Table 2.1. Excluding the 97.5 per cent of water of high salinity contained in the oceans and seas, groundwater accounts for about one third of the freshwater resources of the world (UNESCO, 1999). If the water permanently contained in the polar ice caps and glaciers is also excluded, then groundwater accounts for nearly all of the useable freshwater. Even if consideration is further limited to the most active and accessible groundwater bodies, which were estimated by Lvovitch (1972) at 4 x 10^6 km^3, then they still constitute 95 per cent of the total freshwater. Lakes, swamps, reservoirs and rivers account for 3.5 per cent and soil moisture for 1.5 per cent (Freeze and Cherry, 1979). The dominant role of groundwater resources is clear, their use is fundamental to human life and economic activity, and their proper management and protection are correspondingly vital.

Table 2.1. Estimated water balance of the world (modified from Nace, 1971 and UNESCO, 1999)

Parameter	Surface area (10^6 km^2)	Volume (10^6 km^3)	Volume (%)	Residence time
Oceans and seas	361	1,370	97	~ 4000 years
Groundwater	130	8	0.5	Weeks - 100 000 years
Icecaps and glaciers	17.8	27	2	10-100 000 years
Lakes and reservoirs	1.55	0.13	<0.01	~ 10 years
Soil moisture	130	0.07	<0.01	2 weeks - several years
Atmospheric water	504	0.01	<0.01	~ 10 days
Swamps	<0.1	<0.01	<0.01	1-10 years
River channels	<0.1	<0.01	<0.01	~ 2 weeks
Biospheric water	<0.1	<0.01	<0.01	~ 1 week

The last column of Table 2.1 provides an indication of the range of residence times of water in the various compartments of the hydrological cycle. The great variation in residence times in freshwater bodies is also illustrated in Figure 2.3, which emphasizes the generally slow movement and long residence time of most groundwaters compared to surface waters.

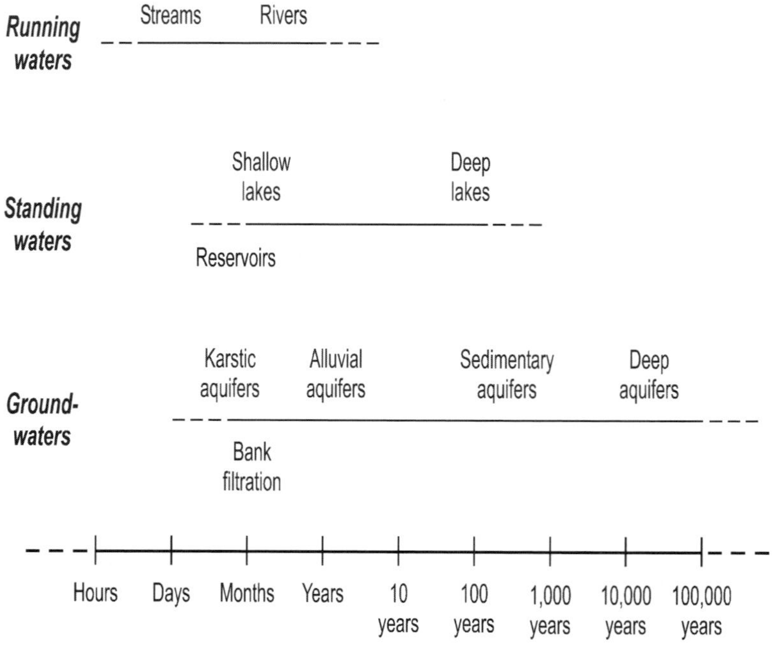

Figure 2.3. Water residence time in inland freshwater bodies (modified from Meybeck *et al.*, 1989)

2.2 GROUNDWATER OCCURRENCE AND MOVEMENT

2.2.1 Groundwater occurrence and storage

Some groundwater occurs in most geological formations because nearly all rocks in the uppermost part of the earth's crust, of whatever type, origin or age, possess openings called pores or voids. Geologists traditionally subdivide rock formations into three classes according to their origins and methods of formation:

Sedimentary rocks are formed by deposition of material, usually under water from lakes, rivers and the sea, and more rarely from the wind. In unconsolidated, granular materials such as sands and gravels, the voids are the spaces between the grains (Figure 2.4A). These may become consolidated physically by compaction and chemically by cementation (Figure 2.4D) to form typical sedimentary rocks such as sandstone, limestone and shale, with much reduced voids between the grains.

Igneous rocks have been formed from molten geological material rising from great depth and cooling to form crystalline rocks either below the ground or at the land surface. The former include rocks such as granites and many volcanic lavas such as basalts. The latter are associated with various types of volcanic eruptions and include lavas and hot ashes. Most igneous rocks are strongly consolidated and, being crystalline, usually have few voids between the grains.

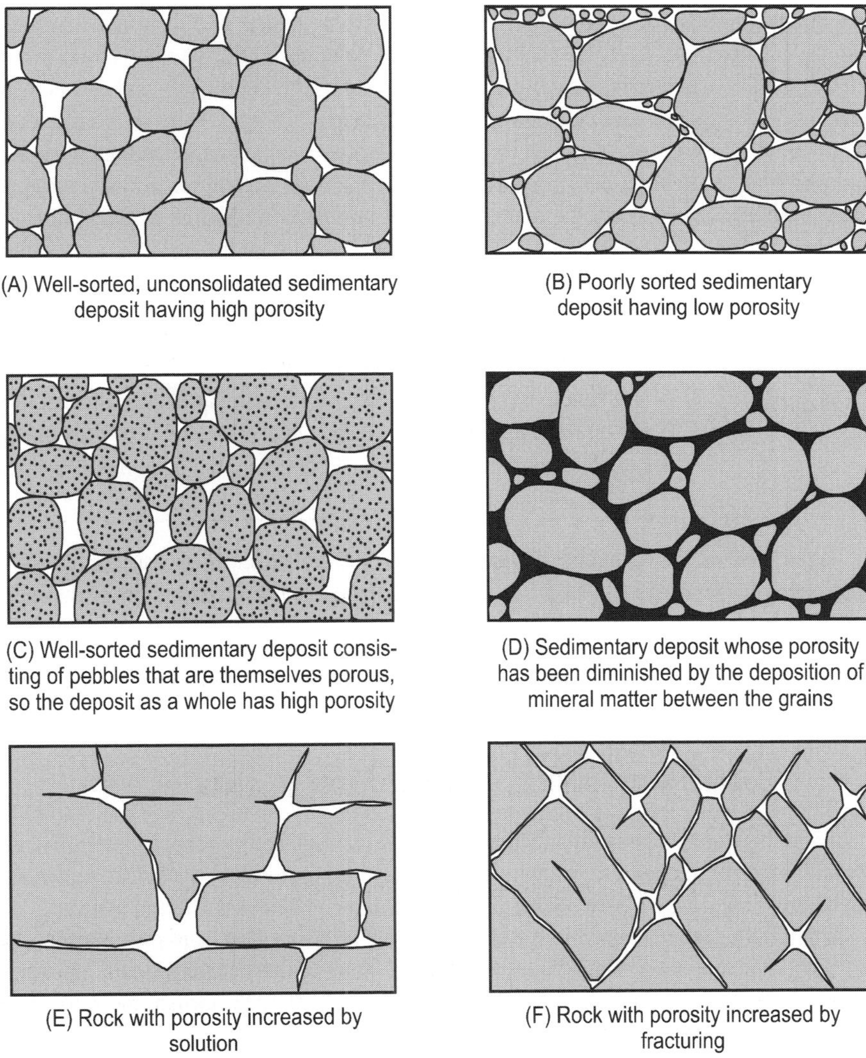

(A) Well-sorted, unconsolidated sedimentary deposit having high porosity

(B) Poorly sorted sedimentary deposit having low porosity

(C) Well-sorted sedimentary deposit consisting of pebbles that are themselves porous, so the deposit as a whole has high porosity

(D) Sedimentary deposit whose porosity has been diminished by the deposition of mineral matter between the grains

(E) Rock with porosity increased by solution

(F) Rock with porosity increased by fracturing

Figure 2.4. Rock texture and porosity of typical aquifer materials (based on Todd, 1980)

Metamorphic rocks have been formed by deep burial, compaction, melting and alteration or re-crystallization of other rocks during periods of intense geological activity.

Metamorphic rocks include gneisses and slates and are also normally consolidated, with few void spaces in the matrix between the grains.

In the more consolidated rocks, such as lavas, gneisses and granites, the only void spaces may be fractures resulting from cooling or stresses due to movement of the earth's crust in the form of folding and faulting. These fractures may be completely closed or have very small and not very extensive or interconnected openings of relatively narrow aperture (Figure 2.4F). Weathering and decomposition of igneous and metamorphic rocks may significantly increase the void spaces in both matrix and fractures. Fractures may be enlarged into open fissures as a result of solution by the flowing groundwater (Figure 2.4E). Limestone, largely made up of calcium carbonate, and evaporates composed of gypsum and other salts, are particularly susceptible to active solution, which can produce the caverns, swallow holes and other characteristic features of karstic aquifers. It is worthwhile becoming aware of the main geological terms, as geological maps are likely to be one of the main sources of information required to characterize a catchment or area of investigation (Chapter 8), but also noting the following important distinction:

NOTE ▶ *A geologist's principal subdivision of rock types is according to origin, whereas hydrogeologists first classify aquifers as unconsolidated or consolidated and hence whether water is stored and moves mainly between the grains of the rock matrix or through fractures.*

The volume of water that can be contained in the rock depends on the proportion of these openings or pores in a given volume of rock, and this is termed porosity of the rock.

DEF ▶ *The porosity of a geological material is the ratio of the volume of the voids to the total volume, expressed as a decimal fraction or percentage.*

Increasing pore space results in higher porosity and greater potential to store water. Typical porosity ranges are shown in Table 2.2 for common geological materials, emphasizing the division between unconsolidated and consolidated referred to above.

Not all of the water contained in fully saturated pore spaces can be abstracted by wells and boreholes and used. Under the influence of gravity when, for example, the water level falls, some of the water drains from the pores but some remains, held by surface tension and molecular effects. The ratio of the water that drains by gravity from an initially saturated rock mass to its own total volume is defined as the specific yield of the material, and typical values are also shown in Table 2.2.

Table 2.2. Porosity and specific yield of geological materials (Freeze and Cherry, 1979; Driscoll, 1986; Domenico and Schwartz, 1998)

Material	Porosity	Specific yield
Unconsolidated sediments		
Gravel	0.25-0.35	0.16-0.23
Coarse sand	0.30-0.45	0.1-0.22
Fine sand	0.26-0.5	0.1-0.25
Silt	0.35-0.5	0.05-0.1
Clay	0.45-0.55	0.01-0.03
Sand and gravel	0.2-0.3	0.1-0.2
Glacial till	0.2-0.3	0.05-0.15
Consolidated sediments		
Sandstone	0.05-0.3	0.03-0.15
Siltstone	0.2-0.4	0.05-0.1
Limestone and dolomite	0.01-0.25	0.005-0.1
Karstic limestone	0.05-0.35	0.02-0.15
Shale	0.01-0.1	0.005-0.05
Igneous and metamorphic rocks		
Vesicular basalt	0.1-0.4	0.05-0.15
Fractured basalt	0.05-0.3	0.02-0.1
Tuff	0.1-0.55	0.05-0.2
Fresh granite and gneiss	0.0001-0.03	<0.001
Weathered granite and gneiss	0.05-0.25	0.005-0.05

Another important way of distinguishing aquifers and the way in which groundwater occurs, when considering both its development and protection, is shown in Figure 2.5. In the figure an unconfined aquifer is one in which the upper limit of the zone in which all the pore spaces are fully saturated, i.e. the water table, is at atmospheric pressure. At any depth below the water table the water pressure is greater than atmospheric, and at any point above, the water pressure is less than atmospheric. In contrast, at greater depths, the effective thickness of an aquifer often extends between two impermeable layers (Figure 2.5).

DEF ▶ *Materials through which water can pass easily are said to be **permeable** and those that scarcely allow water to pass or only with difficulty are described as **impermeable**.*

If the overlying layer has low permeability and restricts the movement of water, then it is known as an aquitard and causes the aquifer beneath to be partially or semi confined. If the overlying layer has such low permeability that it prevents water movement through it, then the aquifer is fully confined. In these situations, at any point in the confined aquifer, the water pressure is greater than atmospheric, because of the elevation of the outcrop receiving recharge. If a borehole is drilled through the confining layer into the aquifer, water rises up the borehole to a level that balances the pressure in the aquifer. An imaginary surface joining the water level in boreholes in a confined aquifer is called the potentiometric surface, which can be above or below the groundwater surface in the overlying unconfined aquifer (Figure 2.5). If the pressure in a confined aquifer is such that the potentiometric surface is above ground level, then a drilled borehole will overflow (Figure 2.5). For a phreatic aquifer, which is the first unconfined aquifer to be formed below the surface, the potentiometric surface and groundwater surface correspond, and this is called the water table.

From the groundwater development point of view, unconfined aquifers are often favoured because their storage properties make them more efficient for exploitation than confined aquifers, and they are likely to be shallower and therefore cheaper to drill into and pump from. On the other hand:

NOTE ▶ *A confined aquifer which has even a modest overlying sequence of less permeable clay strata is likely to be much less vulnerable to pollution than an unconfined aquifer.*

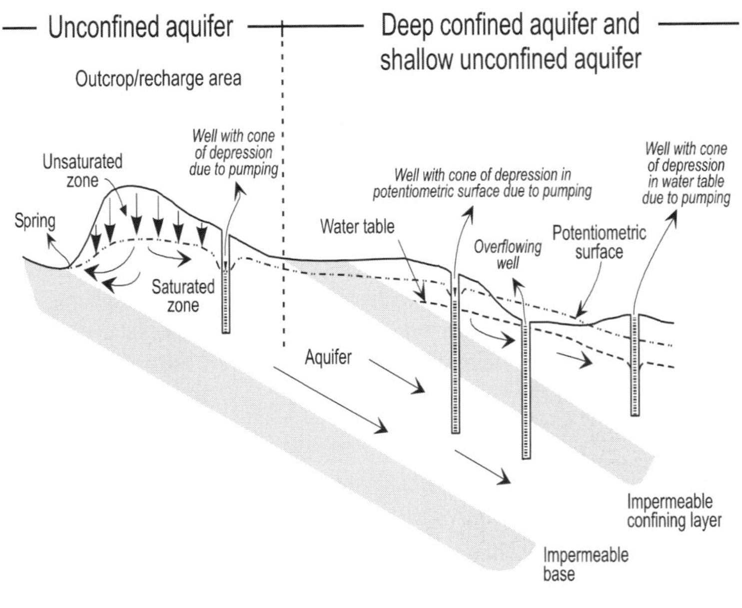

Figure 2.5. Schematic cross-section illustrating confined and unconfined aquifers

2.2.2 Groundwater movement

Groundwater is not usually static but moves slowly through aquifers. However, it needs a source of energy to do so, which is provided by the hydraulic head represented by the height of the water level in an observation well or borehole in the aquifer. The total hydraulic head is made up of two components, the elevation head being the height of the midpoint of the section of the borehole or well that is open to the aquifer, and the pressure head, which is the height of the column of water above this midpoint. The first component thus reflects location and topographic position and the second reflects conditions in the aquifer, including seasonal and longer-term changes in water levels. Hydraulic heads are normally measured with respect to an arbitrary datum, which is often sea level. To obtain a more comprehensive description of these rather difficult concepts, the reader should refer to standard hydrogeological text books such as Freeze and Cherry (1979), Price (1996) or Domenico and Schwartz (1998). For an understanding of groundwater movement for the present purposes, however, it is sufficient to know that groundwater moves from regions of high head to regions of low head.

The flow of groundwater through an aquifer is governed by Darcy's Law, which states that the rate of flow is directly proportional to the hydraulic gradient:

$$Q/A = q = -K(h_1 - h_2)/l = -K\,\Delta h/\Delta l \qquad \text{(Eqn. 2.1)}$$

where Q is the rate of flow through area A under a hydraulic gradient $\Delta h/\Delta l$ which is the difference in hydraulic heads (h_1 - h_2) between two measuring points, and q is the volumetric flow per unit surface area. The direction of groundwater flow in an aquifer is at right angles to lines of equal head. A simple experimental apparatus used to demonstrate Darcy's Law is shown in Figure 2.6, indicating also the elevation and pressure components of hydraulic head referred to above. The equation for Darcy's Law is conventionally written with a minus sign because flow is in the direction of decreasing hydraulic heads.

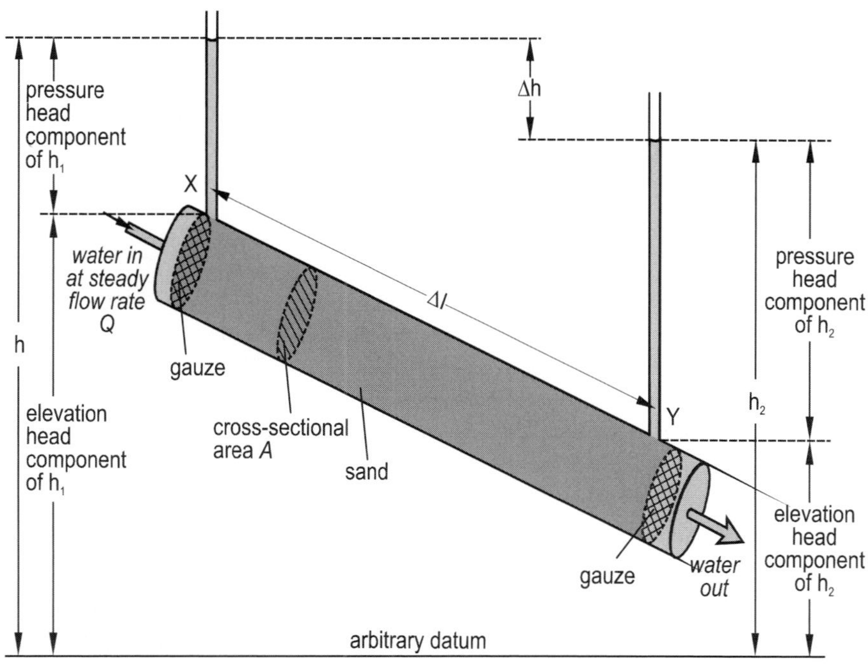

Figure 2.6. Experimental apparatus to demonstrate Darcy's Law (modified from Price, 1996)

The constant of proportionality in the equation, K, has dimensions of length/time because the hydraulic gradient is dimensionless. This parameter is known as hydraulic conductivity, and is a measure of the ease with which water flows through the sand contained in the cylinder in the laboratory experiment or through the various materials that form aquifers and aquicludes. The similarity between Darcy's Law and other important laws of physics governing the flow of both electricity and heat should be noted. The ease with which water can flow through a rock mass depends on a combination of the size of the pores and the degree to which they are interconnected. These features determine the overall permeability of the rock. For clean, granular materials, hydraulic conductivity increases with grain size. Typical ranges of hydraulic conductivity for the main types of geological materials are shown in Figure 2.7.

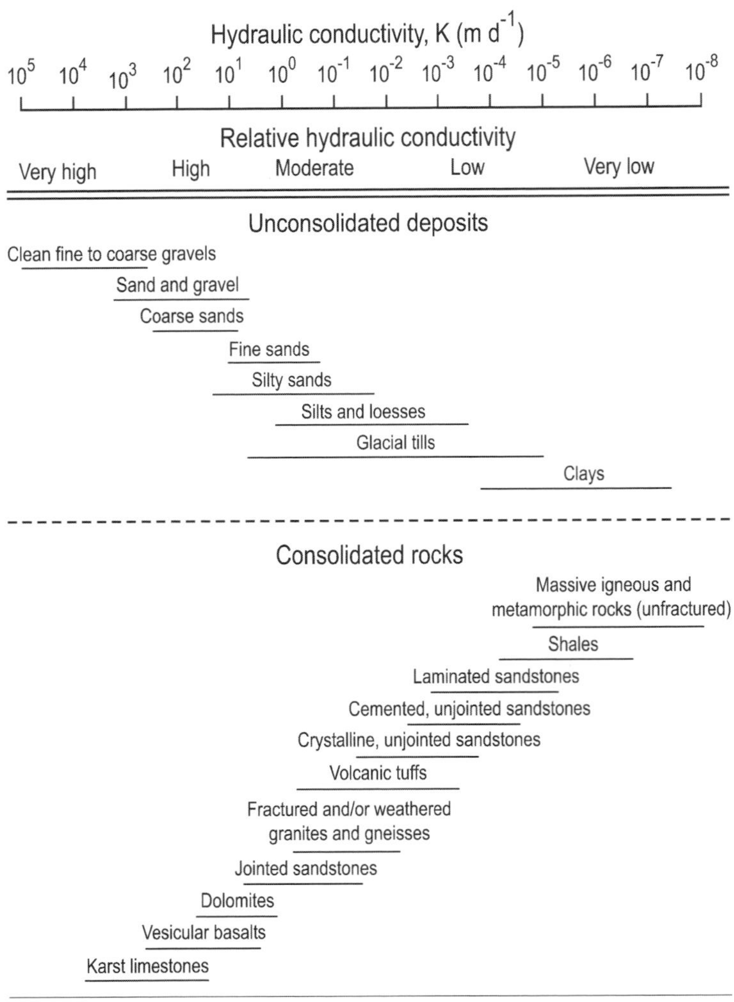

Figure 2.7. Range of hydraulic conductivity (K) values for geological materials (based on Driscoll, 1986 and Todd, 1980)

Darcy's Law can be written in several forms. By substituting q in the equation, it is possible to determine the specific discharge per unit area, if the volumetric flux Q is divided by the full cross-sectional area (A in Figure 2.6). However, this area includes both solids and voids, although clearly flow can only take place through the voids or pore spaces. A more realistic linear pore velocity, v, the volumetric flow rate per area of connected pore space can be calculated if the porosity is known. Thus we can define:

$$v = - q/n = - Ki/n \qquad \text{(Eqn. 2.2)}$$

Where i is conventionally used to represent the hydraulic gradient $\Delta h/\Delta l$. To make this calculation, it is necessary to know the effective or dynamic porosity n_e, which represents the proportion of the total porosity that is involved in groundwater movement. This is difficult to measure, but for unconfined aquifers is probably close to the specific yield values given in Table 2.2.

DEF ▶ *The effective porosity is defined as the proportion of the total volume that consists of interconnected pores able to transmit fluids.*

Thus most materials in which intergranular flow predominates have effective porosities of 0.15 to 0.25, so in these types of aquifers the actual groundwater flow velocity is four to six times the specific discharge. The linear velocity will always be greater than the specific discharge, and increases with decreasing effective porosity. This average velocity in the direction of groundwater flow does not represent the true velocity of water particles through the pore spaces. These microscopic velocities are generally greater because the intergranular flow pathways are irregular and tortuous and longer than the average linear macroscopic pathway. The average linear velocity, *v*, is a key parameter in groundwater protection, as it defines the travel times for water and solutes within aquifers. For unconsolidated granular aquifers, typical natural groundwater flow velocities range from a few mm/d for silts and fine sands to 5-10 m/d for the clean and coarse gravels.

Darcy's Law provides a valid description of the flow of groundwater in most naturally occurring hydrogeological conditions i.e. for fractured rocks as well as granular materials. Fractured rocks are characterized by low porosity and localized high hydraulic conductivity, and very high flow velocities of up to several kilometres per day may result (Orth *et al.*, 1997; US EPA, 1997), especially where a small number of fractures are enlarged by solution (Figure 2.4E). Extensive development of solution in limestone areas can result in karst terrain, which is typified by channels, sinkholes, depressions and caves, into which all traces of surface flow may disappear. Such conditions can be very favourable for groundwater supplies from springs and boreholes, but aquifers of this type are often highly vulnerable to all types of pollution (Malard *et al.*, 1994).

Groundwater flow may occur through the spaces between the grains or through fractures (Figure 2.4) or a combination of the two in, for example, a jointed sandstone or limestone. Hydrogeologists commonly refer to these as dual-porosity aquifers, because they have primary porosity and permeability from the intergranular pores and additional secondary porosity and permeability provided by the fracture systems. The presence of highly fractured rocks should immediately warn of the risk of rapid transport over large distances. The occurrence of potential contaminant sources close to water supplies in such environments should be considered to provide a very high risk of pollution.

The characteristic properties of an aquifer to store and transmit groundwater are normally deduced from the interpretation of pumping tests performed on wells or boreholes (Price, 1996) or by introducing inert tracers into groundwater flow systems and observing their transport (Becker *et al.*, 1998; Käss, 1998). Determinations of aquifer parameters are often difficult and expensive, and information is usually available for at most a few specific locations in an aquifer. However, most geological materials are far from uniform laterally or with depth. As an example, sediments such as river alluvium, deltas and glacial deposits may contain alternating fine and coarse layers, clay lenses, sand channels and many other features and structures which reflect the complex history of deposition. These geological variations mean that aquifers are rarely homogeneous, in which the properties are the same irrespective of position in the aquifer, but more often heterogeneous, with varying properties. Obtaining or selecting aquifer parameters that can apply to and be representative of a whole aquifer or catchment is often, therefore, a difficult task for a hydrogeologist. Describing and quantifying groundwater flow is not as straightforward as a summary text such as this might suggest to the reader, especially in aquifers with complex patterns of fracture flow. However, distinguishing whether intergranular or fracture flow predominates for any aquifer of interest is fundamental to understanding the hydrogeology, which is in turn the basis for developing, managing and protecting groundwater.

NOTE ▶ *Whatever the source of pollution and type of pollutant, understanding the way groundwater occurs and moves is crucial to:*
(1) setting up groundwater protection policies;
(2) establishing water quality monitoring systems;
(3) designing pollution control or aquifer remediation measures.

2.3 GROUNDWATER DISCHARGE AND RECHARGE

It is important to distinguish between infiltration of precipitation and groundwater recharge. Thus looking back at the hydrological cycle in Figure 2.1, when rain falls, some infiltrates into the soil. Much of this moisture is taken up by the roots of plants and is subject to evapotranspiration from the soil zone, and some becomes interflow drainage to streams and rivers. Only a part of the infiltration becomes recharge and moves deeper into the subsurface under gravity, and in arid and semi-arid areas this may be a very small proportion indeed. This distinction becomes very important when considering the estimation of recharge in Chapter 8. Thus conceptually and for estimation purposes:

DEF ▶ ***groundwater recharge*** *should be defined as the downward flow of water reaching the water table and replenishing groundwater resources*

and should be distinguished from infiltration. The latter includes all of the water entering the ground from rainfall or other sources but by no means does all of this become groundwater recharge.

In the subsoil and rock closest to the ground surface, the pore spaces are partly filled with air and partly with water. This was defined as the unsaturated or vadose zone in Figure 2.2, and can vary in depth from nothing to tens of metres. In the unsaturated zone, soil, air and water are in contact and may react with each other, which can be important in the evolution of the hydrochemistry of the water. In the uppermost part of the unsaturated zone, some upward movement occurs in response to seasonal evapotranspiration requirements. Below this, in humid areas, movement in the unsaturated zone is dominantly vertically downwards. The most recent water arriving from the soil displaces downwards the whole column of water already in the unsaturated zone, rather like the movement of a piston, so that the water at the base of the column reaches the water table. For parts of the year, particularly when the weather is dry and no new percolating water passes below the soil, the 'piston' moves very slowly or not at all. In times of heavy rainfall and substantial infiltration, downward movement may be more strongly established.

By sampling the unsaturated zone water to obtain vertical profiles and repeat profiles of tracers such as bromide, nitrate and tritium, average rates of movement of less than 5 m/yr and often less than 1 m/yr have been measured in temperate regions (Wellings, 1984; Geake and Foster, 1989; Barraclough *et al.*, 1994). This means that it could take 20 years or more for infiltrating water to reach a water table 20 m below the ground surface. It is common for the water table to be in the range 10 to 50 m below ground, and the unsaturated zone component of the pollutant pathway can therefore be substantial. In semi-arid regions, recharge can be much less and downward displacement correspondingly very slow (Edmunds and Gaye, 1994). In the most arid areas, the unsaturated zone may only act as a temporary storage reservoir in which water that percolates downwards after occasional heavy rain does not reach the water table but is instead drawn upwards and returns to the atmosphere by evapotranspiration from plants. Residence times in the unsaturated zone thus depend on the thickness and the rate of recharge, and can vary from almost nothing to tens or hundreds of years.

The above applies to aquifers in which downward movement of recharging water takes place only through the intergranular matrix. In fractured and dual porosity aquifer materials, much more rapid, preferential flow to the water table may occur, especially after heavy rainfall. This component of flow can carry pollutants from the ground surface much more quickly, allowing little or no time for attenuation, and such aquifers can be highly vulnerable to pollution.

All subsurface freshwater must have a source of recharge, even if it was long ago. This comes either by direct infiltration of rainfall or snowmelt, or from rivers and lakes. Now that the hydrological cycle has been interfered with as a result of human activities, recharge can also be derived from canals, reservoirs, irrigated land, water mains and sewerage systems in urban areas, mining waste, sewage lagoons, in fact any artificial water body that is in connection with the subsurface. Artificial recharge, which is becoming an increasingly important resource management option, can also introduce water of different origin and quality into aquifers. This of course means that groundwater recharge is not always of the same good quality as infiltrating rainfall, which itself may be contaminated by acid rain or atmospheric acid deposition.

2.4 GROUNDWATER FLOW SYSTEMS

In many aquifers, the hydraulic head reflects the topographic surface of the ground, and groundwater therefore moves from elevated regions where recharge occurs to discharge zones at lower elevations. Thus within the context of the overall cycle shown in Figure 2.1 and the source-pathway-receptor concept, the groundwater flow system (Figure 2.8) is a useful way of describing the physical occurrence, movement and hydrochemical evolution of groundwater.

DEF ▶ *A groundwater flow system is a discrete, closed three-dimensional system containing flow paths from the point at which recharging water enters an aquifer to the topographically lower point at which it leaves the aquifer.*

Infiltration of rainfall on high ground occurs in a recharge area in which the hydraulic head decreases with depth, and net saturated flow is downwards away from the water table and laterally towards areas of lower hydraulic head (A in Figure 2.8). After moving slowly through the aquifer down the hydraulic gradient (B in Figure 2.8), groundwater leaves the aquifer by springs, wetlands, baseflow to rivers (C in Figure 2.8) or discharge to lakes or the oceans. These are known as groundwater discharge areas, and at C (in Figure 2.8) the hydraulic head increases with depth and the net saturated zone flow is upwards towards the water table. In a recharge area, the water table can be at depth, with a considerable thickness of unsaturated zone above it. In a discharge area, the water table is usually at, or very near to, the ground surface. Rivers, canals, lakes and reservoirs may either discharge to or receive recharge from groundwater, and the relationship may change seasonally or over a longer time span or along the course of a single river.

While in many cases groundwater and surface water catchments have more or less the same boundaries, this is not always the case. Seasonally due to recharge, and in response to heavy abstraction, groundwater catchment boundaries may deviate significantly from surface water catchments. In deep aquifers, or sequences of more than one layered aquifer, groundwater recharge may come from great distances and deep groundwater flow may have little relationship to the overlying surface water system. In most cases, however, if there is no information it is a reasonable first estimate to assume surface water and groundwater catchments are similar and that groundwater flow patterns are likely to be a subdued reflection of the surface topography.

In large, deep aquifers, groundwater is likely to move slowly, at rates of a few metres per year, from recharge to discharge area over tens or hundreds of kilometres. This may take hundreds or thousands of years, and typical order-of-magnitude values from time of recharge to point of discharge are indicated in Figure 2.8. Hydrogeologists can confirm these by isotopic dating techniques (Kendal and McDonnel, 1998; Edmunds and Smedley, 2000). In small, shallow aquifers, recharge and discharge areas may be much closer or even adjacent to each other, and residence times can be restricted to a few months or years. In arid and semi-arid regions, groundwater discharge areas are often characterized by poor quality groundwater, particularly with high salinity. Groundwater discharge may be from seepages or salt marshes with distinctive vegetation, known as salinas or playas, in which evapotranspiration at high rates for long periods of time has led to a build-up in salinity.

While the flow system in Figure 2.8 is a useful general illustration, in many cases groundwater does not flow uniformly through the entire thickness of an aquifer, but instead flows predominantly at shallow depths close to the water table (Seiler and Lindner, 1995). In unconsolidated aquifers, both hydraulic conductivity and porosity usually decrease with depth due to consolidation and compaction. In fractured aquifers too, the hydraulic conductivity and porosity provided by the fracture system would be expected to decline with depth as the fractures become less open. This general but variable and not easily predicted decline in groundwater flow properties with depth often restricts the flow and pollutant pathway to the most permeable, near-surface parts of the aquifer. Further, the zone of seasonal water table fluctuation is often where the most active solution of fractures occurs and this helps to enhance the flow dominance of the uppermost part of the saturated aquifer.

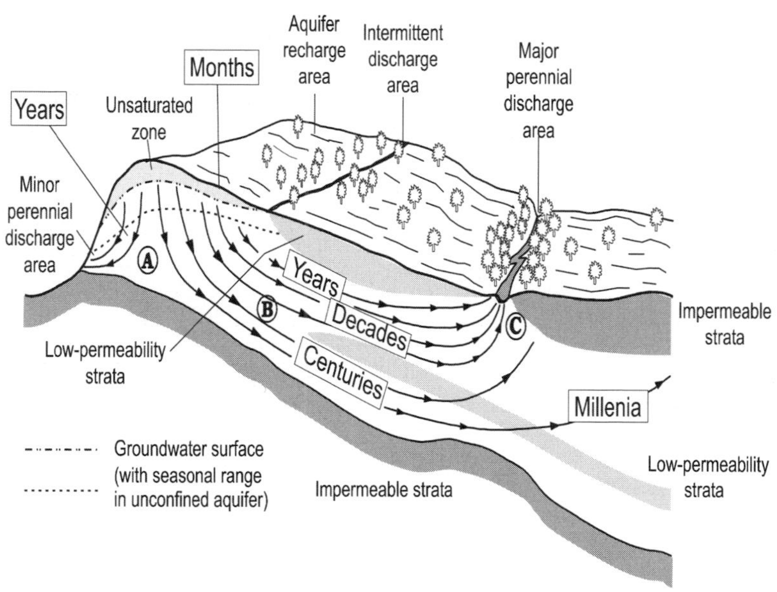

Figure 2.8. Schematic groundwater flow system (modified from Foster *et al.*, 2000)

Given the high porosity values in the upper part of Table 2.2, it can be seen that most types of aquifer, provided they are at least a few metres thick, can contain large volumes of water. Many aquifers are, of course, much thicker, ranging up to several hundred metres. Even in humid areas, recharge comprises only a proportion of the total rainfall and a simple calculation will show that, for typical annual recharge volumes equivalent to tens to a few hundreds of millimetres, the total volume of groundwater in storage in the aquifer is many times larger than the annual recharge. Aquifers are generally, therefore, high storage, low recharge systems with substantial capacity for dilution of incoming pollutants, except in the situations of restricted shallow flow referred to immediately above. In these cases, incoming recharge may be distributed far from evenly through the aquifer, and the resulting groundwater volume available for dilution may be much less than the total storage of the aquifer.

2.5 GEOLOGICAL ENVIRONMENTS AND AQUIFER TYPES

As described above, the natural subsurface geological environment provides the dominant control over the occurrence and movement of groundwater and hence defines which rock types form good aquifers. However, geodiversity and the consequent hydrogeological variability are poorly appreciated by many of those working in water protection and management. The variability both between and within hydrogeological environments can have a profound impact on how aquifers respond to the pressures imposed upon them. Further, if an aquifer is to be protected and managed, it is important to understand the groundwater flow system to be able to assess the susceptibility of the aquifer to these external changes and the types and timescales of the likely responses.

While almost all geological materials contain some water and many different rocks can form useful aquifers, nevertheless it is possible to develop a summary of the most common aquifer types and hydrogeological environments (Table 2.3). This classification is a useful overall basis for helping to identify the major potential concerns for protection and management of groundwater. The general subdivision in Table 2.3 takes into account both the rock type and the geological environment in which the rocks were formed. While such a broad classification is useful, it inevitably involves some simplifications of the true breadth of subsurface geological variation and complexity. The classification shown in Table 2.3, or slight variations of it, proved useful as a basis for discussion of groundwater quality monitoring (Chilton, 1996) and provided the hydrogeological framework within which the management of groundwater in urban areas (Foster *et al.*, 1998) and the development and management of groundwater in rural areas (Foster *et al.*, 2000) can be set. Each of the seven subdivisions is briefly described below.

Major alluvial and coastal plain sediments

The first subdivision (Table 2.3), covers a broad range of materials and lateral and vertical scales. At one end of the scale are extensive sequences of coastal, river and deltaic alluvium, sometimes hundreds of metres thick. These unconsolidated sedimentary deposits form some of the most important aquifers in the world, in which very large volumes of groundwater are stored and from which large quantities of water are pumped for water supply and irrigation. Examples include the Lower Indus and Ganges-Brahmaputra valleys, the Mekong, the Tigris-Euphrates, the north European plain and the Nile valley. Many of the world's largest cities such as Bangkok, Beijing, Cairo, Calcutta, Dhaka, Hanoi, Lima, Madras and Shanghai are located on such deposits and are supplied by groundwater drawn from unconsolidated strata.

These aquifers can cover large areas and contain enormous volumes of water. As an example, aquifers within the unconsolidated sediments underlying the Huang-Hai-Hai Plain of eastern China, which covers an area of 350 000 km^2, provide the potable water requirements for nearly 160 million people and also enough to irrigate some 20 million ha of land. The sediments are of Quaternary age and are typically 200-400 m thick. Groundwater in these sediments can be subdivided into three types: an upper unconfined freshwater zone, a middle saline water zone and a lower confined aquifer. The total volume of groundwater stored exceeds 2 000 000 million m^3, whilst usable groundwater resources have been estimated at more than 49 000 million m^3/a. Other unconsolidated sedimentary aquifers may be much less extensive but can still store sufficient volumes of groundwater to be important sources of water supply. The coastal plain around Jakarta and the Nile valley at Cairo are examples. Smaller but still locally-important aquifers are provided by river valley and coastal plain sediments of more limited lateral extent and depth, and aquifers of much more restricted size and extent may occur in upland river valleys as river terraces.

Aquifers in unconsolidated strata are rarely simple homogeneous systems but typically consist of alternating permeable layers of productive sands and gravels separated by less permeable aquitard layers of clay and silt, reflecting the complex history of deposition. In such sequences, the shallowest aquifer may be the easiest and cheapest to exploit, but is likely to be the most vulnerable to pollution. The presence of aquitards may produce complex groundwater flow patterns, but the permeable horizons may still have a degree of hydraulic continuity, such that pumping from one layer will affect the others, producing significant vertical head gradients and consequent leakage.

The high porosity of unconsolidated sediments, typically in the range 0.25 to 0.35, and the generally low horizontal hydraulic gradients in the major alluvial plains means that groundwater velocities are very low, usually in the range 0.003-0.1 m d^{-1}. These low velocities combined with the significant distances travelled (tens to hundreds of kilometres) indicate that much of the deeper groundwater in thick alluvial sequences is derived from recharge several hundred to several thousand years ago, and the term 'fossil' has sometimes been used to describe deep, old groundwater.

Intermontane alluvial and volcanic systems

Aquifers of this type include some volcanic lavas and pyroclastic rocks, together with alluvial-volcanic and alluvial fan deposits. They are typically associated with rapidly infilled and faulted troughs or basins within mountain regions (Table 2.3). Hydraulic conductivities and porosities are generally high but variable. When combined with the above average rainfall that is often found in the mountainous climatic regimes where many of these environments are found, valuable aquifers occur and are capable of supporting substantial borehole yields. Additional recharge to groundwater often occurs where surface water flowing from the surrounding mountain slopes infiltrates into the highly permeable valley-fill deposits, especially through the alluvial fans and colluvial deposits found on valley margins. Examples of this environment include Mexico City, Guatemala City, San Salvador, Managua and San José in Central America, the Kathmandu Valley in Nepal, Bandung and Yogyakarta in Indonesia, Davao in the Philippines and Sana'a in Yemen. In these mountainous areas, flat land is limited and highly valuable, and is often densely populated. Restrictions on available land for settlement will often result in groundwater abstraction for potable supplies in the basin occurring within densely populated areas, with significant implications for water quality. Furthermore, the concentration of population and the consequent high water demand can result in groundwater abstraction exceeding the safe yield of the aquifer. Long-term decline in groundwater levels and/or contamination of groundwater can result, as for example in Mexico City (NRC, 1995), the Kathmandu Valley (Khadka, 1991) and the Sana'a Valley (Alderwish and Dottridge, 1999).

Consolidated sedimentary aquifers

Important aquifers occur within consolidated sedimentary strata, principally sandstone and limestone (Table 2.3). These can be broadly subdivided into younger, Tertiary formations and older Mesozoic or Palaeozoic formations. Globally, although their distribution is irregular, they are widespread and common being found both in mountain belts such as the Alpine-Himalayan, Andean, Urals and North American cordilleras and in lowlands and plateau areas such as northern Europe and central China.

Sandstones have been formed when sandy marine or continental sediments were buried and compacted to form consolidated rocks. The degree of consolidation generally increases with depth and age of the rocks. Thus the younger, Tertiary sandstones usually retain some degree of primary porosity between the sand grains and are typically of low to moderate permeability. In the older, Mesozoic or Palaeozoic formations with more strongly developed cementing of the grains, the primary porosity may have become largely eliminated. Pre-Tertiary sandstones can range from friable to highly indurated depending on the degree of cementation, and in the latter cases it is the secondary porosity resulting from the development of fractures which can provide adequate permeability and storage for such rocks to form productive aquifers.

Limestones exhibiting solution enhancement of such fractures (called karst) are widespread and can be prolific aquifers, although well yields are highly variable in time and space. For instance, in northern China, karst limestones occupy an area of 800 000 km^2, are typically 300-600 m thick and their groundwater resources have been estimated at 12 800 million m^3 yr^{-1}. In southern China, karst limestones are even more extensive where they cover an area of 1 400 000 km^2, with groundwater resources estimated at 190 000 million m^3 yr^{-1}. Similar, highly permeable limestones occur throughout southern Europe, including along the coast of the Adriatic Sea region from which the karst name derives, in the Middle East and in the USA.

Recent coastal calcareous formations

These formations form important local aquifers. Examples, which include Florida, Jamaica, Cuba, Hispaniola and numerous other islands in the Caribbean, the Yucatan peninsula of Mexico, the Cebu limestone of the Philippines, and the Jaffna limestone in Sri Lanka, provide important sources of potable water for the people living there and for irrigation. Their high to very high permeability derives not only from initially high primary porosities (due to the sedimentary environment of deposition), but also from fractures that have been enhanced by solution. This can produce rapid groundwater movement with velocities frequently in excess of 100 m d^{-1}. The high infiltration capacity of these strata often precludes surface drainage systems and very often groundwater is the only available source of water supply in these environments.

These characteristics have important implications for protecting groundwater quality. Soils can be very thin and water movement from the soil to the water table via fissures is often so rapid that these formations are highly vulnerable to pollution. In addition, being coastal, the aquifers are usually underlain by seawater, often at shallow depths. Excessive abstraction of groundwater, with a consequent lowering of the water table, may induce saline intrusion by lateral movement of the freshwater/seawater interface inland or local upconing and contamination of the fresh groundwater body from below.

Glacial formations

Deposits of glacial and fluvioglacial origin comprise small but locally important aquifers not only in temperate zones of the world but also at altitude in the mountain ranges of the Andes and Himalayas. Ice-transported sediments are commonly unsorted mixtures of all grain sizes from clay to boulders, typically have low permeabilities and act as aquitards or aquicludes. Their geographical distribution is often limited, as they tend to occur in regions of active erosion. In contrast, water-sorted sediments, laid down from glacial melt-waters, include the sands and gravels of kames and eskers, which can form restricted but highly productive aquifer systems. These can sometimes be more extensive, as in the coalescing gravel outwash plains of North America, the eastern Andes and the Himalayas/Pamir/Tienshan cordilleras, or quite narrow and sinuous, as in the glacial channels of the North German Plain and the Great Lakes.

The environment of deposition, from melt-water streams and the upper reaches of braided rivers makes for highly variable lithology. As a result, multiple aquifers are typical, comprising complex systems in which lenses of highly permeable sands and gravels are partly separated vertically and laterally from each other by lower-permeability fine sands, silts and clays. The resultant 'patchy' aquifer can be very productive, but hydraulic continuity between different lenses means that mobile persistent contaminants are able to penetrate to significant depths.

In many glacial areas, the underlying bedrock consists of ancient, hard and unweathered granites and gneisses, which are very unpromising as aquifers. In these terrains, even the small sedimentary aquifers referred to above can provide vital but potentially quite vulnerable water supplies to the small and scattered communities in such regions. Sometimes these aquifers are used for urban supply, either directly by means of boreholes, or as prefilters for high volume riverbank intakes via infiltration galleries or collector wells. Examples include Cincinnati and Lincoln (USA), Berlin and Düsseldorf (Germany), and Vilnius (Lithuania).

Loessic plateau deposits

Fine windblown deposits, called loess, form an important aquifer in China. Although loess is found elsewhere, such as in Argentina and north of the Black Sea, thick deposits are almost entirely restricted to north central China where they cover an area in excess of 600 000 km^2. Of this, some 440 000 km^2 is continuously covered with a thickness of between 100 and 300 m. The loess covers a vast plateau at elevations of between 400 and 2 400 m above sea level. The loess plateau supports a population of 64 million and 7.3 million ha of cultivated land and is dependent on groundwater for domestic water and irrigation in this semi-arid region. The distinctive geomorphological features and geological characteristics produce a complex groundwater system. The loessic plateau aquifers are frequently cut through by gullies and ravines, so that the plateaux form a series of independent water circulation systems. The deposits are generally of low permeability and the presence of palaeo-soils produces a layered aquifer; the deeper zones being partly confined. The water table is often quite deep (30-50 m below surface).

Extensive volcanic terrains

One of the largest and most important areas of volcanic lava flows occurs in the central and western parts of India, where the Deccan basalts cover more than 500 000 km^2. Other extensive volcanic terrains occur in North and Central America, Central and East Africa, and many islands are entirely or predominantly of volcanic origin, such as Hawaii, Iceland, the Canary Islands (Spain) and some of the Caribbean islands. Older lavas such as the Deccan basalts can often be largely impermeable in the rock mass, but younger basalts can provide very large springs. Individual lava flows can be up to 100 m thick, and although the more massive flows are often impermeable, extensive jointing allows water to infiltrate and move through them. The junctions between flows can form highly productive aquifers, because of the cooling cracks and joints, and development of rubble zones caused when the rough surface of the lava is covered by the chilled base of the next flow, weathering and soil in the period of time between successive flows. Extensive lava tubes may be formed where lava drains from beneath a cooled and congealed surface. It is the combination of these features that make the Deccan basalts and other such volcanic rocks important and locally productive aquifers. Other materials are thrown out as volcanic clouds, which sometimes settle as ash deposits or become welded tuffs. The mineralogy and chemistry of the volcanic rocks and their viscosity and gas content determine the precise nature of the volcanic eruptions and resulting rocks. Alternating sequences of ashes and lavas, in which the lavas act as conduits for groundwater flow and the intervening ashes provide the storage, characterize the important aquifer systems of Costa Rica, Nicaragua and El Salvador.

Weathered basement complex

These aquifers are found in ancient crystalline rocks of Precambrian or Lower Palaeozoic age. In sub-Saharan Africa, such rocks cover 40 per cent of the total land area and some 220 million people live on them. Groundwater flow and storage occurs in restricted fractures in the fresh bedrock, but usually more extensively in the superficial weathered layers. The processes of weathering and disaggregation can enhance both porosity and permeability (Chilton and Foster, 1995). Because these ancient rocks occupy stable, continental shield areas, there has been plenty of opportunity for prolonged periods of weathering, and the zone of weathering tends to be better developed and thicker in tropical regions where such processes are more active. As a result, the weathered zone can be as much as 60 m thick, but more commonly in the range of 20 to 30 m. Groundwater velocities in weathered and fractured bedrock aquifers can be very variable.

Even with the beneficial effects of weathering, the volumes of water stored within these aquifers are generally limited, hydraulic conductivities are low, borehole yields modest and groundwater is used mostly for providing potable water supplies for rural communities and small towns, and for small-scale supplementary irrigation. Larger cities located on such formations, such as Kampala in Uganda, may find it difficult to abstract the large quantities of water needed for urban supply and also to dispose of wastewater on-site to the subsurface in a sanitary manner.

Table 2.3. Summary of characteristics of principal hydrogeological environments

Hydrogeological environment	Lithology	Geological description and origin	Groundwater flow regime	Natural groundwater flow rates (m/d)
Major alluvial and coastal plain sediments	Gravels, sands, silts and clays	Unconsolidated deposits of major rivers, deltas and shallow seas, high primary porosity and permeability. Very extensive and thick aquifers	Intergranular	2-10 in gravels, 0.05-1 in sands, 0.001-0.1 in silts
Intermontane alluvial and volcanic systems	Pebbles, gravels, sands and clays, sometimes inter-bedded with lavas and volcanic ashes	Rapid infilling of faulted troughs and basins in mountain regions; unconsolidated, primary porosity and/or permeability usually high for fan sediments and ashes, but lavas and lacustrine deposits are often poor aquifers or confining aquitards. Less extensive but can be thick	Intergranular, fracture in lavas and cemented ashes	0.001-10
Consolidated sedimentary aquifers	Sandstones	Marine or continental deposits buried, compacted and cemented to form consolidated rocks; degree of consolidation generally increases with depth/age of deposition. Primary porosity moderate to poor but secondary porosity introduced by fractures can be significant	Intergranular and fracture	0.001-0.1
	Limestones	Deposited from skeletal material (shell fragments, reefs, reef detritus) in shallow seas and compacted to form consolidated rocks; often have fractures which may be enlarged by solution processes to form characteristic topography, cavities and tunnel systems known as karst	Dominant fracture with variable intergranular component	0.001-0.1 in matrix, up to 1000 in karst fissures
Recent coastal calcareous formations	Limestones and calcareous sands	Usually composed of coral limestones, shellbanks, chemically precipitated oolites and calcareous oozes; often loosely cemented; porosity and permeability can be exceptionally high, especially if fractures are solution-enhanced	Intergranular and fracture	0.01-0.1 in matrix, up to 2000 in karst

Hydrogeological environment	Lithology	Geological description and origin	Groundwater flow regime	Natural groundwater flow rates (m/d)
Glacial formations	Boulders, pebbles, gravels, sands, silts and clays	Ice-transported sediments are commonly unsorted and have low permeability, but water-sorted sediments such as melt-water and outwash deposits often have high porosity and permeability. Can be thin, patchy and shallow	Intergranular	0.001-0.1 in tills, much higher in sands and gravels
Loessic plateau deposits	Silts, fine sands and sandy clays	Usually well-sorted windblown deposits of silt and fine sand, with some sandy clay deposits of secondary fluvial origin; low permeability. Can be extensive and thick but divided into blocks by deep gullies	Intergranular	0.001-0.01
Extensive volcanic terrains	Lavas, tuffs and ashes	Extensive basaltic lava flows or ashes and tuffs from more explosive eruptions. Primary porosity only in ashes and the less welded tuffs and at junctions of lavas. Joints and fractures in lavas. Variable potential, decreasing with age	Fracture with variable intergranular	0.001-10
Weathered basement complex	Crystalline rocks, granites, gneisses, schists	Decomposition of older igneous or metamorphic rocks can produce a weathered mantle of variable thickness, moderate porosity but generally low permeability, underlain by fresher rock which may be fractured; the combination results in a low potential but important very widespread but shallow aquifer system	Dominantly intergranular in weathered zone, fracture below	0.001-0.1

2.6 REFERENCES

Because many of the basic concepts about groundwater occurrence and movement are likely to be new to many of the readers of this monograph, and this introductory discussion may not have done them justice, the following three standard texts can be used to provide additional material if required:

Domenico, P.A. and Schwartz, F.W. (1998) *Physical and Chemical Hydrogeology*, 2nd edn, John Wiley, New York.

Freeze, R.A. and Cherry, J.A. (1979) *Groundwater*, Prentice Hall, Englewood Cliffs, New Jersey.

Price, M. (1996) *Introducing Groundwater*, 2nd edn, Chapman and Hall, London.

Additional references quoted in the chapter:

Alderwish, A.M. and Dottridge, J. (1999) Urban recharge and its influence on groundwater quality in Sana'a, Yemen. In *Groundwater in the Urban Environment: Selected city profiles*, (ed. P.J. Chilton), pp. 85-90, Balkema, Rotterdam.

Barraclough, D., Gardner, C.M.K., Wellings, S.R. and Cooper, J.D. (1994) A tracer investigation into the importance of fissure flow in the unsaturated zone of the British Upper Chalk. *J. Hydrology*, **156**, 459-469.

Becker, M.W., Reimus, P.W. and Vilks, P. (1998) Transport and attenuation of carboxylate-modified latex microspheres in fractured rock, laboratory and field tracer tests. *Ground Water*, **37**(3), 387-395.

Chilton, P.J. (1996) Groundwater. In *Water Quality Assessments: A guide to the use of biota, sediments and water in environmental monitoring*, 2nd edn, (ed. D. Chapman), pp. 413-510, E & FN Spon, London.

Chilton, P.J. and Foster, S.S.D. (1995) Hydrogeological characterization and water supply potential of basement aquifers in tropical Africa. *Hydrogeology J.*, **3**, 36-49.

Driscoll, F.G. (1986) *Groundwater and Wells*, 2nd edn, Johnson Division, St Paul, Minnesota.

Edmunds, W.M. and Gaye, C.B. (1994) Estimating the spatial variability of groundwater recharge in the Sahel using chloride. *J. Hydrology*, **156**, 47-59.

Edmunds, W.M. and Smedley, P.L. (2000) Residence times indicators in groundwater: the East Midlands Triassic sandstone aquifer. *Appl. Geochem.*, **15**, 737-752.

European Community 2000: Directive 2000/60/EC of the European Parliament and of the Council of 23 October 2000 establishing a framework for Community action in the field of water policy, *Official Journal of the European Communities*, **L327,** Brussels.

Foster, S.S.D., Chilton, P.J., Moench, M., Cardy, F. and Schiffler, M. (2000) *Groundwater in Rural Development: facing the challenges of supply and resource sustainability*, World Bank Technical Paper No. 463, Washington, DC.

Foster, S.S.D., Lawrence, A.R. and Morris B.L. (1998) *Groundwater in Urban Development: assessing management needs and formulating policy strategies*, World Bank Technical Paper No. 390, Washington, DC.

Geake, A.K. and Foster, S.S.D. (1989) Sequential isotope and solute profiling of the unsaturated zone of the British Chalk. *Hydrological Sci. J.*, **34**, 79-95.

Ghassemi, F., Jakeman, A.J. and Nix, H.A. (1995) *Stalinization of Land and Water Resources*, University of New South Wales Press, Sydney.

Käss, W. (1998) *Tracing Techniques in Geohydrology*, Balkema, Rotterdam.

Kendal, C. and McDonnel, J.J. (1998) *Isotope Tracers in Catchment Hydrology*, Elsevier, Amsterdam.

Khadka, M.S. (1991) Nepal: Groundwater Quality. In *Groundwater Quality Monitoring in Asia and the Pacific*, UN-ESCAP, Water Resources Series No 70, Bangkok.

Lerner, D.N. (1997) Too much or too little: Recharge in urban areas. In *Groundwater in the Urban Environment: Problems, processes and management*, (eds. P.J. Chilton *et al.*) pp. 41-47, Balkema, Rotterdam.

Lerner, D.N., Issar, A. and Simmers, I. (1990) *Groundwater Recharge: A guide to understanding and estimating natural recharge*, Heisse, Hannover.

Lvovitch, M.I. (1972) World water balance: general report. In IAHS/UNESCO/ WMO *Proc. of Symposium on World Water Balance*, Reading, 1970, pp. 401-415, IAHS, Wallingford.

Malard, F., Reygrobellet, J-L., and Soulie, M. (1994) Transport and retention of faecal bacteria at sewage-polluted fractured rock sites. J. Environ. Qual., 23, 1352-1363.

Meybeck, M., Chapman, D.V. and Helmer, R. (eds.) (1989) Global Freshwater Quality: A first assessment, GEMS Global Environment Monitoring System, WHO/UNEP, Blackwell, Oxford.

Morris, B.L. Lawrence, A.R., Chilton, P.J., Adams, B., Calow, R. and Klinck, B.A. (2003) Groundwater and its Susceptibility to Degradation: A global assessment of the problem and options for management. Early Warning Report Series, RS 03-3, UNEP, Nairobi.

Nace, R.L. (ed) (1971) Scientific Framework of World Water Balance, Technical Papers in Hydrology 7, UNESCO, Paris.

NRC (1995) Mexico City's Water Supply: Improving the outlook for sustainability, NRC, National Academy Press, Washington, DC.

Orth, J.P., Netter, R. and Merkl, G. (1997) Bacterial and chemical contaminant transport tests in a confined karst aquifer (Danube Valley, Swabian Jura, Germany). In *Karst Water and Environmental Impacts*, (eds. G. Gunay and A.I. Johnson) Balkema, Rotterdam.

Seiler, K-P. and Lindner, W. (1995) Near surface and deep groundwater. *J. Hydrology*, **165**, 33-44.

Todd, D.K. (1980) *Groundwater Hydrology*, 2nd edn, John Wiley, New York.

UNESCO (1999) *World Water Resources at the Beginning of the 21st Century*, CD version, UNESCO, Paris.

US EPA (1997) *Guidelines for Wellhead and Spring Protection in Carbonate Rocks*, EPA 904-B-97-0003, Washington, DC.

Wellings, S.R. (1984) Recharge of the upper chalk aquifer at a site in Hampshire, England. *J. Hydrology*, **69**, 275-285.

3

Pathogens: Health relevance, transport and attenuation

S. Pedley, M. Yates, J.F. Schijven, J. West, G. Howard and M. Barrett

This chapter will summarize current knowledge about the distribution of pathogens in groundwater and the factors that control their transport and attenuation. The aim is to provide a level of information and interpretation that will allow public health specialists and water resource managers to estimate risks to the groundwater from microbial contaminants derived from sources described in Section II of this monograph, for example, agricultural and urban sources.

Many factors, some environmental and others linked to the properties of the organism, control the survival and transport of microorganisms in the subsurface. However, it is important to consider that often the factors perceived to be of importance have been studied in isolation using controlled laboratory experiments and the conclusions then extrapolated to predict the fate of pathogens in the environment. This process is known as upscaling and is itself the subject of current research. In contrast, very few studies have attempted to examine the effect of multiple factors interacting in the natural environment. Current knowledge therefore offers a number of guiding principles about the transport and attenuation of pathogens in groundwater, but the complex interaction of factors controlling the fate of pathogens is poorly understood and difficult to predict in some environments.

DEF ▶ ***Microorganisms*** *are microscopic organisms within the categories algae, bacteria, fungi, protozoa, viruses and subviral agents (Singleton and Sainsbury, 1999).*

Pathogens *are any microorganisms which by direct interaction with (infection of) another organism cause disease in that organism (Singleton and Sainsbury, 1999). The strict definition of a pathogen excludes those microorganisms that cause disease indirectly by the synthesis of a toxin that may subsequently be ingested by the victim. Several microorganisms implicated in food poisoning cause disease in this way: Clostridium perfringens and Staphylococcus aureus are examples of this group. Nevertheless, these microorganisms can be pathogens in the strict sense under different conditions.*

3.1 MICROBIAL PATHOGENS AND MICROBIAL INDICATOR ORGANISMS

The ability of a pathogen to inflict damage upon the host is controlled by a combination of factors, in particular the nature of the organism (for example its virulence) and the susceptibility of the host. Several factors combine to determine the susceptibility of the host, including age, nutritional status and immunity. Immunocompromised individuals, for example, are highly susceptible to infection by pathogens, whereas well-nourished young adults are typically less susceptible to infections.

DEF ▶ ***Virulence*** *is the capacity of a pathogen to cause disease, defined broadly in terms of the severity of the symptoms.* ***Infectivity*** *is the ability of a pathogen to become established on or within the tissue of a host.*

Water can be the vehicle for the transmission of many different types of pathogenic microorganism: some being natural aquatic organisms and some being introduced into the water from an infected host. Overall, the pathogens in water that are the main concern to public health originate in the faeces of humans and animals, and establish an infection when contaminated water is consumed by a susceptible host (Boxes 3.1 and 3.2). These are the classical waterborne pathogens (Table 3.1) that are transmitted by the faecal-oral route of infection (Figure 3.1). Waterborne pathogens can be classified into four broad groups according to their chemical, physical and physiological characteristics. Listed in order of increasing functional complexity the groups are viruses, bacteria, protozoa and helminths. In general, the transmission of helminths in groundwater is unlikely, although not impossible, due to the size of the organisms and their eggs. For this reason, and because public health concerns surrounding waterborne disease transmitted through groundwater have concentrated upon the other groups of microorganisms, helminths will not be discussed in this chapter. Within the other groups a large number of microbial pathogens are able to contaminate groundwater (Table 3.1).

Box 3.1. Health impacts of contaminated groundwater in Walkerton, Canada (based on Howard, 2001)

In May 2000, 6 people died and over 2000 others became ill in the small town of Walkerton, Canada, as a result of consuming groundwater contaminated by *E. coli* O157:H7 and *Campylobacter*. Walkerton was almost entirely dependent upon groundwater for domestic supply obtained from production wells varying from 15 to >70 m deep. These wells intercept limestones and dolomites. There was a history of detection of *E. coli* in a number of the wells, but this problem was regarded as being something that could be controlled by routine chlorination of the source water.

The first public health problems (diarrhoea, vomiting) were noted in the days following a particularly violent storm during which 100 mm of rain fell. Initially the problem was suspected as food poisoning and it was not until over a week later that well water was identified as the source and a boil-water alert was issued. *E. coli* O157:H7 and *Campylobacter* spp. were subsequently identified as the cause of the deaths. The source of the organisms was traced to a cattle farm close to one of the shallower production wells. It was initially suggested that storm water runoff conveyed the contaminants to the production wells. However, subsequent evidence indicates that the well contamination causing the outbreak had occurred before the storm event. The precise travel paths enabling the contaminants to enter the production wells (even the deepest of which was contaminated) remain a matter of speculation, although fissure flow in the limestone aquifer is an obvious candidate. It may be significant that there were abandoned production wells in the area that had not been properly sealed. Improperly abandoned boreholes may facilitate the rapid vertical mixing of contaminants entering at or near the surface.

It is clear that the public health impact was a result of the combination of inadequate protection and monitoring of the groundwater resource with a failure in treatment. This is not uncommon, Bramham in the United Kingdom being a further example (Lerner and Barrett, 1996). Clearly reliance on treatment (as a 'single barrier' approach) without other measures such as adequate groundwater protection is a higher risk approach than that of the 'multi barrier'.

The types and numbers of the various pathogens will vary temporally and spatially depending upon the incidence of disease in the community, the known seasonality of human infections, and the characteristics of the aquifer systems. Furthermore, the microbial illnesses and the severity of the disease vary markedly with the organism.

NOTE ▶ *Although some enteric pathogens may circulate within a population all year round, many have a clearly defined seasonal distribution. For example, in temperate zones, the transmission of rotavirus takes place almost exclusively during cold weather.*

Box 3.2. Virus borne outbreaks of gastroenteritis in Wyoming, USA

In February 2001, episodes of acute gastroenteritis were reported to the Wyoming Department of Health among snowmobilers (Anderson *et al.*, 2003). The outbreak was believed to have been caused by noroviruses that could be identified from 8 of 13 stool samples as well as from groundwater samples of one well. A second outbreak of acute gastroenteritis occurred in Wyoming during October 2001 among persons who dined at a tourist saloon (Parshionikar *et al.*, 2003). A norovirus strain (genogroup I, subtype 3) was found in stool samples from three ill persons as well as in the water from the saloon's only well.

Although land use planning, water resource management, water treatment and disinfection are used to control the transmission of waterborne pathogens, the safety of a water source is frequently verified by testing for the presence of microbial parameters. A variety of pathogens may be present, and the different methods that are required to isolate each pathogen prohibit the direct examination of water samples for pathogens on a routine basis. Moreover, pathogens may not be detected due to low levels, or the lack of an appropriate detection method, but they may still be present at a density that represents an unacceptable level of risk. Furthermore, currently unknown and therefore undetectable pathogens may be present. To overcome this difficulty, a separate group of microorganisms is used as an indicator for the potential presence of pathogens. The common descriptive term for this group of organisms is faecal indicator organisms. Gleeson and Gray (1997) have published a thorough review of the application of faecal indicator organisms in water quality monitoring that may be consulted for further information. This group comprises the following:

- total coliform bacteria
- thermotolerant coliform bacteria
- *E. coli*
- faecal streptococci
- bacteriophage

Table 3.1. Pathogenic microorganisms of concern in groundwater (adapted from Macler and Merkle, 2000)

Organism	Associated health effects
Viruses	
Coxsackievirus	Fever, pharyngitis, rash, respiratory disease, diarrhoea, haemorrhagic conjunctivitis, myocarditis, pericarditis, aseptic meningitis, encephalitis, reactive insulin-dependent diabetes, hand, foot and mouth disease
Echovirus	Respiratory disease, aseptic meningitis, rash, fever
Norovirus (formerly Norwalk virus)	Gastroenteritis
Hepatitis A	Fever, nausea, jaundice, liver failure
Hepatitis E	Fever, nausea, jaundice, death
Rotavirus A and C	Gastroenteritis
Enteric adenovirus	Respiratory disease, haemorrhagic conjunctivitis, gastroenteritis
Calicivirus	Gastroenteritis
Astrovirus	Gastroenteritis
Bacteria	
Escherichia coli	Gastroenteritis, Haemolytic Uraemic Syndrome (enterotoxic *E. coli*)
Salmonella spp.	Enterocolitis, endocarditis, meningitis, pericarditis, reactive arthritis, pneumonia
Shigella spp.	Gastroenteritis, dysentery, reactive arthritis
Campylobacter jejuni	Gastroenteritis, Guillain-Barré syndrome
Yersinia spp.	Diarrhoea, reactive arthritis
Legionella spp.	Legionnaire's disease, Pontiac fever
Vibrio cholerae	Cholera
Protozoa	
Cryptosporidium parvum	Diarrhoea
Giardia lamblia	Chronic diarrhoea

NOTE ▶ *The coliform group of bacteria consists of several genera belonging to the family Enterobacteriaceae (Gleeson and Gray, 1997). The relatively limited number of biochemical and physiological attributes used to define the group (including growth at 37°C) means that its members include a heterogeneous mix of bacteria. Although the total coliform group will include bacteria of faecal origin it also includes species that are found in unpolluted environments. Thermotolerant coliforms are those bacteria from within the total coliform group that grow at 44°C. E. coli is a thermotolerant coliform.*

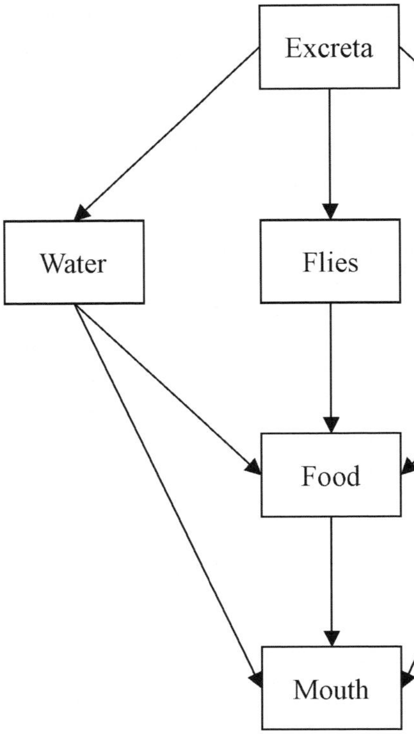

Figure 3.1. Principal elements of the faecal-oral route of disease transmission (Howard *et al.,* 2002)

An ideal microbial indicator of faecal pollution is easily detected, always present in faecal waste, and is more durable in the environment than most enteric pathogens. It should comprise a large percentage of the organisms in faecal waste, exceed the numbers of most enteric pathogens and be roughly proportional to the degree of pollution. Lastly, because indicator organisms should be absent unless faecal contamination is present, they should ideally not be present in drinking-water that is microbially safe for consumption. For many applications, 100 ml of water is used as the standard volume for analysis.

The durability of faecal indicator organisms in the subsurface is an important issue that has a direct bearing on the interpretation of water quality data. Ideally the faecal indicator organism should be removed from the environment at a slower rate than the most durable pathogen so that its potential for dispersal is greater. In practice, this situation will seldom arise. For the purpose of acting as surrogates for pathogens in groundwater tracer studies, other advantageous properties are the ease of preparing high numbers of the indicator and the ease of enumerating the indicator. These properties allow indicators to be used as a tool to identify and quantify removal processes in laboratory and field studies. The most important removal processes are inactivation or die-off, adsorption to the surface of the grains of the porous medium and physical filtration or straining when pore throats are too small to let a microorganism pass. Thus robust indicators are inactivated, adsorbed and strained not more than the pathogens they represent. The removal processes are described in more detail later in this chapter.

It is acknowledged that faecal indicator bacteria do not give absolute resolution to the presence/absence of pathogenic protozoa, bacteria and viruses. These diverse groups of microorganisms have highly variable transport and attenuation characteristics that cannot be wholly represented by the small group of indicator bacteria. However, the density of faecal indicator bacteria does provide a measure of probability of the presence of pathogens.

Faecal indicator bacteria are of limited use for predicting the presence or absence of viruses as viruses often survive significantly longer in groundwater systems. It is not practical to sample and analyse for all pathogenic viruses. As a result bacteriophage – viruses that infect bacteria – from faecal sources are often used to indicate the likely presence or absence of viruses in groundwater. Several groups of bacteriophage have been evaluated as indicator organisms although the focus of attention has been upon the coliphage group of viruses because they can be detected and quantified using relatively simple analytical methods.

Negatively charged bacteriophages, like MS2 and PRD1, have been found to be useful model viruses and have been used in many field and laboratory studies on subsurface transport of viruses (Schijven and Hassanizadeh, 2000). They attach poorly to most soils, and at low temperatures they inactivate at a low rate. PRD1 is relatively insensitive to higher temperatures or extreme pH. These bacteriophages have the same size and shape as many waterborne pathogenic viruses. They are easy to prepare in high numbers and are easy to detect. MS2 is an F-specific RNA bacteriophage. This group of bacteriophages may be naturally present in faecal contaminated water in numbers 10^2 to 10^4 times higher than enteroviruses (Havelaar *et al.*, 1993). However, F-specific bacteriophages are less stable (more temperature sensitive) than somatic coliphages. The latter are usually found in higher number in faecally contaminated water and are therefore also useful indicators.

3.2 DISTRIBUTION OF PATHOGENS AND FAECAL INDICATORS IN GROUNDWATER

Sources of faecal contamination in groundwater are discussed in Section II and potentially include leakage from on-site sanitation systems or sewers (Box 3.3), animal manures (Box 3.1), wastewater or sewage sludge applied in agriculture. The sources can be classified according to their origin. A point source has an identifiable source, such as a leaking septic tank, which may result in a well-defined plume. More difficult to control, and posing a greater risk to groundwater quality, are non-point sources. Non-point sources are larger in scale and produce relatively diffuse pollution originating from either widespread application of contaminated material or many smaller sources. The aggregate of point sources in a leaking sewerage system may, overall, represent a non-point source of contamination to groundwater (usually described as multi-point source of pollution).

Traditionally, hydrogeologists, and many public health scientists, have regarded groundwater as a relatively microbially safe source of drinking-water. Unlike surface waters, which are vulnerable to direct contamination from many sources, groundwater is often shielded from the immediate influence of contamination by the overlying soil and unsaturated zones as described in Chapter 2. In these zones, pathogenic microorganisms have been assumed to be attenuated by the prevailing physical, chemical and biological conditions in the environment. The risk of pathogens being transported into groundwater and producing a threat to public health was, therefore, considered to be low. Consequently, many groundwater sources are used for public supply with a minimum level of treatment, normally chlorination, or with no treatment at all.

The underlying rationale for restricted dispersion in the subsurface has some merit, but it is now known that microbial contamination of groundwater is more widespread than previously believed. Indeed, as was found with chemical contaminant studies, particularly in the 1980s, the more that it is looked for, the more it can be found. Tables 3.2 and 3.3 list some examples of studies that have demonstrated the occurrence of faecal indicators and enteroviruses in groundwater. Although there may be a bias in some of the studies, created by the deliberate selection of vulnerable sites, the data show that a significant percentage, up to 70 per cent in some regions, of groundwater sources contain one or more of the microbial indicators of faecal contamination.

Table 3.2. Occurrence of microbial faecal indicators in groundwater

Organism	Proportion of receptors positive (%)	Study location	Reference
Coliform bacteria	10	USA: 445 public supply wells	Abbaszadegan *et al.*, 1998
Coliphage	21	USA: 444 public supply wells	
Enterococci	9	USA: 355 public supply wells	
Somatic coliphage	50	USA: 30 public water supply wells judged to be vulnerable to faecal contamination	Lieberman *et al.*, 1994
E. coli	50		
Enterococci	70		
Coliform bacteria	40	USA, Montana	Bauder *et al.*, 1991
E. coli	16-24	Canada, Province of Ontario: farmstead domestic wells	Goss *et al.*, 1998
Faecal streptococci	12-24		
E. coli	60	Republic of Moldova, Balatina and Carpini	Melian *et al.*, 1999
Faecal streptococci	50		
Thermotolerant coliforms and faecal streptococci	10-40	Finland: rural wells	Korhonen *et al.*, 1996

Table 3.3 shows the occurrence of pathogenic enteroviruses found in public water supply wells in the USA as defined by the Safe Drinking Water Act (US Government, 1996). Hydrogeological data from these studies were available on aquifer type as part of the study designs and were reviewed for accuracy by hydrogeologists. Limestone (karst), fractured bedrock (igneous and metamorphic rocks) and gravel (formed in high energy depositional environments with little or no sand or other fine grained materials) aquifers are defined as sensitive aquifers under the proposed Ground Water Rule (US EPA,

2000). If a public water supply well draws water from a sensitive aquifer, then the State must find the well sensitive to faecal contamination unless a hydrogeological barrier protects it. A hydrogeological barrier is defined as the physical, biological and chemical factors, singly or in combination, that protect a well from pathogenic organisms. In this proposal, a confining layer is one example of a hydrogeological barrier. If a hydrogeological barrier is present, then the State can nullify the determination that a system is located in a sensitive aquifer. If no suitable hydrogeological barrier exists, then the proposed Ground Water Rule requires the system to conduct faecal indicator source water monitoring.

Table 3.3. Occurrence of enteroviruses in public water supply wells in the USA

Sensitive wells [a]		**Non-sensitive wells**		**Samples per well**	**Average filtered volume (L)**	**Virus type (N_w)** [b]	**Reference**
N^+/N_w	*Positive*	*N^+/N_w*	*Positive*				
3/49	6%	2/10	20%	1	200-1000	Coxsackievirus B5 (1) Echovirus 13 (1) Echovirus 20 (1) Poliovirus 3 (1) [c] Reovirus (3)	Lindsey *et al.*, 2002
0/91	0%	0/0	-	1	1500		Banks and Battigelli, 2002
0/0	-	0[d]/27	0%	1	1500	Rotavirus (1) [d]	Banks *et al.*, 2001
0/92	0%	0/12	0%	1	200-300		Femmer *et al.*, 2000
1/96	1%	0/9	0%	2	200-300	Poliovirus 1 (1) [c]	Davis and Witt, 2000
1/96	1%	0/9	0%	2	200-300	Poliovirus 1 (1) [c]	Davis and Witt, 2000
6/12	50%	1/18	6%	12	6000	Coxsackievirus A7 (1) Coxsackievirus B1 (4) Coxsackievirus B3 (1) Coxsackievirus B4 (5) Coxsackievirus B5 (1) Echovirus 11 (2) Echovirus 15 (4) Echovirus 18 (2) Echovirus 21 (3) Echovirus 24 (2)	Lieberman *et al.*, 2002
0/31	0%	0/79	0%	4	Up to 1500		Doherty *et al.*, 1998

N^+ = number of positive enteroviruses by buffalo-green-monkey tissue culture; N_w = number of wells; a = sensitive wells are wells located in igneous or metamorphic rock aquifers or limestone aquifers or gravel aquifers with very low sand/fine grained content; b = serotyping results to confirm cell culture enteric virus positive samples (serological identification of all these samples was determined by Dan Dahling of the US EPA); c = possible laboratory contamination; d = positive by Rototest assay of RD cell lysate.

Lindsey *et al.* (2002) found enteric viruses in three sensitive aquifers and two that were not sensitive. In Tables 3.2 and 3.3, the data of Lieberman *et al.* (2002) are most meaningful because 12 samples were taken at monthly intervals from each well from 30 sites in the continental USA, the Virgin Islands and Puerto Rico. In addition, the sample volumes used in these studies were significantly larger than routine volumes: average

6000 litres, and maximum even up to 15 000 litres. This significantly increases the probability of detecting virus contamination of a well. Lieberman *et al.* (2002) found enteric viruses primarily in sensitive aquifers (i.e. karst, fractured bedrock and coarse gravel) but only once in a very shallow sand well (5 m deep; see also Dahling, 2002). One reason that as many as 7 of the 30 wells (24 per cent) were contaminated with enterovirus is because almost all wells were selected for enterovirus sampling only if they had a history of total coliform and somatic coliphage occurrence (see Table 3.2). Echovirus 11 was possibly coming from upgradient septic tanks. Viruses found in the karst wells must have travelled a long distance because no potential sources of contamination were near. One of the wells was located in a populated gravel flood plain with a nearby trailer park with septic tanks about 30 m away. In this well water, coxsackieviruses A7, B1 and B4, echoviruses 15, 18, 21 and 24 and reovirus were detected. Coxsackievirus B4 was also found in the water from a well in fractured basalt with septic tanks nearby.

Borchardt *et al.* (2003) studied the incidence of viruses in Wisconsin, USA private household wells located near seepage land application sites or in rural subdivisions served by septic systems. Fifty wells in seven hydrogeological areas were sampled four times over a year, once each season. Of the 50 wells, 4 (8 per cent) were found to be positive for viruses using analytical methods that detect the presence of the viral nucleic acid (reverse transcription PCR). Of these, three wells were positive for hepatitis A virus and the fourth well was positive for rotavirus and norovirus in one sample and for enterovirus in another sample. Culturable enteroviruses were not detected in any of the wells. Virus occurrence could not be associated statistically with faecal indicators (i.e. total coliforms, *E. coli*, faecal enterococci, F-specific RNA bacteriophages).

Clearly, the common perception that groundwater is *per se* a microbially safe source of drinking-water is inaccurate. Whilst usually the microbial contamination of groundwater is likely to be orders of magnitude lower than that of surface waters, it is now apparent that a significant percentage of groundwater sources are contaminated by microorganisms derived from faeces. As shown above, the known presence of infectious pathogenic viruses in some wells, for example, represents an unequivocal message that viruses are mobile in the subsurface, long-lived and capable of causing waterborne illness.

In the same way that surface waters may show rapid changes in the concentration of pathogens, distribution and concentration of pathogens and faecal indicator bacteria in groundwater sources is not static but also demonstrates fluctuation. Temporal and spatial variations are frequently observed that may be linked to seasonal changes in land use and changing weather patterns. For example, fluctuations in the levels of thermotolerant coliforms have been observed in proximity to wastewater irrigation sites in Mexico and in the United Kingdom. The distribution of enteric bacteria and viruses in urban groundwater has been observed to vary both horizontally and vertically (Box 3.3).

Box 3.3. Depth and extent of microbial contamination of groundwater in British urban sandstone aquifers (based on Powell *et al.*, 2000; 2001a; 2001b)

There are few published data on the microbial, and particularly viral, quality of United Kingdom groundwaters. This should not be taken as an indication of a general absence of contamination, rather as a lack of detailed monitoring studies. A number of workers have identified sewage contamination of the Triassic Sandstone aquifers in urban areas of the United Kingdom, derived from leaking sewers.

Powell and co-workers set out to determine the extent and penetration of microbial contaminants in the Triassic Sandstone aquifer underlying Birmingham and Nottingham in the United Kingdom. Five multilevel groundwater monitoring devices were installed into the aquifer, providing a total of some 50 depth-specific sampling points. Viral monitoring was undertaken using a glass wool trap for the concentration of enteric viruses from large volume groundwater samples.

The field data lead to four key findings:

- Sewer leakage-derived microbial contaminants are able to penetrate sandstone aquifers to significant depths (>90 m).
- Human enteric viruses, including pathogenic species are widespread in the aquifer.
- The species of sewage-derived human enteric viruses in groundwater are found to vary temporally, and in parallel with their predicted prevalence in the population. The dominant types found in March and June 2001 were Noroviruses and Coxsackievirus B4 respectively.
- Particular horizons at depth within the sandstone aquifer were found to be rapidly susceptible to microbial contamination (i.e. contaminant distribution is vertically and temporally heterogeneous).
- The Triassic Sandstone aquifer (and, by implication, other similar sandstone aquifers around the world), the second most important in the United Kingdom, is far more vulnerable to microbal contamination than previously assumed. This has public health implications where groundwater is consumed without adequate treatment.

Frequently, the microbial quality of shallow groundwater sources, including springs, will deteriorate after heavy rainfall as surface contamination is washed into the source directly and organisms in the unsaturated zone are mobilized by water percolating through the soil matrix (Box 3.1). Similarly, the relative proportions of indicator bacteria and pathogens, and of bacteria and viruses, will fluctuate such that the dominant species isolated at one time may be absent on subsequent sampling occasions (Box 3.4).

Box 3.4. Temporal fluctuations in microbial contamination of groundwater in Kampala, Uganda (based on Barrett *et al.*, 2000)

Sampling of 15 protected springs was undertaken in areas of high-density and low-density population (peri-urban) in Kampala, Uganda on three successive days (28-30 September 1999). The wet season was in progress, and rainfall occurred on every day of sampling. On the night of 28-29 September, a significantly heavier rainfall event took place. Analysis of the samples for faecal indicators, including thermotolerant coliforms and faecal streptococci, was undertaken. As shown by Figure 3.2, there was a clear pulse reaction of contamination in spring water within 12 hours of the major rainfall event.

A variety of metasedimentary rocks (e.g. quartzites and phyllites) underlie Kampala. Differential weathering has resulted in a pronounced topography with thin weathered mantles of limited extent containing shallow (2-20 m) groundwater flow systems that discharge to valley springs. These are fed by a combination of baseflow and seasonally derived interflow. Many of these springs are used, untreated, by lower income communities with limited access to higher service levels of piped water supply. Clearly the monitoring of protected springs for microbial contaminants in these localized groundwater flow systems must be interpreted in the context of rainfall events.

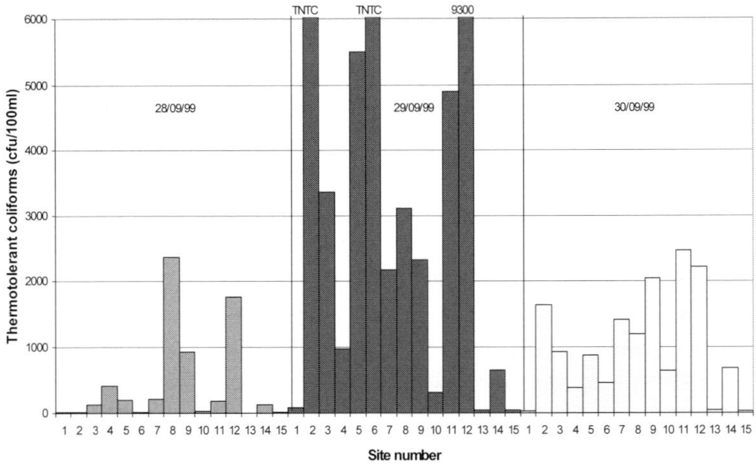

Figure 3.2. Effect of heavy rainfall on the microbial quality of spring water in Kampala, Uganda

3.3 TRANSPORT AND ATTENUATION OF MICROORGANISMS IN THE UNDERGROUND

Some, perhaps many, instances of groundwater receptor contamination will occur by rapid transport pathways accidentally introduced by human intervention and connecting the contamination source to the groundwater abstraction point. Such pathways could include inadequate sanitary completion of springs, wells and boreholes, the presence of a forgotten conduit connecting the source of contamination to the groundwater abstraction point, or voids and fractures in soils that allow direct ingress of contaminated materials. The implementation of management actions to reduce faecal contamination close to the abstraction point or the rehabilitation or improvement of the well or spring is usually sufficient to control access of pathogens to the water source (Section IV).

Rapid transport pathways cannot, however, explain all groundwater source contamination events and it is now widely accepted that the transport of microbial pathogens within groundwater systems is a significant mechanism for waterborne disease transmission. The remainder of this chapter deals with the factors that control the transport and attenuation of pathogens into and through groundwater.

3.3.1 Transport and attenuation of pathogens in the unsaturated zone

Hydrogeological processes in the unsaturated zone are complex and the behaviour of microorganisms is often difficult to predict. Nevertheless, the unsaturated zone can play an important role in retarding (and in some cases eliminating) pathogens and so must be considered when assessing aquifer vulnerability, as described in Chapter 8. Attenuation of pathogens is generally most effective in the uppermost soil layers where biological activity is greatest. The presence of protozoa and other predatory organisms, the rapid changes in soil moisture and temperature, competition from the established microbial community, and the effect of sunlight at the surface combine to reduce the level of pathogens within this zone. The effect of individual environmental factors will be discussed in a later section of this chapter.

The transport of pathogens from the surface into the subsurface requires the presence of moisture. Even during relatively dry periods, soil particles retain sufficient moisture over their surface for pathogens to migrate downwards into the subsurface. Under these conditions the main driving forces will be sedimentation, diffusion and bacterial motility. Within the thin film of moisture the organisms are brought into close contact with the surface of the particle, thus increasing the opportunity for adsorption to the particle surface and further retarding movement. If soil moisture decreases, the strength of the association between the organism and the particle surface will increase to a point where the organism is bound irreversibly to the surface. Passive binding to particle surfaces has been observed with some strains of virus, and it is believed that the strength of the bond can immobilize the virus and contain it at the point of interaction. It is possible that similar interactions occur with other groups of pathogens, but the processes are less well defined. Bacteria, for instance, synthesize extracellular substances that can enhance their attachment to surfaces and promote binding, suggesting that the process involves both passive and active processes. Whether alone, or in combination with the apparently protective effect of adsorption onto surfaces, soil moisture influences the persistence of microorganisms, in particular viruses. In laboratory experiments a soil moisture content of between 10 and 15 per cent was shown to be optimal for the survival of several strains of enteric virus (e.g. Bagdasaryan, 1964; Hurst *et al.,* 1980a; 1980b).

By contrast, an increase in the moisture content of the unsaturated zone may increase the vulnerability of the aquifer to pathogen contamination in two ways: by providing rapid transport pathways and by mobilizing adsorbed organisms. During periods of high recharge, for example during prolonged heavy rain, the intergranular spaces in the unsaturated zone become waterlogged and provide a hydraulic pathway for the rapid transport of pathogens. Where these intergranular spaces expand into fissures the downward migration of pathogens can be extremely rapid. For example, particles ranging in diameter from 0.1-6.0 μm have been found to move through 20 m of unsaturated chalk in less than 3 days by passage through horizontal and vertical fissures (Lawrence *et al.*, 1996). Moreover, the rapid movement of pathogens through fissures limits the potential for attenuation by adsorption to surfaces in the soil matrix.

In the interval between recharge events, the chemistry of the water in the unsaturated zone will change as it equilibrates with the soil matrix. In some soil types, these changes may favour the adsorption of microorganisms to surfaces in the soil matrix. A lowering in the ionic strength or salt content of the surrounding medium, which can occur during a rainfall event, may be sufficient to cause desorption of the organism allowing further migration into the soil. This phenomenon has been observed in laboratory experiments and there is evidence to suggest that it can occur in the field. Furthermore, some workers have noted that the virus particles that have desorbed from the soil surface have a reduced capacity to resorb when the environmental conditions become favourable. The implication of this observation is that virus particles that have been mobilized in the subsurface are unaffected by one of the principal methods of attenuation and are likely, therefore, to be dispersed over a much wider area than would be anticipated.

The size variability of microorganisms (Table 3.4) can, to an extent, control their mobility in the subsurface. Soil and rock pore sizes are also variable and the two ranges are known to overlap (Figure 3.3). Thus in soils that are composed of fine grain particles, typically clayey-silts, the pore space is sufficiently small (<4 μm) to physically prevent the passage of bacterial and protozoal pathogens into the subsurface. This removal process is called physical filtration or straining. Straining has been identified as the principal mechanism for controlling the migration of *Giardia* and *Cryptosporidium* species (*Cryptosporidium* oocysts: 4-6 μm; *Giardia* cysts: 7-14 μm) through these soil types; indeed, experience has shown that up to 99 per cent of *Cryptosporidium* oocysts are retained in the upper layers of the soil. However, the isolation of *Cryptosporidium* and *Giardia* from a small but significant number of groundwater sources in the USA (Hancock *et al.*, 1998) and the United Kingdom indicates that the protective effect of the soil layer is frequently evaded, probably by migration through preferential pathways or bypassing; for example, from sewers that are often located below the soil zone.

Several studies demonstrate a considerable degree of variability between the inactivation or die-off rates of different groups of pathogens, and between inactivation rates of the same organism in different environments. However, as a general rule, enteric viruses persist longer in soils than bacteria. Among the enteric viruses, hepatitis A virus appears to be the most resistant to inactivation in soil (Sobsey *et al.*, 1986) and, in laboratory experiments, shows a lower capacity for adsorption to particle surfaces. The oocysts of *Cryptosporidium* are highly resistant to environmental stress and it has been estimated that they could be detected after 12 months in soil.

Table 3.4. Approximate sizes of selected microorganisms

Class	Microorganism	Size
Virus	Bacteriophage	0.02-0.2 μm diameter
	Poliovirus	0.03 μm diameter
Bacteria	Bacterial spores (*Bacillus*, clostridia)	1 μm
	E. coli	0.5 μm x 1.0 μm x 2.0 μm
	Salmonella typhi	0.6 μm x 0.7 μm x 2.5 μm
	Shigella spp.	0.4 μm x 0.6 μm x 2.5 μm
Protozoa	*Cryptosporidium* oocysts	4.0-6.0 μm diameter
	Giardia	7.0-14.0 μm diameter
	Enteroamoeba histolitica	20-25 μm diameter

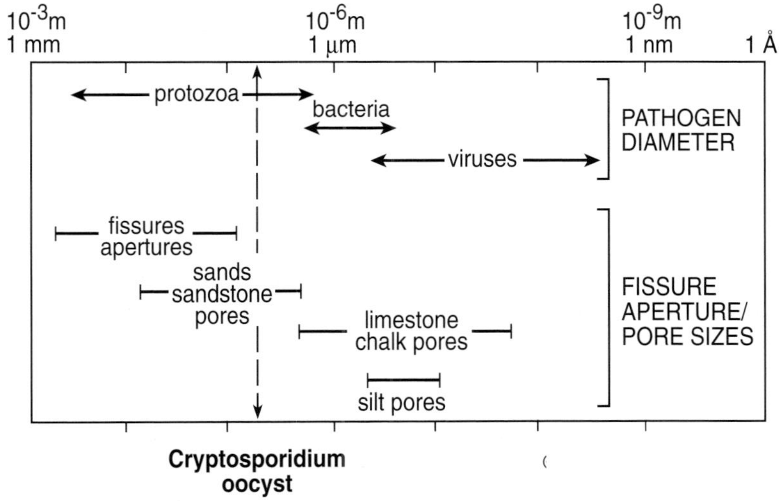

Figure 3.3. Pathogen diameters compared to aquifer matrix diameters (ARGOSS, 2001; British Geological Survey ©NERC)

Some of the factors that contribute to reduced inactivation rates in the unsaturated zone are known (lower temperatures, increased moisture, pH, reduced exposure to sunlight, organic matter and the nature of the pathogen) but the relative contribution of each factor at any field site is difficult to predict, and may be site specific. In these circumstances, the tendency of hydrogeologists and public health microbiologists is to construct general risk assessment models based upon laboratory and field experience. At their most sophisticated, the models comprise computer simulations of pathogen transport (for example Yates and Yates, 1988), but also used are simple tables and diagrams linking risk to the main observable features of the environment. One such model has been constructed by Romero (1972); a second has been developed to accompany guidelines for assessing the risk to groundwater from on-site sanitation (ARGOSS, 2001). However, the problem with such an approach is that only a qualitative indication for risk levels is given without any definition of what is meant by a high or low risk level. The actual useful information from this approach is a rough indication of a relative probability for pathogenic microorganisms to reach groundwater. Table 3.5 shows the different classes of lithology that were defined in decreasing order of ability to limit transport of microorganisms.

Table 3.5. Limitation of the transport of microorganisms by the lithology of the unsaturated zone in decreasing order (top to bottom) (based on ARGOSS, 2001)

Lithology of the unsaturated zone
Fine sand, silt and clay
Weathered basement (soft not consolidated)
Medium clean sand
Coarse sand and gravels
Consolidated rock

Shallow groundwater (<5 m) is assumed to be at the highest probability of contamination irrespective of the lithology of the unsaturated zone. As the depth to the water table increases so the capacity of the unsaturated zone to attenuate microorganisms will also increase, although this will depend upon the composition and structure of the unsaturated zone. For example, fine silts and clays will strongly adsorb bacteria and viruses and also effectively filter out the larger pathogenic microorganisms. Thus the probability of reaching groundwater at greater than five m depth is very low. By contrast, fracture flow through consolidated rock creates a relatively high probability of reaching groundwater even at depths of over 10 m.

In summary, maximizing the residence times in the unsaturated zone has been proposed as the key mechanism for eliminating bacteria and viruses (Lewis *et al.*, 1982) and, in general, this principle is robust. However, there are exceptions, for example:

- The variability in the nature and thickness of the unsaturated zone overlying aquifers means that the residence times may not always be adequate to attenuate all pathogens. In particular, during periods of high recharge, an aquifer may be vulnerable to contamination by pathogens that are transported rapidly through the waterlogged intergranular spaces in the unsaturated zone.
- Where the flow is intergranular within the unsaturated zone there is greater potential for contact with the soil/rock particles and hence greater potential for retention, both sorptive and filtering. However, if excessive loading takes place the filtering effect may lead to a blocking of the pores. The resulting reduction in hydraulic conductivity may reduce the effectiveness of the unsaturated zone to retard contaminants if the clogging forces recharge water into vertical fissures where rapid downward movement can occur.
- The structure of the unsaturated zone is seldom uniform and fissures may exist permanently or develop in any environment when the unsaturated zone dries out. The presence of fissures will always increase the vulnerability of the groundwater to contamination from the surface, and it should be considered that although the soil conditions may facilitate the adsorption and attenuation of pathogens, the existence of bypass channels may offset the protective effect of the soil.

3.3.2 Transport and attenuation of pathogens in the saturated zone

On reaching the saturated zone, microbial contaminants are subject to the same processes of attenuation that are described in Section 3.3.1 but under conditions of natural or artificially induced flow. Thus die-off, adsorption, filtration, predation and dilution all contribute to the attenuation of pathogens in the saturated zone. Due to the heterogeneous nature of aquifer material there may be large variations in hydraulic conductivity and this can significantly influence the movement of microorganisms in the aquifer. Microorganisms are transported in groundwater by advection, dispersion and diffusion, which are defined in Chapter 4. The result is a migration and spreading of the contaminant concentration in space and time. This may result in contamination of increasingly large aquifer volumes as the pollutant moves downgradient. Although the transport of pathogens in some aquifer types can be both rapid and extensive, there are several factors that may attenuate pathogens in groundwater (Table 3.6).

Table 3.6. Factors affecting transport and attenuation of microorganisms in groundwater (adapted from West *et al.*, 1998)

Characteristics of the microorganism	Aquifer/soil (environment) properties
Size	Groundwater flow velocity
Shape	Dispersion
Density	Pore size (intergranular or fracture)
Inactivation rate (die-off)	Kinematic/effective porosity
(Ir)reversible adsorption	Organic carbon content (solid)
Physical filtration	Temperature
	Chemical properties of groundwater (pH, etc.)
	Mineral composition of aquifer/soil material
	Predatory microflora (bacteria, fungi, algae, etc.)
	Moisture content
	Pressure

From the perspective of groundwater management and the estimation of pathogens at the point of abstraction (receptor), highly fractured and karstic aquifers represent a particular problem. As discussed in Chapter 2, groundwater flow through fractured systems may be very rapid, and the potential for microorganisms to be attenuated by interaction with the aquifer matrix is much reduced, although not entirely absent. Consequently, the inactivation rate of the pathogen and the groundwater flow rate will primarily control dispersal in these aquifer systems. Three referenced studies will help to illustrate the potential for rapid pathogen transport in highly fractured aquifers:

- The migration of bacteriophage in a chalk aquifer in the south of England was investigated by Skilton and Wheeler (1988, 1989). They injected three strains of bacteriophage into piezometers that intersected the water table and then collected samples at different sites downgradient to determine the extent of movement. Very high velocities were observed at one site due to the fact that the majority of the water flow is through fissures, fractures, solution openings and cavities. All three phage types were detected 355 m from the injection site approximately 5 hours after introduction. It is noteworthy that viable phage were still being recovered more than 150 days after they were injected into the aquifer.

- Mahler *et al.* (2000) cite the work of Batsch and colleagues who reported the detection of injected bacteria 14 km from the injection site, having been transported at a velocity of about 250 m/h. Mahler's own studies (Mahler *et al.*, 2000) in a karstic aquifer located in the South of France have confirmed the very rapid transport of faecal indicator bacteria in these systems.
- Lee (1993) investigated the contamination of a water supply well by *Giardia* spp. and *Cryptosporidium* spp. in a karstic environment. The karstic nature of the study area provided the potential for rapid infiltration of surface waters to the water table and subsequent transport of the organisms to the well through fractures and fissures. This connection was confirmed by the study. An analysis of particle size revealed that the full range of particle sizes found in the surface waters was not present in the well; there was a cut-off at both high and low ranges. The author concluded that there had been adsorption of smaller particles and straining of larger ones. The size range of the particles that were transported through the system included *Giardia* and *Cryptosporidium*.

These observations have significant implications for the public health risk associated with water abstracted from highly fractured and karstic aquifers. Not only can viral, bacterial and protozoan pathogens be transported rapidly over great distances, but also the groundwater flow pattern between the source and receptor can be very difficult to predict due to the many interconnected fractures in the aquifer. It is possible that well designed tracer studies and groundwater flow models can help to define the potential limits of pathogen dispersion in a highly fractured aquifer; however, with the current uncertainties surrounding pathogen attenuation in groundwater it is prudent to assume that where these aquifer types are exposed to sources of pathogens they are at high risk of contamination over a wide area.

In other aquifer types, the radius of migration from the source of contamination is normally restricted to several tens of meters, or a few hundred meters, depending upon the type of aquifer system and the properties of the organism (Table 3.6). Where the groundwater flow rate is low through unconsolidated sediments the dispersal of pathogens will naturally be limited and, in addition, this type of system offers greater opportunity for the pathogen to interact with the aquifer matrix. Adsorption and physical filtration may then be the major factors controlling pathogen transport.

Despite acting to limit the dispersal of pathogens in the aquifer, interaction with the aquifer matrix may also enhance the survival of the pathogens in the environment. In several cases, adsorption to surfaces in the aquifer (sediment particles and colloids, as well as the aquifer matrix) has been shown to reduce the inactivation rate of both viruses and bacteria. Consequently, although the risk of contamination is contained close to the source, the persistence of pathogens within the zone of contamination may be increased beyond what is predicted from measurements of inactivation in the groundwater.

Schijven and Hassanizadeh (2000) showed that removal of virus in the subsurface often appears to be higher near the source than further away from the source, e.g. within the first 8 m of aquifer passage spores of *Clostridium bifermentans* R5 and bacteriophage MS2 were reduced by 5 $\log_{10}$ and 6 $\log_{10}$ respectively, while in the following 30 m MS2 concentrations were reduced only by 2 $\log_{10}$ and reduction of spore concentrations was negligible. This may be explained by favourable attachment sites that are present in the first meters of transport but rapidly decrease with travel distance or travel time in an exponential fashion, like sites formed by ferric oxyhydroxides. Obviously, predictions of virus removal over larger travel times or distances can be severely overestimated if they are based on removal data from column or field experiments where transport was studied over short times and distances.

Inactivation rates of bacteria and viruses in groundwater vary considerably, not only between the bacteria and virus groups, but also between different strains within each group and between the results of different investigations. Table 3.7 lists inactivation rate coefficients of pathogenic viruses, bacteriophages and bacteria in groundwater. These data are ordered according to microorganism and then according to temperature. Usually inactivation proceeds faster at higher temperatures, although this is highly dependent on the type of microorganism.

Often, inactivation of microorganisms can be described well as a first order rate process, especially under relatively mild conditions like temperatures from 5-20 °C and pH values from 6-8. Under more extreme conditions, the rate of inactivation of, for example viruses is often found to proceed initially at a higher rate followed by a lower rate as if two or more sub-populations exist that differ in stability (see e.g. Hurst *et al.*, 1992). The data given in Table 3.7 are based on the observation, or in some cases assumption, that inactivation or die-off proceeded as a first order process:

$$C_t = C_0 e^{-\mu t} \quad \text{or} \quad \ln\left(\frac{C_t}{C_0}\right) = -\mu t \quad \text{or} \quad \log_{10}\left(\frac{C_t}{C_0}\right) = -\frac{\mu}{2.3}t \qquad \text{(Eqn. 3.1)}$$

where C_t is the remaining concentration of microorganisms after time t, C_0 is the initial concentration at t=0 and μ is the inactivation rate coefficient (T^{-1}). As a means of interpretation, μ is often divided by a factor of 2.3 (equal to the natural logarithm of 10). The inactivation rate coefficient then reflects the number of $\log_{10}$ units per time unit; e.g. a virus decreases in number by 2 $\log_{10}$ (equal to a factor of 100) every 10 days.

From Table 3.7 it can be seen that at common groundwater temperatures of 10-12 °C inactivation rate coefficients for coxsackieviruses B, echoviruses 7, poliovirus 1, hepatitis A virus, but also of bacteriophages ϕX174, MS2 and PRD1 are in the range from 0.01 to 0.04 day^{-1}. This corresponds to a decline in number or concentration of 1 $\log_{10}$ (equal to a factor of 10) every 57 to 230 days. Some studies report an inactivation rate coefficient of zero (e.g. Nasser *et al.*, 1993). This is to be interpreted as no significant inactivation within the time-scale and accuracy of the experiment and is therefore not included in Table 3.7. Given the variation in inactivation rate between microorganisms, inactivation rate coefficients of the more stable microorganisms need to be considered for estimating adequate protection of groundwater wells or removal efficiency of passing microorganisms through soil as a means of treatment.

The data in Table 3.7 make clear that bacterial die-off is both highly dependent on the type of microorganism as well as on temperature. In many cases bacterial inactivation proceeds faster than that of viruses, implying therefore that viruses are more critical for groundwater protection than bacteria.

Table 3.7. Inactivation rate coefficients of pathogenic viruses, bacteriophages and bacteria in groundwater

Microorganism	Temp. (°C)	Other conditions	Inactivation rate coefficient μ (1/day)	Reference
Coxsackievirus A9	10	Sterile	0.019	Matthess *et al.* (1988)
	10		0.027	
	10	Deionized	0.031	
Coxsackievirus B1	10	Sterile	0.012	Matthess *et al.* (1988)
	10		0.019	
	10	Deionized	0.040	
Coxsackievirus B3	3-15		0.49	Keswick *et al.* (1982)
Coxsackievirus B4	5		0.079	Schijven *et al.* (2003)
Coxsackievirus B5	16	1.2 mg/l O2	0.12	Jansons *et al.* (1989a)
	19.4		0.12	Jansons *et al.* (1989b)
Echovirus 1	12		0.24	Yates *et al.* (1985)
	13		0.25	
	17		0.28	
	18		0.35	
	23		0.94	
Echovirus 6	22	0.2 mg/l O2	0.25	Jansons *et al.* (1989a)
Echovirus 7	10	Sterile	0.032	Matthess *et al.* (1988)
	10		0.019	
	10	Deionized	0.038	
Echovirus 11	16	2.3 mg/l O2	0.23	Jansons *et al.* (1989a)
Echovirus 24	16	1.6 mg/l O2	0.12	Jansons *et al.* (1989a)
Hepatitis A virus	10		0.10	Nasser *et al.* (1993)
	20		0.41	
	23	Filtered bottled mineral water	0.038	Biziagos *et al.* (1988)
	25	Sterile	0.082	Sobsey *et al.* (1986)
	25		0.33	
	30		0.054	Nasser *et al.* (1993)
Poliovirus 1	3-15		0.48	Keswick *et al.* (1982)
	4		0.016	Meschke (2001)
	5		0.16	Schijven *et al.* (2003)
	10	Sterile	0.010	Matthess *et al.* (1988)
	10		0.013	
	10	Deionized	0.032	
	10		0.025	Nasser and Oman (1999)
	12		0.18	Yates *et al.* (1985)
	13		0.20	
	14	70 weeks	0.16	Meschke (2001)
	16	5.4 mg/l O2	0.21	Jansons *et al.* (1989a)
	16	0.2 mg/l O2	0.069	
	17		0.19	Yates *et al.* (1985)
	18		0.43	
	20		0.038	Nasser *et al.* (1993)
	22	0.06 mg/l O2	0.16	Jansons *et al.* (1989a)
	22		0.10	Bitton *et al.* (1983)
	23		0.17	Blanc and Nasser (1996)
	23		1.2	Yates *et al.* (1985)
	23	Filtered bottled mineral water	0.044	Biziagos *et al.* (1988)
	24		0.046	Bitton *et al.* (1983)
	25	4 weeks	0.11	Meschke (2001)
	30		0.12	Nasser *et al.* (1993)
Rotavirus	20		0.36	Pancorbo *et al.* (1987)

Microorganism	Temp. (°C)	Other conditions	Inactivation rate coefficient μ (1/day)	Reference
Simian Rotavirus	3-15		0.83	Keswick *et al.*(1982)
	23		0.28	Gerba *et al.* (Undated)
ϕX174	5		0.012	Schijven *et al.* (2002b)
F-specific RNA bacteriophages	10		0.025	Nasser and Oman (1999)
	20		0.0077	Nasser *et al.* (1993)
	30		0.031	Nasser *et al.* (1993)
MS2	2-5		0.030	Schijven *et al.* (1999)
	4		0.037	Meschke (2001)
	4		0.063	Yates *et al.* (1985)
	5		0.064	Schijven *et al.* (1999)
	5		0.082	Schijven *et al.* (2002b)
	7		0.0058-0.10	Yahya *et al.* (1993)
	12	Oxic	0.10	Schijven *et al.* (2000)
	12	Anoxic	0.024	
	12		0.16	Yates *et al.* (1985)
	12		0.065	Yates (1992, unpublished observations)
	13		0.22	Yates *et al.* (1985)
	14	70 weeks	0.45	Meschke (2001)
	17		0.17	Yates *et al.* (1985)
	18		0.19	
	23		0.36	Blanc and Nasser (1996)
	23		0.58-1.3	Yahya *et al.* (1993)
	23		0.73	Yates *et al.* (1985)
	25	4 weeks	0.41	Meschke (2001)
PRD1	5		0.0094	Schijven *et al.* (1999)
	5		0.044	Schijven *et al.* (2002b)
	7		0.010-0.10	Yahya *et al.* (1993)
	12	Oxic	0.054	Schijven *et al.* (2000)
	23		0.035	Blanc and Nasser (1996)
	23		0.12-0.30	Yahya *et al.* (1993)
Bacillus subtilis spores	14	70 weeks	0.1382	Meschke *et al.* (2001)
Cl. perfingens spores	14	70 weeks	0.0714	Meschke *et al.* (2001)
E. coli	12		0.083	Schijven *et al.* (2000)
	14	70 weeks	0.51	Meschke *et al.* (2001)
	20		0.044	Nasser and Oman (1999)
	22		0.36	Bitton *et al.* (1983)
	3-15		0.74	Keswick *et al.* (1982)
	9-13		0.84	McFeters *et al.* (1974)
E. coli O157:H7	20		0.32	Rice (1992)
Faecal coliforms	12-20		0.83	Keswick *et al.* (1982)
Faecal streptococci	22		0.066	Bitton *et al.* (1983)
	3-15		0.53	Keswick *et al.* (1982)
Klebsiella spp.	?		0.031	Dowd and Pillai (1997)
Salmonella spp.	?		0.19	Dowd and Pillai (1997)
Salmonella typhimurium	22		0.30	Bitton *et al.* (1983)
Salmonella typhimurium	9-13		0.50	McFeters *et al.* (1974)
Shigella dysentariae	9-13		1.7	McFeters *et al.* (1974)
Shigella flexeri	9-13		1.4	McFeters *et al.* (1974)
Shigella sonnei	9-13		1.6	McFeters *et al.* (1974)
Vibrio cholerae	9-13		5.3	McFeters *et al.* (1974)

The low inactivation rate of *Klebsiella* spp. in groundwater provides an interesting contrast to the other genera of enteric bacteria. *Klebsiella* is an important member of the Enterobacteriaceae, which are commonly, though inaccurately, referred to as 'coliform bacteria'. Some species of *Klebsiella* are able to grow at 44 °C, which further defines them as thermotolerant (formerly faecal) coliforms, the group that contains *E. coli.* Although *Klebsiella* are found in the bowel and respiratory tract of humans and animals, they can be isolated also from environmental samples. The persistence of *Klebsiella* in the survival experiments may therefore reflect the ability of these organisms to exist in two very distinct environments.

One problem associated with the detection of pathogenic bacteria is that they may become dormant in the environment (Schijven *et al.,* 2002a). In this state, they are viable but non-culturable, which means that the organisms show metabolizing activity even though they cannot be grown on traditional media (Olson, 1993). As an important example, in water the pathogen *E. coli* O157:H7 appears to be able to enter a viable but non-culturable state (Wang and Doyle, 1998). The dormant state prolongs the pathogen survival, and if its pathogenicity is unaltered, this increases the likelihood of a host infection and illness

Some bacteria like Clostridium and Bacillus species can survive for extended periods of time by producing spores. Clostridium spores are especially robust in that their inactivation rate may be considered to be negligible. This property makes it difficult to interpret their removal, e.g. in sand filters or river bank filtration. In column studies by Schijven *et al.* (2003) it was shown that most of the spores of *Cl. perfringens* attach to the grains of sand, however followed by slow detachment, i.e. adsorption was apparently reversible for a large fraction of spores. It was shown that the removal is dependant on the number of spores that accumulate in the porous medium.

At the time of writing, no data are available on inactivation of oocysts of *Cryptosporidium* in groundwater. *Cryptosporidium* oocyst in vitro viability has been measured repeatedly using a surface water matrix and in vitro excystation and/or dye exclusion assay. For example, Heisz *et al.* (1997) report 0.12 day^{-1} (30 °C), and Medema *et al.* (1997) found inactivation rates of 0.023-0.056 day^{-1} (5-15 °C) in river water and no differences in inactivation rates between 5 °C and 15 °C. From the data of Robertson *et al.* (1992) and Chauret *et al.* (1995), inactivation rates of 0.0051 to 0.0062 day^{-1} can be calculated. Chauret *et al.* (1995) further found that inactivation rate was independent of water temperature up to 20 °C.

Based on their small size and longevity in the environment viruses have the highest potential to be transported to, and within, groundwater, and thus they have been the focus of the majority of studies. As a consequence, the discussion in the following chapter has a greater emphasis toward the factors that influence the transport and survival of viruses in the subsurface. From the data that is available, the same factors appear to affect the survival of bacteria; however, some bacteria are able to utilize specific physiological responses to resist environmental stress that are not available to viruses. These responses will be discussed separately at the end of the chapter.

3.3.3 Summary of major factors influencing pathogen transport and attenuation mechanisms in the underground

The mechanisms by which microbial contaminants may undergo transport and attenuation in the saturated and unsaturated zones have been described in Sections 3.3.1 and 3.3.2. There now follows a description of the factors that control the degree of the impact of these mechanisms. The potential for pathogens in manures, faeces and wastewater to contaminate the underlying groundwater is dependent on a number of factors including the physical characteristics of the site (e.g. soil texture), the hydraulic conditions (e.g. wastewater or manure application rates, wetting/drying cycles), the environmental conditions (e.g. rainfall, temperature) at the site and the characteristics of the specific pathogens present in the water. The factors that influence the transport and attenuation of pathogens in the subsurface have been the subject of a number of reviews summarized in Table 3.8 (Vaughn *et al.*, 1983; Yates *et al.*, 1985; Yates and Yates, 1988; Bitton and Harvey, 1992; Robertson and Edberg, 1997; Schijven and Hassanizadeh, 2000). Some of the major factors influencing pathogen transport and attenuation are described in more detail below.

Temperature

Temperature is probably the most important factor influencing the inactivation of bacteria and viruses in the environment. Laboratory studies have demonstrated a negative correlation between water temperature and the survival of coliform bacteria and enteric viruses, although the magnitude of the effect varies between different strains. Roughly, the inactivation rate of viruses may be one order in magnitude higher at 25 °C than at 5 °C (Table 3.7). Similarly this may be the case for enteric bacteria, such as *E. coli*. The influence of temperature on the migration of bacteria and viruses is currently unknown.

Microbial activity

There are several conflicting reports regarding the influence of indigenous populations of microorganisms on the survival of enteric bacteria and viruses, ranging from increasing the rate of inactivation through to having no effect to decreasing the rate of inactivation. Overall, however, the main finding of laboratory studies is that microbial activity in the soil and groundwater increases the inactivation rate of enteric bacteria and viruses. Evaluating the role of microbial activity normally involves a comparison of inactivation rates in sterile and non-sterile environments. Depending on the environmental conditions and the experimental design, virus inactivation is either unaffected, or accelerated in the presence of indigenous bacteria. Hurst *et al.* (1980a) showed that the inactivation rate of two strains of enterovirus was more rapid in non-sterile, aerobic environments than in sterile environments. By contrast, Matthess *et al.* (1988) found no significant difference between the inactivation rates of several viruses in sterile and non-sterile groundwater. Studies with thermotolerant coliforms as a whole, and *E. coli* in particular, have shown that the concentration of the test organism can increase rapidly in sterile environments, but remains static, or is reduced in non-sterile environments (Gerba and McLeod, 1976).

Table 3.8. Influence of major factors on the survival and migration of microorganisms in the subsurface (Vaughn *et al.*, 1983; Yates *et al.*, 1985; Yates and Yates, 1988; Bitton and Harvey, 1992; Robertson and Edberg, 1997; Schijven and Hassanizadeh, 2000)

Factor	**Viruses**		**Bacteria**	
	Influence on survival	Influence on migration	Influence on survival	Influence on migration
Temperature	Persistence is longer at low temperatures	nknown	Persistence is longer at low temperatures	Unknown
Microbial activity	Varies: some viruses are inactivated more readily in the presence of certain micro-organisms, the opposite may also be true, or there may be no effect	Unknown	The presence of indigenous micro-organisms appears to increase the inactivation rate of enteric bacteria; possible synergism with some protozoa may reduce inactivation rates	Unknown
Moisture content	Most viruses survive longer in moist soils and even longer under saturated conditions; unsaturated soil may inactivate viruses at the air-water interface	Virus migration usually increases under saturated flow conditions	Most bacteria survive longer in moist soils than in dry soils	Bacterial migration usually increases under saturated flow conditions
pH	Most enteric viruses are stable over pH range of 3 to 9; however, survival may be prolonged by near neutral pH values	Low pH typically increases virus sorption to soils; high pH causes desorption thereby facilitating greater migration	Most enteric bacteria will survive longer at near neutral pH	Low pH encourages adsorption to soils and the aquifer matrix; the tendency of bacteria to bind to surfaces may reduce detachment at high pH
Salt species and concen-tration	Certain cations may prolong survival depending upon the type of virus	Increasing ionic strength of the surrounding medium generally increases sorption	Unknown	Increasing ionic strength of the surrounding medium generally increases sorption
Association with soil	Association with soil generally increases survival, although attachment to certain mineral surfaces may cause inactivation	Viruses interacting with the soil particles are inhibited from migration through the soil matrix	Adsorption onto solid surfaces reduce inactivation rates; the concentration of bacteria on surfaces may be several orders of magnitude higher than the concentration in the aqueous phase	Bacteria interacting with the soil particles are inhibited from migration through the soil matrix
Soil properties	Probably related to the degree of virus sorption	Greater migration in coarse textured soils; soils with charged surfaces, such as clays, adsorb viruses	Probably related to the degree of bacterial adsorption	Greater migration in coarse textured soils; soils with charged surfaces, such as clays, adsorb bacteria

Factor	Viruses Influence on survival	 Influence on migration	Bacteria Influence on survival	 Influence on migration
Bacteria/ virus type	Different virus types vary in their susceptibility to inactivation by physical, chemical and biological factors	Sorption to soils is related to physico-chemical difference in secondary and tertiary capsid surface structure and amino acid sequence	Different species of bacteria vary in their susceptibility to inactivation by physical, chemical and biological factors	Some species of bacteria are more capable of binding to surfaces; variation may also occur between strains of the same bacterial species
Organic matter	Organic matter may prolong survival by competitively binding at air-water interfaces where inactivation can occur	Soluble organic matter competes with viruses for adsorption on soil particles which may result in increased virus migration	The presence of organic matter may act as a source of nutrients for bacteria, promoting growth and extended survival	Organic matter may condition solid surfaces and promote bacterial adsorption
Hydraulic conditions	Unknown	Virus migration generally increased at higher hydraulic loads and flow rates	Unknown	Bacterial migration generally increased at higher hydraulic loads and flow rates

Soil moisture content

Although some investigators have observed no difference between the inactivation rates of viruses in dried and saturated soils (Lefler and Kott, 1974) the majority of reports have shown that soil moisture content influences the survival of viruses in the subsurface. For example, Hurst *et al.* (1980a) found that moisture content affected the survival of poliovirus in loamy sand. The inactivation rate of poliovirus decreased as the moisture content increased from 5 to 15 per cent. However, further increases in soil moisture content increased the inactivation rate of the virus. It was noted that the inactivation rate peaked near the saturation moisture content of the soil (15-25 per cent), and was slowest at the lowest moisture contents (5-15 per cent). Soil moisture has been reported to influence the fate of bacterial contaminants (Robertson and Edberg, 1997), but the magnitude of the effect, and the value of any correlation has not been described.

pH

The effect of pH on the survival of pathogens in the environment has not been studied extensively and the impact can only be inferred from laboratory investigations of the physiological characteristics of the bacteria and the effect on the structural integrity of viruses. In general, every species of bacteria has a narrow pH range that is optimum for growth. Depending on the normal environment of the organism, the pH requirements can range from highly acidic to highly alkaline: for many human pathogenic bacteria the optimum pH is close to neutral. Despite having a preference for a narrow pH range, most species of bacteria can tolerate a short exposure to a much broader range of pH. Outside these limits, the organisms are rapidly killed. It is likely that pH affects the survival of viruses by altering the structure of the capsid proteins and viral nucleic acid.

Some authors have suggested that pH indirectly influences pathogen survival by controlling adsorption to soil particles and the aquifer matrix. It is the adsorption to surfaces that ultimately reduces the inactivation rate of the pathogens. Generally, bacteria and viruses have negative surface charges generated by the level of ionization of the carboxyl and amine groups that are a major component of surface proteins. As the pH of the medium changes, the ionization of the two groups will change, causing a shift in the net strength and polarity of the surface charge. At a specific pH, which is determined by the molecular structure of the protein, the net charge will be zero; this is termed the isoelectric point of the molecule. The isoelectric point has been determined for many different proteins and for a number of virus strains. At pH values below the isoelectric point a virus will have a net positive charge, whereas the charge will be negative at pH values above the isoelectric point.

Within the pH range of most unpolluted groundwater both the matrix surfaces and the surfaces of the microorganisms carry a net negative charge. Under these conditions the microorganisms will be repelled by most mineral grain surfaces. At low pH values the surface charge on the microorganisms will shift to being net positive, which will favour their adsorption to soils and the aquifer matrix by electrostatic attraction. This hypothesis has been confirmed by several groups of workers for both bacteria and viruses (Sobsey, 1983; Gerba and Bitton, 1984; Bales *et al.,* 1991, 1993; Bitton and Harvey, 1992; Grant *et al.*, 1993; Loveland *et al.,* 1996; Penrod *et al.*, 1996; Redman *et al.*, 1997; Ryan *et al.*, 1999).

There are many complicating factors that can interfere with the mechanism discussed above. One is that a given virus may have more than one isoelectric point and the factors responsible for passage from one form to another are unknown at this time. Other factors, such as cations and humic and fulvic acids, may also influence the net surface charge of the organism.

Whilst changes in pH may affect the mobility of microorganisms in the subsurface, the significance of this factor in any particular aquifer is uncertain. Robertson and Edberg (1997) have noted that the pH of most unpolluted groundwater is generally very stable, and in their experience falls within the near-neutral range of 6.5 to 8.5. There are exceptions, and in those African countries underlain by acidic basement gneisses and granites the pH is likely to be much lower, frequently in the range 5.5 to 6.5. In addition, geological materials comprising most aquifers have a significant buffering capacity that helps to maintain a relatively constant pH. Robertson and Edberg conclude that it is unlikely that significant changes in microbe mobility will occur due to these minor pH changes. This assumption may be valid for many stable groundwater systems but in groundwater that is exposed to contamination from a variety of sources, which may be of unknown and variable quality, for example sewage, pH may emerge as a dominant factor in the mobility of pathogens.

Salt species

The adsorption of microorganisms onto surfaces in the groundwater system has been shown to have two counteracting effects: It reduces the dispersal of the organism in the subsurface, but reduces the inactivation rate of the organism in the affected area. If the prevailing geochemical conditions in the groundwater system create opposing charges on the surface of the organism and the aquifer matrix, adsorption will occur by electrostatic attraction. Frequently, however, these conditions do not exist and the organism and the aquifer matrix each have a negative charge.

The types and concentrations of salts in the environment can have a profound influence on the extent of pathogen transport in the subsurface. Cations (positively charged inorganic species), in particular multivalent cations such as Magnesium (Mg^{2+}) and Calcium (Ca^{2+}), can form a bridge between the solid surface and the organism and significantly enhance adsorption. Clearly, the concentration of the salt is also important, as this will influence the number of sites that are available for binding as well as the number of bridges that can be formed between the two surfaces. Thus the capacity for binding and the strength of the bonds will be affected by the salt concentration. Several studies of virus and bacterial transport through simulated groundwater systems have confirmed this hypothesis (Taylor *et al.,* 1981; Sobsey, 1983; Bitton and Harvey, 1992; Simoni *et al.*, 2000).

A decrease in the salt concentration or ionic strength of the soil water, such as would occur during a rainfall event, can cause desorption of viruses and bacteria from soil particles (Gerba and Bitton, 1984). This phenomenon has been observed in both laboratory and field experiments (Landry *et al.*, 1980; Wellings *et al.,* 1980). Furthermore, there is evidence to suggest that only small changes in the salt concentration can dramatically affect the mobilization of some organisms in groundwater systems (Redman *et al.*, 1999). The implication of this discussion is that salt concentration in the groundwater system may be of greater significance to pathogen transport than pH, although it is important to consider that neither factor will act in isolation.

Organic matter

There are conflicting reports about the influence of organic matter on the survival and transport of microorganisms in the subsurface, with different responses being noted for bacteria and viruses, and for different species and strains within each group. The influence of organic matter on virus survival has not been firmly defined. In some studies it has been found that proteinaceous material present in wastewater may have a protective effect on viruses; however, in other studies no effect has been observed. Whilst similar observations have been made of bacterial survival in the presence of organic matter, there remains an additional concern that enteric bacteria, in particular the major pathogens and faecal indicator organisms, may be able to undergo a certain level of growth in the environment if the conditions are suitable. There is some support for this hypothesis, a few reports have been published demonstrating regrowth of faecal indicator bacteria in organically rich tropical surface waters, but the evidence is still insufficient to confirm that regrowth is a significant issue for most enteric bacteria in groundwater.

Dissolved organic matter has generally been found to decrease virus adsorption by competing for binding sites on soil particles and the aquifer matrix. The consequence of this observation is that organic matter will increase the mobility of viruses in the subsurface (Powelson *et al.*, 1991). However, at relatively low concentrations of organic matter the effect may be reversed, causing increased virus adsorption and significantly reduced mobility in the subsurface (Robertson and Edberg, 1997).

Overall, bacteria may respond differently. Binding to surfaces is a characteristic of the growth cycle of most, if not all species of bacteria. Unlike the passive processes that characterize the attachment of viruses to surfaces, bacterial attachment involves active processes, including the synthesis of extracellular appendages specifically required to stabilize the bacteria-surface interaction. The initiation of this binding is favoured by the formation of a conditioning film of organic molecules deposited on the solid surface (Bitton and Harvey, 1992; Wimpenny, 1996). Thus the presence of organic matter may restrict the dispersal of bacteria in the subsurface but reduce the inactivation rate at the site of attachment.

3.4 REFERENCES

Abbaszadegan, M., Stewart, P., Le Chevallier, M., Rosen, J.S. and Gerba, C.P. (1998) *Occurrence of Viruses in Groundwater in the United States*, American Water Works Service Company, Voorhees, New Jersey, Interim Report.

Anderson, A.D., Heryford, A.G., Sarisky, J.P., Higgins, C., Monroe, S.S., Beard, R.S., Newport, C.M., Cashdollar, C.L., Fout, G.S., Robbins, D.E., Seys, S.A., Musgrave, K.J., Medus, C., Vinjé, J., Bresee, J.S., Mainzer, H.M. and Glass, R.I. (2003) A waterborne outbreak of Norwalk like virus among snowmobilers - Wyoming, 2001. *J Infectious Diseases*, **187**, 303-306.

ARGOSS (2001) *Assessing the risk to groundwater from on-site sanitation*, BGS Commissioned Research Report, Wallingford.

Bagdasaryan, G. (1964) Survival of viruses of the Enterovirus group (Poliomyelitis, Echo, Coxsackie) in soil and on vegetables. *J. Hyg. Epidemiol. Microbiol. and Immunol.*, **8**, 497.

Bales, R.C., Hinkle, S.R., Kroeger, T.W. and Stocking, K. (1991) Bacteriophage adsorption during transport through porous media: Chemical perturbations and reversibility. *Environ. Sci. Technol.*, **25**, 2088-2095.

Bales, R.C., Li, S., Maguire, K.M., Yahya, M.T. and Gerba, C.P. (1993) MS-2 and poliovirus transport in porous media: hydrophobic effects and chemical perturbations. *Water Resour. Res.*, **2**, 957-963.

Banks, W.S.L. and Battigelli, D.A. (2002) *Occurrence and Distribution of Microbiological Contamination and Enteric Viruses in Shallow Ground Water in Baltimore and Harford Counties, Maryland.* Water-Resources Investigations Report 01-4216, Reston, Virginia.

Banks, W.S.L., Klohe, C.A. and Battigelli, D.A. (2001) *Occurrence and Distribution of Enteric Viruses in Shallow Ground Water and Factors Affecting Well Vulnerability to Microbiological Contamination in Worcester and Wicomico Counties, Maryland.* USGS, Water-Resources Investigations Report 01-4147, Reston, Virginia.

Barrett, M.H., Johal, K., Howard, G., Pedley, S. and Nalubega, M. (2000) Sources of faecal contamination of shallow groundwater in Kampala. In *Groundwater: Past achievements and future challenges* (eds. O.T.M. Sililo, S. Appleyard, M. Barrett) pp. 691-696, Balkema, Rotterdam.

Bauder, J.W., White, B.A. and Inskeep, W.P. (1991) Montana extension initiative focuses on private well quality. *J. Soil and Water Conservation*, **46**, 69-74.

Bitton, G. and Harvey, R.W. (1992) Transport of pathogens through soils and aquifers. In *Environmental Microbiology*, (ed. R. Mitchell), pp. 103-124, Wiley-Liss, Inc.

Bitton, G., Farrah, S.R., Ruskin, R.H., Butner, J. and Chou, Y.J. (1983) Survival of pathogenic and indicator organisms in groundwater. *Ground Water*, **21**, 405-410.

Biziagos, E., Passagot, J., Crance, J. and Deloince, R. (1988) Long-term survival of Hepatitis A virus and Poliovirus Type 1 in mineral water. *Appl. Environ. Microbiol.*, **54**, 2705-2710.

Blanc, R. and Nasser, A. (1996) Effect of effluent quality and temperature on the persistence of viruses in soil. *Water Sci. Technol.*, **33**, 237-242.

Borchardt, M.A., Bertz, P.D., Spencer, S.K. and Batigelli, D.A. (2003) Incidence of enteric viruses in groundwater from household wells in Wisconsin. *Appl. Environ. Microbiol.*, **69**, 1172-1180.

Chauret, C., Chen, P., Springthorpe, S. and Sattar, S. (1995) Effect of environmental stressors on the survival of Cryptosporidium oocysts. In *Proc. Water Quality Technology Conference, New Orleans, Louisiana*, pp. 1567-1585. AWWA, Denver, Colorado.

Dahling, D. (2002) An improved filter elution and cell culture as a procedure for evaluating public groundwater systems for culturable enteroviruses. *Water Environ. Res.*, **74**, 654-568.

Davis, J. and Witt, E.C. (2000) *Microbiological and Chemical Quality of Ground Water used as a Source of Public Supply in Southern Missouri- Phase I, May 1997-March 1998.* USGS Water-Resources Investigations Report 00-4038, Reston, Virginia.

Doherty, K. (1998) Status of the New England ground water viral study. In *Proc. AWWA Annual Meeting Dallas Texas, June 23, 1998*, AWWA, Denver.

Dowd, S.E. and Pillai, S.D. (1997) Survival and transport of selected bacterial pathogens and indicator viruses under sandy aquifer conditions. *J Environ. Sci. & Health Part A*, **32**, 2245-2258.

Femmer, S.R. (2000) *Microbiological and Chemical Quality of Ground Water Used as a Source of Public Supply in Southern Missouri-Phase II, April-July, 1998.* USGS Water-Resources Investigations Report 00-4260, USGS, Denver, Colorado.

Gerba, C.P. and Bitton, G. (1984) Microbial pollutants: Their survival and transport pattern to groundwater. In *Groundwater Pollution Microbiology*, (eds. G. Bitton and C.P. Gerba), pp. 65-88, Wiley, New York.

Gerba, C.P. and McLeod, J.S. (1976) Effects of sediments on the survival of *Escherichia coli* in marine waters. *Appl. Environ. Microbiol.*, **32**(1), 114-120.

Gerba, C.P., Kutz, S.M., Bradford, A.W. and DeLeon, R. (Undated) *Rotavirus Survival and Transport in the Subsurface.* Cooperative Agreement Final Report to Steve Hutchson, US EPA, Ada, Oklahoma, CR-813566-0101.

Gleeson, C. and Gray, N. (1997) *The Coliform Index and Waterborne Disease*, 1st edn, E&FN Spon, London.

Goss, M.J., Barry, D.A.J. and Rudolph, D.L. (1998) Contamination of Ontario farmstead domestic wells and its association with agriculture: 1. Results from drinking water wells. *J. Contam. Hydrol.*, **32**, 267-293.

Grant, S.B., List, E.J. and Lidstrom, M.E. (1993) Kinetic analysis of virus adsorption and inactivation in batch experiments. *Water Resour. Res.*, **29**, 2067-2085.

Hancock, C.M., Rose, J.B. and Callahan, M. (1998) Crypto and Giardia in US groundwater. *JAWWA*, **90**(3), 58-61.

Havelaar, A.H., van Olphen, M. and Drost, Y.C. (1993) F-specific RNA bacteriophages are adequate model organisms for enteric viruses in fresh water. *Appl. Environ. Microbiol.*, **59**, 2956-2962.

Heisz, M., Chauret, C., Chen, P., Springthorpe, S. and Sattar, S.A. (1997) In vitro survival of Cryptosporidium oocysts in natural waters. In *Proc. International Symposium on Waterborne Cryptosporidium, March 2-5, 1997, Newport Beach, California*, (eds. C. Fricker, J.L. Clancy and P.A. Rochelle), pp. 71-175, AWWA, Denver, Colorado,.

Howard, K.W.F. (2001) Polluted groundwater - deadly lessons from Walkerton, Ontario, Canada. In *New Approaches to Characterising Groundwater Flow*, (eds. K.P. Seiler and S. Wohnlich), pp. 521-525, Swets and Zeitlinger, Balkema, Rotterdam.

Howard, G., Bogh, C., Goldstein, G., Morgan, J., Prüss, A., Shaw, R. and Teuton, J. (2002). *Healthy villages: A guide for communities and community health workers.* WHO, Geneva.

Hurst, C.J., Gerba, C.P. and Cech, I. (1980a) Effects of environmental variables and soil characteristics on virus survival in soil. *Appl. Environ. Microbiol.*, **40**, 1067-1079.

Hurst, C.J., Gerba, C.P. Lance, J.C. and Rice, R.C. (1980b) Survival of enteroviruses in rapid-infiltration basins during the land application of wastewater. *Appl. Environ. Microbiol.*, **40**,192.

Hurst, C.J., Wild, D.K. and Clark, R.M. (1992) Comparing the accuracy of equation formats for modelling microbial population decay rates. In *Modelling the Metabolic and Physiologic*

Activities of Micro-organisms, (ed. C.J. Hurst), pp. 149-175, John Wiley & Sons. Inc., New York.

Jansons, J., Edmonds, L.W., Speight, B. and Bucens, M.R. (1989a) Movement of viruses after artificial recharge. *Water Res.*, **23**, 293-299.

Jansons, J., Edmonds, L.W., Speight, B. and Bucens, M.R.(1989b) Survival of viruses in groundwater. *Water Res.*, **23**, 301-306.

Keswick, B.H., Gerba, C.P., Secor, S.L., Cech, I. (1982) Survival of enteric viruses and indicator bacteria in groundwater. *J. Environ. Sci. Health*, **A17**, 903-812.

Korhonen, L.K., Niskanen, M., HeinonenTanski, H., Martikainen, P.J., Salonen, L. and Taipalinen, I. (1996) Groundwater quality in wells in central rural Finland: A microbiological and radiochemical survey. *Ambio*, **25**, 343-349.

Landry, E.F., Vaughn, J.M. and Penello, W.P. (1980) Poliovirus retention in 75-cm soil cores after sewage and rainwater application. *Appl. Environ. Microbiol.*, **40**, 1032.

Lawrence, A.R., Williams, A.T., Brewerton, L.T., Barker, J.A., Bird, M. and Ward, R.S. (1996) *Pathogen transport in the Chalk aquifer: An assessment of the risk to chalk groundwater resources*. BGS Technical Report No. WD/96/48, Wallingford.

Lee, J.V. (1993) Contamination of a well by the direct influence of surface water. In *Proc. of the Annual Conference of the American Waterworks Association, San Antonio, Texas*, AWWA, pp. 215-231.

Lefler, E. and Kott, Y. (1974) *Virus Retention and Survival in Sand*, Centre for Research in Water Resources, Austin, Texas.

Lerner, D.N. and Barrett, M.H. (1996) Urban groundwater issues in the United Kingdom. *Hydrogeology J.*, **4**, 80-89.

Lewis, W.J., Foster, S.S.D. and Drasar, B.S. (1982) The risk of groundwater pollution by on-site sanitation in developing countries: A literature review. *Internat. Ref. Centre for Wastes Disposal,* **01**, 82.

Lieberman, R., Shadiz, L.C., Newport, B.S., Frebis, C.P., Safferman, R.S., Stetler, R.E., Lye, D., Fout, G.S. and Dahling, D. (2002) *Microbial Monitoring of Vulnerable Public Ground Water Supplies*. AWWARF, Denver, Colorado.

Lieberman, R.J., Shadix, L.C., Newport, B.S., Crout, S.R., Buescher, S.E., Safferman, R.S., Stetler, R.E., Lye, D., Fout, G.S. and Dahling, D.R. (1994) *Source Water Microbial Quality of Some Vulnerable Public Groundwater Supplies*. AWWA, Denver, Colorado.

Lindsey, B.D., Rasberry, J.S. and Zimmerman, T.M. (2002) *Microbiological Quality of Water from Noncommunity Supply Wells in Carbonate and Crystalline Aquifers of Pennsylvania.* USGS, Water-Resources Investigations Report 01-4268, Reston, Virginia.

Loveland, J.P., Ryan, J.N., Amy, G.L. and Harvey, R.W. (1996) The reversibility of virus attachment to mineral surfaces. *Colloids and Surfaces. A: Physicochemical and Engineering Aspects*, **107**, 205-221.

Macler, B.A. and Merkle, J.C. (2000) Current knowledge on groundwater microbial pathogens and their control. *Hydrogeology J.,* **8,** 29-40.

Mahler, B.J., Personne, J.C., Lods, G.F. and Drogue, C. (2000) Transport of free and particulate-associated bacteria in karst. *J. Hydrology*, **238**, 179-193.

Matthess, G.A., Pekdeger, A. and Schroeter, J. (1988) Persistence and transport of bacteria and viruses in groundwater - a conceptual evaluation. *J. Contam. Hydrol.*, **2**, 171-188.

McFeters, G.A., Bissonnette, G.K., Jezeski, J., Thomson, C.A. and Stuart, D.G. (1974) Comparative survival of indicator bacteria and enteric pathones in well water. *Appl. Microbiol.*, **27**, 823-829.

Medema, G.J., Bahar, M. and Schets, F.M. (1997) Survival of Cryptosporidium parvum, Escherichia coli, faecal enterococci and Clostridium perfringens in river water: Influence of temperature and autochthonous microorganisms. *Water Sci. Technol.*, **35**(11-12), 249-252.

Melian, R., Myrlian, N., Gouriev, A., Moraru, C., and Radstake, F. (1999) Groundwater quality and rural drinking-water supplies in the Republic of Moldova. *Hydrogeology J.*, **7**, 188-196.

Meschke, J.S. (2001) Comparative Adsorption, Persistence, and Mobility of Norwalk virus, Poliovirus type 1, and F+RNA Coliphages in Soil and Groundwater. Ph.D. thesis, School of Public Health, University of North Carolina, Chapel Hill, North Carolina.

Nasser, A., Tchorch, Y. and Fattal, B. (1993) Comparative survival of E. coli, F+bacteriophages, HAV and poliovirus 1 in wastewater and groundwater. *Water Sci. Technol.,* **27**, 401-407.

Nasser, A.M. and Oman, S.D. (1999) Quantitative assessment of the inactivation of pathogenic and indicator viruses in natural water sources. *Water Res.*, **33**, 1748-1752.

Olson, B.H. (1993) Pathogen occurrence in source waters: factors affecting survival and growth. In *Safety of Water Disinfection: Balancing Chemical and Microbial Risks*, (ed. G.F. Craun), ILSI Press, Washington DC.

Pancorbo, O.C., Evanshen, B.G., Campbell, W.F., Lambert, S., Curtis, S.K. and Woolley, T.W. (1987) Infectivity and antigenicity reduction rates of human Rotavirus strain Wa in Fresh Waters. *Appl. Environ. Microbiol.*, **13**, 1803-1811.

Parshionikar, S.U., Willian-True, S., Fout, G.S., Robbins, D.E., Syes, S.A., Cassady, J.D. and Harris, R. (2003) Waterborne outbreak of gastroenteritis associated with a Norovirus. *Appl. Environ. Microbiol.,* **69**(9), 5263-5268.

Penrod, S.L., Olson, T.M. and Grant, S.B. (1996) Deposition kinetics of two viruses in packed beds of quartz granular media. *Langmuir*, **12**, 5576-5587.

Powell, K.L., Barrett, M.H., Pedley, S., Tellam, J.H., Stagg, K.A., Greswell, R.B. and Rivett, M.O., (2000) Enteric virus detection in groundwater using a glasswool trap. In *Groundwater: Past Achievements and Future Challenges*, (ed. O.T.M. Sililo), pp. 813-816, Balkema, Rotterdam.

Powell, K.L., Cronin, A.A., Barrett, M.H., Sellwood, J. and Pedley, S. (2001a) *Distribution of Microbiological Contaminants in Triassic Sandstone Urban Aquifers*. Environment Agency R&D Technical Report P2-255/TR, Bristol.

Powell, K.L., Cronin, A.A., Pedley, S., Barrett, M.H. (2001) Microbiological quality of groundwater in UK urban aquifers: do we know enough? In *Proc. 3rd International conference on Groundwater Quality, 'Groundwater Quality 2001'*, (eds. S. Thornton and S. Oswald), University of Sheffield, June 2001, pp. 236-238.

Powelson, D.K., Simpson, J.R. and Gerba, C.P. (1991) Effects of organic matter on virus transport in unsaturated flow. *Appl. Environ. Microbiol.*, **57**, 2192-2196.

Redman, J.A., Grant, S.B., Olson, T.M., Amkins, J.M., Jackson, L., Castillo, M.S. and Yanko, W.A. (1999) Physiochemical mechanisms responsible for the filtration and mobilisation of filamentous bacteriophage in quartz sand. *Water Res.*, **33**(1), 43-52.

Redman, J.A., Grant, S.B., Olson, T.M., Hardy, M.E. and Estes, M.K. (1997) Filtration of recombinant Norwalk virus particles and bacteriophage MS2 in quartz sand: Importance of electrostatic interactions. *Environ. Sci. Technol.,* **31**, 3378-3383.

Rice, E.W., Johnson, C.H., Wild, D.K. and Reasoner, D.J. (1992) Survival of *Escherichia coli* O157:H7 in drinking-water associated with a waterborne disease outbreak of hemorrhagic colitis. *Letters in Appl. Microbiol.*, **13**, 38-40.

Robertson, J.B. and Edberg, S.C. (1997) Natural protection of spring and well drinking water against surface microbial contamination. I. Hydrogeological Parameters. *Crit. Rev. in Microbiol.*, **23**, 143-178.

Robertson, L.J., Campbell, A.T. and Smith, H.V. (1992) Survival of *Cryptosporidium parvum* oocysts under various environmental pressures. *Appl. Environ. Microbiol.*, **58**, 3494-3500.

Romero, J.C. (1972) The movement of bacteria and viruses through porous media. In *Water Quality in a Stressed Environment*, (ed. W.A. Pettyjohn), pp. 200-224, Burgess Publishing Company, Minnesota.

Ryan, J.N., Elimelech, M., Ard, R.A., Harvey, R.W. and Johnson, P.R. (1999) Bacteriophage PRD1 and silica colloid transport and recovery in an iron oxide-coated sand aquifer. *Environ. Sci. Technol.,* **33**, 63-73.

Schijven, J.F. and Hassanizadeh, S.M. (2000) Removal of viruses by soil passage: Overview of modelling, processes and parameters. *Crit. Rev. Environ. Sci. Technol.*, **30**(1), 49-127.

Schijven, J.F., Hoogenboezem, W., Hassanizadeh, S.M. and Peters, J.H. (1999) Modelling removal of bacteriophages MS2 and PRD1 by dune infiltration at Castricum, the Netherlands. *Water Resour. Res.*, **35**, 1101-1111.

Schijven, J.F., Medema, G., Vogelaar, A.J. and Hassanizadeh, S.M. (2000) Removal of microorganisms by deep well injection. *J Contam. Hydrol.*, **44**, 301-327.

Schijven JF, Berger P, Miettinen I. (2002a) Removal of pathogens, surrogates, indicators and toxins using bank filtration. In *Riverbank Filtration: Improving source water quality*, (eds. R. Chittaranjan, G. Mlin and R.B. Linsky), Kluwer Academic, University of Hawaii.

Schijven, J.F., Hassanizadeh, S.M., de Bruin, H.A.M. (2002b) Two-site kinetic modeling of bacteriophages transport through columns of saturated dune sand. *J Contam. Hydrol.*, **57**, 259-279.

Schijven, J.F., de Bruin, H.A.M., Hassanizadeh, S.M., de Roda Husman, A.M. (2003) Bacteriophages and clostridium spores as indicator organisms for removal of pathogens by passage through saturated dune sand. *Water Res.*, **37**, 2186-2194.

Simoni, S.F., Bosma, T.N.P., Harms, H. and Zehnder, A.J.B. (2000) Bivalent cations increase both the subpopulation of adhering bacteria and their adhesion efficiency in sand columns. *Environ. Sci. Technol.*, **34**(6), 1011-1017.

Singleton, P. and Sainsbury, D. (1999) *Dictionary of Microbiology and Molecular Biology*, 2nd edn, John Wiley & Sons Ltd, New York.

Skilton, H. and Wheeler, D. (1988) Bacteriophage tracer experiments in groundwater. *J. Appl. Bacteriol.*, **65**(5), 387-395.

Skilton, H. and Wheeler, D. (1989) The application of bacteriophage as tracers in chalk aquifer systems. *J. Appl. Bacteriol.*, **66**, 549-557.

Sobsey, M.D. (1983) Transport and fate of viruses in soils. In *Microbial Health Considerations of Soil Disposal of Domestic Wastewaters*, US EPA, Cincinnati, Ohio.

Sobsey, M.D., Shields, P.A., Hauchman, F.H., Hazard, R.L., and Canton, L.W. (1986) Survival and transport of hepatitis A virus in soils, groundwater and wastewater. *Water Sci. Technol.*, **18**, 97-106.

Taylor, D.H., Moore, R.S. and Sturman, L.S. (1981) Influence of pH and electrolyte composition on adsorption of poliovirus by soils and minerals. *Appl. Environ. Microbiol.*, **42**, 976.

US EPA (2000) National Primary Drinking Water Regulations: Ground Water Rule. Proposed Rule. *Federal Register*, **10**, 30194-30274.

US Government (1996) Safe Drinking Water Act, Public Law 104-182.

USGS (1998) *Microbiological Quality of Public Water Supplies in the Ozark Plateaus Aquifer System, Missouri*. Washington DC, USGS, Ref Type: Pamphlet.

Vaughn, J.M., Landry, E.F. and Thomas, M.Z. (1983) Entrainment of viruses from septic tank leach fields through a shallow, sandy soil aquifer. *Appl. Environ. Microbiol.*, **45**, 1474-1480.

Wang, G. and Doyle, M.P. (1998) Survival of enterhemorrhagic *Escherichia coli* O157:H7 in water. *J. Food Protection*, **61**, 662-667.

Wellings, M., Lewis, A.C., Mountain, C.W. and Pierce, L.V. (1975) Demonstration of viruses in groundwater after effluent discharge onto soil. *Appl. Microbiol.*, **29**, 751.

West, J.M., Pedley, S., Baker, S.J., Barrott, L., Morris, B., Storey, A., Ward, R.S. and Barrett, M. (1998) *A Review of the Impact of Microbiological Contaminants in Groundwater*, UK Environment Agency, R and D Technical Report P139, Bristol.

Wimpenny, J.W.T. (1996) Ecological determinants of biofilm formation. *Biofouling*, **10**(1-3), 43-64.

Yahya, M.T., Galsomies, L., Gerba, C.P. and Bales, R.C. (1993) Survival of bacteriophages MS2 and PRD1 in groundwater. *Water Sci. Tech.*, **27**, 409-412.

Yates, M.V. and Yates, S.R. (1988) Modelling microbial fate in the subsurface environment. *Crit. Rev. Environ. Control*, **17**, 307-344.

Yates, M.V., Gerba, C.P. and Kelley, L.M. (1985) Virus persistence in groundwater. *Appl. Environ. Microbiol.*, **49**(4), 778-781.

4

Chemicals: Health relevance, transport and attenuation

M. Rivett, J. Drewes, M. Barrett, J. Chilton, S. Appleyard, H.H. Dieter, D. Wauchope and J. Fastner

The presence of substances in groundwater may be affected by naturally occurring processes as well as by actions directly associated with human activities. Naturally occurring processes such as decomposition of organic material in soils or leaching of mineral deposits can result in increased concentrations of several substances. Those of health concern include fluoride, arsenic, nitrate, selenium, uranium, metals, and radionuclides such as radon. Problems of aesthetic quality and acceptance may be caused by iron, manganese, sulphate, chloride and organic matter.

Sources of groundwater contamination associated with human activities are widespread and include diffuse as well as point source pollution like land application of animal wastes and agrochemicals in agriculture; disposal practices of human excreta and wastes such as leaking sewers or sanitation systems, leakage of waste disposal sites, landfills, underground storage tanks and pipelines; and pollution due to both poor practices and accidental spills in mining, industry, traffic, health care facilities and military sites.

 Protecting Groundwater for Health: Managing the Quality of Drinking-water Sources. Edited by O. Schmoll, G. Howard, J. Chilton and I. Chorus. ISBN: 1843390795. Published by IWA Publishing, London, UK.

The exploitation of petroleum products and the development of industrial chemistry has given rise to a large number of organic chemicals, many of which can be found in the environment. Many organic chemicals are known to have potential human health impacts, and some of these may occur in drinking-water in health-relevant concentrations. In consequence, the list of those for which guideline values and national quality standards have been developed has been continually added to and revised as data on the occurrence of chemicals in water and new toxicological data emerge. Organic chemicals commonly used by industry with known or suspected human health impacts that are often encountered in groundwaters include, for example, aromatic hydrocarbons such as benzene, toluene, ethylbenzene and xylenes (BTEX) as well as volatile chlorinated hydrocarbons such as tetrachloroethene (PCE) and trichloroethene (TCE). A diverse range of pesticides is also found in groundwaters. These are primarily, but not exclusively, ascribed to agricultural activities. Typically pesticide concentrations encountered are low. In some cases they have exceeded regulatory limits for drinking-water supplies, although often the regulatory levels exceeded were often much lower than those based on public health considerations.

This chapter concentrates on the groups of chemical substances that are toxic to humans and have reasonable potential to contaminate drinking-water abstracted from groundwater. It provides foundational knowledge of natural groundwater constituents and anthropogenic groundwater contaminants and discusses their relevance to human health, origin, and transport and attenuation in groundwater systems. The chapter is sub-divided as follows: Section 4.1 provides introductory theory on the transport and attenuation of chemicals in the subsurface; Sections 4.2 to 4.4 focus upon inorganic chemicals – natural inorganic constituents, nitrogen species and metals respectively; Sections 4.5 to 4.8 focus upon organic chemicals including an introductory section on conceptual contaminant models and transport and attenuation theory specific to organic contaminants followed by sections on some organic chemical groups of key concern – aromatic hydrocarbons, chlorinated hydrocarbons and pesticides respectively; finally, the chapter closes with a brief consideration of currently emerging issues (Section 4.9). Further information on the individual chemicals discussed in this chapter is available in the WHO *Guidelines for Drinking-water Quality*, Volume 1 (WHO, 2004a), as well as in detailed background documents on WHO's Water, Sanitation and Health website (http://www.who.int/water_sanitation_health/dwq/chemicals/en/index.html).

4.1 SUBSURFACE TRANSPORT AND ATTENUATION OF CHEMICALS

Understanding of the transport and attenuation of chemicals in the subsurface is fundamental to effective management of risks posed by chemicals and their possible impact on groundwater resources. A risk assessment approach to groundwater protection incorporates the three-stage combination of source, pathway and receptor. All three must be considered and understood to arrive at a balanced view of the risks to health of groundwater users. Informed consideration of the pathway, which in the context of this monograph means transport through the groundwater system, is vital. This not only includes consideration of the general and local hydrogeologic characteristics covered in

Chapters 2 and 8, but also the transport and attenuation of chemicals within that pathway. The latter depend upon the properties of the chemical itself, particularly those properties that control interactions of the chemical with the subsurface regime, a regime that includes not only the host rock and groundwater, but other natural and anthropogenic chemical constituents present as well as microbial life.

Within the overall transport process, attenuation processes may cause movement of the chemical to differ from that of the bulk flowing groundwater, for example dispersion, sorption and chemical or biological degradation of the chemical. Such attenuation processes potentially act to mitigate the impact of chemicals and are a function of both the specific chemical and geologic domain. Indeed, attenuation may vary significantly between individual chemicals and within different geological settings. In recent years natural attenuation of organic contaminants has been increasingly recognized to play an important role in many aquifer systems leading to monitored natural attenuation becoming a recognized remedial strategy to manage risks to groundwater at some contaminated sites (EA, 2000).

This section provides an overview of the key processes that control the transport and attenuation of chemicals in groundwater. Elaboration of some of the more specific attenuation processes is also included in later sections. Further details may be found in the following texts and references therein: Freeze and Cherry (1979), Appelo and Postma (1993), Stumm and Morgan (1996), Domenico and Schwartz (1998), Bedient *et al.* (1999), Fetter (1999) and Schwartz and Zhang (2003).

4.1.1 Natural hydrochemical conditions

It is important to understand at the outset the natural hydrochemical conditions that exist in a given aquifer system, as these provide the baseline from which quality changes caused by human impacts can be determined. The natural hydrochemical conditions may also affect the behaviour of some pollutants. Because groundwater movement is typically slow and residence times long, there is potential for interaction between the water and the rock material through which it passes. The properties of both the water and the material are therefore important, and natural groundwater quality will vary from one rock type to another and within aquifers along groundwater flow paths. Water is essentially a highly polar liquid solvent that will readily dissolve ionic chemical species. Rock material is predominantly inorganic in nature and contact of flowing groundwater with the rock may dissolve inorganic ions into that water, i.e. dissolution of the rock occurs. 'Major ions' present are the anions nitrate, sulphate, chloride and bicarbonate and the cations sodium, potassium, magnesium and calcium. Ions typically present at lower concentration, 'minor ions', include anions such as fluoride and bromide and a wide variety of metal ions that are predominantly cations. Combined, the total inorganic concentration within the water is referred to as the total dissolved solids (TDS).

Natural groundwater quality changes start in the soil, where infiltrating rainfall dissolves carbon dioxide from biological activity in the soil to produce weak carbonic acid that may assist removal of soluble minerals from the underlying rocks, e.g. calcite cements. At the same time, soil organisms consume some of the oxygen that was dissolved in the rainfall. In temperate and humid climates with significant recharge, groundwater moves relatively quickly through the aquifer. Contact time with the rock matrix is short and only readily soluble minerals will be involved in reactions. Groundwater in the outcrop areas of aquifers is likely to be low in overall chemical content, i.e. have low major ion contents and low TDS, with igneous rocks usually having less dissolved constituents than sedimentary rocks (Hem, 1989). In coastal regions, sodium and chloride may exceed calcium, magnesium and bicarbonate and the presence of soluble cement between the grains may allow major ion concentrations to be increased. Groundwaters in carbonate rocks have pH above 7, and mineral contents usually dominated by bicarbonate and calcium.

In many small and shallow aquifers the hydrochemistry does not evolve further. However, the baseline natural quality of groundwater may vary spatially within the same aquifer if the mineral assemblages vary, and also evolves with time as the water moves along groundwater flow lines. If an aquifer dips below a confining layer (Figure 2.5), a sequence of hydrochemical processes occurs with progressive distance downgradient away from the outcrop, including precipitation of some solids when relevant ion concentrations reach saturation levels for a solid mineral phase. These processes have been clearly observed in the United Kingdom, where the geological history is such that all three of the major aquifers exhibit the sequence shown in Figure 4.1, which has been characterized by sampling transects of abstraction boreholes across the aquifers (Edmunds *et al.*, 1987).

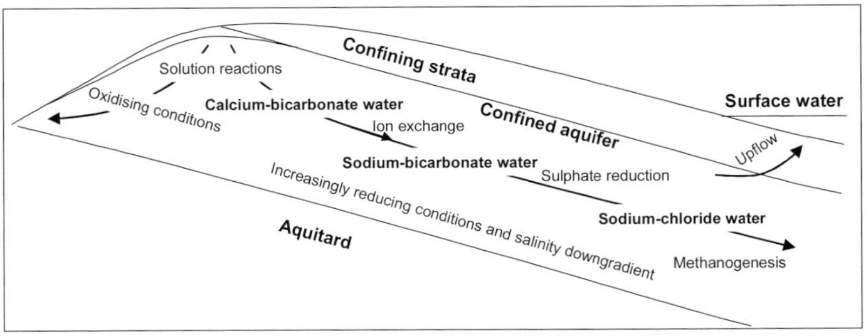

Figure 4.1. Schematic representation of downgradient hydrochemical changes

In the recharge area, oxidizing conditions occur and dissolution of calcium and bicarbonate dominates. As the water continues to move downgradient, further modifications are at first limited. By observing the redox potential (E_h) of abstracted groundwater, a sharp redox barrier is detected beyond the edge of the confining layer, corresponding to the complete exhaustion of dissolved oxygen. Bicarbonate increases and the pH rises until buffering occurs at about 8.3. Sulphate concentrations remain stable in the oxidizing water, but decrease suddenly just beyond the redox boundary due to sulphate reduction. Groundwater becomes steadily more reducing as it flows downgradient, as demonstrated by the presence of sulphide, increase in the solubility of iron and manganese and denitrification of nitrate. After some further kilometres, sodium begins to increase by ion exchange at the expense of calcium, producing a natural softening of the water. Eventually, the available calcium in the water is exhausted, but sodium continues to increase to a level greater than could be achieved purely by cation exchange. As chloride also begins to increase, this marks the point at which recharging water moving slowly down through the aquifer mixes with much older saline water present in the sediments (Figure 4.1). The observed hydrochemical changes can thus be interpreted in terms of oxidation/reduction, ion exchange and mixing processes.

In arid and semi-arid regions, evapotranspiration rates are much higher, recharge is less, flow paths longer and residence times much greater and hence much higher levels of natural mineralization, often dominated by sodium and chloride, can be encountered. Thus the major ion contents and TDS are often high. In some desert regions, even if groundwater can be found it may be so salty (extremely high TDS) as to be undrinkable, and the difficulty of meeting even the most basic domestic requirements can have serious impacts on health and livelihood.

Natural variations in pH and oxygen status are also important and are not restricted to deep environments. Many groundwaters in tropical regions in weathered basement aquifers and alluvial sequences have low pH, and the reducing conditions which prevail can promote the mobilization of metals and other parameters of health significance such as arsenic. Thus prevailing hydrochemical conditions of the groundwater that are naturally present and develop need to be taken into account when: (i) developing schemes for groundwater abstraction for various uses and in protecting groundwater; and (ii) considering the transport and attenuation of additional chemicals entering groundwaters due to human activity.

4.1.2 Conceptual models and attenuation processes

Effective prediction of transport of chemical pollutants through a subsurface groundwater system and associated assessments of risk requires a valid conceptual model of the contaminant migration scenario. The classical contaminant conceptual model is one of a near-surface leachable source zone where chemical contaminant is leached, i.e. dissolved/solubilized, into water infiltrating through the source (Figure 4.2). A dissolved-phase chemical solute plume subsequently emerges in water draining from the base of the contaminant source zone and moves vertically downward through any unsaturated zone present. The dissolved solute plume ultimately penetrates below the water table to subsequently migrate laterally in the flowing groundwater. Many sources, e.g. a landfill, chemical waste lagoon, contaminated industrial site soils, pesticide residues in field soils, may have sufficient chemical mass to enable them to act as long-term generators of dissolved-phase contaminant plumes; potentially such sources can last decades. This will lead to continuous dissolved-phase plumes extending from these sources through the groundwater pathway that grow with time and may ultimately reach distant receptors unless attenuation processes operate. This near-surface leachable source – dissolved-plume conceptual model is the model most frequently invoked and the one to which groundwater vulnerability and protection concepts and groundwater risk-assessment models are most easily applied. It is important to note, however, that the above conceptualization may be too simplified and alternative conceptual models need to be invoked in some cases, most notably for non aqueous phase liquids (NAPLs) as discussed in Section 4.5.

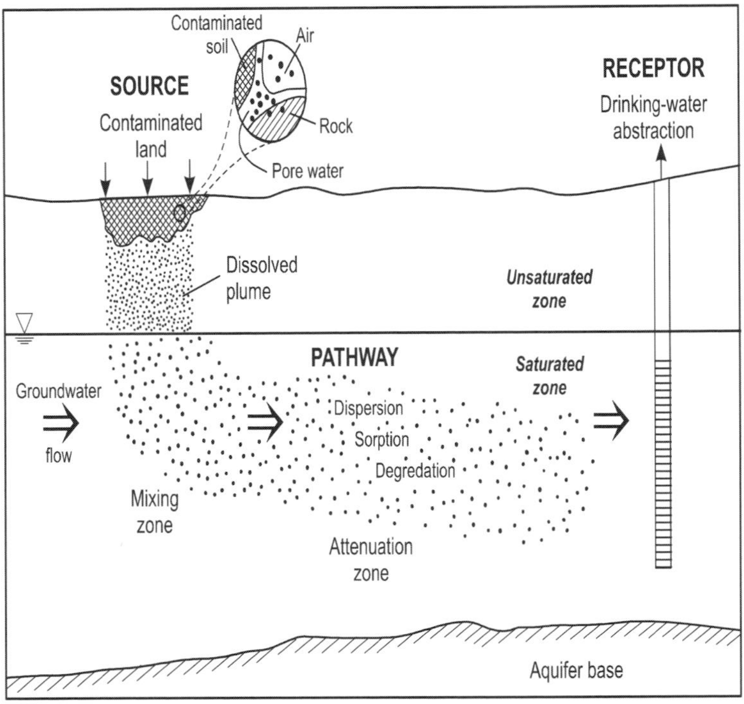

Figure 4.2. Classical contaminant conceptual model

Attenuation processes operative in the groundwater pathway, both for unsaturated and saturated zones, are summarized below and briefly described in the text that follows. Further details may be found in the texts referenced in other sections of this chapter.

DEF ▶

Advection and dispersion

Advection *is the transport of dissolved solute mass present in groundwater due to the bulk flow (movement) of that groundwater. Advection alone (with no dispersion or reactive processes occuring) would cause a non-reactive solute to advect (move) at the mean groundwater pore velocity. All solutes undergo advection, however, reactive solutes are subject to influences by other processes detailed below.*

Molecular diffusion *is the movement of solute ions in the direction of the concentration gradient from high towards low concentrations. It affects all solutes.*

Mechanical dispersion *causes spreading of solute and hence dilution of concentrations, it arises from the tortuosity of the pore channels in a granular aquifer and of the fractures in a consolidated aquifer and the different speeds of groundwater within flow channels of varying width. It affects all solutes.*

Retardation

Sorption *is a process by which chemicals or organisms become attached to soils and/or the geologic rock material (aquifer solids) and removed from the water. Often the sorption process is reversible and solutes desorb and hence dissolved-solute plumes are retarded, rather than solutes being permanently retained by the solids.*

Cation exchange *is the interchange between cations in solution and cations on the surfaces of clay particles or organic colloids.*

Filtration *is a process that affects particulate contaminants (e.g. organic/inorganic colloids or microbes) rather than dissolved solutes. Particles larger than pore throat diameters or fracture apertures are prevented from moving by advection and are therefore attenuated within the soil or rock.*

Reactions and transformations of chemicals

Chemical reactions (abiotic reactions) *are classical chemical reactions that are not mediated by bacteria. They may include reaction processes such as precipitation, hydrolysis, complexation, elimination, substitution, etc., that transform chemicals into other chemicals and potentially alter their phase/state (solid, liquid, gas, dissolved).*

Precipitation *is the removal of ions from solution by the formation of insoluble compounds, i.e. a solid-phase precipitate.*

Hydrolysis *is a process of chemical reaction by the addition of water.*

Complexation *is the reaction process by which compounds are formed in which molecules or ions form coordinate bonds to a metal atom or ion.*

Biodegradation (biotic reactions) *is a reaction process that is facilitated by microbial activity, e.g. by bacteria present in the subsurface. Typically molecules are degraded (broken down) to molecules of a simpler structure that often have lower toxicity.*

Advection

As described in Chapter 2, groundwater moves due to the presence of a hydraulic gradient and may be characterized by the Darcy velocity q (alternatively named the specific discharge). The Darcy velocity may be calculated via Darcy's law and is the product of the geologic media hydraulic conductivity K and the groundwater hydraulic gradient i. The actual mean groundwater pore (linear) velocity of groundwater, henceforth referred to as the 'groundwater velocity' v differs from the Darcy velocity as flow can only occur through the effective porosity n_e of the formation. The groundwater velocity may be quantified by modifying the Darcy equation:

$$v = -Ki / n_e \qquad \text{(Eqn. 4.1)}$$

Advection is the transport of dissolved solutes in groundwater due to the bulk movement of groundwater. The mean advective velocity of non-reactive solutes is equal to the groundwater velocity, v (Eqn. 4.1) and is normally estimated by knowledge of the Equation 4.1 hydrogeological parameters. Occasionally v may be estimated from the mean position of a solute plume, typically within a groundwater tracer test (Mackay *et al.*, 1986). Reactive solutes also advect with the flowing groundwater, however, their velocities are modified due to co-occurrence of attenuation processes.

Dispersion

All reactive and non-reactive solutes will undergo spreading due to dispersion, causing dissolved-phase plumes to broaden both along and perpendicular to the groundwater flow direction (Figure 4.3). Dispersion is most easily observed for 'conservative' non-reactive solutes, such as chloride, as these only undergo advection and dispersion. Dispersion causes mixing of the dissolved-solute plume with uncontaminated water and hence concentration dilution as well as plume spreading. Longitudinal dispersion, spreading in the direction of predominant groundwater flow, is greatest causing solutes to move at greater or less than the mean advective velocity v. Solute spreading is due to mechanical dispersion that can arise at the pore-scale due to (Fetter, 1999): (i) fluids moving faster at pore centres due to less friction; (ii) larger pores allowing faster fluid movement; (iii) routes of varying tortuosity around grains. At a larger scale, macro-dispersion is controlled by the distribution of hydraulic conductivities in the geologic domain; greater geological heterogeneity resulting in greater plume spreading. The above processes cause increasing dispersion with plume travel distance, i.e. dispersion is scale dependent (Gelhar, 1986; Fetter, 1999).

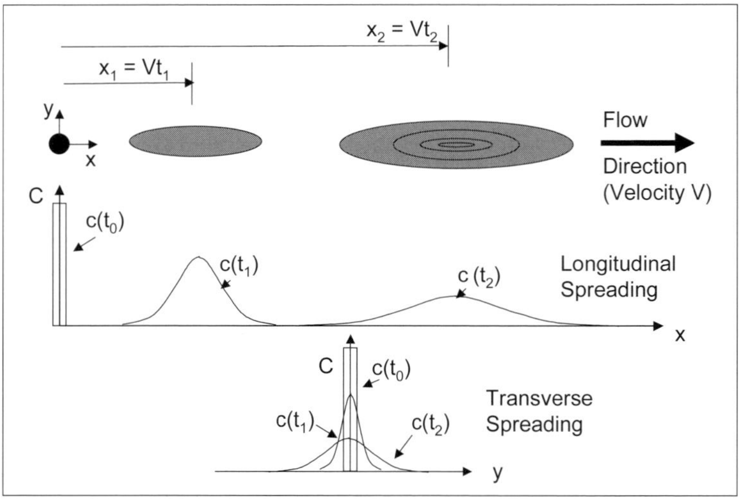

Figure 4.3. Dispersion of a pulse release of dissolved-solute plume

Plume dispersion in other directions is much reduced. Transverse horizontal spreading may arise from flowpath tortuosity and molecular diffusion due to plume chemical-concentration gradients. Transverse vertical spreading occurs for similar reasons, but is generally more restricted due to predominantly near-horizontal layering of geologic strata. Overall, a hydrodynamic dispersion coefficient D is defined for each direction (longitudinal, transverse horizontal, transverse vertical):

$$D = \alpha v + D^* \qquad \text{(Eqn. 4.2)}$$

which is seen to depend upon D*, the solute's effective diffusion coefficient and α the geologic media dispersivity. Dispersion parameters are most reliably obtained from tracer tests or, less reliably, at the larger (>250 m) scale, by model fitting to existing plumes. Collated values have yielded simple empirical relationships to estimate dispersion, e.g. the longitudinal dispersivity is often approximated to be 0.1 (10 per cent) of the mean plume travel distance (Gelhar, 1986). However, such relationships are very approximate.

Retardation

The processes that cause retardation (slowing down) of dissolved-solute plume migration include filtration, sorption and cation exchange. Filtration is a process that affects particulate contaminants (e.g. organic/inorganic colloids or microbes) rather than dissolved solutes, the key focus here. Sorption is a process by which chemicals or organisms become attached to soils and/or the geologic rock material (aquifer solids) and are removed from the water. Often the sorption process is reversible and solutes desorb back into the water phase and hence dissolved-solute plumes are retarded, rather than solutes being permanently retained by the solids. Preferred sorption sites depend upon the chemical solute properties, in general clay strata or organic matter within the geologic solid media are key sorption sites. Such sites may, however, be limited and sorption to other mineral phases, e.g. iron oxyhydroxides, may become important in some cases. Sorption processes normally lead to a Retardation Factor, R_i, being defined that is the ratio of the mean advective velocity (conservative solute velocity) (v) to the mean velocity of the retarded sorbing solute plume (v_i):

$$R_i = v / v_i \qquad \text{(Eqn. 4.3)}$$

Typically R_i is not estimated from Equation 4.3, rather various methods may be used to estimate R_i relating to the specific chemical nature of the sorption interaction and a relevant sorption coefficient (e.g. see Section 4.5.2). Sorption-related processes can be sensitive to the environmental conditions. For example, relatively small pH changes may cause significant changes to the mobilization of metals or perhaps organic contaminants that are themselves acids or bases, e.g. phenols or amines.

Reactions and transformations of chemicals

Many chemicals undergo reaction or transformation in the subsurface environment. In contrast to retardation, contaminants may be removed rather than simply slowed down. Reactions of harmful chemicals to yield benign products prior to arrival at a receptor are the ideal, e.g. many toxic hydrocarbons have potential to biodegrade to simple organic acids (of low health concern and themselves potentially degradable), carbon dioxide (bicarbonate) and water. Transformation often causes a deactivation (lowering) of toxicity. Reactions and/or transformations incorporate processes such as chemical precipitation, complexation, hydrolysis, biodegradation (biotic reactions) and chemical reactions (abiotic reactions).

Chemical *precipitation* and *complexation* are primarily important for the inorganic species. The formation of coordination complexes is typical behaviour of transition metals, which provide the cation or central atom. Ligands include common inorganic anions such as Cl^-, F^-, Br^-, SO_4^{2-}, PO_4^{3-} and CO_3^{2-} as well as organic molecules such as amino acids. Such complexation may facilitate the transport of metals.

Biodegradation is a reaction process mediated by microbial activity (a biotic reaction). Naturally present bacteria may transform the organic molecule to a simpler product, e.g. another organic molecule or even CO_2. Biodegradation has wide applicability to many organic chemicals in a diverse range of subsurface environments. Rates of biodegradation vary widely, some compounds may only degrade very slowly, e.g. high molecular weight polynuclear aromatic hydrocarbons (PAHs) that are relatively recalcitrant (unreactive). Rates are also very dependent upon environmental conditions, including redox, microbial populations present and their activity towards contaminants present.

Abiotic reactions, classic chemical reactions not mediated by bacteria, have been found to be of fairly limited importance in groundwater relative to biodegradation. For example, a few organics, e.g. 1,1,1-trichloroethane (1,1,1-TCA) and some pesticides, may readily undergo reaction with water (hydrolysis), others such as the aromatic hydrocarbon benzene are essentially unreactive to water and a range of other potential chemical reactions.

Potential for attenuation

Potential for attenuation processes to occur varies within the various subsurface zones, i.e. soil, unsaturated and saturated zone. Attenuation processes can be more effective in the soil rather than aquifers due to higher clay contents, organic carbon, microbial populations and replenishable oxygen. This makes the soil a very important first line of defence against groundwater pollution, often termed 'protective layer'. Consideration of the soil and its attenuation properties is a key factor in assessing the vulnerability of groundwater to pollution (Chapter 8). This also means that where the soil is thin or absent the risk of groundwater pollution may be greatly increased. Many human activities that give rise to pollution by-pass the soil completely and introduce pollutants directly into the unsaturated or even saturated zones of aquifers. Examples include landfills, leaking sewers, pit latrines, transportation routes in excavated areas and highway drainage.

4.2 NATURAL INORGANIC CONSTITUENTS

The occurrence of natural constituents in groundwater varies greatly depending on the nature of the aquifer. In general, aquifers in magmatites and metamorphic rocks show lower dissolved contents than in carbonate or sedimentary rocks. The mobility and thus the concentration of nearly all natural groundwater constituents can be significantly influenced by changes of physical and chemical conditions in groundwater through human activities.

Fluoride and arsenic are now recognized as the most serious inorganic contaminants in drinking-water on a worldwide basis. Further natural constituents that can cause a public health risk addressed in this chapter (in alphabetical order) are selenium, radon and uranium. Although nitrate has occasionally been found naturally in health-relevant concentrations, in most cases these are caused anthropogenically, and therefore nitrate is addressed in Section 4.3.

NOTE ▶ *Arsenic, fluoride, selenium, radon and uranium are examples of health-relevant naturally occurring groundwater constituents. Their concentrations in groundwater are strongly dependant on hydrogeological conditions. In some settings, nitrate may naturally occur in health-relevant concentrations.*

4.2.1 Arsenic

Health aspects

The International Agency for Research on Cancer (IARC) has classified arsenic (As) as a Group 1 human carcinogen (IARC, 2001). The health effects of arsenic in drinking-water include skin cancer, internal cancers (lung, bladder, kidney) and peripheral vascular disease (blackfoot disease). Evidence of chronic arsenic poisoning includes melanosis (abnormal black-brown pigmentation of the skin), hyperkeratosis (thickening of the soles of the feet), gangrene and skin and bladder cancer (WHO, 2003a). Arsenic toxicity may not be apparent for some time but the time to appearance of symptoms and the severity of effects will depend on the concentration in the drinking-water, other sources of exposure, dietary habits that may increase arsenic concentrations in staple dishes and a variety of other possible nutritional factors.

While earlier maximum allowable concentrations recommended by WHO for arsenic in drinking-water were higher, in 1993 the provisional WHO guideline value for arsenic in drinking-water was reduced to 10 µg/l based on concerns regarding its carcinogenicity in humans (WHO, 2004a). Regulations in some countries, e.g. the European Union (EU), Japan and the USA follow this guideline value and Australia has established a drinking-water standard for arsenic of 7 µg/l. A number of countries operate at present at a 50 µg/l standard, which corresponds to the provisional WHO guideline value before 1993. Some national authorities are currently seeking to reduce their own limits in line with the WHO guideline value. It is important to realize that the WHO Guidelines emphasize the need for adaptation of national standards to local health priorities, social, cultural, environmental and economic conditions and also advocate progressive improvement that may include interim standards. Furthermore, the WHO Guidelines emphasises the scientific uncertainty of the dose-response curves at low intakes and thus in deriving the guideline value of 10 µg/l. For improving public health benefits, other issues may therefore take priority over upgrading the sensitivity of analytical facilities for detecting lower concentrations or investing in upgrading drinking-water supplies to reduce arsenic concentrations to levels below 50 μg/l.

In recent years both the WHO guideline value and current national standards for arsenic have been found to be frequently exceeded in drinking-water sources. The scale of the arsenic problem in terms of population exposed to high arsenic concentrations is greatest in Bangladesh with between 35 and 77 million people at risk (Smith *et al.*, 2000). However, many other countries are also faced with elevated arsenic concentrations in groundwater, such as Hungary, Chile, Mexico, northeast Canada, the western USA and many countries in South Asia. More detailed information on occurrence and health significance of arsenic can be found in '*Arsenic in Drinking-water*' (WHO, In Press).

Occurrence

Arsenic is an ubiquitous element found in soils and rocks, natural waters and organisms. It occurs naturally in a number of geological environments, but is particularly common in regions of active volcanism where it is present in geothermal fluids and also occurs in sulphide minerals (principally arsenopyrite) precipitated from hydrothermal fluids in metamorphic environments (Hem, 1989). Arsenic may also accumulate in sedimentary environments by being co-precipitated with hydrous iron oxides or as sulphide minerals in anaerobic environments. It is mobilized in the environment through a combination of natural processes such as weathering reactions, biological activity and igneous activity as well as through a range of anthropogenic activities. Of the various routes of exposure to arsenic in the environment, drinking-water probably poses the greatest threat to human health.

Background concentrations of arsenic in groundwater in most countries are less than 10 μg/l. However, surveys performed in arsenic-rich areas showed a very large range, from <0.5 to 5000 μg/l (Smedley and Kinniburgh, 2001). Cases of large scale naturally occurring arsenic in groundwater are mainly restricted to hydrogeological environments characterized by young sediment deposits (often alluvium), and low-lying flat conditions with slow-moving groundwater such as the deltaic areas forming much of Bangladesh. Investigations by WHO in Bangladesh indicate that 20 per cent of 25 000 boreholes tested in that country have arsenic concentrations that exceed 50 μg/l. High concentrations of arsenic in groundwater also occur in regions where oxidation of sulphide minerals (such as arsenopyrite) has occurred (Alaerts *et al.*, 2001).

Arsenic concentration in German groundwater downstream of abandoned waste disposal sites was found to have a mean concentration of 61 μg/l (n = 253 sites) due to arsenic leaching from domestic coal ashes deposited with household wastes. In contrast, the mean arsenic concentration in uncontaminated aquifers is 0.5 μg/l (n = 472 sites) (Kerndorff *et al.*, 1992).

Transport and attenuation

The concentration of arsenic in natural waters is normally controlled by solid-solution interactions, particularly in groundwater where the solid/solution ratio is large. In most soils and aquifers, mineral arsenic interactions are likely to dominate over organic matter-arsenic interactions, although organic matter may interact to some extent through its reactions with the surfaces of minerals (Smedley and Kinniburgh, 2001). One of the best correlations between the concentration of arsenic in sediments and other elements is with iron. These interactions have also been the basis for the use of iron, aluminium and manganese salts in water treatment for arsenic removal.

Arsenic shows a high sensitivity to mobilization at the pH values typically found in groundwater (pH 6.5-8.5) and under both oxidizing and reducing conditions. Arsenic can occur in the environment in several oxidation states (-3, 0, +3 and +5) but in natural waters is mostly found in inorganic oxyanion forms as trivalent arsenite (As(III)) or pentavalent arsenate (As(V)). Redox potential (E_h) and pH are the most important factors controlling arsenic speciation. Relative to the other oxyanion-forming elements, arsenic is among the most problematic in the environment because of its mobility over a wide range of redox conditions (Smedley and Kinniburgh, 2001). Under oxidizing conditions, $H_2AsO_4^-$ is dominant at low pH (less than ~pH 6.9), while at higher pH, $HAsO_4^{2-}$ becomes dominant (H_3AsO_4 and AsO_4^{3-} may be present in extremely acidic and alkaline conditions, respectively). Under reducing conditions at less than ~pH 9.2, the uncharged arsenate(III)-species (H_3AsO_3) will predominate.

Transport is largely controlled by the aquifer conditions, particularly by adsorption on ferric oxohydroxides, humic substances and clays. Arsenic adsorption is most likely to be non-linear, with the rate of adsorption disproportionally decreasing with increasing concentrations in groundwater. This leads to reduced retardation at high concentrations. Since different arsenic species exhibit different retardation behaviour, arsenate (V) and arsenate (III) should travel through an aquifer with different amounts of interactions resulting in different velocities and increased separation along a flow path. This was demonstrated by Gulens *et al.* (in Smedley and Kinniburgh, 2001) using controlled soil-column experiments and various groundwaters. They showed that: (i) As(III) moved five to six times faster than As(V) under oxidizing conditions (at pH 5.7); (ii) with a 'neutral' groundwater (pH 6.9) under oxidizing conditions, As(V) moved much faster than under (i) but was still slower than As(III); (iii) under reducing conditions (at pH 8.3), both As(III) and As(V) moved rapidly through the column; (iv) when the amount of arsenic injected was substantially reduced, the mobility of the As(III) and As(V) was greatly reduced.

There is no process in the subsurface that alters arsenic species beside precipitation and adsorption. If groundwater with elevated arsenic levels is used for drinking-water supply, then treatment should be applied. There has been increasing research into this area and a number of low-cost household treatment technologies are available. Data from studies in Bangladesh suggest that low-cost technologies can remove arsenic to below 0.05 mg/l and sometimes lower (Ahmed *et al.*, 2001). Technologies are also available for system treatment including activated alumina, chemical precipitation and reverse osmosis (for arsenate). However, in some situations, source substitution or mixing is preferable to arsenic removal (Alaerts *et al.*, 2001).

4.2.2 Fluoride

Health aspects

Because fluoride is widely dispersed in the environment, all living organisms are widely exposed to it and tolerate modest amounts. In humans, fluoride has an affinity for accumulating in mineralizing tissues in the body, in young people in bone and teeth, in older people in bone, and incorporation of fluoride into the matrix of teeth during their formation is protective against dental caries.

Health problems associated with the condition known as fluorosis may occur when fluoride concentrations in groundwater exceed 1.5 mg/l: staining of the tooth enamel may become apparent (dental fluorosis) and, with continued exposure, teeth may become extremely brittle. The incidence and severity of dental fluorosis, and the much more serious skeletal fluorosis, depend on a range of factors including the quantity of water drunk and exposure to fluoride from other sources, such as from high fluoride coal in China. Nutritional status may also be important. Estimates based on studies from China and India indicate that for a total intake of 14 mg/day there is a clear excess risk of skeletal adverse effects, and there is suggestive evidence of an increased risk of effects on the skeleton at total fluoride intakes above about 6 mg/day (WHO, 2004b). In its most severe form, this disease is characterized by irregular bone deposits that may cause arthritis and crippling when occurring at joints.

The WHO guideline value for fluoride is 1.5 mg/l since 1984 (WHO, 2004a). The EU maximum admissible concentration for fluoride in drinking-water is 1.5 mg/l. The US Environmental Protection Agency (US EPA) set an enforceable primary maximum contaminant level of 4 mg/l in water systems to prevent crippling skeletal fluorosis. A secondary contaminant level of 2 mg/l was recommended by US EPA to protect against objectionable dental fluorosis. In setting national standards for fluoride, it is particularly important to consider volumes of water intake (which are affected by climatic conditions) and intake of fluoride from other sources (e.g. food, air). Where higher fluoride concentrations occur in groundwater used as drinking-water source, treatment and/or change or mixing with other waste sources containing lower fluoride levels is necessary in order to meet drinking-water standards. In areas with high natural fluoride levels, the public health benefits of investments in the treatment necessary to meet the WHO guideline value may need to be balanced against other priorities for optimising public health benefits.

More detailed information on occurrence and health significance of fluoride can be found in 'Fluorides in Drinking Water' (Bailey *et al.*, In Press).

Occurrence

Fluoride (F^-) naturally occurs in rocks in many geological environments (Hem, 1989) but fluoride concentrations in groundwater are particularly high in groundwater associated with acid volcanic rocks, e.g. in Sudan, Ethiopia, Uganda, Kenya and Tanzania (Bailey *et al.*, In Press). High concentrations of fluoride also occur in some metamorphic and sedimentary rocks that contain significant amounts of fluoride-bearing minerals such as fluorite and apatite. Fluoride in water supply based on groundwater is a problem in a number of countries and over 70 million people worldwide are believed to be at risk of adverse health effects from consumption of water containing high levels of fluoride. India and China have particular problems and estimates suggest up to 60 million are affected in these two countries alone.

Exposure to fluoride from drinking-water depends greatly on natural circumstances. Levels in raw water are normally below 1.5 mg/l, but groundwater has been found to contain >50 mg/l in some areas rich in fluoride-containing minerals. For example, in Kenya, 61 per cent of groundwater samples collected nationally from drinking-water wells exceeded 1 mg/l (Bailey *et al.*, In Press). In general high fluoride concentrations in groundwater show a strong positive correlation with dissolved solids, sodium, and alkalinity, and a strong negative correlation with hardness.

Transport and attenuation

The concentration of fluoride ions in groundwater is driven by the presence of calcium ions and the solubility product of fluorite (CaF_2). In equilibrium, a calcium concentration of 40 mg/l equates to a concentration of 3.2 mg/l fluoride. In groundwater with a high concentration of calcium ions, fluoride concentrations rarely exceed 1 mg/l. Substantially higher fluoride concentrations in groundwater are usually caused by a lack of calcium. During high percolation rates, Flühler *et al.* (1985) observed increased fluoride concentration in the leachate of fluoride-enriched soils due to a limited additional delivery of calcium.

In groundwater with a high pH (>8) and dominated by sodium ions and carbonate species, fluoride concentrations commonly exceed 1 mg/l, and concentrations in excess of 50 mg/l have been recorded in groundwater in South Africa, and in Arizona in the USA (Hem, 1989). Moreover, the fluoride-ion (F^-) can interact with mineral surfaces, but is substituted by hydroxyl-ions at high pH values. Hem (1989) observed a fluoride concentration of 22 mg/l in a caustic thermal groundwater (pH 9.2, 50 °C) in Owyhee County, Idaho. Fluoride ions form strong complexes especially with aluminium, beryllium and iron (III).

4.2.3 Selenium

Health aspects

Selenium is an essential trace element with a physiologically required intake of about 1 µg per kg body weight and day for adults. Deficiencies of selenium in diets may cause a number of health effects, although few reports of clinical signs of deficiency are available. However, the range of concentrations of this element in food and water that provide health benefits appears to be very narrow. When ingested in excess of nutritional requirements in food and drinking-water, selenium can cause a number of acute and chronic health effects including damage to or loss of hair and fingernails, finger deformities, skin lesions, tooth decay and neurological disorders (WHO, 2003b).

Although drinking-water generally accounts for less than 1 per cent of the typical dietary intake of selenium, in some circumstances naturally-occurring concentrations of selenium in groundwater may be sufficiently high to cause health problems. The WHO guideline value for selenium in drinking-water is 0.01 mg/l (WHO, 2004a).

Occurrence

Selenium has similar chemical properties and behaviour to sulphur (Hem, 1989), and is commonly associated with metal sulphide minerals in mineral deposits in a wide range of igneous rocks and with sulphur-rich coal. Sedimentary rocks and overlying soil in some regions may have high background concentrations of selenium. In the western part of the USA, these are associated with uranium and vanadium mineralization in shales and sandstones. In some semi-arid areas in China and India, selenium reaches high concentrations in soil and accumulates in plant tissue. Runoff from irrigated agriculture on seleniferous soil may contain dissolved selenium concentrations of up to 1 mg/l (Hem, 1989), and groundwater in these areas also typically contains high concentrations of leached selenium (Barceloux, 1999). Groundwater concentrations of selenium rarely exceed 1 μg/l (Hem, 1989), but up to 6000 μg/l have been reported (WHO, 2003b), and high concentrations (tens to hundreds of micrograms per litre) may occur in surface water and groundwater near metal-sulphide mine sites.

Selenium concentrations are often particularly high in surface waters and groundwater in coal mining areas where solid wastes and wastewater from coal power stations are disposed to the environment (Barceloux, 1999; US EPA, 2000).

Transport and attenuation

Selenium can exist in nature in four oxidation states: 0 (elemental selenium), -2 (selenide), +4 (selenite) and +6 (selenate). Under oxidizing conditions, the selenium occurs predominantly as selenite (SeO_3^{2-}) and selenate (SeO_4^{2-}) ions in natural waters. These ions have a very high solubility, and can reach very high concentrations in conditions when water is being subjected to high rates of evapotranspiration such as in regions with semi-arid or arid climates. Selenate and selenite minerals can accumulate with sulphates in soils in regions with semi-arid or arid climates.

High concentrations of selenium may also occur in groundwater beneath areas where intense irrigated agriculture flushes selenium compounds through the soil profile, and if groundwater pumping rates are high, the concentration of selenium may be progressively increased by the recycling of salts by the process of pumping, evaporation and recharge of pumped effluent. Consequently, selenium concentrations in shallow groundwater and in drainage from irrigated agriculture on seleniferous soils are often highly toxic to wildlife that ingests the water, as in the widely studied case of the Kesterson National Wildlife Refuge in the San Joaquin Valley of California (NRC, 1989). This water is also potentially toxic to humans who might use shallow groundwater as a drinking-water source, although water contaminated with high selenium concentrations is often too saline for potable use.

Under reducing conditions in groundwater or in marshes, selenium can also be removed from water through co-precipitation with sulphide minerals such as pyrite (FeS_2) or the precipitation of ferroselite ($FeSe_2$); through volatilization as dimethyl selenide or hydrogen selenide, or through the uptake of organo-selenium compounds by plants. Consequently, anaerobic bioreactors or artificial wetlands are being used for selenium removal from water, predominantly to protect receiving environments from the discharge of wastewater contaminated by selenium.

Selenium can be removed from water by adsorption onto iron oxyhydroxide minerals (especially ferrihydrite) and this is one of the preferred water treatment methods. Selenium can also be removed from drinking-water by reverse osmosis and through the use of anion-exchange resins.

4.2.4 Radon

Health aspects

Radon is a radioactive gas emitted from radium, a daughter product of uranium that occurs naturally in rocks and soil. The main health effect of radon is to cause lung cancer. Radon, together with its decay products, emits alpha particles that can damage lung tissue. Although most radon is exhaled before it can do significant damage, its decay products can remain trapped in the respiratory system attached to dust, smoke and other fine particles from the air.

The global average human exposure to radiation from natural sources is 2.4 mSv per year with an average dose from inhalation of radon of 1.2 mSv per year. There are large local variations in this exposure depending on a number of factors, such as height above sea level, the amount and type of radionuclides in the soil, and the amount taken into the body in air, food, and water (WHO, 2004a). Unlike most other naturally occurring groundwater contaminants, most of the health effects of radon in groundwater are considered to be due to its contribution to indoor air quality rather than due to effects caused by direct ingestion of water. UNSCEAR has calculated the average doses from radon in drinking-water as low as 0.025 mSv/year via inhalation and 0.002 mSv/year from ingestion as compared to the inhalation dose from radon in the air of 1.1 mSv/year (UNSCEAR, 2000). The WHO has recommended a reference level of committed effective dose of 0.1 mSv from 1 year's consumption of drinking-water (WHO, 2004a).

Stirring and transferring water from one container to another will liberate dissolved radon. Water that has been left to stand will have reduced radon activity, and boiling will remove radon completely. As a result, it is important that the form of water consumed is taken into account in assessing the dose from ingestion. Moreover, the use of water supplies for other domestic purposes will increase the levels of radon in the air, thus increasing the dose from inhalation. This dose depends markedly on the form of domestic usage and housing construction (NCRP, 1989). The form of water intake, the domestic use of water and the construction of houses vary widely throughout the world. It is therefore not possible to derive an activity concentration for radon in drinking-water that is universally applicable.

WHO recommends implementing controls if the radon concentration of drinking-water exceeds 100 Bq/litre (WHO, 2004a), and the EU likewise recommends assessing the need for protective measures at concentrations above this level (Euratom 2001/928; CEC, 2001).

Occurrence

Radon (Rn) is a naturally occurring, colourless, odourless gaseous element that is soluble in water. It occurs naturally only as a product of the radioactive decay of radium, itself a radioactive decay product of uranium. As is the case for uranium, concentrations of radon are directly related to the local geology, and are particularly high in granitic rocks and pegmatites and sediments with phosphate nodules or heavy mineral sand deposits.

Radon-222 is a frequently encountered radioactive constituent in natural waters and typically exceeds the concentration of other radionuclides, including uranium, thorium and radium, by orders of magnitude. High radon emanation, especially along fracture surfaces, contributes significantly to radon concentrations in groundwater. Data from sampling campaigns indicate that there is a great degree of variability in the radon-222 concentration of samples drawn from any given rock type. The United States Geological Survery (USGS) conducted a study on occurrence of dissolved radon in groundwater in Pennsylvania (Senior, 1998). Findings of this study indicated that rock types with the highest median radon concentrations in groundwater include schist and phyllite (2400 pCi/l) as well as quartzite (2150 pCi/l). The geohydrologic groups with lowest median radon concentrations in ground water include carbonate rocks (540 pCi/l) and other rocks (360 pCi/l). Water from wells in gneiss had a median radon concentration of 1000 pCi/l, and water from wells in Triassic-age sedimentary rocks had a median radon concentration of 1300 pCi/l. Radon concentrations generally do not correlate with well characteristics, the pH of water or concentrations of dissolved major ions and other chemical constituents in the water samples.

Transport and attenuation

The rate of radon's radioactive decay is defined by its half-life, which is the time required for one half of the amount of radon present to break down to form other elements. The half-life of radon is 3.8 days. Several factors probably control the concentration of radon-222 in a water supply. The flux of radon-222 within the ground may be controlled by the radium-226 concentration in the surrounding rocks, the emanation fraction for the radon-222 from the rock matrix and the permeability of the rock to radon-222 movement. For a given flux, the concentration of radon-222 in a water supply would then also be controlled by the ratio of aquifer surface area to volume.

4.2.5 Uranium

Health aspects

Uranium is a heavy metal of toxicological rather than radiological relevance in drinking-water. In particular, it is of concern because of its impact on kidney function following long-term exposure. Because of uncertainties regarding the toxicity of uranium for human beings the WHO has proposed a provisional drinking-water guideline value of 15 μg/l (WHO, 2004a; 2005a). The US EPA maximum contaminant level for uranium in drinking-water is 30 μg/l.

Occurrence

Uranium (U) is widely distributed in the geological environment, but concentrations in groundwater are particularly high in granitic rocks and pegmatites, and locally in some sedimentary rocks like sandstones. Uranium often occurs in oxidizing and sulphate-rich groundwater. There are three naturally occurring isotopes of uranium: ^{234}U (<0.01 per cent), ^{235}U (0.72 per cent), and ^{238}U (99.27 per cent). All three isotopes are equally toxic.

Concentrations of uranium in natural waters usually range between 0.1 and 10 μg/l (Hem, 1989), but are often up to 100 μg/l in groundwater in areas underlain by granitic rocks, and may exceed 1 mg/l near uranium mineral deposits.

Transport and attenuation

The transport of uranium in groundwater varies widely according to the aquifer conditions. In anoxic conditions, uranium is reduced to U(IV) which is relatively insoluble and precipitates. In oxidizing environments, uranium exists mainly as UO_2X_2-(= uranyl)-compounds with U(VI) which is considerably more soluble. Even with the higher solubility of U(VI), transport of U(VI) can be limited as it sorbs strongly to solid surfaces at circum-neutral pH. Very low and very high pH conditions limit sorption as does the presence of certain complexing ligands such as natural organic matter, organic chelating agents and carbonate, all of which can significantly enhance the transport of uranium.

4.3 NITROGEN SPECIES

Ammonia, nitrate and nitrogen containing organic compounds of humic type are the dominating nitrogen compounds in groundwater. Though nitrite is highly toxic, it usually occurs only in very low concentrations in groundwater and these are not relevant to human health. However, nitrite can become relevant from conversion of ammonia or nitrogen in the drinking-water supply system or human body.

NOTE ▶ *Though nitrogen may occur naturally in groundwater, the main sources of groundwater pollution are human activities such as agriculture and sanitation (see Chapters 9 and 10).*

Health aspects

Ammonia in drinking-water is not of direct health relevance, and therefore WHO have not set a health-based guideline value. However, ammonia can compromise disinfection efficiency, can cause the failure of filters for the removal of manganese, and can cause taste and odour problems. Also in distribution systems it can lead to nitrite formation which is of health relevance.

The toxicity of nitrate to humans is mainly attributable to its reduction to nitrite. Nitrite, or nitrate converted to nitrite in the body, causes a chemical reaction that can lead to the induction of methaemoglobinaemia, especially in bottle-fed infants. Methaemoglobin (metHb), normally present at 1-3 per cent in the blood, is the oxidized form of haemoglobin (Hb) and cannot act as an oxygen carrier in the blood. The reduced oxygen transport becomes clinically manifest when the proportion of metHb concentration reaches 5-10 per cent or more of normal Hb values (WHO, 1996a). Nitrate is enzymatically reduced in saliva forming nitrite. Additionally, in infants under one year of age the relatively low acidity in the stomach allows bacteria to form nitrite. Up to 100 per cent of nitrate is reduced to nitrite in infants, as compared to 10 per cent in adults and children over one year of age. When the proportion of metHb reaches 5-10 per cent, the symptoms can include lethargy, shortness of breath and a bluish skin colour ('blue baby syndrome'). Anoxia and death can occur at very high uptakes of nitrite and nitrate from drinking-water.

Methaemoglobinaemia is observed in populations where food for infant formula in prepared with water containing nitrate in excess of around 50 mg/l, but other factors are also involved in disease causation. The risk is enhanced by sewage contamination. This contributes nitrate and renders chemical conditions in the water to be reducing, thus supporting the presence of nitrate reducing bacteria. Moreover, ingestion of microbially contaminated water causes gastroenteritis infection which would also predispose the infant to a nitrate reducing conditions and thereby more nitrite exposure (WHO, 2004a). A review of numerous case studies of water-related infant methaemoglobinaemia in the 1980s indicated high correlation with microbial contamination of the water (US EPA, 1990; WHO, 2005b).

The weight of evidence is strongly against an association between nitrite and nitrate exposure in humans and the risk of cancer (WHO, 2004a). Studies in laboratory animals demonstrate increased tumour incidence only after exposure to extremely high levels of nitrite in the order of 1000 mg/l in drinking-water and simultaneously high levels of nitrosatable precursors (WHO, 1996b). At lower nitrite levels, tumour incidence resembled those of control groups treated with the nitrosatable compound only. On the basis of adequately performed and reported studies, it may be concluded that nitrite itself is not carcinogenic to animals (WHO, 1996a).

Based on methaemoglobinaemia in infants (an acute effect), the WHO has established a guideline value for nitrate ion of 50 mg/l as NO_3^- and a provisional guideline value for nitrite of 3 mg/l as NO_2^- (WHO, 2004a). Because of the possibility of simultaneous occurrence of nitrite and nitrate in drinking-water, the sum of the ratios of the concentrations ($C_{nitrate}$ or $C_{nitrite}$) of each to its guideline value ($GV_{nitrate}$ or $GV_{nitrite}$) should not to exceed one.

Sources and occurrence

Nitrogen is present in human and animal waste in organic form, which may then subsequently be mineralized to inorganic forms. Ammonia (ionized as NH_4^+, non-ionized as NH_3) as well as urea $(NH_2)_2CO$ is a major component of the metabolism of mammals. Ammonia in the environment mainly results from animal feed lots and the use of manures in agriculture (Chapter 9), or from on-site sanitation or leaking sewers (Chapter 10). Thus ammonia in water is often an indicator of sewage pollution. The nitrite ion (NO^-_2) contains nitrogen in a relatively unstable oxidation state. Nitrite does not typically occur in natural waters at significant levels, except temporarily under reducing conditions. Chemical and biological processes can further reduce nitrite to various compounds or oxidize it to nitrate. The nitrate ion (NO^-_3) is the stable form of combined nitrogen for oxygenated systems. Nitrate is one of the major anions in natural waters, but as for ammonia, concentrations can be greatly elevated due to agricultural activities (Chapter 9), and sanitation practices (Chapter 10).

Natural levels of ammonia in ground and surface waters are usually below 0.2 mg/l, and nitrate concentrations in groundwater and surface water typically range between 0-18 mg/l as NO_3^-. Although elevated concentrations of nitrate in groundwater are mostly caused by agricultural activity or sanitation practices, natural nitrate concentrations can also exceed 100 mg/l as NO_3^- as observed in some arid parts of the world such as the Sahel and north Africa (Edmunds and Gaye, 1994) and the arid interior of Australia (Box 4.1).

Box 4.1. Naturally-occurring high nitrate in Australia

High groundwater nitrate concentrations have been observed in the arid interior of Australia, commonly exceeding 45 mg/l, and often exceeding 100 mg/l in groundwater which otherwise meets national and international drinking-water guidelines (Lawrence, 1983; Barnes *et al.*, 1992). The nitrate in this region is partially derived from nitrogen fixing by native vegetation, and by cyanobacteria crusts on soils. Termite mounds appear to be a significant contributory source of the nitrate (Barnes *et al.*, 1992), possibly due to the presence of nitrogen fixing bacteria in many termite species, and the nitrogen-rich secretions used to build the walls of the mounds. Nitrate is leached to the water table in arid Australia after periodic heavy rainfall events, particularly after bush fires that allow soluble nitrate salts to accumulate in soils. Denitrification in these soils appears to be inhibited by low carbon levels.

Transport and attenuation

Ammonium (NH_4^+) shows a high tendency for adsorption to clay minerals, which limits its mobility in the subsurface (saturated and unsaturated zones). In contrast, interactions between minerals and nitrate or nitrite are usually negligible and both ions are mobile in the subsurface.

Under aerobic conditions in the subsurface oxidation of ammonium through nitrite to nitrate by microorganisms is the only process where nitrate is formed in natural systems.

DEF ▶ ***Nitrification*** *is the biological conversion of ammonium through nitrite to nitrate.* ***Denitrification*** *is the biological process of reducing nitrate to ammonia and nitrogen gas.*

Despite the natural high concentrations of nitrate in groundwater in much of inland Australia, there have been no verified cases of MetHb in Aboriginal people (Hearn *et al.*, 1993), who are the main users of groundwater in this part of the country. Because potable quality groundwater is scarce in the interior of Australia, and because the use of water is vital for maintaining hygiene in the region, the National Health and Medical Research Council revised the national water quality guidelines in 1990. The revised guidelines allow the use of groundwater with concentrations of nitrate exceeding 100 mg/l for all non-potable needs, up to 100 mg/l for potable use except for infants under 3 months old, and up to 50 mg/l for infants under 3 months old. Although technologies exist to remove nitrate from drinking-water using microbial denitrification, the equipment is difficult to maintain in remote aboriginal settlements, and it was considered in this case that changing guideline concentrations would produce better health outcomes. These changes were incorporated into the Australian drinking-water guidelines in 1996. The autotrophic conversion of ammonia to nitrite and nitrate (nitrification) requires oxygen. The discharge of ammonia nitrogen into groundwater and its subsequent oxidation can thus seriously reduce the dissolved oxygen content in shallow groundwater, especially where high ammonia loads are applied and re-aeration of the soil is limited.

In the absence of dissolved oxygen (such as in some deep or confined groundwaters), denitrification can occur, driven by denitrifying bacteria. Under fully anaerobic conditions, in an aquifer where predominantly sulphides serve as reduction agents, the microbial oxidation of sulphides into sulphate and simultaneous reduction of nitrate to nitrogen gas can occur which also reduces the nitrate content.

As microbial processes, both nitrification and denitrification are affected by many factors that are of importance to microbial activity. Nitrification and denitrification are optimal at about 25°C and are inhibited at 10°C or less. Other regulating factors are pH and all factors affecting the diffusion of oxygen such as soil density, grain structure, porosity and soil moisture. Warm, moist and well aerated soils provide ideal conditions for nitrification. Denitrification occurs only under anoxic or almost anoxic conditions. Beside the presence of nitrate, the denitrifying bacteria require a carbon source. A soil moisture of more than 80 per cent has been found to be essential for denitrification. Thus in many settings natural attenuation can substantially reduce nitrate concentrations in groundwater over time, but rates of attenuation strongly depend on conditions in the aquifer.

4.4 METALS

The following focuses on those metals which are toxic to humans and which have frequently been observed as groundwater contaminants in connection with human activities and/or have physical and chemical properties which make them potential groundwater contaminants, i.e. cadmium (Cd), lead (Pb), nickel (Ni), chromium (Cr), and copper (Cu).

Health aspects

Cadmium has a high renal toxicity, which is not only due to its mode of action but also to its irreversible accumulation in the kidney. The health based guideline value for cadmium in drinking-water is 3 μg/l (WHO, 2004a).

Lead is a strong neurotoxin in the unborn, newborn and young children. It crosses the placenta easily and is toxic to both the central and the peripheral nervous system, thus causing cognitive and behavioural effects (WHO, 2004a). The threshold of neurotoxicological concern, defined as a group based mean blood lead level, has decreased continually during the last 10 to 20 years, and epidemiological evidence indicates lead levels above 30 μg of lead per litre of blood to be associated with intelligence quotient deficits in children (WHO, 2004a). The use of lead in antiknock and lubricating agents in petrol is being phased out in many countries, thus decreasing this source of contamination. However, a major main source of exposure to lead through water is household plumbing systems, i.e. pipes, fittings, solder and connections from the mains to homes. Dissolution from such materials strongly depends on chemical properties of the drinking-water, with soft, acidic water dissolving the largest amount. The WHO guideline value for lead in drinking-water is 10 μg/l (WHO, 2003c; 2004a).

The significance of Nickel from the health point of view is mainly due to its high allergenic potential. The WHO drinking-water guideline value for the protection of sensitive persons is 20 μg/l (WHO, 2004a).

Chromium can be found in the environment in two valency states, Cr(III) and Cr(VI). The former predominates in soils, whereas the latter occurs exclusively as chromate (CrO_4^{2-}) from anthropogenic sources. Cr(VI) is the form which is of toxicological significance because of its easy uptake into cells together with SO_4^{2-} and PO_4^{2-}. Within cells and during its reduction to Cr(III), the chromate ion represents a considerable genotoxic and clastogenic potential (Costa, 2002). However, since even very high doses of Cr(VI) are subjected to rapid chemical reduction in the upper gastrointestinal tract (Kerger *et al.*, 1997), only negligible amounts of Cr(VI) should reach the blood compartment and other body fluids and organs. The health based guideline value for chromium in drinking-water is 50 μg/l (WHO, 2004a), and while higher concentrations have been reported from some drinking-water supplies, most studies indicate the concentration of chromium in groundwater to be low (WHO, 2003d). Cr(III) in drinking-water may eventually be oxidized to Cr(VI) during its ozonation.

Copper is an essential trace element with an optimal daily oral intake of 1-2 mg per person. Naturally occurring copper concentrations in groundwater are without any health significance and scatter mostly around 20 μg/l. If drinking-water drawn from groundwater contains elevated levels, in most situations corrosion of copper pipes is the primary source. Mean concentrations of more than 2 mg/l could lead to liver cirrhosis in babies if their formula is repeatedly prepared using such water (Zietz *et al.*, 2003). The prevalent endpoint of acute copper toxicity by time, concentration and dose is nausea (Araya *et al.*, 2003). The health based guideline value for copper in drinking-water is 2 mg/l (WHO, 2004a; 2004c).

Sources and occurrence

Metals from activities such as mining, manufacturing industries, metal finishing, wastewater, waste disposal, agriculture and the burning of fossil fuels can reach concentrations in groundwater which are hazardous to human health. Chapter 11 lists industry types together with the metals they commonly emit (see Table 11.2.) Metals are natural constituents in groundwaters, having their origin in weathering and solution of numerous minerals. However, natural concentrations of metals in groundwaters are generally low. Typical concentrations in natural groundwaters are <10 μg/l (copper, nickel), <5 μg/l (lead) or <1 μg/l (cadmium, chromium). Even so, the concentrations can locally increase naturally up to levels which are of toxicological relevance and can exceed drinking-water guidelines, e.g. in aquifers containing high amounts of heavy metal bearing minerals (ore). Metal concentrations in groundwater may be of particular concern where it is directly affected by manufacturing and mining as well as downstream of abandoned waste disposal sites. Another anthropogenic cause of elevated metal concentrations in groundwaters is the acidification of rain and soils by air pollution and the mobilization of metals at lower pH values. This problem predominantly appears in forested areas, because the deposition rates of the acidifying anions sulphur and nitrate from the atmosphere are evidently higher in forests due to the large surface of needles and leafs, and because soils in forests are generally poor in nutrients and have a low neutralization capacity against acids.

Transport and attenuation

Most of the metals of concern occur in groundwater mainly as cations (e.g. Pb^{2+}, Cu^{2+}, Ni^{2+}, Cd^{2+}) which generally become more insoluble as pH increases. At a nearly neutral pH typical for most groundwaters, the solubility of most metal cations is severely limited by precipitation as an oxide, hydroxide, carbonate or phosphate mineral, or more likely by their strong adsorption to hydrous metal oxides, clay or organic matter in the aquifer matrix. The adsorption decreases with decreasing pH. As a consequence, in naturally or anthropogenically acidified groundwaters metals are mobile and can travel long distances. Furthermore, as simple cations there is no microbial or other degradation.

In a soil solution containing a variety of heavy metal cations that tend to adsorb to particle surfaces, there is competition between metals for the available sites. Of several factors that determine this selectivity, ionic potential, which is equal to the charge of an ion over its ionic radius, has a significant effect. Cations with a lower ionic potential tend to release their solvating water molecules more readily so that inner sphere surface complexes can be formed. Selectivity sequences are arranged in order of decreasing ionic radius, which results in increasing ionic potential and decreasing affinity or selectivity for adsorption. As an example the following selectivity sequence of transition elements belonging to group IIb has been determined (Sposito, 1989):

$$Hg^{2+} > Cd^{2+} > Zn^{2+}$$

As a consequence, mercury is the most strongly adsorbed; this being the probable reason for its generally very low concentration occurrence in groundwater.

Metals within the transition group differ in that electron configuration becomes more important than ionic radius in determining selectivity. The relative affinity of some metals belonging to different transition groups is given by:

$$Cu^{2+} > Ni^{2+} > Co^{2+} > Fe^{2+} > Mn^{2+}$$

However, this sequence can be more or less changed in groundwater by naturally occurring complexing agents like fulvic acids which is especially true for copper (Schnitzer and Khan, 1972).

In addition, most oxyanions tend to become less strongly sorbed as the pH increases (Sposito, 1989). Therefore, the oxyanion-forming metals such as chromium are some of the more common trace contaminants in groundwater. Chromium is mobile as stable Cr(VI) oxyanion species under oxidizing conditions, but forms cationic Cr(III) species in reducing environments and hence behaves relatively immobile under these conditions. For example, in contaminated groundwater at industrial and waste disposal sites Chromium occurs as Cr^{3+} and CrO_4^{2-} species, with CrO_4^{2-} being much more toxic but less common than Cr^{3+}. In most aquifers chromium is not very mobile because of precipitation of hydrous chromium(III)oxide. In sulphur-rich, reducing environments, many of the trace metals also form insoluble sulphides (Smedley and Kinniburgh, 2001).

4.5 ORGANIC COMPOUNDS

Organic compounds in groundwater commonly derive from breakdown and leaching of naturally occurring organic material, e.g. from organic-rich soil horizons and organic matter associated with other geologic strata, or human activity, e.g. domestic, agricultural, commercial and industrial activities.

Natural sources will always contribute some organic compounds to groundwater, often at low levels. Natural organic matter comprises water-soluble compounds of a rather complex nature having a broad range of chemical and physical properties. Typically, natural organic matter in groundwater is composed of humic substances (mostly fulvic acids) and non-humic materials, e.g. proteins, carbohydrates, and hydrocarbons (Thurman, 1985; Stevenson, 1994). While natural organic matter is a complex, heterogeneous mixture, it can be characterized according to size, structure, functionality, and reactivity. Natural organic matter can originate from terrestrial sources (allochthonous natural organic matter) and/or algal and bacterial sources within the water (autochthonous natural organic matter). Dissolved organic carbon (DOC) is considered to be a suitable parameter for quantifying organic matter present in groundwater; however, DOC is a bulk organic quality parameter and does not provide specific identification data and may also incorporate organic compounds arising from human activity. Natural organic matter, although considered benign, may still indirectly influence groundwater quality. For example, contaminants may bind to organic-matter colloids allowing their facilitated transport within groundwater, a process proposed (but not proven) to be of most significance for the more highly sorbing organic compounds. Also, routine chlorination of water supplies containing natural organic matter may form disinfection by-products such as trihalomethanes. However, because of their low direct health relevance, natural organic substances are not addressed further herein.

Human activity has released a vast range of anthropogenic organic chemicals, commonly termed 'micro-pollutants', to the environment, some of which may detrimentally impact groundwater quality. This chapter focuses on commercially and industrially derived chemicals which (i) have a high toxicity, (ii) have physical and chemical properties facilitating their occurrence in groundwater and (iii) have been observed to occur frequently as groundwater contaminants. Chapter 11 lists industry types together with substances that may potentially be released to the subsurface from their respective industrial activities. The occurrence of organic pollutants in groundwater is controlled not only by their use intensity and release potential, but also by their physical and chemical properties which influence subsurface transport and attenuation. Discussion of this aspect specific to organic chemicals follows and extends the general concepts covered in Section 4.1.

4.5.1 Conceptual transport models for non aqueous phase liquids

A correct conceptual model of contaminant behaviour is essential for assessing subsurface organic contaminant migration. The classical near-surface leachable source zone – dissolved plume model presented earlier (Section 4.1.2, Figure 4.2) is not applicable for all organic substances. Of key importance is the recognition that organic chemicals have very different affinities for water, ranging from organic compounds that are hydrophilic ("love" water) to organics that are hydrophobic ("fear" water). Such concepts are used below to develop appropriate contaminant conceptual models followed by discussion of specific transport processes applicable within the models developed.

Water is a highly polar solvent, so polar in fact that it develops a hydrogen-bonded structure and will easily dissolve and solvate ionic species. The vast majority of organic compounds are covalent molecules, rather than ionic species, and most have a limited tendency to partition or dissolve into water. Further, many organic compounds found in groundwater are used as liquids, e.g. hydrocarbon fuels or industry solvents. A focus upon organic liquids is hence relevant. Organic compounds that most easily partition or dissolve into water tend to be small molecules, have a polar structure and may hydrogen-bond with water. Typically they have only a few carbon atoms and often contain oxygen. Examples include methanol (CH_3OH) and other short-chain alcohols, e.g. ethanol and propanols that may be used as de-icers, and ketones such as methyl-ethyl-ketone and ethers such as dioxane that are used as industrial solvents. Some compounds are so hydrophilic that they form a single fluid phase with the water and are said to be miscible with the water, e.g. methanol, acetone, dioxane.

Most organic compounds are, however, relatively hydrophobic as they are comparatively large molecules of limited polarity with low hydrogen-bonding potential. Most organic liquids are so hydrophobic that they form a separate organic phase to the water (aqueous) phase. They are immiscible with water and a phase boundary exists between the organic phase and the aqueous phase, with the organic phase generally being referred to as the non aqueous phase liquid (NAPL). When a separate organic NAPL exists it is important to consider the density of the NAPL relative to water as this controls whether the NAPL will be upper or lower phase relative to the water phase. Most hydrocarbon-based organic liquids have a density <1 (g/ml), e.g. benzene is 0.88 and pentane is 0.63 and when in contact with water will be the upper phase and "float" upon the water phase of density 1. Such "light" organic compounds are generally referred to as being LNAPLs .

In contrast, other hydrophobic organics have a relatively high density due to incorporation of dense chlorine (or other halogen) atoms in their structure and for example chlorinated solvents such as trichloroethene (TCE) and 1,1,1-trichloroethane (1,1,1-TCA) and polychlorinated biphenyl (PCB) mixtures have densities in the 1.1 to 1.7 range. Due to their density such organic phases will be the lower phase and 'sink' below the water phase. Such "dense" organic compounds are generally referred to as DNAPLs.

Although hydrophobic, LNAPL and DNAPL organics still have potential for some of their organic molecules to dissolve into the adjacent aqueous phase. The organics are 'sparingly soluble' and will have a finite solubility value in water leading to dissolved concentrations in the water phase. Solubility values achieved by individual organic compounds in water are highly variable between organics and controlled by their relative hydrophobicity. For example, small and/or polar organics have the greatest solubility with for example dichloromethane (DCM) (CH_2Cl_2) being one of the most soluble with a solubility of ca. ~20,000 mg/l, which contrasts with e.g. DDT, a large pesticide molecule that is not easily accommodated in the polar water structure and has a solubility of just about 0.1 mg/l. Similarly benzene, as single aromatic ring hydrocarbon, has a solubility ca. 1,800 mg/l that is much greater than benzo[a]pyrene, a PAH of solubility ca. 0.004 mg/l that is composed of five adjacent aromatic rings.

The above provides fundamental understanding for conceptual models of organic contaminant transport in the subsurface and why specific organic compounds have a tendency to occur or not occur in groundwater. Hydrophilic miscible organics behave similarly to the classical leachable source model (Figure 4.2). In essence, a spill of e.g. a de-icer fluid at surface would migrate as a concentrated organic-aqueous fluid through the unsaturated zone and then migrate laterally in the groundwater as a concentrated dissolved-phase plume. Importantly, hydrophobic immiscible organics, i.e. NAPLs, exhibit very different behaviour. Conceptual models for LNAPL releases and DNAPL releases (Mackay and Cherry, 1989) are shown in Figures 4.4 and 4.5.

NAPLs may migrate as a separate NAPL phase and displace air and water from the pores they invade if they have sufficient head (pressure) to overcome the entry pressure to the pores or fractures. This head is controlled by aspects such as the spill volume and rate and the vertical column of continuous NAPL developed in the subsurface. NAPL migration is also controlled by its density and viscosity. Petrol fuel and chlorinated solvents have viscosities lower than water and more easily migrate in the subsurface; in contrast, PCB oils or coal tar (PAH-based) hydrocarbons may be very viscous and perhaps take years for the NAPL to come to a resting position in the subsurface. Chlorinated solvents such as PCE have high densities and may penetrate to significant depths through aquifer systems in very short time periods. Whereas dissolved pesticides may take years to decades to migrate through a 30 m unsaturated zone, DNAPL may migrate through such a zone in the order of hours to days (Pankow and Cherry, 1996). DNAPLs may penetrate discrete sand horizons and hairline fractures in clays and compromise clay units that are normally an effective barrier to dissolved plume migration.

At the water table, LNAPLs, being less dense than water, will form a floating layer of LNAPL on the water table often slightly elongated in the direction of the water table hydraulic gradient. DNAPL, in contrast may penetrate as a separate immiscible DNAPL below the water table. Predominant movement will be vertically downward due to its density, but some lateral spreading will occur as it encounters lower permeability strata. If spilt in sufficient volume and with sufficient driving head, the DNAPL may penetrate the full aquifer depth to the underlying aquitard/bedrock (Kueper *et al.*, 1993). This should not be assumed to occur in all cases. Migrating NAPL leaves a trail of immobile residual NAPL droplets behind its migration pathway held by capillary forces causing NAPL to spread across an aquifer thickness. DNAPL accumulating on low permeability features, often referred to as pools, is potentially mobile. It may ultimately penetrate that formation due to changes in pressure arising from continued DNAPL spillage, pumping or remediation attempts or via drilling (for boreholes, piling etc) through that layer.

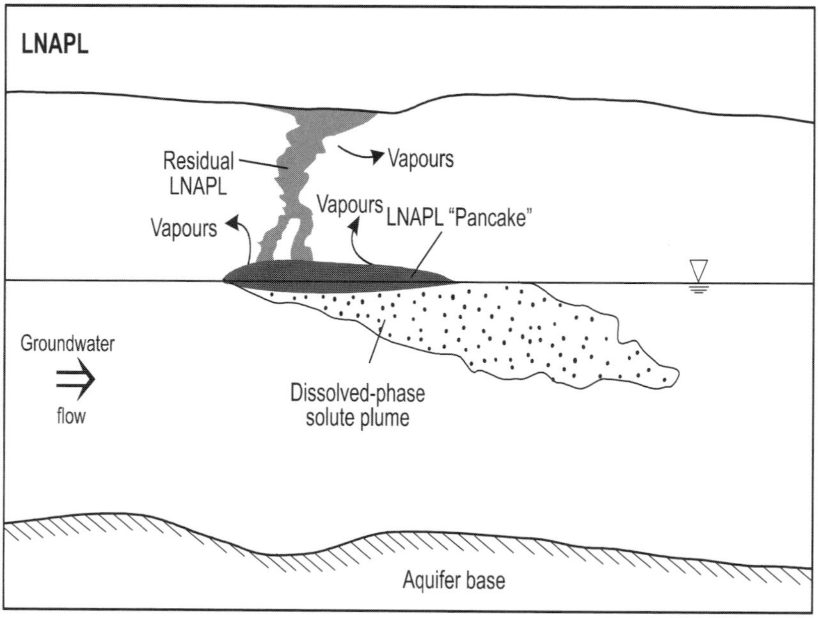

Figure 4.4. Conceptual model of a light non aqueous phase liquid (LNAPL) release

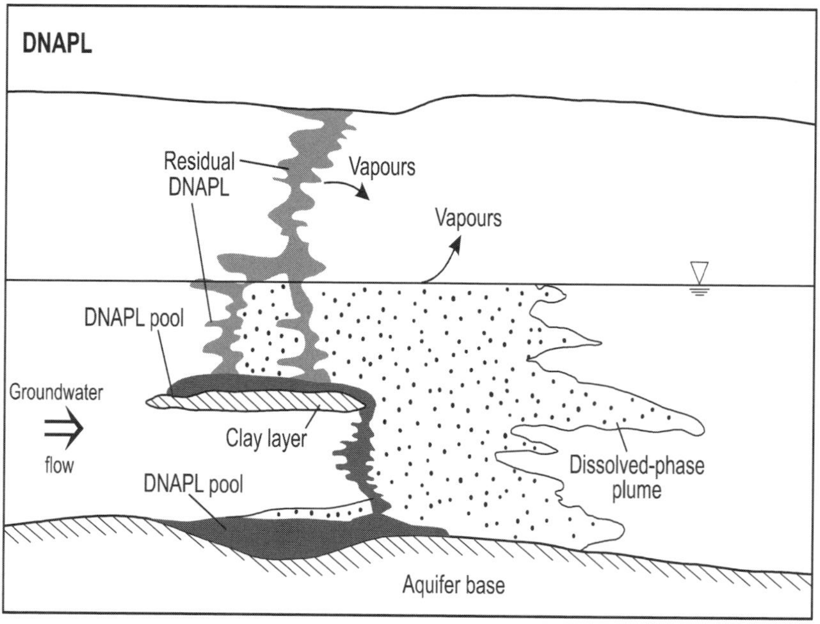

Figure 4.5. Conceptual model of a dense non aqueous phase liquid (DNAPL) release

Often NAPL will remain relatively local to a site, possible exceptions being the migration of LNAPL to a local surface water and perhaps huge NAPL spills, e.g. at a poorly operated oil refinery/distribution facility. Risks posed to groundwater resources and supplies are most often concerned with the migration of the dissolved-phase plume formed by the contact of flowing groundwater with the spilt NAPL. Although the presence of NAPL may impede the flow of groundwater, e.g. in DNAPL pools and central LNAPL body, areas where NAPL residual saturations are lower will still be permeable to water and NAPL dissolution will occur. Often the mass of NAPL is so large and the dissolution (solubilization) of NAPL into water so slow that the entire NAPL body post spill should be regarded as a largely immobile source zone able to continuously generate a dissolved-phase solute plume of organics downgradient for years to decades, even centuries for low-solubility NAPLs. Thus historic spill sites may still constitute major sources of NAPL in the deep subsurface and cause very large dissolved-phase plumes, particularly where dissolved-phase plume attenuation is limited. In general, DNAPLs tend to pose the greatest groundwater threat as they reside deep in groundwater systems and many, being chlorinated, are less susceptible to attenuation. In contrast, LNAPLs are restricted to shallower groundwater-table depths, and are more susceptible to attenuation via biodegradation.

The above provides a basic introduction to NAPLs in groundwater. Much research and field experience has been gained since the pioneering NAPLs research of Schwille (1988) and the reader is referred to Mercer and Cohen (1990) and Pankow and Cherry (1996) and references therein for further details.

4.5.2 General aspects of transport and attenuation of organics

Some of the transport and attenuation processes introduced earlier require specific discussion for organic contaminants. Several physiochemical properties/parameters exert a key control over subsurface organic contaminant migration. A selection of parameters is listed in Table 4.1 for a range of organic chemicals of groundwater-health concern. Values for a specific parameter generally vary over orders of magnitude across the listed chemicals and infer substantial variations in transport and attenuation between organic contaminants. Table 4.1 is not exhaustive: there are many more organic chemicals; values of individual chemical parameters may show significant variability across the literature; and other parameters exist, most notably half-life, that due to their dependency on site conditions display significant variability (see Section 4.5.4 for references to some half-life literature). For more detailed databases and their supporting literature see e.g. US EPA (1996; 1999) and Montgomery (1996).

Table 4.1. Selected physiochemical parameter values for important organic groundwater contaminants at 20-25°C (Mercer and Cohen, 1990; US EPA, 1996; 1999) (see Section 4.5.3 for an explanation of the abbreviations)

Chemical	Density (g/ml)	Absolute viscosity (cP)	Aqueous solubility (mg/l)	Vapour pressure (atm.)	Henry's constant (atm. m^3/mol)	K_{OC} (ml/g)	K_{OW}
Aromatic Hydrocarbons (single aromatic ring)							
Benzene	0.87	0.60	1750	0.13	0.0056	62	130
Toluene	0.86	0.55	535	0.037	0.0064	140	540
Ethylbenzene	0.87	0.68	152	0.0092	0.0064	200	1400
o-Xylene	0.88	0.81	175	0.0087	0.0051	240	890
Chlorinated Hydrocarbons							
DCM	1.34	0.45	20 000	0.48	0.0020	8.8	20
TCM	1.50	0.60	8200	0.20	0.0029	53	93
CTC	1.58	0.97	757	0.12	0.024	152	440
TCE	1.47	0.57	1100	0.076	0.0091	94	240
PCE	1.63	1.93	150	0.024	0.026	265	400
VC	Gas	Gas	2760	3.7	0.027	18	14
1,2-DCA	1.26	0.89	8520	0.084	0.00098	38	30
1,1-DCE	1.22	0.36	2250	0.79	0.034	65	69
cDCE	1.27	0.44	3500	0.27	0.0076	49	5.0
tDCE	1.26	0.40	6300	0.43	0.0066	38	3.0
1,2-DCB	1.30	1.32	100	0.0018	0.0021	379	2790
1,4- DCB	1.28	1.04	73	0.0014	0.0028	616	2580
Others							
Naphthalene	1.16	solid	31	0.00012	0.00048	1190	2290
Anthracene	1.24	solid	0.043	3.6×10^{-8}	0.00007	23 500	35 500
Benzo(a) pyrene	1.35	solid	0.0016	6.4×10^{-12}	0.000001	969 000	1 260 000
PCB-1248	1.44	212	0.054	6.6×10^{-7}	0.0035	437 000	562 000

Volatilization

Although other processes may be enhanced in the unsaturated zone relative to the saturated zone, e.g. biodegradation through the ready availability of oxygen, volatilization is a key process that only occurs in the unsaturated zone. Organic compounds with high vapour pressures (P) (>0.008 atm., i.e. xylene in Table 4.1) are termed (volatile organic compounds (VOCs). The vapour concentration adjacent to a NAPL or organic solid is dictated by its vapour pressure. Although volatilization of subsurface organic contaminants, e.g. NAPL sources, may occur and organic vapours potentially lost to the above ground atmosphere, VOCs are nevertheless very widely encountered in groundwaters; possible reasons include many VOCs are NAPLs of relatively high solubility (S) and low sorbing potential (K_{oc}). Vapours migrate due to diffusion and advection within the air phase and may migrate due to pressure (barometric) and temperature fluctuations, water infiltration and preferential conduit routes (Mendoza *et al.*, 1996). In relation to vapour-phase diffusion, it is emphasized diffusion coefficients in the air phase are ~4 orders of magnitude greater that the water phase. This allows much greater opportunity for lateral (radial) migration of vapour plumes and, due to vapour contact, contamination of underlying groundwaters over a wide area (Rivett, 1995).

Partitioning of dissolved organic solutes between a contaminated water phase and an adjacent air-phase is controlled by the Henry's Law partition coefficient:

$$H=C_A / C_W \quad \text{(Eqn. 4.4)}$$

where H is the Henry's Law constant, C_W is the concentration of an organic compound in water and C_A is its vapour concentration in the gaseous phase. It should be noted that similar concentration units will yield a dimensionless (i.e. unitless) H value, however, the air concentration is often expressed in terms of pressure and hence H values may be quoted with units of the type atmos m^3 mol^{-1}, e.g. for benzene H (dimensionless) is 0.24 and with units H is 5.5×10^{-3} atmos m^3 mol^{-1} (Table 4.1).

Solubilization

As indicated in Section 4.5.1, the differing hydrophobic nature of organic compounds means their solubility in water varies over orders of magnitude (Table 4.1). Solubility values represent maximum concentrations that may be achieved in a dissolved-phase plume. VOC contaminants, e.g. benzene and TCE tend to be small molecules that are moderately soluble, amenable to analysis and hence often detected in groundwater. Although solubilities are relatively low compared to inorganic ions, they may nevertheless achieve concentrations 4-5 orders of magnitude greater than drinking-water standards or guideline values. Larger molecular weight organics will have lower solubilities, hence concentrations e.g. of DDT in groundwater may only reach about 0.1 mg/l (its solubility).Thus DDT at solubility only exceeds the WHO guideline value of 1 µg/l by a factor of approx. 50, hence allowing for dilution and some attenuation, the prospects of exceeding the DDT standard in groundwater are low except in close proximity to a DDT source. DDT's high hydrophobicity and hence high sorption (see below) tends to cause DDT (and other organic chemicals of similar properties, e.g. high molecular weight PCBs and PAHs) to perhaps be more of a soil rather than a groundwater problem.

Sorption

Sorption exerts a key control over the transport of anthropogenic organic contaminants. Organic sorption is a complex topic, a detailed review is provided by Allen-King *et al.* (2002). Sorption is a function of the properties of both the organic solute and aquifer solid. Hydrophobic non-ionic organic contaminants preferentially sorb to the low-polarity components of geosolids, e.g. any organic material present. Sorption is inversely related to organic compound solubility; the more hydrophobic and less soluble an organic solute, the greater its intrinsic potential for sorption to any organic material present in the aquifer solids. Hence hydrophilic organics have negligible sorption, and mild to moderately hydrophobic organics such as the VOCs show limited sorption. In contrast, hydrophobic, high molecular weight, large organics such as PAHs and PCBs of low solubility exhibit high sorption (Table 4.1).

An additional measure of organic compound hydrophobicity often used in sorption research is the octanol-water partition coefficient (K_{ow}) (Table 4.1), that is simply an equilibrium partitioning of the organic solute between the organic octanol phase and water phase. The higher the K_{ow} value, the more hydrophobic, less soluble and more sorptive the organic compound.

The degree of sorption is also controlled by the sorption potential of the sorbate, i.e. the aquifer material that dissolved concentrations in groundwater contact. Frequently sorption is assumed to be at equilibrium and linear with organic solute concentrations, the magnitude of sorption being expressed by the sorption partition coefficient K_d:

$$K_d = C_S / C_W \qquad \text{(Eqn. 4.5)}$$

where C_S is the sorbed concentration. The main sorbing phase for organic solutes is any organic material present in the rock phase originating for example from organic detritus, e.g. humic material, deposited at the time of rock deposition. This organic material is referred to as the fraction of organic carbon (f_{oc}) within the geologic or soil matrix. Although f_{oc} values may be on the order of one per cent or more in organic-rich soil horizons, many aquifers comprise geologic strata with low f_{oc} values, e.g. an f_{oc}~0.02 per cent is recorded for the Borden glaciolacustrine sands, Canada (Rivett and Allen-King, 2003). The f_{oc}, even at such low concentrations, may still be the dominant sorption phase rather than poorly sorbing mineral surfaces. Simultaneous laboratory measurements of f_{oc} and K_d have shown that they are approximately linearly related, with the constant of proportionality being termed the organic-carbon partition coefficient (K_{oc}) of the specific organic solute (Table 4.1). Typically practitioners assessing sorption controls now obtain K_{oc} values from databases (e.g. US EPA, 1996), measure the aquifer f_{oc} (Heron *et al.*, 1997) and estimate the sorption K_d from the relationship:

$$K_d = f_{oc}\, K_{oc} \qquad \text{(Eqn. 4.6)}$$

where f_{oc} is in mass fraction dimensionless units, e.g. expressed as 0.0002 rather than as 0.02 per cent. Hence the greater the f_{oc} of the aquifer deposits (e.g. higher values are often found in shallow soils or recent sub-river, i.e. hyporheic zone), and greater the K_{oc} (Table 4.1), which will increase with solute hydrophobicity, the greater the K_d value, i.e. sorption. Assuming ideal linear equilibrium sorption and calculation of K_d from the above, the retardation factor R_i of organic solute i may be estimated from:

$$R_i = 1 + (\rho/\eta) K_d \qquad \text{(Eqn. 4.7)}$$

where ρ and η are the bulk density and porosity of the porous media respectively.

The above hydrophobic partitioning ideal sorption approach is an approximation of reality; it provides a reasonable first estimate. It should be recognized, however, that non-ideal sorption processes may be significant (Allen-King *et al.*, 2002); these include slow equilibration of sorbed and dissolved phases, dependence of the degree of sorption on dissolved concentration magnitude and the presence of any competing, also sorbing, solutes within multi-contaminant plumes. Further, the nature of the f_{oc}, i.e. ratio of the carbon-hydrogen-oxygen contents, has an important control on the sorption that occurs.

It is often useful to combine some of the above parameters visually to assess how organic contaminants may comparatively behave. Figure 4.6 plots K_{ow}, a measure of compound hydrophobicity that will indicate solubility and sorptive retardation trends against vapour pressure, a measure of volatilization tendency. The figure indicates PAHs are unlikely to volatilize and will undergo high sorption, and this would infer that soil/unsaturated zone solids concentrations of PAHs could often be high (frequently the case) and perhaps there is relatively limited development of PAH plumes to groundwater (often relatively small plumes are encountered). The chlorinated hydrocarbons, in contrast, are volatile but of low sorption potential. It is likely they would vaporize (and potentially be a vapour hazard to receptors at the soil surface) and also leach to groundwater leaving low concentrations in soils and unsaturated samples (quite often the case).

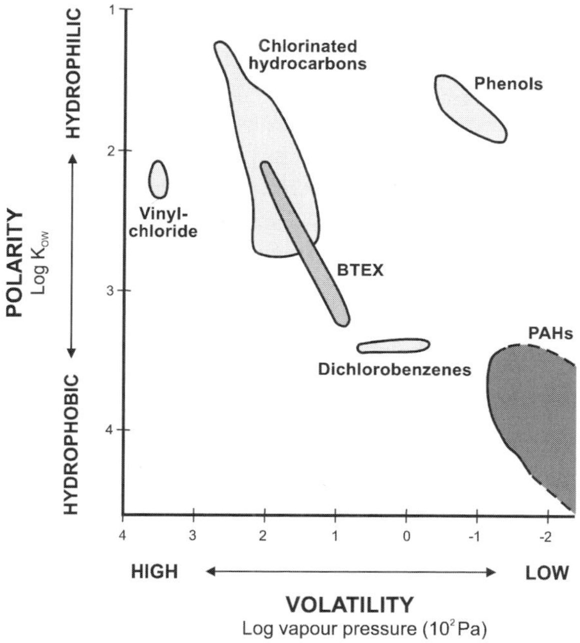

Figure 4.6. Polarity-volatility diagram for selected organic contaminants

Chemical reactions

Although there are a multitude of possible chemical reactions (abiotic reactions, i.e. not mediated by bacteria), reactions of low-concentration organics in a water-based environment tend to be fairly limited. Perhaps the most common reaction is that of slightly positively charged organic compounds ($C^{\delta+}$) with negatively charged (nucleophilic) species, such as HS^-, OH^- or water. The latter is a hydrolysis reaction. Reactive organic solutes tend to be organic halides, particularly brominated compounds and to a lesser extent chlorinated compounds. An example of a chemical reaction is that of the chlorinated solvent 1,1,1-TCA which was commonly used to degrease metals and circuit boards (as a less toxic replacement to TCE, before concerns were raised about its ozone-depletion potential). TCA in water will either undergo an elimination reaction to yield 1,1-DCE, or alternatively a sequential hydrolysis reaction replacing all the chlorine atoms as chloride to yield ethanoic acid. Interestingly it may also biodegrade to predominantly form a different product, 1,1,-DCA (Klecka *et al.*, 1990). Further information on chemical reactions in water may be found in Schwarzenbach *et al.* (1993).

Biodegradation

Bacteria degrade organic contaminants to simpler, often less toxic, products. Biodegradation is perceived to be the primary attenuation process that may mitigate dissolved-plume impacts to receptors by organic chemicals. monitored natural attenuation (MNA) remedial strategies generally have their main focus upon demonstration of occurrence of biodegradation (Wiedemeier *et al*., 1999). This not only entails monitoring the disappearance of the organic contaminant, but also the appearance of intermediate organic contaminants that may themselves persist or be further biodegraded, ideally to benign inorganic products, e.g. water, carbon dioxide, chloride.

Monitoring of the inorganic hydrochemistry is also a key requirement to assessing biodegradation occurrence. Sites may be initially aerobic/oxic (containing oxygen), and under these conditions biodegradation of many contaminants is often the most rapid. Dissolved oxygen concentrations in groundwater are usually low, maximally ~10 mg/l. Such levels can easily be depleted by even low to moderate levels of organic contamination present and are not easily renewed as dispersive mixing in groundwaters to allow oxygen re-entry is typically low. Other electron acceptors, for example sulphate, nitrate, iron and manganese, are then used to allow biodegradation to continue under anaerobic conditions. Finally site conditions may become so reducing that biodegradation occurs under methanogenic conditions.

Specific examples of biodegradation are included in the sections that follow. In general, most hydrocarbon-based compounds and most oxygenated-organics are relatively biodegradable under a wide range of conditions, and natural attenuation of such plumes often significant. Chlorinated (halogenated) compounds are generally less biodegradable but evidence has increasingly shown that they do biodegrade under appropriate redox-bacterial conditions. A sequence of reactions under varying redox conditions may be required to allow complete biodegradation to benign products. This means that for some contaminants and sites full biodegradation to benign products is difficult and there may be persistence of both the original contaminants and their intermediate degradation products, both of which may have toxicity.

4.5.3 Organic chemicals of major concern in groundwater

Consideration of the above transport and attenuation processes, together with data on organic chemical toxicity, use of chemicals and associated potential for release to the subsurface, and actual chemical occurrence in groundwater data, enables identification of groups of organic chemicals, as well as individual chemicals, thought to be of major concern in groundwater. Two organic chemical groups of key concern include:

- aromatic hydrocarbons: benzene, toluene, ethylbenzene and xylenes (BTEX);
- chlorinated hydrocarbons (aliphatic and aromatic): dichloromethane (DCM), trichloromethane (TCM, also known as chloroform), tetrachloromethane (also known as carbon tetrachloride, CTC), trichloroethene (TCE), tetrachloroethene (PCE, also known as perchloroethylene), vinyl chloride (VC), 1,2-dichloroethane (1,2-DCA), 1,1-dichloroethene (1,1-DCE), 1,2-dichloroethene (cis and trans isomers, cDCE and tDCE respectively), 1,2 dichlorobenzene (1,2-DCB) and 1,4 dichlorobenzene (1,4-DCB).

Properties of the above compounds are included in Table 4.1.

Figure 4.7 presents data based upon 250 sites in Germany and 500 sites in the USA (Kerndorff *et al.*, 1992) and indicates the prevalence of the above compounds in groundwater. A third (predominantly) organic chemical group of key concern in groundwater is pesticides.

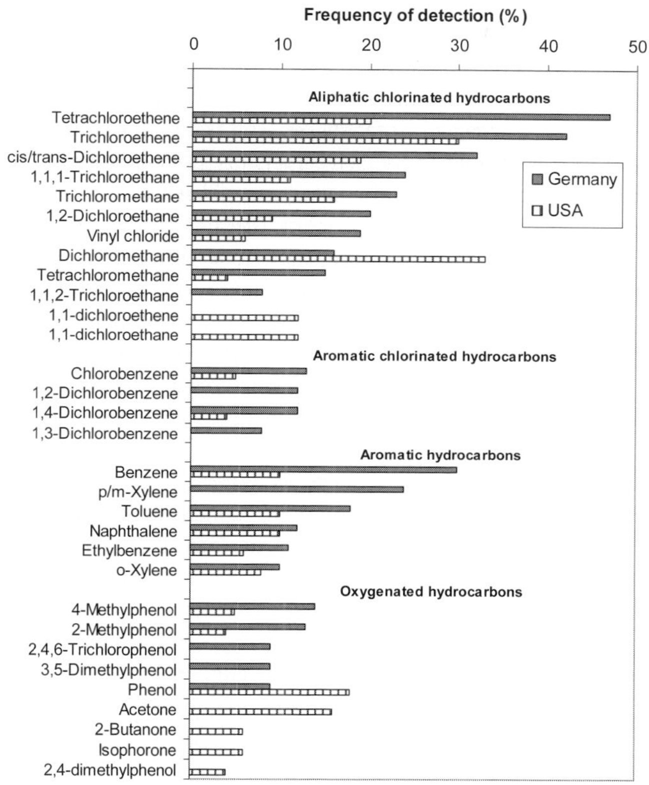

Figure 4.7. The 25 most frequently detected organic groundwater contaminants at hazardous waste sites in Germany (250 sites) and the USA (500 sites) (based on concentrations $\geq 1 \mu g/l$) (adapted from Kerndorff *et al.*, 1992)

Of the many organic chemicals that may potentially contaminate groundwater, these three groups have perhaps received the most attention from both the groundwater practitioner and research communities during the 1980s and 1990s. The following Sections 4.5.4, 4.5.5 and 4.6 are hence devoted to these three groups.

The focus upon the above three groups does not preclude the potential importance of other organic contaminants in groundwater. Although compounds may perhaps not occur frequently due to restricted use within specialized industries, some compounds may have low natural attenuation (NA) properties and potentially develop extensive plumes. Other chemical groups may, in contrast, have received widespread industrial use, but were perhaps thought (sometimes mistakenly) to pose a much lower risk to groundwater due to high NA properties. Examples of the latter may include PAHs and PCBs that are both briefly discussed below.

PAHs are a component of creosotes and coal tars frequently associated with former gasworks and coal carbonization (coking) works (Johansen *et al.*, 1997). They are a diverse class of compounds of natural and anthropogenic origin, some of which show carcinogenic properties. WHO has derived a guideline value of 0.7 μg/l for benzo(a)pyrene because it may be released from coal tar coatings of drinking-water distribution pipes. Most PAHs have extremely low solubilities in water and have a high tendency to adsorb to the organic matrix of soils and sediments, particularly the higher molecular mass, higher-ring PAHs, e.g. the 5-ring benzo(a)pyrene. Thus, they are generally not found in water in notable concentrations, and human exposure is mostly through food prepared at high temperatures and air (particularly from open fires) (WHO, 2004a). Countering the processes that may serve to attenuate groundwater impacts, however, are: (i) creosote and coal tars may occur as a DNAPL (density is composition dependent, but may be around 1.05) and slowly migrate as a DNAPL deep into the subsurface potentially penetrating the water table; and (ii) higher molecular mass, higher ring-member PAHs are much more resistant to biodegradation and hence dissolved plumes, although slow to develop, may persist and grow over decades (King and Barker, 1999).

PCBs are a class of stable compounds, each containing a biphenyl nucleus (two linked benzene rings) with two or more substituent chlorine atoms. PCBs are produced industrially as complex mixtures that often contain between 40 and 60 different chlorinated biphenyls. Similar to PAHs, most PCBs are of low solubility in water and sorptive and hence dissolved-phase plumes in groundwater tend not to be large. However, PCB oils, historically used in electrical transformer facilities, are DNAPLs and may potentially penetrate deep into aquifer systems. Dissolved PCBs are generally slow to biodegrade and hence PCBs, like PAHs, may serve as long-term sources of groundwater contamination.

An emergent groundwater issue in the 1990s relating to hydrocarbon fuels has been the use of oxygenates, particularly methyl tertiary-butyl ether (MTBE) and methanol or ethanol within fuels (Squillace *et al.*, 1996). Use of MTBE has been most significant in the USA with MTBE first used in gasoline at greater than 10 per cent by volume in 1992. MTBE has a strong taste and odour and is likely to impair drinking-water quality at concentrations in the 0.01-0.1 mg/l range. However, because of its low relevance to human health, MTBE is not further discussed in this chapter; see WHO (2005c) for further information on MTBE.

4.5.4 Aromatic hydrocarbons (BTEX)

Health aspects

Mononuclear (single-ring) aromatic hydrocarbons such as BTEX are amongst the most common groundwater contaminants (Figure 4.7) and the main aromatic faction of many hydrocarbon fuels. The key compound of health relevance within the BTEX group is benzene, a proven carcinogen in humans. The mechanism or metabolic form by which it exerts its action (haematological changes, including leukaemia) is not clear. WHO (2004a) has established a guideline value for benzene in drinking-water of 10 μg/l, corresponding to a lifespan risk of 10^{-5} to contract cancer by exposure to benzene via drinking-water. Alkylated benzenes are much less toxic than benzene, and correspondingly their WHO guideline values are 700 μg/l for toluene, 300 μg/L for ethylbenzene and 500 μg/l for xylene (WHO, 2004a), whereas no health-based guideline values are given for trimethylbenzenes and butylbenzene. However, some of them may be perceived by odour and/or taste at only a few micrograms per litre.

Sources and occurrence

The aromatic hydrocarbons BTEX are the primary contaminants of concern associated with point-sources of fuels and fuel-related contamination originating from petroleum production, refining and wholesale and retail distribution (service stations) of petroleum products (Newell *et al.*, 1995). They are also used as solvents and raw materials in chemical production. Spills and accidental releases of gasoline (petrol), kerosene and diesel are common sources of their occurrence in the environment. The German-USA groundwater survey data in Figure 4.7 indicate BTEX was relatively common, benzene being the most prominent.

Transport and attenuation

Most petroleum products are LNAPLs and hence the Figure 4.4 conceptual model applies. BTEX components typically comprise just a few per cent of the LNAPL fuel. BTEX concentrations dissolving in groundwater near fuel sources are reduced from their pure-phase solubility values (as individual component solubilized concentrations form NAPL mixtures depend upon the mass (strictly mole) faction of that component in the NAPL). BTEX-aromatics, being the most soluble hydrocarbons, are still the main risk driver for groundwater at hydrocarbon-contaminated sites, as solubilized concentrations and mobility of other alkane (branched and straight-chain) and PAH hydrocarbons are much lower. Since the early 1990s, it has been recognized that natural attenuation (NA) of BTEX is highly significant at most sites due to the high biodegradability of BTEX under a range of conditions. Monitored natural attenuation, i.e. monitoring of the growth, stability and eventual decline of dissolved-phase BTEX plumes, has indeed become a viable and cost-effective remediation option at many sites rather than active remedial measures such as pump-and-treat (McAllister and Chiang, 1994).

Studies that have examined dissolved-phase hydrocarbon plume lengths from over 600 hydrocarbon-release sites in the USA are particularly instructive on the potential for NA applicability (Newell and Connor, 1998; Wiedemeier *et al.*, 1999 and references therein). Of the 604 plumes evaluated, 86 per cent were less than 300 feet (~100 m) long with only 2 per cent of plumes greater than 900 feet. One of the studies that examined 271 plumes indicated only 8 per cent of these plumes were still growing, 59 per cent of plumes were approximately stable as mass being dissolved from the source was balanced by mass being depleted by attenuation (biodegradation), and 33 per cent of plumes were shrinking as source mass inputs declined or biodegradation of contaminants perhaps became increasingly efficient with time. These studies hence provide a strong rationale for occurrence of NA across a variety of site conditions. Under the vast majority of circumstances the potential for impacts of hydrocarbon plumes is limited to distances of a few hundred meters from source zones. Thus although hydrocarbon sources can be numerous, hydrocarbon and BTEX impacts are likely to remain local to those sources zones. However, although dissolved-phase plume may extend to relatively short distances from source areas, LNAPL source zones (i.e. pancake-like of hydrocarbon) themselves can on occasion be very extensive, for example Albu *et al.* (2002) depict zones of LNAPL extending over 5 km around oil refinery sites near Ploiesti, Romania.

Much insight into the importance of biodegradation and associated controlling factors has been obtained in plume studies. A controlled injection of dissolved-phase benzene, toluene and xylene at the Borden site, Canada (Barker *et al.*, 1987) showed complete benzene, toluene and xylene biodegradation by just over 400 days with only benzene persisting beyond 270 days. This study, and many real spill sites indicate BTEX are readily degraded when dissolved oxygen is present in groundwater. Under anaerobic conditions, rates of biodegradation of remaining hydrocarbon was governed by both the rate of oxygen re-entry across the contaminant plume fringe and rates of alternative anaerobic biodegradation pathways using less efficient electron acceptors such as nitrate, sulphate, and iron(III). Many other field sites have demonstrated the importance of anaerobic processes, primarily through changes in the groundwater geochemistry. For example, the Plattsburgh, Hill and Patrick Air Force Bases in the USA (Wiedemeier *et al.*, 1999) indicate development of depleted dissolved oxygen, nitrate and sulphate coincident with the BTEX plumes as these electron acceptors are consumed in the oxidation of the BTEX. pH declines as well as production of Fe(II) and methane arising from methanogenic activity were also evident where the most reducing conditions prevailed.

A myriad of aerobic and anaerobic BTEX biodegradation rates are available from the literature, e.g. data and references within Wiedemeier *et al.* (1999) and Noble and Morgan (2002). Rates are typically expressed as a first order rate constant or an equivalent half-life. Table 4.2 summarizes half-life data provided in the review by Noble and Morgan (2002) for BTEX, naphthalene and some chlorinated hydrocarbons. These chemicals represent the most studied groundwater contaminants in relation to biodegradation. The table subdivides rate data between laboratory and field studies that are in turn subdivided to aerobic and anaerobic conditions. Although the work of Noble and Morgan is reasonably comprehensive and based upon many citations, it should be noted that Table 4.2 aims to be illustrative rather than comprehensive of all literature available on biodegradation rates for the chemicals listed; biodegradation is an active area of study worldwide and half-life data continue to be published.

Rates selected for risk assessment modelling at other sites (where field data are insufficient to determine rates) need to be used with care as modelling results, and hence plume attenuation predicted and any site risk-based remediation standards computed, are very sensitive to degradation mass-loss parameters selected. Rates may vary significantly for individual compounds that may be a reflection of rates being lab-based or field based and the particular aerobic-anaerobic site conditions. This is apparent from examination of Table 4.2. Also, field biodegradation rates may be derived from a localized point measurement, or more often a rate predicted from whole plume behaviour that will average varying rates and different biodegradation processes and aerobic/anaerobic conditions occurring throughout the plume.

The laboratory half-lives in Table 4.2 are generally shorter than the equivalent field-based values, i.e. plumes are apparently more persistent in the field presumably as a reflection of field conditions, e.g. supply of electron acceptors being less optimal than can be achieved in a laboratory. In general, BTEX degradation rates will be lower under anaerobic conditions and plume may be persistent and indeed attain significant lengths where groundwater is naturally anaerobic, e.g. confined aquifer conditions. This is reasonably demonstrated by the Table 4.2 field data, but less so by the laboratory-based data; the latter is in part from the limited studies undertaken (at least reported) for some of the chemicals. A cautionary approach is warranted to the application to sites of the half-life data provided in Table 4.2 (and elsewhere); clearly the range in half-life values for a specific contaminant is large and for some contaminants and conditions insufficient data exist to yield a reliable average and range. Some Table 4.2 entries are based on a single study, in some cases no data are provided. The latter may be a reflection of a genuine lack of data or else biodegradation not being effective under the specific conditions (e.g. aerobic biodegradation of PCE).

Table 4.2. Summary of biodegradation half-life data (at 10°C) for important organic groundwater contaminants (adapted from Noble and Morgan, 2002)

Chemical	Aerobic - laboratory data			Aerobic - field study data			Anaerobic - laboratory data			Anaerobic - field study data		
	No. of studies (n)	Mean half-life (days)	Range in half-life (days)	No. of studies (n)	Mean half-life (days)	Range in half-life (days)	No. of studies (n)	Mean half-life (days)	Range in half-life (days)	No. of studies (n)	Mean half-life (days)	Range in half-life (days)
Aromatic Hydrocarbons												
Benzene	18	34	3-200	6	220	15-490	11	79	20-200	13	502	85-2000
Toluene	5	120	5-320	3	120	15-178	15	64	5-320	9	257	10-660
Ethylbenzene	No data provided			No data provided			4	239	53-548	4	519	238-693
Xylenes	1	11	11-11	3	58	1-123	5	63	30-155	8	489	72-800
Naphthalene	11	138	10-400	No data provided			No data provided			No data provided		
Chlorinated Hydrocarbons												
PCE	No data provided			No data provided			6	23	4-62	4	1600	4-3600
TCE	No data provided			No data provided			8	43	3-99	12	1460	3-6600
DCE	5	2	0.5-3	3	39	12-56	1	280	280-280	9	4060	42-16860
VC	7	14	84	1	10	10-10	5	81	24-124	5	948	506-1265
DCA	No data provided			No data provided			5	157	16-340	2	1430	460-2400

4.5.5 Chlorinated hydrocarbons

Health aspects

A number of aliphatic and aromatic chlorinated hydrocarbons are of health significance because of their toxicity and occurrence in drinking-water, particularly as in groundwater, their concentrations do not decrease rapidly through volatilisation, anaerobic degradation is slow, and in consequence contaminants may persist for some time.

WHO (2004) gives guideline values for dichloromethane (DCM), trihalomethanes chloromethane (e.g. chloroform which may be generated as by-products of disinfection), tetrachloromethane (also known as carbon tetrachloride, CTC), trichloroethene (TCE), tetrachloroethene (PCE, also known as perchloroethylene), vinyl chloride (VC), 1,2-dichloroethane (1,2-DCA), 1,1-dichloroethene (1,1-DCE), 1,2-dichloroethene (cis and trans isomers, cDCE and tDCE respectively), 1,2 dichlorobenzene (1,2-DCB) and 1,4 dichlorobenzene (1,4-DCB) (see Table 4.1 for selected physiochemical parameter values). Generally, among the saturated chlorinated compounds, the 1,1- halogenated ones (e.g. 1,1-dichloroethane) are of less health concern than the 1,2-halogenated ones (e.g. 1,2-DCA), since they are metabolized differently. This is also true for higher halogenated compounds (e.g. trichloroethanes).

Dichloromethane (DCM, also known as methylene chloride) is of low acute toxicity, and current evidence suggests that it is not a genotoxic carcinogen. The WHO drinking-water guideline value for DCM is 20 µg/l, based on hepatotoxic effects observed in rats (WHO, 2004a). Carbon tetrachloride (CTC) has a WHO guideline value of 4 µg/l based on its liver toxicity, and from carcinogenic effects observed in laboratory animals it is classified as possibly carcinogenic to humans. 1,2-dichloroethane (1,2-DCA) is potentially genotoxic and a proven carcinogen in experimental animals with a WHO guideline value of 30 µg/L (WHO, 2003d; 2004a).

Trichloroethene (TCE) and tetrachloroethene (PCE) may degrade to the more toxic vinyl chloride. The provisional WHO guideline value for TCE is 70 µg/l based on liver effects in mice (WHO, 2004a; 2005d). PCE causes nervous disorders at high dose, whereas at lower doses kidney and liver damage have been reported. It is classified as possible human carcinogen with overall evidence indicating that it is not genotoxic (WHO, 2003e), and its WHO guideline value for drinking-water is 40 µg/l (WHO, 2004a). Vinyl chloride (VC) is genotoxic and carcinogenic in experimental animals as well as in humans. Administered orally to experimental animals, it produced cancer at a variety of sites (WHO, 2004d). Its WHO drinking-water guideline value is 0.3 µg/l (WHO, 2004a).

1,1-Dichloroethene (1.1-DCE) is a weak in vitro-mutagen and not classifiable as to its carcinogenicity to humans. It is a central nervous system depressant and may cause liver and kidney toxicity. The WHO guideline value is 30 µg/l (WHO, 2004a; 2005e). Among the two isomers of 1,2-dichloroethene, the cis-form is detected more frequently and at higher concentrations than the trans-form as a water contaminant, since the former is the main anaerobic metabolite of TCE and PCE. As such it may indicate as well the presence of vinyl chloride, the next anaerobic breakdown product, which is not only much more toxic than all higher chlorinated ethenes but also a genotoxic human carcinogen (see above). In contrast, both 1,2-DCEs do not seem to be genotoxic and there is no information on their carcinogenic potential. The WHO drinking-water guideline value for each of the 2 isomers is 50 µg/l (WHO, 2003f; 2004a).

Dichlorobenzenes are the least toxic of this group of contaminants. Their health based guideline values in drinking-water are 300 and 1000 µg/l for 1,2- and 1,4-DCB, respectively. These exceed their odour threshold range of 0.3-30 µg/l by far (WHO, 2003g; 2004a).

Sources and occurrence
Chlorinated hydrocarbons are employed in a variety of industrial activities, including almost any facility where degreasing, e.g. of metals, circuit boards, textiles (dry cleaning) and animal/leather hides, metal stripping, chemical manufacturing, pesticide production or other activities where chlorinated solvents, cleaners, dry cleaning fluids, paint removers are used (Chapter 11).

In many industrialized countries, chlorinated hydrocarbons are the most frequently detected groundwater contaminants at hazardous waste sites (Kerndorff *et al.*, 1992; Plumb, 1992; NRC, 1994). This is highlighted by the German-USA survey data in Figure 4.7. TCE and PCE together with their principal metabolites cDCE and VC have been the most frequently detected chlorinated hydrocarbons at the investigated sites. Point source release of chlorinated hydrocarbons to groundwater is anticipated to be the main source of groundwater contamination. Complex mixtures of chlorinated hydrocarbons may arise from leakages at hazardous waste disposal sites where many solvent types may have been disposed. In contrast, spills at industrial manufacturing/processing sites may well comprise liquid chlorinated hydrocarbon as a DNAPL with a high proportion of a single chlorinated hydrocarbon component. A multitude of point sources exist in many urban areas due to the diversity and frequency of chlorinated hydrocarbon users.

Examples of regional chlorinated hydrocarbon contamination within aquifers underlying urban towns and cities emerged during the 1980s. Many groundwater supplies or monitoring wells were contaminated in some instances, particularly by TCE and to a lesser extent PCE. Examples include Milan, Italy (Cavallero *et al.*, 1985); the New Jersey coastal plain aquifer, USA (Fusillo, 1985) and Birmingham, United Kingdom (Rivett *et al.*, 1990); the latter example is described in Box 4.2.

Transport and attenuation
Many of the chlorinated hydrocarbons will have entered the subsurface in the DNAPL form and may reside to significant depths within aquifers (Pankow and Cherry, 1996). They typically have low to medium water solubility (in the range of 0.2-20 g/l; Table 4.1). Dissolution of DNAPL sources is expected to be slow taking years to decades, particularly from long lengths of residual DNAPL pools that have invaded or diffused into low-permeability strata. Dissolved-phase plumes of chlorinated hydrocarbons can be very extensive, for example Mackay and Cherry (1989) depict several plumes in the km-scale and Jackson (1998) several plumes in the alluvial aquifers of the southwestern USA that are around 10 km in length. Some plumes have lead to high profile court cases and set legal precedents on apportioning liability for historic contamination events, e.g. the near 2-km PCE plume that caused contamination of the

Sawston public water supply borehole in Cambridgeshire, United Kingdom (Ashley, 1998).

Box 4.2. Chlorinated-hydrocarbon contamination of groundwater in Birmingham, United Kingdom (based on Rivett *et al.*, 1990; Rivett *et al.*, 2005)

Birmingham is the second largest city in the United Kingdom and has a long history of manufacturing, particularly in metal-related industries. Groundwater samples were taken during the late 1980s from 59 abstraction boreholes typically screened over 100 m in the Triassic Sandstone aquifer underlying the city. Chlorinated solvents were found to be widespread, particularly TCE detected in 78 per cent of abstraction boreholes with over 40 per cent of the sampled boreholes showing concentrations over 30 µg/l to a maximum of 5500 µg/l. The majority of highly contaminated abstractions were located in solvent-user sites, predominantly metals-related industry. The predominance of TCE was ascribed to its main United Kingdom use within metal cleaning applications since the 1930s. PCE was less evident as it has generally only been used for dry cleaning in the United Kingdom since 1950s. Lower TCA occurrence was ascribed to its much later introduction in the United Kingdom starting about 1965 as a less toxic replacement to TCE. Greatest groundwater contamination occurred in the Tame valley area that was hydrogeologically vulnerable due to low depths to groundwater and limited aquifer protection by low permeability drift. Moderate contamination was present in other less vulnerable areas of the unconfined aquifer with least contamination evident in the Mercia Mudstone confined aquifer.

The aquifer was re-visited a decade later during the late 1990s. Declines in industrial use of groundwater meant only 36 abstractions were active and available for sampling, of these 26 were from the 1980s survey. Overall contamination detected was less and attributed to most of the new boreholes being located in industry areas where solvent use appeared limited. Also, many of the former highly contaminated abstractions had ceased operation due to industry closure. The latter was of some concern as contamination previously inadvertently captured by such abstractions was now able to more freely migrate into the wider aquifer. Comparison of the 26 abstractions common to both surveys indicated contamination at individual boreholes was at similar or greater concentrations in the more recent survey compared to the decade-earlier survey. These levels are unlikely to be due to major ongoing contamination, rather, it is reasonably assumed that incidences of new contamination will have declined over the decade as industry has become much more environmentally aware. The sustained level of contamination was hence ascribed to persistent sources of chlorinated solvents, likely DNAPL sources at depth. These will have remained unaffected by remedial works implemented at many sites to date because under a land redevelopment focused agenda these predominantly focused upon shallow soil and groundwater problems.

Under aerobic conditions, biodegradation of solvents such as TCE and PCE can be limited to non-existent and may account for the extensive plume examples noted above. Sorption is often limited too, particularly for the less hydrophobic compounds where compound solubility exceeds 1g/l (Table 4.1). A controlled emplacement of a DNAPL chlorinated solvent source in the Borden aquifer research site, Canada, resulted in TCM and TCE plumes exhibiting near conservative behaviour with retardation factors in the range of 1.0-1.2 and no evidence of biodegradation for these solvents and also PCE that was more retarded at about 1.6 (Rivett *et al.*, 2001; Rivett and Allen-King, 2003). Dispersion of these plumes, although moderate in this relatively homogeneous sand aquifer, nevertheless produced leading plume contours at concentrations in the range of drinking-water standards that had travelled toward 100 per cent further than the mean advection (groundwater) velocity.

In contrast, other sites have shown significant natural attenuation of chlorinated hydrocarbons due to biodegradation activity. The most well known biodegradation pathways are those involving the sequential reductive dechlorination of chlorinated hydrocarbons where lesser chlorinated organics, chloride and ultimately hydrocarbons such as ethane or ethane, are formed (Vogel *et al.*, 1987), e.g. PCE is transformed to TCE to cDCE (usually the predominant isomer) to VC to ethene. On average chlorinated hydrocarbon plumes are significantly longer than the aforementioned BTEX plumes. For example, Newell *et al.* (1990) reported a median length of 1000 feet (about 300 m) for chlorinated ethene (PCE, TCE, DCE, VC) plumes (88 sites sampled).

Biodegradation of chlorinated hydrocarbons has proven to be relatively complicated with five possible degradation processes (Wiedemeier *et al.*, 1999). Most chlorinated compounds have been observed to biodegrade by three or four of these processes, only DCE and VC may biodegrade via all five processes. Under anaerobic or low oxygen conditions degradation processes include (i) dehalorespiration, in which the chlorinated hydrocarbon is used as the electron acceptor and effectively respired, (ii) direct anaerobic oxidation and (iii) anaerobic co-metabolism. Under aerobic conditions, further processes are (iv) direct aerobic oxidation and (v) aerobic co-metabolism. Direct processes involve the chlorinated hydrocarbon being used as the primary growth substrate. Dehalorespiration and co-metabolism both require an alternative primary growth substrate to be present. That primary substrate is normally a relatively biodegradable substrate and may include anthropogenic carbon such as BTEX contamination. Alternatively, anaerobic conditions may be driven by high levels of naturally occurring carbon acting as the substrate, a primary example being wetland sediments and sub riverbed deposits, e.g. Lorah and Olsen (1999) observed TCE and 1,1,2,2-PCE dechlorinations in the former.

Due to the complexity of biodegradation processes outlined, there is a wide divergence in reported biodegradation rates of chlorinated hydrocarbons (Wiedemeier *et al.*, 1999; Noble and Morgan, 2002; Table 4.2). This is clearly illustrated by the Table 4.2 half-life data for the more common chlorinated hydrocarbon groundwater contaminants, e.g. DCE field-based half life data vary from just 42 days to nearly 17 000 days. Also, Table 4.2 indicates laboratory half-life data are generally much shorter (by 1-2 orders of magnitude) than equivalent field data, e.g. TCE data under anaerobic conditions indicate a laboratory half life mean of 43 days compared to a field mean of

1460 days. This is perhaps ascribed to the fact that optimal anaerobic reducing conditions can be achieved in the laboratory for the whole sample, whereas in the field such anaerobic conditions may in fact only occur in localized portions of a plume. Table 4.2 emphasizes the sensitivity of half life to aerobic and anaerobic conditions and that much longer half life values may occur for chlorinated hydrocarbons relative to the aromatics. The above strongly endorses the need to recognize that literature half-life data have very significant uncertainty when applied in a predictive manner to sites elsewhere. The unfortunate reality is that most sites require individual case-by-case assessment to allow effective prediction of natural attenuation rates.

4.6 PESTICIDES

Pesticides represent a wide range of compounds used mostly as insecticides, herbicides, and fungicides. Formerly a small number of classes of chemicals included most pesticides, i.e. organochlorines, organophosphates, carbamates, phenoxyacetic acids and triazine herbicides. However, modern pesticides include other types of chemicals, and therefore such a classification is of more limited use for descriptive purposes. Many of the historically used pesticides, such as the organochlorines, are however environmentally persistent and may pose a long-term groundwater problem.

Health aspects

In general, health effects associated with pesticides are specific for each chemical. This is reflected in their different WHO guideline values for drinking-water quality (see WHO, 2004a) and in the wide range of acceptable daily intake values derived by the Food and Agricultural Organization (FAO) for exposure through food (resulting from pesticide uses on crops; FAO, 2004). Most health effect studies are conducted using single compounds, little is known about effects associated with pesticide mixtures. Health effects from acute (short-term and high level) or chronic (long-term and low-level) exposure include liver and kidney damage, major interference with nervous, immune and reproductive system functions, birth defects and cancer. In most cases the risk from food contaminated by unduly high levels of pesticides is likely to be more significant than that posed by pesticide levels in drinking-water.

Chronic exposure associated with pesticides has declined in Europe and North America as many of the more persistent herbicides (such as chlordane, DDT, dieldrin, endrin, heptachlor, γ-HCH and toxaphene) have been restricted or phased out (Barnard *et al.*, 1997). They have been replaced with less persistent and more species-specific toxicants. While acute toxicities have often increased, some important new biologically-derived insecticides have very low mammalian toxicity.

Sources and occurrence

Pesticides are intentionally applied to protect crops in agriculture (Chapter 9) as well as to control pests and unwanted vegetation in gardens, buildings, railway tracks, forests and roadsides (Chapter 13). They may be accidentally released from production sites (Chapter 11) or, more often, transported away from their site of application in water, air or dust. Pesticides can reach groundwater after accidental spills or excessive application in geologically sensitive settings, from contamination of poorly sealed wells by surface runoff after intensive rains following field application and from storage or production sites. Though some organochlorine insecticides have been banned or are subject to severe restrictions in many countries, in several developing countries production and use of, for example, DDT has continued because of its relatively inexpensive production and its high efficacy against mosquitoes in malaria control. Generally the dilemma of the low cost and high efficacy of persistent pesticides versus their long term health and environmental effects remains a contentious global issue which has been addressed by a global convention (see Box 9.5).

As sampling has become more extensive and monitoring programmes developed, increasing numbers of pesticide compounds are being detected in groundwater. A major study, the National Pesticide Survey, conducted by the US EPA in the late 1980s detected 46 pesticides in groundwater in 26 states originating from normal agricultural practice (Williams *et al.*, 1988) but with low frequency and usually below health-based standards. Pesticides detected in more than five states were alachlor, aldicarb, atrazine, cyanazine, metolachlor and simazine. More recently, extensive sampling within the USGS National Water Quality Assessment programme has confirmed the widespread occurrence of pesticides in both surface water resources and groundwater, but generally at concentrations below their respective allowable maximum contaminant levels (Kolpin *et al.*, 2000). The newer work has, however, shown that pesticides, especially insecticides, are also reaching water resources in urban and suburban areas, including residential sources. This work has also demonstrated widespread detection of pesticide metabolites, often at concentrations exceeding the parent compound, and for which there may not be adequate toxicity data to establish their health significance.

This picture is largely confirmed by monitoring efforts in Europe. Herbicides which are widely used in cereal cultivation, such as MCPP and isoproturon are detected in the countries of northern Europe (Spliid and Køppen, 1998) with carbamates perhaps more common further south. Most detected groundwater pesticide concentrations were in the range 0.1 to 10 μg/l. Concentrations significantly above this range can probably be attributed to local point source contamination from poor disposal practices, or from non-agricultural usage such as on railways.

Because of high analysis costs, much less monitoring has been undertaken in low-income countries and data from tropical regions are scarce. However, atrazine residues from its use in sugar cane cultivation were widely observed in groundwater in Barbados and carbofuran was detected in shallow groundwater beneath irrigated vegetable cultivation in Sri Lanka (Chilton *et al.*, 1998). Elsewhere, presence of organochlorines in groundwater reflects their highly persistent nature and perhaps continuing usage even when banned (Matin *et al.*, 1996).

Transport and attenuation

The mobility and persistence of pesticides in the environment are well understood because admission of a new pesticide for the market requires a series of standardized laboratory and field experiments.

The overall likelihood of a pesticide to be a groundwater pollutant is dependent both on its persistence and its soil sorption. Table 4.3 lists pesticides used in agriculture for which WHO has derived health-based drinking-water guideline values. It provides a classification of their leaching and runoff potential based on their physical-chemical characteristics, i.e. their persistence (characterized by soil half-lives) and soil organic carbon sorption (K_{OC}). Other pesticides used for public health purposes (e.g. DDT, chlorpyrifos and pyriproxyfen in malaria control) and wood conservation (e.g. pentachlorophenol or PCP) are not listed in Table 4.3.

As for the organic contaminants discussed in Sections 4.1.2 and 4.5.2, soil organic matter content, clay content and permeability all affect the potential for pesticides to leach through soils. In general, soils with moderate-to-high organic matter and clay content will absorb pesticides onto soil particles, making them less available for leaching, and moderate or low permeability soils allow less water infiltration.

A wide range of pesticide soil sorption K_d values (as defined earlier in Section 4.5.2) exist. DDT, for example, has a K_d value roughly 20 000 times as high as that for aldicarb and 1500 times as high as that for atrazine. This explains why aldicarb and atrazine have been found in groundwater in agricultural areas while DDT has not.

There are several processes by which pesticide may be degraded. Exposure to sunlight may cause photolysis before they leach into soils. Hydrolysis, the degradation of a chemical in reaction with water that may occur at surface, in the soil zone and underlying groundwater; the longer the hydrolysis half-life the more probable it will enter groundwater. Biodegradation, i.e. enzymatic reactions driven by microorganisms, will occur at greatest rates in microbially active soil. Chlorinated pesticides and triazine herbicides are the most resistant to biodegradation and may persist for years following application. Although the mobility of some organochlorine insecticides is limited by their high hydrophobicity (Table 4.3), their persistence is mirrored in the accumulation in fatty tissues in animals, including fish and humans, mostly from pesticides in surface-water food chains. Organic phosphorus pesticides tend to hydrolyse rather quickly at pH values above neutral, thus losing their toxic properties. However, under dry conditions some have been observed to persist for many months (Graham-Bryce, 1981). Carbamates are noted for their high susceptibility to degradation (Williams *et al.*, 1988).

Higher water solubility does not necessarily correlate with a lower degree of persistence, but highly sorbed pesticides tend to be more persistent. The biodegradability of pesticides depends on their molecular structure and soil half-lives can vary between a couple of days to years (Table 4.3). Quite long half lives can occur once pesticides leave the soil and reach the less biologically active zones of aquifers (Lavu *et al.,* 1996; Chilton *et al.*, 2000). It should be noted that although many pesticide half-lives, have been determined for soils (Table 4.3); use of such half-lives to predict aquifer behaviour may cause misleadingly optimistic attenuation estimates.

Several other pesticide concerns remain. Little is known about the fate of pesticides in tropical environments, most published data are from registration trials in temperate regions. For all pesticides there is potential for incomplete transformation of the parent compound into metabolites which may also be more or less toxic (Sawyer *et al.*, 1994) and may themselves be persistent enough to be detected in groundwater. When pesticides do get into groundwater, cleanup of the contamination is usually prohibitively costly and often may not be practically feasible. The contamination can last many years and spread over a large area before dilution and degradation eventually reduce the pesticide concentrations.

Table 4.3. Classification of leaching potential for agricultural pesticides for which WHO has derived guideline values (data from the US Department of Agriculture Natural Resources Conservation Service and Agricultural Research Service)

Common name	CAS-No.	WHO GV (mg/l)	Water solubility at 20-25 °C	Soil half-life (days)	K_{OC} (ml/g)	Leaching potential	Solution runoff potential	Adsorbed runoff potential
Alachlor	15972-60-8	20	240	15	170	Medium	Medium	Low
Aldicarb	116-06-3	10	6000	30	30	High	Medium	Low
Aldrin	309-00-2	0.03	0.027	365	5000	Low	Medium	High
Dieldrin	60-57-1	0.03	0.200	1000*	12000	Very low	Medium	High
Atrazine	1912-24-9	2	33	60	100	High	High	Medium
Carbofuran	1563-66-2	7	351	50	22	High	High	Medium
Chlordane	57-74-9	0.2	0.060	350*	20000	Very low	Medium	High
Chlorotoluron	15545-48-9	30	74	35	350	Medium	High	Low
Cyanazine	21725-46-2	0.6	170	14	190	Medium	Medium	Low
2,4-D	94-75-7	30	890	10	20	Medium	Medium	Low
2,4-DB	94-82-6	90	46	5	440	Low	High	Low
1,2-Dibromo-3-chloro-propane	96-12-8	1	1000	180	70	High	High	Medium
1,2-Dibromoethane (ethylene dibromide)	106-93-4	0.4 (P)	4300	100*	34	High	High	Medium
1,2-Dichloropropane	78-87-5	40 (P)	2700	700*	50	High	High	Medium
1,3-Dichloropropene	542-75-6	20	2250	10	32	Medium	Medium	Low
Dimethoate	60-51-5	6	39800	7	20	Medium	Medium	Low
Endrin	72-20-8	0.6	0.230	4300	10000	Low	Medium	High
2,4,5-TP	93-72-1	0.09	140	21	300	Medium	Medium	Low
HCB	118-74-1	None[1]	0.005	1000*	50000	Very low	Medium	High
Isoproturon	34123-59-6	9	700	21	130	Medium	Medium	Low
γ-HCH	58-89-9	None[2]	7	400	1100	Medium	High	High
MCPA	2039-46-5	2	866000*	25	20	High	Medium	Low
MCPP	708-51-90	None[2]	660000*	21	20	High	Medium	Low
Methoxychlor	72-43-5	20	0.100	120	80000	Very low	Medium	High
Metolachlor	51218-45-2	10	530	90	200	High	High	Medium
Molinate	2212-67-1	6	970	21	190	Medium	Medium	Low
Pendimethalin	40487-42-1	20	0.275	90	5000	Low	Medium	High
Simazine	122-34-9	2	6	60	130	High	High	Medium
2,4,5-T	93-76-5	9	278	30	80	High	Medium	Low
Terbuthylazine	5915-41-3	7	9	45	200	High	High	Medium
Trifluralin	1582-09-8	20	0.300	60	8000	Low	Medium	High

* Estimated value; [1] occurs in drinking-water at concentrations well below those causing toxic effects; [2] unlikely to occur in drinking-water

4.7 EMERGING ISSUES

4.7.1 Pharmaceuticals

There is increasing concern about micropollutants originating from pharmaceuticals and active ingredients in personal care products excreted by people as complex mixtures into wastewater systems (Kümmerer, 2004). There are a number of routes through which pharmaceuticals can impact groundwater, but primarily the sources are both untreated and treated sewage. There is also evidence that substances of pharmaceutical origin are not completely eliminated during wastewater treatment or biodegraded in the environment (Daughton and Ternes, 1999; Drewes and Shore, 2001).

Health aspects

Current knowledge on the health effects of pharmaceutically active compounds at concentration levels found in groundwater samples, which are several orders of magnitude lower than concentrations which would be therapeutically active, indicates that there are no effects on human health reasonably to be expected from this source of exposure (Ternes, 2001). However, there is an ongoing debate on how comprehensive health effect data from short-term high dose exposure during diagnosis and treatment should be extrapolated to long-term low dose exposure during drinking-water consumption. Moreover, the problem of correctly assessing the risk from unexpected environmental (underground) and technical metabolites (from oxidative drinking-water treatment) is not resolved. Investigation on the fate of pharmaceutically active substances in drinking-water unit operations and processes is in progress in numerous research studies e.g. in Europe, Australia, Japan, and North America. Additionally, proposals for risk assessment procedures have been suggested (Montforts, 2001).

Transport and attenuation

A lack of knowledge still persists regarding the fate of pharmaceuticals during travel through the subsurface. Findings of recent studies indicate that travel through the subsurface can substantially attenuate the majority of pharmaceutically active compounds where surface water or domestic wastewater is used for groundwater recharge. However, where groundwater is influenced by surface water, such as artificial recharge, polar pharmaceutically active compounds such as clofibric acid (blood-lipid regulating agent), carbamazepine and primidone (antiepileptic drugs) and iodinated X-ray contrast agent can migrate through the subsurface and have been detected in groundwater samples in Germany and the USA (Heberer *et al.*, 1998; Kuehn and Mueller, 2000; Drewes *et al.*, 2001).

4.7.2 Endocrine disrupting compounds

There has been increasing public concern about various environmental contaminants which mimic estrogens and other sex-hormones and hence interfere with endogenous endocrine systems, with potential adverse effects on human health. A global assessment of the state of knowledge on endocrine disrupters was published by the International Programme on Chemical Safety (Damstra *et al.*, 2002).

More than 70 000 chemicals are discussed with respect to endocrine disruptive potential (Bradley and Zacharewski, 1998). These compounds represent both synthetic chemicals produced industrially (such as cleaners, pesticides, food additives, birth control pills, cosmetics) and naturally occurring compounds (such as steroidal hormones, plant-produced estrogens, herbal supplements and metals). Whilst endocrine disrupting compounds (EDCs) are largely organic compounds, it should be noted that some inorganic substances such as metals are also suspected of endocrine disrupting effects. Although their potential occurrence in drinking-water has been the subject of public attention and discussion in some countries, it is important to note that human exposure is chiefly through food.

The steroidal sex hormones estradiol, estrone and testosterone are a class of hormonally active agents of particular interest because they are naturally excreted into the environment from human and animal sources as well as extensively used as pharmaceuticals (e.g. birth control pills). Of the numerous synthetic chemicals that have been implicated as endocrine disrupters, many are no longer used in commerce in many countries, such as some organochlorine pesticides (e.g. DDT, endosulphan, dieldrin, and toxaphene), and PCBs. Other hormonally active compounds, such as various phenolics and phthalates, continue to be used in a variety of industrial applications worldwide (NRC, 1999). Alkylphenol is a biological metabolite of alkylphenol polyethoxylates commonly used in a variety of industrial, agricultural and household applications as non-ionic surfactants. Alkylphenol and compounds are both believed to be endocrine disrupters (Lye *et al.*, 1999). Another synthetic chemical that has measurable hormonal activity is Bisphenol A used as a chemical intermediate for numerous industrial products including polymers, resins, dyes and flame retardants.

Health effects of hormonally active compounds are based on binding of these compounds on steroid hormone receptors which control fundamental mechanisms of gene regulation. The disruption of this process can result mainly in reproductive changes. Developmental defects, neurobehavioral abnormalities, immunological deficits, carcinogenesis and ecologic effects can also be induced (NRC, 1999).

Determining the risk of EDCs to humans is difficult because exposure to these agents has not been routinely monitored, and effects that might be attributed to background concentrations could be complicated by endogenous hormones, pharmacological estrogens (e.g. hormonal contraceptives), and naturally occurring hormonally active agents (e.g. phytoestrogens) that are ubiquitous in the environment (NRC, 1999). Although it is clear that exposures to EDCs at high concentrations can affect human health, the extent of harm caused by exposure to these compounds in concentrations that are commonly found in groundwater is debated (NRC, 1999). Generally, natural and synthetic steroidal hormones are several thousand times more potent than industrial chemicals, pesticides and metals (Khan and Ongerth, 2000).

The WHO has not yet specifically proposed any guidelines for the occurrence of EDCs in drinking-water. However, some of the organochlorine compounds are regulated as pesticides.

The relevance to human health of EDCs occurring in water is currently uncertain. Their occurrence in groundwater is linked to the release of sewage, manure, or spill of specific synthetic chemicals into the environment. The specific processes used in wastewater treatment facilities play a key role in the introduction of EDCs into surface water and groundwater (Drewes and Shore, 2001). The transport of EDCs to groundwater depends on their hydrophobicity and degradability. The majority of highly potent compounds such as steroids are hydrophobic and degradable. Degradation rates of EDC compounds depend on temperature, soil characteristics and their molecular weights (IUPAC, 2003).

The potential risk related to an uptake of individual EDCs present in wastewater affected groundwater by humans does not appear to be very significant. The small data set about the fate of EDCs (such as natural and synthetic hormones, surfactants and pesticides) during percolation through the soil and aquifer and the lack of toxicity data on long-term exposure of low concentrations makes it at present impossible to finally assess the impact of EDCs in groundwater on human health. However, contaminated groundwater may be impacted by a mix of different compounds, which could additively impose endocrine disrupting effects.

4.8 REFERENCES

Ahmed, F.M., Ali, M.A. and Adeel, Z. (2001) *Technologies for Arsenic Removal From Drinking Water*, BUET, Dhaka.

Alaerts, G.J., Khouri, N. and Kabir, B. (2001) Strategies to mitigate arsenic contamination in water supply. In *United Nations Synthesis Report on Arsenic in Drinking Water*, www.who.int/water_sanitation_health/dwq/en/arsenicun8.pdf (accessed April 29. 2005).

Albu, M.A., Morris, L.M., Nash. H. and Rivett, M.O. (2002) Hydrocarbon contamination of groundwater at Ploiesti, Romania. In: *Sustainable Groundwater Development* (eds. K.M. Hiscock, M.O. Rivett and R.M. Davison), pp. 293-301, Geological Society Special Publication No. 193, London.

Allen-King, R.M., Ball, W.P. and Grathwohl, P. (2002) New modeling paradigms for the sorption of hydrophobic organic chemicals to heterogeneous carbonaceous matter in soils, sediments, and rocks. *Advances in Water Resour.*, **25**, 985-1016.

Appelo, C.A.J. and Postma, D. (1993) *Geochemistry, Groundwater and Pollution*, Balkema, Rotterdam.

Araya, M., Chen, B., Klevay, L.M., Strain, J.J., Johnson, L., Robson, P., Shi, W., Nielsen, F., Zhu, H., Olivares, M., Pizarro, F., Haber, L.T. (2003) Confirmation of an acute no-observed-adverse-effect and low-observed-adverse-effect level for copper in bottled drinking water in a multi-site international study. *Regul. Toxicol. Pharmacol.* **38**, 389-399.

Ashley, R.P. (1998) Foreseeability of environmental hazards: the implications of the Cambridge Water Company case. In: *Groundwater Contaminants and their Migration*, (ed. J. Mather), pp. 23-26, Special Publication 128, Geological Society, London.

Bailey, K., Chilton, J., Dahi, E, Lennon, M. and Jackson, P (In Press) *Fluoride in Drinking Water.* WHO Drinking-water Series. IWA Publishing, London.

Barceloux, D.G. (1999) Selenium. *J. Toxicol. Clin. Toxicol.,* **37**(2), 145-172.

Barker, J.F., Patrick, G.C. and Major, D. (1987) Natural attenuation of aromatic hydrocarbons in a shallow sand aquifer. *Ground Water Monitor Review* **7**(4), 64-71.

Barnard, C., Daberkow, S., Padgitt, M., Smith, M.E. and Uri, N.D. (1997) Alternative measures of pesticide use. *Sci. Total Environ.*, **203**, 229-244.

Barnes, C.J. Jacobson, G. and Smith, G.D. (1992) The origin of high-nitrate groundwaters in the Australian arid zone. *J. Hydrology*, **137**, 181-197.

Bedient, P.B., Rifai, H.S. and Newell, C. (1999) *Ground Water Contamination: Transport and Remediation*, 2nd edn, Prentice-Hall, New Jersey.

Bradley, E.G. and Zacharewski, T.R. (1998) Exoestrogens: mechanisms of action and strategies for identification and assessment. *Environ. Toxicol. Contam.*, **17**, 49-57.

Cavallero, A., Corradi, C., De Felice, G. and Grassi, P. (1985) Underground water pollution in Milan by industrial chlorinated organic compounds. In *Effects of Landuse Upon Freshwaters*, (ed. J.F. de L.G. Solbe), pp. 68-84, Ellis Harwood, Chichester.

CEC (2001) Commission Recommendation on the protection of the public against exposure to radon in drinking water supplies. Document number C (2001) 4580. *Official Journal of the European Communities,* L 344/85.

Chilton, P.J., Lawrence, A.R. and Stuart, M.E. (1998) Pesticides in groundwater: some preliminary results from recent research in temperate and tropical environments. In *Groundwater Contaminants and Their Migration,* (ed. J. Mather), pp. 333-345, Special Publication 128, Geological Society, London.

Chilton, P.J., Stuart, M.E., Lawrence, A.R., Gooddy, D.C., Williams, R.J. and Johnson, A.C. (2000) Assessing pesticide pollution of groundwater: current knowledge and remaining gaps. In *Groundwater: Past Achievements and Future Challenges,* (ed. O.T.N. Sililo), pp. 17-26, Balkema, Rotterdam.

Costa, M. (2002) Potential hazards of hexavalent chromate in our drinking water. *Toxicol. Appl. Pharmacol.*, **188**, 1-5.

Damstra, T., Barlow, S., Bergman, A., Kavlock, R. and van der Kraak, G. (2002) *Global Assessment of the State-of-the-Science of Endocrine Disruptors*. International Programme on Chemical Safety. http://www.who.int/ipcs/publications/new_issues/endocrine_disruptors/en/index.html

Daughton, C. and Ternes, T. (1999) Pharmaceuticals and personal care products in the environment: agents of subtle change? *EHP,* **107**, suppl. 6, 907-938.

Domenico, P.A. and Schwartz, R.W. (1998) *Physical and Chemical Hydrogeology*, 2nd edn, John Wiley & Sons, New York.

Drewes, J.E. and Shore, L.S. (2001) Concerns about pharmaceuticals in water reuse, groundwater recharge, and animal waste. In *Pharmaceuticals and Personal Care Products in the Environment*, (eds. Ch. Daughton and T.L. Jones-Lepp), pp. 206-228, American Chemical Society Symposium Series No. 791, Washington, DC.

Drewes, J.E., Fox, P. and Jekel, M. (2001) Occurrence of iodinated X-ray contrast media in domestic effluents and their fate during indirect potable reuse. *J. Environ. Sci. Health Part A,* **36A**(9), 1633-1645.

EA (2000) *Guidance on the Assessment and Monitoring of Natural Attenuation of Contaminants in Groundwater*. R&D Publication 95, Environment Agency, Bristol.

Edmunds, W.M. and Gaye, C.B. (1994) Estimating the spatial variability of groundwater recharge in the Sahel using chloride. *J. Hydrology*, **156**, 47-59.

Edmunds, W.M., Cook, J.M., Darling, W.G., Kinniburgh, D.G., Miles, D.L., Bath, A.H., Morgan-Jones, M. and Andrews, J.N. (1987) Baseline geochemical conditions in the Chalk aquifer, Berkshire UK: a basis for groundwater quality management. *Appl. Geochem.*, **2**(2), 252-274.

FAO (2004) Pesticide residues in food - 2004. Report of the Joint Meeting of the FAO Panel of Experts on Pesticide Residues in Food and the Environment and the WHO Core Assessment Group on Pesticide Residues. FAO Plant Production and Protection Paper 178, Food and Agricultural Organization, Rome, Italy.

Fetter, C.W. (1999) *Contaminant Hydrogeology*, 2nd edn, Macmillan, New York.

Flühler, H.P., Ferlin, P., Selim, H.M., and Schulin R. (1985) Transport von Fluorid, Bromid und Chlorid in Bodensäulen und in einem natürlich gelagertem Boden. *Z. Dtsch. Geol. Gesellschaft*, **136**, 375-383.

Freeze, R.A. and Cherry, J.A. (1979) *Groundwater*, Prentice-Hall, Englewood Cliffs, New Jersey.

Fusillo, T.V., Hochreiter J.J. and Lord, D.G. (1985) Distribution of volatile organic compounds in a New Jersey coastal plain aquifer. *Ground Water,* **23**(3), 354-360.

Gelhar, L.W. (1986) Stochastic subsurface hydrology from theory to applications. *Water Resour. Res.*, **22**(9), 135S-145S.

Graham-Bryce, I.J. (1981) The behaviour of pesticides in soil. In *The Chemistry of Soil Processes*, (eds. D.J. Greenland and M.H.B. Hayes), pp. 621-671, John Wiley & Sons Ltd., Sussex.

Hearn, W., Henderson, G., Houston, S., Wadw, A. and Walker, B. (1993) Water supply and Aboriginal and Torres Strait Ilsander health: an overview. *AGSO J. Austral. Geology and Geophysics*, **14**(2/3), 135-146.

Heberer, T., Schmidt-Bäumler K. and Stan H–J. (1998) Occurrence and distribution of organic contaminants in the aquatic system in Berlin. Part I: Drug residues and other polar contaminants in Berlin surface and groundwater. *Acta Hydrochim. Hydrobiol.* **26**, 272-278.

Hem, J.D. (1989) *Study and Interpretation of the Chemical Characteristics of Natural Water*, 3rd ed, USGS, Water Supply Paper 2254.

Heron, G., Barcelona, M.J., Anderson, M.L. and Christensen, T.H. (1997) Determination of nonvolatile organic carbon in aquifer solids after carbonate removal by sulphurous acid. *Ground Water*, **35**, 6-11.

IARC (2001) *Overall Evaluations of Carcinogenicity to Humans*, http://www-cie.iarc.fr/monoeval/crthall.html (accessed April 29. 2005).

IUPAC (2003) Endocrine Disrupters in the Environment. IUPAC Technical Report, *Pure and Applied Chemistry*, **75**(5), 631-681.

Jackson, R.E. (1998) The migration, dissolution and fate of chlorinated solvents in the urbanized alluvial valleys of the southwestern USA. *Hydrogeology J.*, **6**(1), 144-155.

Johansen, S.S., Hansen, A.B., Mosbaek, H. and Arvin, E. (1997) Characterization of heteroaromatic and other organic compounds in groundwater at creosote-contaminated sites in Denmark. *Groundwater Mon. Remediation*, **17**(2), 106-115.

Kerger, B.D., Finley, B.L., Corbett, G.E., Dodge, D.G. and Paustenbach, D.J. (1997) Ingestion of chromium(VI) in drinking water by human volunteers: absorption, distribution, and excretion of single and repeated doses. *J. Toxicol. Environ. Health*, **50**, 67-95.

Kerndorff, H., Schleyer, R., Milde, G. and Plumb, R. H. (1992) Geochemistry of groundwater pollutants at German waste disposal sites. In *Groundwater Contamination and Analysis at Hazardous Waste Sites*, (eds. S. Lesage and R.E. Jackson), Marcel Dekker, Inc. New York, Basel, Hong Kong.

Khan, S. and Ongerth, J. (2000) *Hormonally Active Agents in Domestic Wastewater Treatment – A Literature Review.* Report prepared for CGE Australia.

King, M.W.G. and Barker, J.F. (1999) Migration and natural fate of a coal tar creosote plume 1. Overview and plume development. *J. Contam. Hydrol.*, **39**, 249-279.

Klecka, G.M., Gonsior, S.J. and Markham, D.A. (1990) Biological transformations of 1,1,1-trichloroethane in subsurface soils and groundwater. *Environ. Sci. Technol.*, **9**, 1437-1451.

Kolpin, D.W., Barbash, J.E. and Gilliom, R.J. (2000) Pesticides in groundwater of the United States, 1992-96. *Ground Water*, **38**(6), 858-863.

Kuehn, W. and Mueller, U. (2000) Riverbank filtration – an overview. *JAWA*, **92**(12), 60-69.

Kueper, B.H., Redman, D., Starr, R.C., Reitsma, S. and Mah, M. (1993) A field experiment to study the behavior of tetrachloroethylene below the water table: spatial distribution of residual and pooled DNAPL. *Ground Water*, **31**(5), 756-766.

Kümmerer, K. (2004) Ecotoxicological evaluation of pharmaceuticals. In *Pharmaceuticals in the Environment: Sources, occurrence, fate and effects*, 2nd ed, Springer, Heidelberg, New York.

Lavy, T.L., Mattice, J.D., Massey, J.H., Skulman, B.W. and Senseman, S.A. (1996) Long-term in situ leaching and degradation of six herbicides aged in subsoils. *J. Environ. Qual.*, **25**, 1268-1279.

Lawrence, C.R. (1983) *Nitrate-rich groundwaters in Australia.* Australian Water Resources Council Technical Paper No. 79, Australian Government Publishing Service, Canberra.

Lorah, M.M. and Olsen, L.D. (1999) Natural attenuation of chlorinated volatile organic compounds in a freshwater tidal wetland: field evidence of anaerobic biodegradation. *Water Resour. Res.*, **35**, 3811-3827.

Lye, C.M., Frid, C.L.J., Gill, M.E., Cooper, D.W. and Jones, D.M. (1999) Estrogenic alkylphenols in fish tissues, sediments, and waters from the UK Tyne and Tees Estuaries. *Environ. Sci. Technol.*, **33**(7), 1009-1014.

Mackay, D.M. and Cherry, J.A. (1989) Groundwater contamination: pump-and-treat remediation. *Environ. Sci. Technol.*, **23**(6), 630-636.

Mackay, D.M., Freyberg, D.L., Roberts, P.V. and Cherry, J.A. (1986) A natural gradient experiment on solute transport in a sand aquifer, 1. Approach and overview of plume movement. *Water Resour. Res.*, **22**(13), 2017-2029.

Matin, M.A., Malek, M.A., Amin, M.R., Rahman, S., Khatoon, J., Rahman, M., Aminuddin, M.. McLachlan, M.S., Horstmann, M. and Hinkel, M. (1996) Polychlorinated dibenso-p-dioxins and dibenzofurans in sewage sludge: sources and fate following sludge application to land. *Sci. Total Environ.*, **185**, 109-123.

McAllister, P.M. and Chiang, C.Y. (1994) A practical approach to evaluating natural attenuation of contaminants in ground water. *Ground Water Monitoring & Remediation*, **14**(2), 161-173.

Mendoza, C.A., Johnson, R.L. and Gillham, R.W. (1996) Vapor migration in the vadose zone. In *Dense Chlorinated Solvents and Other DNAPLs in Groundwater*, (eds. J.F. Pankow and J.A. Cherry), pp. 179-201, Waterloo Press, Portland, Oregon.

Mercer, J.W. and Cohen, R.M. (1990) A review of immiscible fluids in the subsurface: Properties, models, characterization and remediation. *J. Contam. Hydrol.*, **6**, 107-163.

Montforts, M. (2001) Regulatory and methodological aspects concerning the risk assessment for medicinal products; need for research. In *Pharmaceuticals in the Environment: Sources, Occurrence, Fate and Effects*, (ed. K. Kümmerer), pp. 159-174, Springer, Heidelberg, New York.

Montgomery, J.H. (1996) *Groundwater Chemicals Desk Reference*, 2nd edn, CRC Press, Inc., Florida.

NCRP (1989) *Control of Radon in Houses: Recommendations of the NCRP*. National Council on Radiation Protection Report No. 103, Bethesda, Maryland.

Newell, C.J. and Connor, J.A. (1998) *Characteristics of Dissolved Hydrocarbon Plumes: Results from four studies*, API technical transfer bulletin, American Petroleum Institute, Washington DC.

Newell, C.J., Hopkins, L.P. and Bedient, P.B. (1990) A hydrogeologic database for ground-water modelling. *Ground Water*, **28**(5), 703-714.

Newell, C.J., Acree, S.D., Ross, R.R. and Huling, S.G. (1995) *Light Non Aqueous Phase Liquids*, US EPA, Ada, Oklahoma.

Noble, P. and Morgan, P. (2002) *The Effects of Contaminant Concentration on the Potential For Natural Attenuation*. EA (for England & Wales) R&D Technical Report P2-228/TR. Bristol.

NRC (1989) *Irrigation-induced water quality problems*, National Academy Press, National Research Council, Washington, DC.

NRC (1994) *Alternatives of Ground Water Cleanup*, National Academy Press, National Research Council, Washington, DC.

NRC (1999) *Hormonally Active Agents in the Environment*, National Academy Press, National Research Council, Washington, DC.

Pankow, J.F. and Cherry, J.A. (1996) *Dense Chlorinated Solvents and Other DNAPLs in Groundwater*, Waterloo Press, Portland, Oregon.

Plumb, R.H. (1992) The importance of volatile organic compounds as a disposal site monitoring parameter. In *Groundwater Contamination and Analysis at Hazardous Waste Sites*, (eds. S. Lesage and R.E. Jackson), pp. 173-197, Marcel Dekker, Inc. New York, Basel, Hong Kong.

Rivett, M.O. (1995) Soil-gas signatures from volatile chlorinated solvents: Borden field experiments. *Ground Water*, **33**(1), 84-98.

Rivett, M.O. and Allen-King, R.M. (2003) A controlled field experiment on groundwater contamination by a multicomponent DNAPL: dissolved-plume retardation. *J. Contam. Hydrol.*, **66**, 117-146

Rivett, M.O., Lerner, D.N., Lloyd, J.W. and Clark, L. (1990) Organic contamination of the Birmingham aquifer. *UK J. Hydrology*, **113**, 307-323.

Rivett, M.O., Feenstra, S. and Cherry J.A. (2001) A controlled field experiment on groundwater contamination by a multicomponent DNAPL: creation of the emplaced-source and overview of dissolved plume development. *J. Contam. Hydrol.*, **49**, 111-149.

Rivett, M.O., Shepherd, K.A., Keeys, L. and Brennan, A.E., (2005) Chlorinated Solvents in the Birmingham Aquifer, UK: 1986 – 2001. *Quarterly Journal of Engineering Geology & Hydrogeology*, 38(4), 337-350.

Sawyer, C.N., McCarty, P.L. and Parkin, G.F. (1994) *Chemistry for Environmental Engineering*, 4th edn, McGraw-Hill, New York.

Schnitzer, M. and Khan, S.U. (1972) *Humic Substances in the Environment*, Marcel Dekker, Inc., New York.

Schwartz, F.W. and Zhang, H. (2003) *Fundamentals of Groundwater*, John Wiley & Sons, Inc., New York.

Schwarzenbach, R.P., Gschwend, P.M. and Imboden, D.M. (1993) *Environmental Organic Chemistry*, John Wiley & Sons, Inc, New Jersey.

Schwille, F. (1988) *Dense Chlorinated Solvents in Porous and Fractured Media*, (ed. J.F. Pankow), Lewis Publishers, Chelsea, Michigan.

Senior, L.A. (1998) *Radon-222 in the Ground Water of Chester County, Pennsylvania:* USGS Water-Resources Investigations Report 98-4169.

Smedley, P.L. and Kinniburgh, D.G. (2001) Source and behaviour of arsenic in natural waters. In *United Nations Synthesis Report on Arsenic in Drinking Water*, www.who.int/water_sanitation_health/dwq/en/arsenicun1.pdf (accessed April 29, 2005).

Smith, A.H., Lingas, E.O. and Rahman, M. (2000) Contamination of drinking-water by arsenic in Bangladesh: a public health emergency. *Bulletin of the WHO*, **78**, 1093-1103.

Spliid, N.H. and Køppen, B. (1998) Occurrence of pesticides in Danish shallow groundwater. *Chemosphere*, **37**(7), 1307-1316.

Sposito, G. (1989) *The Chemistry of Soils*, OUP, New York.

Squillace, P.J., Zogorski, J.S., Wilber, W.G. and Price, C.V. (1996) Preliminary assessment of the occurrences and possible sources of MTBE in groundwater in the United States, 1993-1994. *Environ. Sci. Technol.*, **30**(5), 1721-1730.

Stevenson, F.J. (1994) Structural basis of humic substances. In *Humus Chemistry: Genesis, Composition, Reactions*, (ed. P.J. Stevenson), 2nd edn, pp. 285-302, John Wiley & Sons, New York.

Stumm, W. and Morgan, J.J. (1996) *Aquatic Chemistry*, 3rd edn, John Wiley & Sons, New York.

Ternes, T. (2001) Pharmaceuticals as New Emerging Environmental Contaminants: A Survey. In *Proc. Second International Conference on Pharmaceuticals and Endocrine Disrupting Chemicals in Water, Minneapolis, Minnesota,* pp. 1-9.

Thurman, E.M. (1985) *Organic Geochemistry of Natural Waters: Developments in Biogeochemistry*, Martinus Nijhoff/Dr. W. Junk Publishers, Dordrecht.

UNSCEAR (2000) *Sources, Effects and Risks of Ionizing Radiation*, UN, New York.

US EPA (1996) *Soil Screening Guidance*, Office of Emergency and Remedial Response, EPA/540/R95/128. http://www.epa.gov/superfund/resources/soil/index.htm (accessed August 11, 2005).

US EPA (1999) *Screening Level Ecological Risk Assessment Protocol for Hazardous Waste Combustion Facilities*, vol. 2, Appendix A, EPA530-D-99-001B. http://www.epa.gov/combustion/eco-risk/volume2/appx-a.pdf (accessed August 11, 2005).

US EPA (2000) *Summary of EPA Coal Combustion Waste Damage Cases Involving Sand and Gravel Mines/Pits/Operations*, http://www.epa.gov/epaoswer/other/fossil/ sandgrav.pdf (accessed April 29, 2005).

Vogel, T.M., Criddle, C.S. and McCarty, P.L. (1987) Transformations of halogenated aliphatic compounds. *Environ. Sci. Technol.*, **21**, 722-736.

Wiedemeier, T.H., Rifai, H.S., Newell, C.J. and Wilson, J.T. (1999) *Natural Attenuation of Fuels and Chlorinated Solvents in the Subsurface*, John Wiley & Sons, New York.

Williams, W.M., Holden, P. W., Parsons, D.W. and Lorber, M.W. (1988) *Pesticides in Ground Water Data Base*, Interim Report, US EPA, Office of Pesticide Programs, Washington, DC.

WHO (1993) *Guidelines for Drinking-water Quality: Recommendations*, 2nd edn, vol. 1, World Health Organization, Geneva.

WHO (1996a) *Guidelines for Drinking-water Quality: Health criteria and other supporting information*, 2nd edn, vol 2, World Health Organization, Geneva.

WHO (1996b) *Toxicological Evaluation of Certain Food Additives and Contaminants*. Prepared by the 44th Meeting of the Joint FAO/WHO Expert Committee on Food Additives (JECFA), Geneva.

WHO (2003a) *Arsenic in Drinking-water: Background Document for development of WHO Guidelines for Drinking-water Quality*. World Health Organization, Geneva, http://www.who.int/water_sanitation_health/dwq/chemicals/arsenic.pdf (accessed September 8, 2005).

WHO (2003b) *Selenium in Drinking-water: Background Document for development of WHO Guidelines for Drinking-water Quality*. World Health Organization, Geneva, http://www.who.int/water_sanitation_health/dwq/chemicals/selenium.pdf (accessed September 8, 2005).

WHO (2003c) *Lead in Drinking-water: Background Document for development of WHO Guidelines for Drinking-water Quality*. World Health Organization, Geneva, http://www.who.int/water_sanitation_health/dwq/chemicals/lead.pdf (accessed September 8, 2005).

WHO (2003d) *Chromium in Drinking-water: Background Document for development of WHO Guidelines for Drinking-water Quality*. World Health Organization, Geneva, http://www.who.int/water_sanitation_health/dwq/chemicals/chromium.pdf (accessed September 8, 2005).

WHO (2003e) *Vinyl Chloride in Drinking-water: Background Document for development of WHO Guidelines for Drinking-water Quality*. World Health Organization, Geneva, http://www.who.int/water_sanitation_health/dwq/chemicals/vinylchloride.pdf (accessed September 8, 2005).

WHO (2003f) *1,2-Dichloroethene in Drinking-water: Background Document for development of WHO Guidelines for Drinking-water Quality*. World Health Organization, Geneva, http://www.who.int/water_sanitation_health/dwq/1,2-Dichloroethene.pdf (accessed September 8, 2005).

WHO (2003g) *Dichlorobenzenes in Drinking-water: Background Document for development of WHO Guidelines for Drinking-water Quality*. World Health Organization, Geneva http://www.who.int/water_sanitation_health/dwq/chemicals/dichlorobenzenes.pdf (accessed September 8, 2005).

WHO (2004a) *Guidelines for Drinking-water Quality: Recommendations*. 3rd edn, vol. 1, Geneva.

WHO (2004b) *Fluoride in Drinking-water: Background Document for development of WHO Guidelines for Drinking-water Quality*. World Health Organization, Geneva, http://www.who.int/water_sanitation_health/dwq/chemicals/fluoride.pdf (accessed September 8, 2005)..

WHO (2004c) *Copper in Drinking-water: Background Document for development of WHO Guidelines for Drinking-water Quality*. World Health Organization, Geneva, http://www.who.int/water_sanitation_health/dwq/chemicals/copper.pdf (accessed September 8, 2005).

WHO (2004d) *Vinyl Chloride in Drinking-water: Background Document for development of WHO Guidelines for Drinking-water Quality*. World Health Organization, Geneva http://www.who.int/water_sanitation_health/dwq/chemicals/vinylchloride.pdf (accessed September 8, 2005).

WHO (2005a) *Uranium Drinking-water: Background Document for development of WHO Guidelines for Drinking-water Quality.* World Health Organization, Geneva, http://www.who.int/water_sanitation_health/dwq/chemicals/uranium290605.pdf (accessed September 8, 2005).

WHO (2005b) *Nitrate in Drinking-water: Background Document for development of WHO Guidelines for Drinking-water Quality.* World Health Organization, Geneva, http://www.who.int/water_sanitation_health/dwq/chemicals/en/index.html (accessed September 8, 2005).

WHO (2005c) *Methyl tertiary-butyl ether (MTBE) in Drinking-water: Background Document for development of WHO Guidelines for Drinking-water Quality.* World Health Organization, Geneva, http://www.who.int/water_sanitation_health/dwq/chemicals/MTBE200605.pdf (accessed September 8, 2005).

WHO (2005d) *Trichloroethene in Drinking-water: Background Document for development of WHO Guidelines for Drinking-water Quality.* World Health Organization, Geneva, http://www.who.int/water_sanitation_health/dwq/chemicals/trichloroethenemay05.doc (accessed September 8, 2005).

WHO (2005e) *Dichloroethene in Drinking-water: Background Document for development of WHO Guidelines for Drinking-water Quality.* World Health Organization, Geneva, http://www.who.int/water_sanitation_health/dwq/chemicals/11dichloroethenefinal.pdf (accessed September 8, 2005).

WHO (In Press) *Arsenic in Drinking-water*, WHO Drinking-water Quality Series. IWA Publishing, London.

Zietz, B., Dieter, H.H., Lakomek, M., Schneider, H., Kessler-Gaedtke, B. and Dunkelberg, H. (2003) Epidemiological Investigation on chronic copper toxicity to children exposed via the public drinking water supply. *Sci. Tot. Environ.*, **302**, 127-144.

5

Socioeconomic, institutional and legal aspects in groundwater assessment and protection

G. Howard, P. Chave, P. Bakir and B. Hoque

The socioeconomic conditions in areas where groundwater is used or recharge occurs are critical to the development of groundwater protection measures. The protection of groundwater may be expensive and disruptive to the inhabitants of the land above the aquifers. Socioeconomic conditions play an important role in determining the likely contaminant loads and types of contaminant that affect the groundwater. It is also critical in determining what types of intervention are possible, how such interventions will be implemented and what resources (including human) will be required and are available.

Institutional and legal issues are also critical in determining the success or failure of groundwater protection policies and strategies. Weak institutions and poor institutional frameworks are commonly identified with poor implementation of water policy (World Bank, 1993; WELL, 1998). The development of groundwater policies and strategies must therefore provide adequate consideration of the appropriate institutional arrangements and consider how the needs of all stakeholders can be incorporated into the policy. The latter demands that there are effective processes of public consultation and participation in policy and strategy development. Legislation is also vital to support effective groundwater protection. Not only should the law be supportive of groundwater protection, effective implementation of the law is required to ensure groundwater protection objectives are met (Caponera, 1992; Foster *et al.*, 1992).

The purpose of this chapter is to discuss some of the key socioeconomic, institutional and legal issues that are important to consider in groundwater protection. In Chapter 7, the types of information and methods of data collection will be discussed; Chapter 20 outlines how socioeconomic issues and the institutional and legal frameworks can be addressed in developing groundwater protection.

5.1 SOCIOECONOMIC STATUS: ISSUES OF POVERTY AND WEALTH

Socioeconomic status is a measure of the wealth of individuals, households and communities and reflects their assets as well as the ability of households to obtain goods and services. Socioeconomic status is of importance when considering the level of investment in groundwater management which individuals, communities and societies are willing and able to make.

Socioeconomic conditions influence the capacity for different groups to protect their environment. For instance, in some communities short-term priorities for resource exploitation override the need for resource protection necessary to secure a long-term livelihood, despite the recognition in the communities of the need for such protection.

The poor are usually at greater risk from the adverse effects of poor resource management and it is essential that their needs be properly addressed when developing groundwater strategies. Critical to this approach is to avoid disadvantage for the poor caused by the implementation of groundwater protection policies and strategies. Such disadvantage may occur, for instance, because agricultural use of land is restricted in order to protect groundwater, which may result in reduced incomes and decreased security for poor farmers. Consideration must be given to compensation, financial support, the creation of alternative employment opportunities or provision of new land when no restrictions apply. However, the latter is often difficult to implement and should only be considered where there is strong evidence from consultations that such an approach is acceptable to the communities affected and that the proposed land for relocation is at least the same quality as the land being left.

The implementation of groundwater protection measures will often have important implications for the livelihoods of the households affected and this applies in all countries. For instance, significant changes in land use regulations in developed countries will also have a profound impact on the users of land, water and other resources. Changes may have positive or negative impacts on some or all of the components of livelihoods. The implications of such impacts in terms of compensation, social services and environmental protection should be taken into account when reaching a decision about what and how land use regulations are applied. The population affected must be fully consulted and be willing to accept any restrictions as part of the process of establishing protection norms. Box 5.1 outlines the problems faced in parts of Germany in relation to controlling groundwater contamination when families facing financial hardship are able to access alternative sources of water.

Box 5.1. Socioeconomic factors and illegal use of private wells in some rural areas of Germany

After the political changes in Germany in 1990, the connection of rural areas in eastern Germany to central water supplies was rapidly developed. This was particularly urgent in some mountainous areas of Erzgebirge, Thuringia, as the supply from individual wells was highly unsatisfactory because aquifers in fractured bedrock fell dry at intervals. They were also vulnerable to short-circuiting with surface water and sewage. Connection to central supplies of high quality and reliable quantity in the 1990s was therefore warmly welcomed by the population. Individual wells were abandoned and sometimes illegally misused as undrained sewage pits. As local aquifers were no longer needed for feeding household wells, their protection was no longer perceived as a priority.

However, the introduction of cost recovery for drinking-water significantly increased prices within a few years. At the same time, unemployment rates were very high and available work often poorly paid. As a result, many households struggled financially. This made individual supplies attractive again, and a large number were re-activated illegally. The German Drinking-water Ordinance requires annual monitoring of private wells, which typically involves partial or total cost coverage for the analyses by the well owner, but numerous households connected to a central supply avoided these costs by not registering their re-activated wells for surveillance. Their use only became known because metering showed large numbers of households that were not using any water from the central supply.

Ensuing public health concerns include the high rate of household wells with microbial contamination (up to 60 per cent in one survey) and a high risk of unintentional cross-connection between self-built piping from the household well and the public supply, potentially contaminating the public supply. Furthermore, because of increasing stagnation in the mains due to reduced flow, the costs of the central supply have further increased as flushing is required more often to prevent microbial re-growth. This further reduces the attractiveness of the central supply and encourages greater use of the re-activated wells.

When actions are required by specific communities to protect groundwater resources, consideration must be given to the incentives that may be required and how these can be provided. It should be noted that in many cases such incentives do not refer to direct monetary compensation packages but could, for example, address improved security of tenure for poor farmers in order to promote reduced pollution loads derived from agriculture.

The socioeconomic status of communities is likely to influence the type of interventions that will be feasible for groundwater protection. For instance, in low-income communities in developing countries with shallow groundwater, the use of pit latrines may not be the preferred technical solution for excreta disposal, as they lead to an increased risk of contamination. However, alternative technologies may be too expensive for the majority of the population to sustain. In this case, some degree of contamination of groundwater may be tolerated in order to reduce a greater health risk caused by the lack of excreta disposal. In urban areas, if contamination is deemed unacceptable, then it is often more cost-effective to provide an alternative (often piped) water supply that uses water from a more distant and protected water source (Franceys *et al.,* 1992). In rural areas it is often more difficult to implement such solutions and the use of an alternative sanitation technology may need to be considered and potentially subsidized.

Similar situations may occur in developed countries where balances need to be made between protection of the groundwater resource and sanitation provision. In rural communities, septic tanks may be used where shallow groundwater is tapped for domestic supply thus representing a risk to the quality of the groundwater source. Off-site methods may not be feasible because of the cost of operation and maintenance. In this case, it is likely to be more cost effective to treat the water or use an alternative source of water rather than attempt to change the sanitation technology.

5.1.1 Livelihood concepts

The concept of livelihoods is now used in many countries when considering the nature of poverty. The basic concept of a livelihoods approach is that the ability of households and communities to sustain and improve their livelihood relies on income, assets, capabilities and their vulnerability. This approach also takes into account gender, environmental sustainability and cultural norms in defining sustainable livelihoods. Chambers and Conway (1991) provided definitions of sustainable livelihoods in relation to both environmental sustainability and social sustainability. DFID (2003) has combined these to define a sustainable livelihood:

> *'A livelihood is sustainable when it can cope with and recover from stresses and shocks and maintain or enhance its capabilities and assets both now and in the future, while not undermining the natural resource base.'*

Stresses and shocks refer to events or changes in assets, income or vulnerability that put pressure on the livelihood. For instance, sudden loss of employment or large increases in price of basic goods result in a shock or stress to the livelihood. Equally, a poor harvest or sudden change in allowed land use may affect the asset base of a community. The onset of a significant health problem may increase vulnerability (infection with human immunodeficiency virus (HIV) being a good example) of an individual such that previous levels of health protection are no longer adequate.

Water and health are both considered as assets within this framework and the degree to which households or communities have access to these assets and their resilience to shocks and stresses are fundamental components of securing a sustainable livelihood. The livelihood approach also encompasses concepts of vulnerability and environmental sustainability when considering poverty.

Vulnerability is composed of risks that are shared by a community (sometimes called exposure), which includes lack of access to a specific water resource, and those risks unique to an individual (often termed susceptibility) such as HIV infection. Vulnerability may be physical, social or political (Nichol, 2000). In the context of groundwater protection, physical vulnerability may refer to the increased risk of contamination from inappropriate land use. Social vulnerability arises from marginalization of parts of a community within the larger community or society and factors such as gender-specific restrictions to assets or decision-making. In relation to groundwater, this may result in marginalization of women in decision-making regarding groundwater development, management and protection. Political vulnerability typically relates to the capability of communities to be engaged in wider decision-making processes in relation to resource access and management.

The livelihood approach ensures that the sustainability of natural resources and the environment is given an important place in the understanding of poverty. This may be given greater priority in rural areas where livelihood may depend on sustainable use of natural resources (Tamuno *et al.*, 2003). However, natural resources always remain an important component of livelihoods, as sustainability is defined in terms of a livelihood that does not degrade the asset base. This has implications for both rural and urban dwellers. For instance, the protection and sustainable use of groundwater has important implications for urban households that rely on groundwater for domestic supply, as both deterioration and protection of the resource may increase water costs and affect the livelihood of the users.

5.1.2 Source of livelihoods

Understanding the source of livelihood of communities that utilize and potentially pollute groundwater sources is important. Different means of sustaining a livelihood will result in different types of pollution. Where commercial farming is the principal source of livelihood, groundwater may be vulnerable to pollution derived from agrochemicals such as fertilizers and pesticides (Chapter 9). Where irrigation is practised, contamination is likely to increase because many irrigation systems are inefficient, resulting in significant volumes of water infiltrating the aquifer. The water used for irrigation is frequently under-priced and this tends to reinforce the inefficient use of water (World Bank, 1993). However, where irrigation is essential for growing crops, the development of groundwater protection strategies will have to take this into account and compensation packages and alternative irrigation practices (such as drip irrigation) promoted (Chapter 21).

Small subsistence or near subsistence farming may make relatively little use of agrochemicals or irrigation, but their use may be significant in countries where there are government subsidies on agricultural production. In this case, it may be more appropriate to remove the subsidy on agrochemical use than to try to regulate application in particular areas.

Where agrochemical use derives from private purchase, the groundwater strategy will have to consider the capacity of regulatory bodies to develop and deliver incentives to reduce or change applications and the cost of inspection and monitoring. Where there is widespread small-scale private use of agrochemicals, it will be important to consider targeting those areas where groundwater is at greatest risk from pollution, rather than trying to implement broad measures.

The situation with large commercial farming may be simpler to regulate, as there will be a smaller number of people to deal with. Where direct actions are taken to change land use to an economically less productive use, the land-owner would usually expect compensation and this would have to reflect their overall economic loss. However, in many cases, the restrictions may actually apply more directly to applications of agrochemicals on a seasonal basis, which would not require the same level of economic recompense. It may, however, require systems of monitoring to ensure compliance.

Where the majority of the population derive an income from small-scale agriculture, groundwater protection may be more difficult to regulate as there will be many more farmers whose needs must be addressed. Any compensation packages that are developed in such situations may have a lower per capita outlay than larger farms but it is likely to result in a higher per ha cost. This will increase overall direct costs of the protection strategy. As noted above, alternative incentives may need to be developed in some situations. These may be related to land tenure, but also include aspects such as providing more secure markets for produce or providing improved extension programmes as a way of off-setting economic losses.

5.2 POPULATION AND POPULATION DENSITY

Increasing population and population density can increase the risk to groundwater from pollution and unsustainable abstraction. Balancing the needs for protection of resources against demands from rapidly increasing populations is a key element in groundwater protection. Population growth often provides an impetus for improving protection strategies as the need to secure and conserve high-quality water resources for domestic supply becomes increasingly important. This can provide a strong argument for the need to protect groundwater against pollution.

It should be noted, however, that the protection of particular groundwater resources is also dependent on whether it is considered a key source of domestic water in the long-term. In some cases, other resources (either surface water or more remote groundwater) can satisfy demands for water and the threatened groundwater will not form a key part of the water resources used for supply. This is common in wetter countries where urban groundwater has been abandoned. In other situations, typically much drier counties, alternatives may not exist and groundwater resources will therefore need to be protected.

5.3 COMMUNITY PARTICIPATION AND CONSULTATION

Protection of groundwater resources is a public concern and a public responsibility and therefore requires public participation. Participation can be defined as a process through which all stakeholders influence and share control over development and environmental initiatives and the decisions and resources which affect them. The principle of public participation and consultation is found in developed and developing countries.

The Regional Environmental Centre for Central and Eastern Europe states that: '*The most fundamental interest that must be addressed in the process of public participation is the basic right of individuals to have a say in matters affecting their lives... The basic right to participate in decisions affecting oneself... applies in circumstances where the rights and interests may be less recognizable [such as right to have a clean environment]... Taking into account the users' interests should actively involve the users themselves.*' (REC, 1995).

The American Waterworks Association Policy Statement on Public Involvement states that: '*Involving the public in decision making... is... important because many drinking water issues, including adequacy of supply, water quality, rates and conservation, are not only technical issues, they are also social, political, personal health, and economic issues. As such, they are best resolved through a process of meaningful dialogue with concerned parties and the public.*' (Kusel, 1998; AWWA, 1995).

The World Bank Policy Research Working Paper states that: '*Recent evidence from Asia, Latin America and North America suggests that neighbouring communities can have a powerful influence on factories' environmental performance... where formal regulators are present, communities use the political process to influence the tightness of enforcement. Where formal regulators are absent or ineffective, 'informal regulation' is implemented through community groups or NGOs.*' (Afsah and Benoit-Wheeler, 1996).

Community or public participation and consultation are important aspects of resource management as successful implementation is commonly dependent on broad agreement with the objectives and in some cases active public participation in programmes, to ensure these objectives are met. Although the general public in most countries is aware that pollution of surface water is caused by mismanagement of waste and inappropriate land use, awareness is more limited when it comes to groundwater, which is often considered 'pure' and clean. This may present particular challenges to ensuring commitment and participation by the public in protecting groundwater resources.

The role of communities may be critical to promoting improved protection, but the nature of the role that they will play may vary. In many situations, communities are consulted but play limited practical roles in the implementation of groundwater protection strategies. In other cases, communities are expected to play an active role in the design, planning and implementation of groundwater protection.

It is important to be clear about the differences in two of the principal approaches to community involvement: consultation and participation. Consultation is a process of discussion with stakeholders about proposed actions or strategy and is geared towards obtaining the opinion from each stakeholder about these and to review the options that are available. However, it may not mean that the agency undertaking the consultation is bound by the outcome of these discussions and usually does not imply a responsibility for action by the community.

Participation is a set of processes where communities and individuals play an active role in the design, planning and implementation of programmes of water resource development or protection. This often implies that the agency and the community have responsibilities for ensuring agreed actions are performed. It is therefore a more long-term and proactive process than consultation. However, for successful participation there must be effective consultation and therefore the two processes are often combined.

5.3.1 Consultation

It is essential that there is proper consultation with stakeholders in the development of policy and implementation of groundwater protection plans. A key activity in the initial stages of policy development is to ensure that the views and needs of different stakeholders are properly reviewed and incorporated into the policy being developed as far as possible. The stakeholders should also have an opportunity to comment on the policy and strategies developed to ensure that these reflect a position of agreement among key stakeholding groups.

Consultation should bring in the views of Government, affected interest groups and the views of the broader society. Therefore various consultation exercises may need to be undertaken to ensure that the views of all concerned and in particular those groups whose livelihood may be directly affected are collected and concerns addressed. Very often, these groups are those most directly affected by water resource management through lack of access to safe drinking-water supplies, contamination of water sources and limited water for irrigation. In order for policy to be effectively implemented, it is important that there is general support for the overall policy and strategy framework within the country. This is an ongoing process and not something that is engaged in only at the start of policy development. It should be seen as a necessary process which supports the development and implementation of resource management policy and strategy.

Perceptions and cultural values attached to water are also important to understand in the context of groundwater quality management and protection. Many of these concepts provide a foundation upon which to build effective protection strategies as they attach important religious or cultural values on the protection of the groundwater. Examples include some aboriginal beliefs about the origins and sacred nature of water in Australia. In other examples traditional beliefs may hamper the development of groundwater protection strategies. For instance in Uganda beliefs about the use of certain springs by ancestral spirits prevented action being taken to improve water sources.

5.3.2 Participation

In wealthier industrialized countries although public participation occurs, the emphasis tends to be on consultation in the development of the underlying principles, policies and plans that define the development of environmental protection. In most cases, groundwater protection strategies are implemented by local or central government with systems of land use restriction, compensation and appeal processes operating. Specific activities required will often result in specific negotiated agreements with individual land-owners.

By contrast, in developing countries, the development of groundwater protection plans and implementation of protection measures is likely to require the direct involvement of large numbers of people and communities. Many of the tasks that will be required can only be undertaken by local people taking responsibility themselves to enforce protection measures, although this means that communities need support to develop effective capacity. The development of community management committees or users organizations is an important component in promoting effective resource use (Subramanian *et al.*, 1997).

By understanding these issues, appropriate strategies and plans can be developed that identify key stakeholders, where responsibilities lie and what role is expected to be undertaken by the community. Such decisions will be arrived at partly through stakeholder consultation. However, during the initial stages of the development of the policies and strategies, it is important to collect information about communities in order to be able to provide direction for subsequent discussions.

5.4 LAND TENURE AND PROPERTY RIGHTS

Land tenure and property rights are an important consideration when planning interventions to protect groundwater resources as they will directly influence the scope and depth of consultation and negotiation regarding land use. They may also influence what type of intervention is possible and the nature of any regulations that will need to be developed.

One aspect of land tenure of particular importance is the degree to which ownership of land confers rights of ownership and use of underlying resources. In many countries, ownership of land may confer automatic rights to exploit, although these are increasingly subject to licensing and permitting procedures, but ownership resides with the Government. In these cases, controls over abstraction and land use may be easier to implement and monitor.

In other countries, resource ownership has historically resided with the land owner, although this is being revised in many countries. In the Sultanate of Oman, for example, private ownership of water was abolished by a Royal decree in 1988 and a centrally regulated system of water management introduced with an associated well permit system (Government of Oman, 1995).

Revisions to land laws may require significant transitional periods. For instance, the Spanish Royal Decree of 1986 (No 849) considers underground waters to be in the public domain and licences to abstract are required. Public ownership is however subject to the right of landowners to carry out activities on their land but these must not interfere with groundwater quality. In order to avoid opposition to the transition from private ownership to public resource, the Act gave extensive protection to existing rights owners, and complete transition will not occur until 75 years have elapsed.

Land tenure is often complicated and there are many different forms of rights including customary rights to land, private freehold ownership and publicly owned land, with many different variants (Payne, 1997). In addition to issues of ownership, the nature of tenancy arrangements varies and there are further groups who lack any form of *de jure* right to abode, but which may have a variety of *de facto* rights (Hardoy and Satterthwaite, 1989). There are also a significant number of people who have no rights and no security of tenure. The sections below review some key forms of tenure and discuss their implications in relation to groundwater management.

5.4.1 Private land ownership

Private land-ownership is common in many parts of the world and refers to situations where individuals own land, for instance through freehold arrangements. This may be complicated where land is subsequently let to third parties, a common arrangement in European agricultural areas.

This form of land ownership has particular consequences for the development of groundwater protection strategies. The large number of land-owners or tenants may make the process of consultation more cumbersome as the numbers of people involved may increase the time it takes to collect and synthesize local opinions and a greater range of views may need to be accommodated. Such patterns of land ownership will also often result in compensation packages being developed to offset loss of earnings resulting from restrictions placed on land use. However, where such tenure is in place, it may be easier to define a legal framework that can be transparent in its operation and where compulsory purchase or mandatory development controls can be enforced.

5.4.2 Customary land rights

This is common throughout much the developing world and reflects a situation where rights to land are held by a community, although ownership is retained by an individual or the Government. An example of such an arrangement is ‘common’ land within a village where all residents have the right to graze their livestock. In some parts of the world, this may be expanded into communal ownership of land.

Customary rights imply that decisions relating to the use of the land require agreement with all those with rights to use the land, which may result in a more difficult decision-making process when establishing protection strategies. However, customary rights may already implicitly or explicitly restrict activities acceptable on the ‘common’ land, for instance by proscribing activities that would restrict the full enjoyment of rights by others.

Communal rights to land can also offer benefits in terms of discussions with communities regarding actions required to protect water resources. Firstly, the impact of poor groundwater management is likely to be felt directly by the community as in many cases they may be using the sources being polluted from land that is used by the community. Secondly, it introduces the broader concept of public goods that may be easier to accept when restrictions apply across a community rather than to specific individuals. Thirdly, management and protection strategies can be designed to respond to the demands of the community. Where communities have been active participants in strategy development, they will be able to provide a degree of self-policing which may ultimately prove more effective than outside inspection.

5.4.3 Publicly owned land

Publicly owned land is land owned by a Government for all its population. Examples of publicly owned land include national parks where the land is held for the nation, even though some of the land may be let to individual farmers. In some European countries, the catchment areas of major sources of water are purchased specifically for the purpose of protecting the quality of the water source, particularly where it is used for domestic purposes.

As most groundwater protection policies and plans are implemented by Government bodies, publicly-owned land is the most amenable to restriction of land use, but will still require a process of public consultation during policy development. Particular issues that are likely to need resolution will be changes in allowable use of land where part of the land is let in long-term tenancy to farmers or where there is a public right of access. In the former case, changes may need to be phased or supported by compensation packages, whilst in the latter case, broad consultation should be undertaken to ensure that there is public acceptance of the need for such restrictions. Where public access rights are maintained within areas where there are restrictions, it is essential that appropriate services (such as public toilets) are provided to reduce the potential for release of contaminants into the groundwater.

5.4.4 Informal settlements

Informal settlements – situations where land tenure is unclear and where rights are limited – represent particular problems for groundwater protection. In many cases the residents of such settlements have little or no resources and are vulnerable both to exploitation and to ill-heath derived from contamination in the environment. At the same time, such settlements may become a major source of pollution for groundwater resources as they typically lack basic sanitation, solid waste disposal or surface water drainage. Where water supplies are also lacking use is likely to be made of shallow groundwater systems, potentially leading to direct impacts on the health of the community.

Enforcement of land use restrictions in informal areas is unlikely to be successful as they are illegal and unlicensed settlements. However, simply trying to remove such settlements is not only highly discriminatory against the poor, but it is unlikely to be effective and will result in simply shifting the problem and not resolving it. In these situations, it is more appropriate to identify ways of working with community groups to make improvements in environmental health that reduces health risks.

5.5 VALUING AND COSTING GROUNDWATER PROTECTION

An important approach to protection of groundwater is to put an economic and social value on groundwater resources. This value should take into account the direct and indirect cost of protecting the resources as a function of the direct compensation costs (if any) and lost opportunity costs from other, potentially more productive, uses of the land. This should be balanced through placing a value on the aquifer in relation to its importance in supporting economic growth. The latter should consider the current value of groundwater to different industries and the value of each industry to the overall economy. It should also include the incremental marginal costs caused by increased treatment costs (either derived from use of alternative sources or due to pollution of groundwater) and increased abstraction costs derived from exploitation of deeper resources due to contamination of shallow groundwater.

Most environmental protection activities will result in some increase in the cost of production and distribution of drinking-water and more generally in terms of overall environmental protection. For instance, there may be a requirement to pay compensation to existing land-users or to purchase land in drinking-water catchment areas. Within this debate it is important to obtain the views of the public (perhaps represented by consumer groups) on their willingness to pay for such improvements and to assess whether this will be sufficient to off-set costs, discounted over an appropriate period where necessary. Unless there is a willingness by the public (or specific water consumers) to pay the costs of protection, it may be very difficult to sustain intervention strategies. This is discussed further in Chapter 20.

In addition to a direct balancing of economic costs, it is important that social aspects such as the access to safe water supply and the burdens placed upon poor families from having to walk long distances to collect poor quality water should also be assigned a value. One way of doing this is to calculate the likely public health burden derived from poor access to water supply. This may also include a factoring in of the numbers of people whose welfare depends on the continued exploitation of groundwater, whether for domestic use or in agriculture or industry. An example of an approach to valuing groundwater protection is shown in Box 5.2.

Box 5.2. Putting a value on groundwater: Managua (based on Scharp *et al.*, 1997)

The city of Managua in Nicaragua is dependent on groundwater for domestic water supplies and therefore groundwater protection is a priority. Work undertaken by the Sustainable Use of Water Resources project developed a methodology to assign a groundwater protection value to groundwater sources as an input to groundwater protection planning. The project used four criteria: available quantity, groundwater quality, present or planned use and sensitivity to changes in groundwater level. These criteria were based on the economic valuation of water resources in relation to current use, option for future use and environmental significance.

A protection value was calculated based on scores calculated for each criteria. The scores for quantity, quality and sensitivity to changes in groundwater level were used to define the protection value, whilst the present or planned use criteria was overlain on a final map to indicate current and planned abstraction. The authors note that the protection value was a relative measure based on five classes. The data was compiled into a map on a Geographical Information System (GIS) platform that allowed abstraction to be overlain on the protection value. The authors concluded that the areas where there was currently greatest use corresponded to the areas where the protection value was highest. They also concluded that the approach provided a simple and effective tool to assist planners to develop groundwater protection plans.

5.6 SETTING GOALS AND OBJECTIVES – HOW MUCH WILL BE PROTECTED?

The goals and objectives of groundwater protection programmes must be determined before appropriate choices can be taken. Where they existed, in the past, water resource management and environmental protection agencies often made decisions on goals and objectives exclusively. More recently, however, it has been increasingly recognized that planning agencies, local government authorities, key industry groups and the general community need to be consulted. If the inherent conflicts in land use controls are to be resolved, then the understanding of the resource should be accompanied by an appreciation of its value to the community and of the potential impacts of specific land uses on groundwater quality. The community as a whole should decide what needs to be protected and how much protection it can afford. The introduction of obligatory environmental impact studies in Chile, for example, has included the whole community of affected interests into the decision-making process for the first time (Garcés, 2000). Of course, there are still many countries where little or no protection is afforded.

Protection can either extend across an entire aquifer or be restricted to important recharge areas, or capture zones, for specific water supply wells. The question of how much protection is needed or desired depends on the characteristics of the resource, the degree to which it is used, as well as other community social and economic goals. Alternative macro-protection land use management policies include:

No degradation. The maintenance of the quality of groundwater at no worse than existing levels. Generally, such a policy would only be applied to vital resources, typically a resource that provides the sole source of drinking-water. For practical reasons it can only be applied to groundwater resources in undeveloped areas, or areas of very low intensity development. Further land development will normally be excluded from the designated area

Limited or controlled degradation. Such a policy acknowledges that existing or proposed land uses will cause a deterioration of groundwater quality, but strives to maintain the quality above certain specified limits. This policy normally involves controlling the density and types of land development, and the prescription of specific management practices for activities that can affect groundwater quality.

Differential protection. Differential protection policies allow for combinations of no-degradation and limited degradation. Land use management practices normally result in a combination of exclusion and restriction. Such differential protection policies allow the development of different protection objectives taking into account factors such as present and potential uses of water resources, the tenure, zoning and uses of land in the locality, and the desires of the communities involved.

Conflicts in land use management for groundwater quality protection. Restrictions on land use for groundwater protection will always have an economic cost, and decisions must be made about how to minimize these costs while maximizing protection. For example, any limitation on the type and amount of industrial or urban development will have a cost. Consequently, the setting of land use controls for groundwater protection can be very controversial. Obviously, landowners have an expectation that their land can be used freely in its economic highest value use. The wider community interest, on the other hand, can require that groundwater should not be put at risk of pollution. Those responsible, therefore, usually have the difficult task of trying to balance the optimum protection of groundwater resources with the economic interests of the owners of the overlying land surface.

There may be many potential complications involved in land use management, for instance in trying to control problems such as the salinization of groundwater due to irrigation return waters. There may be extreme examples of conflicts, as for instance in the Doon Valley in Uttar Pradesh, India where limestone quarrying is physically destroying the aquifer (Shaman, 1996). Many different types of land use may have to be restricted to protect the quality of groundwater. It is not simply a question of limitations on activities involving toxic materials or the disposal of sewage. Run off from urban areas can be a serious contaminant and it and other sources of diffuse contamination, particularly from agriculture, are those that are best controlled through land use management policies. However, it is with controls over diffuse sources that most conflicts will tend to arise. The problem is particularly acute in rapid growing cities and in those cities in poorer countries where there is reliance on water supplies from shallow aquifers and disposal of excreta in situ.

5.7 INSTITUTIONAL ISSUES

The development of groundwater protection strategies and policies requires effective institutions responsible for the planning, implementation and management of groundwater in the country. In a great number of cases, the failure to protect groundwater resources results not from a lack of appropriate legislation, but because of the poor enforcement of existing regulations. This frequently reflects both weaknesses in the overall institutional framework for groundwater protection and weaknesses within key institutions themselves. Part of the weakness often noted is that the institutions dealing with health, water resource or water supply fail to collaborate to define groundwater protection needs.

Where groundwater policies do not exist or are in need of revision, it is important that a lead institution should be identified for policy direction, and the co-operation of other relevant organizations in decision making should be sought. Generally the lead organization is placed at the central government level. Even in countries where local public participation in water management is high, such as parts of the USA, local water management plans must be consistent with national water quality management objectives and plans.

It is essential that the different roles and responsibilities of different agencies working in the water sector are clearly defined and that one agency is charged with the responsibility to develop, implement and enforce a groundwater protection plan. It is important that the institution identified does not have any conflicts of interest that will compromise its ability to work independently. It is usually preferred that a water resources management body is established that is involved in the approval of development of water sources and the control of the quality of the resource, but is not directly involved in water source development.

Institutional mandates must reflect the aims and objectives of each institution with respect to their roles and responsibilities within the sector (Alaerts, 1997). One of the consequences of this is to consider carefully the scale and scope of activities. For instance, water and wastewater service provision is often most effectively performed when decision-making is devolved to decentralized bodies such as municipalities, water companies or water users associations. National central bodies may still retain some responsibility for policy or strategic development, but may have little influence on operational matters. This implies that when considering roles and responsibilities, the level of action (national policy or local implementation) needs to be considered as well as the area over which the institution has a mandate for action.

The regulation and control of groundwater quality requires a somewhat different approach, although the principles of decentralized operation are still valid. However, although implementation of regulatory activities may be decentralized, there is a need for a strong national institution capable of providing the overall policy and strategic guidance for groundwater protection.

5.8 LEGAL FRAMEWORK

The protection of groundwater requires an adequate legal framework (Caponera, 1992; Soulsby *et al.*, 1999). As governments move towards the strategic management of the country's water resources, it is often necessary to replace basic common law and property rights with statutory provisions regulating the use, development and protection of water (Caponera, 1992). Legal issues related to water ownership, the means used to control abstraction and polluting activities, and the enforcement of such legislation become important. The framework must be supported by appropriate institutions that are capable of implementing the policies and enforcing the relevant laws and regulations, and these organizations must also have the necessary legal status and powers. The willingness to enforce compliance with pollution control measures and whether regulatory frameworks create incentives for potential polluters to comply are critical in ensuring effective regulation (Lane *et al.*, 1999). Within the general considerations of the scope of environmental legislation, the legitimate demands of economic development must be considered to ensure that a sensible balance is struck between the two (Lane *et al.*, 1999).

Legal frameworks in place or developed for water protection deal often with many other issues apart from groundwater. It is possible to identify the shortcuts that relate solely to groundwater, but it is not usually possible to change the laws so as to concentrate only on groundwater quality. The use of more general laws must therefore be accepted, and the specific legislative provisions that may be applicable to the groundwater situation should be identified and used as appropriate, working within the framework of all the provisions of the relevant legislation.

Legislative reform may be required in order to achieve the objectives of groundwater protection. This may involve the revision of existing legislation to encompass these policy objectives or the development of new legislation geared towards groundwater. The approach adopted depends in large part on the nature of existing legislation, the ease with which this may be updated (bearing in mind that existing legislation may deal with broader issues and updating may be time consuming) and the importance of groundwater in the national water resources. Furthermore, as noted by Foster *et al.* (1992), legislative reform will only be effective where political will exists to ensure implementation. These issues are discussed further in Chapter 20.

5.9 REFERENCES

Afsah, S-L. and Benoit-Wheeler, D. (1996) *Controlling Industrial Pollution: A new paradigm.* World Bank Policy Research Working Paper WPS1672.

Alaerts, G. (1997) Institutional arrangements. In *Water Pollution Control*, (eds. R. Helmer and I. Hespanhol), pp. 219-244, E&F Spon, London.

AWWA (1995) *AWWA government affairs: public involvement policy.* www.awwa.org (accessed April 29, 2005).

Caponera, D.E. (1992) *Principles of Water Law and Administration*, Balkema, Rotterdam.

Chambers, R. and Conway, G. (1991) *Sustainable Rural Livelihoods: Practical Concepts for the 21st Century*, IDS Discussion Paper 296, Brighton.

DFID (2003) *Sustainable livelihoods guidance sheets.* www.livelihoods.org/info/info-guidancesheets.html (accessed April 29, 2005).

Foster, S., Adams, B., Morales, M. and Tenjo, S. (1992) *Estrategias par la protección de aguas subterráneas*, Centro Panamericano de Ingeniería Sanitaria y Ciencias del Ambiente, Lima.

Franceys, R., Pickford, J. and Reed, R. (1992) *A Guide to the Development of On-site Sanitation.* World Health Organization, Geneva.

Garcés, J.A. (2000) *El Sistema de Evaluación de Impacto Ambental y el Sector Sanitario*, Revista AIDIS, No. 25, Chile.

Government of Oman (1995) *Water Resources of the Sultanate of Oman, Ministry of Water Resources*, Oman.

Hardoy, J.E. and Satterthwaite, D. (1989) *Squatter Citizen: Life in the Urban Third World*, Earthscan Publications, London.

Kusel, M. (1998) Perspective. *American Waterworks Association Mainstream*, **42**(3), 5.

Lane, S.N., Richards, K.S., Sinha, S. and Wy, S. (1999) Screw the lid even tighter? Water pollution and the enforcement of environmental regulation in less developed countries. In *Water Quality: Processes and Policy*, (eds. S.T. Trudgill, D.E. Walling and B.W. Webb), pp. 233-255, John Wiley & Sons, Chichester.

Nichol, A. (2000) *Adopting a Sustainable Livelihoods Approach to Water Projects: Policy and Practical Implications*, Working Paper No. 133, ODI, London.

Payne, G. (1997) *Urban Land Tenure and Property Rights in Developing Countries: A review*, IT publications and ODA, London.

REC (1995) *Status of Public Participation Practices in Environmental Decision making in Central and Eastern Europe Case Studies of Albania, Bulgaria, the Czech Republic, Croatia, Estonia, Hungary, Latvia, Lithuania, FYR Macedonia, Poland, Romania, the Slovak Republic and Slovenia.* Report of the Regional Environmental Centre for Central and Eastern Europe, Budapest, http://www.rec.org/REC/Publications/PPstatus/cover. html (accessed April 29, 2005).

Scharp, C., Alveteg, T., Johansson, P-O. and Caldera, M. (1997) Assigning a groundwater protection value: methodology development. In *Groundwater in the Urban Environment: Problems, Processes and Management*, (ed. J. Chilton), pp. 659-664, Balkema, Rotterdam.

Shaman, D. (1996) *India's Pollution Regulatory Structure and Background*, NIPR Papers, World Bank, Washington.

Soulsby, C., Chen, M. and Malcolm, R. (1999) Towards groundwater protection in the UK: problems of integrating science and policy. In *Water Quality: Processes and policy*, (eds. S.T. Trudgill, D.E. Walling and B.W. Webb), pp. 151-179, John Wiley & Sons, Chichester.

Subramanian, A., Jaggannathan, N.V. and Meinzen-Dick, R. (1997) *User organizations for Sustainable Water Services,* World Bank Technical Paper No. 354, The World Bank, Washington DC.

Tamuno, P.B.L., Howard, G. and Smith, M.D. (2003) Ecological Risk Assessment (ERA): How appropriate for developing countries? *Environmental Infomatics Archives*, **1**, 204-212.

WELL (1998) *Guidance Manual on Water Supply and Sanitation Programmes*, WEDC, Loughborough.

World Bank (1993) *Water Resources Management,* World Bank Policy Paper, World Bank, Washington, DC.

Section II

Understanding the drinking-water catchment

6

Collecting information for characterizing the catchment and assessing pollution potential

I. Chorus, J. Chilton and O. Schmoll

Sufficient information is key for adequately assessing both the groundwater pollution potential and safety of a drinking-water source. This assessment requires a variety of information, particularly relating to the hydrogeological setting (i.e. aquifer vulnerability), socioeconomic conditions and the range of anthropogenic activities present in the catchment which potentially release pollutants. The establishment of an information inventory is therefore a central tool for developing a sound understanding of potential pollution sources and the likelihood with which pollutants may reach groundwater in concentrations hazardous to human health. The type of information required for assessing groundwater pollution potential is highlighted by the checklists at the end of Chapters 7-13. This chapter addresses general aspects of establishing an information inventory, focusing on the following three questions:

- Which types of information are useful or necessary, and where can they be obtained?
- Who needs to work together to collate the information needed for a comprehensive characterization of the drinking-water catchment area?

- What quality of information is necessary to assess groundwater pollution potential, prioritize hazards and decide on management options?

A key element when compiling an information inventory is to establish adequate data management. This is crucial for viable assessment approaches, regardless of the level of sophistication of data and information that are available and are being used. A well maintained information inventory also supports transparency in the chain from assessment to subsequent management decisions, and is therefore helpful for ensuring acceptance and relevance to all involved parties.

6.1 TYPES OF INFORMATION AND ACCESS TO IT

Information needed for assessing the groundwater pollution potential, i.e. the likelihood that disease agents such as pathogens or chemicals reach groundwater (Chapter 14), can be gained from four types of sources:

- site and catchment inspections;
- public consultation, i.e. communication with the local population;
- collating existing data;
- targeted hydrogeological field surveys (e.g. for aquifer vulnerability mapping as discussed in Chapter 8), and groundwater quality screening or monitoring programmes involving laboratory analyses.

For the most effective use of available resources it is helpful to begin with an initial assessment identifying which information is available. From this, it is often possible to prioritize additional information needed as a basis for making decisions. Such an initial assessment will begin with site inspection and careful evaluation of the available data. Further steps may follow, involving different levels of field surveys including laboratory analysis of samples.

Generally, financial resources and institutional capacity significantly influence both access to different types of information as well as scope and extent of the information collection process. Institutional capacities may be very limited with respect to analytical laboratory equipment or with respect to staff, or both, and this does not necessarily correlate with economic potential. In some settings, financial resources and institutional capacity may allow for regular groundwater monitoring with sophisticated equipment and analytical methods, including modelling of groundwater flows and loading. In contrast, in other settings available options may be limited to visual information gleaned from site or catchment inspection and informal information from communication with the population in the drinking-water catchment.

Assessment strategies that are mainly based on high-tech monitoring easily neglect site or catchment inspection as an important first-hand information sources whereas communities with limited resources can successfully perform pollution potential assessments on the basis of information gleaned from site inspection and public communication. Some countries with excellent laboratory capacity face increasing staff reductions, and site inspection has often been the first activity compromised by structural changes in responsible authorities since they are labour intensive. Other countries with lower labour costs may have more staff capacity for conducting site inspection – in spite of substantially less developed laboratory capacity – and thus have a high potential for good situation assessments, if the staff are adequately trained.

6.1.1 Site and catchment inspection

Comprehensive site and catchment inspections provide highly valuable first hand information about the area of interest, and in many cases they uncover self-evident pollution problems in a catchment. Inspections often provide an important early warning function as they can indicate the potential of future groundwater pollution which would only be detected by monitoring after the fact.

Sanitary inspections are specifically designed to provide an overview of the status of drinking-water source contamination risks. They identify probable causes of failures when drinking-water contamination is found (Howard, 2002a), and may include, for example, an assessment of the proximity of polluting activities in relation to drinking-water abstraction points and of the condition of wellheads. A more detailed description of sanitary inspection is given in Chapter 18, and further guidance can be obtained from, for example, Lloyd and Helmer (1991), WHO (1997) and Howard (2002a, 2002b).

Inspections on a broader, catchment oriented scale target the collection of information for the characterization of visible activities which may potentially pollute groundwater and of hydrogeological conditions which determine aquifer pollution vulnerability. The checklists at the end of Chapters 7-13 provide examples of aspects to consider when performing site or catchment inspections. Inspections need to be repeated periodically, as conditions and activities in the catchment are not static and change over time as new development occurs in the area, or may be important only seasonally.

Inspections can often gain more information if performed in collaboration with those responsible for, or operating, activities in the catchment which may potentially contaminate groundwater (e.g. farmers, operators of waste disposal sites or sanitation infrastructure, etc.).

6.1.2 Consulting the public

Public consultation and involvement is best undertaken whilst collecting information about groundwater. This is important to establish a basis of open dialogue and mutual trust as well as ownership which is crucial for the implementation of management responses. At this stage, however, public consultation is also important as a potential source of information. People in the community can report how their livelihoods relate to water use and water quality, which sources and wells are used for which purpose and in what amounts, and pressures that have an impact on water uses and on pollution. This helps assess the economic and social values placed upon groundwater in a given community and understand reasons for quality and quantity problems (see also Chapters 5 and 7).

Understanding the community's perception of issues, e.g. land tenure rights in relation to groundwater use, or performance of government institutions, may be an important source of information for developing successful management responses. Other issues for which this may be important are the perception of quality, e.g. confidence in public supply and perhaps how this compares to confidence in water from private wells, as well as willingness to pay for better quality of water.

As discussed in Chapter 7, in undertaking community consultation, it is important to ensure that all stakeholder groups are included, and to avoid potential bias in the findings by concentrating on particularly vociferous groups that may have an unrepresentatively negative or positive view of certain issues.

Community consultation is generally an important source of 'informal local knowledge'. Targeted interviews, questionnaires, telephone or door-to-door surveys are common tools for establishing communication with the local population and thus accessing their knowledge. A wide range of information about technical groundwater issues may be gleaned from interviewing community members, e.g. on water levels in wells, patterns of rainfall and inundations, changes in vegetation cover that might indicate changes in groundwater levels or seasonal patterns relating to groundwater. Communication with the local population may also be a good – though usually incomplete – source of information on potential pollution from human activities (including illicit activities), as discussions can help identify the reasons for which adequate control measures are lacking. Information gleaned from local knowledge may be particularly valuable where those operating polluting activities are reluctant to provide required information or deny access to operation sites for inspection. Consultation can also help identify the numbers and locations of private wells and to map these in relation to public supplies and to centres of population. Further, the community may be able to actively support groundwater monitoring or even conduct elements of it.

In order to maintain open dialogue and trust, it is important to inform the communities of the results of the catchment assessment, e.g. by providing copies of relevant reports and assessments and/or conducting meetings for oral reporting and discussion.

6.1.3 Evaluating existing data

Careful evaluation of existing data may prove to be a good basis for assessing groundwater pollution potential. In most settings, current and historical data are likely to be available although often compiled for specific purposes other than groundwater assessments, i.e. for research, land use planning, environmental impact assessments or registration and licensing of commercial activities. Data sources that can be evaluated vary widely but include:

- statistical data (e.g. population, water usage, agrochemical usage, economic activities, land ownership and use, health and epidemiology);
- recorded data from ongoing monitoring programmes (e.g. drinking-water or groundwater quality, drinking-water abstraction, meteorology);
- hydrogeological and geographical information (e.g. area, geologic, vulnerability or land use maps, air photos, satellite images);
- published studies (e.g. from earlier surveys, programmes, inspections, environmental impact assessments).

Water quality data may exist from various types of groundwater surveys or ongoing monitoring programmes. Such data are a highly valuable support for the assessment of pollution. Water quality data for delivered drinking-water are most frequently available and can be helpful for assessing aquifer pollution as contaminants found in drinking-water are likely to have their source in the catchment area. Drinking-water quality cannot, however, always be taken as indicative of the quality of groundwater within the aquifer. This is because samples for drinking-water quality monitoring may be taken in the distribution system, which may be after treatment to remove pollutants and/or after mixing of water from more than one source. These data should be used with care, but can provide useful information for many parameters if these potential constraints are taken into account. In particular, data for natural constituents can provide information on the origin and movement of the water and time series data are important for recognizing potential trends of increasing or decreasing pollutant loading.

Generally, when using data from programmes targeted at purposes other than groundwater pollution potential assessments, their limitations must be kept in mind. For example, existing monitoring programmes may be limited in scope and scale (e.g. in selection of sampling sites, parameters tested or sampling frequency), and thus may not provide satisfactory temporal and spatial coverage in relation to the requirements of a comprehensive pollution potential assessment. However, even if information about specific pollutants is lacking, general hydrochemical information can be valuable for assessing groundwater flow patterns and residence times, as baseline quality from which the effects of human impacts may be distinguished, as well as to provide vital information about naturally-occurring groundwater constituents such as fluoride and arsenic. General hydrochemical data can also provide useful information about the likely behaviour of some pollutants (e.g. the mobility of metals), as referred to in Chapter 4.

Data are available from various governmental and non-governmental bodies and thus may be widely scattered. They include, but are not restricted to:

- public authorities or agencies at different administrative levels (i.e. local, regional, national) and different responsibilities such as:
- health, environment, water management or geology;
- planning, permission or licensing bodies for commercial enterprises (e.g. industry, mining), sanitation, and traffic;
- water suppliers and wastewater agencies;
- health care facilities;
- university departments and other research institutions;
- NGOs
- local community initiatives (e.g. in water supply, sanitation, environment or agriculture);
- statistical bureaus;
- scientific literature or archives;
- aid or development organizations;
- professional associations (e.g. of farmers, industries).

Personnel and time demands on 'data mining' may be substantial: the authorities and agencies possessing valuable databases are often scattered across different administrative levels. Also, more often than not, data are likely to be available by administrative areas, rather than by catchments. Thus for the purpose of an assessment of a given catchment, it may be necessary to seek contact with authorities in several different administrative units.

Much of the data will be unpublished, available only in raw form, perhaps not electronically, and possibly even classified as confidential. Distinction between useful and unimportant data is not always self-evident. Considerable effort may thus be needed to extract, process and categorize useful information. The challenge is to organize and critically assess such data, and to document their sources, including suspected uncertainties about data quality. A further challenge is balancing time and effort needed to obtain data against their usefulness.

Data mining may require more than one iteration of filling information gaps by targeted checking for availability of further data that initially were not readily accessible. This is often the case when data are scattered between different public authorities, water suppliers or universities that produce them. Putting further effort into data mining by revisiting the step of searching for information on activities in the catchment area on a more detailed and targeted level may substantially improve the information inventory or even close critical gaps.

6.1.4 Generating new water quality data

In the context of groundwater protection, the term monitoring is usually understood to imply water quality monitoring, and is defined and sub-divided into categories in rather different ways by different authors. Groundwater monitoring tends to be taken for granted in most guidance documents on groundwater protection, e.g. as an 'additional and essential component' (Foster *et al.*, 2002). However, in practice groundwater monitoring programmes are often implicitly geared towards scientific investigation per se, and objectives are often not clearly defined and stated. Adriaanse and Lindgaard-Jørgensen (1997) point out the importance of establishing meaningful objectives for water monitoring programmes in order to avoid the data-rich, but information poor syndrome. For providing a basis for management decisions, monitoring programmes need to be tailored to the information needs in the specific setting. Often, the most important information needs for the objective of controlling groundwater quality in drinking-water catchments may not be gained from groundwater monitoring, but rather from monitoring potentially polluting activities and the implementation of measures to prevent them from releasing contaminants to the subsurface.

Water quality monitoring

Regular groundwater quality monitoring programmes have been established in many countries worldwide. Generally, they provide valuable information on groundwater chemistry, groundwater levels, seasonal quality patterns and other trends over time. Thus they are a supportive component for sound understanding of the hydrological environment, for developing conceptual models of groundwater systems or for mapping groundwater vulnerability. However, as discussed above, it is important to realize that comprehensive groundwater quality monitoring is not an essential prerequisite for assessing groundwater pollution potential in a given drinking-water catchment (Chapter 14). Particularly (though not exclusively) in settings where implementation of regular groundwater monitoring programmes is not feasible (e.g. due to limited financial resources or institutional capacities) or is only just beginning, investigations such as site and catchment inspections and the evaluation of already existing data (Sections 6.1.1-6.1.3) are also highly valuable information sources on which the assessment can build.

In many settings, targeted water quality surveys or screening programmes for selected parameters and/or specific purposes, possibly repeated at more extensive time intervals, may prove sufficient to fill the crucial information gaps and thus confirm and improve pollution potential assessments, e.g. to investigate the extent and severity of suspected pollution 'hot spots'. In addition, regular monitoring over a certain period of time – often one to two years – may be important to reduce uncertainty of the assessment, particularly in settings where important groundwater parameters show seasonal patterns and trends over time.

Groundwater monitoring in this context can have different objectives:

- Understanding natural groundwater quality (i.e. by installing observation wells in pristine areas). Chemical parameters typically chosen for this objective may include conductivity, pH, redox conditions and natural groundwater constituents including those potentially hazardous to human health, e.g. arsenic and fluoride.
- Provision of data needed for developing groundwater flow models. In this case, monitoring would include tracers or indicators of water movement, e.g. chloride or temperature.
- Characterization of the current level of aquifer pollution. For this purpose, observation wells would be installed in areas of the aquifer considered to be representative of the human activities as well as of the hydrogeological conditions. Parameters for analyses would be selected in relation to contaminants expected from ongoing or previous activities.

Where actual contamination of a drinking-water source is occurring, targeted analysis of the raw water quality will usually identify the pollutant and its concentration. This will then aid in tracking potential pollutant sources and in defining control measures and/or drinking-water treatment requirements. If a pollutant is suspected, but not present in drinking-water wells, water samples from observation wells between the pollution source and the drinking-water well can be used to track movement of the pollutant. In many cases, such observation points are critical in risk assessment of the pollution potential from a known source. Where sufficiently comprehensive programmes can be implemented, they may actually provide quantitative data on microbial and/or chemical contaminants in groundwater to be used in quantitative risk assessments.

Further guidance on specifying information needs in the context of planning monitoring programmes is given by Bartram and Balance (1996), Chapman (1996), UNECE (1996, 2000), Adriaanse and Lindgaard-Jørgensen (1997) and Timmerman and Mulder (1999), provide general guidance and overview on water quality monitoring. Technical aspects of constructing observation wells, sampling and analyses are beyond the scope of this book, and readers are referred to Nielsen (1991), Lapham *et al.* (1997) and Boulding and Ginn (2003). Because novel sampling and analytical methods are always being developed to improve the monitoring and characterization of subsurface environmental quality, printed material can quickly become dated, and the more technically interested reader should look at the websites of organizations such as US Environmental Protection Agency (US EPA), the American Society for Testing Materials, and US Geological Survey (USGS).

Other types of monitoring

For the purpose of this book, the use and understanding of the term monitoring goes beyond groundwater quality monitoring discussed above. The following monitoring categories, which subsume other technical monitoring categories proposed elsewhere, are used in this book, particularly in the context of Water Safety Plans (WSPs) (Chapter 17):

- Operational monitoring for process control does not focus on measuring groundwater quality but is a planned series of observations or measurements to timely quantify efficacy and changes in performance of a control measure that is established to control the occurrence of pathogens or chemicals in groundwater (for more details see Chapter 17 and for examples Chapters 20-25).
- Monitoring for verification is the application of methods, procedures, tests and other evaluations in addition to operational monitoring to determine compliance with and efficacy of the WSP or the groundwater management system, respectively (for more details see Chapter 17 and for examples Chapters 20-25).

6.2 THE NEED FOR COLLABORATION

Groundwater assessments are typically undertaken by hydrogeologists, potentially supported by sanitary engineers and/or environmental scientists. These qualifications are indeed essential for understanding groundwater flow and potential contaminant transport. For understanding the potential for contaminant loads to the aquifer, broader competence is needed, preferably in a team that integrates hydrogeological knowledge and an understanding of the potentially polluting human activities (e.g. from agriculture, industry, sanitation, etc.).

Also, as discussed above in Section 6.1.3, existing information on activities potentially impacting on groundwater quality is likely to be broadly scattered. For example, while health or environmental authorities may be responsible for assessing the risk of groundwater pollution, information on human activity in the catchment may be available from diverse government bodies responsible for health, commerce, statistics, traffic, tourism, agriculture, mining, etc. Government agencies with executive and policy functions for the environment and groundwater may have different types of information from local government.

Teamwork and intersectoral collaboration is vital for groundwater assessments. This includes cooperation between all institutions that can contribute from various angles of access to information. The task of collecting and evaluating relevant information will in many cases be the responsibility of a core team of public health, groundwater and environmental experts from public authorities, often supported by water suppliers. However, this core team can be expanded by external experts from the scientific community, representatives of the public, and possibly stakeholders from the catchment. Health authorities may take the initiative in assembling such teams and/or in leading them. The example from India provided in Box 20.2 shows how competence for groundwater pollution potential assessments as well as for developing management options can be built by training employees of government agencies.

Water suppliers can successfully take the initiative in establishing interdisciplinary teams for identifying groundwater pollution problems and implementing protection strategies for their specific supply setting. This is mirrored by the WSP approach in which the formation of an interdisciplinary team is a key requirement (Chapter 16). Box 6.1 provides an example of such an approach in an urban setting and Box 16.1 shows how farmers were involved specifically to control nitrate contamination in a rural setting.

Box 6.1. Collaboration of water supply, sanitation and public authorities in Berlin, Germany: the Hygiene Commission of the Berlin Waterworks

Members of The Hygiene Commission of the Berlin Waterworks include:

- a staff member responsible for resource protection, the technical director and the laboratory director of the waterworks;
- a staff member of the municipal public health authority;
- a staff member of the municipal environmental authority;
- a staff member of the municipal forestry authority;
- a staff member of the local police department;
- invited experts, e.g. from federal agencies.

Tabling information. All of these members report to the Commission any changes that occurred in the water supply and its catchment since the last meeting. For example, the water supply representatives report on issues regarding water treatment and distribution as well as protection zones. The public authority members report any illegal activities in the catchment observed and/or reported, such as illicit waste disposal, construction activity or traffic, newly identified historic waste sites, and any conspicuous changes observed in the catchment. Reporting also includes public complaints about water quality, odour from sewage treatment works, or observations on recipient water bodies.

Discussing solutions to problems. The Commission evaluates all reports and discusses proposed solutions. This includes administrational measures of public authorities such as issuing permits for construction of facilities, new or changed systems for wastewater collection and treatment, as well as direct regulatory measures of the local police directorate such as removing illegally parked vehicles, closing roads to traffic and posting signs and notices.

Targets of the Hygiene Commission are to:

- inform all members about current issues regarding operation of the water supply and sanitation system as well as catchment protection;
- identify pollution potential for management of catchment and supply system;
- coordinate management responses of public authorities from different sectors and develop intersectorally harmonized strategies for sustainable provision of safe drinking-water;
- coordinate contingency planning.

In many settings worldwide, the key challenge will be to build motivation for collaboration. This may be particularly difficult in situations where, for example, data on contaminant loading to aquifers are likely to be available only from the polluters themselves, who have no primary interest in control and potential restriction of their activity. Creating ownership, for example through involvement of such stakeholders in drinking-water resource protection teams (Box 6.1), may, therefore, be crucial. Motivation and commitment for collaboration are likely to increase if all players fully understand both their individual contribution to, and benefits from, efforts in protecting groundwater as a drinking-water source.

Accounting for value judgements

Assessments of the potential for groundwater contamination typically contain a major component of expert judgement rather than being fully quantitative, and they often include substantial uncertainty (Section 6.3). This has two major implications for collecting and compiling information:

- Hazard assessments themselves, but also the information on which they are based, are subject to bias by the knowledge background and the value judgments of the experts involved. It is important to understand implicit value judgements that may drive experts' assessments, as they are likely to play a role in selecting questions asked, choosing which data are important, assessing data and drawing conclusions. This can be dealt with more appropriately if value judgements are understood, made transparent and explicitly acknowledged.
- Teamwork may be encumbered by implicit value judgements due to differences in scientific background and perception of issues. Both for effective collaboration and for making uncertainties of pollution potential assessments transparent, it is important to be aware of differences in value judgements. In her comprehensive discussion of this aspect of evaluating information for decision-making, Harding (1998) makes the point that 'even though controversies are typically seen as disputes over "facts", in most environmental disputes it is the clash between people's value positions which fuels debate, rather than a disagreement over the "facts"' (page 61). This applies both to disputes within expert teams and with the general public and is therefore also important when organizing public involvement in the assessment and management of groundwater resources.

For the perception of pollution hazards in groundwater, the general shift to greener or more environmentalist ideology has become relevant in many countries. Environmental protection arguments are often quoted in the context of protecting human health, even where this connection is scientifically unsubstantiated. Perception of health hazards from different classes of substances tends to be linked to the 'reputation' the substance has in the media. For example, traces of pesticides in drinking-water may be seen as poisons and therefore as objectionable even at orders of magnitude below health-based guideline values. Specifically for groundwater, the knowledge of the often insurmountable difficulties of removing pollutants from the subsurface has produced strong environmental arguments for protecting aquifers from the impact of human activity.

Often, in assessing information for protecting drinking-water catchments, it will be important to explicitly differentiate between priorities for protecting the environment and priorities for protecting human health. Showing respect for the ethical positions and principles of others may facilitate making motivations transparent. For example, explicitly acknowledging and respecting the position of a team member who considers any groundwater pollution unethical and incompatible with the target of sustainable environmental management may be the basis for attaining his or her support in prioritizing pollutants for the purpose of keeping drinking-water safe for human health.

6.3 SUFFICIENCY AND QUALITY OF INFORMATION – DEALING WITH UNCERTAINTY

Uncertainty is likely to be an issue in all elements of catchment analysis and groundwater pollution potential assessment. The information base on hydrogeological conditions and on the scale and range of human activities in the catchment are only rarely sufficiently comprehensive to allow a fully quantitative determination or prediction of the pollution potential or contaminant concentrations. For example,

- data about the population's water needs may be incomplete, e.g. with respect to the amount of groundwater abstracted from private wells, possibly illegally;
- information on human activities in the catchment is likely to be incomplete, for potential loads from on-site sanitation and/or leaky sewers as well as for the range and amounts of hazardous substances used and potentially released to the underground in diverse enterprises, agriculture and traffic;
- the meaning of data gained from groundwater monitoring is dependent on the representative selection of sites for observation wells, and costs for their installation constrain the number of sites that can be sampled.

One consequence of this uncertainty is that the target in most cases is a qualitative or semi-quantitative determination of pollution potential, as discussed in Chapter 14 as a basis for risk assessment.

Understanding uncertainty of the information available

The concept of uncertainty can be classified in different ways. Harding (1998) proposes the following categories that are relevant for assessing groundwater contamination of a catchment:

- Risk is the most quantifiable and measurable type of uncertainty. It implies that the parameters driving groundwater quality are basically understood. Though the concentration of a contaminant may not be fully quantified or predicted, structured analysis of mechanisms and probabilities allows prediction of the likelihood that it occurs or will occur in concentrations relevant to human health.
- Uncertainty implies that the parameters driving the processes are understood, but not sufficiently well to assess the probability of pollution occurring.
- Ignorance means that not even the most important parameters are understood, i.e. 'we don't know what we don't know'.

A key issue for making management decisions is to understand how sound the available information base is for assessing aquifer groundwater pollution, i.e. the amount of uncertainty involved. This is important both for avoiding wasting resources by poorly informed decisions on management options which later turn out to be ineffective, and for avoiding undue postponement of urgent measures in situations where in fact the information base is already sufficient for making adequate decisions. The question to ask is 'how much do we need to know to assess hazards for groundwater quality, and how much uncertainty can we tolerate?' Aspects of information quality in relation to taking management decisions are discussed in Chapter 15. In the following, the aspects of documenting the quality and reliability of information during its collection are covered together with options for identifying and filling gaps.

Systematic documentation of the information inventory and the information sources, including indication of information quality and reliability, is a useful tool for identifying critical information gaps. Earmarking uncertainties in the database or in the assessment of its quality helps to guide further work on the data inventory. This can be achieved by a list of the factors which probably affect the data but are not well known (e.g. heterogeneity of the subsurface environment as a factor affecting data on estimated flow rates, or illegal manure application as a factor affecting nitrate loads). Such documentation may prove important for acquiring funding for targeted programmes to close gaps.

A next step for assessing sufficiency and quality of information as well as of getting improved information on missing, inconsistent or unverified data is to assign a level of confidence to all data sets in the form of an uncertainty score. Low values of uncertainty indicate data of high reliability, whilst high uncertainty scores indicate estimates or unverifiable information. No data (i.e. a gap) should be given the highest uncertainty score. As discussed in WHO (1982) and by Foster and Hirata (1988), assessments of data quality will often be somewhat subjective. This is endorsed and encouraged, provided that subjectivity is made transparent in the report. As assessment progresses, the aim should be to reduce the uncertainty scores by infilling with more accurate and reliable data. The use of uncertainty scores will aid in prioritization of data acquisition. When the data are used collectively, a summed or weighted uncertainty score can be used to indicate the relative confidence in the interpretation. An indication of where uncertainties are most relevant for assessing potential pollution can often be gleaned from targeted inspection of the respective sites and activities in the catchment.

Reducing uncertainty

An effective option for filling information gaps is the design of specific programmes that may prove to be quite feasible if they are properly targeted. For example, where potential pollution sources are known from assessments of activities in the catchment area, but the scale of the activity with regard to the likelihood of the pollutants actually reaching the aquifer is poorly understood or cannot be assessed, a survey of pollutant occurrence in groundwater may prove possible even with limited budgets. Such surveys may be small-scale and well-targeted to address the specific pollutants and limited to localities critical for water supply. For example, in Tajikistan the authorities responsible for drinking-water quality needed information on pesticide levels as a basis for making a decision whether to use shallow aquifers as drinking-water sources. In the face of data showing a substantial decline in the application of pesticides, a small initial survey of pesticide concentrations in the aquifers that were envisaged for providing supplies helped decide which ones could actually be used, and indeed this showed pesticide pollution to be much less of a problem than had been assumed from the historic knowledge of high rates formerly applied.

Hydrogeological information gaps are often perceived as intimidating. However, here also, targeting investigations towards specific questions may narrow down the effort required and thus make programmes feasible. Examples are the use of groundwater temperature to indicate ingress of surface water (applicable in settings with seasonal patterns of surface water temperature), the use of electrical conductivity as a simple indicator of saline intrusion or the use of substances characteristic for sewage (e.g. detergents, ammonia, caffeine) as tracers for sewage ingress.

Another aspect of filling information gaps is checking whether further data may indeed be available, though perhaps initially not readily accessible, as discussed in Section 6.1.

Decisions on investments into improving the information base will depend on the consequences of uncertainty. For example, if the assessment of aquifer contamination is highly uncertain (in ignorance of some polluting activity or because of poor understanding of its vulnerability) and a large number of people use it as drinking-water source, the consequence might be severe, e.g. high incidence of waterborne disease. This would be a strong rationale for improving the information base. Vice versa, consequences in terms of public health would be minor if the population is connected to a central supply using an alternative water source, and improving public health would not be a reason to invest in improving the database.

6.4 SUMMARY – HOW TO PROCEED

Figure 6.1 outlines principal planning steps that may be taken to glean sufficient information for characterizing a drinking-water catchment as a basis for assessing groundwater pollution potential (Chapter 14). Taking the initiative for compiling a catchment specific information inventory may be the role of local or regional public authorities either in charge of surveillance of drinking-water supplies or of overall catchment management (e.g. environmental authorities or departments), but it may also be taken by a water supplier in the context of protecting the supply's catchment and/or developing a WSP.

The first step for those initiating a programme for catchment characterization and building an information inventory, both a basis for the assessment of groundwater pollution potential, is to outline the scope of the process. This particularly requires a preliminary understanding of:

- What area to assess, i.e. the delineation of the catchment of the water supply or well field. In many settings, some hydrogeological information is available in relation to the water supply. Where this is lacking, the surface profile gleaned from a topographic map and/or a first visit to the site may provide a preliminary indication.
- The type of information expected to be available (e.g. on socioeconomic aspects, groundwater vulnerability and anthropogenic activities). This can be achieved by compiling and reviewing a first overview of the information sources accessible and the stakeholders involved.
- Key issues to investigate within the process of catchment characterization.

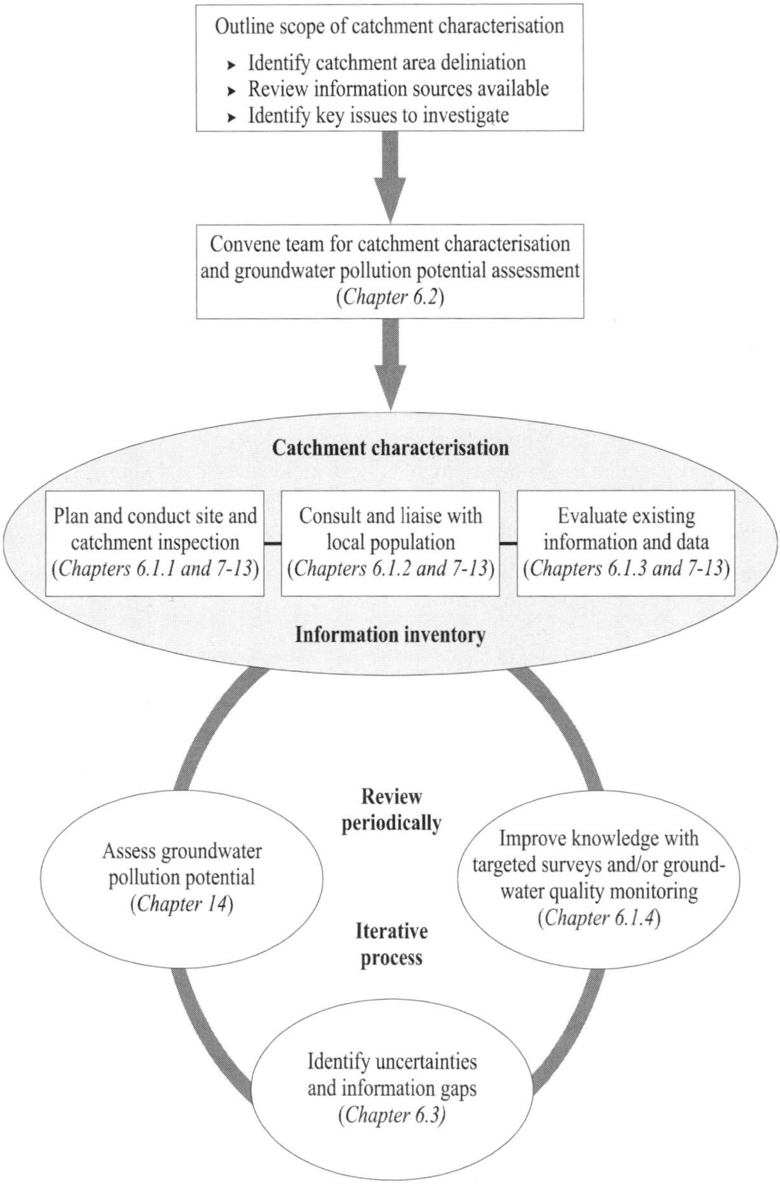

Figure 6.1. Principal planning steps for collecting information

Planning and designing the actual investigation requires defining its targets in consideration of the financial, institutional and personnel resources available. For more detailed guidance on how to design a groundwater quality assessment, see UNECE (2000).

The second step is to identify expertise and key players who need to be involved in catchment investigation and pollution potential assessment, and to convene a team for conducting both. As discussed in Section 6.2, it is useful to involve public authorities from the sectors important in the catchment as well as experts, particularly in the fields of hydrogeology and public health. Often, it is useful to strive for public participation in this early planning phase by including members of communities affected or special interest groups. Supportive involvement of stakeholders from potentially polluting activities may be particularly useful for obtaining sensitive information, e.g. on pollutant loads.

The third step is to carry out the investigation. For gathering comprehensive information, it may be useful to split the team into working-groups which focus on three different activities that may be conducted in parallel. These include:

- Site and catchment inspection (Section 6.1.1) to record and map all features potentially relevant to groundwater pollution, for example settlements and water use practices (Chapter 7), topographic features (such as sinkholes and abandoned wells) that could facilitate rapid transport of pollutants to groundwater (Chapter 8), and human activities that might lead to groundwater pollution (Chapters 9-13). Depending on the scope of catchment characterization, this can include recording of hydrogeological information for assessing groundwater vulnerability, or the latter would be a separate activity conducted by a team of hydrogeologists. Templates for checklists to assist collecting this information at the end of Chapters 7-13 assist inspection and would be tailored to the specific setting. Notes should be included on uncertainties and information gaps perceived as potentially relevant to the assessment (Section 6.3).
- Consult and liase with the local population as a source of information, particularly to glean local knowledge about, for example, economic or cultural values placed upon groundwater, community perceptions on the use and protection of groundwater resources, or information of a more technical nature, e.g. on potential pollution sources, location and use of wells, etc. (Section 6.1.2).
- Evaluation of existing information and data (Section 6.1.3) already available from previous programmes and investigations into groundwater and drinking-water quality, aquifer characteristics which define its vulnerability, and anthropogenic activities in the catchment. This information amends the checklists from catchment inspection.

Step three is followed by an iterative process as shown in Figure 6.1. It involves the following elements, the order of which will depend on the situation and the approach preferred by the team:

- Drafting a groundwater pollution potential assessment as described in Chapter 14.
- Identifying uncertainties and information gaps. The evaluation of information available may reveal that the understanding of catchment characteristics is not sufficient, or information from different sources is inconsistent, and therefore the information inventory needs to be improved for adequately assessing groundwater pollution potential as a decision basis for developing management responses to protect the drinking-water catchment.
- Improving knowledge. Some of the identified information gaps may be fairly readily closed by revisiting specific sites in the catchment for more thorough inspection, or by requesting information from, for example, operators of farms, enterprises or other activities. If uncertainty is too large to make decisions, information gaps need to be closed with specifically targeted groundwater

surveys and or regular groundwater quality monitoring. In many cases the need for this can be limited to selected localities and/or polluting activities in the catchment (Section 6.1.3).

As settlements and human activities in the catchment develop and change, so will pollution potential. Further, the impact of management responses needs to be assessed. In consequence, the topicality of information inventories and pollution potential assessments needs to be reviewed and repeated periodically, and improvement of the assessment is an iterative process. The experience and knowledge gained over time will serve to reduce the uncertainty of the assessment and increase the safety of the water supply in the long term.

6.5 REFERENCES

Adriaanse, M. and Lindgaard-Jørgensen, P. (1997) Information systems. In *Water Pollution Control*, (eds. R. Helmer and I. Hespanhol), pp. 245-274, UNEP, Water Supply & Sanitation Control and WHO, E & FN Spon, London.

Bartram, J. and Balance, R. (1996) *Water Quality Monitoring*, UNEP and WHO, E & FN Spon, London.

Boulding, J.R. and Ginn, J.S. (2003) *Practical Handbook of Soil, Vadose Zone and Groundwater Contamination Assessment, Prevention and Remediation*, 2nd edn, CRC Press, Florida.

Chapman, D. (1996) *Water Quality Assessments: A guide to the use of biota, sediments and water in environmental monitoring*, 2nd edn, UNEP, UNESCO and WHO, E&FN Spon, London.

Foster, S. and Hirata, R. (1988) *Groundwater Pollution Risk Assessment*. CEPIS Regional Program on the prevention and control of groundwater pollution, PAHO/WHO, Lima.

Foster, S., Hirata, R., Gomes, D., D'Elia, M. and Paris, M. (2002) *Groundwater Quality Protection – A guide for water utilities, municipal authorities, and environmental agencies*, The World Bank, Washington, DC.

Harding, R. (ed.) (1998) *Environmental Decision-Making*. The Federation Press, Sydney.

Howard, A.G. (2002a) *Water Supply Surveillance – A Reference Manual*, WEDC, Loughborough University.

Howard, A.G. (2002b): *Water Supply Surveillance – A Practical Guide*, WEDC, Loughborough University.

Lapham, W.L., Wilde, F.D. and Koterba, M.T. (1997) *Guidelines and Standard Procedures For Studies of Groundwater Quality: Selection and installation of wells, and supporting documentation*, Water Resources Investigations Report 96-4233, USGS, Reston, Virginia.

Lloyd, B. and Helmer, R. (1991) *Surveillance of Drinking Water Quality in Rural Areas*. Longman, London.

Nielsen, D.M. (ed) (1991) *Practical Handbook of Groundwater Monitoring*. Lewis Publishers, Chelsea, Michigan.

Timmerman, J.G. and Mulder, W.H. (1999) Information needs as the basis for monitoring. *European Water Man.*, **2**(2), 41-45.

UNECE (1996) *Guidelines on Monitoring and Assessment of Transboundary Rivers*. RIZA, Lelystad.

UNECE (2000) *Guidelines on Monitoring and Assessment of Transboundary Groundwaters*. RIZA, Lelystad.

WHO (1982) Rapid assessment of sources of air, water, and land pollution. *World Health Organization Offset Publication*, **62**, 1-113.

WHO (1997) Guidelines for Drinking-water Quality, Surveillance and Control of Community Supplies. 2nd edn, vol . 3, World Health Organization, Geneva.

7

Characterization of the socioeconomic, institutional and legal setting

G. Howard, P. Chave, P. Bakir and B. Hoque

Building on Chapter 5, the purpose of this chapter is to review how data may be collected on socioeconomic conditions, institutional and legal frameworks and valuing of groundwater protection that may affect groundwater protection policies and strategies. The chapter is a review of some of the tools that can be used to collect data and a short check-list is provided for readers to identify the types of information that should be collected.

7.1 DEFINING SOCIOECONOMIC STATUS

How socioeconomic status is defined is important as it has implications with regard to the impact of policy on livelihoods, support provided to households to help cope with adverse economic consequences of land use restrictions, and to determine whether and what compensation may be offered to households. There are a number of different ways in which socioeconomic status can be assessed. The selection of the means of defining socioeconomic status and in particular which households are poor, depends on how such information will be used. Many countries define national ‘poverty lines’ that represent a benchmark level of income associated with poverty. This level of income would usually represent a level below which a set of basic goods and services can no longer be afforded (Satterthwaite, 1997).

In an international context, many organizations apply a measure of absolute monetary poverty, which is usually taken at being less than US$ 1 per day per capita (World Bank, 2002). In many cases, this figure is refined based on purchasing power parity (PPP) which is adjusted to the value of the dollar in 1993. The use of PPP is an attempt to reflect that costs of living as well as incomes vary between countries (World Resources Institute, 1996). The PPP uses a standard ‘shopping basket’ of goods and services and calculates how many units of the national income are required to purchase the shopping basket contents in comparison to the cost in US$ in an ‘average’ country (defined as being the average costs from all countries included in international comparisons). Gross domestic product is adjusted in light of these differential costs and typically lowers the gross domestic product of wealthy countries and raises that of poorer countries (World Resources Institute, 1996).

The numbers of people living in absolute poverty in developing countries remains high and about 1.2 billion people globally live on less than US$ 1 per day (World Bank, 2002). Another level of monetary poverty is sometimes applied to middle-income countries (e.g. in Central and Eastern Europe and Latin America) of US$ 2 per capita per day as a more meaningful description of poverty in these countries (World Bank, 2002).

Although monetary factors are important, the usefulness of absolute monetary values to describe poverty is questioned by many workers who argue that relative poverty is a better measure as it reflects the inequalities within a society. Relative poverty is often deemed to be more influential in determining access to goods and services than measures of income (Hardoy and Satterthwaite, 1989; Stephens *et al.*, 1997).

Monetary definitions of poverty are also criticized because poverty lines are often set too low in comparison to costs of living and because they place too great an emphasis on income as a determinant of poverty (Satterthwaite, 1997). Income is often difficult to precisely gauge and security of income may be at least as important as its value in obtaining services. Furthermore, the means by which households obtain basic goods and services is often complex and is not necessarily reliant on the cash economy (Bigsten and Kayizzi-Mugerwa, 1992; Moser, 1995; Rakodi, 1995; Wratten, 1995). It should also be noted that the poverty line approach takes the view that the lack of services is primarily a consequence rather than a cause of poverty.

An alternative view on poverty is to consider it as a complex set of social and economic relationships (Moser, 1995; Satterthwaite, 1997). Such an approach defines poverty as being dependent on many factors that influence the ability of a household to access goods and services and their ability to fully participate in the society. An example of an approach defining poverty in relation to access to basic goods and services is provided by the UNDP Human Poverty Index (UNDP, 1999). This index incorporates aspects such as access to water and sanitation, education and health care services, as being a defining feature of poverty rather than simply a consequence of poverty.

Socioeconomic indices

Another approach that has been used in defining socioeconomic status is through the use of indices based on a set of factors that reflect the standard of living of the household. These indices include aspects such as housing quality or numbers of people living in a dwelling (Rakodi, 1995; Satterthwaite, 1997). This information can take the form of a quantitative index based on data from a census or a qualitative assessment built upon community perceptions and utilizing a range of participatory techniques. Such approaches have been used in developed and developing countries as a mechanism to define vulnerability and disadvantage in relation to access to health care and other services (Jarman, 1984; Stephens *et al.*, 1997). These approaches have proven effective in identifying priority areas and vulnerable populations and therefore in targeting resources at those of greatest risk (Howard, 2002).

Socioeconomic indices can be used at a variety of levels, including national, international and city/town (Townsend *et al.*, 1992; Stephens *et al.*, 1997). When defining these indices, it is important that the variables selected are those that are deemed to be sensitive to changes in relative wealth or status and where there is more than one condition that may be found within a country. Commonly used variables include employment type, roof material, house type and level of education. These variables may also include possession of consumable durables (such as televisions, radios, cars and bikes). It is important when selecting variables to also consider the age of data, how frequently these data will be collected and the rapidity with which ownership of a 'variable' by a household may change. For instance, in emerging economies the ownership of consumer durables may provide a good indication of relative wealth. However, if data on ownership of items are only collected once every five to ten years and sales of items is rapidly increasing, the inclusion of such variables within a socioeconomic index would have limited value as the data would rapidly become inaccurate.

The variables selected are typically weighted to reflect their sensitivity to socioeconomic change. Thus variables deemed to be responsive to changes in wealth are given a higher weight than those deemed to be important indicators but less sensitive to change. For example, in the application of an index in Uganda, roofing material was given a high weight and main source of livelihood a much lower weight as the former was considered to be more sensitive to changes in wealth. Within each variable a number of conditions is defined (for instance different types of roofing material) and these are allocated a score that shows the socioeconomic status that the condition represents. Commonly, negative scores are allocated to conditions indicating poverty and positive scores to conditions indicating wealth, with zero representing a medium level of socioeconomic status.

Selection of variables, their weighting and condition scores depends in large part on expert judgement. It is important to ensure that the expert judgement is drawn from a range of people and a useful method to employ is the Delphi method, which allows repeated consultation with different groups of experts building on the conclusions of each group when initiating consultation with the next group (Stephens *et al.,* 1997).

Qualitative approaches to defining poverty

All the above approaches rely to a certain extent on quantitative measures of poverty and this aspect has been criticized, because it confers external judgements on the social condition of people (Chambers, 1989). As such approaches use standardized measures, which are usually determined by professionals, they allow little opportunity for communities to describe their own circumstances in relation to their needs and perceptions. As a result, there is often little depth to the description of poverty and this may limit the understanding of the problems and difficulties faced by the poor and the relative priority accorded by poor people to those problems. A consequence of this approach can be that interventions to reduce poverty do not address the underlying causes of poverty – such as lack of influence on decision-making – and therefore make at best superficial changes in socioeconomic status while delivering little fundamental community improvement.

One approach to overcoming such problems is to use more qualitative techniques to describe and identify poverty which allow greater depth to the definition of poverty and place greater emphasis on the needs and perceptions of the poor themselves in defining priorities (Wratten, 1995). Such approaches have an inherent advantage over quantitative methods as they allow the people affected by poverty to define what it means to be poor, what assets they hold that could be used to reduce poverty and what external support is required. Many poverty assessments now utilize participatory approaches, which allow much greater opportunity for those affected by poverty to directly engage in defining ways of increasing assets, reducing vulnerability and achieving more sustainable use of the natural resource base. This approach also has merit because it has a big impact on capacity building for the community through allowing:

- community identification of issues;
- community identification of its own preferred approaches for addressing the issues;
- community monitoring to ensure that the approaches selected are working to deliver the required outcomes.

Such approaches have been used at national levels in poverty assessment as well as small-scale community-orientated approaches to water resources management. However, it is likely that for large-scale water resource management strategies both quantitative and qualitative data may be required (Wratten, 1995). This approach may lead to application of quantitative data being used to identify areas where poverty is greatest or where natural resources are most vulnerable, with qualitative assessments used to define appropriate local priorities and interventions.

Livelihoods analysis

Livelihoods analysis has emerged as a new paradigm for assessing the needs of communities and the impacts of interventions by considering how the members of the community ensure a sustainable livelihood. This approach provides a conceptual framework in which to understand communities and the range of assets they posses and the threats they may face in sustaining livelihoods. It therefore places community needs, capabilities and vulnerabilities into the core of understanding socioeconomic conditions.

A set of tools has been developed for undertaking analysis of livelihoods which addresses environmental sustainability as one of the core aspects. The guidance notes on livelihoods prepared by DFID (2003) provide a wide range of quantitative and qualitative tools that can be used to analyse livelihoods. Because there is a wide range of data that may be collected and for each type of data a number of different tools that can be used, they are not discussed in detail here.

Data will often be collected through participatory assessment using tools such as wealth ranking, transect walks, seasonal calendars, resource maps and a range of different methods of interview and group discussion. Data may also be collected from sample surveys looking at specific aspects of the livelihood, for instance access to water sources or economic data. Data from secondary sources (often government or NGO databases) are also often analysed and included within the overall analysis of livelihoods.

Livelihoods analysis usually involves both macro and micro-level assessments and for data to be collected on a wide range of factors that will influence livelihoods. The detail of different methods is not reviewed here as each is underpinned by an extensive literature. However, the most critical aspect of livelihoods analysis is the need to keep livelihood considerations as the primary reason as to why the data are collected, rather than the application of particular methods of data collection.

During the livelihoods analysis, and when considering the likely type and extent of pollution that can be derived from different areas, it is important to collect data on the major source of livelihood and consider the impact that this may have on groundwater. This approach would typically emerge from the assessments on environment and vulnerability, but may also draw information from other aspects, for instance assessment of social capital where governments subsidize access to agrochemicals.

Population and land tenure

Information on population density allows both the need for groundwater protection to be defined and should also provide information as to what measures (if any) can be taken in particular areas. It will also help in making a case for protection of less densely populated areas by highlighting potentially negative impacts from more densely populated settlements. Overall, population growth and numbers of people relying on groundwater also help in shaping a policy and strategy that meets long and short-term needs. Population density is relatively easy to calculate when reliable census and cadastral data are available. Alternatively, qualitative estimates based on ranking may also be used.

The degree to which land tenure systems confer rights over the ownership of groundwater resources should be established. In many countries, ownership of land confers rights to exploit resources on or underlying the land, although this may be limited with regard to how much resource may be removed, over what time period and to within what depth below the ground surface automatic rights to abstract extend. In many other countries, ownership of land does not confer automatic rights to exploit resources underlying the surface and any resource, including groundwater, is under public ownership and exploitation is a public rather than private decision. The nature of rights to exploit resources underground will influence how effectively protection strategies can be implemented. This last point is important as rights are increasingly being unpacked from a piece of land and are now being assigned to aspects such as biodiversity, the land itself, water, other resources, erosion credits, salinity credits. All these factors are recognized as having a significant impact on water quality and quantity and environmental integrity.

Assessing the nature and security of land tenure will help determine what approaches and incentives will need to be available for different communities and may help in directing the overall approach to groundwater resource management. Much of this information would emerge in an analysis of livelihoods as tenure systems will have important asset and vulnerability implications.

7.2 INSTITUTIONAL AND STAKEHOLDER ANALYSIS

When characterizing current capacity and future development of groundwater protection the different institutions and stakeholders, their roles and responsibilities should all be identified and reviewed. This approach would typically be undertaken through a stakeholder analysis that attempts to identify all those organizations, agencies and departments (governmental and non-governmental) that have an interest in the use, management or protection of the groundwater resource. A key element of this process is to assess whether the current institutional responsibilities are supportive of the development of effective institutional arrangements.

A common problem is that institutional responsibilities are highly fragmented with multiple organizations taking some responsibility for either provision of services or control of groundwater. These responsibilities commonly overlap, have inherent internal conflicts of interest and frequent inter-institutional conflicts. Therefore, a key component for the institutional analysis is to define what roles different organizations play, the degree to which these are consistent both with their internal and external environment and what rationalizations are required to make institutional relationships more effective. For example, a process for achieving integrated water cycle management has been developed in New South Wales, Australia. Integrated water cycle management has been particularly successful as it includes key government agency and community stakeholder workshops at the beginning and throughout the study process. This approach ensures that the agencies are not only cognisant of the issues pertinent to the study area, and therefore impacts on the water resource(s), but also that the acts, policies, regulations and issues which are administered by the agencies are made transparent and accommodated as part of the integrated approach (Schneider *et al.*, 2003).

This process supports the identification of the appropriate institution to lead groundwater protection and to outline its relations with other institutions and stakeholders. It also helps to define what further strengthening and capacity building is required in order to support the lead institution to develop and implement an effective groundwater protection strategy.

A review should also be undertaken of the roles, responsibilities and interests in water safety of external stakeholders. This should address statutory roles, all aspects of regulation (financial, safety, environment), involvement in capital and operational investment, roles in specific circumstances (for instance epidemics) and interest groups. For each stakeholder, the relative influence each has over policy, investment, regulation and operations should be noted as a means of identifying how each stakeholder interacts with the supplier. In undertaking this exercise, it is important to identify all stakeholders and not solely concentrate on those deemed powerful or influential. For instance, consumers may not be powerful, but are the most important stakeholders.

Developing a matrix as a result of the institutional analysis is a useful mechanism to summarize the data and to gain a clear idea of which organizations are responsible to different activities. Table 7.1 provides a summary of an institutional analysis in relation to legislation and regulation for establishing WSPs.

7.2.1 The government environment

Government departments and agencies have a key role to play in the development of groundwater management strategies as they usually develop policy and strategic plans. It is important that both national and local government roles are analysed as both may exert significant influence on the actual implementation of any groundwater strategy.

The policy environment within which each institution operates should be analysed and the mandate and jurisdiction that each holds should be identified and clearly defined. Key questions at this stage include:

- What level of autonomy does each institution have?
- What level of decision-making are they invested with?
- To what degree do they have to refer to other institutions in order to implement their plans and strategies?
- Do they control their own budgets?
- What proportion of this budget derives from local and what proportion from national grants?

These questions help define the relative influence of each institution in the practical application of the policy framework into definable strategies on the ground. For instance, if a department is largely autonomous with its own funding base, it tends to have a significant influence on how policy is translated into action. By contrast, weak departments that have limited autonomy and little budgetary control are likely to have little direct influence on actions related to policy.

It is important to assess whether there are overlapping institutional responsibilities between several government departments and levels of government. It is not uncommon to find that several national government departments have some mandate on the development of groundwater or land use for different purposes and have different policy objectives on which they have to act. Rationalizing the institutional responsibilities is often the critical first step in developing sustainable groundwater management policies and strategies. Resistance to change and responsibility is common. Potential conflicts between different institutions should be identified and addressed from the outset.

Table 7.1. Summary of an institutional analysis in relation to legislation and regulation within the context of establishing Water Safety Plans

Activity	Ministry of Water	Ministry of Health	Independent regulator	Water utility	Bureau of Standards	Communities	Independent consultants	Government chemist
Legislate for water quality regulatory framework	Work with relevant government departments to have a legal framework strengthened	Participate in legislation processes	Participate in legislation processes	Participate in legislation processes	In consultation with other stakeholders, establish National Drinking Water Quality Standards	Feed into the legislative process		
Review and revise health based targets	Participate in the review and revision process	Take leading role in the review and revision process	Participate in the review and revision process	Participate in the review and revision process		Feed into the review and revision process		Participate in the review and revision process
Appoint independent water quality auditors through open tendering	Participate in the process of appointing auditors	Appoint chief water quality inspector, who leads the process of recruiting Auditors	Participate in the process of appointing auditors and fund audit operations				Competitive bidding	Participate in the process of appointing auditors

Activity	Ministry of Water	Ministry of Health	Independent regulator	Water utility	Bureau of Standards	Communities	Independent consultants	Government chemist
Regular inspection of utility labs and other facilities		Provide personnel support for inspection	Receive reports	Avail facilities for inspection			Carry out inspection	
Regular review of WSPs		Provide manpower support for inspection	Receive reports	Present WSPs		Participate in review process	Lead role in review process	
Strengthen the national system database	Provide historical data	Provide historical data and participate in process	Receive water quality management information	Provide water quality monitoring and surveillance data			Receive, analyse and process data	
Regular system audits		Provide man-power support for audits	Receive reports and send out feed back	Receive feedback		Receive feedback	Carry out audits Carry out independent sample analysis	

7.2.2 The non-governmental sector

NGOs include farmers groups, industry groups, Chambers of Commerce and environmental groups that have a stake in groundwater management and use. It is important to identify these players, review their potential contribution to the development of groundwater policy and collect their views on the need for groundwater management. This is important information which feeds into the policy development framework.

When assessing the current and potential roles of the non-government sector, it is important to understand whose views each organization represents, the role the organization has in civic society and the influence the organization exerts on policy and public opinion.

7.2.3 Governance

The style and means by which decisions are actually made are important considerations within a review of institutional, policy and legal frameworks. The degree to which outlined processes of decision-making and enforcement of legislation are followed by the institutions responsible ultimately determines whether these processes are effective. It is therefore important to evaluate the decision-making process when undertaking a situation analysis.

Evaluating governance can be politically sensitive, although this should not prevent it being undertaken. The key aspect of assessing governance is to determine to what extent requirements under existing legislation are met and whether procedures and criteria for decision-making are documented and followed. This assessment can be done by reviewing acts, policies and operating procedures to assess what should be done, and through consultation with key stakeholders to obtain their perceptions of how processes are followed. Where appropriate, case law may also be reviewed.

Consultation with key stakeholders can be undertaken using a range of techniques, including in-depth interviews, focus-group discussions and semi-structured interviews. In undertaking consultation on these issues, it is important to ensure that all stakeholder groups are included. It is important to avoid potential bias in the findings of such exercises by only concentrating on particular groups that may have an unrepresentatively negative or positive view. Important stakeholder groups will include Government agencies with executive and policy functions for the environment and groundwater, local government, pressure and consumer groups, agriculture and potential polluters. Within these groups it is also important to obtain copies of relevant reports and assessments of the implementation of protection laws and regulations, as this may provide a clear idea of the overall performance.

The review of enforcement actions and results of cases brought to prosecution is also important in determining the extent to which procedures and processes are being followed. This review will require some kind of value judgement being made on whether these (in terms of both quantity and content) constitute showing good governance or whether they indicate a failure in governance and unacceptable circumventing of procedures. Assessing the seriousness of single deviations and the accumulated deviances from documented procedure are likely to be included.

7.3 MANAGING STAKEHOLDER DISCUSSIONS – LEVELLING THE PLAYING FIELD

It is important to ensure that all relevant stakeholders are able to participate in the development process: '*different stakeholders have different levels of power, different interests, and different resources. Arrangements are needed to "level the playing field" and enable different stakeholders to interact on an equitable and genuinely collaborative basis*' (World Bank, 1996).

Most activities involving groundwater protection involve some technical aspects – assessment, design, construction and/or operation and maintenance. Public participation needs to be co-ordinated with the technical aspects. In addition, the technical aspects need to be carried out professionally and efficiently otherwise the public participation will not ensure mutual trust and support between stakeholders. Although proponents of public participation often blame project failure on the fact that a community was not consulted, there are many cases where failure is due to faulty design of technical aspects. There is no faster way to kill stakeholder support for a project than to include them in all the planning, design, etc., and then to find that the actual design is faulty.

Public participation takes place in all countries of the world, in some form or another. The level of participation varies from country to country, project to project, but many of the lessons learned remain the same. In developing countries, the participating hygiene and sanitation transformation (PHAST) approach has been used as a methodology of participatory learning, which builds on people's innate ability to address and resolve their own problems. PHAST was developed specifically to help communities manage their water supplies and control sanitation-related diseases. Examples of how this has been used to promote source protection is given in Box 7.1.

7.4 DEVELOPING PUBLIC PARTICIPATION

Human settlements are made up of communities and within each community, a number of social structures may be found which represent the interests of the community as it deals with the external world. Whilst community structures and participation are often discussed in the context of developing countries, similar structures often exist in developed countries and in both cases existing community organizations should be incorporated within an assessment of the socioeconomic factors of importance for groundwater protection.

Public participation is useful in improving the outcomes of strategy development, even if it comes too little or too late. However, the ideal scenario is for stakeholders to be included right from the start of project conceptualization and planning. Better yet, the idea comes from the stakeholders themselves, with the role of the outsiders being only one of advisors and catalysts to assist the process.

Box 7.1. PHAST and the children of Dhlabane (based on Breslin, 2000)

Dhlabane is an isolated community in rural KwaZulu/Natal with limited water supply options. Half of Dhlabane is serviced by a reticulated water system while the other half relies on unprotected springs. A series of water quality tests was administered and analysed by children at a local secondary school. The results suggested that water quality was poor at all collection points in the community. Children participated in a range of PHAST exercises that helped them clarify the links between poor water quality and health using a sorting exercise with pictures that depicted activities that could be interpreted as either beneficial or detrimental to water source protection. Children suggested that factors contributing to poor water quality in Dhlabane were:

- open-field defecation near community water sources;
- poor animal management near community water sources;
- poor waste disposal practices;
- infrequent hand washing.

Children were then introduced to a planning exercise story with a gap. The first picture depicted a community with poor water quality, while the second picture was of a community that enjoyed access to clean water supplies. Children were asked how Dhlabane could move to the improved situation depicted.

The children then developed a water quality plan for Dhlabane. They decided to protect a spring that was servicing the school. They developed a plan showing where the pipes would go, what they would feed into, and how the spring could be protected from human and animal contamination. They suggested that Dhlabane residents should provide the labour, but that help was required to fine-tune the design.

The spring was constructed under the supervision and guidance of a local engineering firm that was active in the area and also had invested into child-to-child programmes in the area. The children's model was followed, although some technical modifications were made. Through the process, the children not only took on responsibility for their water quality, but also managed the process of change in their village.

A fence was put up to protect the spring from contamination (as suggested by the children), and students continue to monitor water quality at the source as part of a school course. Results to date have suggested that water quality has been improved and sustained. The children have also developed plays highlighting the linkages between water quality and poor health, which has also raised awareness..

The earlier that stakeholders are involved, the greater chance that they will support changes and be active partners in the process. There is a recognized sequence towards successful community participation as set out below (Oakley, 1989):

(1) Initial contact by change-agents; participant observation and assessment by change-agent.

(2) Group identification and analysis of problems.
(3) Development and strengthening of community structures; emergence of appropriate organizations, identification of local cadres.
(4) Widespread community awareness of causes of problems; awareness of community ability to resolve problems.
(5) Leadership training, briefing, community education.
(6) Concrete group action; programme management.
(7) Networking; making outside contacts; building alliances.
(8) Self-evaluation; adjusting strategies; expansion.
(9) Stabilization; autonomy; functioning alone.

There is no magic formula for the process of participation, and many good references are available which describe the various techniques and examples (Narayan, 1993; Nagy *et al.*, 1994; Yacoob and Whiteford, 1995; World Bank, 1996). The most important aspect of participation is establishing open dialogue and mutual trust, which can only be achieved by understanding the current level of knowledge, attitudes and practices of the various stakeholders. The US EPA has developed seven rules for successful communication with the public concerning groundwater contamination, shown in Box 7.2.

Box 7.2. Seven cardinal rules for communication on groundwater contamination (based on Chun and Den, 1999)

1. Accept and involve the public as a legitimate partner in the issue
2. Plan carefully and evaluate your performance as a communicaton
3. Listen to the public's feelings (active listening)
4. Be honest, open, frank, kind and respectful
5. Coordinate and collaborate with other credible sources
6. Meet the needs of the media
7. Speak clearly with compassion

In developing a communication plan and information needs, the example of the WaterCOM approach, developed for use in water and wastewater projects, can be used (Chemonics International, 2000). Key organizations, groups and individuals are identified and analysed in terms of their potential roles in policy formulation, policy implementation, management of water resources and actual implementation of water, wastewater and/or irrigation projects. These groups cover all levels, from the individual citizen to the top policy makers. Once these groups and individuals are identified, three types of communication flow are established:

- UP from water users to policy makers and senior managers, e.g. on needs, willingness to pay for services, willingness to support water conservation and pollution prevention policies.
- DOWN from policy makers and senior managers to water users, e.g. explanation of a situation and scale of the water shortage problem, technical

considerations, government priorities for assistance to different areas, input needed from the users.

- ACROSS between groups, e.g. between Ministries of Agriculture and Water, NGOs and government, technical groups and NGOs. Communication between government ministries needs to take place to reduce confusion and conflicts, which can lead to distrust in government and therefore lack of support for water conservation and pollution prevention programmes.

Once established, these communication links are used throughout a project to continue dialogue and fulfilment of commitments made by government and other service agencies.

7.5 ANALYSIS OF LAND USE AND GROUNDWATER USE FOR POLICY DEVELOPMENT

In the development of policies, the core elements in groundwater management are the functions and uses of the groundwater, the problems and threats to it, and the impact of possible measures that are proposed to deal with the problems (UNECE, 2000; Schneider *et al.*, 2003). Measures can include investigations, risk analysis, remediation, control of polluting activities or land uses. When developing management strategies the following need to be identified:

- the boundaries of the aquifers and their relation to surface waters and associated ecosystems (Chapter 8);
- human uses (Chapters 9-13) and ecological functions of the groundwaters;
- pressures which have an impact on the uses;
- management targets which can be implemented within a specified timescale.

Some functions have an impact on other important functions, which may not be directly related to health or to groundwater, but which must be taken into account when protection policies are being developed because they affect the success of protection measures. For example, cultural or legal problems may be barriers to the use of some possible protective measures and these must be recognized at the outset. However, if it can be shown that the chosen control measure ensures the best use of the resource, can minimize extractions from the environment and is best suited to the community for which it is intended, then it is not out of the question to lobby for changes in statutory or other potentially limiting barriers connected (with the appropriate controls) to the reticulated supply.

If land use planning and management is to be an effective tool for groundwater protection, a good understanding of the distribution and quality of the groundwater resource is necessary. Effective land use management means that sufficient investigations should be undertaken to define groundwater flow patterns and interactions between aquifers, identify recharge and discharge areas and recognize spatial variations in groundwater quality (see Chapter 8). Vulnerability mapping will identify areas that are particularly susceptible to contamination (see Chapter 8). Modelling techniques can be used to predict the movement of contaminants in aquifers, estimate the impact of land use changes in drinking-water catchments, and predict the effects of particular types of developments on groundwater quality. Models can also help define the maximum density of particular types of development compatible with specific groundwater quality objectives.

It is important in this context that spatial variations in groundwater quality, both those caused by natural processes and those caused by existing land use, are identified. Information is also required about the distribution of land use and its impact on groundwater quality. Some of this information is common to all uses and can be obtained from studies carried out elsewhere. However, there are often local peculiarities in climate, hydrogeology or in the way a particular activity is undertaken that will require site-specific investigations (see Chapters 9-13). Local problems can be obscured by scale, so detailed information is often required (Rudnitski, 1998). Moreover, it is not only present economic activities that affect the environment, and so as well as information about the distribution of existing land uses, information about the distribution of past uses that may have affected groundwater quality should also be investigated if it has not been recorded. Also intrinsic to water resource use is the changing consumption patterns and growth of the community and how this will impact on the resource base. Not only does information need to be gathered on community demographics and growth, but also on system analysis including factors such as unaccounted for water (system leakages) and where water can be accessed from other sources, such as rainwater, storm water and effluent, to supplement the potable water supply. Knowing how water can be best targeted for its intended use can facilitate the sustainable use of the community's and the environment's water resources.

7.5.1 Importance of groundwater for domestic supply

One of the major issues to be addressed when developing groundwater protection strategies is the level of groundwater use within the country, to what types of use groundwater is put and the long-term strategies for groundwater development within the country. In industrialized countries, use of groundwater is often easy to calculate as access to domestic water supply direct to the homes of the population is effectively universal and the proportion of this water that is derived from groundwater is then easily calculated. A similar situation would also be true for agricultural use of groundwater. However, in some countries, a lack of knowledge of water resources has resulted in over allocation of licences and this fact may cause problems in terms not only of the resource itself but also in terms of compensation as water licences may need to be bought back by the state.

In developing countries, the use of groundwater is likely to be more difficult to accurately calculate. Access to water supplies piped to the home may be restricted to a relatively small proportion of the population, such as the wealthy elite. The remaining population must develop a variety of different strategies to secure a water supply (Howard, 2002). In some countries, people who lack access to a water supply piped to the home will use communal sources, such as public taps or small groundwater sources such as protected springs, tubewells and dug wells (Gelinas *et al.*, 1996; Rahman, 1996; Howard *et al.*, 1999). Levels of use of such supplies may be significant, for instance in Kampala, the capital city of Uganda, it was estimated that protected springs were used for part or all of the domestic water collection by over 60 per cent of poor households living in areas where springs were found. In a town in the east of the country, this figure went up to over 70 per cent. However in both cases it was noted that most households used several water sources (Howard *et al.*, 2002).

In other countries, whilst multiple source use may be common, the use of water from particular sources may be restricted to specific uses, based on the perception of the users on the quality of that water (Madanat and Humplick, 1993). As the protection of groundwater should take into account the use for which the water is intended, it is therefore important to understand water use patterns. This requires a water usage study to be undertaken in at least a sample of communities in order to understand the importance of groundwater.

7.5.2 Private supplies

In many countries, a significant proportion of the population will construct a private water supply and these are often shallow tubewells or dug wells. These are found in both developed and developing countries and the quality of the water and its surveillance are often unsatisfactory. Where access to piped water supply is limited, the total number of private supplies may be very high and greatly exceed the number of public supplies.

For example, in many large Asian cities by far the majority of the wells and boreholes sunk are owned by individuals and thus any estimated use of groundwater would need to take private as well as public tubewells into account. Private supplies are often not adequately controlled. In a number of large cities (e.g. Bangkok, Jakarta, Manila and Dhaka) there are very large numbers of shallow private wells that are largely uncontrolled and unlicensed, are poorly constructed and are rarely monitored (Foster and Lawrence, 1995). When assessments of quality are made, these often indicate poor and deteriorating quality of the water and increasing health risks. In these situations, control of licensed large water supplies operated by utilities or municipalities is of limited benefit if the private supplies are not also adequately controlled (Foster and Lawrence, 1995).

Similar situations are also found in rural areas and the lack of information and control of such small private supplies represents a more significant challenge to the protection of groundwater than where exploitation is through a relatively small number of high volume production sources. It is important to identify the numbers and location of private wells and to map these in relation to public supplies and to centres of population. It is also important to establish some programme of monitoring of these supplies and what response will be made to evidence of poor quality.

The use of private supplies is also widespread in rural areas in Europe, North America, Australia and New Zealand and in many cases the quality of such supplies is problematic. Quality problems include microbial and chemical hazards and control of the quality of such supplies still represents a significant challenge in many countries. In parts of Central and Eastern Europe, there have been very large increases in the numbers of private water supplies since the early 1990s, at a time when the ability of national and local authorities to monitor and advise owners on improvements has declined.

In developing a response to identified contamination, the likely importance of private wells providing a route for contamination of the wider aquifer, thus potentially affecting public supplies, should be considered as there is evidence in some situations of this causing significant quality problems (Rojas *et al.*, 1995). There may therefore be a case for closure of existing private wells and/or prohibiting development of new wells as a means of protecting the public supply. However, such approaches are only feasible if viable alternative supplies are provided to the population: simply prohibiting private supplies in situations where the public supply is inadequate will not be effective and would be likely to lead to greater public health risks. Furthermore, prohibition of such sources (particularly in rural areas) is unlikely to be fully effective and will be expensive and complicated to enforce.

7.5.3 Long-term sector plans

Long-term water supply development plans are also important to consider. In particular it is important to collect information on how shallow and deep groundwater will be developed in the long term. For instance, some contamination of shallow groundwaters may be tolerated if the long-term plan is a water supply based on deeper groundwater or surface water. In the case of deeper groundwater, however, the degree of hydraulic continuity between shallow and deep groundwater should be taken into consideration. This will help to avoid the creation of a long-term resource quality problem that will not be easy to resolve.

When collecting information for a situational analysis, it is important to determine what economic and social values should be placed upon the groundwater taking into account the issues raised above. In Chapter 17 examples are provided of how groundwater protection strategies can take into account the economic value placed upon groundwater through practical measures.

7.6 VALUING GROUNDWATER PROTECTION

Valuing of groundwater and estimating the costs of protection measures must take into account several factors in order to gain a clear understanding of the comparative costs of different levels and methods of protection that can be applied. This should address whether it is more effective to purchase land in the drinking-water catchment or whether changes in existing land users' practice is more effective. The purchase of the land in the catchment will represent a large single capital payment, which can either be recovered or written off (depending on the nature of body purchasing the land and their accounting obligations) or defrayed over several years. Where the approach is to change existing land use, costs will be incurred on an annual basis if compensation is provided to land users. In some cases this may be covered in land rents charged for the use of the land, but where this is not covered the incremental costs must be either absorbed into the overall costs of water production or written off. Where these costs are incorporated into the water production costs they will typically be passed, at least in part, to the consumers of the water produced.

The protection of groundwater will usually incur some costs to the public through taxation, water tariffs or costs of goods that would otherwise have been produced on the land where activities are to be controlled. Therefore, it is important that there is consultation with the public to develop a coherent overview regarding their willingness to pay for groundwater protection. Willingness to pay may be critical in defining what level of protection will be implemented.

Willingness to pay for environmental improvements is often undertaken through the use of contingent valuation methodologies (CVM) (Mitchell and Carson, 1989; McGranahan *et al.*, 1997). In CVM, an attempt is made to provide a monetary value to non-financial resources or public goods and to elicit from participants what they would pay for environmental improvements or protection. CVM were originally applied in the 1960s but were only later more widely used in the evaluation of the willingness to pay for public goods. This has included increasing application within environmental improvements in and provision of services such as water supply (Briscoe and Garn, 1995; McGranahan *et al.,* 1997; Wedgewood and Sansom, 2003).

There are a number of different approaches to using CVM, but common to all is that participants are asked to state a value they would be willing to pay for a (as yet) hypothetical level of protection or service (Wedgewood and Sansom, 2003). Some approaches present participants with a specified amount for a public good and they find out whether the participant would be willing to pay that sum for the good. This is a relatively simplistic approach and may have limited value for groundwater protection, where it may be more important to provide participants with a broader range of levels of protection and costs in order to determine the optimum willingness to pay. More sophisticated CVM approaches involve bidding games which allow the researcher to raise the stakes in terms of costs and in levels of protection as a means of identifying both the upper limits of willingness to pay and the limits of protection desired. The advantage of bidding games is that it allows much greater flexibility in the options that can be offered to participants and an iterative approach can be used to define optimum interventions matched against willingness to pay. However, it is more complex to analyse as different participants will identify different levels of willingness to pay.

Some other approaches to CVM try to establish lower and upper limits of willingness to pay, which may utilize bidding game processes or may simply ask people to state the least and most they would be willing to pay for groundwater protection. The specific methods are not detailed here and the choice of method will be determined by the range of options available, resource to undertake studies and the participation of the public.

CVM studies may also be subject to a variety of biases and these must be addressed during the design stage. Wedgewood and Sansom (2003) note that these include:

- bias introduced by participants deliberately under-stating their willingness to pay as they perceive that this may deliver lower-cost solutions (strategic bias);
- bias introduced because the participants do not understand or believe the hypothetical scenarios (hypothetical bias);
- bias introduced because there is insufficient quantity and/or quality of information (information bias);
- bias introduced because the starting point in relation to the costs is excessively high or low (starting point bias);
- bias introduced by the interviewer or because the respondent attempts to guess the 'right' answer (interviewer/compliance bias);
- bias introduced by the nature of payment method (payment method bias).

All these biases may be overcome through good design, but their potential should not be ignored. In addition to potential for bias in the results, a sound statistical basis for the survey design is preferred. This can use a range of techniques, from simple random sampling to cluster sampling, with the technique being determined by the size, distribution and nature of the population being studied and the degree to which these have common traits among sub-groups.

7.7 CHECKLIST

NOTE ▶ *The following checklist outlines the information needed for characterizing the socioeconomic, institutional and legal setting in the drinking-water catchment area. It is neither complete nor designed as a template for direct use and needs to be adapted for local conditions.*

ⓘ **What are the socioeconomic issues in the drinking-water catchment area?**

- ✓ Analyse socioeconomic status in regions or communities in areas likely to be affected by groundwater protection strategies
- ✓ Check whether there are particular groups that are poor or vulnerable: where are such groups found and will they be further disadvantaged by the groundwater protection strategy?
- ✓ Identify main sources of livelihood in protected areas
- ✓ Assess economic impact of protection measures on livelihood in protected areas
- ✓ Check compensation requirements in protected areas: costs and compensation delivery mechanisms
- ✓ Compile information on the population residing in protected areas and density of population
- ✓ Evaluate short and long-term projected demands for water for drinking and for other sectors
- ✓ …

ⓘ **What is the level of service provision in the drinking-water catchment area?**

- ✓ Estimate the proportion of the population currently having access to public water supplies (divide this into categories based on source type, service level, functional status and use of sources by population)
- ✓ Compile information on the number of private supplies present: where are they found and what type of technology is used?
- ✓ Assess the condition and quality of private supplies

- ✓ Check whether plans exist for future development of groundwater for drinking and domestic supply
- ✓ …

What are the community characteristics in relation to participation and consultation in the drinking-water catchment area?

- ✓ Identify the type of communities living in protected areas: rural, peri-urban, urban
- ✓ Estimate the number of communities living in the protected areas
- ✓ Evaluate social structures existing in the communities and tradition of community management of resources
- ✓ Analyse experience and demand for public participation in the country/region
- ✓ Check consultation methods commonly used (will these provide the information required?)
- ✓ …

What are the land tenure and property rights in the drinking-water catchment area?

- ✓ Identify ownership of groundwater and rights of exploitation for land owners
- ✓ Estimate number of private land-owners in protected areas
- ✓ Estimate proportion of land under customary ownership in protected areas
- ✓ Estimate proportion of land publicly owned in protected areas
- ✓ Compile information on number, size and type of informal settlements in protected areas
- ✓ …

What is the basis for valuing and costing groundwater protection measures?

- ✓ Estimate economic value of groundwater resources
- ✓ Evaluate social and cultural values of groundwater resources
- ✓ Estimate costs of protection strategies
- ✓ Assess current water charges and increases likely to result from protection measures
- ✓ Check for subsidies currently applied to any users of groundwater
- ✓ Identify particular political constraints to cost-recovery
- ✓ …

What are the institutional structures and needs in the drinking-water catchment area?

- ✓ Identify government bodies involved in water resources management
- ✓ Identify government bodies responsible for development of groundwater
- ✓ Identify government bodies responsible for protection of groundwater
- ✓ Identify government bodies that govern activities that may pollute groundwater
- ✓ Identify departments having the strongest claim for a groundwater protection mandate
- ✓ Assess how rationalizations will be made to the institutional framework
- ✓ Identify NGOs or community groups having an interest in groundwater
- ✓ Assess NGOs' opinion and influences they exert
- ✓ …

Documentation and visualization of information on the socioeconomic and institutional setting.

- ✓ Compile summarizing report and consolidate information from checklist points above
- ✓ Map population and settlement structure as well as water supply service structure (use GIS if possible)
- ✓ …

7.8 REFERENCES

Bigsten, A. and Kayizzi-Mugerwa, S. (1992) Adaptation and distress in the urban economy: a study of Kampala households. *World Dev.*, **20**(10), 1423-1441.

Breslin, E.D. (2000) Protecting drinking water: water quality testing and PHAST in South Africa. In *Water, Sanitation and Health*, (eds. I. Chorus, U. Ringelband, G. Schlag and O. Schmoll), pp. 89-93, IWA, London.

Briscoe, J. and Garn, H.A. (1995) Financing water supply and sanitation under Agenda 21. *Natural Resour. Forum*, **19**(1), 59-70.

Chambers, R. (1989) Editorial introduction: vulnerability, coping and policy. *IDS Bull.*, **20**(2), 1-7.

Chemonics International web site 2000. WaterCOM. www.chemonics.com (accessed April 29, 2005).

Chun, A. and Den, A.R. (1999) A key to decision making: a challenging workshop on how to communicate, listen and work with our public. In *Presenter's Manual for: Superfund Risk Assessment and How You Can Help*. http://www.epa.gov/oswer/riskassessment/pdf/vdmanual.pdf (accessed August 11, 2005).

DFID (2003) *Sustainable Livelihoods Guidance Sheets*. www.livelihoods.org (accessed April 29, 2005).

Foster, S.S.D. and Lawrence, A.R. (1995) Groundwater quality in Asia: an overview of trends and concerns. *UN-ESCAP Water Resour. J. Series C*, **184**, 97-110.

Gelinas, Y., Randall, H., Robidoux, L. and Schmit, J-P. (1996) Well water survey in two districts of Conakry (republic of Guinea) and comparison with the piped city water. *Water Resour.*, **30**(9), 2017-2026.

Hardoy, J.E. and Satterthwaite, D. (1989) *Squatter Citizen: Life in the Urban Third World*, Earthscan Publications, London.

Howard, G. (ed) (2002) Water Supply Surveillance – A Reference Manual, WEDC, Loughborough University.

Howard, G., Bartram, J.K. and Luyima, P.G. (1999) Small water supplies in urban areas of developing countries. In Providing Safe Drinking Water in Small Systems: Technology, operations and economics, (eds. J.A. Cotruvo, G.F. Craun and N. Hearne), pp. 83-93, Lewis Publishers, Washington DC.

Howard, G., Teuton, J., Luyima, P. and Odongo, R. (2002) Water usage patterns in low-income urban communities in Uganda: implications for surveillance. *Internat. J. Environ. Health Res.*, **12**(1), 63-73.

Jarman, B. (1984) Underprivileged areas: validation and distribution of scores. *BMJ*, **289**, 1587-1592.

Madanat, S. and Humplick, F. (1993) A model of household choice of water supply systems in developing countries. *Water Resour. Res.*, **29**(5), 1353-1358.

McGranahan, G., Leitman, J. and Surjadi, C. (1997) *Understanding Environmental Problems in Disadvantaged Neighbourhoods: Broad spectrum surveys, participatory appraisal and contingent valuation*, Urban Management Programme Working Paper 16 Stockholm Environment Institute, Stockholm.

Mitchell, R.C. and Carson, R.T. (1989) *Using Surveys to Value Public Goods: The contingent valuation method*, Resources for the Future, Washington, DC.

Moser, C.O.N. (1995) Urban social policy and poverty reduction. *Environ. and Urbanization*, **7**(1), 159-171.

Nagy, M.T., Bowman, M., Dusik, J., Jendroska, J., Stec, S., van der Zwiep, K. and Zlinsky, J. (eds) (1994) *Manual on Public Participation in Environmental Decision making: Current Practice and Future Possibilities in Central and Eastern Europe*, REC, Budapest.

Narayan, D. (1993) *Participatory evaluation: tools for managing change in water and sanitation*. The World Bank, Washington, DC.

Oakley, P. (1989) *Community Involvement in Health Development: An Examination of Critical Issues*. World Health Organization, Geneva.

Rahman, A. (1996) Groundwater as source of contamination for water supply in rapidly growing megacities of Asia: Case of Karachi, Pakistan. *Water Sci. Technol.*, **34**(7-8), 285-292.

Rakodi, C. (1995) Poverty lines or household strategies? A review of conceptual issues in the study of urban poverty. *Habitat Internat.*, **19**(4), 407-426.

Rojas, R., Howard, G. and Bartram, J. (1995) Groundwater quality and water supply in Lima, Peru. In *Groundwater Quality, AGID Report No. 16*, (eds. H. Nash and G.J.H. McCall), pp. 159-167, Chapman and Hall, London.

Rudnitski, K. (1998) *Land Use Planning For Sustainable Groundwater Management in the Port Perry Region, Problems in Land Resource Science*. University of Guelph, Guelph, Ontario.

Satterthwaite, D. (1997) Urban poverty: reconsidering its scale and nature. *IDS Bull.*, **28**(2), 9-23.

Schneider, P., Davison, A.D., Langdon, A., Freeman, G., Essery, C., Beatty, R. and Toop, P. (2003) Integrated water cycle planning for NSW towns. *Water Sci. Technol.*, **47**(7-8), 87-94.

Stephens, C., Akerman, M., Avle, S., Maia, P.B., Campanario, P., Doe, B. and Tetteh, D. (1997) Urban equity and urban health: using existing data to understand inequalities in health and environment in Accra, Ghana and Sao Paulo, Brazil. *Environ. and Urbanisation*, **9**(1), 181-202.

Townsend, P., Whitehead, M. and Davidson, N. (1992) *Inequalities in Health*. Penguin Books, Harmondsworth.

UNDP (1999) *Human Development Report 1999*, Oxford University Press, New York.

UNECE (2000) *Guidelines on Monitoring and Assessment of Transboundary Groundwaters*, RIZA, Inst for Inland Water Management and Wastewater Treatment, United Nations Economic Commission for Europe Task Force for Monitoring and Assessment, Lelystad.

Wedgewood, A. and Sansom, K. (2003) *Willingness-To-Pay Surveys – A Streamlined Approach: Guidance Notes for Small Town Water Supplies.* WEDC, Loughborough University.

World Bank (1996) *Participation Sourcebook*, Washington, DC.

World Bank (2002) *World Development Report 2003: Sustainable Development in a Dynamic World*, OUP, New York.

World Resources Institute (1996) *World Resources: A Guide to the Global Environment: The Urban Environment*, OUP, New York.

Wratten, E. (1995) Conceptualizing urban poverty. *Environ. and Urbanization*, **7**(1), 11-36.

Yacoob, M.B.E. and Whiteford, L. (1995) *Public Participation in Urban Environmental Management.* WASH Reprint, Technical Report No. 90, Environmental Health Project, Washington, DC.

8

Assessment of aquifer pollution vulnerability and susceptibility to the impacts of abstraction

J. Chilton

Information about subsurface conditions is needed for the area of investigation, which may be a complete catchment, the outcrop or recharge area of an aquifer or the part of the aquifer contributing water to individual public supply sources or wellfields. The latter is often referred to as the capture zone, and represents the size of the area from which adequate recharge is obtained to balance the total amount of water abstracted. The information required is that which will enable assessments to be made of both the vulnerability of the aquifer to pollution and its susceptibility to the impacts of heavy or even excessive abstraction of groundwater. This chapter first defines aquifer vulnerability and describes how it is assessed, and reviews the range of information types that are likely to be needed about the hydrogeological conditions to enable this to be done. The chapter also provides some general guidance on where such information might be found. While this information is mostly of a physical geographical and geological nature, it can also include land use, as this is often closely linked to or determined by geographical factors. An obvious example of this would be mining, the presence or absence of which is clearly determined by the geology. The chapter provides a brief summary of the ways in which abstraction can have negative consequences for groundwater. As with the other chapters in this section, a checklist is provided at the end.

NOTE ▶ *Hydrogeological conditions which determine aquifer pollution potential vary greatly. They therefore need to be analysed specifically for the conditions in a given setting. The information in this chapter supports hazard analysis in the context of developing a Water Safety Plan for a given water supply (Chapter 16).*

8.1 DEFINING, CHARACTERIZING AND MAPPING GROUNDWATER VULNERABILITY

8.1.1 Vulnerability of groundwater to pollution

In view of the importance of groundwater for potable supplies, it might be expected that aquifer protection to prevent groundwater quality deterioration would have received due attention. However, even in and around urban and industrial areas where many actual or potential sources of pollution are located, aquifer protection has, until relatively recently, not been given adequate consideration. One important reason for this lack of consideration is that groundwater flow and pollutant transport are neither readily observed nor easily measured. These are generally slow processes in the subsurface, and there is widespread ignorance and indeed complacency about the risk of groundwater pollution amongst administrators and planners with responsibility for managing land and water resources. In the long term, however, protection of groundwater resources is of direct practical importance because, once pollution of groundwater has been allowed to occur, the scale and persistence of such pollution makes restoration technically difficult and costly. In taking care of the quality of groundwater, as in many other things, prevention is better than cure.

Natural attenuation capacity varies widely according to geological and soil conditions. Instead of applying controls on possible polluting activities everywhere, it is more cost-effective and provides less severe constraints on economic development if the degree of control is varied according to attenuation capacity. This is the general principle underlying the concept of aquifer vulnerability, and the need for mapping vulnerability distribution (Foster *et al.*, 2002).

Given the complexity of the factors governing pollutant pathways and transport of pollutants to aquifers, it might appear that hydrogeological conditions are just too complicated for vulnerability to be mapped, and that each polluting activity or pollutant should be treated separately. While it is clear that general vulnerability to a universal contaminant cannot really be valid, nevertheless trying to define vulnerability separately for specific pollutants is unlikely to achieve either adequate coverage or universal acceptance, and would have data requirements that are unrealistic in terms of human and financial resources.

A logical approach to assessing the likelihood of groundwater pollution is to think of it as the interaction between the pollutant load that is, or might be, applied to the subsurface environment as a result of human activity and the pollution vulnerability, which is determined by the characteristics of the strata separating the aquifer from the land surface.

In these terms, vulnerability is a function of the ease of access to the saturated aquifer for water and pollutants, and the attenuation capacity of the soil and geological strata between the pollution source and the groundwater. Information needs concerning possible pollutant loads and sources are dealt with in Chapters 9 to 13, and the general guiding principles of aquifer vulnerability are covered below. Firstly, however, some words of caution need to be born in mind in relation to the applicability of the above approach (Foster and Hirata, 1988; NRC, 1993):

NOTE ▶ *All aquifers are vulnerable to persistent, mobile pollutants in the long term.*

Less vulnerable aquifers are not easily polluted, but once polluted they are more difficult to restore.

Uncertainty is inherent in all pollution vulnerability assessments.

If complex assessment systems are developed, obvious factors may be obscured, and subtle differences may become indistinguishable.

The term pollution vulnerability refers to the intrinsic characteristics of an aquifer that represent its sensitivity to being adversely affected by an imposed contaminant load. It is, in effect, the inverse of the pollution assimilation capacity of the receiving water in river quality management, but with the difference that aquifers usually have at least some overlying strata that can provide additional protection. If such a scheme is adopted, it is possible to have high vulnerability but no pollution risk, because there is no pollution loading, or vice versa. Both are quite consistent in practice. Moreover, the contaminant load can be removed, controlled or modified, but the aquifer vulnerability, which depends on the intrinsic properties of the subsurface, cannot.

8.1.2 Defining aquifer pollution vulnerability

The concept of groundwater vulnerability is derived from the assumption that the physical environment may provide some degree of protection of groundwater against natural and human impacts, especially with regard to pollutants entering the subsurface environment. The term 'vulnerability of groundwater to contamination' was probably first introduced in France in the late 1960s (Albinet and Margat, 1970). The general intention was to show that the protection provided by the natural environment varied from place to place. This would be done by describing in map form the degree of vulnerability of groundwater to pollution as a function of the hydrogeological conditions. Thus the fundamental principle of groundwater vulnerability is that some land areas are more vulnerable to pollution than others, and the goal of a vulnerability map is to subdivide an area accordingly. The differentiation between mapped units was considered arbitrary because the maps showed the vulnerability of certain areas relative to others, and did not represent absolute values. The maps, however, would provide information from which land use and associated human activities could be planned and/or controlled as an integral part of an overall policy of groundwater protection at national, sub-national (province or state) or catchment scale.

Although the general concept has been in use for more than thirty years, there is not really a generally accepted definition of the term. The historical evolution of the concept of vulnerability was reviewed by Vrba and Zaporozec (1994). Hydrogeologists have debated in particular whether vulnerability should be determined in a general way for all pollutants, or specifically for individual or groups of pollutants. The following is considered to adequately define the concept of vulnerability:

DEF ▶ ***Vulnerability*** *comprises the intrinsic properties of the strata separating a saturated aquifer from the land surface which determine the sensitivity of that aquifer to being adversely affected by pollution loads applied at the land surface.*

Vrba and Zaporozec (1994) recognized that there could be more than one type of vulnerability: intrinsic (or natural) which was defined purely as a function of hydrogeological factors, and specific for those users who wished to prepare and use maps related to specific pollutants, for example agricultural nitrate, pesticides, or atmospheric deposition. It would be more scientifically robust to evaluate vulnerability for each pollutant or class of pollutant or group of polluting activities separately (Anderson and Gosk, 1987), especially for such diverse pollutants and activities as those listed above, or unsewered sanitation and wastewater use, for example. There is, however, unlikely to be adequate data or human resources to achieve this. Development of a generally recognized and accepted definition of vulnerability does not, however, imply that there should be a standardized approach to its mapping. Hydrogeological environments and user requirements in terms of scales are too diverse to be dealt with in a standardized way, but it is important to agree on a common basis – the definition of vulnerability – before considering approaches to the assessment of such diversified conditions.

Representation of the vulnerability of groundwater to pollution by means of maps has become an important tool by which hydrogeologists can assist the planning community. However, the inevitable limitations of such maps need to be explained to the users by the groundwater specialists who prepare them. These limitations come from the conceptual distinction between intrinsic and specific maps referred to above, from the simplifications imposed by the scale of heterogeneity of soils and aquifers compared to the scale of the map, and from deficiencies in the data available for whatever method of depicting vulnerability is adopted. Given an appreciation of these limitations, vulnerability maps have been demonstrated to play a useful part in groundwater protection (NRC, 1993; Vrba and Zaporozec, 1994).

8.1.3 Classifying aquifer vulnerability

Vulnerability assessment involves evaluating likely travel times from the ground surface to the water table, or to the aquifer in the case of confined conditions. The greater the travel time, the more potential there is for pollutant attenuation by the processes outlined in Chapters 3 and 4. Aquifer vulnerability can be subdivided simply into five broad classes (Table 8.1). Extreme vulnerability is associated with aquifers having a high density of open fractures and with shallow water tables, which offer little chance for pollutant attenuation.

Table 8.1. Broad classification of aquifer vulnerability (based on Foster *et al.*, 2002)

Vulnerability class	Definition
Extreme	Vulnerable to most water pollutants with relatively rapid impact in many pollution scenarios
High	Vulnerable to many pollutants, except those highly absorbed and/or readily transformed, in many pollution scenarios
Moderate	Vulnerable to some pollutants, but only when continuously discharged or leached
Low	Only vulnerable to the most persistent pollutants in the long term, when continuously and widely discharged or leached
Negligible	Confining beds are present and prevent any significant vertical groundwater flow

Thus for preliminary assessment purposes, it is instructive to note that the hydrogeological environments described in Chapter 2 differ greatly in the time taken for recharge entering at the land surface to reach the water table or potentiometric surface of the aquifer (Table 8.2). Table 8.1 also indicates the likely vulnerability class for each environment, and the general vulnerability of some common soils and rocks is summarized in Figure 8.1, in which the arrows indicate increasing vulnerability, and the three classes used in this earlier attempt at classification roughly correspond to the three middle classes in Table 8.1.

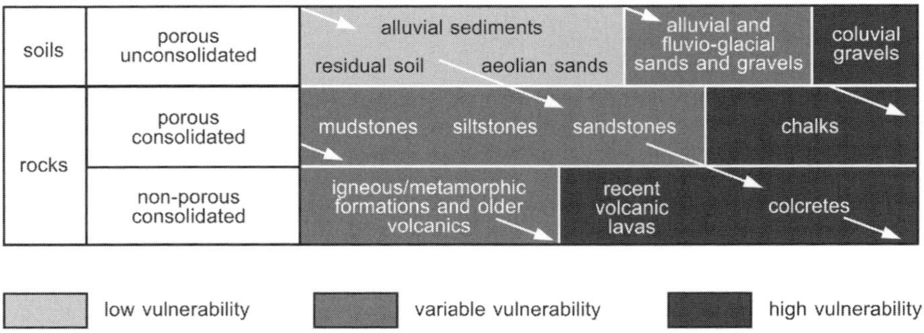

Figure 8.1. Vulnerability of soils and rocks to groundwater pollution (modified from Lewis *et al.*, 1980)

Table 8.2. Hydrogeological environments and their associated groundwater pollution vulnerability (based on Morris *et al.*, 2003)

Hydrogeological environment		Typical travel times to water-table	Attenuation potential of aquifer	Pollution vulnerability
Alluvial and coastal plain sediments	unconfined semi-confined	weeks-months years-decades	high-moderate High	moderate low
Intermontane valley-fill and volcanic systems	unconfined semi-confined	months-years years-decades	moderate moderate	moderate moderate-low
Consolidated sedimentary aquifers	porous sandstones	weeks-years	high	moderate-high
	karstic limestones	days-weeks	low	extreme
Coastal limestones	unconfined	days-weeks	low-moderate	high-extreme
Glacial deposits	unconfined	weeks-years	moderate-low	high-moderate
Extensive volcanic sequences	lavas ash/lava sequences	days-months months-years	low-moderate high	high-extreme low
Weathered basement	unconfined semi-confined	days-weeks weeks-years	low moderate	high-extreme moderate
Loessic plateaux	unconfined	days-months	low-moderate	moderate-high

Unsaturated zone travel time and aquifer residence time are important factors in any aquifer assessment because they affect the ability of the aquifer in question to protect against pollution. For instance, a residence period of a month or so is adequate to eliminate most bacterial pathogens (Chapter 3). Spillages of more intractable pollutants such as petrol or other fuels, and other organic compounds can, given time, undergo significant degradation in-situ by an aquifer's indigenous microbial population (Chapter 4).

The soil zone is usually regarded as a principal factors in the assessment of groundwater vulnerability and the first line of defence against pollution. The main properties of soils that relate to vulnerability to pollution are discussed in Section 8.2.4 below. The soil layer is usually continuous, but the spatial variability of its physical, chemical and biological properties can be very great, and generalizations of soil parameters have to be undertaken with some care. Because of its potential to attenuate a range of pollutants, it plays a critical role when considering specific vulnerability to diffuse sources of pollution such as agricultural fertilizers, pesticides and acid deposition. Not all soil profiles and underlying materials are equally effective in attenuating pollutants, and the degree of attenuation will vary widely with the types of pollutant and polluting process in any given environment.

The soil has a particularly important position amongst vulnerability factors because the soil itself is vulnerable. The soil's function as a natural protective filter can be damaged rather easily by such routine activities as cultivation and tillage, irrigation, compaction and drainage. Human activities at the land surface can greatly modify the existing natural mechanisms of groundwater recharge and introduce new ones, changing the rate, frequency and quality of groundwater recharge. This is especially the case in arid and semi-arid regions where there may be relatively little and infrequent natural recharge, but also applies to more humid regions. Understanding these mechanisms and diagnosing such changes are critical, and the use of soil properties in vulnerability assessment should always take into consideration whether the soils in the area of interest are in their natural state. Further, there are many potentially polluting human activities in which the soil is removed or otherwise by-passed and for these the component of protection provided by the soil does not apply.

Below the soil, the unsaturated zone is very important in protecting the underlying groundwater, especially where soils are thin and/or poorly developed. The character of the unsaturated zone and its potential attenuation capacity then determine decisively the degree of groundwater vulnerability. The main unsaturated zone properties that are important in vulnerability assessment are the thickness, lithology and vertical hydraulic conductivity of the materials. The thickness depends on the depth to the water table, which can vary significantly due to local topography and also fluctuates seasonally, and both these have to be taken into account when determining thickness. The importance of hydraulic conductivity, its distribution and its role in determining groundwater flow rates should be particularly emphasized. Porosity, storage properties, and groundwater flow direction may also be important, and another supplementary parameter in some types of aquifers and circumstances may be the depth and degree of weathering of the upper part of the unsaturated zone.

Degree of confinement is also an important factor in determining vulnerability. Concern about groundwater pollution relates primarily to unconfined aquifers, especially where the unsaturated zone is thin because the water table is at shallow depth. Significant risk of pollution may also occur in semi-confined aquifers, if the confining aquitards are relatively thin and permeable. Groundwater supplies drawn from deeper and more fully confined aquifers will normally be affected only by the most persistent pollutants and in the long term. Data requirements are summarized in Table 8.3 below.

Table 8.3 Data requirements for principal factors contributing to vulnerability assessments (modified from Foster *et al.*, 2002)

Component of vulnerability	Ideally required	Normal availability and source
Hydraulic inaccessibility (ease and speed of water movement)	Degree of aquifer confinement (incl. partial or semi-confined)	Simple division between confined and unconfined from geological maps
	Soil thickness and permeability	Soil classes in map form and their accompanying descriptions
	Depth to groundwater table or potentiometric surface in map form (gives thickness of unsaturated zone)	Varying amounts of water table data from individual wells
	Unsaturated zone moisture content and vertical hydraulic conductivity of strata in unsaturated zone or confining layers	Little or no data: typical values inferred from existing studies or literature
Attenuation capacity	Mineral and organic matter content of soil, and its thickness	Soil classes in map form and their accompanying descriptions
	Grain size and/or fracture distribution of strata in unsaturated zone or confining layers	General distinction between intergranular and fracture flow from geological maps
	Mineralogy of strata in unsaturated zone or confining beds, including organic content	Maybe found in descriptive memoirs or reports accompanying maps, or from existing studies or literature

Additional attributes that can be considered as of secondary importance to those in Table 8.3 include topography, surface water features and the nature of geological formations beneath the aquifer of interest. Some of these may have only local significance. Topography influences the location of recharge, soil development, properties and thickness and local groundwater flow. The interaction between surface water and groundwater, i.e. in which direction water is moving, may be important locally.

8.1.4 Mapping aquifer vulnerability

A vulnerability map shows in a more or less subjective way the capacity of the subsurface environment to protect groundwater. Like all derivative or interpreted maps, it is somewhat subjective because it must meet the requirements of the user. The maps should provide the user with the most accurate and informative assessment of the sensitivity to impacts, allowing comparison between aquifers and between different locations and different parts of the same aquifer. Preparation of the maps usually involves combining or overlaying several thematic maps of selected physical factors that have been chosen to depict vulnerability. These are discussed below, but have been grouped by Vrba and Zaporozec (1994) into those associated with:

- the hydrogeological framework – the characteristics of the soils and underlying geological materials;
- the groundwater flow system – the direction and speed of groundwater movement;
- the climate – the amount and type of recharge.

A further general consideration is that vulnerability maps are normally prepared from existing information only, without the collection of new field data.

A number of approaches to the assessment and mapping of vulnerability have been developed, using varying combinations of the soil, unsaturated and saturated zone factors outlined above. These can be considered in three groups:

- hydrogeological setting methods
- parametric methods
- analogical relation and numerical model methods.

All methods are briefly explained below and described in more detail by Vrba and Zaporozec (1994). The ways they are incorporated into groundwater protection strategies are discussed in Chapter 17.

The hydrogeological setting methods use the comparison of a subject area to criteria that are judged to represent vulnerable conditions in other areas. A hierarchical system of two or usually more classes is established to cover the range of vulnerability. These widely used methods evaluate the vulnerability of hydrogeological complexes and settings, using an overlay cartographic method (Albinet and Margat, 1970). These methods produce universal systems suitable for large areas and a variety of hydrogeological conditions, and therefore lend themselves to the production of maps of large land areas, including national coverage.

The second group, parametric methods, can be further subdivided into matrix, rating or point count systems, although the overall approach is the same. Factors judged to be representative of vulnerability are selected, and each has a range that is divided into discrete hierarchical intervals (e.g. 0-5 m, 5-10 m, 10-20 m depth to water), and each interval is assigned a value reflecting its relative degree of vulnerability. Matrix systems are based on a limited number of factors for example two to four classes of vulnerability of soil and aquifer types, two or three intervals of depth to water. Vulnerability mapping of the United Kingdom follows this approach (Adams and Foster, 1992). The Jordan case study presented in Section 8.1.5 is an example of a rating system. Rating systems are largely derived from the work of Le Grand (1983). A fixed range is given to any parameter considered necessary and adequate for vulnerability assessment. The range is divided according to the variation interval of each parameter and the sum of the ratings for each parameter provides the vulnerability assessment for any point or area. The range of final scores is divided into segments representing relative vulnerability. Some rating systems use primarily soil parameters, others hydrogeological factors, but they are generic and not specific to any pollutant.

An index-based parametric method developed by Foster and Hirata (1988) has been applied in several Latin American countries (Foster *et al.*, 2002). The GOD system uses three generally available or readily estimated parameters, the degree of **G**roundwater hydraulic confinement, geological nature of the **O**verlying strata and **D**epth to groundwater. Each is rated on a vulnerability scale of 0 (lowest) to 1 (highest), and multiplied together to provide an overall index of pollution vulnerability. This method has recently been further developed and adapted to take account of the soil component of vulnerability (Foster *et al.*, 2002). Another approach specifically for karstic areas is the EPIK method developed by Doerfliger and Zwahlen (1998) and discussed by Daly *et al.* (2001).

Point count systems are a further development in parametric approaches in which a weighting factor or multiplier is added to represent the importance of each parameter in the vulnerability assessment. The ratings for each interval are multiplied by the weight for the parameter and the products summed to obtain a final numerical score, which is higher for greater vulnerability. However, a potential problem with this approach is the breaking down of the final scores into classes of relative vulnerability. One of the best known point count systems is DRASTIC, developed by the US EPA (Aller *et al.*, 1987; US EPA, 1992), which has been widely tested (Kalinski *et al.*, 1994; Rosen, 1994). The method employs seven hydrogeological factors to develop an index of vulnerability:

- Depth to water table
- net Recharge
- Aquifer media
- Soil media
- Topography (slope)
- Impact on the vadose zone
- hydraulic Conductivity

An index is generated by applying a weight to each hydrogeological factor that is represented numerically. As the hydrogeological factors vary spatially, the DRASTIC index provides a systematic way of mapping the relative vulnerability of groundwater to contamination and can be readily incorporated into a GIS (Kim and Hamm, 1999; Shahid, 2000). However, the large number of parameters included means that data requirements are invariably difficult to meet. Further, the large number of variables factored into the final index number may mean that critical parameters may be subdued by other parameters having little or no bearing on vulnerability in that particular setting. Some DRASTIC parameters, such as aquifer and soil media and hydraulic conductivity, are not fully independent but interact with each other.

Whichever system is used, the primary sources of data for assessing aquifer vulnerability are soil and geological maps and cross-sections, data or maps of depth to groundwater, supplemented by information from existing hydrogeological investigations that can provide additional information on subsurface transport and attenuation properties (Table 8.3).

Overall, allowing for the cautionary words at the beginning of this discussion, the concept of groundwater vulnerability has become both broadly accepted and widely used (NRC, 1993; Vrba and Zaporozec, 1994). Vulnerability maps should not be used to assess hazards where pollutants are discharged deeper into the subsurface, for example leaking tanks and landfills, or for spillages of DNAPLs. A further note of warning is that for most methods the resulting assessment of vulnerability applies only to the aquifer closest to the ground surface if there is more than one aquifer in a vertical sequence. While this is often the most important for local water supply, deeper aquifers may also be exploited. At first sight, such aquifers may appear to be more than adequately protected from pollution, but they may in fact be vulnerable to downward leakage of pollutants, which can be induced by pumping from the deeper horizons, or to pollutants moving laterally from a more remote source.

8.1.5 Case study: Groundwater vulnerability mapping in the Irbid area in Northern Jordan

The method applied

The method established by the State Geological Surveys of Germany (Hölting *et al.*, 1995) for the preparation of groundwater vulnerability maps uses a rating system for the properties of the unsaturated zone. The degree of vulnerability is specified according to the protective effectiveness (the ability of the cover above an aquifer to protect the groundwater) of the soil cover down to a depth of one metre (the average rooting depth), and the rock cover (the unsaturated zone). The following parameters are considered for the assessment of the overall protective effectiveness: effective field capacity of the soil, percolation rate factor, rock type and thickness of the rock cover above the saturated aquifer. Additional positive weightings are given for perched aquifer systems, and for aquifers under strong hydraulic pressure and upward flow conditions.

The process of calculating the overall protective effectiveness for a large area is complex and requires the use of Geographical Information System (GIS) software. Hölting *et al.* (1995) distinguished five classes of overall protective effectiveness of the soil and rock cover (Table 8.4). The higher the total number of points, the longer the approximate residence time for water percolating through the unsaturated zone and in consequence the greater the overall protective effectiveness.

The Irbid area was selected to test vulnerability mapping for the first time in Jordan, and this method was selected since it allows assessment of groundwater vulnerability over large areas based on existing data, i.e. at low cost and in a short amount of time.

Table 8.4. Classes of overall protective effectiveness of soil and rock cover and corresponding water residence time in the unsaturated zone (based on Hölting *et al.*, 1995)

Overall protective effectiveness	Total number of points	Approximate residence time in the unsaturated zone
Very high	≥4000	>25 years
High	>2000- 4000	10-25 years
Moderate	>1000-2000	3-10 years
Low	>500-1000	several months to about 3 years
Very low	≤500	a few days to about one year, in karstic rocks often less

Features of the study area

The study area is characterized by altitudes varying from more than 1100 m above sea level in the Ajlun Mountains in the south to more than 200 m below sea level in the Yarmouk valley in the north. Towards the Yarmouk River and the Jordan valley, the wadis are steeply incised and slopes exceeding 30 per cent are common. Part of the study area is intensively cultivated and industrial development is expected to increase rapidly in the future. The climate is semi-arid with annual rainfall ranging from less than 200 mm in the east to more than 500 mm in the area west of Irbid.

The main aquifer of the Jordanian Highlands is the moderately to highly fractured and moderately karstified A7/B2 (limestone) aquifer with a total thickness of 300-500 m (Figure 8.2). In the western and northern directions, the A7/B2 aquifer is covered by the predominantly marly B3 aquitard with a thickness increasing from some 100 m in the east to more than 500 m towards the Jordan and Yarmouk valleys. In the northern half of the study area the moderately fractured but almost unkarstified B4/B5 (limestone) constitutes the uppermost aquifer overlying the B3 aquitard (Figure 8.2). Its thickness may exceed 200 m.

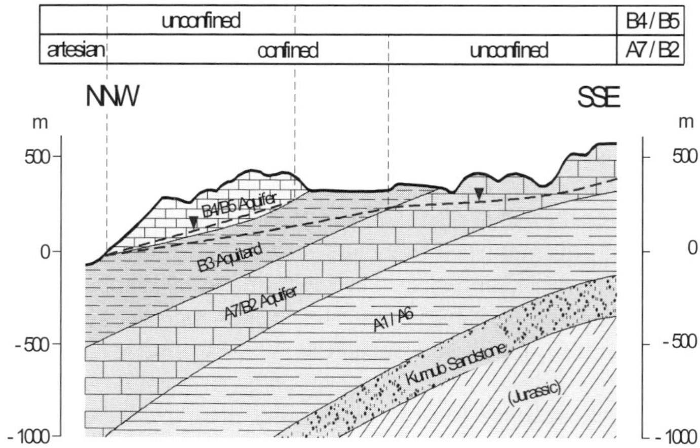

Figure 8.2. Schematic hydrogeological cross section through the study area (adapted from Margane *et al.*, 1999)

In areas where the A7/B2 aquifer is protected by the overlying B3 aquitard, observed groundwater nitrate concentrations are usually below 15 mg/l and often below 1 mg/l. At a few sites of uncovered A7/B2 aquifer, nitrate concentrations above 80 mg/l indicate anthropogenic contamination. In the B4/B5 aquifer, however, many springs cannot be used for public water supply because of chemical or bacteriological contamination. At some intensively cultivated sites nitrate concentrations exceed 100 mg/l.

The resulting vulnerability map

Figure 8.3 shows the resulting groundwater vulnerability map of the Irbid area. The protective effectiveness of the soil and rock cover above the saturated B4/B5 aquifer has been classified as low to very low. In the main wadis and in the areas where the groundwater is close to the ground surface, the vulnerability of the groundwater is extremely high. Protective effectiveness is classified as moderate only on the high plateaus between the deeply incised wadis running towards the Yarmouk River in the north. Vulnerability of the A7/B2 aquifer is especially high in areas where an effective soil cover is missing, the groundwater table is comparatively shallow and the aquifer is unconfined. Areas of medium vulnerability are widely distributed on the outcrop areas of the A7/B2 aquifer in the southern part of the mapped area.

Further north and west, where the A7/B2 aquifer is overlain by the predominantly marly B3 aquitard and well developed soils, the protective effectiveness of the soil and rock cover has been classified as high and, in the areas where the groundwater is confined, as very high. The Yarmouk Valley in the extreme northern part of the study area, where the B4/B5 unit has been eroded and the highly confined A7/B2 aquifer forms the uppermost aquifer, also belongs in this category. Associated mapping of potential groundwater pollution sources in the same area was also part of the study (Margane *et al.*, 1999).

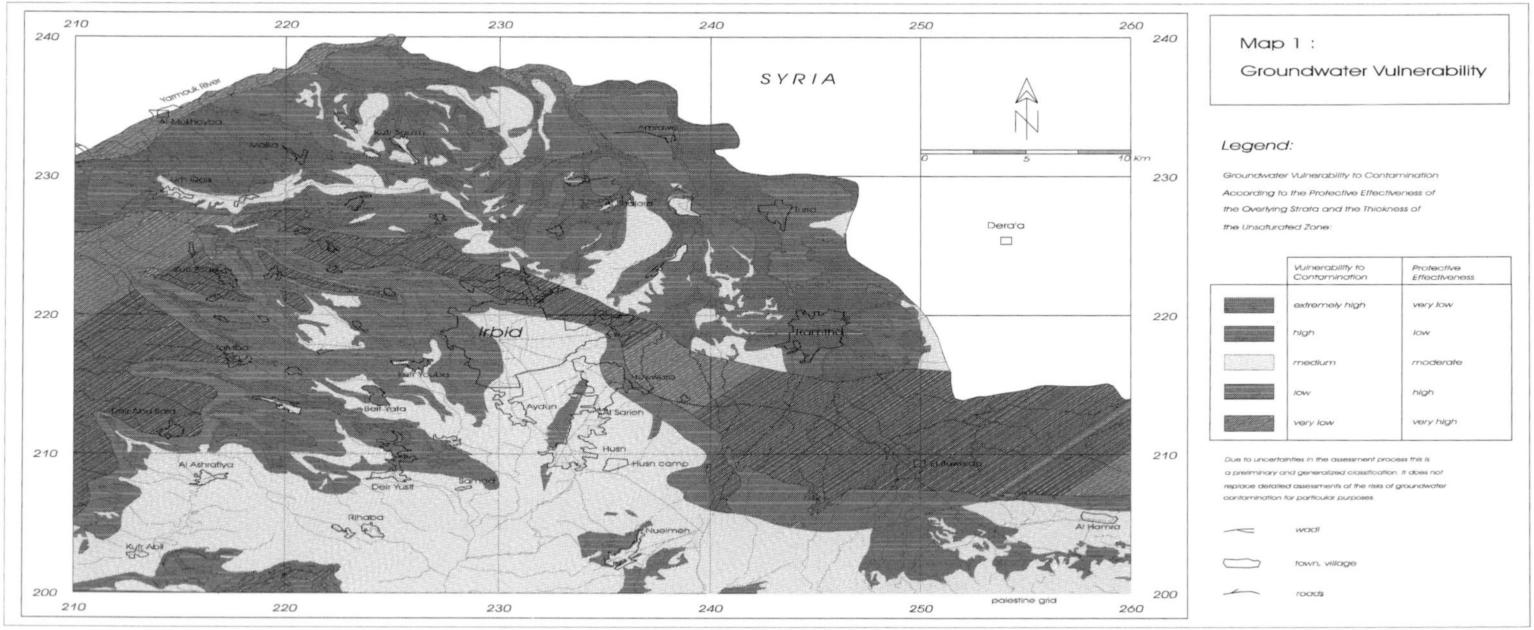

Figure 8.3. Groundwater vulnerability map of the Irbid area, northern Jordan (adapted from Margane *et al*., 1999)

8.2 INFORMATION NEEDS AND DATA SOURCES FOR VULNERABILITY ASSESSMENT

8.2.1 Regional geological and hydrogeological setting

The first important step in characterizing the physical environment for groundwater protection is to define the principal features of the regional geology so that the dominant aquifer types and hydrogeological settings outlined in Chapter 2 can be identified. The best sources of information from which the aquifer types can be defined are geological maps, which in most countries are produced by the national geological survey organization. Where these are published, printed and sold, they provide a cheap and usually easy to obtain source of this basic but vital information. However, these maps, which are often accompanied by descriptive memoirs, are prepared by, and mainly for, geologists, and this is the reason that at least some knowledge of the technical terms is required, as mentioned in Chapter 2. The associated descriptive notes or memoirs can be very useful as they usually include cross-sections showing the geometry, structure, dip and orientation of the various geological formations, from which the first indications of the likely groundwater flow system can be obtained. In some countries, such memoirs also include a general summary of the hydrogeology and groundwater development of the area covered.

An alternative and often better source of basic information about the groundwater conditions in an area is a hydrogeological map, if such exists. Their production and usage has been promoted for many years by UNESCO, which produced a universal legend to assist in the preparation of hydrogeological maps that could be easily compared with each other (UNESCO, 1970). As a result, national and regional maps showing the distribution of productive aquifers and less productive lower permeability materials now exist for much of the world. The maps distinguish between aquifers in which groundwater flow is dominantly intergranular and those in which it is dominantly through fractures, and also indicate the distribution of the main lithological types. Groundwater level contours provide a general indication of flow directions. The maps can also, therefore, be used as a source of information from which to develop a conceptual model of the groundwater flow regime. The principal groundwater supply sources – boreholes, springs or wellfields are sometimes shown. Groundwater quality information is usually restricted to indications of general groundwater salinity. Struckmeier and Margat (1995) provide a comprehensive list of hydrogeological maps, together with guidance on map preparation and a revised standard legend.

Issues of scale of information availability in relation to the scale of interest can be important. National geological mapping is often undertaken in the field at scales ranging from 1:10 000 to 1:50 000, and the final maps are usually produced at 1:50 000 to 1:100 000, or even broader scales for large countries which are being mapped for the first time. Maps of such scale can provide information about the distribution of the main rock types and may give a preliminary indication of their nature as either granular or fractured aquifers. The maps may, however, be rather scientific and technical in the names and ages of the rock types depicted, and some help and interpretation from a geologist or hydrogeologist will probably be required. Additional information and first-hand knowledge and experience can often be obtained from local technical staff of, for example, the water utility operating the groundwater supply sources which require protection. Hydrogeological maps may have more simplified depictions of the geological units and are often prepared at even broader scales, 1:250 000 being common. With increasing usage of digital technology, it may be possible to obtain digital geological or hydrogeological maps to be used as layers in a GIS approach to depicting groundwater vulnerability and planning groundwater protection.

8.2.2 Groundwater flow systems

Having identified the overall hydrogeological setting and determined the lithology and geometry – the extent and depth – of the relevant aquifers in the area of interest or catchment, the next step is to develop a broad conceptual model of the groundwater flow system. In relation to groundwater protection, this means understanding where groundwater recharge occurs, how it moves and where it discharges, as this forms the hydrogeological basis for the source-pathway-receptor approach to considering pollution threats.

Figures 8.4 and 8.5 provide examples of conceptual groundwater models. The first example shows a sedimentary sequence from the north east of the United Kingdom in which several aquifers and intervening clay layers are dipping eastwards towards the coast. The main chalk aquifer, the minor aquifers of the Carstone and Roach Formation immediately below it and the Spilsby Sandstone (Figure 8.4) are confined. The general scarcity of boreholes in the area means that there are few groundwater level measurements and the positions of the respective piezometric levels are uncertain, as show in Figure 8.4. A component of groundwater discharge from the chalk occurs as springs at the buried cliff, and it is assumed that there is a component of vertical flow between the aquifers through the intervening clay strata.

In contrast to the largely natural groundwater flow system in Figure 8.4, Figure 8.5 shows a highly-developed shallow aquifer beneath the city of Merida in Mexico, which is characterized by very high hydraulic conductivity, low hydraulic gradient and water table within a few meters of the ground surface. The aquifer is an unconfined karstic limestone, with little or no soil cover, providing little scope for attenuation of pollutants, and hence is highly vulnerable to pollution. Rainfall infiltration is supplemented by additional water from leaking mains, wastewater from unsewered sanitation and storm drainage, more than doubling the annual recharge (Morris *et al.*, 2003). As a result, microbial pollution of the shallow aquifer is widespread (Morris *et al.*, 2003), and there is a danger that the increased recharge within the city boundary could change the shallow hydraulic gradient and allow pollution to move towards the public supply wellfields in the neighboring peri-urban and rural areas (Figure 8.5).

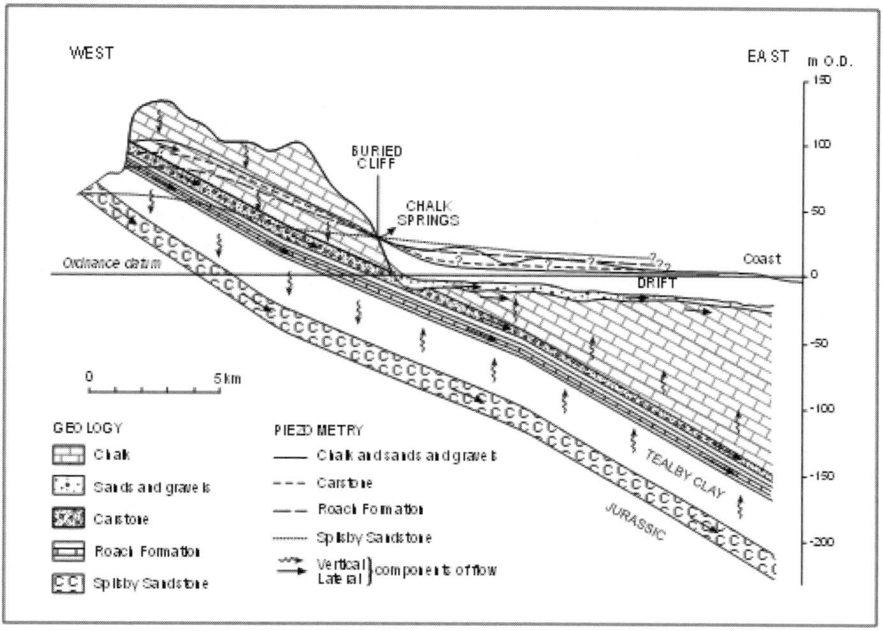

Figure 8.4. Conceptual model of natural groundwater flow system in eastern England (modified from Groundwater Development Consultants, 1989)

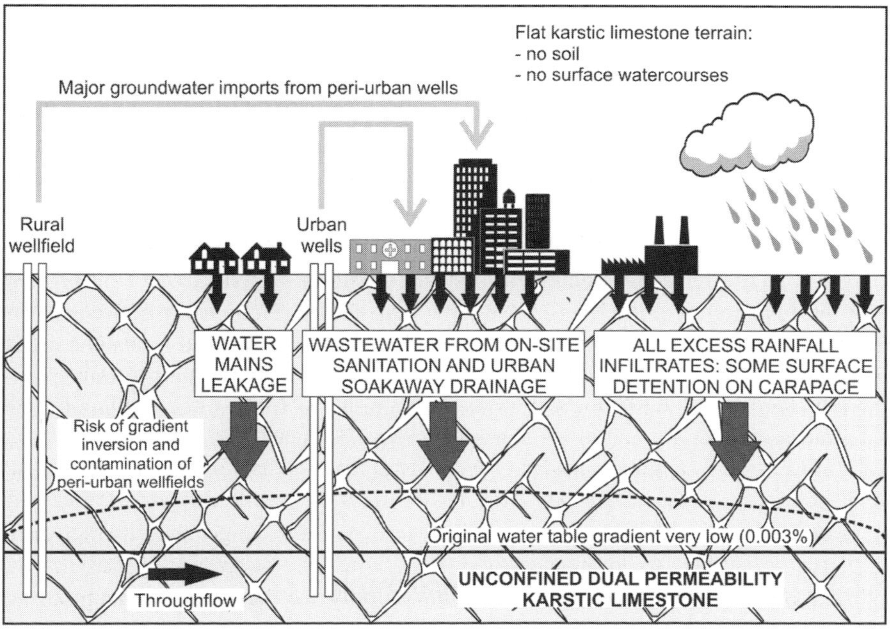

Figure 8.5. Conceptual model of groundwater system Merida, Mexico (modified from Morris *et al.*, 1994)

The geological history and structure of the region provides some of the most important information, especially about the nature of the boundaries of the aquifers. The geological age and stratigraphic sequence define the vertical distribution of aquifers and aquitards, and the folding history determines the way in which aquifer sequences are tilted and dipping. Geological faults with significant displacement can bring permeable and impermeable materials adjacent to each other and reduce or prevent lateral groundwater flow. The geological structure may in fact help to determine the limits of the catchment or recharge area. Understanding of the aquifer boundaries and their hydraulic nature should be part of the development of a three-dimensional conceptual model of the groundwater flow system.

As introduced in Chapter 2, natural groundwater flow directions generally reflect the land surface, and movement is usually from topographically high recharge areas to lower areas of groundwater discharge (Figure 2.7). While this generalization holds for many cases, the occurrence of separate local and regional groundwater flow systems operating at different scales and depths (Figure 8.4) may mean that groundwater flow can be in opposite directions at different depths, and sometimes contrary to the topographic gradient. Also, human activities can affect the flow system by providing additional recharge sources and by abstraction of groundwater modifying or even reversing groundwater flow directions (Figure 8.5).

In many situations, there will be scope for interactions between surface water and groundwater. Thus rivers and lakes may be either losing or gaining, i.e. water can move from the surface water body downwards into the ground, or groundwater may be discharged into the river. To make matters more complicated, the direction of water movement may be different along the length of the river or lake, depending on the topographic relationship and hydraulic connection and gradient between surface water and groundwater. Thus in some places where the stream or riverbed is cut down into a shallow aquifer, groundwater will flow towards the river and augment surface water discharge. Elsewhere, a stream or canal may be located above an aquifer and separated from it by either a significant unsaturated zone or impermeable materials. Slow infiltration of water from the river to the aquifer could occur. Further, the direction of groundwater movement may be reversed at different times of the year as the relationship between river levels and groundwater table can vary seasonally. Only in karstic limestone areas is surface water largely absent, as rainfall that does not evaporate infiltrates, and there is hardly any runoff. Indeed, extreme examples of surface water/groundwater relationships changing along a river course are seen where streams flowing over relatively impermeable materials cross a geological boundary onto karstic limestones and disappear completely.

8.2.3 Physical geography and topography

Having established the regional geological setting and consequent hydrogeological conditions and flow regime as outlined above, some other general physical features of the study area are also important. Useful information can often be gained from the topographic maps of the national survey, commonly at scales of 1:50 000 or 1:25 000, and from driving around and looking at the area, and this should always be done. One of the main features directly linked to geology that is of interest from the point of view of this monograph is the occurrence of minerals and the related mining activities, which are discussed in Chapter 11.

Geology and present or past climate interact to define the topography and geomorphology, the hills, mountains, valleys, rivers and lakes and other physical expressions of surface landforms. Steepness of slopes helps define runoff to rivers and concentrates and localizes recharge to groundwater. The configuration of the drainage system defines individual catchments and sub-catchments, and may help to indicate whether there is likely to be close interaction between surface water and groundwater systems. Limestone terrains are, for example, often characterized by a lack of surface drainage, and this often shows on topographic maps. The presence of springs, caves, swallow holes, often marked by their own symbols, also provides an indication of limestone and rapid conduit flow systems. Even the names of villages, farms and natural features such as hills and rivers can provide useful information about the area.

The interaction in turn between physical geomorphology and climate defines soil conditions and fertility, and hence land use and human activities, population density and distribution. As well as controlling runoff and recharge as mentioned above, steepness of slopes also plays a key role in land use – steep slopes may be unsuitable for both cultivation and human settlement. Topographic maps can indicate the main features of land use such as forests, orchards, artificial drainage, irrigated farming, glasshouse cultivation, nature reserves and other protected areas, but actual visits and looking may be required to determine the type of crops grown, cultivation regimes and livestock farming and to provide the more detailed information specified in Chapter 9. The patterns of rural, periurban and urban settlement, and transport infrastructure such as roads, motorways, railways and airfields are also apparent from topographic maps. This overview will, however, need to be supplemented by visual inspection and specific surveys to see the types of industries and their age and degree of activity, as described in Chapter 11.

8.2.4 Characteristics of the soil

The soil is the uppermost layer of the earth's crust and is the product of complex interactions between climate, living organisms, parent material and topography. Soils develop through the accumulation of unconsolidated mineral grains from the physical and chemical weathering of rock fragments and the addition of organic material from vegetation. Soil is defined and described in many ways, which differ according to the interests and requirements of the user. For the purposes of groundwater protection, it can be considered as the weathered zone into which plants will root and which experiences seasonal changes in moisture content, temperature and gaseous composition. In temperate regions, it is generally 1-2 m thick and in tropical regions can exceed 5 m. It should already be clear from

Chapters 3 and 4 that the soil is an important factor in groundwater protection, because it is the most chemically and biologically active part of the subsurface environment.

The characteristics of the soil at any particular location and time depend on five main groups of factors that have helped to produce it (Palmer *et al.*, 1995):

- physical and chemical constitution of the parent material;
- past and present climate;
- relief and hydrology;
- length of time during which soil forming processes have operated;
- the ecosystem, including the modifying effects of man's activities.

To provide a consistent and systematic basis for differentiating the characteristics and properties of soils, soil scientists develop classifications which group soils that behave in similar, and therefore predictable ways. Maps of soil types (usually called Soil Series) are based largely on the following observable or measurable criteria (Palmer and Lewis, 1998):

- texture of the whole soil profile;
- soil water regime – depth to and duration of waterlogging in a soil;
- substrate type – the underlying geological material from which the soil has developed;
- organic matter content.

In the United Kingdom, this approach defines the 725 Soil Series used to produce a national soil map. Given the complexity of interactions that are possible between the five groups of factors listed above, it is clear that the resulting spatial distribution of soil series within the landscape can be very complex. It may be difficult to map the variations adequately even at a scale of 1:10 000. For maps at a scale of 1:100 000 or smaller, which may be the chosen publication scale in many instances, soil series which are so intricately mixed within the landscape that they cannot be represented separately may be grouped together into soil associations. These usually reflect the same parent material (and hence the same underlying aquifer), but differ in characteristics related to texture, relief and hydrological conditions.

So that soils can contribute to the assessment of groundwater vulnerability, these series have then been classified according to their potential for allowing pollutants at the ground surface to be leached into underlying aquifers. The classification is based on knowledge of those physical and chemical properties routinely measured during soil surveys. These properties include texture, stoniness, organic matter content, presence of raw peaty topsoils and low permeability layers, and soil water regime, and they will determine the soil's tendency to encourage lateral movement of pollutants, speed of downward pollutant movement and capacity for attenuation and degradation of pollutants by the processes outlined in Chapters 3 and 4. Derivation of the resulting leaching potential classification is described by Palmer *et al.* (1995) and summarized in Table 8.5.

The nature of the soil is important in two other respects. Firstly soils have a direct influence on land use, especially in conjunction with climate, and help to determine the distribution of human activities. The deepest and most fertile soils are generally used for cultivation with or without irrigation, and poorer soils for forestry, grazing and wildlife conservation. Soil type hence influences the distribution of potential pollutants. Secondly, soil properties that affect leaching potential also have a bearing on the mechanisms and amounts of recharge to groundwater.

Table 8.5. Soil leaching potential classification for groundwater vulnerability (based on Palmer and Lewis, 1998)

High soil leaching potential (four sub classes, with H1 having the highest potential)	
H1 Soils with groundwater at shallow depth Soils with rock, rock-rubble or gravel at shallow depth Undrained lowland peat soils	**H3** Sandy soils with moderate topsoil organic matter content Soils with rock, rock-rubble or gravel at relatively shallow depth
H2 Sandy soils with low topsoil organic matter content	**HU** Soils in urban environments and sites of restored mineral workings where H1 leaching potential is assumed until proven otherwise
Intermediate soil leaching potential (two sub-classes: one for mineral soils / one for peaty soils or humose mineral soils)	
I1 Deep loamy and clayey mineral soils unaffected by marked seasonal waterlogging	**I2** Drained peat with humified topsoil Soils with a humose topsoil
Low soil leaching potential (no sub-classes)	
L Soils with a slowly permeable layer restricting downward water movement Upland soils with a raw peaty topsoil	

8.3 ESTIMATING GROUNDWATER RECHARGE

In characterizing catchments or aquifers for protection, understanding how and where recharge occurs is necessary for three principal reasons:

- the relationship between the amount of recharge and the amount of abstraction defines the land area subject to or receiving the recharge that needs to be protected;
- the locations and processes of recharge and their relationship to potential sources of pollution help to determine pollutant loads;
- the relationship between the amount of recharge and the amount of abstraction helps to define the susceptibility of the aquifer to the effects of excessive pumping.

The distinction between the last two is important. Thus in relation to the objective of groundwater protection, it may often be more critical to identify locations, mechanisms and speed of recharge rather than total volumes. General estimates of total recharge volumes are needed to help define catchments and to estimate diffuse pollution loads. A greater degree of effort is required to make estimates that are as precise and reliable as possible for groundwater resources management. Recharge estimation can be technically difficult and costly.

8.3.1 Recharge components and processes

Recharge of groundwater may occur naturally from precipitation, rivers or lakes and/or from a whole range of man's activities such as irrigation and urbanization. Further, an important way of categorizing recharge is to consider it as direct, localized or indirect (Simmers, 1997). The first is defined as water that is in excess of soil moisture deficits and evapotranspiration and which is added to the groundwater reservoir by direct vertical percolation through the unsaturated zone. The second is an intermediate form of recharge that results from percolation to the water table following surface or near-surface movement and subsequent collection and ponding in low-lying areas and in fractured zones as a result of small-scale topographic or geological variability. Indirect recharge is percolation to the water table through the beds of rivers, lakes and canals (Figure 8.6).

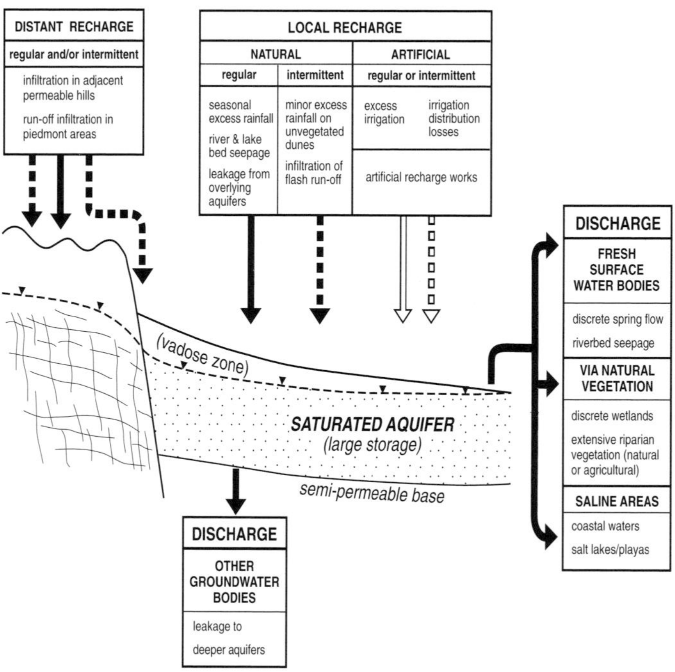

Figure 8.6. Components of groundwater recharge (Foster *et al.*, 2000)

While important conceptually, these distinctions are in practice a simplification of complex natural environments in which both may occur. However, comprehensive reviews of the subject by Lerner *et al.* (1990) and Simmers (1997) concluded that the following general guidelines were evident from the literature:

- recharge occurs, albeit to a limited degree, even in the most arid environments although increasing aridity will be characterized by a decreasing net downward flux and greater time variability;
- direct recharge is likely to become less important and indirect recharge more important with increasing aridity;
- estimates of direct recharge are likely to be more readily derived than those of localized or indirect recharge.

These generalizations certainly show that successful estimation of groundwater recharge depends on first identifying the probable recharge mechanisms and the important features influencing recharge, and secondly on selecting an estimation method which is suitable for the environment. Even with this understanding, recharge estimation remains one of the most difficult tasks for the groundwater specialist and, in many circumstances, groundwater recharge has proved much more difficult to measure than other components of the hydrological cycle.

8.3.2 Methods for estimating recharge

While comprehensive technical guidance on the estimation of recharge is outside the scope of this monograph, nevertheless the features that good methods of recharge estimates should have, and the most likely sources of error can be summarized. Simmers (1997) identified four general sources of error:

- *Adopting an incorrect conceptual model.* This is the most common and serious source of error, and arises when the groundwater flow system and recharge processes are not fully understood or the simplifying assumptions made are too great or unsound. It is important on the one hand to take account of all of the natural and artificial sources of recharge and on the other hand to avoid double accounting of any sources.
- *Neglecting spatial and temporal variability.* A particular rainfall amount may not cause recharge if it falls at low intensity during times of high evapotranspiration, but the same amount could produce recharge if it occurred with high intensity when evapotranspiration was low. Major errors can arise if temporal variation is not taken proper account of by using monthly, annual or longer-term average data. Recharge estimates over long periods should be obtained from the sum of values over shorter periods – soil moisture balances based on monthly data may indicate no recharge, especially in arid and semi-arid areas, whereas daily time steps will often show that recharge can occur. The high degree of spatial heterogeneity of soils and aquifers will also limit the degree to which estimates at one location can be applied regionally.
- *Measurement error.* This is governed by the equipment used and operator skills and, for those parameters that are readily measured such has rainfall, is unlikely to be as important as either of the two above.

- *Calculation errors.* These can usually be avoided by taking care and by checking that the units in which the various parameters used are either compatible or correctly converted.

Further discussion of these sources of error is given in Lerner *et al.* (1990). There are five features that should be looked for in a recharge estimation method (Lerner *et al.*, 1990; Simmers, 1997):

- The method should explicitly account for the water that does not become recharge.
- Most methods rely on knowledge of the processes that convert source water into recharge and of the flow mechanisms for that water. Good methods should reveal if the conceptual model underlying the method is correct.
- The method should have low errors associated with it and should not be sensitive to parameters which are difficult to measure or to estimate accurately.
- The method should be easy to use.
- Methods utilizing readily and widely available data, such as rainfall are more useful than those requiring specialized observations.

The applicability of a number of recharge methods is shown in Table 8.6. The time in the last column of the table refers to the typical period over which data are needed to apply the method. For those methods that are event, season or yearly based, the actual data for parameters used may be needed at time steps ranging from hourly through daily and weekly to monthly, depending on the hydrologic setting and the precision required. Large volumes of existing data for things like rainfall, potential evapotranspiration and river flow may need to be collected from the appropriate authorities.

Table 8.6. Direct techniques for recharge estimation (modified from Foster *et al.*, 2000)

Technique	Applicability	Cost range	Time
Soil water balance from hydro-meteorological data	D(L)	+[1]	ESYH
Hydrological data interpretation			
water table fluctuations	D(L)	+[1]	YH
differential stream/canal flow	L	+[1]	E
Chemical and isotopic analyses from saturated zone	D and L	+-++[2]	HG
Chemical and isotopic profiling of unsaturated zone	D[a,b]	++-+++[2]	HG
Soil physics measurements	D[a]	+++	SY

D/L: diffuse(direct)/localized(indirect) distribution of recharge; a: only suitable for relatively uniform soil profiles; b: not appropriate for irrigated agricultural areas; +, ++, +++: approximate relative range of costs; 1: excluding construction and operation of basic data collection network, which is assumed to exist already; 2: isotopic analyses increase the costs substantially; E: event; S: season; Y: year; H: hydrological time; G: geological time.

Most of the methods listed in Table 8.6 have some but not all of these features, and it is therefore desirable to apply and compare a number of independent approaches. The applicability and potential accuracy of these methods depends on the superficial environment of geology, geomorphology, soils and the climatic and hydrological regime. These factors determine the spatial variability of recharge processes, the distribution and extent of runoff, and the characteristics of the vegetation cover, particularly whether it is natural or agricultural and whether the latter is with or without irrigation. An initial indication of how variable the recharge is likely to be can be gained from the overview of the physical conditions of the area of interest outlined above. It is important also to appreciate that average groundwater recharge is not necessarily constant over time, as changes in land use, irrigation infrastructure (such as construction and operation of canals), irrigation technology and cropping regimes can produce significant changes in recharge rates and also affect the quality of the infiltrating recharge water, as shown in Chapter 9.

Two other major human activities can greatly influence recharge processes. Firstly, the covering of the land by impermeable surfaces in urban areas reduces direct infiltration, can lower evaporation and increase and accelerate surface runoff. Depending on the specific arrangements for storm drainage, overall groundwater recharge may be increased or decreased (Foster *et al.*, 1998). Leaking water mains, on-site sanitation and leaking sewers also contribute urban recharge of differing water quality, and the preliminary physical appraisal of the catchment referred to in Chapter 6 will indicate whether a more detailed consideration of urban recharge processes, as given in Lerner *et al.* (1990) and Foster *et al.* (1998) is required to support the assessment of urban impacts in Chapters 10 to 12.

Secondly, because of ever-rising water demands and increasing scarcity of freshwater resources, artificial recharge is becoming more widely used as a component of overall resource management to maintain resources, restore groundwater levels or prevent saline intrusion (Chapter 19). Various sources of water are used in artificial recharge schemes, including rainfall harvesting and collection, river water, mains water, groundwater and reclaimed wastewater, and a wide range of technologies is employed. In general, artificial recharge may improve groundwater quality as well as increasing the available storage in an aquifer. Reclaimed wastewater, however, may contain chemical or microbial pollutants that are poorly characterized but could impact human health. Primary, secondary or tertiary treatment progressively removes these pollutants, provided the treatment facilities are of sufficient capacity, properly designed and correctly operated. Reclaimed water has been used to augment groundwater supplies in the southern USA and in Israel for many years. These are well monitored, high technology installations, but there is increasing interest in the use of untreated or partially treated wastewater to augment scarce water resources, either by direct infiltration or through irrigated cultivation (Chapter 9), and an understanding of the likely pollution threats is therefore required.

8.4 NATURAL HYDROCHEMICAL AND GEOCHEMICAL ENVIRONMENTS

Natural quality varies from one rock type to another, and also within aquifers along groundwater flow paths, as described in Chapter 4. It is important for those with responsibility for protecting groundwater quality to be aware of the geological environments in which naturally-occurring substances are likely to exceed drinking-water criteria so that groundwater is properly tested for these substances and, if necessary, adequately treated to ensure that water is safe for potable use.

The potential for a naturally occurring chemical constituent to pose a threat to public health from drinking-water depends on the distribution of the constituent in the environment, and on the extent to which the physical and chemical environmental conditions ensure that the constituent has a high solubility, and remains soluble. The geology of an area fundamentally controls the distribution of chemicals in the environment, as particular chemical constituents are generally associated with particular rock types. Very high concentrations can occur in rocks associated with specifically mineralized areas (Chapter 11). Climate also plays an important role in controlling the way that rocks are broken down, and climatic factors influence soil forming processes and the extent to which specific constituents are either concentrated in soil profiles, or are leached into rivers or groundwater.

The geological environments from which the most important health-related chemicals are derived are shown in Table 8.7. Further information about the most important individual chemicals is provided in Chapter 4.

Table 8.7. Environmental factors affecting the distribution of naturally occurring toxic chemicals in water and soil

Geological setting[1]	Climate	Possible health-related constituents in soil and water
Felsic igneous rocks (e.g. granites, pegmatites)	Humid, arid	As, Ba, B, Mo, F, Pb, Rn, U; concentrations of B, F, U likely to be higher in drier areas
Alkaline igneous and volcanic rocks	Humid, arid	As, Ba, B, Mo, F, Pb, Rn, U
Mafic and ultramafic igneous and volcanic rocks	Humid, arid	Co, Cr, Ni, SO_4^{2-}
Contact metamorphic rocks	Humid, arid	Mo, U
Iron-rich sedimentary rocks (e.g. feruginous sandstones, siltstones)	Mainly arid	As, Co, Ni, Se
Manganese rich sedimentary rocks	Mainly arid	As, Ba, Co, Mo, Ni
Phosphorus-rich sedimentary rocks (limestones, mudstones, siltstones)	Mainly arid	Mo, Pb, F, U
Black shales	Humid, arid	As, Mo, Ni, Pb, Sb
Sulphide mineralization	Humid, arid	Al, As, Co, Cd, Cr, Pb, Mo, Ni, Sb, Se
Gold mineralization	Humid, arid	As, CN, Hg
Alluvial plains, mainly in coastal areas	Humid, arid	As, Co, Cd, Cr, Pb, Mo, Ni, Sb, Se
All	Arid	NO_3; high concentrations may occur where there are leguminous plants (e.g. Acacia species, Box 4.1)
All	Humid	I; very low concentrations occur in areas of very high rainfall or very high relief

[1]: Geological association of inorganic constituents based on data presented by Rose *et al.* (1979).

8.5 CHARACTERIZING GROUNDWATER ABSTRACTION

The scale of groundwater abstraction and methods of abstraction used are also important factors in assessing groundwater pollution and protecting groundwater quality. In relation to the former, abstraction sources may create pathways for groundwater pollution, either directly via the borehole or well itself or through the aquifer because heavy and prolonged pumping can modify natural groundwater flow regimes. In relation to the latter, the type, scale and numbers of groundwater abstraction sources have a bearing on the way in which groundwater protection measures can be implemented. Further, the condition of wellheads will influence the potential for direct contamination of drinking-water during abstraction, as well as the potential for ingress of pollutants into the aquifer. The situation assessment should therefore include an appraisal of the condition of the wellhead and its surroundings.

8.5.1 Groundwater abstraction types

Groundwater abstraction takes many forms and employs a range of techniques. The use of traditional open dug wells goes back thousands of years, drawing water by hand or using animal power, and there are many parts of the developing world where these remain important supply sources. They are cheap, relatively easy to construct in unconsolidated aquifers and simple to maintain, and therefore remain popular in programmes in which community involvement is strongly promoted. They are, however, highly sensitive to pollution being directly introduced into the open top of the well or through the ground immediately around the well. This direct pollution can be greatly reduced by the use of proper sanitary seals and aprons around the well (Chapter 18), by covering them and by installing hand pumps. Drilled boreholes take many forms from very simple, narrow diameter holes for hand pumps producing less than 0.5 l/s to shafts up to 1 m across from which tens of litres per second can be abstracted for urban supply or irrigation (Driscoll, 1986), or wellfields of individual boreholes all connected up to a major supply, such as those of the Great Man Made River project in Libya. Groundwater is also abstracted from protected springs and from galleries such as the ancient qanats of the Middle East.

Clearly, large-scale groundwater abstraction requires major investment in drilling, borehole materials, pumps and power supplies and the associated pipelines and tanks. The loss of a major supply could be a significant problem to the operator if groundwater pollution were severe enough to render the water unusable. Treatment to remove pollutants, or the location and development of an alternative supply, both of which could be very costly, would be required, emphasizing the benefits of prevention by protecting groundwater.

8.5.2 Groundwater abstraction and pollutant pathways

Large and prolonged abstraction can modify groundwater flow rates and directions by reducing or reversing hydraulic gradients and producing cones of depression in the water table or piezometric surface around pumping wells and wellfields. These hydraulic changes can in turn create new pollutant pathways or modify existing ones. In multi-layered aquifer systems in urban areas, the uppermost zones have usually been developed first for groundwater supply, often with many shallow, relatively small boreholes and wells. These are typically privately owned and used for domestic, industrial and commercial supply, and often unregistered and uncontrolled. As cities have grown, the uppermost aquifer is also used, either deliberately or accidentally, as a receptor for urban waste and the shallow groundwater becomes more and more polluted, sometimes to the extent that it becomes unusable. Municipal authorities and other larger groundwater users consequently drill into deeper aquifers in search of better quality groundwater, and the increasing abstraction from depth can induce downward movement of polluted groundwater, threatening these deeper supplies, as described in Section 8.6.

In some circumstances, the very act of constructing wells or boreholes may in itself encourage groundwater pollution by puncturing protective layers above or between aquifers. There are examples both of boreholes permitting downward movement of polluted groundwater from shallow to deeper aquifers, and of deep boreholes penetrating confining layers and allowing naturally saline groundwater to move up into aquifers containing high quality water. Sometimes even observation boreholes for measuring groundwater levels or taking groundwater samples, if constructed without proper understanding of the three-dimensional hydrogeological conditions, can allow this to happen. Abandoned boreholes can, therefore, remain as a potential short-circuiting route for pollutants, and consideration may need to be given to sealing them to try to restore the protective layer. If disused or abandoned boreholes are used, perhaps covertly and illegally, for effluent disposal then their passive short-circuiting role can become an active one as a pollution source, releasing pollutants directly into what may be the most permeable and productive part of the aquifer. Backfilling and sealing may then be urgently required.

Within the situation analysis it is also necessary to assess wellhead protection/sanitary completion. It is important to bear in mind the source-pathway-receptor model introduced in Section I, as pollution at wellheads may require a number of factors to be present. These include hazards (i.e. sources of pollutants) and pathways (which often reflect specific problems with the infrastructure). In addition, it may also be useful to assess indirect or contributing factors (Howard, 2002). These do not either directly cause contamination or offer a pathway for the contamination to enter the groundwater source, but may contribute to the development of a pathway or lead to build up of contamination within the immediate vicinity of the borehole. Examples include aspects such as fencing around the groundwater source, allowing animals to gain access close to the source, lack of drainage to divert contaminated surface water from the wellhead area and deterioration in the engineering works at the wellhead (Table 8.8).

Table 8.8. Examples of pathways and contributing factors for microbial contamination (modified from ARGOSS, 2001)

Factors	Conditions facilitating pollutant ingress
Hazards that may cause contamination through direct ingress (hazard factors)	Open-air defecation Stagnant surface water uphill of the source Waste or refuse dumps Animal faecal matter stored above ground
Pathways for contaminants to enter the source (pathway factors)	Cracked lining Lack of seal on top of rising main Lack of cover on well Cracked or damaged apron or pump-house floor Lack of head wall on well Faulty masonry on spring protection works Eroded backfill or catchment area Rope and bucket used to withdraw water
Contributing factors to contamination (indirect factors)	Lack of fencing Lack of lockable pump-house Lack of adequate diversion drainage to remove surface water Lack of drainage to remove wastewater Animal access to source

8.5.3 Abstraction types and groundwater protection

The ways in which groundwater is abstracted need to be considered when thinking about its protection. Scale and distribution are particularly important. Taking the example of the United Kingdom, three large consolidated aquifers provide half or more of the public supplies in the south, centre and east of the country from several hundred high yielding boreholes. The United Kingdom's national approach to protecting groundwater supplies (Adams and Foster, 1992) designates zones from which the recharge is derived around these supplies, in which certain potentially polluting activities are prohibited or controlled. This and other strategies for groundwater protection are discussed in more detail in Chapter 17. While this approach is sometimes hydrogeologically problematic, given the complex local groundwater flow systems that are often encountered in aquifers, it is logistically and institutionally reasonable to establish protection zones around a relatively small number of large abstractions. This protection approach has been applied widely in Europe and North America to large groundwater supplies, both in the form of individual wells or boreholes and as wellfields – small groups of closely-spaced boreholes all drawing water from the same aquifer and feeding into a common pipeline to convey the water to where it is being used.

The situation would be very different where groundwater is drawn from a large number of much smaller groundwater supplies widely dispersed over the aquifer such as, for example, in India and Bangladesh and large parts of Africa. Not only is it more difficult to implement protection zoning on this broad scale, the shallow aquifers used for supply may be highly vulnerable to pollution, and many of the boreholes or wells may be privately owned. While this situation may occur in either rural or urban environments, protecting small groundwater supplies from the wide range of potential polluting activities in urban areas is especially problematic. Shallow hand dug wells are often particularly difficult to protect. The conventional approach of protection zoning cannot be easily applied, and the best strategy may be to ensure careful borehole or well siting in relation to pollution sources, together with good construction practice with adequate sanitary seals. This may be backed up a by a policy of resource protection, in which the whole aquifer outcrop, rather than defined zones around individual supplies, is subject to some degree of control of likely pollution sources. In practice, however, any groundwater protection measures may be difficult to implement for large numbers of small dispersed, perhaps private and usually unregistered supplies. Siting supplies to avoid pollution sources (and vice versa) and adequate sanitary protection should always be seen as important lines of defence in protecting health.

At the opposite end of the spectrum of groundwater supplies, in karstic areas large springs are often used. Because of the rapid groundwater flow and response times in karstic aquifers, these may be especially vulnerable to pollution. In addition, karst springs may be supported by groundwater recharge from large catchments that are notoriously difficult to define.

NOTE ▶ *Approaches to protecting groundwater quality need to be matched not only to the hydrogeological situation, but also to the types and scales of groundwater abstraction.*

8.6 SUSCEPTIBILITY OF GROUNDWATER RESOURCES TO DEGRADATION

8.6.1 Scope and scale of resource degradation

As a component of the overall sustainable management of water resources, i.e. the quantity of water available for use, is largely outside the scope of this monograph, but it is nevertheless necessary to comment on the potential impacts of groundwater usage on the overall resource situation. This is because heavy and prolonged groundwater abstraction and poor management of groundwater resources can have negative consequences for groundwater quality. These consequences can be severe and are often difficult and costly to manage and remedy. Also, lack of water availability for domestic uses can have severe public health consequences, and in such circumstances health authorities may need to ensure their needs are taken account of in overall water resources management. Approaches to managing water resources are dealt with in Chapter 19.

The availability of groundwater in an area affects current and future drinking-water quality either by the direct effects on groundwater quality or indirectly by changing the degree to which groundwater is available for use. For example, lack of future availability of groundwater may force communities to use surface water that is more contaminated or more difficult to access. Thus there is a need to interact with those outside the health sector involved in the management of groundwater to address all aspects of water management. It is also important to recognize that domestic use of groundwater for drinking-water and household purposes is but one use of groundwater. In many areas, withdrawals for other water-use sectors, particularly agriculture, but also possibly industrial uses, may far exceed withdrawals for domestic use. Worldwide, about 70 per cent of total water use (surface water and groundwater) is for agriculture, 22 per cent for industry, and 8 per cent for domestic purposes (Bowden, 2002). Heavy groundwater use for irrigation is commonly a major cause of groundwater level declines. The linkages between water quality and water quantity suggest that monitoring programmes for each should be integrated. Greater attention is needed to the long-term value of water-level data collected as part of water-quality monitoring and to the potential synergies between water quality and water level monitoring networks (Taylor and Alley, 2002).

Traditionally, surface water and groundwater have been treated as separate water resources, so that one could be utilized independently without affecting the other. With increased utilization of water resources, has come greater recognition of surface water and groundwater as fundamentally interconnected (Winter *et al.*, 1998). Depletion of one resource eventually results in depletion of the other, and likewise, contamination of one resource can contaminate the other. This recognition of groundwater and surface water as a single resource has increased the imperative for those involved in development of groundwater to interact with those involved with surface water, and vice versa. This is true from both water quantity and water quality perspectives.

Over-exploitation of groundwater resources or individual aquifers by uncontrolled, excessive abstraction, while often not precisely definable in scientific terms is nevertheless an emotive term when used at the political or institutional level (Foster *et al.*, 2000). At this level, concern relates more to the consequences of excessive abstraction rather than to the volumes of water themselves, although the latter need to be regularly monitored to determine whether control measures are proving effective. Many of the changes in response to groundwater pumping are subtle, and they may occur over long periods of time. Development of groundwater resources increasingly requires a more complete understanding of the effects of abstractions on groundwater systems (Alley *et al.*, 2002).

The impacts of excessive abstraction range from the often reversible interference with springs and other wells to much less reversible degradation caused by ingress of saline or polluted water and land subsidence (Foster, 1992;

Alley *et al.*, 1999; Custodio, 2002). While declines in groundwater levels and reduced spring flow or baseflow may be reversible in humid areas, groundwater mining in arid regions can be virtually irreversible (in the absence of artificial recharge) on any practical time scale. The cumulative effects of pumping can cause significant and unanticipated consequences when not properly considered in management plans. The possible consequences of large abstractions are summarized in Table 8.9 and in the text below, and the susceptibility of the broad classes of hydrogeological environments defined in Chapter 2 to some of these effects is shown in Table 8.10.

Table 8.9. Consequences of excessive groundwater abstraction (adapted from Foster *et al.*, 2000)

Consequences of excessive abstraction		Factors affecting susceptibility
Reversible interference	pumping lifts/costs increase borehole yield reduction spring flow/base flow reduction	aquifer response characteristic drawdown to productive horizon aquifer storage characteristic
Reversible/ irreversible	phreatophytic vegetation stress (both natural and agricultural) aquifer compaction/transmissivity reduction	depth to groundwater table aquifer compressibility
Irreversible deterioration	saline water intrusion ingress of polluted water (from shallow aquifer, river or canal) land subsidence and related effects	proximity of saline or polluted water vertical compressibility of overlying/interbedded aquitards

The reversible effects in the upper part of Table 8.9 result when an excessive number of boreholes or wells are constructed and heavily pumped, especially if the uppermost part of the subsurface geological sequence provides the most productive aquifer. The two effects in the middle of the table, vegetation stress and aquifer compaction with reduction in transmissivity, may be reversible or irreversible, depending on the hydrogeological conditions and the length of time over which the excessive abstraction has been established. The most serious effects, which are irreversible or nearly so, include the mechanisms of quality deterioration given at the bottom of Table 8.9.

8.6.2 Deterioration of groundwater quality

Groundwater abstraction can affect water quality in several ways. Perhaps best known is saltwater intrusion in coastal areas resulting from large withdrawals of groundwater. Some inland aquifers also experience similar problems, where withdrawal of good-quality water from the upper parts of aquifers allows underlying saline water to move upward and degrade water quality. Changes in the quality of water as a result of abstractions also can occur as water levels decline and the pumped water originates from different parts of the aquifer system, usually because of the downward movement of polluted water in response to pumping. Likewise, declining water levels can result in changing oxidation and other hydrochemical conditions, mobilizing or precipitating different chemical constituents. In addition, surface water can be drawn into the aquifer with poorer water quality or of a chemical composition that mobilizes naturally occurring chemicals.

There are numerous examples of saline intrusion where heavy groundwater abstraction from productive coastal limestone or alluvial aquifers for urban, industrial or agricultural usage has produced serious intrusion of saline water into these aquifers, often stretching far inland. Under natural conditions in coastal aquifers, fresh water derived from recharge overlies saline water in such a way that flow takes place towards the sea (Figure 8.7A). Their relative densities govern the position of the boundary between fresh and saline water. Under simplifying assumptions (homogeneous aquifer properties and no vertical gradients in heads), the depth of the interface below sea level can be assumed to be about 40 times the height of the fresh groundwater table above sea level. This is known as the Ghyben-Herzberg relationship, and is described in most hydrogeology textbooks (Freeze and Cherry, 1979; Domenico and Schwartz, 1998). On small islands such as those of the Caribbean and the Florida Keys the result is a lens-shaped body of freshwater which can be difficult to exploit without causing quality deterioration.

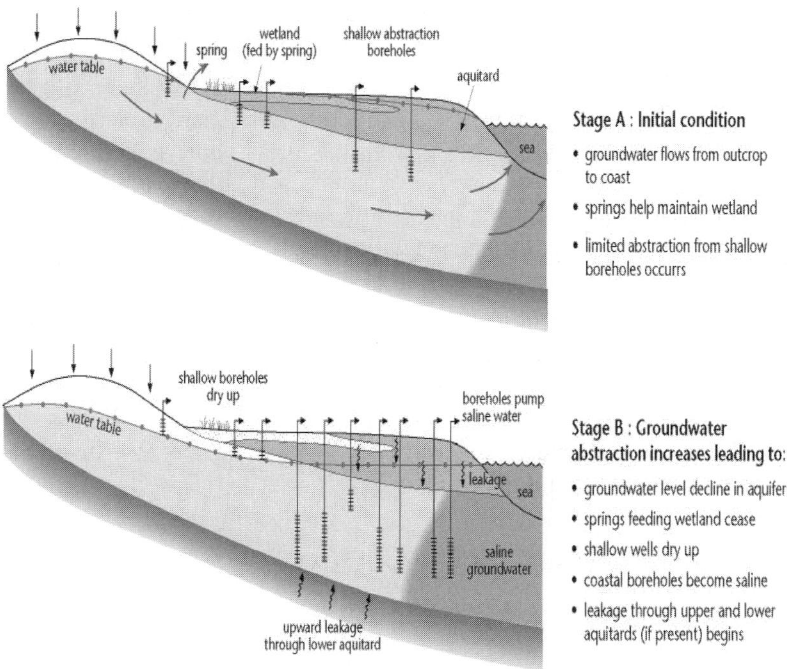

Figure 8.7. Impact of abstraction on a coastal aquifer system (Morris *et al.*, 2003; British Geological Survey ©NERC)

When pumping disturbs the natural conditions (Figure 8.7B), the consequent lowering of the freshwater table results in a corresponding movement of the freshwater-salt water interface which, given the relationship described above, is in the ratio of 40:1, i.e. 1 m of water level decline produces about 40 m of upward movement of the interface. In practice the interface is not usually sharp and uniform, but affected by dispersion, and in detail the three-dimensional picture of saline intrusion can be particularly complicated where rapid movement along fractures and fissures permits a complex fingering of saline water far inland. Given that coastal plains are often densely populated and agriculturally productive regions, it is not surprising that groundwater abstraction has grown rapidly. The resulting serious saline intrusion problems have been encountered in many aquifers, e.g. on the Mediterranean Coast of Spain, in the Netherlands, Mexico, Florida, Cyprus and some of the Caribbean islands. Even far from coastal regions, upconing of saline water from deeper aquifers can be caused by heavy groundwater abstraction from freshwater zones above.

Induced downward movement of pollutants is widely observed in many rapidly developing urban areas underlain by multi-layered aquifer sequences in which the establishment of large-scale abstraction from deeper aquifer horizons imposes or accentuates downward hydraulic gradients and induces accelerated downward movement of polluted water from shallower aquifers. In the city of Santa Cruz in Bolivia, for example, most private boreholes supplying water for industries, small businesses and private domestic use draw groundwater from less than 90 m depth, whereas the public supply is largely dawn from deeper aquifers between 90 and 350 m below ground. The shallow groundwater has become polluted (Figure 8.8) from poor waste disposal practices, and abstraction has induced downward movement of polluted water. This general situation is rather widespread, but the degree of protection of the deeper aquifers and likely timescale of any deterioration in the quality of the groundwater abstracted from them depends on the local hydrogeological situation, and will certainly need specific assessment.

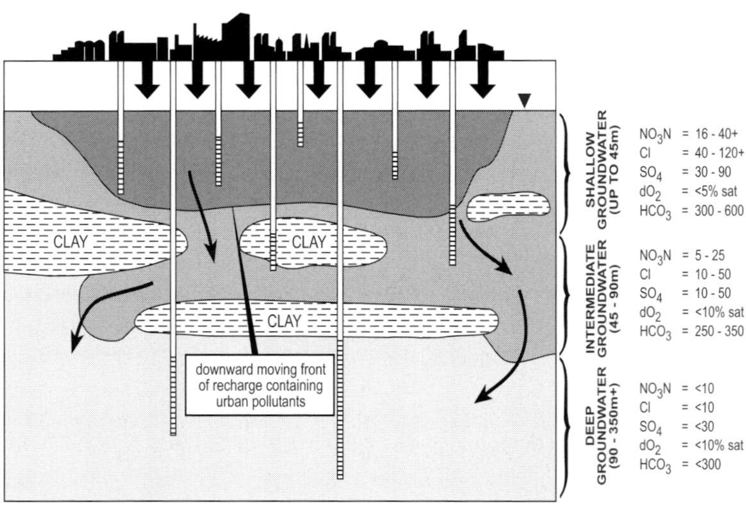

Figure 8.8. Schematic cross-section illustrating downward pollutant migration induced by pumping in a multi-aquifer sequence, Santa Cruz, Bolivia (modified from Morris *et al.*, 1994)

The second situation mentioned above, i.e. changes in the oxidation-reduction potential of the groundwater resulting in the mobilization of metals is illustrated by the situation in Hat Yai, Thailand. The city obtains about 50 per cent of its water supply from private boreholes drawing from an aquifer below a semi-confining layer consisting of about 30 m of silts. Seepage of organic-rich wastewaters from collection canals into the upper part of the underlying alluvial aquifer sequence has produced strongly reducing groundwater. This permits the release of naturally-occurring iron and manganese from the sediments and allows the build up of troublesome concentrations of ammonium, which are being drawn down to the abstraction boreholes (Figure 8.9).

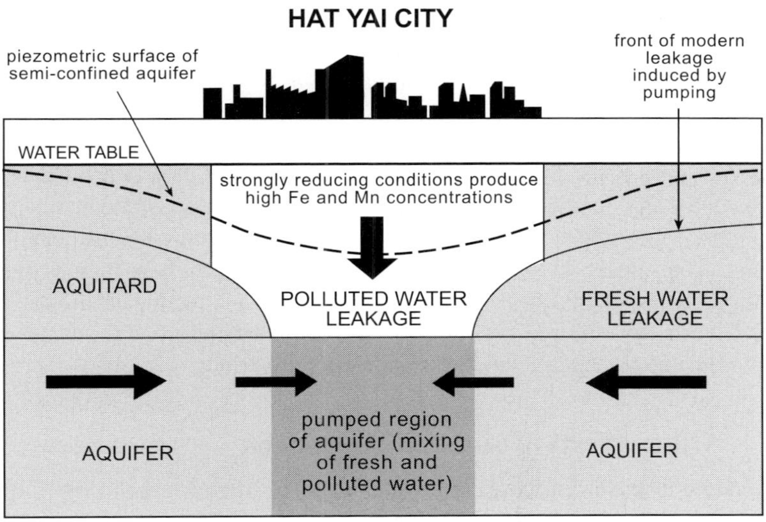

Figure 8.9. Groundwater quality degradation induced by pumping, Hat Yai, Thailand (Foster *et al.*, 1998)

This general picture of urban public supplies being vulnerable to polluted shallow groundwater is probably a widespread problem, and cities such as Nottingham (United Kingdom), Lahore, Karachi (Pakistan) and New Delhi (India) are either experiencing such pollution already or it can be anticipated in the future. The degree of protection afforded by any intervening lower-permeability strata, and the likely timescale of any impacts on deeper groundwater require specific investigations in each case. Similar problems can also be experienced where municipal water supplies are drawn from neighbouring rural areas in which urban wastewater is used for irrigation.

The linkages between groundwater quantity and quality management are well illustrated by the region west of Lake Michigan in the USA, an area that includes the major cities of Chicago and Milwaukee and hosts a population in excess of 12 million people. A number of converging issues have placed increasing pressure in recent years on groundwater resource for these communities, most of which rely heavily on a regional sandstone aquifer. Pumping from this aquifer has resulted in a drawdown cone that extends throughout much of the region, with water-level declines in excess of 300 m measured at some locations. This regional drawdown cone is one of the largest in the USA. Although partial recovery has taken place in some areas through reduction in withdrawals, concern continues about the drawdowns, as well as their water quality implications. For example, the presence of high levels of arsenic in the upper part of the sandstone aquifer in some areas has been attributed to the oxidation of minerals in the newly unsaturated deposits at the top of the sandstone aquifer. Likewise, proposals for artificial recharge to store Lake Michigan water in the aquifer have been hampered by the detection of arsenic in the recovered water. Drawdown in the sandstone aquifer has also coincided with increases in the concentration of total dissolved solids (TDS) in much of the aquifer from upcoming of saline water and leakage from shale beds. Radium concentrations generally show a direct correlation with the concentration of TDS, and it is anticipated that increases in TDS associated with drawdown in the sandstone aquifer may result in an increase in radium concentrations. To properly address these multi-faceted issues, cooperative efforts are needed on a region wide basis to examine various approaches such as strategically shifting water supply to surface water or to other aquifers, optimizing the location and pumping from wells so as to minimize drawdown problems across pumping centres, installation of deeper casing for new private wells or use of deeper 'cluster' wells for multiple households, further investigation of the water-quality effects of artificial recharge, and various treatment options.

8.6.3 Other effects of excessive abstraction

As the depth to water increases, the water must be lifted higher to reach the land surface, and as the lift distance increases, greater energy is required to drive the pump. Depending on the use of the water and the cost of energy, it may no longer be economically feasible to use water for a given purpose. Furthermore, with declining water levels, well yields will decline, possibly below usable rates. In extreme cases, groundwater levels may fall below the bottom of existing pumps, necessitating the expense of lowering the pump, deepening the well, or drilling a deeper replacement well.

In many environments, surface-water and groundwater systems are intimately linked. Groundwater abstraction can reduce spring flow, or alter how water moves between an aquifer and streams, lakes, or wetlands. The decrease in contribution to surface water may occur either by intercepting groundwater flow that discharged into a surface water body under natural conditions, or by increasing the rate of water movement from surface water into an aquifer.

Although several different earth processes can cause land subsidence, a considerable amount is caused by groundwater withdrawals. For example, more than 80 per cent of the land subsidence in the USA is related to the withdrawal of groundwater (Galloway *et al.*, 1999). Geologic conditions most susceptible to subsidence are the existence of compressible clay and silt layers or rocks that are relatively soluble, such as limestone, dolomite or evaporite deposits.

8.6.4 Impacts of abstraction and hydrogeological environments

All groundwater abstraction results in some decline in water levels in the aquifer over a certain area. Some reduction is often necessary and desirable since improved land drainage is often a side effect or even an objective of the pumping. A degree of induced seasonal water level decline may also be desirable for creating subsurface storage to receive the high rates of wet season recharge. If, however, the overall abstraction rate in the area of interest, or in the aquifer as a whole, exceeds the long-term average rate of replenishment, then there will be continuous decline in groundwater levels and mining of aquifer storage is the result. The presence or absence, and the relative severity of the effects of excessive groundwater exploitation are highly dependent on hydrogeological environment, as shown in Table 8.10.

Table 8.10. Susceptibility of hydrogeological environments to adverse effects of excessive abstraction

Hydrogeological environment	**Type of side-effect**		
	Saline intrusion or upcoming	Land subsidence	Induced pollution
Major alluvial and coastal plain sediments			
coastal	√√	√√	√√
inland	√	√	√√
Intermontane alluvium and volcanics			
with lacustrine deposits	√√	√√	√
without lacustrine deposits	√	√	√√
with permeable lavas/breccias	√√	-	√√
without permeable lavas/breccias	√	-	√
Consolidated sedimentary aquifers	√√	-	√
Recent coastal calcareous formations	√√	-	√
Glacial deposits	√	√	√
Loessic plateau deposits	-	√	-
Weathered basement complex	-	-	√

√√: major effects; √: occurrences known; -: not applicable or rare.

8.7 CHECKLIST

NOTE ▶ *The following checklist outlines information needed for assessing aquifer pollution vulnerability and susceptibility to the impacts of abstraction in the drinking-water catchment area. It supports hazard analysis in the context of developing a Water Safety Plan (Chapter 16). It is neither complete nor designed as a template for direct use but needs to be adapted for local conditions.*

What is the main hydrogeological environment in the recharge area?

- ✓ Establish the dominant geology and the major rock types
- ✓ From the geological description and lithology, check whether intergranular or fracture flow is predominant
- ✓ Check whether a hydrogeological map is available which gives this information
- ✓ …

Is the groundwater flow regime already known, or can a conceptual model of the regional flow system be developed?

- ✓ Identify the vertical and lateral aquifer boundaries
- ✓ Check whether the aquifer is single or multi-layered
- ✓ Check whether the groundwater is confined or unconfined
- ✓ Check whether there are major faults which could form hydraulic barriers
- ✓ Check whether there are rivers, lakes or other surface waters which could interact with groundwater
- ✓ Identify recharge and discharge areas
- ✓ Establish whether there is hydraulic connection with and groundwater discharge to the sea
- ✓ …

What are the physical conditions at the land surface?

- ✓ Define the topography and slopes
- ✓ Estimate annual rainfall, its seasonal pattern and variability from year to year
- ✓ Identify the main land use types

- ✓ Identify roads, railways, airports and other major infrastructure
- ✓ …

ⓘ **What are the soil conditions?**

- ✓ Check whether soil maps of the area are available
- ✓ Identify the main soil types
- ✓ Assess the leaching potentials of the various soil types
- ✓ …

ⓘ **What are the main sources of recharge to groundwater?**

- ✓ Analyse whether there is direct recharge by infiltration from rainfall
- ✓ Evaluate whether the topography is such as to provide scope for localized recharge
- ✓ Check whether there are sources and routes for indirect recharge
- ✓ Identify man-made modifications to the natural recharge regime (e.g. canals, irrigated fields, urban storm water infiltration, leaking water mains, leaking sewers)
- ✓ Identify artificial recharge facilities
- ✓ Check whether identified recharge sources can be quantified
- ✓ Evaluate whether identified recharge sources can affect groundwater quality: is their quality known?
- ✓ Analyse whether the identified recharge sources are changing significantly with time in either quantity or quality
- ✓ …

ⓘ **What is the vulnerability of groundwater to pollution?**

- ✓ Check whether groundwater vulnerability has already been characterized and mapped at an appropriate scale
- ✓ If not, check whether a broad classification from the geology does mean that it should be formally assessed
- ✓ Evaluate whether there is sufficient available information to assess groundwater vulnerability
- ✓ Based on the available information, assess whether a suitable approach can be developed
- ✓ …

What are the natural baseline hydrochemical conditions in the recharge area?

- ✓ Analyse underlying geology and assess whether it is likely to give rise to natural groundwater quality constraints
- ✓ Check availability of monitoring data from which the baseline quality can be established
- ✓ If not, obtain a preliminary idea of baseline quality from existing literature
- ✓ Check whether the hydrochemical conditions are oxidizing or reducing
- ✓ Evaluate whether the conditions change along groundwater flow lines or with depth
- ✓ Evaluate whether there are elevated constituents of the natural groundwater quality that affect its required uses
- ✓ …

What is the groundwater abstraction in the study area?

- ✓ Compile information on the location of groundwater abstraction
- ✓ Compile information on techniques employed for groundwater abstraction
- ✓ Check adequacy of wellhead protection measures, wellhead construction and maintenance as well as sanitary seals used (Chapter 18)
- ✓ Evaluate whether there is a multi-layered aquifer sequence with abstraction from different levels
- ✓ Assess the potential for shallow polluted groundwater to be induced downwards
- ✓ Identify occurrence and location of abandoned wells or boreholes which could act as pollutant pathways
- ✓ …

Is the groundwater susceptible to resource degradation?

- ✓ Assess whether groundwater abstraction is increasing and likely to continue increasing
- ✓ Evaluate whether groundwater abstraction exceeds average recharge
- ✓ Identify observable signs of persistent decline in groundwater levels
- ✓ …

ⓘ **Documentation and visualization of information on aquifer pollution vulnerability and impacts of abstraction**

- ✓ Compile summarizing report and consolidate information from checklist points above
- ✓ …

8.8 REFERENCES

Adams, B. and Foster, S.S.D. (1992) Land-surface zoning for groundwater protection. *J. Institution of Water & Environ. Man.*, **6**, 312-320.

Albinet, M. and Margat, J. (1970) Cartographie de la vulnerabilité à la pollution des nappes d'eau souterraine. *Bulletin BRGM 2nd Series*, **3**(4), 13-22.

Aller, L., Bennett, T., Lehr, J.H., Pett, R.J. and Hackett, G. (1987) *DRASTIC: A Standardised System For Evaluating Groundwater Pollution Potential Using Hydrogeologic Settings*. EPA/600/2-85/018, US EPA, Ada, Oklahoma.

Alley, W.M., Reilly, T.E. and Franke, O.L. (1999) *Sustainability of Ground-water Resources*. USGS, Circular 1186. http://pubs.water.usgs.gov./circ1186/ (accessed April 29, 2005).

Alley, W.M., Healy, R.W., LaBaugh, J.W. and Reilly, T.E. (2002) Flow and storage in groundwater systems. *Science*, **296**, 1985-1990.

Andersen, L.J. and Gosk, E. (1987) Applicability of vulnerability maps. *TNO Committee for Hydrological Research: Proceedings and information*, **38**, 321-332, Delft.

ARGOSS (2001) *Guidelines For Assessing the Risk to Groundwater From On-site Sanitation*. BGS Commissioned Report, CR/01/142, Keyworth.

Bowden, R. (2002) *Water Supply: Our Impact on the Planet*, Hodder Wayland, London.

Chapman, D. (ed) (1996) *Water Quality Assessments*, 2nd edn, E & F N Spon, London.

Custodio, E. (2002) Aquifer overexploitation: what does it mean? *Hydrogeology J.*, **10**, 254-277.

Daly, D., Dassargues, A., Drew, D., Dunne, N., Goldschneider, N., Neale, I., Popescu, C. and Zwahlen, F. (2001) Main concepts of the European approach for karst groundwater resource assessment and mapping. *Hydrogeology J.*, **10**, 340-345.

Doerfliger, N. and Zwahlen, F. (1998) *Practical Guide to Groundwater Vulnerability Mapping in Karstic Regions*. Swiss Agency for Environment, Forest and Landscape, Berne.

Domenico, P.A. and Schwartz, F.W. (1998) *Physical and Chemical Hydrogeology*, 2nd edn, John Wiley & Sons, New York.

Driscoll, F.G. (1986) *Groundwater and Wells*, 2nd edn, Johnson Division, St Paul, Minnesota.

Foster, S.S.D. (1992) Unsustainable development and irrational exploitation of groundwater resources in developing nations - an overview. In *Selected Papers on Aquifer Overexploitation*, vol. 3, pp. 321-336, Heise, Hannover.

Foster, S.S.D. and Hirata, R. (1988) *Groundwater Pollution Risk Assessment*. Pan American Centre for Sanitary Engineering and Environmental Sciences, Lima.

Foster, S., Chilton, J., Moench, M., Cardy, F. and Schiffler, M. (2000) *Groundwater in Rural Development: Facing the challenges of supply and resource sustainability*. World Bank Technical Paper 463, Washington, DC.

Foster, S., Lawrence, A. and Morris, B. (1998) *Groundwater in Urban Development: Assessing management needs and formulating policy strategies*. World Bank Technical Paper 390, Washington, DC.

Foster, S.S.D., Hirata, R, Gomes, D., D'Elia, M. and Paris, M. (2002) *Groundwater Quality Protection : A guide for water utilities, municipal authorities and environment agencies*. The World Bank, Washington, DC.

Freeze, R.A. and Cherry, J.A. (1979) *Groundwater*. Prentice Hall, Englewood Cliffs, New Jersey.

Galloway, D.L., Jones, D.R. and Ingebritsen, S.E. (1999) *Land Subsidence in the United States*. USGS, Circular 1182. http://pubs.water.usgs.gov/circ1182/ (accessed April 29, 2005).

Groundwater Development Consultants (1989) *Spilsby Sandstone Investigation*. CDP 3R-56. Final Report to Anglian Water, Lincoln.

Hölting, B., Haertlé, T., Hohberger, K-H., Nachtigall, K.H., Villinger, E., Weinzierl, W. and Wrobel, J-P. (1995) Konzept zur Ermittlung der Schutzfunktion der Grundwasserüberdeckung. *Geol. Jb.*, **C-63**, 5-24. [translated into English by the Federal Institute for Geosciences and Natural Resources: *Concept for the Determination of the Protective Effectiveness of the Cover above the Groundwater against Pollution*. – Ad-hoc Working Group on Hydrogeology, Hanover]

Howard, G. (ed) (2002) *Urban Water Supply Surveillance - A Reference Manual*. WEDC, Loughborough University.

Kalinski, R.J., Kelly, W.E., Bogardi, I., Ehrman, R.L. and Yamamoto, P.D. (1994) Correlation between DRASTIC vulnerabilities and incidents of VOC contamination of municipal wells in Nebraska. *Ground Water*, **32**(1), 31-34.

Kim, Y.J. and Hamm, S.Y. (1999) Assessment of the potential for groundwater contamination using the DRASTIC/EGIS technique, Cheonju area, South Korea. *Hydrogeology J.*, **7**, 227-235.

Le Grand, H.E. (1983) *A Standardised System For Evaluating Waste-Disposal Sites*. National Water Well Association, Dublin, Ohio.

Lerner, D.N., Issar, A. and Simmers, I. (1990) Groundwater recharge; a guide to understanding and estimating natural recharge. Internat. Assoc. Hydrogeologists, 8, 345.

Lewis W.J., Foster S.S.D. and Drasar B.S. (1980) Risk of groundwater pollution by on-site sanitation in developing countries: a literature review. IRCWD Report 01/82

Margane, A., Hobler, M. and Subah, A. (1999) Mapping of groundwater vulnerability and hazards to groundwater in the Irbid Area, N. Jordan. Z. Angew. Geol., 45, 175-187.

Morris, B.L., Lawrence, A.R. and Stuart M.E. (1994) The Impact of Urbanization on Groundwater Quality – Project Summary Report. Technical Report WC/94/56, BGS, Keyworth.

Morris, B.L. Lawrence, A.R., Chilton, P.J., Adams, B., Calow, R. and Klinck, B.A. (2003) Groundwater and its Susceptibility to Degradation: A global assessment of the problem and options for management. Early Warning Report Series, RS 03-3, UNEP, Nairobi.

NRC (1993) Groundwater Vulnerability Assessment: Contamination potential under *conditions of uncertainty*. NRC, National Academy Press, Washington, DC.

Palmer, R.C. and Lewis, M.A. (1998) Assessment of groundwater vulnerability in England and Wales. In *Groundwater Pollution, Aquifer Recharge and Vulnerability*, pp. 191-198, Geological Society Special Publication 130, London.

Palmer, R.C., Holman, I., Robins, N.S. and Lewis, M.A. (1995) *Guide to Groundwater Vulnerability Mapping in England and Wales*. National Rivers Authority, HMSO, London.

Rose, A.W., Hawkes, H.E. and Webb, J.S. (1979) *Geochemistry in Mineral Exploration*. Academic Press, London.

Rosen, L. (1994) A study of the DRASTIC methodology with emphasis on Swedish conditions. *Ground Water*, **32**(2), 278-285.

Shahid, S. (2000) A study of groundwater pollution vulnerability using DRASTIC/GIS, West Bengal, India. *J. Environ. Hydrol.*, **8**, 1-9.

Simmers, I. (ed) (1997) Recharge of phreatic aquifers in (semi-) arid areas. *Internat. Assoc. Hydrogeologists*, **19**, 277.

Struckmeier, W.F. and Margat, J. (1995) Hydrogeological maps: a guide and a standard legend. *Internat. Assoc. Hydrogeologists*, **17**, 177.

Taylor, C.J. and Alley, W.M. (2002) *Ground-water-Level Monitoring and the Importance of Long-term Water-level Data*. USGS, Circular 1217, http://pubs.water.usgs.gov/circ/circ1217/ (accessed April 29, 2005).

UNESCO (1970) *International Legend For Hydrogeological Maps*. UNESCO/IASH/IAH/IGS, Paris.

US EPA (1992) *Handbook: Ground Water and Wellhead Protection*. US EPA Report No. EPA/625/R-94/001, Washington, DC.

Vrba, J. and Zaporozec, A. (eds) (1994) Guidebook on mapping groundwater vulnerability. *Internat. Assoc. Hydrogeologists*, **16**, 131.

Winter, T.C., Harvey, J.W., Franke, O.L. and Alley, W.M. (1998) *Ground Water and Surface Water- A Single Resource*. USGS, Circular 1139. http://pubs.water.usgs.gov/circ/circ1139/ (accessed April 29, 2005).

9

Agriculture: Potential hazards and information needs

S. Appleyard and O. Schmoll

Agriculture has only received serious attention as a source of groundwater contamination in the last few decades because of the intense focus on industrial and urban pollution problems in many developed countries. However, agricultural practices are often significant sources of health-relevant groundwater pollution. Nitrate contamination can be found in many parts of the world mainly due to the large land area used for agriculture and the usage of chemical fertilizers and animal manures to enhance crop yields. The use of animal manures and the production and disposal of wastes from livestock can also contaminate groundwater with pathogens. A wide range of pesticides used in agriculture to control weeds, insects, nematodes, and fungi in crops, can pollute groundwater. Further, land clearing for agriculture can also lead to groundwater quality problems due to changes in hydrological conditions.

Contamination of groundwater by agriculture can cause serious health problems in rural and urban populations that depend on groundwater for water supply. The discharge of polluted groundwater into wetlands, rivers, estuaries and the coastal environment can contribute to toxic algal blooms in these water bodies that can also cause health problems. These problems have progressively increased over the last few decades with the general intensification of agriculture to feed the world's growing population.

 Protecting Groundwater for Health: Managing the Quality of Drinking-water Sources. Edited by O. Schmoll, G. Howard, J. Chilton and I. Chorus. ISBN: 1843390795. Published by IWA Publishing, London, UK.

In many parts of the world, more than 40 per cent of the land surface is used for agricultural production, and in very densely populated countries, the proportion of agricultural land is often greater than 70 per cent of the land surface. An increasing proportion of the world's population is moving from rural areas to large urban centres. In both developing and developed countries, horticulture and market gardening are increasingly being carried out on vacant land within cities and in peri-urban areas. Urban agriculture is extremely important for impoverished urban dwellers in low income countries, as it provides a measure of food security when there is little disposable income to purchase food, and urban agriculture typically provides more than half of the urban household's food needs in many Asian cities.

Agricultural practices vary enormously throughout the world due to variations in climate and soil types, population density and traditional and modern methods of cultivation. However, there are a number of common agricultural activities that frequently are significant sources of groundwater pollution. These are presented here together with guidance on how to compile the information needed for situation assessment.

NOTE ▶ *Agricultural practices and the environment in which they take place vary greatly. Health hazards arising from agriculture and their potential to pollute groundwater therefore need to be analysed specifically for the conditions in a given setting. The information in this chapter supports hazard analysis in the context of developing a Water Safety Plan for a given water supply (Chapter 16). Options for controlling these risks are introduced in Chapter 21.*

9.1 USE OF MANURES AND FERTILIZERS

Animal manures and human excrement have probably been used as a source of nutrients to enhance crop yields since agriculture commenced more than 4000 years ago. They are still widely used as fertilizers in the developing world, where chemical fertilizers are often expensive and not widely available. Animal manures are also still used in the developed world to reduce fertilizer costs and as a disposal method for animal wastes, particularly from intensive animal rearing. They are also being re-introduced to agriculture as alternatives to chemical fertilizers through the increased popularity of organic farming methods. Chemical fertilizers in contrast have only been available for the last 100 years, and only widely available for the last 50 years.

Nitrate contamination

Among the nutrients, nitrogen species in manures and fertilizers are the chemicals of greatest concern as groundwater pollutants, and among them particularly nitrate because of the high solubility of most nitrate salts and because of its potential health effects in drinking-water.

Livestock manures mainly consist of organic matter, nutrients (nitrogen, phosphorus and potassium) and trace elements. When managed correctly, nutrients in livestock can be a valuable resource. The nutrient content of manures varies considerably depending on animal species, manure moisture, feeding methods and nutritional conditions, but is generally much higher in poultry than other livestock. Table 9.1 shows the mass of nitrogen, phosphorus and potassium contained in excreta for a variety of animals.

Table 9.1. Typical nitrogen (N), phosphorus (P) and potassium (K) content in solid and liquid manure (adapted from US EPA, 2000)

Type of stock	Total N	Total P	Total K
Solid manure	(kg/t)	(kg/t)	(kg/t)
Dairy cattle	2-7	0.7-2	0.7-5
Beef cattle	6-21	1-3	4-9
Swine	8	1-2	2-3
Poultry	9-32	3-12	4-19
Sheep	12	1-2	9-10
Horse	19	0.7	5
Liquid manure	($kg/10^3$ l)	($kg/10^3$ l)	($kg/10^3$ l)
Dairy cattle	0.5-4	0.2-1	0.5-3
Beef cattle	0.5-5	0.5-1	0.5-3
Poultry	8-10	2-3	3-10
Swine	0.5-4	0.1-1	0.4-2

Nitrogen occurs in both inorganic and organic chemical species in manure. The ammonium form (NH_4^+) originates from urea nitrogen in the urine, and the more stable organic form largely originates from faeces. The nitrogen content of manures varies considerably depending on its age and on how it is handled and stored, and is generally much lower in well aged manures due to the volatilization of ammonia (NH_3) into the atmosphere. During storage either in open or closed systems ammonia losses occur in a range of 5-40 per cent (White and Sharpley, 1996).

Losses due to ammonia volatilization can occur very rapidly, and up to 20 per cent of nitrogen can be lost within 4 days of fresh manures being applied to the surface of soils. If manures are ploughed into the soil, ammonium is either directly available to the plants or converts to another plant available form, nitrate nitrogen (NO_3^-), and the losses may be reduced to about 5 per cent (White and Sharpley, 1996). Figure 9.1 summarizes the chemical behaviour of nitrogenous material applied to soils in manure and fertilizers.

Once manure is applied to a soil as solids or in slurries, a number of chemical and microbial processes can occur depending on chemical conditions within the soil. Organic matter containing nitrogen is progressively broken down by microbial activity releasing ammonium ions that are generally absorbed by clay particles in the soil. In well aerated soils, nitrifying bacteria oxidize ammonium to nitrate, and if there is abundant organic matter within the soil profile and anaerobic conditions, denitrifying bacteria can convert nitrate to nitrogen gas which diffuses through the soil back into the atmosphere. Plant uptake of nitrate and ammonium removes nitrogen from the soil profile. Organic nitrogen is largely unavailable to plants until microbial activity in soil releases ammonium from the organic matter (Figure 9.1; also see Chapter 4.3).

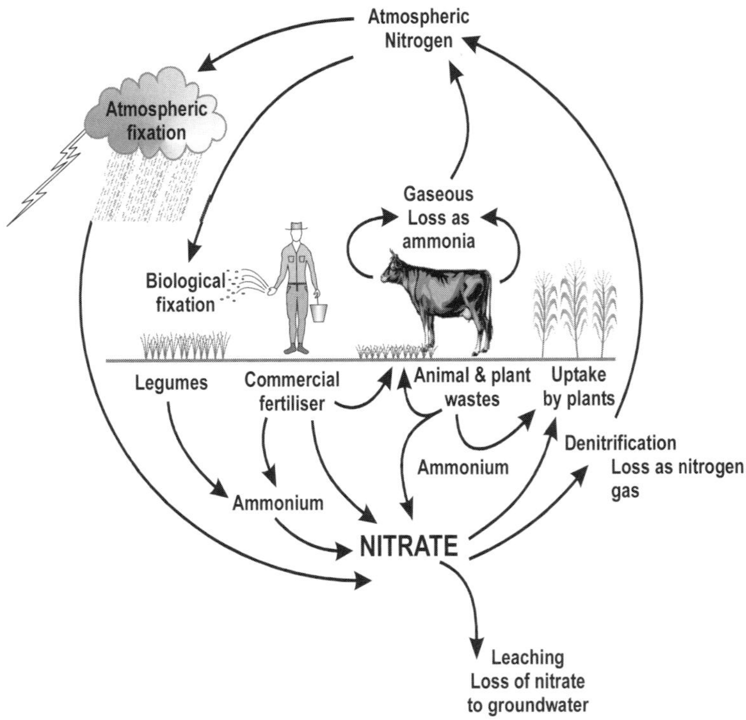

Figure 9.1. The physical and chemical behaviour of nitrogen in an agricultural catchment

Chemical fertilizers consist of inorganic salts of nitrogen, phosphorus, potassium and sulphur with the addition of some trace metals necessary for healthy plant growth. Generally, nitrogen is present in inorganic fertilizers in the form of soluble salts of ammonium and nitrate. Nitrogen in fertilizer is generally more available for plant uptake than from manures, but is also more easily leached into groundwater if used in excess. Slow release fertilizers greatly reduce the risk of nutrient leaching, but these are often too expensive for widespread use in agriculture. Table 9.2 shows the typical chemical composition of common inorganic fertilizers or constituents of fertilizer formulations.

The amount of nitrogen applied in manure and fertilizer varies greatly depending on a range of factors. Ideally, application rates should be adapted to the cropping system used, plant uptake rates, soil contents of nutrient fraction available to the crop and climate and soil conditions in the drinking-water catchment area (see Chapter 21). However, in many situations rates of application exceed crop uptake rates. This is especially the case in areas with high livestock densities or intensive livestock farming where volumes of animal waste produced often exceed local demand for use as fertilizer. In these areas, application practices for manures may be driven by the need of getting rid of them rather than their use as a nutrient source. This problem tends to occur where other means of managing the large amounts of animal wastes are difficult and transportation costs for a wider distribution of manure to areas where soils show fertilizer deficits are unattractive or prohibitive. Moreover, application of chemical fertilizers and manures can also substantially exceed recommended rates due to the lack of knowledge or information of farmers on the factors which determine good application practices, or under the influence of 'advisers' or aggressive marketing strategies that try to convince farmers that crop yields will be bigger and better by using excessive fertilizers. In both cases, behaviour of single farmers can become the key factor in whether or not nitrogen pollution of groundwater is likely to occur.

Table 9.2. Chemical composition of common inorganic fertilizers or constituents of fertilizer formulations (adapted from US EPA, 2000)

Common name	Chemical formula	N (%)	P_2O_5 (%)	K_2O (%)
Ammonium nitrate	NH_4NO_3	34	0	0
Ammonium sulphate	$(NH_4)_2SO_4$	21	0	0
Ammonium nitrate-urea	$NH_4NO_3 + (NH_2)_2CO$	32	0	0
Anhydrous ammonia	NH_3	82	0	0
Aqua ammonia	NH_4OH	20	0	0
Urea	$(NH_2)_2CO$	46	0	0
Superphosphate	$Ca(H_2PO_4)_2$	0	20-46	0
Monoammonium phosphate	$NH_4H_2PO_4$	13	52	0
Diammonium phosphate	$(NH_4)_2HPO_4$	18	46	0
Urea-ammonium phosphate	$(NH_2)_2CO + (NH_4)_2HPO_4$	28	28	0
Potassium chloride	KCl	0	0	60
Monopotassium phosphate	KH_2PO_4	0	50	40
Potassium nitrate	KNO_3	13	0	45
Potassium sulphate	K_2SO_4	0	0	50

Often more than 200 kg/ha of nitrogen is applied for intensive production of non-leguminous crops, and Salama *et al.* (1999) reported even more than 500 kg/ha on some horticultural crops in Malaysia. Loading rates on fertilized improved grazed pasture may exceed 400 kg/ha of nitrogen due to the combined effects of fertilizer use and high stock densities in fields (Sumner and McLaughlin, 1996). Nitrogen application in excess of 140 kg/ha on coarse sandy soils, or in excess of 200 kg/ha on loamy soils, have caused nitrate concentrations in groundwater to exceed drinking-water criteria in irrigated horticulture in Perth, Western Australia (Water and Rivers Commission, 1996). Although these loadings do not seem high in comparison with crop nitrogen requirements which typically vary between 50 and 400 kg/ha (USGS, 1999), the rate and timing of individual applications of fertilizer is critical in determining how much nitrogen leaches past the root zone and into groundwater.

Much of the ammonia in animal urine may be oxidized to nitrate, and the nitrification of localized patches of urine in soil can cause significant contamination of groundwater by nitrate (Close *et al.*, 2001). The amount of mineralized nitrogen in urine patches on grazed pasture may be up 600 kg N/ha (Ball and Ryden, 1984) and can greatly exceed the capacity of the pasture to take up the nitrate, leading to leaching to groundwater. The problem is exacerbated by the uneven distribution of urine patches in pasture, with high concentrations of mineralized nitrogen in soil often occurring near watering points or stock yards. For example, Ruz-Jerez *et al.* (1994) estimated that for a rotationally grazed clover-ryegrass pasture in New Zealand that was fixing 144 kg/ha of nitrogen yearly, only about 10 per cent of the area was affected at any time by urine patches, but these patches contributed about 55 per cent of the nitrate leached from the pasture.

In addition to the application rate of manure and fertilizer, the factors that significantly determine the extent to which nitrate is leached from soils are the physical and chemical properties of the soil, climate, land use and whether or not irrigation is used. The soils most vulnerable to nitrate leaching are sandy soils with a low organic matter content. Nitrogen compounds applied to these soils are readily nitrified to form nitrate, and the high permeability of the soils may allow nitrate ions to be rapidly leached, allowing limited opportunity for plant uptake or denitrification (Box 9.1). The poor nature of these soils often encourages farmers to add excess manure and fertilizer to get reasonable yields, particularly in horticultural areas. Nitrogen is retained more effectively in loamy soils containing large amounts of organic carbon, but nevertheless nitrate leaching still occurs in this type of soil.

Box 9.1. Sandy soils and nitrate concentrations in groundwaters

Work by Pionke *et al.* (1990) estimated that farmers applied between four and seven times the amount of nitrogen in poultry manure and fertilizer than could be taken up by crops in areas used for irrigated horticulture on sandy soils in Western Australia. Nitrate concentrations in leachate below the root zone of the crops were up to 200 mg N/l. Concentrations in groundwater beneath the crops commonly ranged between 10 and 70 mg N/l (Pionke *et al.*, 1990; Lantzke, 1999), compared to concentrations of less than 1 mg N/l in groundwater beneath uncleared native vegetation.

Weil *et al.* (1990) found that groundwater in sandy soils in Maryland in the USA, an area used for irrigated maize cropping, contained between 10 and 20 mg N/l and 20 and 30 mg N/l from fertilizer and manure use respectively.

Nitrate leaching is also strongly influenced by climatic factors. In climates with strongly seasonal rainfall, nitrate concentrations in groundwater may vary seasonally, peaking after the onset of the rainy season when infiltrating water flushes nitrate out of the vadose zone into groundwater. In climates with cold winters where the ground freezes or there is snow cover, maximum nitrate leaching often takes place during the spring thaw (Box 9.2).

Land use and vegetation cover have a strong effect on nitrate leaching. Less nitrate is generally leached under permanent grassland than under either arable land or ploughed grassland (DOE, 1986). Tilling and leaving agricultural land fallow increases the risk of nitrate contamination caused by the mineralization of nitrogen-rich organic material in the soil profile. Keeping land fallow can accentuate nitrate leaching by up to nine-fold from pasture, and by a factor of two from cropped land (Juergens-Gschwind, 1989).

The use of artificial irrigation in pasture and cropping increases the risk of nitrate being leached to the water table due to increases in infiltration rates of water through the soil profile. This may cause nitrogen leaching rates to more than double. For instance, annual nitrogen leaching rates from unimproved (i.e. not irrigated) dairy pasture in New Zealand are typically 10-25 kg/ha, whereas leaching rates from irrigated pasture areas are

65-70 kg/ha (Burden, 1986). Excessive irrigation in semi-arid or arid areas can further increase nitrate and other dissolved salt concentrations through evapotranspiration of water in the soil (Romijn, 1986). This may also cause water near irrigation areas to become too saline for potable use.

Box 9.2. Rainfall and nitrate concentrations in groundwaters

In the United Kingdom nitrate leaching is least in spring and summer due to crop uptake and is greatest during autumn and winter when there is little uptake and soils are saturated with water. Annual rates of nitrogen leaching from arable land in the United Kingdom range from 40-120 kg/ha, with a weighted mean of about 50 kg/ha (Addiscott and Gold, 1994).

In tropical regions with monsoonal climates, leaching of nitrogen from soil profiles increases at the onset of the wet season. In a study of the impacts of agriculture on groundwater quality in Sri Lanka, Lawrence and Kumppnarachi (1986) found that nitrate concentrations in groundwater beneath rice paddies and other horticultural areas typically increased from an average of 10 to 25 mg N/l in the dry season to more than 40 mg N/l at the onset of the wet season. Concentrations in groundwater then progressively declined to dry season levels. The high nitrogen concentrations in groundwater were due to both the high rate at which nutrients were applied to crops, and due to the fact that several crops could be grown in a year in the tropical climate.

Pathogen contamination

In addition to being a potential source of nitrate contamination, manure can also contaminate groundwater with a variety of pathogens that can affect human health, either through the ingestion of unwashed crops, or through the ingestion of polluted groundwater used as a source of drinking-water.

Animal manure can contain large numbers of pathogenic organisms such as bacteria, viruses, protozoa and helminths that can cause human disease, i.e. up to 10^6 pathogens per gram of faeces (Gannon *et al.,* 2004). However many species excreted by farm animals do not cause disease in humans. Pathogens from manure that may cause disease from drinking contaminated groundwater are shown in Table 9.3.

The number of viable pathogens in manure can be greatly reduced by storing manure before use. Survival times of disease-causing bacteria and protozoa are greatly affected by ambient temperature (Table 9.4). Viruses may become dormant and can persist for long periods in manure. For example, the infectious avian influenza virus can survive in water for 207 days at 17 °C (Brown and Alexander, 1998), and rotaviruses are stable in faeces for 7 to 9 months (Goss *et al.*, 2001). However, the longevity of some viruses can be reduced by the presence of predatory bacteria. The longevity of pathogens can be further reduced by either aerobically composting or drying manure, and most pathogens die within a week if manure is treated in this manner provided that the temperature in the compost pile reaches 55 °C (Goss *et al.*, 2001) (see Chapter 21 for details).

Table 9.3. Examples of human pathogens potentially present in manure (Playford and Leech, 1977; Addis *et al*., 1999; Goss *et al*., 2001; Gannon *et al*., 2004; WHO, 2004)

Pathogenic organism	Main source
Bacteria	
E. coli O157:H7	Livestock excrement (particularly cattle and sheep and, to a lesser extend, goats, pigs and chickens) (see also example in Box 9.3)
Leptospira species	Pig urine
Yersinia enterocolitica	Pig excrement
Campylobacter species	Poultry, pig and cattle excrement (see also example in Box 9.3)
Listeria moncytogenes	Animal excrement
Salmonella species	Wild animal and livestock excrement (incl. poultry, cattle, pigs, sheep)
Viruses	
Hepatitis E virus	Livestock excrement (particularly pigs, as well as cattle and goats)
Protozoa	
Cryptosporidium parvum	Livestock excrement (particularly young animals)
Giardia lamblia	Livestock excrement

Table 9.4. Survival of potentially pathogenic bacteria and protozoa in manure at various temperatures (based on Goss *et al*., 2001)

Organism	Survival time (days)		
	Frozen	5 °C	30 °C
E. coli	>100	>100	10
E. coli O157:H7		70	49
Salmonella	>150	150	28
Campylobacter	50	21	7
Giardia	<1	7	7
Cryptosporidium	>300	50	28

Soils generally provide an effective barrier against pathogens reaching the water table, and short die-off time of most pathogens in the sub-surface ensures that the number of viable organisms reaching groundwater are low (Chapter 3). Most cases of waterborne disease from groundwater consumption are caused by viruses and bacteria as protozoa and helminths are too large to be transmitted far through the pore spaces between soil particles (although some occurrences of groundwater contamination by the protozoa *Giardia* and *Cryptosporidium* have been recorded). Most cases of waterborne disease from wells where there is a thick soil cover are due to the faulty construction of head works, or to the use of manure near sinkholes, abandoned wells or other features that will allow water and contaminated material direct, rapid access to the water table (Figure 9.2).

Contamination of groundwater by pathogens is often a significant issue in areas where there is a thin or no soil cover over fractured rock or karstic limestone (Box 9.3; Figure 9.2). In these terrains, pathogens can be rapidly carried through preferential flow paths into groundwater with little or no attenuation (Chapters 2 and 3). Pore spaces in karstic aquifers may even allow large organisms like protozoa to be transmitted through the aquifer to water supply wells.

Box 9.3. Microbial quality of water in karst aquifers and in badly constructed wells

Studies of bacterial contamination in springs in a limestone terrain in Ireland (Thorn and Coxon, 1992) indicated contamination with coliform bacteria in spring water samples to be derived from dairy cattle. Detected numbers of microorganisms vary greatly with time from 0 to 300 cfu/100 ml of water within a two hour period. The bacterial quality of water is usually at its worst after heavy rainfall, and water from some springs was estimated to have travelled about 1 km from recharge areas within a 12 to 18 hour period after heavy rain.

One of the best documented outbreaks of disease caused by groundwater contamination by manure occurred in the small Canadian town of Walkerton in May 2000. Seven people died and more than 2300 people became seriously ill when pathogens from manure spread on a nearby farm was washed in surface runoff into a badly constructed and poorly monitored well used as a water supply for the town. The disease was caused by the virulent strain of *E. coli* O157:H7 and by *Campylobacter jejuni*. A Parliamentary Enquiry into the incident was held in 2001, and an initial report published (O'Connor, 2001).

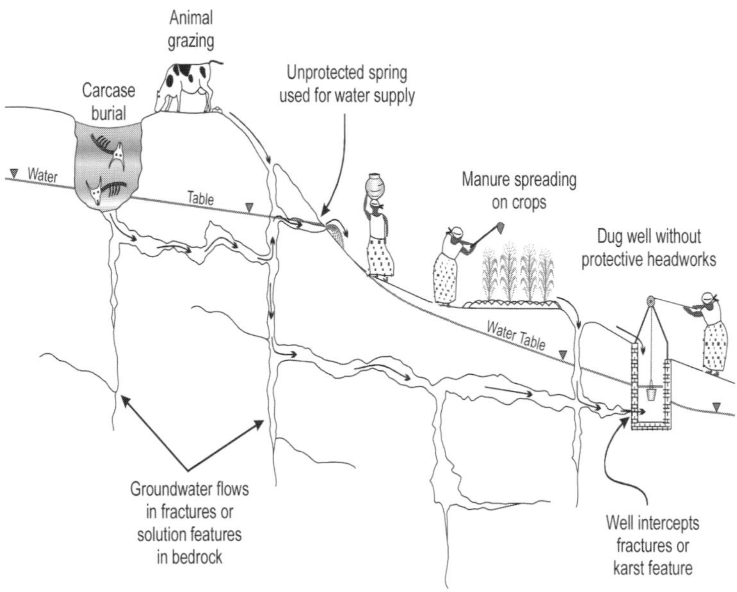

Figure 9.2. Potential pathways for groundwater contamination by pathogens in an agricultural area

Contamination of groundwater by pathogens is also an issue in shallow fractured rock aquifers and in volcanic aquifers. Large cave systems may also be formed in some volcanic terrains as a result of the degassing of volatiles dissolved in molten lava flows, and the interface between individual lava flows may contain interconnected voids that may allow groundwater to transmit contaminants over long distances. Therefore volcanic aquifers show some of the hydrogeological characteristics of karst aquifers, and share the same extremely high vulnerability to groundwater contamination.

Karst-like features can also develop in tropical or subtropical regions with lateritic soils. Voids can form in lateritic duricrusts due to the erosion of soft, poorly consolidated clays in an otherwise cemented ferruginous matrix. This can give rise to a spongy network of interconnected voids which may be at least several millimetres in diameter, and which can rapidly transmit water and pathogens (Box 9.4).

Box 9.4. Voids in lateritic soils

In Brazil, voids in lateritic soils appear to be caused by termite activity (Mendonça *et al.*, 1994). Where intensive urban development takes place, increased point sources of recharge such as storm water soakwells and sewage disposal systems, coupled with intensive groundwater abstraction, can erode these voids to form sinkholes and caves which may be 5 m or more in diameter and several hundred metres long. These so called 'pseudosinkholes' may cause groundwater contamination problems of a similar nature to limestone karst aquifers. Contamination of shallow groundwater by pathogens may become a problem in countries like Brazil where tropical forests overlying laterites are being rapidly cleared and developed for agriculture and urban land use.

9.2 DISPOSAL OF ANIMAL CARCASSES

The disposal of animal carcasses by burial on farms or in landfill sites may pose public health problems if burial sites are located near wells or tubewells used as a source of drinking-water (see Figure 9.2). Generally, the risks are similar than those posed by the excessive use of animal manures, including contaminating groundwater with nitrates and pathogens such as verotoxin-producing *E. coli*, *Campylobacter*, *Salmonella*, *Cryptosporidium* and *Giardia*.

The risks of groundwater contamination are greatest when very large numbers of animals may be destroyed and buried to control the spread of animal diseases in agricultural areas. Of particular concern to public health are epidemics of animal disease that may also cause disease in humans, especially where the disease-causing agent is persistent in soil and groundwater. The mass burial of animal carcasses infected by such a disease-causing agent may pose a risk to nearby groundwater supplies, although the risks are often reduced by burning of carcases prior to burial (UK Department of Health, 2001).

One group of disease-causing agents of concern in this regard are prions, particularly the prion that causes the disease BSE (Mad Cow disease) in cattle. Ingestion of the prion can also infect humans and cause the disease variant Creutzfeldt-Jakob disease (CJD) in humans. Prions are inanimate disease-causing agents. It is hypothesized that they are distorted forms of proteins naturally present in neural as well as many other body tissues of animals (Gannon, 2004). In contrast to beef products, the BSE agent is likely to be highly dispersed in water and therefore the daily intake of infectious prions is likely to be very low by nature (Gannon, 2004). The proteins are highly resistant to physical and chemical agents, such as heat, ultraviolet light and oxidants (e.g. chlorine) and may pass through water treatment plants (Gannon, 2004).

Prions behave as particulates in soil and groundwater, and the movement of these proteins in geological media is likely to be affected by the same processes that influence the behaviour of other particulate infectious agents such as bacteria and viruses (UK EA, 2000). The prion protein has both hydrophilic and hydrophobic domains, and this property limits mobility in groundwater and promotes adhesion to hydrophobic particles (Gannon, 2004). However, it is uncertain whether significant amounts of prions are moved by filtration in soils, and thus the risk of groundwater contamination is uncertain (UK Department of Health, 2001). Therefore the risk of groundwater contamination is most likely when:

- prion-contaminate carcass are disposed of in areas where there is little or no soil cover, and fractures or karstic features provide a direct conduit between land surface and groundwater;
- overland flow transports material from carcasses in fields or from prion-contaminated animal-based fertilizer (blood and bone fertilizer) directly into poorly constructed wells or tubewells.

9.3 ANIMAL FEEDLOTS

Stocking densities on agricultural land have progressively increased in many parts of the world due to an increasing demand for animal products and improvements in agricultural techniques. This has culminated in the development of animal feedlots, where animals are maintained in pens in a controlled environment. Feedlots may either be open-air facilities, or completely enclosed within large buildings. Typically feedlots are used for beef and pork production, and for poultry, meat products and eggs. Dairies are similar to feedlots in that a large number of cows are gathered together for milking, although they may be allowed to run free range in between milking events.

The large number of stock housed in animal feedlots generates great amount of wastes that can become an unintended, but nonetheless substantial, point sources of widespread groundwater pollution if not managed properly. For example, beef feedlots may contain 500 or more steers, and a typical 450 kg steer will produce up to 30 kg of solid and liquid wastes each day. The 1996 National Water Quality Inventory carried out by the US EPA (1997) found that agriculture in general was the leading cause of water pollution in the USA and that 20 per cent of the problems were due to animal feedlots alone.

The major sources of pollution from feedlots are manure, animal carcasses, process wastewater (e.g. dairy wastes), feed, and bedding materials. The contaminants of highest concern for groundwater are nutrients, particularly nitrogen and pathogens. Wastewater from feedlots may also contain growth hormones and pharmaceuticals (e.g. antibiotics) used to accelerate the growth of livestock. Table 9.5 shows the typical nutrient content of wastes from feedlots.

Table 9.5. Nutrient content of wastes from feedlots (based on Goss *et al.*, 2001)

Waste source	Dry matter (%)	Nitrogen (%)	Phosphorus (%)	Potassium (%)	NH_4-N (mg/l)
Beef cattle					
Solid	18-63	0.4-1.0	0.1-0.2	0.3-1.0	30-1050
Liquid	1-13	0.1-0.5	0.02-0.2	0.1-0.2	700-2100
Pig					
Solid	17-51	0.8-1.8	0.4-1.2	0.2-1.2	1700-4000
Liquid	1-13	0.2-0.8	0.05-0.4	0.1-0.4	1500-5450
Poultry					
Solid	16-90	0.9-3.2	0.4-1.4	0.4-1.6	3221-6450
Liquid	0.5-12	0.2-0.9	0.02-0.4	0.01-0.4	900-6250

Groundwater contamination problems from feedlots are mostly from wastewater generated by the washdown of the feedlots, and storm water runoff from manure stockpiles and other wastes. If feedlot pens are constructed on bare soil without impermeable floors, leachate for manure and urine can percolate into groundwater. Moreover, if liquid wastes and washdown are not drained into lined retention and treatment ponds but are allowed to discharge directly to the environment, feedlots may become significant point-sources of groundwater contamination. This is particularly critical for feedlots sited in areas with sandy soils or high groundwater tables, and if feedlots are located in flood-prone areas or next to a sinkhole, abandoned well or other feature that will allow direct access to the water table. For example, the Western Australian policy on feedlots states that there is a high risk of groundwater contamination from feedlots occurring if the water table is less than 1.5 m below the surface, particularly if soils are sandy and if they are sited in flood-prone areas with more than 1 in 100 year flooding frequency, or on land sloping at more than 5 per cent (1 in 200 grade), as this makes wastewater management in the feedlot extremely difficult, and indirectly increases the risk of contamination (Agriculture WA *et al.*, 2000). The Government British Columbia, Canada, requires that feedlots and manure storages are not located within 30 m of wells used for water supply, on land with a slope of more than 4 per cent, and that there is sufficient land area to allow for manure application in a sustainable manner that will not cause groundwater pollution (BC MAFF, 2000).

Well-managed feedlots usually have wastewater retention and treatment ponds to contain feedlot effluent. However, feedlots commonly leak if pond liners have not been properly constructed or are not well maintained: a study of feedlot wastewater ponds in North Carolina, USA, found that 50 per cent of the ponds leaked (Centner and Risse, 1999).

Typically, there is a much higher amount of waste generated, and therefore a greater risk of groundwater contamination occurring, in uncovered open-air facilities than in enclosed feedlots with roofs. This is particularly the case if treatment and/or storage ponds do not have sufficient storage capacity to store water from intense rainfall events. Treated pond effluent is commonly used to irrigate pasture (Section 9.4), but treatment ponds have to be designed to store water for the wettest period of the year when there is little opportunity to dispose of effluent through land irrigation.

Manure collected in feedlots will commonly be applied as fertilizer on cultivated fields and may become diffuse source of groundwater pollution, particularly in areas where feedlots significantly increase stock densities in relation to application options for manure within economically viable distances. Application of manures is discussed in the previous Section 9.1.

9.4 USE OF WASTEWATER AND SEWAGE SLUDGE ON LAND AND IN AQUACULTURE

There has been an increasing interest in the use of wastewater in agriculture over the last few decades due to increased demand for fresh water. Population growth, increased per capita use of water, and the demands of industry and of the agricultural sector have all put pressure on limited fresh water resources. The use of wastewater has been successful for irrigation of a wide array of crops, and increases in crop yields from 10-30 per cent have been reported. In addition, the use of treated wastewater for irrigation and industrial purposes can be a strategy to increase the amount of fresh water available for domestic use, and to improve the quality of river waters used for abstraction of drinking-water (by reducing the disposal of effluent into rivers).

Sewage sludge (or biosolids) are the organic solids derived from municipal sewage and septic tank treatment processes (Chapter 10). In many countries, this material is being disposed of to landfill or offshore, but these practices are increasingly being seen as environmentally unacceptable, and an increasing proportion of this material is being used as a source of nutrients and as a soil amendment in many agricultural areas.

Municipal wastewater and sewage sludge contain considerable amounts of nitrogen and phosphorus. In sewage sludge, nitrogen content ranges between less than 0.1 and 18 per cent (dry weight), and nitrogen concentrations in secondary effluents of wastewaters range between 10 and 30 mg/l (US EPA, 1996). Used at appropriate application rates, these are a valuable resource. However, in regions with a high population density which produce large volumes of sewage wastes, excessive use on agricultural land can pollute groundwater, particularly with nitrate. As with manures and fertilizers, whether or not irrigation of wastewaters and land application of sludges leads to groundwater contamination depends strongly on physical conditions in the drinking-water catchment area, and on the criteria on which rates and timing of application are based (for more detailed information see Section 9.1).

The variety and concentrations of pathogens contained in sewage sludges and wastewater derived from sewage and agriculture strongly depends on treatment and storage practices, which impact on their die-off rates. Viruses in particular accumulate in sewage sludge. For example, Cliver (1987) reported 2400-115 000 pfu/l in primary sludge and 5000 pfu/l in secondary sludge. Poorly or untreated sludge and wastewater can cause significant health effects. The most significant health risk from wastewater use is the consumption of food directly irrigated with the effluent or the direct contact with contaminated wastes (e.g. by field workers or by children playing in irrigation channels). Consequently, the WHO has developed guidelines for the application of wastewater intended for use in agriculture to protect human health (WHO, 2005).

In groundwater, wastewater irrigation and spreading of sludges in agriculture – like use of manures – can cause serious pathogen contamination, particularly in areas with high vulnerability (e.g. high water table, thin soil cover over fractured rock or karstic limestone) and features that allow rapid movement of pathogens in the subsurface (Chapter 8 and Section 9.1). For example, Moore *et al.* (1981) has found virus particles in groundwater up to 27 m below sites irrigated with sewage wastewater, and Jorgensen and Lund (1985) found enteroviruses 3 m below a forest site used for sludge application. However, Liu (1982) found that over a period of 4 years of heavy sludge application to farmland, 92-98 per cent of the bacteria were inactivated by the soil.

Use of sludges which are treated either by composting or by other disinfection methods generally poses a lower risk of groundwater contamination due to greatly reduced numbers of pathogens. Risks to groundwater will depend on both good agricultural practices in use of sludge (e.g. maintaining a suitable buffer zone between areas used for application and water supply wells – Chapter 21) and good sanitation practices in sludge treatment.

Chemical pollutants may be a further concern if wastewater or sewage sludge used in agriculture originates from treatment plants receiving substantial amounts of industrial effluent, or if specific household chemicals are widely used. However, these contaminants generally accumulate in the soil profile (Chapter 4) and are strongly bound by organic matter in the sludge. In most cases, the predominant health concerns are the uptake of these chemicals in crops used for human consumption rather than their role as groundwater pollutants. However, these contaminants are very persistent in soils and may take many decades to degrade into harmless by-products. If excessive amounts of wastes containing these chemicals are applied to soil, there is a risk that the normal biological degradation processes in the soil could break down, and that leaching into groundwater could take place. Therefore, detailed investigations of local soil properties are required if contaminated sludges and wastewater are applied on a long term basis in a particular area to ensure that groundwater contamination does not occur.

Wastewater use in aquaculture

Aquaculture is a possible reuse strategy. Fish raised in wastewater-fed ponds are an important source of protein for many millions of people, particularly in countries in Asia where the fertilization of fish ponds with human wastes has been practised for several thousand years. Today, at least two-thirds of the world yield of farmed fish comes from ponds fertilized in this way. China alone produces 60 per cent of the world's farmed fish in only 27 per cent of the world's area of fish ponds. The largest wastewater-fed aquaculture project in the world is the Kolkata wetland system. This consists of a 3000 ha area of constructed fish ponds which are fed with 550 000 m^3 of untreated wastewater each day. The wetlands produce about 13 000 tonnes of fish each year (mainly Carp and Tilapia) which are supplied to fish markets in central Kolkata and are consumed more widely in the region.

The use of wastewater and human excreta in aquaculture poses a great threat to groundwater quality. This is because fish ponds are often unlined and in direct contact with the water table, and there is no soil profile to allow the die-off of pathogens or the removal of chemical constituents from the effluent. For this reason, there are significant health risks (particularly from pathogens) from using groundwater from wells constructed near fish ponds filled with untreated wastewater and fertilized with raw excrement.

9.5 USE OF PESTICIDES

Many pesticides and degradation products are toxic at low concentrations and have the potential to cause health effects if groundwater used for water supply is polluted with these chemicals.

In general, the progression of pesticide development has moved from highly toxic, persistent and bioaccumulating pesticides such as DDT, to pesticides that degrade rapidly in the environment and are less toxic to non-target organisms. Many of the older pesticides are now banned in many countries due to their health and environmental effects (see Box 9.5).

There are a large number of pesticides for control of insects (insecticides), weeds (herbicides), fungus (fungicides), nematodes (nematicides) and mites (acaricides) currently used in agriculture throughout the world. However, the largest usage tends to be associated with a relatively small number of pesticides.

Table 9.6 lists the world's 35 major crops (on a harvested area basis according to FAO, 2002 statistics) and assigns the known, major uses for pesticides for which WHO has set drinking-water guideline values. In addition to the major uses listed in Table 9.6 there will also be many minor uses of each pesticide which are not stated in the literature because the manufacturers consider this use to be insignificant. In general, similar suites of pesticides are used on different crops in a related group. For example, many more sorghum pesticides may be used on millet than is shown in the table and similar pesticides are likely to be used on chickpeas as cow peas. Table 9.6 is also based on approved or recommended uses for each pesticide. In countries where pesticide use is less well regulated, farmers and growers might use a much greater range of pesticides on each crop than is shown, particularly if the pesticides do not cause unacceptable damage to the crop. Many insecticides in particular could be used on a much wider range of crops than is shown, without damage to the crops (unlike herbicides). It is possible that most of the insecticides shown could be used on most of the crops shown.

Pesticides use creates a risk of a diffuse groundwater pollution. Whether or not pesticides reach groundwater depends on the chemical and physical properties of the active ingredient (see Chapter 4.6), on the local hydrogeological conditions and soil characteristics, and on the manner in which these chemicals are applied to crops. Box 9.6 shows some examples of groundwater contamination by pesticides.

Generally, the usage of pesticides should follow codes of good practice. Application patterns and rates need to be based on the recommendations given by the producer or/and the criteria developed by licensing authorities. The way that pesticides are applied depends on the target pest and on the scale of the agricultural operation. Formulations to kill nematodes, fungal infections or insects (systemic insecticides) are usually applied directly to the soil as solid granules, solutions or sprays. Most other pesticides are applied as foliar sprays, either from hand-held spraying equipment for small plots, or from vehicles or aircraft for commercial scale agricultural operations.

Box 9.5. Stockholm Convention (based on UNDP, 2001)

Organochlorine pesticides belong to a group of organic compounds known as persistent organic pollutants (POPs) which are considered to pose such a significant threat to human health and the environment that there is a major international effort to remove these chemicals from use. The Stockholm Convention on Persistent Organic Pollutants was adopted in May 2001. Its objective is to protect human health and the environment from POPs. Convention parties will be required to take actions to reduce or eliminate POPs releases and ultimately eliminate the production of these chemicals. The convention identifies 12 initial POPs of global concern which include 9 organochlorine pesticides. All substances are characterized by adverse effects on human health and environment, high persistence, and the potential for bio-accumulation and long-range environmental transport.

Seven of the listed POPs are produced mainly for use as insecticides – Aldrin, Chlordane, Dieldrin, Endrin, Heptachlor, Mirex and Toxaphene. These have been mainly applied in agriculture. In many countries, all seven insecticides are already banned or are subject to severe restrictions. However, the Convention will lead to the use of these pesticides being phased out and banned. This will include the prohibition of their production and use, and bans on the import and export of these chemicals.

Dichlorodiphenyltrichloroethane (DDT) is also an insecticide. It was extensively used against insect pests on a variety of agricultural crops (e.g. cotton). Another important use has been in combating vector born diseases such as malaria. While the Convention requires the phasing out of all agricultural DDT uses, production and use will be permitted for disease vector control under specific circumstances (e.g. areas where malaria is endemic).

Hexachlorobenze (HCB) has mainly been used as a fungicide for seed treatment, or as solvent in other pesticide applications. Under the POPs Convention, the use of HCB will also be phased out.

Despite international efforts to remove these chemicals from use, large amounts of organochlorine pesticides are stored throughout the world and continue to be traded on the black market. As it is likely that usage of these will continue in many countries until stockpiles are depleted, it is important that water suppliers continue to test for the presence of organochlorine pesticides in groundwater supplies in agricultural areas used as a source of drinking-water.

Lack of knowledge or of training in the usage of pesticides may lead to over-application of these chemicals in an inappropriate way (Figure 9.3). Improper usage of pesticides such as the use of inappropriate spraying equipment, or practices like preventative spraying instead of scheduling application to crop needs and avoiding spraying prior to heavy rainfall, can significantly increase the pollution risk.

Table 9.6. Known pesticide uses in selected crops (based on Page and Thomson, 1990; Tomlin, 2000; FAO, 2002; WHO, 2004)

Crop	Insecticides	Herbicides
Cereals		
Barley	aldrin*; dieldrin*; dimethoate; γ-HCH	chlorotoluron; cyanazine; 2,4-D; 2,4-DB; dichlorprop; isoproturon; MCPA; MCPP; pendimethalin
Maize	aldicarb; aldrin*; dieldrin*; carbofuran; γ-HCH	alachlor; atrazine; cyanazine; 2,4-D; 2,4,5-TP; metalochlor; pendimethalin; simazine; terbuthylazine
Millet	1,3-dichloropropene	2,4-D
Oats	aldrin*; dieldrin*; dimethoate; γ-HCH	cyanazine; 2,4-D; 2,4-DB; dichlorprop; MCPA; MCPP; pendimethalin
Rice (paddy)	carbofuran; dimethoate	2,4-D; MCPA; molinate; pendimethalin
Rye	dimethoate; γ-HCH	chlorotoluron; 2,4-D; 2,4-DB; dichlorprop; isoproturon; MCPA; MCPP; pendimethalin
Sorghum	aldicarb; carbofuran	atrazine; 2,4-D; metalochlor; pendimethalin; terbuthylazine
Wheat	aldrin*; dieldrin*; dimethoate; γ-HCH	chlorotoluron; cyanazine; 2,4-D; 2,4-DB; dichlorprop; isoproturon; MCPA; MCPP; pendimethalin
Fibre crops		
Flax fibre and tow	1,3-dichloropropene	MCPA; trifluralin
Seed cotton	aldicarb; aldrin*; dieldrin*; carbofuran; dimethoate; endrin*	alachlor; cyanazine; metalochlor; pendimethalin; trifluralin
Fruits		
Apples	1,2-dichloropropane*; 1,3-dichloropropene; dimethoate; γ-HCH; methoxychlor*	2,4-D; pendimethalin; simazine
Bananas	aldicarb; carbofuran	simazine
Citrus fruits	aldicarb; dimethoate	metolachlor; pendimethalin; simazine; trifluralin
Grapes	1,2-dichloropropane*; 1,3-dichloropropene; carbofuran; dimethoate; γ-HCH; methoxychlor*	2,4-D; metolachlor; pendimethalin; simazine; terbuthylazine; trifluralin
Oilcrops		
Coconuts	-	2,4,5-T
Groundnuts in shell	aldicarb; carbofuran; 1,2-dichloropropane; 1,3-dichloropropene; ethylene dibromide	alachlor; 2,4-DB; metalochlor; pendimethalin; trifluralin
Oil palm fruit		simazine; 2,4,5-T; terbuthylazine
Olives	1,2-dichloropropane*; 1,3-dichloropropene	simazine; terbuthylazine
Rapeseed	carbofuran; γ-HCH	alachlor; cyanazine; simazine; trifluralin

Crop	Insecticides	Herbicides
Soybeans	aldicarb; carbofuran	alachlor; cyanazine; 2,4-DB; metalochlor; pendimethalin; trifluralin
Sunflower seed	carbofuran; γ-HCH	alachlor; metalochlor; pendimethalin; trifluralin
Pulses		
Beans (dry)	aldicarb; 1,2-dichloropropane*; 1,3-dichloropropene; dimethoate; γ-HCH; mehoxychlor*	cyanazine; pendimethalin; simazine; terbuthylazine; trifluralin
Chick-peas	dimethoate	cyanazine; MCPA; pendimethalin; simazine; terbuthylazine; trifluralin
Cow peas (dry)	1,2-dichloropropane*; 1,3-dichloropropene; dimethoate; methoxychlor*	cyanazine; MCPA; pendimethalin; simazine; terbuthylazine; trifluralin
Roots and tubers		
Cassava	dimethoate	metolachlor; pendimethalin
Potatoes	aldicarb; aldrin*; carbofuran; 1;2-dichloropropane; 1,3-dichloropropene; dieldrin; endrin*; ethylene dibromide; dimethoate	cyanazine; MCPA; metalochlor; pendimethalin; terbuthylazine
Sweet potatoes	aldicarb; 1,2-dichloropropane; 1,3-dichloropropene; methoxychlor*	-
Sugar crops		
Sugar beet	aldicarb; carbofuran; 1,2-dichloropropane; 1,3-dichloropropene; ethylene dibromide; dimethoate; γ-HCH	metalochlor; trifluralin
Sugar cane	aldicarb; carbofuran	alachlor; atrazine; cyanazine; 2,4-D; TP; metalochlor; simazine; terbuthylazine; trifluralin
Vegetables		
Cabbages		
	1,2-dichloropropane*; 1,3-dichloropropene; dimethoate; γ-HCH; methoxychlor*	metolachlor; trifluralin
Onions	1,2-dichloropropane*; 1,3-dichloropropene; γ-HCH	trifluralin
Tomatoes	1,2-dichloropropane*; 1,3-dichloropropene; dimethoate; γ-HCH; methoxychlor*	trifluralin
Other crops		
Cocoa beans	dimethoate; γ-HCH	simazine; terbuthylazine
Coffee (green)	aldicarb; carbofuran; dimethoate	simazine; terbuthylazine
Natural rubber	-	simazine; 2,4,5-T; terbuthylazine

* superseded pesticides (materials believed to be no longer manufactured or marketed for crop protection use).

Box 9.6. Examples of groundwater contamination by pesticides

A nationwide study of more than 1000 shallow wells and springs in the USA found that one or more pesticides were detected in more than half of the samples collected (Kolpin *et al.*, 1998). 95 per cent of the pesticide detections were at concentrations less than 1 μg/l. The most commonly detected pesticide was atrazine, reflecting both the widespread use of this chemical in agriculture, and its chemical stability and mobility in soil and groundwater. The frequency of pesticide detections was highest in agricultural areas where there was intensive use of these chemicals. Dieldrin was also commonly detected, despite its ban for agricultural use in the USA since the mid 1970s.

Some 40 pesticides were detected in drinking-water supplies in the United Kingdom (England and Wales) each year between 1989 and 1996. Most frequently detected pesticides were the herbicides isoproturon, simazine, atrazine, chlorotoluron MCPP and diuron. The percentage failing the drinking-water standard of 0.1 μg/l fell from 2.8 in 1991 to 0.2 per cent in 1996 (DWI, 1992, 1996, 1997).

Brazil has become the third largest user of pesticides in the world, only exceeded by France and the USA (Andreoli 1993). Contamination of surface water and groundwater by pesticide residues is widespread, and 100 per cent of water samples from springs in the Pirapó sub-basin contained pesticide residues.

A pesticide spill at a mixing site into a storm water soakwell in Western Australia contaminated a nearby irrigation well with concentrations of fenamiphos and atrazine exceeding 1000 and 2000 μg/l respectively (Appleyard, 1995). The concentration of fenamiphos was high enough to be toxic on prolonged skin contact, and would have caused severe health effects if the water had been used for drinking. Ten years after the spill, contamination is still present in toxic levels in groundwater at a distance of more than 300 m from the spill site. Investigations at other pesticide mixing sites in Western Australia (Appleyard *et al.*, 1997) detected a range of pesticides in groundwater including atrazine, chlorphyrifos, diazinon, dimethoate, fenamiphos, maldison, aldrin, chlordane, DDT, dieldrin and heptachlor. The most commonly detected pesticide in groundwater was diazinon, which was detected in 63 of the 78 groundwater samples collected at concentrations ranging between 0.1 and 4 μg/l.

Pesticides have a wide range of physical and chemical properties. As discussed in Chapter 4.6, the extent to which pesticides can be leached into groundwater through normal agricultural use depends on a number of factors. These include the degree to which pesticides are adsorbed onto organic matter in soils, the degree to which they are volatilized from the soil, the rate at which they degrade within the soil environment, their solubility in water, and the amount of water percolating through the soil profile available to mobilize a specific pesticide (Figure 9.3). Table 4.2 in Chapter 4.6 summarizes the extent to which the major groups of pesticides used in agriculture are retained or degraded in soils.

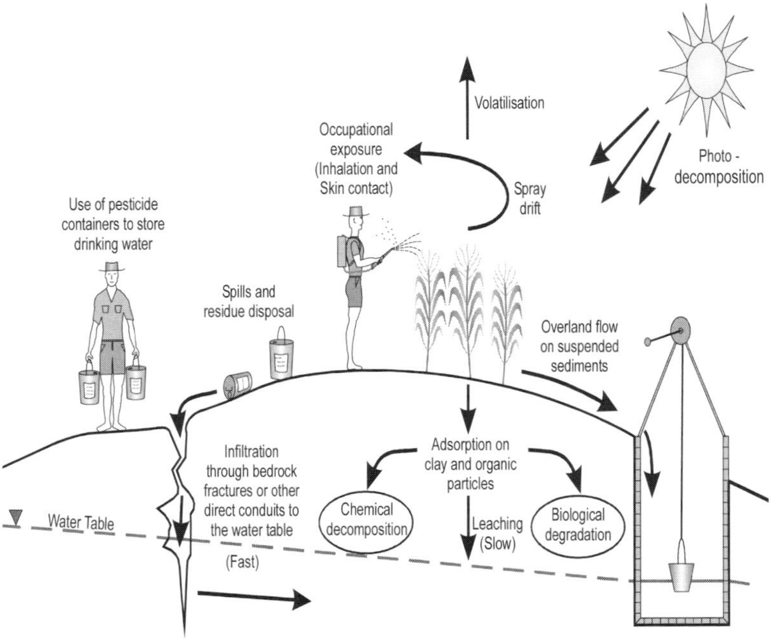

Figure 9.3. Transport of pesticides to groundwater, degradation behaviour and possible pathways that can cause human health problems

Low level contamination of groundwater by pesticides (concentrations less than 1 μg/l) commonly occurs in agricultural areas where the water table is shallow, and where soils are sandy and contain little or no organic matter. High concentrations of pesticides in groundwater (greater than 5 μg/l) are much more likely to result from the percolation of contaminated runoff into natural and man-made pathways through the soil profile than through their normal agricultural use (DeMartinis and Cooper, 1994). Natural preferred pathways due to vulnerable hydrogeology include karst features, fissured or fractured bedrock exposed at the surface, fissured clays, and deep animal burrows (Figure 9.3).

The most common man-made causes of severe groundwater pollution by pesticides is the infiltration of contaminated runoff directly into poorly constructed or abandoned water-supply wells, soak wells and other man-made conduits. This takes place either through the open casing, if the well is not properly capped, or through the annular space around the casing if the well has not been properly constructed. In particular, practices such as mixing of pesticides and cleaning of pesticide spraying equipment on bare soil, or the disposal of pesticide residues and empty drums near to man-made or natural features that allow rapid infiltration of contaminated runoff, pose a significant risk to groundwater. Drainage wells used to drain excess water in some irrigation areas are another common point of entry of pesticides into groundwater (DeMartinis and Cooper, 1994). Important point sources of pesticide pollution in groundwater are dip sites used to treat sheep and other livestock for ectoparasites (Hadfield and Smith, 1997), and spills or poor handling practice at sites used to mix or store pesticides before use. Moreover, there are large stockpiles of persistent pesticides in many countries, and if storage facilities are poorly designed or constructed (e.g. not sealed, covered or lockable) they may leak and cause pollution of waters. In this context, abandoned stockpiles (for example as a result of storage of older batches of pesticides or of civil war) are of particular concern as those normally are not subject to any means of control. Generally, all point sources can produce extremely high concentrations of pesticides that may persist in groundwater for long periods with little or no degradation.

Generally, there is much less information on the extent and severity of groundwater contamination by pesticides in developing countries. Pesticide usage in low-income countries is extremely variable, varying from almost no usage in large parts of Africa and Asia that rely on traditional and subsistence agriculture, to very high application rates in plantations in Central and South America.

9.6 IRRIGATION AND DRAINAGE

Irrigation is widely used in agriculture as a means of making otherwise unproductive land suitable for agriculture, or for substantially increasing agricultural productivity. However, this activity can affect groundwater quality by altering the water and salt balance in soil profiles, which in turn can change the physical and chemical characteristics of the soil. The large volumes of water used in irrigation – typically between 5000 and 15 000 cubic m per year per ha (Romijn, 1986) – allows solutes in irrigated soil profiles to be readily leached and affect groundwater quality.

Water can be applied to crops or pasture by surface irrigation (often by land flooding), by sub-surface irrigation, or by overhead irrigation through sprinklers, drip or trickle systems. Growing plants selectively uptake nutrients and some solutes, such as calcium and magnesium ions, from the irrigation water and water in the soil leaving residual soil water that is progressively enriched in some ions, particularly sodium. The solute content in the residual soil water cannot be allowed to increase too high, or else the changes in osmotic pressure across plant roots will inhibit the ability of plants to take up moisture and nutrients, so salts are generally allowed to leach to groundwater by applying excess irrigation water. However, even minor increases in the salinity of groundwater can affect the viability of growing some highly salt-sensitive crops like some vegetable species.

The problems can be compounded if irrigation water contains a high proportion of sodium ions. Sodium ion concentrations in residual water in these soils can accumulate to the point that ion-exchange sites on clays in the soil become saturated with sodium, changing the physical properties of the soil. Soils with excess sodium often become structureless and impermeable to water and air, and may not support plant growth. Sodicity problems in soils can also develop if the irrigation water contains high concentrations of bicarbonate or carbonate ions which causes calcium and magnesium carbonates to precipitate within the soil profile. Under these circumstances, residual soil water again becomes preferentially enriched in sodium ions.

As well as creating salinity and sodicity problems, irrigation can introduce a number of contaminants into groundwater that can affect human health. The large amount of water used in irrigation makes the risk of nitrate and pesticide leaching greater in irrigated than non-irrigated areas. In areas where soils contain significant concentrations of selenium, like parts of the USA and the Indian subcontinent, the infiltration of irrigation water can leach selenium and locally contaminate groundwater with this element. Water in soils beneath irrigation areas is often alkaline, and can leach significant amounts of fluoride that naturally occurs in soils at high concentrations in many areas, and is also introduced into soils as a contaminant in phosphatic fertilizers (Kolaja *et al.*, 1986).

Water is often drained from land to maintain or increase the agricultural productivity, but this activity can also have adverse impacts on groundwater quality. The major impacts include increasing groundwater salinity and nitrate concentrations and, in some areas, increasing sulphate, iron and heavy metal concentrations due to the acidification of soils caused by drainage.

Drainage systems often consist of a series of ditches or drainage pipes in or around individual fields to control the position of the water table, connected to collector and main drains designed to move the drained water to a convenient discharge area. In irrigation areas, water collected in drains can have a salinity up to ten times that of the applied irrigation water (Romijn, 1986), and in semi arid or arid areas, the salinity can be further increased by evaporation as water is moved in open channels away from the irrigated fields. Leakage of water from drainage channels can contaminate groundwater at some distance from irrigated areas with salt and nitrate and, in some areas, with other contaminants like selenium and fluoride.

Reclamation of waterlogged peaty soils by drainage causes land subsidence, firstly due to the shrinkage and compaction of the peat, and then due to the oxidation of the organic matter. It is estimated that drainage of peaty soils over the last 900 years in the Netherlands has caused land subsidence of about 2 m, of which 85 per cent is due to the oxidation of the peat (Romijn, 1986). The oxidation releases substantial amounts of nitrogen bound up in organic form in the peat, and nitrate concentrations in groundwater near drained peat areas often greatly exceed drinking-water guidelines. Drained peat soils are often grazed by a high density of livestock which can further increase nitrate concentrations in groundwater. The rate of oxidation of organic matter in peat is much greater in subtropical or tropical climates than in temperate areas, and the peat loss caused by oxidation can exceed 5 cm per year in some areas (Romijn, 1986).

The drainage of soils containing pyrite and other sulphide minerals can further cause severe environmental and health problems due to the release of sulphuric acid and toxic levels of metals caused by the oxidation of sulphides. These so called ASS were first recognized in the Netherlands more than 250 years ago, and although localized problems occur in parts of Europe, these soils are particularly widespread in some coastal parts of Asia, Australia, Africa and Latin America, and worldwide cover an area of more than one million square kilometres (ARMCANZ, 2000). It has been estimated conservatively that Australia alone has over 40 000 km^2 of acid sulphate soils, containing in excess of one billion tonnes of pyrite. If these soils were fully drained and developed for agriculture, the total amount of sulphuric acid released would be about 1.6 billion tonnes.

Groundwater beneath ASS areas can become contaminated with high concentrations of arsenic and heavy metals. The health status of farming communities living in ASS areas in Asia is often poor as they rely on metal rich acidic water for water supply (ARMCANZ, 2000). There is also evidence that certain species of disease-carrying mosquito actively seek acid drainage for breeding, compounding health effects in these areas.

9.7 CHECKLIST

NOTE ▶ *The following checklist outlines information needed for characterizing agricultural activities in the drinking-water catchment area. It supports hazard analysis in the context of developing a Water Safety Plan (Chapter 16). It is neither complete nor designed as a template for direct use but needs to be specially adapted for local conditions. The analysis of the potential of groundwater pollution from human activity requires combining the checklist below with information about socioeconomic conditions (Chapter 7), aquifer pollution vulnerability (Chapter 8), and other specific polluting activities in the catchment area (Chapters 10-13).*

ⓘ **What are the agricultural land use characteristics in the drinking-water catchment area?**

- ✓ Determine the proportion of land covered by agriculture
- ✓ Compile information on the types of agriculture (e.g. pasture land, arable land, irrigated or drained agriculture, horticulture, market gardening)
- ✓ Identify the main crops cultivated (including change over time)
- ✓ Compile information on the location and spatial distribution of agricultural land and different cultivation types
- ✓ Evaluate information on the historical evolution of land use patterns
- ✓ ...

ⓘ **Is manure applied in the drinking-water catchment area?**

- ✓ Estimate livestock densities, animal species and amount of manures produced
- ✓ Characterize storage conditions and handling practices for manures and estimate nitrogen volatilization and pathogen die-off rates
- ✓ Estimate composition of manures: presence of pathogens, nitrogen content, presence of veterinary pharmaceuticals

- ✓ Estimate quantity of manures applied and concentrations or loads of potential groundwater pollutants (e.g. pathogens, nitrogen, pharmaceuticals), including their change over time
- ✓ Evaluate patterns of manure application:
  - Assess adequacy of application rates: check whether criteria are based on (a) nutrient budgets and crop uptake rates, and/or (b) the need of getting rid of manure in areas with high livestock densities or intensive livestock farming
  - Assess timing of application in relation to hydrological events and to seasonal aspects (e.g. presence/absence of vegetation cover, frozen ground)
  - Assess adequacy of spreading methods
  - Assess adequacy of irrigation practices (if employed)
- ✓ Evaluate manure application practices in relation to aquifer vulnerability and physical conditions in the drinking-water catchment area (e.g. water table, soil, hydrogeology), and features that allow direct access of disease agents to the water table (e.g. abandoned wells, voids): consider checklist for Chapter 8
- ✓ …

Are animal carcasses being buried in the drinking-water catchment area?

- ✓ Determine the location and density of carcass burial sites
- ✓ Determine whether animals were culled because of a disease outbreak, and identify the disease
- ✓ …

Are fertilizers applied in the drinking-water catchment area?

- ✓ Characterize types and products of fertilizers used (slow or fast release)
- ✓ Check composition of fertilizers (e.g. nitrogen content)
- ✓ Estimate quantity of fertilizers applied and concentrations or loads of potential groundwater pollutants (e.g. nitrogen), including their change over time
- ✓ Evaluate patterns of fertilizer application: consider checklist for manure application (see above) for adequacy of application rates, timing, spreading methods and irrigation practices
- ✓ Evaluate fertilizer application practices in relation to aquifer vulnerability and physical conditions in the drinking-water catchment area (e.g. water table, soil, hydrogeology), and natural or man-made features that allow direct access of nitrogen to the water table (e.g. abandoned wells, voids): consider checklist for Chapter 8
- ✓ …

ⓘ **Are feedlots operated in the drinking-water catchment area?**

- ✓ Define locations and size (stock numbers) of feedlots
- ✓ Evaluate adequacy of siting and operation in relation to vulnerability and physical conditions in the drinking-water catchment area (e.g. water table, soil, hydrogeology), and features that allow direct access of disease agents to the water table (e.g. abandoned wells, voids): consider checklist for Chapter 8
- ✓ Assess adequacy of design, construction, condition, operation and maintenance (e.g. sealing, lining, open-air or closed facilities)
- ✓ Characterize wastes generated (e.g. manure, process wastewater, feed and bedding materials, silage)
- ✓ Evaluate availability, storage capacity, treatment efficiency and adequacy of design, construction, condition and maintenance of wastewater treatment facilities
- ✓ Check and assess disposal practices for treated or non-treated wastewater (e.g. irrigation): consider checklist for use of wastewater (see below)
- ✓ Estimate quantity of wastes generated and concentrations or loads of potential groundwater pollutants disposed (e.g. pathogens, nitrogen, pharmaceuticals), including their change over time
- ✓ Check and assess disposal practices for manures: consider checklist for manure application (see above)
- ✓ …

ⓘ **Is sewage sludge or wastewater used in the drinking-water catchment area?**

- ✓ Estimate composition of sludges and treated or non-treated wastewaters (e.g. pathogens, nitrogen, household and/or industrial chemicals): consider checklist for Chapter 10
- ✓ Evaluate adequacy of sludge treatment (e.g. composting) and/or storage time before land application
- ✓ Estimate quantity of sewage sludge or wastewater applied and concentrations or loads of potential groundwater pollutants (e.g. pathogens, nitrogen, pharmaceuticals), including their change over time
- ✓ Evaluate patterns of land application: consider checklist for manure application (see above) for adequacy of application rates, timing, spreading methods and irrigation practices
- ✓ Evaluate sludge and wastewater application practices in relation to aquifer vulnerability and physical conditions in the drinking-water catchment area (e.g. water table, soil, hydrogeology), and natural or man-made features that allow direct access of disease agents to the water table (e.g. abandoned wells, voids): consider checklist for Chapter 8
- ✓ Check whether fishponds are operated, define their locations

- ✓ Evaluate adequacy of siting and operation of fishponds in relation to vulnerability and physical conditions in the drinking-water catchment area (e.g. water table, soil, hydrogeology): consider checklist for Chapter 8
- ✓ Assess adequacy of design, construction, condition, operation and maintenance of fishponds (e.g. sealing, lining)
- ✓ ...

ⓘ **Are pesticides used in the drinking-water catchment area?**

- ✓ Check which pesticides are used and which active ingredients they contain
- ✓ Check whether there is indication for illegal use of banned pesticides
- ✓ Assess adequacy of siting, design, construction and practices of handling and mixing sites as well as storage facilities
- ✓ Check the location of dip sites for livestock treatment, and assess adequacy of practices employed
- ✓ Check whether there is indication of inadequate disposal practices of residues, surplus pesticides or drums
- ✓ Check whether there is indication of abandoned pesticide stocks
- ✓ Estimate quantity of pesticides applied as well as concentrations or loads leached to groundwater, including their change over time
- ✓ Assess patterns of pesticide application:
  - Assess adequacy of application rates: check whether criteria are based on (a) recommendations of producer and/or licensing authorities, and/or (b) the need to get rid of surplus pesticides, and/or (c) preventative spraying practice
  - Assess timing of application in relation to hydrological events, seasonal aspects (e.g. presence/absence of vegetation cover, frozen ground), and crop needs
  - Assess adequacy of spreading methods
  - Assess adequacy of irrigation practices (if employed)
- ✓ Evaluate pesticide application practices in relation to aquifer vulnerability and physical conditions in the drinking-water catchment area (e.g. water table, soil, hydrogeology), and natural or man-made features that allow direct access of pesticides to the water table (e.g. abandoned wells, voids): consider checklist for Chapter 8
- ✓ ...

ⓘ **Are irrigation and drainage practiced in the drinking-water catchment area?**

- ✓ Determine the scale to which irrigation and drainage is practised
- ✓ Compile information on irrigation and drainage techniques employed

- ✓ Assess irrigation methods used in relation to amounts reaching groundwater
- ✓ Check chemical composition of irrigation water and whether admixture of agrochemicals to drainage water is practised
- ✓ Evaluate irrigation and drainage practices in relation to aquifer vulnerability and physical conditions in the drinking-water catchment area (e.g. water table, soil, hydrogeology): consider checklist for Chapter 8
- ✓ …

ⓘ **Are hazardous events likely to increase groundwater pollution potential?**

- ✓ Evaluate whether and how storm water events would enhance transport of pollutants to the aquifer
- ✓ Evaluate which spills and accidents are likely to cause groundwater pollution
- ✓ …

ⓘ **Is drinking-water abstracted in proximity to agricultural activity?**

- ✓ Assess distance between agricultural activity and drinking-water abstraction
- ✓ Check adequacy of wellhead protection measures, wellhead construction and maintenance as well as sanitary seals used (see Chapter 18) to prevent ingress of contaminants from agriculture
- ✓ …

ⓘ **Are groundwater quality data available to indicate pollution from agricultural activity?**

- ✓ Compile historic data from the area of interest, e.g. from local or regional surveys, research projects or previous monitoring programmes
- ✓ Check need and options for implementation of new or expanded monitoring programmes likely to detect contamination from agricultural activities
- ✓ …

ⓘ **What regulatory framework exists for agricultural activity?**

- ✓ Compile information on national, regional, local, or catchment area specific legislation, regulations, recommendations, voluntary cooperation agreements, or common codes of good practices on the use, restrictions, ban, prohibition of substances in agriculture

- ✓ Check whether the regulatory framework adequately addresses environmental and specifically groundwater protection
- ✓ Identify known gaps and weaknesses which may encourage specific pollution problems
- ✓ …

Documentation and visualization of information on agricultural practices.

- ✓ Compile summarizing report and consolidate information from checklist points above
- ✓ Compile summary of types and amounts of substances produced, intentionally applied on land, non-intentionally released into soils or generated as wastes and which are potentially hazardous if they leach into the aquifer
- ✓ Map spatial distribution of general land use and specific agricultural uses (use GIS if possible)
- ✓ Map locations of fish ponds, feedlots, storage facilities (e.g. for pesticides, manures, sewage sludges), mixing, cleaning, livestock treatment sites and groundwater abstraction points (use GIS if possible)
- ✓ …

9.8 REFERENCES

Addiscott, T.M. and Gold, A.J. (1994) Challenges and implications of changes in nitrate limits for natural or potable water. In *Proc. of the XXV Congress of IAH 'Water Down Under'*, pp. 273-278, Adelaide.

Addis, P.B., Blaha, T., Crooker, B., Diez, F., Feirtag, J., Goyal, S., Greaves, I., Hathaway, M., Janni, K., Kirkhorn, S., Moon, R., Morse, D.E., Phillips, C., Reneau, J., Shutske, J. and Wells, S. (1999) *Generic Environmental Impact Statement on Animal Agriculture: A summary of the literature related to the effects of animal agriculture on human health*. Report produced by the University of Minnesota.

Agriculture W.A. (Western Australia) (2000) *Guidelines for the Environmental Management of Beef Cattle Feedlots in Western Australia*. Department of Environmental Protection, Health Department of WA, WRiversc and Western Australian Lot Feeders Association, Government of WA, South Perth.

Andreoli, C.V. (1993) The influence of agriculture on water quality. In *Proc. FAO Expert Consultation 'Prevention of Water Pollution by Agriculture and Related Activities'*, Santiago, Chile, 1992, pp. 53-65, Water Report 1, FAO, Rome.

Appleyard, S.J. (1995) Investigation of ground water contamination by fenamiphos and atrazine in a residential area: source and distribution of contamination. *Ground Water Mon. Remed.*, **15**(4), 110-114.

Appleyard, S.J., Manning, P. and Thorpe, P. (1997) Pest control depots as sources of groundwater contamination in Perth, Western Australia. *Land Contam. Remed.,* **5**(4), 299-305.

ARMCANZ (2000) *National Strategy for the Management of Coastal Acid Sulfate Soils*. Agriculture and Resource Management Council of Australia and New Zealand Discussion Paper, Tasmania.

Ball, P.R. and Ryden, J.C. (1984) Nitrogen relationships in intensively managed temperate grasslands. *Plant Soil*, **76**, 23-33.

Bauld, J. (1994) Groundwater quality in irrigation areas in Australia: interactions of agriculture and hydrogeology. In *Proc. of the XXV Congress of the International Association of Hydrogeologists 'Management to Sustain Shallow Groundwater Systems'*, pp. 423-432, Adelaide.

BC MAFF (2000) *Environmental Guidelines for Beef Producers*, http://www.agf.gov.bc.ca/resmgmt/fppa/pubs/environ/beef/beeftoc.htm (accessed April 29, 2005).

Brown, I.H. and Alexander, D.J. (1998) Influenza. In *Zoonoses: Biology, clinical practice and public health control*, (eds. Lord Soulsby and D.I. Simpson), pp. 365-386, OUP.

Burden, R.J. (1986) Elevated levels of nitrate in ground water beneath intensively grazed pastureland in New Zealand. In *Impact of Agricultural Activities on Ground Water* (eds. J. Vrba and E. Romijn), pp. 263-270, IAH Monograph vol 5, International Association of Hydrogeologists/UNESCO.

Centner, T.J. and Risse, M. (1999) US regulations regarding diffuse animal waste pollution. In *Proc. of an International Conference of Diffuse Pollution, 'Solutions and Innovations'*, pp. 352-361, Perth.

Cliver, D.O. (1987) Fate of viruses during sludge processing. In *Human Viruses in Sediments, Sludges and Soils*, (eds. V.C. Rao and J.L. Melnick), pp. 111-127, CRC Press, Boca Raton, Florida.

Close, M.E., Rosen, M.R. and Smith, V.R. (2001) Fate and transport of nitrate and pesticides in New Zealand's aquifers In *Groundwaters of New Zealand*, (eds. M.R. Rosen and P.A. White), pp. 185-220, NZ Hydrological Society Inc., Wellington.

DeMartinis, J.M. and Cooper, S.C. (1994) Natural and man-made modes of entry (of pesticides) in agronomic areas. In *Mechanisms of Pesticide Movement into Ground Water*, (eds. R.C. Honeycutt and D.J. Schabacker), pp. 165-175, Lewis Publishers, Boca Raton, Florida.

DOE (1986) *Nitrate in Water, A Report of the Nitrate Coordination Group*. Pollution Paper No. 26, Her Majesty's Stationery Office, London.

DWI (1992; 1996; 1997) *Nitrate, Pesticides and Lead Series* – 1989 and 1990; 1991 to 1994; 1995 and 1996, Drinking Water Inspectorate, London.

FAO (2002) *Statistical online database on agricultural production-crops primary*, http://apps.fao.org/subscriber/page/collections?subset=agriculture (accessed April 29, 2005).

Gannon, V.P.J. (2004) Control of zoonotic waterborne pathogens in animal reservoirs. In *Waterborne Zoonoses: Identification, causes and control*, (eds. J.A. Cotruvo, A. Dufour, G. Rees, J. Bartram, R. Carr, D.O. Clover, G.F. Craun, R. Frayer and V.P.J. Gannon), pp. 380-408, IWA Publishing, London.

Gannon, V.P.J., Humenik, F., Rice, M. Cicmanec, J.L., Smith Jr, J.E. and Carr, R. (2004) Control of zoonotic pathogens in animal wastes. In *Waterborne Zoonoses: Identification, causes and control*, (eds. J.A. Cotruvo, A. Dufour, G. Rees, J. Bartram, R. Carr, D.O. Clover, G.F. Craun, R. Frayer and V.P.J. Gannon), pp. 409-425, IWA Publishing, London.

Goss, M.J., Rollins, K.S., McEwan, K., Shaw, J.R. and Lammers-Helps, H. (2001) *The Management of Manure in Ontario with Respect to Water Quality*. Report produced by the University of Guelph for the Walkerton Inquiry, http://www.walkertoninquiry.com/part2info/commissuepapers (accessed April 29, 2005).

Hadfield, J. and Smith, D. (1997) Pesticide contamination of groundwater: an investigation in the Waikato region, New Zealand. In *Proc. of the 24th Australasian Hydrology and Water Resources Symposium 'Wai Whenua'*, pp. 246-251, Auckland.

Jorgensen, P.H. and Lund. E. (1985) Detection and stability of enteric viruses and indicator bacteria in sludge, soil and groundwater. *Water Sci. Technol.*, **17**, 185-195.

Juergens-Gschwind, S. (1989) Groundwater nitrates in other developed countries (Europe): Relationships to land use patterns. In *Nitrogen Management and Ground Water Protection*, (ed. R.F. Follet), pp. 75-138, Elsevier, Amsterdam.

Kolaja, V., Vrba, J. and Zwirnmann, K.H. (1986) Control and management of agricultural impact on groundwater. In *Impact of Agricultural Activities on Ground Water*, (eds. J. Vrba and E. Romijn), pp. 197-229, International Association of Hydrogeologists Monograph No 5, Verlag Heinz Heise.

Kolpin, D.W., Barbash, J.E. and Gilliom, R.J. (1998) Occurrence of pesticides in shallow ground water of the United States: initial results from the National Water Quality Assessment Program. *Environ. Sci. Technol.*, **32**, 558-566.

Lantzke, N.C. (1999) Phosphorus and nitrate loss from horticulture on the Swan Coastal Plain. In *Proc. of an International Conference of Diffuse Pollution, 'Solutions and Innovations'*, pp. 157-166, Perth.

Lawrence, A.R. and Kumppnarachi, (1986) *Impact of Agriculture on Groundwater Quality in Kalpitiya, Sri Lanka*. BGS, Report WD/03/86/20, Wallingford.

Liu, D. (1982) The effect of sewage sludge land disposal on the microbiological quality of groundwater. *Water Res.*, **16**, 957-961.

Mendonça, A.F., Pires, A.C.B. and Barros, J.G.C. (1994) Pseudosinkhole occurrences in Brasilia, Brazil. *Environ. Geol.*, **23**, 36-40.

Moore, B.E., Sagik, B.P. and Sorber, C.A. (1981) Viral transport to ground water at a wastewater land application site. *J. Water Pollution Control Foundation*, **53**, 1492-1502.

O'Connor, The Hon. D.R. (2001) *Report of the Inquiry into the Walkerton Incident*. Report published by the Ontario Ministry of the Attorney General.

Page, B.G. and Thomson, W.T. (1990) *The Insecticide, Herbicide, Fungicide Quick Guide*. Thomson Publications, Fresno, California.

Pionke, H.B., Sharma, M.L. and Hirschberg, K.-J. (1990) Impact of irrigated horticulture on nitrate concentrations in groundwater. *Agri., Ecosystems and Environ.*, **32**, 119-132.

Playford, P.E. and Leech, R.E. (1977) *Geology and Hydrology of Rottenest Island*. Geological Survey of Western Australia, Report 6.

Romijn, E. (1986) Influence of irrigation and drainage. In *Impact of Agricultural Activities on Ground Water*, (eds. J. Vrba and E. Romijn), pp.133-148, IAH Monograph vol 5, International Association of Hydrogeologists/UNESCO.

Ruz-Jerez, B.E., White, R.E. and Ball, P.R. (1994) Long-term measurement of denitrification in three contrasting pastures grazed by sheep. *Soil Biol. Biochem.*, **26**, 29-39.

Salama, R., Pollock, D. and Byrne, J. (1999) Experimental and modelling results of pesticide leaching in agricultural systems: examples from Thailand, Malaysia and Western Australia. In *Proc. of an International Conference of Diffuse Pollution, 'Solutions and Innovations'*, pp. 454-462, Perth.

Sumner, M.E. and McLaughlin, M.J. (1996) Adverse impacts of agriculture on soil, water and food quality. In *Contaminants and the Soil Environment in the Australasia-Pacific Region*, (eds. R. Naidu, R.S. Kookana, D.P. Oliver, S. Rogers and M.J. McLaughlin), pp. 125-182, Kluwer Academic, Dordrecht.

Thorn, R.H. and Coxon, C.E. (1992) Hydrogeological aspects of bacterial contamination of some Western Ireland karstic limestone aquifers. *Environ. Geol. and Water Sci.*, **20**(1), 65-72.

Tomlin, C.D.S. (2000) *The Pesticide Manual*, 12th edn, The British Crop Protection Council, Alton, Hampshire.

UK Department of Health (2001) *A Rapid Qualitative Assessment of Possible Risks to Public Health From Current Foot and Mouth Disposal Options*. Department of Health technical report, http://www.doh.gov.uk/fmdguidance/disposalrisk assessment.htm (accessed April 29, 2005).

UK Environment Agency (2000) Witness statement made to the UK BSE Enquiry, http://www.bseinquiry.gov.uk/files/ws/s490.pdf (accessed April 29, 2005).

UNDP (2001) *UNDP-GEF Persistent Organic Pollutants (POPs) Resource Kit*, GEF, New York.

US EPA (1996) *Use of Reclaimed Water and Sludge in Food Crop Production*. Report produced by the Office of Wastewater Management. http://www.epa.gov/OW-OWM.html/mtb/biosolids/biolib.htm (accessed August 11, 2005).

US EPA (1997) *National Water Quality Inventory: 1996 Report to Congress*. US EPA Report EPA841-R-97-008. http://www.epa.gov/305b/96report/index.html (accessed August 11, 2005).

US EPA (2000) *National Management Measures to Control Nonpoint Source Pollution From Agriculture*. Report produced by the Office of Water, US EPA. Contract No 68-C99-249, Washington, DC.

USGS (1999) *Fertilizers-Sustaining Global Food Supplies*, Fact sheet published by the USGS, http://pubs.usgs.gov/factsheet/fs155-99/fs155-99.html (accessed April 29, 2005).

Water and Rivers Commission of Western Australia (1996) *Guidelines for Irrigation of Vegetated Land With Nutrient-Rich Wastewaters*, Government of Western Australia, Perth.

Weil, R.R., Weismiller, R.A. and Turner, R.S. (1990) Nitrate contamination of groundwater under irrigated coastal plain soils. *J. Environ. Qual.*, **19**, 441-448.

White, R.E. and Sharpley, A.N. (1996) The fate of non-metal contaminants in the soil environment. In *Contaminants and the Soil Environment in the Australasia-Pacific Region*, (eds. R. Naidu, R.S. Kookana, D.P. Oliver, S. Rogers and M.J. McLaughlin), pp. 29-68, Kluwer Press, Dordrecht.

WHO (2004) *Guidelines for Drinking-water Quality: Recommendations, Volume 1, 3rd edition*, World Health Organization, Geneva.

WHO (2005) *Guidelines for the Safe Use of Wastewater, Excreta and Greywater. Volume 2: Wastewater use in Agriculture*, World Health Organization, Geneva.

10

Human excreta and sanitation: Potential hazards and information needs

G. Howard, J. Jahnel, F.H. Frimmel, D. McChesney, B. Reed, J. Schijven and E. Braun-Howland

In 2002 some 2.6 billion people (almost half of the world's population) did not have access to basic sanitation, based on the definitions given in Table 10.1 below (WHO and UNICEF, 2004). The population lacking sanitation is more than twice as many people as those lacking access to an 'improved' water supply. Increases in sanitation coverage have been achieved during the last decade, but they have essentially done little more than kept pace with population increase. It is in the rural areas of developing countries that access to sanitation remains most limited, in particular African and South Asian countries have very low rates of rural sanitation access. Current estimates suggest that access to improved sanitation has not increased above approximately half of the population of developing countries (WHO and UNICEF, 2004).

The lack of adequate sanitation is a key contributing factor to the ongoing high rates of diarrhoeal disease noted in developing countries. Improvement in sanitation has been consistently identified as being an important intervention to improve health (Esrey *et al.*, 1991; Esrey, 1996).

 Protecting Groundwater for Health: Managing the Quality of Drinking-water Sources. Edited by O. Schmoll, G. Howard, J. Chilton and I. Chorus. ISBN: 1843390795. Published by IWA Publishing, London, UK.

Table 10.1. Definitions of improved and unimproved sanitation (WHO and UNICEF, 2004)

Improved technologies	**Unimproved technologies**
Connection to a public sewer	Service or bucket latrines
Connection to a septic system	(where excreta are manually removed)
Pour-flush latrine	Public latrines
Simple pit latrine	Latrines with an open pit
Ventilated improved pit latrine	

Improvements in sanitation continue to lag behind the needs of the population and the provision of water services for a number of reasons. The water and sanitation sector has traditionally focused more on the provision of water services than sanitation facilities. This is partly due to the differing domains in which such facilities operate and the greater attractiveness of community as opposed to household interventions. As the provision of sanitation primarily functions at a household level, it has attracted less support from governments and donors more interested in community interventions, despite the gains delivered to the broader community by sanitation improvements. The demand for sanitation by households and communities has also sometimes been limited when there are other competing demands for environmental improvements (Briscoe, 1996).

There are numerous low-cost technologies that may be used to improve access, mainly using simple on-site disposal methods. The potential for such sanitation facilities for contaminating groundwater used for drinking supply is well recognized (Lewis *et al.*, 1982; ARGOSS, 2001), and it is strongly related to the hydrogeological and demographic characteristics of the settings in which they are applied. There is often a need to balance these risks and to accept some degradation in water quality where the health gains from improved sanitation outweigh potential risks of contamination from sanitation. This is most obvious in relation to chemical contamination, but also applies to microbial quality. For example, where potential contamination of shallow groundwater from on-site facilities is not a health threat because other water sources are used for drinking or because the filtration efficiency of the soil is likely to reduce pathogens, health gains from excreta disposal will outweigh potential risks from groundwater contamination through sanitation. However, such degradation should be prevented as far as feasible in order to avoid compromising future as well as present use of such groundwaters.

Contamination of groundwater by sanitation systems also occurs in settings where centralized sewerage systems are widely in place, as is common in most industrialized countries. They are often poorly maintained and leaks contaminate groundwater with pathogens and a diverse array of household and industrial chemicals. If sewage is treated in treatment plants and discharged into surface waters, in a number of settings persistent substances then reach groundwater, particularly where artificial recharge or bank filtration are practised. Further, in high income countries, groundwater contamination from decentralized on-site sanitation systems is also common, and may be due to inadequate design and maintenance. These typically occur in proximity to private wells and may become a hazard to such decentralized drinking-water supplies.

NOTE ▶ *Sanitation practices and the environment in which they take place vary greatly. Health hazards arising from sanitation practices and their potential to pollute groundwater needs to be analysed specifically for the conditions in a given setting. This chapter supports hazard analysis in the context of developing a WSP for a given water supply (Chapter 16).Options for controlling these risks are introduced in Chapter 22.*

10.1 CONTAMINANTS OF CONCERN FROM SANITATION SYSTEMS

10.1.1 Pathogens

Table 3.1 in Chapter 3 provides an overview of those viruses, bacteria, and in some settings protozoa, which are of concern in groundwater affected by human excreta. When assessing the risk of pathogen occurrence in groundwater, it is important to bear in mind that pathogens may be transmitted via a number of routes other than ingestion from water, including direct contact with excreta, food, flies or from aerosols emanating from excreted wastes. In developed countries, because of their infectivity, small size, persistence and low adsorption to solid surfaces, viruses can be regarded as the most critical microorganisms with respect to groundwater contamination and the related health risks. In developing countries, viral exposures may be much greater through other routes, notably related to poor hygiene and sanitation and the residual risk presented by viruses in groundwater is very low in comparison. Bacterial contamination of groundwater in these situations remains common and prevention of this may take precedence.

The presence of pathogens derived from human faeces in groundwater requires that faecal material leaching into the sub-surface contains pathogens excreted by infected individuals. Predicting whether pathogens may be in the population is difficult as the outward health of an individual cannot be taken as determinant of their status as pathogen reservoirs. An asymptomatic individual may harbour pathogenic organisms and serve as a carrier of disease. Therefore, because the majority of pathogens that affect human health are derived from human faeces, sanitation facilities should be sited, designed, operated and maintained on the assumption that excreta will contain pathogens.

Table 10.2 below provides data on the microbial content of untreated sewage entering into two sewage treatment works in the Netherlands and waste stabilization ponds in Brazil. The figures all refer to organisms per litre and the figures for bacterial content from Brazil have been adjusted from those reported in relation to numbers per 100 ml.

Table 10.2. Geometric mean concentration of selected microorganisms per litre in untreated sewage from two wastewater treatment works in the Netherlands (Hoogenboezem *et al.*, 2001) and waste stabilization ponds in Brazil (Oragui *et al.*, 1987)

Organism	Rotterdam Kralingseveer	Amsterdam Westpoort	Waste stabilization ponds Brazil
Cryptosporidium	540	4650	-
Giardia	1220	21 300	-
Sulphite-reducing clostridia	6.2×10^5	7.9×10^5	-
Clostridium perfringens	6.0×10^5	5.4×10^5	5×10^5
Thermotolerant coliforms	9.4×10^7	1.6×10^8	2×10^8
Faecal streptococci	3.6×10^6	1.6×10^7	3×10^7
Campylobactera	-	-	700
Salmonellae	-	-	200
F-specific RNA bacteriophage	2.2×10^6	4.3×10^6	-
Enterovirus	34	190	1×10^4
Reovirus	69	370	-
Rotavirus	-	-	800

10.1.2 Chemical contaminants

Where dry on-site latrines are used and no other wastes are disposed into the on-site system, contaminants are expected to be derived wholly from excreta. The major risk will therefore be from nitrate contamination. Nitrate is formed by the sequential, microbially-catalysed oxidation of ammonia to nitrite and then to nitrate (Cantor, 1997). Most nitrogen is excreted as urea, which readily degrades to ammonium. The person specific nitrogen load daily excreted amounts to 11-12 g (Hamm, 1991). With respect to manure, storage times and conditions will affect ammonium losses due to volatilization (Chapter 9). Most relevant for groundwater, microbial oxidation may convert ammonium to nitrate, which is conserved in oxidizing subsurface environments. As nitrate is highly soluble in water and very mobile it readily poses a risk to groundwater (Chapter 4). However, it should be noted that in reducing conditions nitrogen remains in a reduced form (ammonia and nitrite) and this has been noted in groundwater underlying several cities dependent on groundwater (Lawrence *et al.*, 1997).

Wet sanitation systems, especially those which serve household waste water as well as excreta, likely contain a more complex mix of chemicals including those derived from household use such as laundry detergents. Sewage that has centralized management, and serves both residential and industrial users, is likely to contain a complex mixture of organic and inorganic chemicals mainly used in manufacturing and processing. Industrial contaminants will depend on the types of industry in the catchment of the sewerage system, and their mixture in sewage is typically very variable and complex (see e.g. Burston *et al.*, 1993; Bishop *et al.*, 1998). If released to the subsurface, their persistence and mobility determines the potential to contaminate groundwater (Chapter 4).

The concentrations of dissolved constituents in sewage depend both on the household consumption of water and the relative proportion of industrial effluent in relation to domestic sewage in municipal sewers. The average composition of organic and inorganic substances found in domestic sewage is shown below in Table 10.3.

Sewage composition shows diurnal patterns as domestic water use changes. It may also change over larger time intervals due to changes in industrial effluent inputs, as can be seen in the example of the raw sewage composition in Nottingham in Table 10.4.

Table 10.3. Average composition of domestic sewage (Koppe and Stozek, 1986; Klopp, 1999)

Parameter	**Concentration (mg/l)**	**Parameter**	**Concentration (mg/l)**
Carbohydrates	95	Magnesium	15
Fats	100	Zinc	0.2
Proteins	115	Manganese	0.15
Detergents	43	Copper	0.15
Phosphorus	10	Lead	0.1
Sulphur	46	Nickel	0.04
Chloride	50	Chromium	0.03
Boron	2	Tin	0.015
Sodium	80	Silver	0.01
Potassium	19	Cadmiun	<1
Calcium	70	Mercury	<0.1

Table 10.4. Filtered raw sewage composition in Nottingham (based on Barrett *et al.*, 1997)

Parameter	**Sample date**			
	2 Feb 1995	15 Aug 1995	1 Feb 1996	13 June 1996
$CaCO_3$ (mg/l)		162		282
Cl (mg/l)	322	353	386	200
NH_4-N (mg/l)	25.7	40.6	22.2	30.0
NO_2-N (mg/l)	<0.1	0.1	<0.1	<0.1
NO_3-N (mg/l)	<0.3	<0.3	<0.1	<0.3
Ca (mg/l)	86	76	99	63
Mg (mg/l)	33	30	41	21
Na (mg/l)	226	289	170	157
K (mg/l)	17.5	24.5	13	18
SO_4 (mg/l)	135	78	178	69
PO_4-P (mg/l)	6.3	10.7	4.0	6.7
Ag (μg/l)		40		
B (μg/l)	1050	1700	980	1300
Pb (μg/l)	<10	41	<10	
Hg (μg/l)	0.4	0.5	<0.1	8.0
Cr (μg/l)	<5.0	0.04	9	
Cu (μg/l)	<20	130	25	
Ni (μg/l)	9	0.15	20	
Zn (μg/l)	177	290	95	
Cd (μg/l)	2.1	<1.0	<1.0	
TTC(cfu/100 ml)	200	2 086 000	58 000	>500 000
TON (mg/l)		24.2		14.1
TCM (μg/l)	6	Trace	0	0
Toluene (μg/l)	14	138	215	50
TeCE (μg/l)	11	10	28	63
Decane	Not quantified	0	0	0
Oxylene	0	0	Trace	0
TCE	0	0	0	Trace
1,4-DCB	0	0	0	Trace

This example is typical of bulk sewage from a medium-sized, moderately industrial city and highlights the diversity and variability of substances that can potentially impair groundwater quality, where such sewage leaks into groundwater. The greatest risk posed to groundwater under such settings is often from leaking sewers (see Section 10.2.3).

The prevalence of pharmaceutical chemicals and personal care products (PCPs) is increasingly observed in water supplies (Doughton and Ternes, 1999). Sources of PCPs and pharmaceuticals can be from manufacturers, medical facilities or personal use. Pharmaceuticals may be excreted by patients and can reach the aquatic environment via waste water and potentially enter groundwater (see also Chapter 4.7.1).

10.2 TYPES OF SANITATION AND THEIR POTENTIAL TO CONTAMINATE GROUNDWATER

This section provides an overview of widely practised methods of excreta disposal. There are numerous excreta disposal methods, which range from low-cost options suitable for low-income communities to expensive methods involving several stages of collection and treatment. In general terms, excreta disposal methods fall into two broad categories – on-site and off-site systems. On-site systems are point sources and therefore will be expected to exert the greatest impact on the groundwater in their vicinity, although where there are large numbers of on-site systems the overall impact may be widespread. There are a number of references that provide detailed descriptions of design, engineering, construction, use, and maintenance of various on-site sanitation systems (Metcalf and Eddy, 1991; Franceys *et al.*, 1992; Mara, 1996). Off-site systems represent more diffuse pollution and risks to groundwater may be found along the sewer lines, at the treatment works and from the final effluent discharged to the environment. The discussion in this section provides a summary of the available knowledge of the potential of these technologies to contaminate groundwater and the type of information needed to assess this potential.

10.2.1 Open air defecation

WHO and UNICEF (2004) estimated that in 2002 more than one-third of the global population (ca. 40 per cent) still lack access to basic sanitation facilities. The unserved population primarily resides in lower-income countries in Africa and South Asia and within rural areas. The lack of adequate sanitation for half of the Earth's population indicates that open-air defecation is practised by millions of people. Open-air defecation is generally found in developing countries, particularly in low-income rural and peri-urban communities (Pedley and Howard, 1997).

The risks of groundwater contamination from open-air defecation are variable and largely depend on local conditions, including groundwater use for drinking-water supply. Pathogenic microorganisms in faeces may contaminate groundwater or spring abstraction points by leaching through soils into shallow groundwater or springs, flowing into outcropping or shallow rock fractures, seeping into pits or low areas (recharge zones) or runoff to surface water, with secondary transport to connected aquifers. Several studies in developing countries have shown that open-air defecation in urban areas may actually be the primary source of faecal matter washed into poorly maintained water sources (Gelinas *et al.*, 1996; ARGOSS, 2002). In rural areas of developing countries, the concentrations of microbial contamination is often seen to be highest at the onset of the rains and then declines during the remaining wet season and into the dry season. The build-up of faecal matter that is readily washed into groundwater sources may provide a plausible explanation of this phenomenon. Risks of chemical contamination from open-air defecation are generally lower (Gelinas *et al.*, 1996). Open-air defecation is also associated with the transmission of helminths such as hookworm, although this occurs via contact with contaminated soils rather than groundwater (Cairncross and Feacham, 1993).

Even where the risk is reduced by decreasing the potential for defecation near wells or springs, open-air defecation always represents a major risk to public health and the use of safe alternatives should be encouraged. The provision of improved sanitation would always be preferred, although in some cases, burial of faecal material several centimetres below the ground surface may be promoted as an interim measure. The risk to groundwater from such burial is likely to be limited in low-density settlements, but would potentially represent a more significant risk in urban areas.

10.2.2 On-site sanitation

The general principles of design and operation of on-site sanitation systems are discussed by Franceys *et al.* (1992) and the risk posed to groundwater has been the subject of several major reviews and research (Lewis *et al.*, 1982; van Ryneveld and Fourie, 1997; ARGOSS, 2002; see Box 10.1 below).

On-site systems are generally those in which excreta and anal cleansing materials are deposited directly into some sort of container, most commonly a subsurface excavation or tank.

Common to all forms of on-site sanitation is that part of the decomposition process is performed on-site. In some cases, the sludge will be fully decomposed within the pit, whilst in some others periodic desludging and treatment of the waste is required. The risk of such a contamination from the collection of waste at a single point depends largely on the design of the facility, but has been noted as being significant (Lewis *et al.*, 1982; van Ryneveld and Fourie, 1997; ARGOSS, 2002). On-site wet systems also typically require some form of soakaway to dispose of excess effluent and this may increase risks from both pathogens and nitrate (ARGOSS, 2001).

On-site systems may be low-cost options such as various forms of pit latrine, or high-cost options such as septic tanks that provide a similar level of service to sewerage. They may be 'wet' systems, where water is used to flush the waste into a tank or pit, or dry systems using little or no water. Wet systems may require very limited water (for instance a pour-flush latrine) or require large volumes of water consistent with water piped inside the house with multiple faucets and which can receive all wastewater produced by the household.

Contained systems, such as vault latrines, cess pits and composting latrines are an intermediate stage between on- and off-site options, with varying amounts of decomposition occurring on-site, but final treatment and disposal of the excreta occurring elsewhere. Collection is in batches rather than continuous.

Box 10.1. Case Study: Urbanization's Impacts in Santa Cruz, Bolivia (based on BGS, 1995)

Santa Cruz is located on the eastern plains of Bolivia about 25 km from the easternmost edge of the Andean Cordillera. Population growth in Santa Cruz is about 4 per cent per year, one of the greatest growth rates in the Americas. Though the city is largely urban, industrial and commercial development are experiencing rapid growth. The city is laid out in a radial pattern along the banks of the River Pirai. A surface ridge bisects Santa Cruz, effectively forming a surface water drainage divide.

Groundwater provides all of the potable water supply in Santa Cruz. The aquifer comprises variable alluvial beds of sand, clay, sandy clay lenses and gravels. The alluvium in the southern part of Santa Cruz is dominated by medium to coarse grained gravels and sands. In the east, northeast and northern parts of the city, beds tend to be thicker and contain an increased proportion of clays. Groundwater flow has been described as southwest to northeast. Groundwater levels vary from essentially at the ground surface to 15 m below ground surface beneath the water divide ridge bisecting the city.

The population density in Santa Cruz is relatively low, at only about 45 persons/ha on average, and less than 30 persons per ha in outer areas. The inner city has densities in excess of 120 persons per ha, and the south eastern portion of the city has the greatest density of more than 150 persons per ha. Growth projections are strongest in the south eastern portions of the city.

Water supplies in developed areas are provided by a diverse consortium of municipal, collaborative and private entities. In less developed areas, progress is being made to consolidate water supply into the formal system. However, on the southeast edge of the city, water supply varies in sophistication from dug wells to shallow hand pump fitted wells to deeper boreholes supplying local networks.

The rapid development in Santa Cruz and the vulnerability of the aquifer which supplies drinking-water is cause for concern. The BGS conducted a study to measure the effects of urbanization on the aquifer.

The results showed that urbanization is resulting in negative impacts to Santa Cruz's groundwater. Much of the water that is used in the city is expected to be returned to groundwater as recharge from on-site sanitation – used over about 85 per cent of the city's area, and leaking sewers. While for the most part not a current health concern, concentrations of thermotolerant coliforms, nitrate, chloride and manganese are increasing in shallow groundwater and are beginning to leach into deeper groundwater. In fact, most of the shallow dug wells contained high counts of thermotolerant coliforms. Conditions are worse in areas of high urbanization, with nitrate and manganese currently exceeding WHO guideline values for drinking-water quality in shallow wells.

In on-site systems, the solid part of the waste undergoes anaerobic decomposition within the pit and the contaminants in the effluent are removed through attenuation, die-off, predation and dilution as it travels through the underlying soil, or from soakaways and drain fields. Natural attenuation processes tend to act most effectively in the upper soil level and because so many on-site sanitation systems bypass this layer, this may increase risks of groundwater contamination. This may be a particular problem with designs that allow for direct infiltration of effluent through the sides and base of the pit, particularly where hydraulic loads are high. However research suggests that within a relatively short period of time a biologically active layer forms around the active layers of the pit (i.e. those receiving faecal material) and forms a mat of gelatinous material of bacteria and fungi. Previous research suggests that within three months this layer inhibits movement of faecal bacteria and within seven months the presence of bacteria is largely restricted to the latrine (Caldwell and Parr, 1937). Other studies point to the limited penetration of bacteria to within a travel time equivalent to five to seven days and lateral breakthrough generally limited to within five meters (Subrahmanyan and Bharaskan, 1980; Chidavaenzi *et al.*, 2000).

The biologically active layer appears to work in two key ways. Firstly, the presence of predatory microorganisms within the biologically active layer allows for permanent removal of some pathogens. Secondly the nature of the mat reduces the porosity of the soil matrix – essentially clogging the soil pores – thereby allowing an increased period for attenuation. In a properly sited, designed and operated system, the mat can also provide a protective barrier between waste and groundwater by effectively maintaining an unsaturated zone between the clogged pores and groundwater. This would appear to work in dry systems but is unlikely to be as effective when wet systems are used as the hydraulic load may be sufficient to overload the natural removal of microorganisms and may extend the saturated zone between the pit and the water table (ARGOSS, 2001).

It has been noted that the formation of a biologically active layer varies with subsurface nature. Some formations allow much greater breakthrough (presumably because the nature of the pores does not allow such ready formation of biologically active layers). Viruses in particular are not so readily retarded. Fissures in the sub-surface, hydraulic overloading of the system or a high groundwater table can result in system failure and contamination of groundwater or breakthrough. For instance, a study by Pyle *et al.* (1979) showed that bacteria could travel as far as 920 m where flow rates were high.

The biologically active layer appears to have a far more limited impact on chemical pollution, particularly nitrate. In most shallow groundwaters nitrate concentrations are likely to increase with the use of on-site sanitation. In some groundwater environments, this risk is mitigated because of reducing conditions that promote reduction of nitrate to ammonia. This may lead to very limited nitrate concentrations in groundwater despite a high density of pit latrines (ARGOSS, 2002). Oxidation reactions in shallow groundwater where organic loading is high (as would be typically found under urban areas) may enhance denitrification by using nitrate as further oxidizing agent after oxygen has been consumed (Lawrence *et al.*, 1997). However, in lower density urban settlements and in rural areas, oxidizing conditions are more likely to be found, and denitrification does not take place, allowing nitrate from on-site sanitation to accumulate. This potentially affects long-term availability of groundwater as a source of drinking-water (ARGOSS, 2002).

Box 10.2. Nitrate contamination of groundwater in areas with pit latrines in Francistown, Botswana

A study carried out on nitrate contamination of groundwater in populated areas of Botswana in 1976 found elevated concentrations of up to 80 mg/l in wells in several major villages (Hutton *et al.*, 1976). The phenomenon was related to the occurrence of pit latrines and septic tanks used for sanitation. In 2000 the Geological Survey of Botswana in co-operation with the Federal Institute of Geosciences and Natural Resources, Germany, carried out a study on groundwater contamination in the city of Francistown.

Francistown is located in the semi-arid region of eastern Botswana at the confluence of the seasonal Tati and Ntshe rivers. Rainfall averages around 500 mm/a. Born during the late 19th century as a gold mining town, the city's rapid economic development particularly since the 1970s has turned Francistown into the second largest city in Botswana with a population of approximately 80 000 inhabitants.

The water demand of the town used to be met entirely by groundwater resources locally available from shallow alluvial and fractured volcanic rock aquifers. The overlying soils are rarely thicker than 0.5 m. In the 1970s it was found that groundwater produced from the city's public wells contained elevated concentrations of nitrate. In addition the growing water demand could no longer be met by the groundwater resources. For these reasons public water supply was shifted to surface water from the new Shashe dam in 1982, which was originally constructed for mining purposes 30 km from Francistown.

The rapid population growth led to an extensive development of pit latrines, which also serve for domestic wastewater disposal. A centralized sewerage system covering approximately 50 per cent of the city area started to operate some years ago, now discharging to a wastewater treatment plant. Recently the sewerage system was extended to the remaining city area. However, connection to the sewerage system is voluntary, and the use of pit latrines and – to a lesser extent – septic tanks is currently still the main means of wastewater discharge in the newly connected areas.

The recent groundwater quality survey sampled 47 public and private wells within and around Francistown. Analyses showed that nitrate concentrations well above the maximum allowable limit for drinking-water in Botswana of 45 mg/l were frequent within the city area, often reaching values between 100 and 300 mg/l (Figure 10.1). Some of these wells had already been sampled in the mid 1970s, and comparison with the old extremes of 80 mg/l shows that nitrate levels had significantly increased. Groundwater sampled in the recent survey from wells situated in remote areas outside the city contained considerably less nitrate, in most cases below 40 mg/l, which indicates that the cause of nitrate contamination is likely to be anthropogenic.

To gain knowledge on the possible causes of contamination, potential point and non-point hazards to groundwater pollution such as areas with pit latrines, intensive farming, mining activity, cemeteries, abattoirs etc. were mapped. In addition flow directions of groundwater were inferred from the construction of a groundwater contour map. Combining the results of the nitrate analyses with this information showed that nitrate concentrations in fact maximize in areas with pit latrines. Also, not a single borehole lying in or close to such areas was found to have a nitrate concentration below 100 mg/l. This relation between the location of areas with pit latrines and the occurrence of nitrate contamination can be seen in Figure 10.1, which shows a part of the project area in the south eastern part of Francistown.

However, nitrate concentrations seem to quickly decrease with distance from a contamination source, whether this occurs by denitrification or dilution. The July 2000 nitrate contamination of groundwater might have been aggravated by extraordinarily high rainfall in the range of 2000 mm in February and March 2000.

The findings support the conception that the use of pit latrines can cause serious nitrate contamination in groundwater. Nitrate contamination is promoted when households are connected to water supply but not to sewerage, as this causes an increase of percolation of waste water from the pit latrines to the groundwater. An improvement is likely when all households of Francistown are connected to the centralized sewerage system.

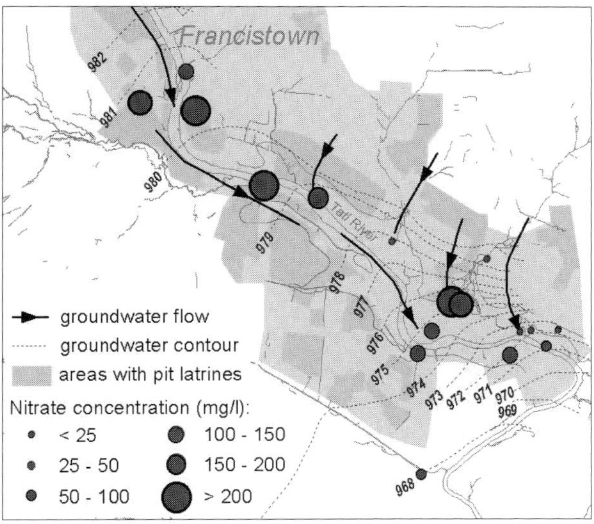

Figure 10.1. Nitrate concentration in groundwater in southeastern Francistown, eastern Botswana, in July 2000

Pit and trench latrines

Pit and trench latrines are widely used for direct disposal of excreta due to the simplicity of design, ease of construction and low cost. Examples of pit latrine designs include simple pit latrines, ventilated improved pit latrines, pour-flush latrines, raised pit latrines, and composting latrines (Franceys *et al.*, 1992). Aqua privies are discussed in conjunction with septic systems below.

Some pit latrine designs incorporate a twin-pit design, which provides several benefits (Pickford, 1995). Additional time is afforded for decomposition of the waste in the pit not being used, rendering a well-composted, odourless product which is relatively easy to handle when the pit is emptied. Where two pits are used, they are often dug to shallower depths, thereby increasing the depth of the unsaturated soil zone and providing more time for attenuation and, consequently, enhanced protection of groundwater. However, the use of twin pits requires effective user operation and this cannot always be assured, leading to overloading and risks incurred during emptying.

The risks from pit latrines are primarily derived from the increased risk of accumulation of faecal material in one place that may allow subsurface leaching of contaminants. The Francistown case study given in Box 10.2 highlights the potential for nitrate contamination from pit latrines and shows how situation assessment can document this by mapping.

As noted above, the presence of biologically active layers reduces the risk of subsurface leaching caused by the accumulation of faecal material. It has been suggested that ensuring that there are two meters of sandy or loamy soil below the base of the pit will reduce the risk of microbial contamination of groundwater (Franceys *et al.*, 1992). The depth of the soil in the pit or trench in relation to the seasonal high groundwater table will indicate the seasonal risk of contamination.

Pits constructed into the water table are almost always a substantial cause for concern. In general, if pit or trench latrines are constructed so that they are safe from flooding and sufficient depth between the base of the pit and the groundwater is maintained, risks of microbial pollution are reduced.

Septic tanks and aqua privies

Septic systems provide a similar level of service as sewers as they can be linked to water closets located within a house. In unsewered cities, such as many in North America, septic tanks are the main method of sewage disposal (Lerner, 1996). While some large, community-scale septic systems are in use, most accommodate waste loads from a single dwelling to a few dwellings. Aqua privies are essentially limited to single or a few dwellings. The amount of space necessary for a septic system's drainage field, described below, can limit the number of individuals or dwellings served by septic systems.

Septic tanks and aqua privies operate by initial deposition of excreta into an impermeable tank with overflow of excess liquid into a soakaway or drain field. In some cities, such as Hanoi in Vietnam, the effluent enters the surface water drainage system. In both technologies, the sludge is retained under water and this must be maintained to reduce offensive odours. Septic tanks are usually located at a distance from the toilet and water is used to flush excreta into the tank. Typically water volumes required are similar to those used in sewerage and most household wastewater may be discharged into the septic tank. The tank in an aqua privy is located just below or adjacent to the toilet. Water requirements are often lower than for septic tank systems, but the tank requires periodic addition of water to ensure a water seal is maintained.

Inside the tank of septic systems and aqua privies, solids settle out and are deposited on the tank bottom; a scum forms a crust on the surface. As the tank fills with liquid, the overflow is channelled out of the tank. A variety of options are available for the liquid once it leaves the tank. Many systems connect to a lateral permeable pipe or series of parallel permeable pipes. A method which allows for both filtering and treatment by soil microorganisms is to disperse the liquid into a shallow soil drainage field by means of a shallow buried permeable pipe or parallel series of pipes. Other options include piping the overflow to a soakaway, to a sand filtration unit or pit, or to a central sewer.

Microorganisms inhabiting the tank and drainage field of a properly operated, well maintained septic system can degrade carbonaceous, nitrogenous, and microbial waste constituents. Carbonaceous constituents may be completely degraded, however, organic nitrogen and ammonia are likely to be oxidized to nitrate before leachate reaches the groundwater in oxidizing environments. Nitrate contamination is common in areas where septic system density is moderate to high. Where the environment is reducing (i.e. anaerobic), microbial action may transform nitrogen to nitrogen gas.

Pathogen destruction via predation, attenuation and thermophilic or natural die off occurs in the tank and drainage field, but may be incomplete especially for viruses. This may result from high flow rates reducing the period of contact for predation and attenuation and promoting breakthrough, and from low clay content which reduces the potential for adsorption (Scandura and Sobsey, 1997). Both aerobic and anaerobic systems may preferentially destroy bacterial pathogens incompatible with the associated environments, although this will have limited impact on facultative microbes such as *E. coli*, which survive in both aerobic and anaerobic environments.

Disease outbreaks associated with inadequately sited, inadequately maintained, overloaded and malfunctioning septic tanks have been documented (Craun, 1984; 1985) and an example is provided in Box 10.3. When assessing the risk from septic tanks and aqua privies, it is important to bear in mind that there are two distinct components that must be managed. The tanks containing the sludge must be impermeable and properly maintained. They therefore require periodic inspection, which is most easily performed immediately after emptying. Furthermore, the disposal of the sludge is important and the benefits of good design and operation of a septic tank or aqua privy will be undermined if subsequent disposal of the waste is poorly managed.

In addition, is it important that the drain fields or soakaway are properly located and designed, taking into account infiltration rates of the soil, depth to groundwater, groundwater velocity and direction and distance to the nearest groundwater source used for supply of drinking-water. Septic tank use is viable in areas where soils contain relatively high concentrations of organic matter and infiltration rates are 10-50 l/m^2 per day, although this is dependent on the distance to the nearest groundwater source and depth to water table (Franceys *et al.*, 1992). It is important to bear in mind that the drain field will eventually become clogged and a new site developed (Mara, 1996). There should always be a minimum distance to the water table beneath the base of trenches or seepage pits, for instance the US Public Health Service (1965) recommends a minimum of four feet (i.e. 1.2 m).

Box 10.3. A severe outbreak of waterborne disease – an on-site treatment system suspected of contaminating a well (based on CDC, 1999)

The investigation by the New York State Department of Health into the 1999 outbreak associated with attendance at the Washington County Fair indicated that the outbreak may have resulted from contamination of the Fair's Well 6 by a dormitory septic system on the fairgrounds.

A total of 781 people were affected with either confirmed or suspected cases in the outbreak. Of these, 127 cases of *E. coli* infection and 45 cases *Campylobacter jejuni* were confirmed, with 2 deaths and 71 people hospitalized, of which 14 developed haemolytic uraemic syndrome, a severe complication of *E. coli* O157:H7 infection that can lead to kidney failure.

A case-control study concluded that consumption of beverages sold by vendors supplied with water from Well 6 was a key risk factor for patients with culture-confirmed illness. *E. coli* O157:H7 was found in samples from Well 6, which supplied unchlorinated water to some vendors, and water distribution pipes leading from Well 6. *E. coli* O157:H7 was found in the septic system of the dormitory building. The discharge area (seepage pit) of that septic system was approximately 36 feet from Well 6 and tests showed a hydraulic connection, at the time of the tests, from the septic system to Well 6. The source of the *E. coli* O157:H7 in the dormitory septic system is unknown and tests did not identify *Campylobacter* in samples from the dormitory septic system or Well 6.

Tests did not demonstrate a connection from the manure storage area near the Youth Cattle Barn or the presence of *E. coli* O157:H7 in samples taken from that area. However, because exact environmental conditions (including drought followed by rain) present at the time of the Fair could not be replicated and because manure was removed daily, it may never be known if manure-contaminated water percolated from the manure storage area to Well 6.

Contained systems

Contained systems, also referred to as 'cartage systems', are impermeable vessels used to collect excreted wastes. Bucket latrines are contained systems in their simplest form. Excreta in a bucket latrine, commonly referred to as night soil, must be emptied routinely. Bucket latrines represent a major health risk, particularly to the collectors and their use should not be promoted. WHO has developed guidelines for the safe use of excreta in agriculture (WHO, 2005).

The risk of groundwater contamination from bucket latrines may be limited and will certainly be far less than occupational hazards of collectors and transmission of pathogens via flies for users. If bucket latrines are used, the risk to groundwater sources used for drinking-water will be reduced by disposal in lined pits or trenches, or in a sewer if available. This should largely limit the potential for leaching of pathogens and nitrate into groundwater. Cartage to areas far removed or downgradient from groundwater supplies affords more options, with unlined pits or trenches as disposal options for a lesser risk to drinking-water. Dispersed burial reduces the potential for significant groundwater contamination and provides the opportunity for reducing health impacts due to direct contact and insect or vermin attraction, as well as permitting soil bacteria to degrade excreta. The presence of a confining soil layer will determine whether or not leaching into groundwater occurs. Disposal of night soil into surface waters may not only contaminate these waters but also groundwater if the aquifer is under the influence of surface water.

Impermeable tanks are an alternative system that primarily provide a holding area for excreta. Wastes must be removed periodically, generally by hand or by pumping. Wastes are then transported to a centralized treatment system. Such latrines are often referred to as vault latrines or cess pits and some decomposition will occur during storage, although nonetheless the waste must be periodically removed. If the container is intact, no risk to groundwater exists other than that from waste which might seep into soils during emptying or spills during transport. However, maintaining a completely impermeable container may be difficult. Furthermore, in some settings deliberate drilling into the container is practised, often illegally, which allows liquids to leak to the subsurface thus reducing the amount of wastes to be collected and transported to centralized treatment systems, and therefore saving disposal costs for the owner of the container. Though hard to identify, this practise may pose an increased risk to groundwater pollution with pathogens where hydrogeological conditions favour their transport.

Composting technologies

Composting utilizes bacteria, fungi, and other microorganisms to degrade organic waste materials. Composting may be performed in anaerobic or aerobic regimes. As treatment is effected, temperatures in the degrading waste increase due to microbial activity. As the temperature increases, a succession of microorganisms, progressing from mesophilic to thermophilic, inhabit the compost until temperatures rise beyond the ability of organisms to survive. When properly managed, composting temperature can be controlled to optimize degradation, often through the introduction of air and mixing. Introduction of air also minimizes odour production. Odour problems are more predominant in anaerobic systems which generally require additional maintenance and collection of methane off-gas. Control of moisture is important in any composting system to optimize degradation rate. At the end of the composting process, temperature can be allowed to increase to effect pathogen destruction.

Latrines that employ composting as a treatment process are more advantageous, in many respects, than alternative treatment or disposal processes. Modern designs generally require minimal maintenance, destroy waste products and pathogens biologically and produce a by-product that may be used as a soil amendment or fill material. Some designs include the need to segregate urine from faeces (i.e. urine diversion). Disposal of the urine can result in contamination of surface water or groundwater, if not controlled.

A functioning composting process will protect groundwater from contamination with pathogens and will contribute to denitrification, thus reducing nitrate loads. The potential for contamination arises from incomplete composting processes in conjunction with poor attenuation in sub-surface, following the same pathways and with the same risk factors as for other on–site systems. Equally, contamination with household chemicals and personal care products will contaminate the compost product with those components which are not readily biodegradable.

There is increasing interest in the use of ecological sanitation that is designed to maximize the recycling and use of nutrients from excreta and to minimize environmental discharges. However, such technologies are not risk free in relation to groundwater, although the restriction on hydraulic loads can be expected to significantly reduce the risk, and the use of sealed containers may also reduce risks. Risk assessment approaches for urine diversion technologies should take into account the potential for accidental leakage and spills and also consider the end-use and disposal of the treated waste.

Although systems which incorporate composting in a centralized facility are not in general use, the concept holds promise in light of designs more widely applied to composting municipal sludge. They would locally focus potential risks to groundwater. Thus as in other centralized systems, they require increased attention in situation assessment, design and surveillance.

10.2.3 Off-site sanitation: Sewerage and centralized treatment

Off-site systems are forms of sewerage which transport excreta through sewer pipes using water. They only transport faecal matter away from the household and do not involve on-site decomposition to a significant degree. Conventional sewerage systems usually transport excreta and wastewater to centralized treatment plants. They are the norm in urban areas of most developed countries and use large diameter pipes with significant hydraulic gradients to ensure continuous suspension of solids. Conventional sewers usually require significant volumes of water to transport the waste and therefore require high levels of reliable water supply service. Sewerage systems need to be maintained regularly to prevent leakage. Leaking sewers are likely to represent a significant risk to groundwater where centralized wastewater collection is practised. Leaking sewers are therefore described in a separate section below.

There are two forms of modified sewerage which use lower volumes of water and are found in Latin American and Asia (Mara, 1996). Simplified sewers (sometimes called condominial sewers) require typically lower quantities of water. Research in Brazil has shown that simplified sewers can cause suspension of solids at relatively low velocities and are more efficient than conventional sewers where flows are intermittent and solids are moved along the sewer line through a process of repeated deposition and re-suspension (Mara, 1996). However, to do this, sufficient numbers of people must be connected to ensure the necessary level of flow is maintained. Small-bore sewers are another form of modified sewerage that carry effluent and have an inceptor tank to remove solids at the household level (Reed, 1995; Mara, 1996). They therefore work as a mixture of both on-site and off-site sanitation, although there is usually little decomposition in the inceptor tanks, which require periodic desludging, and disposal has been noted to be often poorly managed (Reed, 1995).

In some settings, sewerage systems include rainwater drainage from roads and other paved surfaces, and these overflow periodically when water volumes from precipitation are beyond their capacity, thus leading to an overflow of a mixture of excreta and surface runoff that is commonly discharged to surface or marine waters with no or mechanical treatment.

The use of off-site methods requires treatment of wastes prior to their final disposal to prevent health-relevant contaminants in the effluent from reaching water intended for human use. In almost all cases, the final discharge of treated wastewaters and a significant proportion of untreated wastewater is to surface or marine waters. Risks to groundwater are therefore often dependent on the nature of the relationship between ground and surface water, in particular whether groundwater is recharged by surface water or whether groundwater provides baseflow to surface water bodies. This relationship may not be static and seasonal changes are common (Foster and Hirata, 1988).

Centralized wastewater treatment and storage facilities can bestow significant benefits to communities. Processes can be combined to optimize treatment of physical, chemical, and biological constituents in a waste stream. Pathogen removal and destruction vary between different types of sewage treatment technology, but may be very significant. The quality of the effluent required will not usually depend on meeting drinking-water quality requirements in groundwater sources, but rather relate to surface water use for abstracting drinking-water, recreation, or use of wastewater in agriculture and aquaculture (Bartram and Rees, 2000; Fewtrell and Bartram, 2001).

Centralized wastewater treatment produces substantial amounts of sludges – essentially the biomass that remains after biological treatment of wastewater. These must be disposed of or re-used as a soil amendment, fill, landfill cover, or other beneficial uses. Application to land may represent a significant risk to groundwater where poorly designed and operated. Composting either aerobically or anaerobically is a viable option for treating and thermally disinfecting wastewater sludges, which may then be put to use as soil amendments, backfill and landfill cover without inducing risks due to pathogens.

Where wastewater includes effluent from industries and dispersed small enterprises, it is likely to be contaminated with an array of often unknown chemical substances. Distribution of such sludges in the environment may be a substantial non-point source of contamination, some of which may reach groundwater. This in some cases is mitigated where application is on the land surface, as many contaminants will be removed through adsorption, sequestration and complexing with organic material. Situation assessments of centralized wastewater treatment systems need to include sludge quality and disposal.

Sewage treatment

Various options are available for successfully treating waste flows at centralized facilities. These include trickling filters, activated sludge treatment, oxidation ditches, sequencing batch reactors, constructed wetlands, irrigation fields and sand filtration (Dinges, 1982; Metcalf and Eddy, 1991; US EPA, 1996; Lens *et al.*, 2001). Advanced technologies include filtration, e.g. through membranes, for pathogen removal, wastewater disinfection, nitrification and denitrification, as well as phosphorus removal either biologically or by chemical precipitation using alum, iron salts and lime.

Wastewater treatment facilities can impose inadvertent risks to groundwater. Spills, leaks and overflows, either accidental or occurring during rain events, can enter the ground at or nearby the facility. Raw or non-disinfected wastewater which directly enters or migrates to soils can percolate to groundwater. Large events, or those associated with flooding, also pose risks to groundwater via entry through recharge areas, excavations, abandoned wells, trenches or pits, and by leaking around or into the well itself. The location of wastewater treatment and effluent discharge in relation to a potable groundwater source are important considerations for situation assessments.

Where treated wastewater effluent is used for groundwater recharge, determination of travel times for the recharged groundwater is imperative to ensure that adequate time exists for pathogen attenuation to occur. The use of recharge basins or other techniques of application to the soil surface has the distinct advantage of being able to utilize natural soils above the natural groundwater table for additional treatment and natural filtering. Retention times in basins may also be engineered to be long enough to permit significant pathogen destruction.

The use of waste stabilization ponds as a form of treatment has also become widely used in many parts of the world. When designed and operated properly the use of waste stabilization ponds has been shown to be effective in a variety of settings from arid to humid tropics and have proved to produce effluent of high quality (Pearson *et al.*, 1987; WHO, 1987; Horan, 1990; Mara *et al.*, 1992; Mara, 1996).

There are three principal types of ponds: facultative, anaerobic and maturation. In some cases only a single pond (usually a facultative pond) may be employed, whilst in other as series of ponds may be used. Anaerobic ponds are often used prior to facultative ponds to ensure adequate BOD reduction and sludge decomposition. Such ponds would typically be used where septic tank wastes are disposed of into the treatment works. Maturation ponds are used where good effluent quality is required and these have been shown to produce effluent of very high quality (Mara *et al.*, 1992).

Whether leaching from waste stabilization ponds represents a risk to groundwater depends on the quality of construction, the depth of the unsaturated zone, and whether this is altered by the pond, and depth to the water table. If ponds are unlined, the unsaturated zone below the ponds is low, or materials below the pond are highly permeable, then the risk of leaching may increase. Some workers note that in well-designed ponds, the risk to underlying groundwater from both chemical and microbial contaminants will be limited (WHO, 1987; Foster and Gomes, 1989).

Leaking sewers

Leaking sewers have been shown to be a significant source of groundwater pollution in numerous urban settings of the world though information on the full extent of the potential hazard confronting groundwater is rather poor. Evidence of sewer-related groundwater contamination incidents is known from Bolivia, United Kingdom, Germany, Ireland, Israel, Sweden and the USA (BGS, 1995; Misstear and Bishop, 1997; Bishop *et al.*, 1998).

A number of documented outbreaks have occurred that can be linked to leaking of sewers and subsequent ingress into water supplies, including sources such as wells and distribution systems (Hejkal *et al.*, 1982; D'Antonio *et al.*, 1985; Swerdlow *et al.*, 1992; Bergmire-Sweat *et al.*, 1998). The most dramatic incident was reported from Haifa, Israel in 1985, where leakage from a broken sewer in an adjacent village resulted in epidemics of typhoid and dysentery, with 6000 people affected and one fatality. In the United Kingdom a serious outbreak of disease occurred in 1980 at Bramham in Yorkshire, resulting in 3000 cases of gastroenteritis (Short, 1988; Lerner and Barrett,

1996). In Naas, Ireland leakage from a sewer led to 4000 cases of gastroenteritis (Bishop *et al.*, 1998). Many studies cite groundwater quality data as evidence of sewer-related contamination (Burston *et al.*, 1993). These all indicate the need to address sewer proximity to groundwater sources and condition as important in situation assessments.

Two of the most detailed recent studies of sewer leakage in the United Kingdom have been carried out in Coventry (Burston *et al.*, 1993) and Nottingham (Barrett *et al.*, 1997; 1999a). The Nottingham case study given in Box 10.4 shows the extent to which leaky sewers can compromise groundwater quality with a range of contaminants, including microbial indicators. Yang *et al.* (1999) estimated that about five per cent of recharge of groundwater in Nottingham in the United Kingdom came from sewer leakage, although noted that the confidence intervals on this estimation were very wide. This is in line with estimates derived from an international review by Misstear *et al.* (1996), although these authors noted that estimates of the proportion of sewers that were defective are much higher.

In Germany, the extent of leakage from sewers was estimated to be about 15 per cent according to a poll which registered 17 per cent of the public sewerage system (Dyk and Lohaus, 1998). Eiswirth and Hötzl (1997) estimated that several 100 million m^3 wastewater leak every year from damaged sewers in Germany. Härig and Mull (1992) calculated the extent of exfiltration to the aquifer below the city of Hannover, Germany to be 6.5 million m^3/year and the infiltrated water to 20 million m^3/year.

In earthquake-prone countries, loss rates are significantly higher, for example up to 60 per cent were estimated for parts of Lima, Peru (Lerner, 1996).

There is far less available information regarding the risks posed by modified sewerage (as discussed above in Section 10.2.2), in part because their use is limited. However, there is significant potential for modified sewers to increase risks both through leaks and by their mode of operation. The use of interceptor tanks in small-bore sewers causes similar potential risks as on-site sanitation, although as contained systems this risk may be limited. The shallow depth of most small-bore sewers increases the likelihood of breakage particularly where these are close to roads and may therefore lead to infiltration of contaminated effluent into the sub-surface. The design of the sewer also means surcharging will be more frequent than with conventional sewers. Similar problems may occur with shallow sewers.

The quality of sewage will depend on the source. Sewers draining industrial areas will reflect the waste being disposed of in the factories. This may vary in quality and quantity depending on the activities. Washing down a factory floor may result in large quantities of water with inert suspended solids, but five minutes later there may be a much smaller volume of water from washing equipment that is contaminated with complex chemicals.

Sewers draining residential areas will have a more consistent load, but people dispose of a wide range of wastes down sewers. Oils, grease, household chemicals and faeces may be expected, but chemical spills can occur anywhere. Even surface water cannot be regarded as uncontaminated. Unless it is running off a clean surface, rainfall will pick up dust, spilt oil, air deposits (such as particulate matter from car exhausts) and these can lead to pollution.

Box 10.4. Leaky sewers compromising groundwater safety in Nottingham

Cities may dramatically change recharge pathways and quantities (see e.g. Lerner *et al.*, 1982; Lerner, 1986; 1990). This is highlighted by a study of groundwater quality in Nottingham, UK conducted from 1994 to 2001.

The setting. Public water supply to the Nottingham area is from several reservoirs fed by boreholes, only one of which is located within the city itself. Chlorination is carried out at the borehole sites prior to pumping to the reservoir. The borehole located within the city is currently used only in drought situations due to concerns regarding quality. Limited blending of sources is carried out before filling some reservoirs to ensure nitrate limits are not exceeded: As well as the mains supply, some industrial sites and hospitals have private boreholes. Groundwater is mainly found in the Triassic Sherwood Sandstone Group, which varies in thickness from zero in the west to over 150 m in the north. It is confined to the east and south by the Mercia Mudstone Group, and overlain in the valleys of the rivers Trent and Leen by alluvium. Regional groundwater flow is to the south and east, discharging into large boreholes and the two rivers. In much of the urban area there is little drift cover and the aquifer is unconfined and vulnerable to contamination.

Study results. Analysis showed degradation of the inorganic quality of the urban aquifer in comparison to the surrounding unconfined rural aquifer. Nitrate concentrations frequently exceeded 50 mg/l. Within industrial sites contamination by chlorinated solvents was widespread, TeCE being the most common contaminant and exceeding drinking-water limits in 50 per cent of samples. A survey of shallow groundwater in a residential district for nitrogen isotopes and TTC showed all the sampled shallow piezometers at the water table to be contaminated by sewage. Contamination is concentrated at specific horizons (preferential flow paths), but these exist throughout the aquifer thickness. Vertical and temporal distributions of microbial contaminants were found to be far more variable than those of chemical solutes, reflecting different source and transport characteristics.

It is not possible to quantify the groundwater recharges of the city directly. There are three main sources of recharge in Nottingham (precipitation, mains and sewers). This study simultaneously calibrated four flux balance models for groundwater and three conservative chemical species (chloride, sulphate and total nitrogen). A groundwater flow model and three solute transport models were constructed and calibrated against hydrographs and all available measurements of solute concentrations dating back to the 19th century.

The study area was divided into six zones, and the study period of 1850-1995 was divided into 13 periods to represent the spatial and temporal variations in recharge, pumping and solute loads. At 700 mm of rainfall per year, the average rates of recharge in the urban area for the most recent period were calculated to be: 211 mm per year total recharge, 138 mm per year mains water leakage, 64 mm per year precipitation, and 9 mm per year sewer leakage. There is uncertainty associated with these estimates due to the scarcity of hydrological data and limited historical data on quality. On a broad scale, sewage is found to be the major threat to the quality of urban Triassic Sandstone aquifers. The study highlights the vulnerability of sandstone aquifers to microbial contamination and the challenge to the delineation of wellhead protection areas.

The variation in sewage quality (see also Table 10.4) will lead to varying impacts from sewage leakage and may result in periodic discharges of highly contaminated sewage into the environment. Assessing sewage quality and likely variations will help inform control strategies to ensure that areas of particular high-risk are noted and that actions are prioritized in these areas.

As pollutants are transformed within sewers, it has proven useful to classify them accordingly into primary, secondary and tertiary products. Primary pollutants include microbes and ammonia as well as substances used in bleaching, cleansing and disinfection. Detergents, solvents, fertilizers and salts also increase concentrations of bulk organics reflected by parameters such as dissolved organic carbon (DOC) and absorbable organic halides. Secondary pollution effects arise from subsequent reactions, such as lack of dissolved oxygen, formation of carbon dioxide, decrease in redox potential and changes in electrical conductivity due to microbial degradation of wastewater compounds (Eiswirth *et al.*, 1995). Tertiary pollution effects are changes of specific water constituents through biological and chemical reactions such as ammonification, nitrification, denitrification, desulphurization and mobilization processes (Eiswirth and Hötzl, 1997).

In discussing interactions between groundwater and sewers, some confusion may prevail with terminology. A sewer engineer will refer to infiltration as water coming *into* the sewer. A groundwater specialist will refer to infiltration as water going into the ground. This may be water that is exfiltrated from the sewer. The problems of infiltration and exfiltration may occur alternately where sewers are at a level between maximum and minimum groundwater table positions. Water table fluctuations may cause reversals of infiltration and exfiltration with consequent potential for groundwater contamination.

Sewer managers are normally concerned with infiltration of groundwater into sewers. Too much water entering the sewer will result in elevated amounts of wastewater that needs to be treated at the treatment works. This may overload the capacity of the sewers and reduce the efficiency of wastewater treatment plants (Härig and Mull, 1992). For example, in Plittersdorf, Germany nearly 52 per cent of the total wastewater discharge is infiltrated groundwater (Eiswirth and Hötzl, 1997). Infiltrated water can also bring silt into the pipe blocking it. A steady stream of groundwater entering the pipe may wash away the surrounding soil, causing settlement and damage to the pipe. An assessment of infiltration can indicate the condition of the pipe.

More important for groundwater quality, leaks in sewers may cause exfiltration from the sewer into the groundwater. If sewers are situated above the zone of fluctuation of the groundwater level, the impact of exfiltration from damaged sewers to groundwater is controlled by aquifer vulnerability, depending on permeability of the material between a leaking sewer and the aquifer (Misstear *et al.*, 1996; Bishop *et al.*, 1998). The pollution of groundwater with exfiltrated harmful chemicals from wastewater typically results in long-term damage that can only be rectified over very long periods of time and with considerable technical and financial effort.

In order to understand the mechanisms of groundwater pollution from sewers and methods of monitoring and mitigation, some background information is required on the construction of sewers. Sewers are the pipes that form a sewerage system that convey sewage or wastewater. They can be made from a variety of materials such as concrete, cast iron, plastic or, especially for older sewers, brick. The sewer pipes are laid in a trench as shown in Figure 10.2 below.

The trench is filled with a granular material. This bedding and surround protects and supports the pipe, makes it easier to compact the material in the trench and thus avoid settlement of the ground above the pipe and also makes it easier to lay the pipes so they are in the correct position. Pipes are typically manufactured in short lengths to make it easier to transport and lay them, but this does mean they have to be joined. Some metal pipes are bolted together, some plastic pipes can be heated and welded to form a joint, but many cast iron and concrete pipes have a spigot and socket joint. The end of one pipe can be pushed into a socket of the next pipe. This allows for some movement and allows the pipes to be laid at a slight angle to each other in a gentle curve.

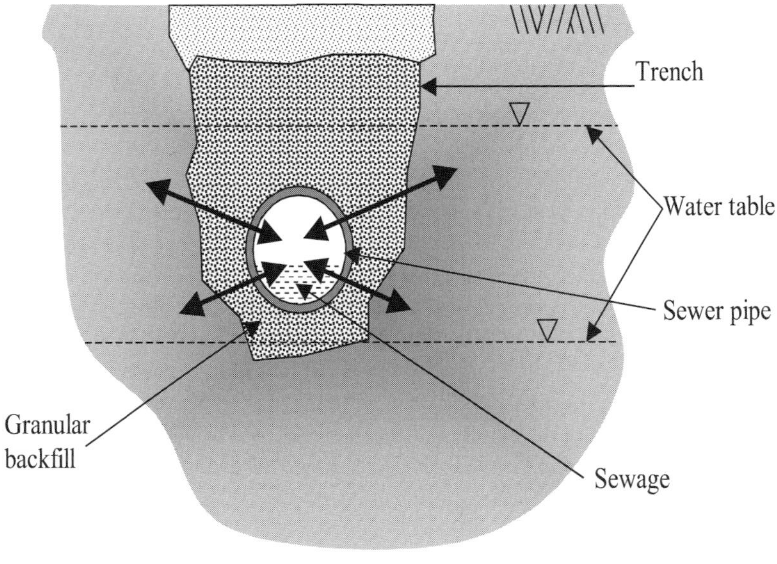

Figure 10.2. Sewer design

Infiltration and exfiltration can occur via a variety of routes as shown in Figure 10.3 below. Loose joints (1), displaced joints (2), and cracks in the structure of the pipe (3) all allow water into and sewage out of the pipe. Other routes include poor junctions between pipe branches (4), leaking manholes (6) and at the points where pipes enter structures (7) or as inflow, directly along pipes (for example pipes leading to abandoned or future connections) (5). These are shown in Figure 10.4 below.

Physical damage is caused when pipes settle relative to other pipes or structures. Chemical damage can occur when hydrogen sulphide from anaerobic decay of organic matter, oxygen and bacteria combine to form sulphuric acid that can attack the soffit (roof) of the sewer and damage materials such as concrete.

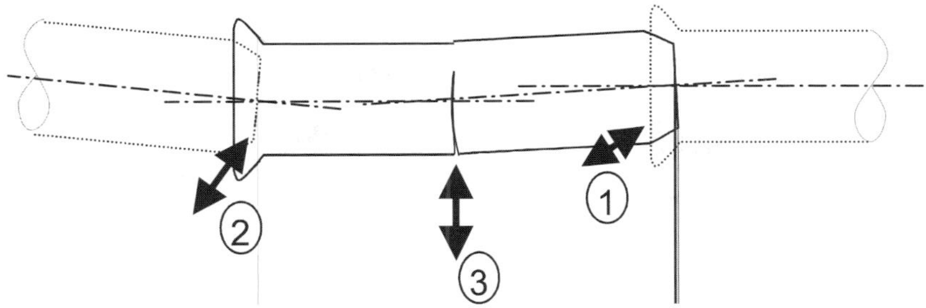

Figure 10.3. Points of leakage from sewers

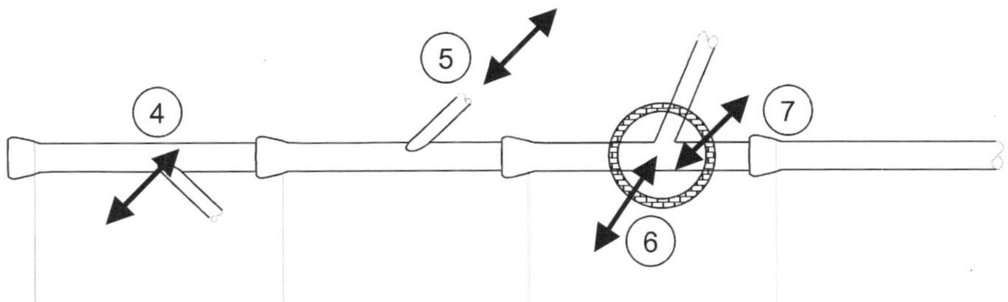

Figure 10.4. Further leakage points from sewers

Leakage from sewers and other damage also occurs due to mechanical malfunctions such as open or displaced joints, deformation, collapse and blockage. From an extensive survey, Bishop *et al.* (1998) showed a steady rise in the rate of failure as sewers age, the most common cause of sewer failure being joint fracturing related to the practice of using rigid joints, while deterioration of pipe material did not play an important role. Tree root ingress, rodent activity (Battersby and Pond, 1997) and damage from subsurface work on other utilities may also cause leaks.

Factors that influence leakage of sewers include (Aigner *et al.*, 1998):

- type of subsoil;
- height of groundwater level above sewers;
- standard of workmanship in laying pipes;
- type of pipe joint, number of joints and pipe size;
- total length of sewer (including house connections);
- number and size of manholes and inspection chambers;
- rainfall (daily and seasonal);
- age of system.

Whether the water is moving into or out of the sewer will depend on the relative head of the water. A sewer will only leak if it is below the water table or if it is overloaded or surcharged. The pipe trench can then act as a soakaway. The sewage can flow through the granular backfill, away from the leaking crack or joint and this can increase the rate the wastewater can infiltrate into the ground. Where wastewater has to be pumped uphill, pressure sewers will have a positive head, so more attention needs to paid to ensure they do not leak.

Sewers can be graded according to their structural condition (WRc, 1994). Using methods such as close-circuit television, the pipes are examined for cracks and deformation. Sewers graded in the most severe category may collapse. Leaking sewers that are not necessarily in danger of structural failure will have a lower priority for action. However, groundwater pollution can arise from both the long term diffuse leakage from pipes as well as the sudden collapse of a sewer. Most utilities conduct structural assessments only for important large sewers, as planned replacement of these is preferable to having to react to a large hole opening up in a road. The collapse of smaller sewers has less of an economic impact and so they are not routinely monitored, yet their length can contribute significantly to diffuse pollution and their shallowness lead to structural damage. In the United Kingdom about 75 per cent of the sewage network is regarded as structurally non-critical (Read and Vickridge, 1997). House connections constructed by non-specialists can also provide a source of groundwater contamination that would not be recorded by a structural survey.

Pollution routes do not only occur through leakage into aquifers. Blockages or hydraulic overloading (due to storm water or increased foul flows) can cause the sewer to become full and eventually flood. This can be through lifting manholes, backing up into people's houses or through specifically designed combined sewer overflows (CSOs). Pumping stations also require emergency overflows in case of complete failure of the pumps. The design should ensure that sewage flooding occurs in an area with minimal social, economic and environmental impact, rather than flooding homes, public areas or environmentally sensitive zones.

Storm water management

Current practice separates foul sewage from surface water runoff in many urban areas. However in older urban areas or areas with poor regulation and enforcement, sewers can carry both surface and foul water. Storm water drainage systems may collect effluent from septic tanks, misconnections from foul sewers, excreta and other wastes washed from the ground surface and flood water contaminated by inundated pit latrines. As storm sewers are 'officially' for rainwater, they are not designed to cater for faecally contaminated material. Open drains and lack of treatment before discharge to surface or groundwaters can provide a route for pollutant transmission.

10.3 ASSESSING THE RISKS TO GROUNDWATER

10.3.1 Assessing risk from on-site sanitation

Understanding the hydrogeological environment and the siting of on-site sanitation is as important as knowing the specific design of the facility, and the two interact to define the level of risk. Of particular concern is whether the natural attenuation of pathogens will be effective. The use of pit latrines, for instance, is often of particular concern as they by-pass the major attenuating layer of the soil. Although the development of biologically active layers around the pit reduces breakthrough, this cannot be wholly relied upon in all situations and periodic overloading may occur.

The assessment of risks to groundwater from on-site sanitation should take into account the hydraulic load, the depth to the water table, the nature of the groundwater (whether oxidizing or reducing) and the time taken for water to travel from the pollutant source to the groundwater abstraction point. The lithology of the unsaturated zone will also be important in relation to the potential for attenuation. For chemicals, and to a lesser extent pathogens, the density of population may also be important in assessing whether significant contamination will occur. Relatively simple approaches are available for assessing such risks using only limited hydrogeological data (ARGOSS, 2001). However, it is important to ensure that faecal material cannot enter the drinking-water source through other means. This can be determined through targeted assessments.

10.3.2 Assessing the risks to groundwater from sewerage

Assessing risks of groundwater pollution from sewerage systems has in the past not attracted much attention and assumptions regarding the expected attenuation of microbial contaminants resulted in little concerted effort to define the extent and nature of groundwater contamination (Reynolds and Barrett, 2003). Powell *et al.* (2003) demonstrated that microbial contaminants (both bacterial and viral) derived from sewage can penetrate to depths of up to 90 m in some aquifers. These included indicator organisms and enteric viruses including Norovirus and Coxsackie B virus.

In assessing the risks of contamination from sewerage, due consideration must be given to the nature of sub-surface below the sewer including the depth of unsaturated zone, the lithology (and likely attenuation potential), the depth of groundwater and the hydraulic loading that could be derived from leaking sewers. In particular, estimating the volume of exfiltration from sewers into groundwater will assist in determining the risk to groundwater, although this is a complex task. Several methods have been developed to quantify the water exchange between groundwater and sewers. Härig and Mull (1992) used budgets of wastewater flow streams, a calibrated groundwater flow model and the detection of sewage indicators. Yang *et al.* (1999) developed a model supplemented by solute budgets of chloride, sulphate, and total nitrogen for the estimation of recharge to groundwater in Nottingham.

In addition, assessments can be made of the state and age of the sewer infrastructure as a guide to the risk of sewer leakage. Older sewers will be more likely to leak as the likelihood of breakage increases because of age and often because of construction techniques that increase vulnerability to breaking. However, whilst more recent sewers may be less prone to breaking, construction techniques may increase the likelihood of leakage paths in the surrounds around the pipes. Reynolds and Barrett (2003) suggest that assessing risks to groundwater from sewer leakage should be based on the frequency of sewer breaks noted, age and methods of construction, grading of material surrounding the pipe and groundwater level. Active monitoring programmes may also assist in identifying the extent and nature of contamination from sewers.

10.4 ANALYTICAL INDICATION OF HUMAN EXCRETA AND SEWAGE IN GROUNDWATER

Identifying pollutant sources from sewage is often a major task due to the multitude of potential sources and pollutants in an urban environment and detection of leaks in sewers may be difficult. Indicator organisms such as *E. coli*, faecal streptococci and bacteriophages remain useful in detecting general faecal contamination although it may be difficult to precisely identify the source of contamination. Less robust organisms such as sulphite-reducing bifidobacteria may be useful as species can be identified that are unique to human faeces.

Barrett *et al.* (1999a) tried to use chemical marker species which can be used to indicate groundwater recharge from sewage. Most useful major ions are chloride, sulphate and individual nitrogen species (Härig and Mull, 1992; Eiswirth *et al.*, 1995), while cation ratios can change due to ion exchange processes (Trauth and Xanthopoulos, 1997). Potential markers for sewage include ingredients of detergents like phosphate, boron, ethylendiamine tetraacetic acid (EDTA), optical brighteners and d-limonene. Constraints on the value of these substances as markers are the variable composition of detergents so that these compounds may not always be present. In addition, boron and phosphate are not unique to sewage, and their occurrence and mobility in groundwater is influenced by pH. Elevated groundwater concentrations of phosphorous may result from overloading soil adsorption capacities from waste treatment systems (i.e. indicating wastewater ingress), but may arise from agricultural use as fertilizer (Day, 2001).

Stable nitrogen isotopes and microbial parameters are further tools. However due to the die-off of microorganisms and the mixing and fractionation processes affecting the nitrogen isotopes both parameters are not absolute indicators and may be difficult to interpret. The study described by Barrett *et al.* (1999a) shows that ideal marker species are rare because most groundwater constituents are present in more than one potential source of recharge. There is a need for a multi-component approach rather than using individual markers.

The introduction of sewage to an aquifer may cause significant changes in the chemical quality of groundwater. Observed effects may include depletion of dissolved oxygen, lowering of pH, increases in DOC, chemical oxygen demand (COD), biological oxygen demand (BOD) and conductivity as well as decrease of redox potential (Rivers *et al.*, 1996; BGS, 1997; Barrett *et al.*, 1999a). A further important effect of organic matter is that it blocks sites for attachment on the surface of grains of the porous medium, thereby reducing attenuation of microorganisms. Clogging will also eventually occur to block flow paths.

Nitrate is frequently used as a marker of sewage input, as it is derived from the microbial oxidation of excreted ammonia in soils, and is generally conserved in groundwater. However, nitrate may be derived from a number of other sources (e.g. fertilizer or manure application). A more reliable tool is the ratio of nitrate to chloride. High nitrate to chloride ratios are indicative of faecal origin, although the precise ratio will depend upon population density and leaching to groundwater (Morris *et al.*, 1994). A drawback is that this ratio may vary seasonally, particularly in shallow groundwater. Nitrate levels have been shown to decrease through dilution during the early part of the rainy season and then to subsequently increase in shallow groundwater in Kampala, Uganda (Barrett *et al.*, 1999b; ARGOSS, 2002). Isotopic nitrogen ratios have demonstrated promise in distinguishing between various sources of nitrogen inputs, thereby providing a useful tool for assessing sources of nitrate pollution (Rivers *et al.*, 1996; Barrett *et al.*, 1999a). Ammonia may be indicative of very recent sewage contamination of shallow groundwater, but is likely to be rapidly oxidized to nitrate under typical conditions in shallow, unconfined aquifers.

A promising marker of sewage is 1-aminopropanone, which is present in human urine and which is not produced significantly by other natural processes. Caffeine may be a non-adsorbed, conservative indicator of sewage inputs, but it not be readily detectable in groundwater (Stroud, 2001). Other potential chemical indicators of sewage contamination in groundwater include trace metals, faecal sterols (e.g. coprostanol), sodium dodecyl sulphate and sodium tripolyphosphate (Ashbolt *et al.*, 2001; Barrett *et al.*, 1999a)

10.5 CHECKLIST

NOTE ▶ *The following checklist outlines information needed for characterizing sanitation practices in the drinking-water catchment area. It supports hazard analysis in the context of developing a Water Safety Plan (Chapter 16). It is neither complete nor designed as a template for direct use but needs to be specially adapted for local conditions. The analysis of the potential of groundwater pollution from human activity requires combining the checklist below with information about socioeconomic conditions (Chapter 7), aquifer pollution vulnerability (Chapter 8) and other specific polluting activities in the catchment area (Chapters 9 and 11-13).*

Is on-site sanitation practised in the drinking-water catchment area?

- ✓ Compile inventory on coverage with different types of on-site and/or off-site sanitation systems (including change over time)
- ✓ Assess size and proportion of population using on-site sanitation
- ✓ Estimate quantity of excreta disposed and loadings of pathogens, nitrate and other chemicals
- ✓ Evaluate adequacy of design, construction, condition and maintenance of on-site systems in relation to aquifer vulnerability and physical conditions in the catchment area (e.g. water table, soil, hydrogeology): consider checklist for Chapter 8
- ✓ Analyse population awareness regarding the need for protecting their groundwater sources through adequate design, construction, and maintenance of on-site systems

- ✓ For trench or pit latrines: assess siting in relation to groundwater levels, vulnerability to flooding, routines for excrement removal and inspection of liner integrity (and access of potential disease vectors such as insects and rodents for differentiation between them and drinking-water as cause of disease)
- ✓ For septic tank systems: assess siting both of tanks and drainage fields in relation to groundwater levels, vulnerability to flooding, adequacy of routines of sludge removal, tank inspection
- ✓ For contained or cartage systems: assess adequacy of collection, transportation and disposal practices in relation to groundwater sources
- ✓ For composting latrines or central systems: assess efficacy of the composting process as well as siting in relation to groundwater levels, vulnerability and to flooding
- ✓ …

Are centralized sewage treatment facilities located in the drinking-water catchment area?

- ✓ Check structure of services (e.g. percentage of population and areas of the settlement connected to storm water sewers, foul sewers and/or combined systems), and estimate wastewater volume per capita
- ✓ Evaluate adequacy of design, construction, condition and maintenance of treatment and sewage systems in relation to aquifer vulnerability and physical conditions in the in the drinking-water catchment area: consider checklist for Chapter 8
- ✓ Assess siting of treatment facilities in relation to groundwater, integrity of containment, susceptibility of facilities to flooding
- ✓ Assess practices for re-use of treated wastewater irrigation, aquifer recharge, fish ponds or other purposes
- ✓ Assess practices of human excreta or sludge re-use and/or disposal, e.g. land application: consider checklist for Chapter 9
- ✓ Evaluate the potential for contamination of sewage (and sludge arising from its treatment) with industrial chemicals, particularly persistent and toxic substances from an inventory of commercial activities in the catchment of the facility and licenses for connection to the system: consider checklist for Chapter 11
- ✓ In some settings, conduct microbiological analyses of raw sewage and effluent to assess treatment performance for pathogen elimination
- ✓ In some settings, particularly with reuse of effluent or sludge, conduct chemical analyses of concentrations of substances that potentially could contaminate groundwater in effluents and/or sludge
- ✓ …

ⓘ **Are there sewers in the drinking-water catchment area that may leak into groundwater?**

Note: In many settings, assessing the risk of groundwater contamination from leaky sewers may be effectively combined with assessing the risk of direct ingress of sewage leaking out of sewers into the central drinking-water distribution system.

- ✓ Check depth of sewers in relation to groundwater table (for assessing likelihood of exfiltration and infiltration)
- ✓ Compile registered information on design, material and age of the sewer system
- ✓ Check whether regular sewer inspections are carried out (e.g. visual or close-circuit television)
- ✓ Compile inventory of licensed industrial and commercial discharges into the sewer system
- ✓ Compile inventory of medical care facilities connected to the system
- ✓ Compile information on land use and historic waste deposits that may indicate unregistered connections to the sewerage system or potential infiltration through leaks
- ✓ Compile information from laboratory analyses of groundwater samples taken in the vicinity of sewers (marker species, e.g. stable nitrogen isotopes, multi-component analyses in relation to known sewage constituents)
- ✓ Check for indication of leaks from budgets of wastewater flow streams and groundwater flow models
- ✓ …

ⓘ **Are hazardous events likely to increase groundwater pollution potential?**

- ✓ Evaluate whether and how storm water events would enhance transport of pollutants to the aquifer
- ✓ Evaluate which spills and accidents are likely to cause groundwater pollution
- ✓ …

ⓘ **Is drinking-water abstracted in proximity to sanitation facilities?**

- ✓ Assess distance between sanitation facilities and drinking-water abstraction
- ✓ Check adequacy of wellhead protection measures, wellhead construction and maintenance as well as sanitary seals used (see Chapter 18) to prevent ingress of contaminants from excreta disposal practices
- ✓ …

Are groundwater quality data available to indicate pollution from sanitation?

- ✓ Compile historic data from the area of interest, e.g. from local or regional surveys, research projects or previous monitoring programmes
- ✓ Check need and options for implementation of new or expanded monitoring programmes likely to detect contamination from sanitation
- ✓ …

What regulatory framework exists for sanitation?

- ✓ Compile information on national, regional, local or catchment area specific legislation, regulations, recommendations or common codes of good practices on siting, construction, operation and maintenance of sanitation facilities
- ✓ Check whether the regulatory framework adequately addresses environmental and specifically groundwater protection
- ✓ Identify gaps and weaknesses known which may encourage specific pollution problems
- ✓ …

Documentation and visualization of information on sanitation practices.

- ✓ Compile summarizing report and consolidate information from checklist points above
- ✓ Compile summary of types and amounts of wastewater and sludges generated, and of disease agents which are potentially hazardous if they leach into the aquifer
- ✓ Map locations of settlements and inventory sanitation facilities (use GIS if possible)
- ✓ …

10.6 REFERENCES

Aigner, C.M, Armstrong, R.J. and Butler, D. (1998) *Dry Weather Flow in Sewers*, CIRIA Report 177, London.

ARGOSS (2001) *Guidelines for Assessing the Risk to Groundwater From On-site Sanitation*. BGS Commissioned Report CR/01/42, BGS, Wallingford.

ARGOSS (2002) *Assessing Risk to Groundwater From On-site Sanitation: Scientific review and case studies*. BGS Commissioned Report CR/02/079N, BGS, Wallingford.

Ashbolt, N.J., Grabow, W. and Snozzi, M. (2001) Indicators of microbial water quality. In *Water Quality: Guidelines, standards and health – assessment of risk and risk management for water-related infectious disease*, (eds. L. Fewtrell and J. Bartram), pp. 289-315, IWA, London.

Barrett, M.H., Hiscock, K.M., Pedley, S., Lerner, D.N. and Tellam, J.H. (1997) The use of marker species to establish the impact of the city of Nottingham (UK) on the quantity and quality of its underlying groundwater. *Groundwater in the Urban Environ.*, **1**, 85-90.

Barrett, M.H., Hiscock, K.M, Pedley, S., Lerner, D.N., Tellam, J.H. and French, M.J. (1999a) Marker species for identifying urban groundwater recharge sources: a review and case study in Nottingham, UK. *Water Res.*, **33**(14), 3083-3097.

Barrett, M.H., Nalubega, M. and Pedley, S. (1999b) On-site sanitation systems and urban aquifer systems in Uganda. *Waterlines*, **17**(4), 10-13.

Bartram, J. and Rees, G. (2000) *Monitoring Bathing Waters: A practical guide to the design and implementation of assessments and monitoring programmes*. E&F Spon, London.

Battersby, S. and Pond, K. (1997) *Rat Infestations and Underground Drainage - Report of a Survey of Local Authorities*. Robens Institute for Public and Environmental Health, Guildford.

Bergmire-Sweat, D., Wilson, K., Marengo, L., Lee, Y.M., MacKenzie, W.R., Morgan, J. Von Alt, K., Bennett, T., Tsang, V.C.W. and Furness, B. (1999) Cryptosporidiosis in Brush Creek: describing the epidemiology and causes of a large outbreak in Texas, 1998. In *Proc. International Conference on Emerging Infectious Diseases, Milwaukee, WI*, AWWA, Denver, Colorado.

Bishop, P.K., Misstear, B.D., White, M. and Harding, N.J. (1998) Impacts of sewers on groundwater quality. *Water & Environ. Man.*, **12**(3), 216-223.

Briscoe, J. (1996) Financing water and sanitation services: the old and new challenges. *Water Supply*, **14**(3/4), 1-17.

BGS (1995) *Impact of Urbanisation on Groundwater in Santa Cruz, Bolivia: Final report.* International Division, British Geological Survey, Nottingham.

BGS (1997) *The Study of Pollution Risk to Deep Groundwaters from Urban Wastewaters: Project summary report*. International Division, British Geological Survey, Nottingham.

Burston, M.W., Nazari, M.M., Bishop, P.K. and Lerner, D.N. (1993) Pollution of groundwater in the Coventry region (UK) by chlorinated hydrocarbon solvents. *J. Hydrology,* **149**, 137-161.

Cairncross, S. and Feachem, R. (1993) *Environmental Health Engineering in the Tropics: An introductory text*, 2nd edn, John Wiley & Sons, Chichester.

Caldwell, E.L. and Parr, L.W. (1937) Groundwater pollution and the bored hole latrine. *J. Infectious Diseases*, **61**, 148-183.

Cantor, L.W. (1997) *Nitrates in Groundwater*. Lewis Publishers/CRC Press, New York.

CDC (1999) Outbreak of *Escherichia coli* O157:H7 and *Camplyobacter* among attendees of the Washington County Fair – New York 1999. *Morbidity and Mortality Weekly Report*, **48**(36), 803-804.

Chidavaenzi, M., Bradley, M., Jere, M. and Nhandara, C. (2000) Pit latrine effluent infiltration into groundwater: the Epworth case study. In *Water, Sanitation and Health,* (eds. I. Chorus, U. Ringelband, G. Schalg and O. Schmoll), pp. 171-178, IWA, London.

Craun, G.F. (1984) Health aspects of groundwater pollution. In *Groundwater Pollution*, (eds. G. Bitton and C. Gerba), pp. 135-179, John Wiley & Sons, New York.

Craun, G,F. (1985) A summary of waterborne illness transmitted through contaminated groundwater. *J. Environ. Health*, **48**, 122-127.

D'Antonio, RG., Winn, RE., Taylor, J.P., Gustafson, T.L., Current, W.L., Rhodes, M.M. Gary, W. and Zajac, R.A.. (1985) A waterborne outbreak of Cryptosporidiosis in normal hosts. *Annals of Int. Med.*, **103**, 886-888.

Day, L.D. (2001) *Phosphorous Impacts from Onsite Septic Systems to Surface Waters in the Cannonsville Reservoir Basin, NY*. Delaware County Soil and Water Conservation District, Walton, New York.

Dinges, R. (1982) *Natural Systems for Water Pollution Control*. Van Nostrand Reinhold, New York.

Doughton, C.G., and Thomas A.T. (1999) Pharmaceuticals and personal care products in the environment: agents of subtle change? *EHP*, **107**(6), 907-938.

Dyk, C. and Lohaus, J. (1998) Der Zustand der Kanalisation in der Bundesrepublik - Ergebnisse der ATV-Umfrage 1997. *Korrespondenz Abwasser*, **45**, 865-874.

Eiswirth, M. and Hötzl, H. (1997) The impact of leaking sewers on urban groundwater. *Groundwater in the Urban Environ.*, **1**, 399-404.

Eiswirth, M., Hötzl, H., Lazar, C. and Merkler, G-P. (1995) Detection of contaminant transport from damaged sewerage systems and leaky landfills. In *Groundwater Quality: Remediation and protection*, pp. 337-346, IAHS 225.

Esrey, S. (1996) Water, waste and wellbeing: the results of a multicountry study. *Am. J. Epidem.*, **143**(6), 608-623.

Esrey, S.A., Potash, J.B., Roberts, L. and Shiff, C. (1991) Effects of improved water supply and sanitation on ascariasis, diarrhoea, dracunculiasis, hookworm infection, schistosomiasis, and trachoma. *Bulletin of the WHO*, **69**(5), 609-621.

Fewtrell, L. and Bartram, J. (2001) *Water Quality: Guidelines, standards and health*. IWA Publishing, London.

Foster S.S.D and Hirata R. (1988) *Groundwater Pollution Risk Assessment – A Methodology Using Available Data*. CEPIS, Lima.

Foster, S.S.D. and Gomes, DC. (1989) *Groundwater Quality Monitoring: An Appraisal of Practices and Costs*. CEPIS, Lima.

Franceys, R., Pickford, J. and Reed, R. (1992) *A Guide to the Development of On-Site Sanitation*. WHO, Geneva.

Gelinas, Y., Randall, H., Robidoux, L. and Schmit, J-P. (1996) Well water survey in two districts of Conakry (Republic of Guinea) and comparison with the piped city water. *Water Resour.*, **30**(9), 2017-2026.

Hamm, A. (ed.) (1991) *Studie über Wirkungen und Qualitätsziele von Nährstoffen in Fließgewässern*. Academia Verlag, Sankt Augustin.

Härig, F. and Mull, R. (1992) Undichte Kanalisationssysteme - Einfluß auf das Grundwasser. *GWF Wasser/Abwasser*, **133**(4), 196-200.

Hejkal, T.W., Keswick, B., LaBelle, R.L., Gerba, C.P., Sanchez, Y., Dressman, G., Hafkin, B. and Melnick, J.L. (1982) Viruses in a community water supply associated with an outbreak of gastroenteritis and infectious hepatitis. *JAWWA*, **74**, 318-321.

Hoogenboezem, W., Ketelaars, H.A.M., Medema, G.J., Rijs, G. and Schijven, J. (2001) *Cryptosporidium and Giardia: Occurrence in sewage, manure and surface water*. RIWA/RIVM/RIZA-report, Amsterdam.

Horan, N.J. (1990) *Biological Wastewater Treatment Systems*. John Wiley & Sons, Chichester.

Hutton, L.G., Lewis, W.J. and Skinner, A.C. (1976) *A Report on Nitrate Contamination of Groundwater in Some Populated Areas of Botswana*. Report No. BGSD/8/76, Department of Geological Survey, Lobatse.

Klopp, R. (1999) Kommunales Abwasser und seine Behandlung. In *Wasser und Gewässer*, (ed. F.H. Frimmel), pp. 369-444, Spektrum Akademischer Verlag, Heidelberg.

Koppe, P. and Stozek, A. (1986) *Kommunales Abwasser*. Vulkan Verlag, Essen.

Lawrence, A.R., Morris, B.L., Gooddy, D.C., Calow, R. and Baird, M.J. (1997) *The Study of the Pollution Risk to Deep Groundwater From Urban Wastewaters: Project summary report*, BGS, Technical Report WC/97/15, Wallingford.

Lens, P., Zeeman, G. and Lettinga, G. (eds.) (2001) *Decentralised Sanitation and Reuse*. IWA Publishing.

Lerner, D.N. (1986) Leaking pipes recharge groundwater. *Ground Water*, **24**(5), 654-662.

Lerner, D.N. (1990) Groundwater recharge in urban areas. *Atmospheric Environ.*, **24**(1), 29-33.

Lerner, D.N. (1996) Urban groundwater an asset for the sustainable city. *European Water Control*, **6**, 43-51.

Lerner, D.N. and Barrett, M.H. (1996) Urban groundwater issues in the United Kingdom. *Hydrogeology J.*, **4**(1), 80-89.

Lerner, D.N., Dellow, D.J., Lloyd, J.W. and Mansell-Moullin, M. (1982) Groundwater studies for Lima, Peru. In *Proc. symposium on Optimal Allocation of Water Resources, Exeter, July 1982*, pp. 17-35, IAHS Publication 135.

Lewis, J.W., Stephen, S.D.F., Bohumil, S. and Drasar, B.S. (1982) *The Risk of Groundwater Pollution by On-Site Sanitation in Developing Countries*. International Reference Centre for Wastes Disposal (IRCWD), Duebendorf.

Mara, D.D. (1996) *Low-Cost Urban Sanitation*. John Wiley & Sons, Chichester.

Mara, D.D., Alabaster, G.P., Pearson, H.W. and Mills, S.W. (1992) *Waste Stabilisation Ponds: A design manual for Eastern Africa*. Lagoon Technology, Leeds.

Metcalf and Eddy, Inc. (1991) *Wastewater Engineering: Treatment, disposal, and reuse*, 3rd edn, (revised by G. Tchobanoglous and F.L. Burton), McGraw-Hill, New York.

Misstear, B.D. and Bishop, P.K. (1997) Groundwater contaminations from sewers: experience from Britain and Ireland. *Groundwater in the Urban Environ.*, **1**, 491-496.

Misstear, B., White, M., Bishop, P. and Anderson, G. (1996) *Reliability of Sewers in Environmentally Vulnerable Areas*. Construction Industry Research and Information Association (CIRIA) Project Report 44, London.

Morris, B.L., Lawrence, A.R., Stuart, M.E. (1994) *The Impact of Urbanisation on Groundwater Quality: Project summary report*, BGS Technical Report WC/94/56, Wallingford.

Oragui, J.I., Curtis, T.P., Silva, S.A. and Mara, D.D. (1987) The removal of excreted bacteria and viruses in waste stabilisation ponds in northeast Brazil. *Water Res.,* **27**(4), 727-728.

Pearson, H.W., Mara, D.D. and Bartone, C.R. (1987) Guidelines for the minimum evaluation of the performance of full-scale waste stabilisation pond systems. *Water Res.*, **21**(9), 1067-1075.

Pedley, S. and Howard, G. (1997) The public health implications of microbiological contamination of groundwater. *Q. J. Eng. Geol.*, **30**, 179-188.

Pickford J. (1995) *Low-Cost Sanitation: A review of practical experience*. IT Publications, London.

Powell, K.L., Taylor, R.G., Cronin, A.A., Barrett, M.H., Pedley, S., Sellwood, J., Trowsdale, S.A. and Lerner, D.N. (2003) Microbial contamination of two urban sandstone groundwater aquifers in the UK. *Water Res.*, **37**(2), 339-352.

Pyle, B.H., Sinton, L.W., Noonana, M.J. and McNabb, J.F. (1979) The movement of micro-organisms in groundwater. In *Proc. New Zealand Institution of Engineers Technical Group on Water, Groundwater Symposium*, Victoria University, Wellington.

Read, G.F. and Vickridge, I.A. (eds.) (1997) *Sewers: Rehabilitation and new construction.* Arnold, London, New York.

Reed, R.A. (1995) *Sustainable Sewerage: A guide for community schemes*. IT Publications, London.

Reynolds, J.H. and Barrett, M.H. (2003) A review of the effects of sewer leakage on groundwater quality. *JCIWEM*, **17**(1), 34-39.

Rivers, C.N., Barrett, M.H., Hiscock, K.M , Dennis, P.F., Feast, N.A. and Lerner, D.N (1996) Use of isotopes to identify nitrogen contamination of the Sherwood sandstone aquifer beneath the City of Nottingham, United Kingdom. *Hydrogeology J.*, **4**(1), 90-102.

Scandura, J.E. and Sonsey, M.D. (1997) Viral and bacterial contamination of groundwater from on-site treatment systems. *Water Sci. Technol.*, **35**(11-12), 141-146.

Short, C.S. (1988) The Bramham incident 1980 - an outbreak of water borne disease. *J. Instit. Water & Environ. Man.*, **2**, 383-390.

Stroud Water Research Center (2001) *New York City Watersheds Year One Progress Report.* SWRC, Pennsylvania.

Subrahmanyan, K. and Bharaskan, T.R. (1980) Pollution of groundwater from borehole latrines. *Indian Medical Gazette*, **85**(9), 418-423.

Swerdlow, D.L., Woodruff, B.A., Brady, R.C., Griffin, P.M., Tippen, S., Donnel Jr. H., Geldreich, E., Payne, B.J., Meyer Jr. A., Wells, J.G., Greene, K.D., Bright, M., Bean, N.H. and Blake, P.A. (1992) A waterborne outbreak in Missouri of *Escherichia coli* O157:H7 associated with bloody diarrhoea and death. *Annals Int. Med.*, **117** (10), 812–819.

Trauth, R. and Xanthopoulos, C. (1997) Non-point pollution of urban groundwater. *Water Res.*, **31**, 2711-2718.

US EPA (1996) *Wastewater Treatment: Alternatives to Septic Systems*. EPA/909-K-96-001, Region 9 Drinking Water Program, San Francisco.

US Public Health Service (1965) *Manual of Septic Tank Practice*. USPHS, Washington, DC.

Van der Ryn, S. (1995) *The Toilet Papers: Recycling waste and conserving water*. Ecological Design Press, Sausalito, CA.

Van Ryneveld, M.B. and Fourie, A.B. (1997) A strategy for evaluating the environmental impact of on-site sanitation systems. *Water SA*, **23**(4), 279-291.

WHO (1987) *Wastewater Stablization Ponds: Principles of planning and practice*. World Health Organization-Eastern Mediterranean Regional Office, Alexandria.

WHO and UNICEF (2004) *Meeting the MDG Drinking Water and Sanitation Target: A mid-term assessment of progress*. World Health Organization, Geneva, and UNICEF, New York.

WHO (2005) *Guidelines for the Safe Use of Wastewater, Excreta and Greywater. Volume 4: Excreta and Greywater Use in Agriculture*, World Health Organization, Geneva.

WRc (1994) *Sewerage Rehabilitation Manual*, 3rd ed. UK Water Research Centre/Water Authorities Association, Swindon.

Yang, Y., Lerner, D.N., Barrett, M.H. and Tellam, J.H. (1999) Quantification of groundwater recharge in the City of Nottingham, UK. *Environ. Geol.*, **38**(3), 183-198.

11

Industry, mining and military sites: Potential hazards and information needs

C. Teaf, B. Merkel, H.-M. Mulisch, M. Kuperberg and E. Wcislo

Many similarities exist between aspects of site characterization and facility evaluation that are applicable for industrial and mining activities as well as military facilities, as a result of the wide range of activities conducted there. This includes use of fuels, solvents and other chemicals, as well as large storage volumes for some raw materials and wastes. Many elements of planning and preventive measures can be developed based on knowledge of site attributes, processes that are conducted, waste management procedures and site closure activities. With the possible exception of high explosives and ammunition, a large number of potential organic and inorganic groundwater impacts for these types of facilities are coincident.

As illustrations of the principles and strategies for site characterization, this chapter provides case studies. Some of the elements of these case studies are immediately evident, while others emphasize the need for careful collection and interpretation of the data. While they focus on groundwater contamination by one type of human activity,

 Protecting Groundwater for Health: Managing the Quality of Drinking-water Sources. Edited by O. Schmoll, G. Howard, J. Chilton and I. Chorus. ISBN: 1843390795. Published by IWA Publishing, London, UK.

they also highlight that settings are often influenced by a more complex mixture of multiple pollution sources to be identified in situation assessment.

NOTE ▶ *Industrial, mining and military activities as well as the environment in which they take place vary greatly. Health hazards arising from industrial, mining and military activities and their potential to pollute groundwater therefore needs to be analysed specifically for the conditions in a given setting. The information in this chapter supports hazard analysis in the context of developing a Water Safety Plan for a given water supply (Chapter 16). Options for controlling these risks are introduced in Chapter 23.*

11.1 INDUSTRIAL ACTIVITIES

Industrial activities in groundwater recharge zones have significant potential to affect large areas of local or regional groundwater as a result of normal operations (e.g. waste disposal, materials storage) as well as short term adverse events (e.g. spills, leaks). Activities defined as industrial may include a wide variety of large scale or small scale commercial, public, governmental or military facilities that are engaged in manufacturing, chemical processing, power generation or ancillary services (US EPA, 1999). In many instances, the aggregate effect of several small local industrial facilities, or even the effects from a single small facility, have severely affected groundwater quality, with impacts on drinking-water supplies.

Although it is an important consideration, apparent industrial facility size may not be the sole or even the principal determinant of risk to groundwater. The degree to which a facility poses risks to aquifers can be related to many factors, including:

- the specific industrial processes and chemicals in use;
- the age and size of the facility;
- corporate 'housekeeping' or environmental management practices;
- local geological and hydrological characteristics.

In addition to contaminant release issues, industrial activities in drinking-water catchment areas may exert other non-chemical influences which change the vertical or horizontal flow regime of contaminants (e.g. changes in recharge inflow quantity or percolation rate), or which serve to reduce the overall capacity of the recharge area (e.g. groundwater withdrawal). Thus as discussed further in Chapter 23, the most effective preventive or management strategies for drinking-water catchments from an engineering and cost perspective are those which seek to eliminate, minimize or carefully control potential contaminant sources (Zektser *et al.*, 1995; Berg *et al.*, 1999).

11.1.1 Types of industrial facilities and potential impacts to groundwater

Many industrial facilities employ practices that historically have been associated with groundwater contamination, such as the production, treatment or handling of metals, petroleum, paints and coatings, rubber and plastics, electrical components, pharmaceuticals, pesticides, non-chlorinated and chlorinated solvents, paper, inks and dyes, fabrics, adhesives, fertilizers, wood preservatives, laundry/dry cleaning and explosives. In addition, complex facilities may have a significant component of vehicular traffic, power production, water withdrawal/treatment and grounds maintenance, all of which may be associated with problems related to non-process use of fuels and lubricants. As a result of the extreme industrial diversity in many regions of the world, it is important to identify the types and sizes, as well as numbers of facilities, that may be potential contributors to groundwater pollution. Table 11.1 illustrates categories of industrial plant processes and ancillary processes that may be potential groundwater pollution sources.

Table 11.1. Typical industrial plant processes and ancillary processes

Industrial plant processes	Ancillary processes
Transfer and storage of raw materials	Transfer, storage and use of fuels and lubricants
Production process	Power production
Storage and management of waste products	Water withdrawal and treatment
Storage, transfer and transportation of final product	Run-off management
	Grounds maintenance employing fertilizers and pesticides

Recognition of potential classes of groundwater contaminants for specific types of industries serves as an aid in the development of appropriate plans for the monitoring of ongoing activities (US EPA, 1999), as well as guiding the investigation of impacts that may be associated with past site operations (see Chapter 23).

Table 11.2 presents a number of examples of chemicals commonly associated with important industrial processes that historically have caused groundwater contamination. Substances marked bold are both toxic to humans and frequently found in groundwater. Their behaviour and attenuation in groundwater is discussed in more detail in Chapter 4.

Some of the substances listed in Table 11.2 are raw materials or production intermediates for chemical manufacturing, while some are typical of waste streams from the indicated industrial process. Yet others, however, may not be released from the facility, but rather are formed as natural degradation phenomena once the release of a parent chemical has occurred (Canter and Knox, 1987). For example, dichloroethene (DCE) or vinyl chloride (VC) may in rare instances be released directly but, more commonly, they are detected in groundwater as breakdown products in anaerobic conditions related to the release of perchloroethylene/tetrachloroethene (PCE) or trichloroethene (TCE) in the presence of specific subsurface bacterial populations. Similar changes may be observed in the progressive, sequential reductive dehalogenation of polychlorinated biphenyls (PCBs), which can result in the formation of lower chlorinated homologues from the parent PCBs. Presence of the new analytes, e.g. in groundwater samples, may create uncertainty regarding the source of the contaminants and, hence, may complicate decisions concerning the appropriate response measures.

Table 11.2. Potential groundwater contaminants from common industrial operations (substances marked bold are both toxic to humans and frequently found in groundwater)

Industry type or industrial process	Representative potential groundwater contaminants
Adhesives	acrylates, aluminium, **chlorinated solvents**, formaldehyde, isocyanates, mineral spirits, naphthalene, phenol, phthalates, **toluene**
Electrical components	acids, aluminium, arsenic, beryllium, **cadmium**, caustics, **chlorinated solvents**, cyanides, **lead**, mercury, **nickel**, **selenium**
Explosives	ethyl acetate, HMX, methanol, **nitrobenzenes**, nitroglycerine, **nitrotoluenes**, pentaerythritol tetranitrate, RDX, tetrazene, tetryl, 1,3-DNB
Fabrics	acetic acid, acetone, acrylates, ammonia, **chlorinated solvents**, **copper**, formaldehyde, naphthalene, **nickel**, phthalates
Fertilizers	ammonia, **arsenic**, chlorides, **lead**, phosphates, potassium, **nitrate**, sulphur
Foods and beverages	chlorine, chlorine dioxide, **nitrate**/nitrite, **pesticides**, biogenic amines, methane, dioxins, general organic wastes
Inks and dyes	acrylates, ammonia, anthraquinones, **arsenic**, benzidine, **cadmium**, chlorinated solvents, chromium, ethyl acetate, hexane, **nickel**, oxalic acid, phenol, phthalates, **toluene**
Laundry and dry-cleaning	calcium hypochlorite, **DCE**, **PCE**, Stoddard solvent, **TCE**, **VC**
Metals production and fabrication	acids, **arsenic**, beryllium, **cadmium**, **chlorinated solvents**, **chromium**, **lead**, mercury, mineral oils, napthtalene, **nickel**, sulphur
Solvents (chlorinated)	carbon tetrachloride/tetrachloromethane **(CTC)**, chlorofluoroethanes, **DCE**, methylene chloride, **PCE**, **TCE**, **VC**, 1,1,1-trichloethane
Solvents (non chlorinated)	acetates, alcohols, **benzene**, **ethylbenzene**, ketones, napthtalene, **toluene**, **xylene**
Paints and coatings	acetates, acrylates, alcohols, aluminium, **cadmium**, **chlorinated solvents**, chromium, cyanides, glycol ethers, ketones, **lead**, **mercury**, methylene chloride, mineral spirits, **nickel**, phthalates, styrene, terpenes, **toluene**, 1,4-dioxane
Paper manufacturing	acrylates, **chlorinated solvents**, dioxins, mercury, phenols, styrene, sulphur
Pesticides	**arsenic**, carbamates, **chlorinated insecticides**, cyanides, **ethylbenzene**, **lead**, naphthalene, organophosphates, phenols, phthalates, **toluene**, **xylene**
Petroleum refining	alkanes, **benzene**, **ethylbenzene**, **nickel**, polynuclear aromatic hydrocarbon (PAHs), napthtalene, sulphur, **toluene**, **xylene**
Pharmaceuticals	alcohols, benzoates, bismuth, dyes, glycols, mercury, mineral spirits, sulphur
Rubber and plastics	acrylonitrile, antimony, **benzene**, butadiene, **cadmium**, **chloroform**, **chromium**, **DCE**, **lead**, phenols, phthalates, styrene, **VC**
Wood preserving	ammonia, **arsenic**, **chromium**, **copper**, creosote, dioxins, pentachlorophenol (PCP), phenol, tri-n-butyltin oxide

In the example of both the chlorinated solvents and PCBs, the result may be that Chemical A is disappearing, when in parallel the concentration of the corresponding Chemical B is increasing. This seeming benefit of the reducing concentration of Chemical A can mask marked increases in the significance of potential risk when the degradation product (Chemical B) is more toxicologically potent than the parent molecule (e.g. VC >> DCE).

11.1.2 Types of industrial practices potentially impacting on groundwater quality

Virtually any aspect of industrial operations has the potential to release chemicals, though some processes are more likely than others to be of consequence when considering the vulnerability of groundwater. Organic and inorganic contaminants may reach groundwater most readily as a result of discharge to the ground surface and subsequent leaching through and from soils, or through subsurface releases from tanks, ponds, underground pipelines, injection wells, and similar structures (Canter and Knox, 1987; US EPA, 1999). Problems and characteristics of contamination that are related to individual chemicals may be compounded by events such as fires or explosions which often cause major changes in the chemical structures, chemical properties and distribution of industrial releases.

The withdrawal of groundwater, though not a waste-related or discharge-related matter, may dramatically affect the subsurface movement and distribution of chemicals, especially if the withdrawal is of large volume such as for single pass cooling water. In many parts of the world, permits are required for such withdrawal. Therefore, permits may represent one source of valuable information during reviews of potential industrial impacts in a drinking-water catchment area. Similarly, discharge permits may provide a valuable source of data on potential contamination origins.

Releases of small or large magnitude to surface soils may occur from raw material piles, aboveground tanks, drums and other containers, as well as from process leaks that may occur within the plant, and which may subsequently reach the ground from improper routing of wash waters or process overflow volumes. Industrial material handling operations, including incoming and outgoing shipments, as well as in-house transfers of materials, are often associated with episodic releases of chemicals over time (Canter and Knox, 1987). While each individual event is not necessarily significant, the cumulative effects can be severe.

Factors such as ground cover type (e.g. paved vs. unpaved), interceptor drains, local precipitation rate, soil types and aquifer vulnerability (see Chapter 8) as well as water solubility, vapour pressure, soil microbial activity and the mobility of the chemicals of interest (see Chapter 4) will influence how rapidly and in what form they move toward groundwater through the soil column once released at the surface. Non-production related activities on a site, such as grounds maintenance, may represent potential groundwater impacts from use of chemicals such as fertilizers, pesticides and herbicides (US EPA, 1999). Table 11.3 illustrates a number of acute and chronic release possibilities that have the potential to contaminate groundwater.

Table 11.3. Potential release points and mechanisms

Chronic releases	Acute releases
Direct discharge: ground surface or surface water	Explosion
Subsurface discharge: injection wells	Fire
Leaks: tanks, pipes or impoundments	Catastrophic failure: storage site or transfer system
Transfer loss: pipelines, transfer points, storage facilities	
Non-process activities: herbicides, fertilizers, pesticides	

Once released, low water solubility and strong binding behaviour cause some materials to move slowly in the subsurface environment, in comparison to substances that are highly water soluble and that do not attach to soil particles (see Chapter 4). In addition, high vapour pressure indicates that a chemical will favour volatilization, and spilled materials may be lost to air from water or soils, as opposed to leaching to groundwater. Local and regional meteorology will exert effects on whether or to what extent these or other airborne materials may be subject to later atmospheric 'washout' by precipitation, and subsequent re-deposition on the ground in complexed form, which then is available for future soil leaching processes that may contaminate groundwater.

Subsurface releases often represent the most direct pathway by which industrial contaminants may reach groundwater. These occur most commonly as a result of storage or disposal of liquids to pits, ponds, basins and underground tanks (Canter and Knox, 1987), as highlighted in Box 11.1. Such structures often are designed to act principally as evaporation or holding structures; however, as a practical matter, those that are not lined with clay or synthetic materials frequently exhibit a percolation component through the floor and walls of the structure, or through cracks in theoretically impervious tank materials (e.g. concrete, metal). The type, age, burial depth of the structure, soil type, proximity to (or contact with) the groundwater interface, the care with which it was constructed or installed, and the regularity of maintenance procedures, all are important influences on the likelihood that such holding structures may serve as sources to long term groundwater contamination potentially to be addressed in situation assessment.

Subsurface releases also may be caused by leakage from underground pipes at connections and valve locations, or as a result of rupture related to pressurization, corrosion and mechanical damage. Such releases frequently go unnoticed and over time may contribute to significant subsurface contamination. The oil refinery case study in Box 11.1 is an example of this type of situation. The frequency, duration and volume of such events, as well as the mobility and toxicity characteristics of the materials that are released will determine the potential risks within drinking-water catchment areas. Some important factors to consider with regard to storage structures are shown in Table 11.4.

Box 11.1. Groundwater pollution with aromatic hydrocarbons and metals caused by a petroleum refinery site in Czechowice, Poland

The process of refining hydrocarbons carries with it potential problems of raw materials transport, handling/storage in large volumes, and chemical production processes. These activities represent points at which chemical substances may be lost to the environment as a result of leaks, spills or other short and long term events. Many thousands of such facilities globally have been the source of local or regional groundwater contamination, particularly in the case of the more water soluble, aromatic hydrocarbon components (e.g. benzene, toluene) and of some historically common additives (e.g. lead).

This phenomenon is well-illustrated by the case of a 100-year old refinery located in an urban industrialized area of Poland. Capacity has more than doubled from early production rates of 40 000 tons of paraffinic crude oil a year producing gasoline, engine oil and fuel oil, as well as specialty oil products. Disposal from by-products of the historical sulphuric acid-based oil refining resulted in the deposition of more than 140 000 tons of acidic petroleum sludges in a series of open, unlined waste lagoons.

The refinery site is underlain predominantly by silty sands, interspersed with several thin discontinuous subsurface clay layers that do little to retard vertical movement of contaminants. Groundwater was located at about 10 m below the ground level in most areas on-site. There is a nearby water supply well, used for commercial and industrial purposes, where increasing levels of petroleum substances have been observed in recent years. Nearby residences are connected to a public water supply system.

A comprehensive refinery site investigation was conducted to assess the extent, degree and potential migration of site contamination, focusing on several principal indicator chemicals, including BTEX, PAHs and heavy metals. These were selected on the basis of their concentrations, mobilities and toxicological properties, as well as their known linkage to historical facility operations. These typically are 'sentinel compounds' for the evaluation of potential risks at facilities such as the Polish site.

Soil and groundwater sampling data indicated broadly variable contaminant levels at the refinery site, with a definite 'hot spot' within the large lagoon area. Groundwater was found to be heavily impacted mainly by benzene and toluene, though these substances often were at low levels in the lagoon sludges, due to their volatility and the long residence time of sludges in the lagoon. It was concluded from the observed distribution of contamination that the lagoons represent at least localized long-term sources to groundwater for hydrocarbons and some metals, largely limited by the viscosity and low water solubility of their contents. However, more recent plumes of volatile chemicals (e.g. benzene, toluene) are likely a result of ongoing refinery operations, probably related to pipeline leaks, spills associated with product transfer and product losses in other areas of the site. It was recommended that remedial action should be undertaken at the refinery site, strategies be implemented for the prevention of releases and that efforts be initiated to contain the expanding groundwater plume (see Chapter 23, Box 23.1).

Table 11.4. Information needs for storage vessels

Type	Tank, lagoon, pit, pipeline
Age	Years in service, planned life span
Contents	pH, water content, corrosivity
Construction material	Native soils, concrete, metal, clay, plastic
Containment	Type, volume and security of secondary containment structures
Location	Above/below ground surface
	Proximity to groundwater
	Location and nature of pipes and valves

As an example, some metal manufacturing and finishing processes generate large volumes of liquid, semisolid and solid wastes that historically have required at least some element of on-site storage and/or disposal. These same facilities often have extensive above ground and underground piping systems that may be sources of groundwater contaminants. The historical use of pits, ponds, lagoons and tanks to store oil or hydrocarbon wastes, as well as solvent-contaminated washwaters and acidic (low pH) or caustic (high pH) sludges, has resulted in many instances of broad scale aquifer contamination. Such contamination includes water-soluble substances of health significance (e.g. arsenic, lead, mercury, chlorinated solvents, fuel components, acidic solutions), as well as those with minimal solubility (e.g. PAHs and PCBs). The large facility size and long operational time frames for many smelters and metal production plants pose specific concerns in terms of clearly identifying and characterizing releases, as well as in terms of implementing effective containment or remedial measures to address the problems.

The use of injection wells for the purpose of liquid industrial waste disposal has the capacity to introduce large volumes of chemical constituents, often of poorly understood composition, into deep groundwater. Well type, construction and integrity, as well as injection depth, chemical composition and duration/volume of injection events all will influence the likelihood that an injection well serves as a source of groundwater contamination.

Impacted streams and rivers are often overlooked as potential sources of groundwater contamination, though they may serve as significant contributors to local groundwater quality if the surface water body recharges local groundwater. Thus the detailed understanding of local and regional surface water quality and/or quantity may play a role in assessing the impact of industrial facilities in areas where upstream discharges to or withdrawal of river volumes affects the downstream recharge characteristics.

Aside from the industrial releases themselves, environmental transport of contaminants from soil to groundwater or within groundwater may be enhanced greatly by the presence of conditions which act to mobilize otherwise recalcitrant chemicals. For example, many organic substances may be bound well to soil if they are present alone, but may become quite mobile if they are present concurrently with another chemical that acts as a cosolvent (e.g. fuels mobilize organic chemical residues in soil). Similarly, mobility of a number of metals in soils (e.g. lead) is dramatically enhanced by low pH conditions in the soil or in local precipitation (Mather *et al.*, 1998). Thus site-specific geological, physicochemical, and land cover or land use considerations often are the dominant features in determining the likelihood that a facility may contribute to groundwater contamination.

Box 11.2. Groundwater pollution with chlorinated solvents caused by leather tanning industry in the United Kingdom

As with refinery sites, large and small chlorinated solvent sites around the world have been associated with groundwater contamination. This is as a result of their historical widespread use for degreasing, metals cleaning, textile treatments and other applications. Although these solvents exhibit comparatively low water solubility, their environmental behaviour and their ability to act as dense non-aqueous phase liquids (DNAPLs) (see Chapter 4) often cause disproportionate problems in developing engineered remediation solutions. In addition, many countries have established quite restrictive water quality protection criteria for chlorinated solvents (e.g. TCE, PCE) or potential environmental degradation products (e.g. VC). A case which has elements reminiscent of many others involved the Cambridge Water Company and several local tanneries in the United Kingdom during the 1950s through to the 1990s.

TCE and PCE are among the most common chlorinated solvents encountered, and were used in the leather tanning process. On-site handling practices, as well as spills and other releases, caused soil contamination at this industrial site. The complex geology in the area (multilayered Chalk composition) complicated several efforts to model the contaminant flow in the vertical and horizontal direction. However, it was concluded that the releases likely occurred in the early years of chlorinated solvent usage at the facility (i.e. the late 1950s). Discovery of contamination of a local water supply well in the early 1980s triggered an extensive investigation by the local Water Authority and the British Geological Survey (BGS), which ultimately demonstrated that significant contamination was broadly distributed in the area at concentrations exceeding 1000 micrograms per litre. Despite conversion of the local water supply well to a pump-and-treat recovery well (which recovered over 3600 litres of PCE in 5 years), a substantial quantity was unrecoverable, as is often the case with the chlorinated solvents.

Although there is a tendency to focus on large industries as most likely to cause large groundwater impacts, the judicial actions surrounding this case emphasized the potential for contributions to local groundwater pollution by many small industries in an area, as well as the valuable benefits of planning and proper chemical handling, as opposed to attempting remedial actions decades after the release has occurred. Of course this observation can be made for other industries as well including, for example, textile operations, tanneries, motor vehicle fuel stations, electroplating shops, etc.

Implementation of rigorous management practices at individual facilities may provide an excellent organizational structure for maintaining good control of raw materials and wastes that have the potential to contaminate groundwater. Recycling, waste minimization and good materials balance accounting have the potential to reduce energy requirements, transportation requirements, chemical demands, water demand and waste disposal needs (see Chapter 23). Situation assessment therefore needs to identify the extent to which such practices are operating.

Furthermore while in many regions of the world practices in production, transport and containment of hazardous chemicals has substantially improved during the past two decades, historical contamination may be substantial. The case study in Box 11.2 shows how discovery of contamination in drinking-water may lead to detection of large-scale contamination of historic origin, particularly also from a high number of small-scale enterprises.

Effective documentation of the hazards that may be posed to groundwater by any particular facility will be a function of the ability to show that good management practices and responsible chemical stewardship are in place and working (Ekmekci and Gunay, 1997). Further evaluation of the existing impacts to groundwater (or lack thereof) can be assisted by access to historical groundwater data for the facility, adjacent facilities or the region in which the facility is located. Based upon that information, it should be possible to expand existing programmes of monitoring or develop new programmes to more effectively detect industrial contamination.

11.2 MINING ACTIVITIES

A number of activities associated with mining operations have a significant potential to pollute groundwater resources. With the exception of some deep mining areas, mines tend to be at higher elevation in the catchment areas where rocks are closer to the surface. Thus impacts from these mining activities may affect downgradient groundwater resources as well.

The broad term mining includes both open pit (surface mines) and underground mines, as well as oil and gas mining (via wells), solution mining, in situ leaching (ISL), heat (geothermal) mining and even gas hydrate mining or ocean dredging. Open pit mining includes not only ore and lignite mining, but also excavation of gravel, sand and clay for the construction industry. While the latter may be less intrusive, they can result in severe environmental effects if not properly managed. Historically, mining was chiefly conducted as underground mining following veins of e.g. ore and coal, whereas more recently the availability of large machinery has promoted a tendency towards open pit mining. Today, criteria determining which option is preferred include economic considerations such as costs for labor (which is substantially more intensive in deep mining) and investment into machinery (which is high for open pit mining), as well as aspects of occupational safety (deep mining tends to be more hazardous) and the acceptability of sacrificing large areas for devastation as open mine pits. ISL is a high-tech option chiefly employed for mining copper and uranium.

Impacts on groundwater quality from mining operations include but are not restricted to:

- mobilization of metals and metalloids due to low pH values in acid mine drainage;
- leaching of substances from rock formations ISL;

- leaching from inadequately designed or operated mining waste dumps or tailing piles (i.e. overburden soil and rock);
- activities directly linked to mining operations, often in their direct vicinity, such as inappropriate usage, handling, storage or spillage of chemicals employed in ore treatment, underground or surface traffic, heavy mining machinery, workshop or refining work operations as well as wood strut preservation in underground mines.

Figure 11.1 provides a general overview of activities associated with the operation of underground mines.

Mining activities may directly impact on groundwater quantity. Both open pit mines and underground mines are often associated with groundwater withdrawal which creates a cone of depression during the operating lifetime of the mine. Thus the unsaturated zone (zone of aeration) is significantly enlarged, leaving rocks and sediments exposed to oxygen for a long time, which may cause the oxidation of sulphides and other minerals. This phenomenon also may occur with 'heaps' or tailing piles from milling sites where minerals can be oxidized.

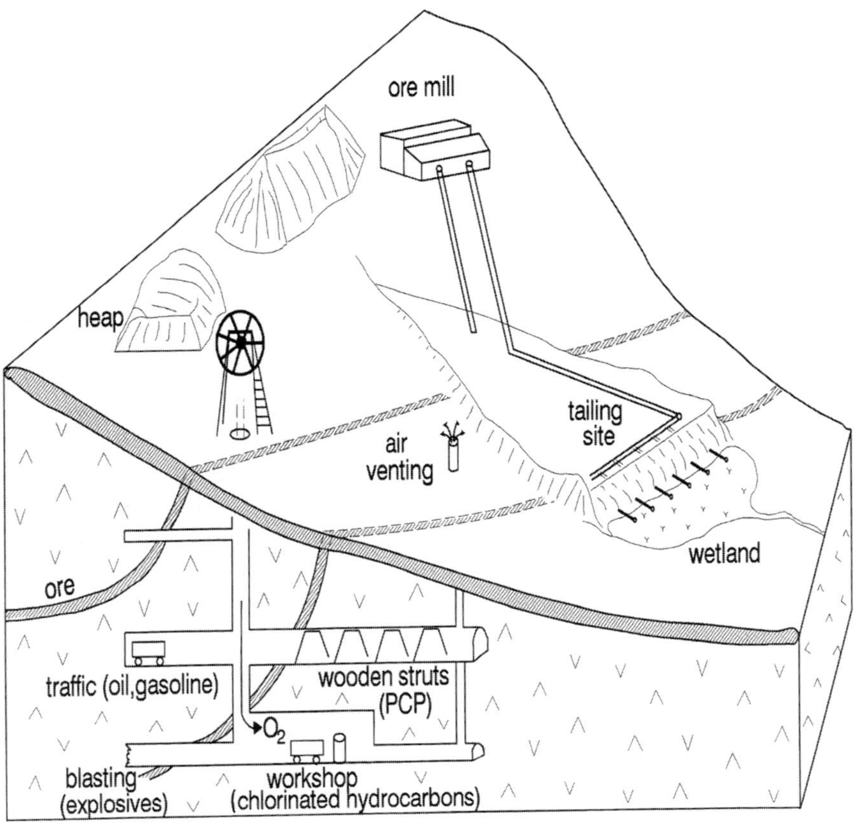

Figure 11.1. Sources of groundwater contamination linked to mining activities

Additional and different problems occur when mines are closed down, as discussed in more detail in Sections 11.2.2 and 11.2.3. Open pit mines may be left open or refilled with waste rock, tailings, industrial by-products and/or they become landfills for municipal waste. Cessation of groundwater withdrawal that is associated with normal mine operations typically leads to a rise in groundwater levels, which may form a lake or may infiltrate the backfill materials and flood former mine shafts. As underground mining changes rock permeability significantly, it is unlikely that natural groundwater conditions will recover to previous levels after mine closure. Mining at the West Rand Goldfields in South Africa, for example, created links through underground penetration of original dykes which originally had separated different groundwater reservoirs in dolomitic areas.

Rewatering after closure of the mines may be a very complicated process with great uncertainty as to the eventual outcome to groundwater quality. Re-establishment of local natural flow conditions can allow polluted waters to contaminate hitherto pollution-free areas. To protect the aquifer against negative impacts after mine closure, a sustainable groundwater management system in the vicinity of abandoned mines is required (see Chapter 23.2). Thus situation assessment in settings with mine closure would address how well such processes are controlled, which often has not been the case in the past.

The scale of mining activities is an important factor to the potential for groundwater pollution since small-scale mining activities are more difficult to monitor and control. On the other hand, large-scale mining operations typically have a greater impact on local groundwater resources. Many countries have environmental legislation to address impacts from mining operations or to guide closure and reclamation of individual newer mines; however old mines often are not considered. There is a lack of definition for controls on mine water quality and mine water in excess of maximum contamination levels being allowed to spill into groundwater or surface water in many instances.

11.2.1 Operation of mines: Chemical processes and potential impacts to groundwater

Of all the processes which occur during mining activities, sulphide oxidation is one of the most severe pollution problems where sulphide minerals (e.g. pyrite) occur geologically. It can lead to acid mine drainage that is often enriched with metals such as iron, aluminum, arsenic, cadmium, lead, mercury and uranium. In this respect, the Boshan case study given in Box 11.3 is typical for groundwater contamination through mining, though contaminants from industrial activity more or less strongly linked to mining were found as well.

Sulphide oxidation happens when sulphidic minerals (e.g. pyrite) are exposed to air and water. This process is complex because it involves chemical, microbiological, and electrochemical reactions. The rate of oxidation is controlled by a number of parameters including water pH, partial pressure of oxygen, mineral surfaces and the presence or absence of bacteria.

Geologically, sulphidic minerals (e.g. pyrite) are formed in a reducing environment when sulphate and iron or other inorganics are supplied by water in the presence of decomposable organic matter. Depending upon the boundary conditions and time, the formation of sulphide can be quite variable resulting in different crystal structures of sulphides and pyrite (e.g. framboid, polyframboid, conglomerates and massive octahedron). A limited supply of organic matter in marine sediments is assumed to result in low sulphate reducing rates, and thus formation, of framboidal pyrite (Evangelou, 1995).

Framboidal pyrite is believed to be the major contributor to acid mine drainage due to its large surface area and resulting rapid rates of reactivity. Pyrite in waste piles and tailings is of finer grain and therefore much more reactive than forms that may be present in the original bedrock (Langmuir, 1997). In the absence of buffering material, such drainage is extremely acidic with pH values approaching zero and even negative pH values (Nordstrom *et al.*, 2000).

Box 11.3. Mining and industrial contamination of groundwater in Boshan, Shandong Province, China

Boshan is located in the centre of Shandong Province in China, and is an important mining, industrial and manufacturing centre known particularly for the production of ceramics. Although Boshan only has a population of a few hundred thousand people, the surrounding District is densely populated and industrialized, close to the industrial centre of Zibo City, which has a population in excess of 3.5 million. Like many towns and cities in densely populated regions, it is difficult to distinguish a clear boundary between the urban centre of Boshan and the surrounding semi-rural or rural areas, which in effect form an extensive peri-urban region.

The City of Boshan and the Boshan District are totally dependent on groundwater for water supplies, because surface streams have become seriously contaminated through the disposal of mine water, sewage and industrial wastes. Currently more than 80 000 m^3 per day is pumped from the Tianjinwan well-field located to the east of Boshan. Another well field, the Liangzhuang well field, was abandoned in 1986 due to contamination.

There is widespread contamination of the limestone aquifers near Boshan, which has greatly reduced the amount of groundwater that is available for potable supplies. Possibly the single largest source of contamination in the District is acid mine drainage containing high concentrations of dissolved iron and sulphate from nearby coal mines. Currently about 40 000 m^3 per day of acid mine water is drained from mines, and this has caused progressive increases in the sulphate concentration in local groundwater, with concentrations commonly exceeding 500 mg/l.

In addition to pollution from mining, a major pathway for groundwater pollution in Boshan is seepage from recharge basins and surface streams, particularly from the Xiaofu Stream that passes through urban areas of Boshan. These are heavily contaminated with industrial wastewater containing a variety of chemicals including sulphate, petroleum hydrocarbons, phenols, cyanide, arsenic and other metals.

Sulphate and iron are the most common inorganic contaminants in acid mine water. In addition, chloride, sodium and potassium are increased significantly if halite is present. Several metals and metalloids of public health importance that are associated with increased concentrations due to mining activities include arsenic, manganese, lead, cadmium, nickel, copper, zinc, aluminum, mercury and uranium. For selected elements, Table 11.5 shows ranges of typical background concentrations in relation to increased concentration ranges induced through mining and thus helps in assessing whether concentrations found in groundwater potentially affected by mining might indeed be elevated through this activity.

Uranium, radium, radon and thorium are radioactive elements which are encountered not only in uranium mining, but also in metal ore, lignite and coal mining, where they commonly occur in increased concentrations. In the case of ISL mining, chemicals (e.g. acids or alkaline brines) are pumped into injection wells in large quantities and may remain to some extent in the underground. Oil and gas exploration may be associated with the presence of groundwater and brines with elevated concentrations of potentially hazardous elements (e.g. boron, lithium, selenium, arsenic, bromine, barium and thallium). Mining of evaporites (e.g. halite) is of special concern owing to the extremely high solubility of the salt. Waste rock piles from halite mining commonly contain large amounts of easily dissolved salt. Therefore, groundwater and surface water bodies are commonly impacted by salty waste water during active mine operations.

Table 11.5. Element concentrations from oilfield groundwater and from metal mining in comparison to WHO guidelines values and ranges of background concentrations observed in groundwater under natural conditions (adapted from Merkel and Sperling, 1998; WHO, 2004)

Element	**Oilfield groundwater** *(mg/l)*	**Metal mining**	**Background (extreme values)**	**WHO guideline value** *(µg/l)*
Arsenic	0.05-0.8	0.1-50	0.0001-0.01 (>1.0)	0.01 (P)
Barium	20-180	1-90	0.005-0.1	0.7
Boron	120-400	No data	0.005-0.070	0.5 (P)
Cadmium	0.001-0.10	0.001-0.4	<0.001-0.005 (0.07)	0.003
Lead	0.001-0.03	0.01-1.5	<0.001-0.01	0.01
Manganese	2-30	0.1-220	0.02-8.0	0.4
Mercury	No data	0.01-0.5	0.00001-0.0005	0.001
Nickel	0.001-0.5	0.001-100	0.001-0.17	0.02 (P)
Selenium	0.5-4.0	1-30	0.0001-0.14	0.01
Uranium	<0.001	0.1-200	<0.001-0.02 (>0.1)	0.015 (P)

P = provisional

Different chemicals, some of health relevance, are used for ore treatment which is commonly conducted close to the mine to reduce transport costs. Cyanide is used, for example, during the extraction of gold. Another problem is the fixation of unwanted by-products, for example in the case of uranium ore treatment where radium is fixed by adding barium chloride to the tailing water in order to precipitate the solid solution mineral Ba(Ra)-sulphate in the tailings. These precipitates may represent long term concerns as potential sources of groundwater contaminants.

During mining activities, precipitation of mineral phases termed secondary minerals may occur. In addition to clay minerals being a common weathering by-product, some minerals such as $KAl_3(SO_4)_2(OH)_6$ (Alunite) and $KFe_3(SO_4)_2(OH)_6$ (Jarosite), can form solid solution complexes with Fe^{3+} and Al^{3+} by evaporation in deep mines and at depth in saturated tailings or in heaps under acid sulphate conditions. Their solubility is relatively low under pH conditions greater than 3. Also, gypsum can be formed in the presence of sulphate and calcium and melanterite ($FeSO_4 \cdot 7H_2O$) in the presence of iron. Most secondary minerals have low solubility which is important in the case of mine flooding because it limits the rate of their re-solution. Thus secondary minerals may act as temporary or permanent sinks for groundwater contaminants making their investigation important.

Large quantities of hydrocarbonsare used in the mining industry for trucks and heavy mining machinery, and chlorinated hydrocarbons may be used for equipment cleaning purposes. PCP, γ-HCH and creosote are commonly used for the preservation of wood which may be of special concern in deep mines using wood struts. When a mine is closed down quantities of hydrocarbons and chlorinated hydrocarbons may be left in the subsurface or in surface contaminant areas and may act as long-term sources for potential contamination. Wood treatment agents like PCP will be leached slowly from remaining wood struts and may contaminate the groundwater in the mine vicinity over a long period. During open pit mining (e.g. lignite mining) spills of fuel and oil may occur regularly. Also, pipelines from oil and gas fields often run over hundreds or thousands of kilometres and leaks and spills can occur (see Chapter 13). Further, explosives from mining activities were suspected to be the chief source of nitrate contamination at the Orapa diamond mine in Botswana (Box 11.4).

Box 11.4. Diamond mining as potential nitrate source in groundwater at the Orapa diamond mine in Botswana

Mining of the kimberlite pipe at Orapa began in 1971. This is the second largest kimberlite pipe in the world in terms of area, covering 117 ha. Mining operation is in a conventional open pit with the pit bottom now at approximately 110 m below surface.

Although no contamination of the groundwater is to be expected from diamond mining, a study by the Debswana Diamond Company (Pty) Ltd revealed elevated levels of nitrate (generally >50 mg/l), in particular for wellfield 2 and 5 (Mokokwe, 1999). In addition, the available data revealed a strong spatial and temporal variability of the observed nitrate levels. The latter was surprising since the main aquifer, the Ntane sandstone, is largely confined and features water of an old age. Thus one would expect nitrate levels to be rather uniform and of natural origin.

A potential anthropogenic source of nitrate, even if only at a limited scale, is the 240 t of ammonium nitrate-based explosives that are used in the pit monthly. Although most of the explosives will be converted to nitrogen and other gases, and be vaporized during the explosion, pollution of the groundwater in the overlying Kalahari Group sediments may occur, in particular through leachates from the slimes and slurry dams (Tredoux, 2000). Because high nitrate levels constitute a health risk to infants, the Federal Institute for Geosciences and Natural Resources, Germany and the Technical University of Berlin and the Debswana Diamond Company (Pty) Ltd carried out a survey in Orapa in January 2000 of 60 boreholes and wells.

The results of this survey again highlighted that nitrate concentrations in Orapa almost always exceeded the WHO guideline level (Figure 11.2). The highest concentration of 199 mg/l was found to the east of the old mining dumps. Concentrations decreased in the direction of the surrounding production boreholes which indicates that this very high nitrate concentration is caused by leachate from the old refuse dumps.

In contrast, groundwater samples from the observation boreholes around the new township landfill site displayed neither significant ion concentration nor changes of ion ratios. Hence, this waste disposal site does not seem to be a source of groundwater pollution yet.

Nitrate was also high at a borehole north of the Orapa Township, probably due to ingress of excreta from households that are not connected to the sewage system. This case highlights the complexity of potential sources of nitrogen.

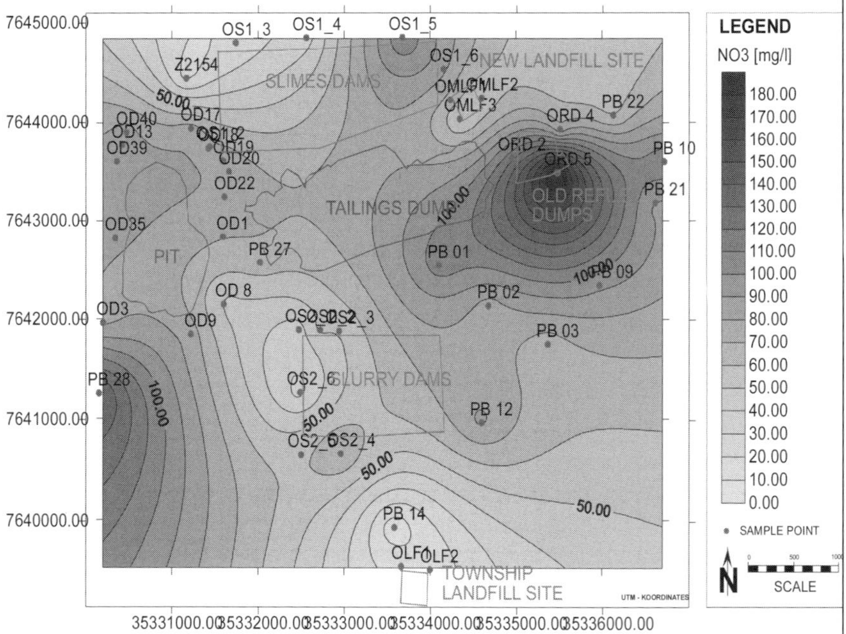

Figure 11.2. Nitrate distribution at the Orapa diamond mine and at wellfield 5

11.2.2 Closure of deep mines

Deep mining is often conducted in hard rock environments with a small degree of total porosity. In the large majority of settings, their operation requires continuous dewatering of the mine shafts. Thus the recovery rate to refill a cone of depression caused by long term dewatering operations can be very slow. However, if the bedrock is more porous, as in the case of sandstone with both fracture flow and pore flow (Chapter 2), the artificial and natural fracture cavities will refill quickly, while the pores will refill much more slowly. This may cause entrapment of air in some parts of the mine. This entrapped air can be removed only by diffusion over a long period of time, until the whole mine area is saturated.

Effective flooding of deep mines will vary according to local conditions. The simplest way is to switch off the pumps used for dewatering. Groundwater levels will then recover at rates dependent upon the hydrogeology of the area. However, due to the higher permeability of adits (drainage tunnels) and shafts, water levels and flow directions are unlikely to recover to natural pre-mining conditions (Figure 11.3). When the cone of depression is refilled, groundwater will resume flow towards the natural drainage and may transport potential contaminants in a downgradient direction. This process of mine closure is referred to as uncontrolled flooding.

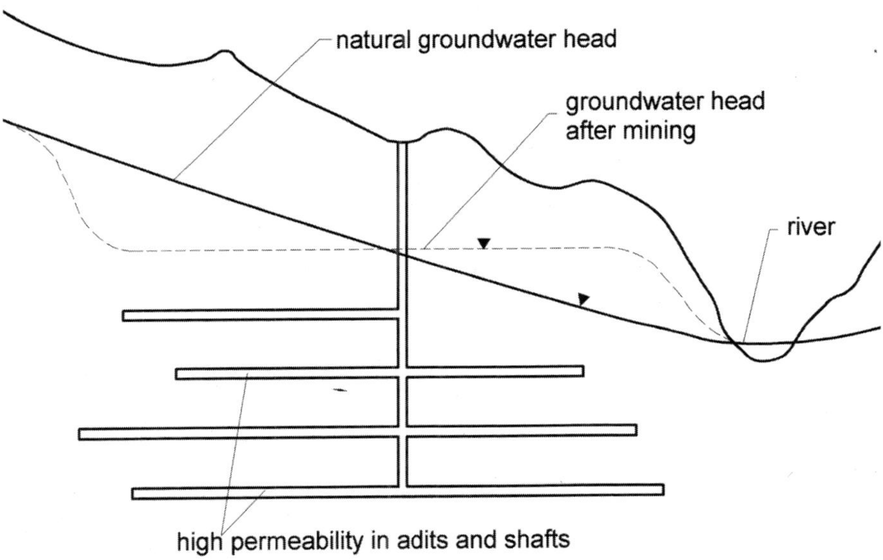

Figure 11.3. Changes in groundwater hydraulic heads due to increased conductivity caused by adits and shafts in mined areas

The rapid solution of minerals on first contact with water ('first flush') often results in a maximum of inorganic contaminants during the initial flushing of the mine at closure. Groundwater quality from a flooded mine may recover to background concentrations after some years or concentrations may remain increased for decades or centuries if pyrite oxidation is still taking place in the unsaturated zone. Some of the secondary minerals that have been precipitated during the mine operational time are also dissolved during the mine flooding process. Thus groundwater often contains significantly increased concentrations of various contaminants at the very beginning of groundwater recovery; however, they may decrease with time due to dissolution of secondary minerals like gypsum or melanterite ($FeSO_4$).

It is quite common in mountainous areas to manage groundwater withdrawal passively from an active mine by means of drainage tunnels known as adits. Construction of such tunnels is expensive; however they are comparatively inexpensive to operate (no need for pumps, no electrical power consumption). If such tunnels are not sealed by concrete dams at or during mine closure, groundwater drainage will continue as long as the tunnel remains open. Thus groundwater levels will not return to pre-mining conditions. In consequence, pyrite oxidation may continue to take place until all sulphide minerals are consumed. Low pH and elevated metal concentrations associated with pyrite oxidation may make the water unsafe for drinking purposes both locally and downgradient.

11.2.3 Closure of open pit mines

Closing surface mines is related to the refilling of excavations with overburden and the recovery of groundwater levels. If other wastes (e.g. industrial wastes, municipal wastes) are deposited together with waste rock during mine closure, additional contamination problems may occur (Chapter 12). Waste rock which was backfilled into the open pit mine may have a high potential to produce acid mine drainage due to oxygen contact over long time periods and thus the formation of secondary minerals which can be easily dissolved with the recovering water table.

Depending on the amount of ore or coal/lignite mined, it is quite common that the deficit volume of an open pit mine is filled largely with groundwater, forming a lake. This lake formation may result in a deformation of local groundwater elevations. At the in-flowing (upgradient) end of the lake the depth-to-groundwater elevation will be increased and will be decreased at the out-flowing lakeshore. If the lake is elongated this impact is more severe.

Whereas during mine operation dewatering chiefly transports contaminants into surface waters, post-mining lakes often become highly acidic and contain high concentrations of metals. As they are integral elements of the local groundwater system, these pollutants will be transported downgradient and impact the aquifer. This is a large-scale quality problem in the Lausitz Region of Germany, as highlighted in Box 11.5.

In arid and semiarid areas with low flow groundwater conditions, which are linked with a low gradient of the groundwater table, any open pit mine lake which forms will be subject to intensive evaporation from the open water surface. Thus the mine lake actually may become a depression in the regional groundwater system even without downgradient outflow. Consequently, the salinity of groundwater or concentrations of other inorganic constituents may increase with time, making the resource unusable for drinking-water or other purposes.

Box 11.5. Consequences of the closure of open pit mines in the Lausitz region, Germany

When groundwater withdrawal occurs over long periods, many mining areas suffer from the effects of the export of groundwater long after mine closure, as illustrated by the Lausitz lignite open pit mining district in Germany. Since the beginning of the 20th century lignite mining took place in an area of approximately 2100 km^2. Groundwater withdrawal initially was conducted at a rate of 2 to 3 m^3/s, increasing to a maximum of 33 m^3/s in 1989. The water was largely pumped to the rivers Spree and Schwarze Elster. This dewatering activity accumulated to a groundwater deficit of 13 billion m^3 in 1990, at which time lignite mining was reduced dramatically following the unification of western and eastern Germany. Natural recharge is not locally sufficient to replace these massive deficits within a short time and measures have been set in operation to ensure a minimum base flow in the rivers. Thus a river catchment and groundwater management system was implemented, which is expected to remain in operation for at least two to three decades, until nearly natural conditions have been re-established in groundwater and surface water levels. The process of filling the open pit lakes created by the mining activities with river water has the potential to result in infiltration into oxidized waste rock piles, thereby creating strong potential for development of acid mine drainage. An acidic plume can already be observed migrating downstream of these lakes. The concern is that even after two to three decades when the groundwater deficit is reversed the interconnected lakes will result in severe problems of acid groundwater. Further, these pH impacts, with associated toxic metals, are likely to affect the local lakes and rivers, rendering them also unusable for abstracting drinking-water for several decades at minimum.

11.2.4 Predicting post-mining groundwater quality

Mine operators and local authorities need to understand water quality both during and after mining operations. A first step in prediction is to consider water quality at operating or abandoned mines in similar geological and hydrogeological conditions. A second step is the analysis of rock samples from the site to determine both alkaline-producing potential and acid potential (Brady and Cravotta, 1992). In addition to such static tests, kinetic leaching tests have been developed as measures of water quality effects; however, long term field verification is lacking for these kinetic tests. In some cases, the prediction of post-mining hydrochemical conditions can be accomplished by the use of regression analyses.

In order to understand complex chemical processes and/or to predict post-mining water quality, a variety of hydrogeochemical models are available, such as PhreeqC2, Phrqpitz or EQ3/6 (Plummer *et al.*, 1988; Wolery, 1992; Parkhurst and Appelo, 1999). Where acid neutralization reactions can have a strong effect on the transport of dissolved metals, models based on the coupled solute-transport/hydrogeochemical mass-transfer like MINTRAN (Walter *et al.*, 1994) or TREAC are preferable. They predict pH buffering sequences and metal attenuation mechanisms that are similar to those observed at field sites.

11.3 MILITARY FACILITIES AND ACTIVITIES

Regional and international conflicts or military occupations, coupled with day-to-day operations of military bases and support facilities, have resulted in environmental degradation and long-term contamination of soil and groundwater at both former and active military sites. An example of severe contamination from day-to-day operations is given in the Valcunai case study in Box 11.6. For assessing impact on groundwater, a distinction can often be drawn between direct military actions or conflicts on one hand and the (often longer term) use history of military sites and manufactories on the other. Depending upon the nature of the site, contamination to be addressed by situation assessment will include non-ordnance-related chemicals that are associated with typical ancillary military operations (e.g. fuels, pest control chemicals and municipal wastes). Because military bases often resemble towns or small cities in their breadth of activities (Teaf, 1995), many of the considerations that are presented in Chapters 10, 11.1 and 12 regarding municipal and industrial risks to groundwater recharge areas also are relevant here.

Box 11.6. Impacts on groundwater quality by a former military base in Lithuania

The Valcunai Oil Product base, which is located 14 km south of Vilnius, was a site of underground storage of light oils and rocket propellants (nitrogen tetroxide) from 1963 until 1993. Over 33 000 m^3 of storage in leaky underground tanks was in service (27 500 m^3 for light oils and over 3000 m^3 for rocket fuels). A three year monitoring study (Seirys and Marcinonis, 1999) determined that groundwater has been impacted both by pure oil product (3000 to 3500 m^2 at a depth of 5.5-7.5 m), chlorinated solvents (e.g. TCE, PCE, carbon tetrachloride) as well as dissolved aqueous phase constituents from oils and other organics and inorganics (nitrates, metals). While a glacial till confining unit of approximately 60 m thickness separates the shallow aquifer from the productive deeper aquifer beneath the oil storage areas, this confining unit is absent beneath the rocket fuel storage areas, rendering the deep aquifer very vulnerable. This aquifer supplies the Pagairai Wellfield, the largest potable water supply for the City of Vilnius. Subsurface ravines representing historical glacial features act as drainage channels for groundwater. Contamination has been observed to be mobile both vertically and horizontally in groundwater. The study reported contaminated shallow groundwater to be discharging to the Rudamina River approximately 0.8 km distant. Thus the Valcunai site represents an example of complex subsurface characteristics with high levels of contamination affecting soils, groundwater and surface water, all of which illustrate the difficult technical aspects of remediation. Removal of free phase hydrocarbons is addressed with extraction wells and oil separation units treating very large volumes of groundwater. The estimated goal of recovery is in the range of 400 to 500 m^3 per two years.

Groundwater pollution resulting from deployment of explosives in military conflict includes the impact of damage or destruction of industry, traffic facilities and municipal sewage that may lead to pollutant release. This is similar to spills in 'extreme events' as discussed in Chapters 10, 11.1 and 12, but may additionally include combustion products and their transformation products. Further aspects of groundwater pollution through military conflict are physical damage to water supply infrastructure, and the direct contamination of water by residues of many types of explosives, for most of which health and environmental impacts as well as their behaviour in groundwater are poorly understood. The discussion below focuses on the potential for groundwater contamination released from warfare agents' production and military operation sites.

In historical review, as with many other human activities potentially polluting groundwater, the scale of deployment and the variety of warfare agents increased dramatically during the 20th century. The additional use of chemical warfare (CW) agents introduced a potential for environmental damage. After conflicts, entire armament production plants and military facilities have been dismantled or destroyed and ordnance buried or dumped without any environmental safety precautions.

The groundwater pollution potential from military operations became apparent and subject of scientific research only in the early 1990s. The results disproved the often-expressed hope that many of the military chemicals which are classified as dangerous would quickly be degraded in soil to non-hazardous concentrations (Mulisch *et al.*, 1999a; 1999b).

The specific difficulty in assessing the potential for chemical impairments to groundwater from military sites is that substances used are often subject to secrecy, while their identification in a historical review of site-specific activities would greatly support situation assessment. Only if the substances to be expected are known can their hazard potential be determined for a given site on the basis of substance-specific data. Investigations of the complex biophysical, chemical and biochemical transfer processes as well as of microbial metabolism of organics further supplement the basis for a prognosis of the likely groundwater impacts in the recharge area, including contaminant fate, distribution, bioavailability and degradation in the subsurface (Mulisch *et al.*, 1996; 2000). Historical research to identify military chemical production sites and the areas in which these products were used can substantially help to identify possible risks to groundwater in a region.

The description of potential groundwater contaminant problems, as given in the following section, follows the basic categorization of sites into armament production sites (manufactories) and areas/sites at which the products are used (military operation sites). The latter may include base facilities as well as deployment areas (e.g. shooting ranges, test sites). Potential control and remediation measures for settings in which military sites are suspected sources of groundwater contamination are similar to those for industrial sites and are discussed together with these in Chapter 23.

11.3.1 Potential groundwater contaminants from military production sites

Both active and abandoned military production and manufactory sites may comprise explosives and powder factories, plants for the production of CW and smoke agents, armament filling plants and munitions factories, as well sites at which munitions were stored, disposed of or buried. For production-related reasons (e.g. need for large volumes of groundwater), these armament complexes are often located in areas that are rich in groundwater resources, which potentially elevates risks of large-scale pollution. In particular, contamination has been reported from sites where loading/unloading and filling operations took place, as well as cleaning and maintenance work on machinery or the cleaning/refurbishment of containers and ammunition. Wastewater from cleaning operations was generally highly contaminated by explosives, such as trinitrotoluene (TNT) isomers. Blasting operations can result in widespread contamination by explosives especially at large plants. Areas of suspected contamination also include, in particular, sites at which residues of explosives and off-specification batches of munitions were burned or buried.

The groups of products generally referred to as military warfare agents mainly comprise explosives and a number of chemical agents. Of the categories of explosives, the high-brisance (very powerful) explosives are of greatest interest, largely because they typically are safer to handle on a regular basis and have very high detonation velocities. Major representatives are 2,4,6-TNT, 2,4 or 2,6-DNT, 1,3-DNB, hexogen (cyclotrimethylenetrinitramine) and picric acid (2,4,6-trinitrophenol), but also nitropenta and tetryl. To detonate the high-brisance explosives, it is necessary to ignite them with highly sensitive initiating explosives (e.g. nitropenta and tetryl as well as lead azide, mercury fulminate, thallium azide, and tetrazene). Other important explosives include cyclotrimethylenetetranitramine (RDX), cyclotetramethylenetetranitramine (HMX), ammonium picrate, ammonium nitrate, nitroguanidine, nitroglycerin and dinitrophenols.

Several of the military explosives are of more dominant interest than others because of the large volume of their use, their potential to migrate to groundwater, persistence and toxicity characteristics. For example, 2,4,6-TNT has been very widely used as filling for bombs, mines and shells, readily dissolves in water and can move to groundwater with ease, is persistent (though microbial degradation to aminodinitrotoluolene occurs), and has a high degree of toxicity. It also is classified by US EPA as a possible human carcinogen. Dinitrotoluene (DNT) is mobile in groundwater, is scarcely oxidized biochemically, and not hydrolysed under environmental conditions. Many examples of TNT and DNT contamination have been identified in European and USA military facilities or support industries. Tetryl slowly hydrolyses to picric acid, which does not degrade biochemically under aerobic conditions and only slowly to picramic acid under anaerobic conditions.

Some military ordnance chemicals (e.g. amines, nitro compounds and nitroso compounds) are of interest also because their degradation products (e.g. nitrates) represent significant groundwater contamination sources. Table 11.6 identifies a number of the most common explosive or other military substances with notes on their environmental and health concerns.

Table 11.6. Health-relevant military ordnance chemicals in groundwater (ATSDR, 1995-2001; NIOSH, 1997; US EPA, 2002)

Chemical	Potential migration	Non-cancer toxicity	Cancer potential
Explosives			
2,4,6-TNT	High	High	Yes
2,4 or 2,6-DNT33	High	High	Yes
Nitroglycerin	High	High	Unknown
Dinitrophenols	High/moderate	High	Unknown
RDX	High	Moderate	Yes
HMX	High	Moderate	Unknown
Tetryl	High	Moderate	Unknown
1,3-DNB	Moderate	Moderate	Unknown
CW agents			
Phosgene	High	High	No
HD	High	High	No
Organophosphates (e.g. sarin)	High	High	No
Hydrocyanic acid	High	High	No

The human health relevance of most of these compounds is a result of the specific metabolic transformations (mostly reductions) of their nitrogroup(s) by the intestinal microorganisms and by hepatic metabolism. Many intermediates are strongly electrophilic compounds. They have the potential to damage DNA either directly or by disturbing its regulation and expression by epigenetic mechanisms up to the protein and cellular level. High intakes of some compounds are also acutely dangerous by inhibiting the transport of oxygen by oxidizing Hb to metHb. The in vivo toxicologic database, especially for some environmental and microbial metabolites, is very poor.

Chemical weapons

About 70 different chemicals were used or stockpiled as CW agents during the 20th century. Now banned worldwide, CW agents can still be found at former manufactories or at clandestine production and storage facilities. They may be categorized according to their main effects on the human organism, and include but are not restricted to blood agents (e.g. hydrocyanic acid), nerve agents (e.g. sarin), skin agents (e.g. mustard gas; HD), or respiratory agents (e.g. phosgene).

CW agents are frequently called 'war gases' though this historic term is no longer correct, since many effective CW agents are liquids or solids and only gases in specialized circumstances. As with other chemicals, the relevance of CW agents to potential groundwater contamination depends primarily on their water solubility and their hydrolytic stability (see Chapter 4). Table 11.6 gives examples of warfare agents in relation to their potential to migrate in groundwater.

11.3.2 Potential groundwater contaminants from military operation sites

For assessing pollution potential of military bases and operation sites, both the military-specific substances discussed above and contaminants from other activities on the site may be relevant. In times of peace, warfare agents are used only for training purposes, though such activities may pose contamination risks for groundwater as well. Abandoned sites are categorized according to their main original uses (e.g. training grounds, barracks, facilities for maintenance of technical equipment, airfields and missile sites). Even at locations that, based on their use history, initially are not suspected of encompassing contaminated areas (e.g. administrative buildings, dwellings), considerable contamination including that extending into the groundwater has been found in many cases, and this contamination can threaten the local drinking-water supply if it occurs in recharge areas (Teaf, 1995).

In contrast, training grounds can be expected to exhibit widespread contamination from munitions, as a result of non-point inputs caused directly by military activities (e.g. bomb-dropping grounds, shooting ranges for tanks). Likewise refueling activities in the field or at base support facilities have caused widespread contamination of soil and groundwater by hydrocarbons. Contamination from point sources arises within individual operational areas such as areas in which military equipment is maintained or cleaned, burning sites, or as a result of exceptional events (hazardous release incidents). All of these activities have resulted in groundwater contamination with metals, solvents, hydrocarbons, explosives and other substances.

Many substances used in the military sector are also used in the civilian sector, including petroleum products from fuel depots for motor vehicles and aircraft. According to results obtained from numerous exploratory investigations into contaminated military sites, contamination by petroleum hydrocarbons ranks first, in quantitative terms. Other contaminants found are mainly organic solvents (e.g. chlorinated hydrocarbons) such as were used in large quantities in maintenance facilities and tank-washing installations. Plant treatment, agricultural and pest control products were used in large quantities as defoliants or to keep strategic or militarily sensitive areas free of vegetation or nuisance insects. Large residential facilities (e.g. base housing and daily military operations) also generated the equivalent of municipal waste which, if improperly disposed (see Chapter 12), has the potential to affect groundwater (Herndon *et al.*, 1995).

At training grounds known as ABC facilities (training in defense against Atomic, Biological and Chemical weapons), contamination has occurred mainly from the use of CW agents, from decontamination activities and from the use of incendiaries and smoke agents. Original warfare agents were used for training purposes at these facilities only in very small quantities. Today, CW agents may be found buried on training grounds or in disposal areas at active as well as closed military bases.

On many military properties, illegal and/or uncharacterized waste dumps were established, so that no information is available about the chemical composition of the waste dumped or the length of time the dumps were in use. The special conditions that typically govern the establishment of landfills, and the associated use restrictions (e.g. restrictions on abstraction of drinking-water in the catchment area), tend to have been disregarded by military forces. The variability and site-specificity of these conditions has made chemical characterization, site investigation and groundwater pollution assessment difficult (Herndon *et al.*, 1995).

11.4 CHECKLIST

NOTE ▶ *The following checklist outlines information needed for characterizing industrial, mining and military activities in the drinking-water catchment area. It supports hazard analysis in the context of developing a Water Safety Plan (Chapter 16). It is neither complete nor designed as a template for direct use but needs to be specially adapted for local conditions. The analysis of the potential of groundwater pollution from human activity requires combining the checklist below with information about socioeconomic conditions (Chapter 7), aquifer pollution vulnerability (Chapter 8), and other specific polluting activities in the catchment area (Chapters 9-10 and 12-13).*

ⓘ **Are active or abandoned industrial, mining and military sites located in the drinking-water catchment area?**

- ✓ Compile inventory of registered large-scale facilities and operations, and check their locations
- ✓ Compile inventory of small-scale enterprises, production sites, mining sites or military operations, and check their locations
- ✓ List operations at these sites for
  - Industry: processes employed and goods produced
  - Mining: products mined (e.g. ore, coal, lignite, gravel, sand) and mine type (e.g. deep or open pit mining, ISL)
  - Military: type (e.g. ordinance testing, troop training, logistic support)
- ✓ Compile inventory of abandoned industrial sites, mines, military facilities and disposal areas that may still be leaching pollutants to groundwater
- ✓ Check data about past accidents (e.g. fires, explosions, spillages) which may have left potential 'hot spots' on historic or active facilities
- ✓ ...

ⓘ **What kind and which amounts of materials are used, transported and stored at individual facilities?**

- ✓ Compile inventory of raw materials needed for production or operation at individual industry, mining or military facilities (including potentially hazardous degradation products if known)

- ✓ Classify site-related goods and materials according to their potential hazard to groundwater
- ✓ Compile inventory of permits for discharging effluents to soils, water bodies, injection wells (including predisposal treatment if known)
- ✓ Compile information on transportation to and from the facility, i.e. on raw materials, ore, potentially hazardous products and wastes
- ✓ Compile inventory of number, size, type, age and materials held in pipelines, storage ponds, lagoons and tanks for liquids, with particular consideration given to subsurface structures
- ✓ Check for indication of episodic releases accumulating contaminants over time
- ✓ Estimate amount of groundwater withdrawn by industries, mines or military sites, including uses if known (e.g. process water, cooling water)
- ✓ …

ⓘ **Are the individual facilities in good condition and safe locations?**

- ✓ Check existence of containment structures for storage, production and transportation of hazardous goods and materials (e.g. pipelines, storage ponds, lagoons, tanks for liquids)
- ✓ Evaluate siting, design, construction and technical condition of individual structures and facilities in relation to aquifer vulnerability and physical conditions in the catchment area (e.g. water table, soil, hydrogeology): consider checklist for Chapter 8
- ✓ For mining: Evaluate location of heaps and tailings in relation to aquifer vulnerability
- ✓ Check type of grounds maintenance and use of chemicals (e.g. herbicides, explosives, pesticides, fertilizers, combustible hydrocarbons)
- ✓ …

ⓘ **Are good management practices implemented at individual sites and facilities to protect groundwater?**

Note: See Chapter 23 for the information background for these items.

- ✓ Check availability and implementation of environmental management concepts, and whether there are audits for best management practice and operational precautions in relation to groundwater protection
- ✓ Check closure plans and maintenance of decommissioned sites:
  - For industrial and military sites: adequate dismantling of facilities and removal of potential groundwater contaminants from sites
  - For mining: adequate management of acid mine drainage to prevent acidification and mobilization of metals
- ✓ Check availability and implementation of emergency response plans, particularly in relation to groundwater protection
- ✓ Check availability and implementation of waste management concepts

- ✓ Check whether there is accounting for materials brought in, materials processed, wastes requiring disposal and long term closure procedures
- ✓ Evaluate operation and management practices at individual facilities in relation to aquifer vulnerability and physical conditions in the catchment area (e.g. water table, soil, hydrogeology): consider checklist for Chapter 8
- ✓ …

Are side effects of production processes also relevant to groundwater contamination?

- ✓ Identify vehicular traffic, power production, water withdrawal/treatment, and grounds maintenance
- ✓ Evaluate emission of substances that act as cosolvents (e.g. fuel, acids) and are likely to mobilize other hazardous chemicals
- ✓ Identify construction activities on industrial, mining or military sites that may physically affect the aquifer or cause contaminant emissions
- ✓ …

Are hazardous events likely to increase groundwater pollution potential?

- ✓ Evaluate whether and how storm water events would enhance transport of pollutants to the aquifer
- ✓ Evaluate which spills and accidents are likely to cause groundwater pollution
- ✓ …

Is drinking-water abstracted in proximity to industry, manufacturing, mining or military sites?

- ✓ Assess distance between such sites and drinking-water abstraction (see Chapter 8)
- ✓ Check adequacy of wellhead protection measures, wellhead construction and maintenance as well as sanitary seals used (see Chapter 18) to prevent ingress of contaminants from production, mining or military sites
- ✓ …

Are groundwater quality data available to indicate pollution from industrial, mining or military activities?

- ✓ Compile historic data from the areas and facilities of interest, e.g. from local or regional surveys, research projects or previous monitoring programmes

- ✓ Check need and options for implementation of new or expanded monitoring programmes likely to detect contamination from industrial, mining or military operations
- ✓ …

ⓘ **What regulatory framework exists for industrial, mining and military activities?**

- ✓ Compile information on national, regional, local, or catchment area specific legislation, regulations, recommendations, voluntary agreements or common codes of good practices on siting, construction, operation, maintenance of sites, and on restrictions, ban or prohibition of substances produced, processed or generated as wastes
- ✓ Check whether the regulatory framework adequately addresses environmental and specifically groundwater protection
- ✓ Identify gaps and weaknesses known which may encourage specific pollution problems
- ✓ …

ⓘ **Documentation and visualization of information on practices at industry, manufacturing, mining and military sites and operations.**

- ✓ Compile summarizing report and consolidate information from checklist points above
- ✓ Compile summary of types and amounts of substances produced, processed or generated as wastes and which are potentially hazardous if they leach into the aquifer
- ✓ Map industrial and mining production sites and military facilities (in-use and abandoned), preferably including suspected 'hot spots' of contamination (use GIS if possible)
- ✓ …

11.5 REFERENCES

ATSDR (1995-2001) *Toxicological Profiles*. Agency for Toxic Substances and Disease Registry, U/S. Public Health Service, Atlanta, Georgia.

Berg, R.C., Curry, B. Brandon, B. and Olshansky, R. (1999) Tools for groundwater protection planning: An example from McHenry County, IL, USA. *Environ. Man.*, **23**, 321-331.

Brady K.B.C. and Cravotta C.A. (1992) Acid-base accounting: an improved method of interpreting overburden chemistry to predict quality of coal-mine drainage. In *Proc. 13th Annual West Virginia Surface Mine Drainage Task Force Symposium, WV Mining and Reclamation Association*, pp. 13-21.

Canter, L.W. and R.C. Knox. (1987) *Ground Water Quality Protection*, Lewis Publishers, Chelsea, Michigan.

Ekmekci, M. and G. Gunay. (1997) Role of public awareness in groundwater protection. *Environ. Geol.*, **30**,1/2.

Evangelou V.P. (1995) *Pyrite Oxidation and its Control*, CRC Press Boca Raton, New York, London, Tokyo.

Herndon, R.C., Richter, P.I., Moerlins, J.E., Kuperberg, J.M. and Biczo I.L. (eds.) (1995) *Clean-up of Former Soviet Military Installations*, Springer-Verlag, Berlin.

Langmuir D. (1997) *Aqueous Environmental Geochemistry*, Prentice Hall, Upper Saddle River, New Jersey.

Mather, J., Banks, D., Dumpleton, S. and Fermor M. (eds.) (1998) *Groundwater Contaminants and their Migration*. Geological Society, London, Special Publications, 128.

Merkel B. and Sperling B. (1998) Hydrogeochemische Stoffsysteme II. *DVWK Schriftenreihe*, **117**.

Mokokwe, K. (1999) *Occurrence of Groundwater with High Nitrate Content - Orapa Well Fields*. Inception Report, Department of Geological Survey (DGS), Lobatse.

Mulisch, H-M., Winter, W. and Grohmann, A. (1996) Assessment of the hazards to drinking water from former military sites. *J. Water SRT – Aqua,* **45**(2), 72-75.

Mulisch, H.M., Winter, W. and Schörling, H. (1999a) Assessment of the Hazards to Drinking Water from Former Military Sites (Advanced Study). In *Society of Environmental Toxicology and Chemistry, 9th Annual Meeting of SETAC-Europe 25-29 May 1999: Quality of life and environment in cultured landscapes*, p. 229, Leipzig.

Mulisch, H-M., Winter, W. and de Lima, S. (1999b) Drinking water protection in the associated field of former properties of the west group of the troops of the Soviet Army (WGT). *TEXTE*, **77**, 99, Umweltbundesamt, Berlin.

Mulisch, H.-M., Winter, W., Dieter, H. H. and Kerndorff, H. (2000) Kriterienkatalog zur Beurteilung von Gefahren in Einzugsgebieten von Trinkwassergewinnungsanlagen am Beispiel militärischen Altlasten. *WaBoLu-Hefte*, **6**, 99, Umweltbundesamt.

NIOSH (1997) *Pocket Guide to Chemical Hazards*, National Institute for Occupational Safety and Health, Washington, DC.

Nordstrom D.K., Alpers C.N., Ptacek C.J. and Blowes D.W. (2000) Negative pH and extremely acid mine waters from Iron Mountain, California. *Environ. Sci. Technol.*, **34**(2), 254-258.

Parkhurst D.L. and Appelo C.A.J. (1999) *User's Guide to PHREEQC (Version 2) - A Computer Program for Speciation, Batch-Reaction, One-Dimensional Transport, and Inverse Geochemical Calculations.* Water Res. Invest. Report 99-4259.

Plummer L.N., Parkhurst D.L., Fleming G.W. and Dukle S.A. (1988) *A Computer Program Incorporating Pitzer's Equations for Calculation of Geochemical Reactions in Brines*. USGS. Water Res. Invest. Report 88-4153, Reston, Virginia.

Seirys, N. and Marcinonis, A. (1999) Underground pollution at a former military installation in Valcunai (Lithuania). In *Bioavailability or Organic Xenobiotics in the Environment*, (eds. P. Bayeve *et al.*), pp. 57-72, Kluwer Academic Publishers, Dordrecht.

Teaf, C.M. (1995) Human health and environmental risks associated with contaminated military sites. In *Clean-up of Former Soviet Military Installations*, (eds. R.C. Herndon *et al.*), Springer-Verlag, Berlin.

Tredoux, G. (2000) *Report on a Visit to Botswana Regarding Nitrate in Groundwater*. Research proposals for the Orapa area and elsewhere. Department of Geological Survey (DGS), Lobatse.

US EPA (1999) *Guide for Industrial Waste Management: Ground-water modeling technical background document*. EPA530-R-99-002, Washington, DC.

US EPA (2002) *Integrated Risk Information System (IRIS) Database*. US EPA, Washington, DC.

Walter, A.L., Frind, E.O., Blowes, D.W., Ptacek, C.J. and Molson, J.W. (1994) Modeling of multicomponent reactive transport in groundwater, 1, Model development and evaluation. *Water Resources Res.,* **30,** 3137-3148.

Wolery T.J. (1992) EQ3/6, a software package for geochemical modeling of aqueous Systems. Package overview and installation guide (Version 7.0). UCRL_MA-110662-PT-I, Lawrence Livermore National Laboratory Livermore.

Zektser, I.S., Belousova A.P. and Dudov, V.Y. (1995) Regional assessment and mapping of groundwater vulnerability to contamination. *Environ. Geol.*, **25**, 225-231.

12

Waste disposal and landfill: Potential hazards and information needs

R. Taylor and A. Allen

Solid wastes, the subject of this chapter, are mainly disposed of to landfill, because landfill is the simplest, cheapest and most cost-effective method of disposing of waste (Barrett and Lawlor, 1995). In most low to medium income developing nations, almost 100 per cent of generated waste goes to landfill. Even in many developed countries, most solid waste is landfilled. For instance, within the European Union, although policies of reduction, reuse and diversion from landfill are strongly promoted, more than half of the member states send in excess of 75 per cent of their waste to landfill (e.g. Ireland 92 per cent), and in 1999 landfill was still by far the main waste disposal option (EEA, 2003). Furthermore, although the proportion of waste to landfill may in future decrease, the total volume of municipal solid waste (MSW) being produced is still increasing significantly, in excess of 3 per cent per annum for many developed nations (Douglas, 1992). Landfill is therefore likely to remain a relevant source of groundwater contamination for the foreseeable future (Allen, 2001).

Solid waste composition, rate of generation and methods of treatment and disposal vary considerably throughout the world and largely determine the potential of waste to impair groundwater quality. The purpose of this chapter is to outline the risk that waste disposal presents to groundwater quality and the information that is required to assess this risk.

NOTE ▶ *Waste disposal and landfill activities and the environment in which they are placed vary greatly. Health hazards arising from waste disposal and landfill and their potential to pollute groundwater therefore needs to be analysed specifically for the conditions in a given setting. The information in this chapter supports hazard analysis in the context of developing a Water Safety Plan for a given water supply (Chapter 16).*

12.1 TYPES OF SOLID WASTE

Wastes generated by the full extent of human activities range from relatively innocuous substances such as food and paper waste to toxic substances such as paint, batteries, asbestos, healthcare waste, sewage sludge derived from wastewater treatment and, as an extreme example, high-level (radioactive) waste in the form of spent nuclear fuel rods. Numerous classifications of solid wastes have been proposed (e.g. Tchobanoglous *et al.*, 1993; Ali *et al.*, 1999), and the following represents a simple classification of waste into broad categories according to its origin and risk to human and environmental health:

- household waste;
- MSW;
- commercial and non-hazardous industrial wastes;
- hazardous (toxic) industrial wastes;
- construction and demolition (C&D) waste;
- health care wastes – waste generated in health care facilities (e.g. hospitals, medical research facilities);
- human and animal wastes;
- incinerator wastes.

Household waste represents waste generated in the home often collected by municipal waste collection services. MSW includes also shop and office waste, food waste from restaurants, etc., also collected by municipal waste collection systems, plus waste derived from street cleaning, and green (organic) waste generated in parks and gardens.

Storage of waste in a disposal facility serves to minimize the effects of waste on the environment. This is achieved by restricting any effluent derived from the waste to a single location, where emissions can be controlled. If control is lacking or inadequate, disposal facilities may become point sources of groundwater contamination. In many regions, centralized waste disposal has historically occurred by landfilling, wherein local quarries and gravel pits have been filled with waste because, in many cases, they simply constituted an appropriately sized hole in the ground. Such locations typically offered little protection against contamination of adjacent groundwater supplies. Legislation designed to protect usable groundwater has helped to reduce the incidence of this practice in many high to middle income countries (e.g. US EPA, 1974; CEC, 1980; NRA, 1992). Modern waste management practices involve disposal of waste in specially sited and engineered sites known as sanitary landfills (see Chapter 24).

Waste accepted in municipal waste landfills in developed countries would normally consist of MSWs, plus commercial and non-hazardous industrial wastes and C&D waste. There is a tendency in many countries for C&D waste, usually regarded as inert, to be buried on the construction site where it is generated. However, since downward percolating rainwater may leach heavy metals from C&D waste, recent waste regulations in some developed countries require all C&D waste to be disposed of in landfills.

Hazardous and non-hazardous wastes are differentiated in the waste management legislation of many countries. A range of legal definitions exist for hazardous waste, but it can generally be thought of as waste or a combination of wastes with the capacity to impair human health or the environment due to its quantity, concentration or physical, chemical or infectious characteristics when improperly used, treated, stored, transported or disposed. In many countries, hazardous (toxic) industrial wastes (both organic and inorganic), solid incinerator residues, bottom and fly ash are disposed in special hazardous waste landfills, and specialized disposal or incineration may also be practiced for healthcare wastes (see Box 12.1).

Box 12.1. Health care and research facilities

Health-care facilities can contaminate groundwater through wastes and wastewater containing infectious pathogens, e.g. from contaminated blood or infectious body parts. Health-care facilities may also release various pharmaceuticals, diagnostics (e.g. radiochemicals) and disinfectants depending on the kinds of medical examination being conducted and local practices for handling these substances. These include, but are not restricted to the following:

- cytostatic agents applied in cancer therapy;
- antibiotics;
- disinfectants for surface, instrument and skin disinfection;
- heavy metals such as platinum from excretion by patients treated with the cytostatic agents, mercury from preservatives, disinfectants, diuretic agents, amalgam separators;
- adsorbable organic halogenes from solvents, disinfectants, cleaners and drugs containing chlorine, as well as iodized X-ray contrast media.

Research institutions may use solvents and other potentially hazardous chemicals and radiochemicals (e.g. mutagenic substances used in molecular biology). Also, organisms used in production and research, especially pathogenic bacteria and viruses, as well as genetically modified organisms, are utilized.

For health-care and research facilities, situation assessment should be based on an inventory of substances used or produced, and of processes, which have the potential to emit hazardous organisms or chemicals. Such assessments should necessarily cover storage (containment), handling and disposal practices (e.g. disinfection of wastes). Additionally, assessment should address how effectively these practices are being implemented and the ultimate destination of disposal (e.g. local dump, sewage mains), as this will determine the nature and magnitude of the risk to groundwater.

In many low to medium income parts of the world, where uncontrolled open dumps are common, all waste tends to be dumped together, regardless of its origins or its hazardous nature. A specific characteristic of leachate from hazardous industrial waste is that it may be toxic to the bacteria naturally present and thus delay biodegradation of organic substances in leachate.

Human and animal wastes are usually not disposed of in landfills, although animal carcasses and waste from abattoirs may in some countries be disposed of in dumps and landfills. Human corpses are not generally regarded as waste, but they degrade in a similar way to other organic waste, and also produce leachate in significant quantities. The majority of corpses are buried in cemeteries (see Box 12.2), although a significant proportion are cremated (incinerated), the proportion varying from country to country depending on the proportions of different religious groups in the population and their funeral rites. The main health concern with human and animal wastes is the high concentrations of pathogenic organisms associated with this type of waste, and the potential it has to spread disease.

Box 12.2. Cemeteries

In many regions burials are concentrated into relatively small areas, such as municipal cemeteries, where each body introduces a heavy burden of organic, inorganic and biological parameters into the subsurface. Hydrogeological factors have historically not been taken into account when locating cemeteries and the potential impact of cemeteries on groundwater quality has not been considered.

Animal and human remains, although not considered a waste product, represent a risk to local groundwater because of the proliferation of microorganisms that occurs during the process of corpse decomposition (Pacheco *et al*., 1991). There are more bacteria in a human body than human cells. Many of the bacteria are harmless saprophytes that benefit the host (e.g. by synthesizing vitamins or by metabolizing toxic waste products). However, some will be pathogenic or have the potential to be pathogenic. In addition, the human body is host to a variety of different viruses, fungi and protozoa that may cause disease if transmitted to a susceptible person. Most pathogens will remain viable for a period of time after the host dies. In most cases long-term survival of the pathogen is unlikely, but notable exceptions have generated concerns during investigations of burial sites. The examination of graves containing the remains of smallpox, cholera, anthrax and plague victims, as well as victims of the 1918 influenza pandemic, have been subject to rigorous controls to prevent the potential dispersal of the pathogen from the burial site.

One of the main agents in decomposition (putrefaction) is *Clostridium perfringens.* These bacteria spread along blood vessels causing haemolysis, proteolysis and gas formation in blood and other tissues. The liquids produced through putrefaction contain a high density of microorganisms. Very few studies have been carried out on the microbiology of human putrefaction. Corry (1978), published a catalogue of bacterial species that have been isolated from human cadavers; some of the species listed are pathogenic. These liquids can migrate down into the water table, particularly as coffins and caskets are not water tight and are liable to decay. Microbial contaminants that may result from the decomposition of cadavers include *Staphylococcus* spp., *Bacillus* spp., *Enterobacteriacae* spp., faecal streptococci, *Clostridium* spp., *Helicobacter pylori*, enteroviruses, rotavirus, calicivirus, and F-specific RNA phage.

Spongberg and Becks (2000) list potential chemicals that can be released from cemeteries. These include arsenic and mercury (embalming and burial practices), formaldehyde (embalming, varnishes, sealers and preservatives) as well as lead, zinc and copper (coffins). Spongberg and Becks (2000) also discuss investigations in Ohio where increases in zinc, copper and lead in the soil at a large cemetery were observed. Significant increases in arsenic were thought to indicate contamination from embalming fluids or wood preservatives.

There are several historical accounts of pollution of water wells in the vicinity of cemeteries (e.g. Teale, 1881), but few recent studies of the microbial impact of cemeteries on groundwater (West *et al.*, 1998). An analysis of groundwater quality beneath an active cemetery in the United Kingdom provided evidence that confirms the risk to groundwater, although no pathogens or viruses were isolated. The impacts on groundwater of three cemeteries in Sao Paulo and Santos, Brazil have been monitored by Pacheco *et al.* (1991), by installing piezometers throughout each of the cemetery sites. One cemetery is situated on Tertiary sediments, 4-12 m above the water table, one is on weathered granite, 4-9 m above the water table, and the third is on Quaternary sandy marine sediments, 0.6-2.2 m above the water table. Contamination of the piezometers by faecal coliforms, faecal streptococci and sulphite reducing clostridia was found to be widespread throughout all of the cemeteries. Thus assessing groundwater pollution potential clearly needs to include the potential for pathogens from cemeteries, particularly from large cemeteries.

Although the above considerations are valid for long-term permanent situations, it is generally accepted that corpses in situations of disaster do not constitute a major health hazard. When a disaster strikes a community, authorities should prioritise their actions to attend to the injured and the displaced. The risk of deaths and epidemics due to unattended dead bodies is far smaller than the risk of deaths and diseases due to lack of sufficient food, shelter, drinking water, sanitation and basic medical care. Evidence obtained from emergency operations would indicate that in the majority of the cases the dead bodies do not pose an appreciable risk for public health in areas where there are no endemic diseases (Üçisik and Rushbrook, 1998; Western, 2004).

The rate at which waste is generated corresponds roughly with levels of income. In high income countries of Europe and North America between 500 and 750 kg of solid waste are produced per person per year (OECD, 1997). In contrast, urban populations in most low income countries, for example in Nigeria and Côte d'Ivoire, generate between 100 and 200 kg of solid waste per person per year (Attahi, 1999; Onibokun and Kumuyi, 1999). Despite this lower rate, rapid urbanization, particularly in low income developing countries, has left little space for disposal of the increasing amounts of waste material being generated in urban settings (Sangodoyin, 1993). As a result, uncontrolled disposal (i.e. fly tipping) is rife in many countries, and is a diffuse source of groundwater contamination.

12.2 WASTE STORAGE, TREATMENT AND DISPOSAL SITES

The processes of storage, collection, transport, treatment and disposal of wastes all have the potential to pollute the environment and particularly groundwater due to uncontrolled migration of fluids (leachate) derived from the wastes. In addition to the potential for groundwater pollution at sites where wastes are produced and stored prior to collection, sites associated with the treatment and disposal of wastes, where leachate may be generated include:

- landfills (both controlled as sanitary landfill or uncontrolled as open dumps)
- scrap-yards
- cemeteries
- waste collection and processing facilities
- composting facilities.

For situation assessment, landfills are most readily identified with the pollution of groundwater by waste-derived liquids. However, any site where waste is concentrated, processed (e.g. recycled) and stored even for a short period of time may be a potential point source of groundwater contamination. Such processing facilities are often not well regulated or licensed and frequently occur in urban or semi-urban settings, where local water supply points may be impacted by these activities. An inventory of these locations, the types of waste handled and management processes for waste products will aid in the assessment of the polluting capability of such sites.

For situation assessment, a critical criterion in estimating potential groundwater pollution from waste disposal is the siting of all of the above mentioned waste treatment and disposal facilities, particularly sanitary landfills and open dumps (discussed in Sections 12.3.2 and 12.3.3). Most modern landfills in high to medium income countries require licenses to operate (see Chapter 24.2), and must be engineered to prevent groundwater pollution. This generally involves lining the site with an artificial lining system, but liners leak and degrade with time (Chapter 24.3). Even if the site is well engineered and managed with an artificial lining system installed and even if or the waste materials are inert, leachate, which may have the potential to pollute groundwater, will be produced. It is therefore essential to assess the capacity of the underlying geology to protect groundwater in the event of liner failure. The likelihood of disposed wastes polluting groundwater depends on the thickness of the unsaturated zone and the attenuation capacity of the overburden (i.e. any loose unconsolidated material which overlies solid bedrock) underlying the site, and also on the total and effective precipitation at the site, since the quantity and concentration of leachate generated is a function of the access of water to the waste. Thus the potential for pollution of groundwater will be least at sites carefully selected to take advantage of the most favourable geological/ hydrogeological conditions.

Historic landfills (dumps) were generally not subject to the regulations governing modern landfills, and were usually sited for convenience, such as the presence of a pre-existing hole into which the waste could be deposited. The general assumption that an aftercare period of 30 years is adequate to allow for degradation of waste to an inert state, is now being questioned, with recent studies (Hjelmar *et al.*, 1995; Wall and Zeiss, 1995; Kruempelbeck and Ehlrig, 1999; Röhrs *et al.*, 2000; Fourie and Morris, 2003) suggesting that waste may remain active for many decades and even hundreds of years, particularly under moisture-deficient conditions. This includes not only landfills from regions where evaporation exceeds precipitation, but also all lined and capped landfills employing the concept of dry entombment of waste.

In the past, hazardous and non-hazardous wastes were not distinguished so that hazardous substances may be stored in all of these landfills. For situation assessment, it is important to locate all waste disposal sites in the drinking-water catchment, including both currently operating landfills, and historic dumps, now closed and covered over (see Chapter 24). Assessment of all landfills, but in particular historic landfills, should include age and type of waste, underlying geology, most importantly type and thickness of overburden and thickness of the unsaturated zone. The state of degradation of the waste can be ascertained by analysing the leachate and landfill gases generated, as degradation of waste follows a distinctive pattern manifested in well-known and documented compositional variations in liquid and gaseous emissions. All of this, together with the proximity of all of these sites to sources of drinking-water, can determine the threat to public health posed by waste disposal.

12.3 FACTORS GOVERNING CONTAMINATION OF GROUNDWATER BY DISPOSAL OF WASTE

Waste deposited in landfills or in refuse dumps immediately becomes part of the prevailing hydrological system. Fluids derived from rainfall, snowmelt and groundwater, together with liquids generated by the waste itself through processes of hydrolysis and solubilization, brought about by a whole series of complex biochemical reactions during degradation of organic wastes, percolate through the deposit and mobilize other components within the waste. The resulting leachate, subsequently migrates from the landfill or dump and has the potential to contaminate local groundwater either through direct infiltration on site or by infiltration of leachate-laden runoff off-site. The risk posed to groundwater-fed drinking-water sources by waste disposal in landfills or dumps can be considered in terms of three controls:

- waste composition and loading
- leachate production
- leachate migration – attenuation and dilution.

12.3.1 Waste composition and loading

The composition and volume of disposed wastes vary nationally and regionally in relation to the local human activities, and the quantity and type of products that communities consume (Table 12.1). Discarded waste in lower income areas is typically rich in food-related waste, i.e. organic (carbon-rich) substances (Table 12.1). Although such waste is not in itself toxic, decomposition of organic matter can alter the physicochemical quality of groundwater and enhance the mobility of hazardous chemicals including metals and solvents (see Section 12.3.2). The proportion of manufactured (e.g. paper) and potentially hazardous (e.g. textiles, metals, plastics) wastes increases in relation to income and degree of industrialization (Table 12.1), and waste disposal leachate from highly industrialized settings may contain a wide range of anthropogenic contaminants (see Section 12.3.2). The types of hazardous substances likely to occur in discarded waste may be assessed from the types of industry, small-scale enterprise and other human activity of a particular area.

Table 12.1. Solid-waste generation and composition from selected regions in the world (OECD, 1993, 1997; Attahi, 1999; Lusugga Kironde 1999; Onibokun and Kumuyi, 1999)

Location	**Rate (kg/pers/year)**	**Composition (%)**						
		Paper	Food	Plastics	Glass	Metals	Textiles	Other
China	285	3	60	4	1	0	2	-
Denmark	520	30	37	7	6	3	17	-
France	560	30	25	10	12	6	17	-
Iran	324	8	74	5	3	1	2	-
Mexico	320	14	52	4	6	3	20	-
Poland	290	10	38	10	12	8	23	-
USA	730	38	23	9	7	8	16	-
Abidjan (Côte d'Ivoire)	211	4	63	5	1	1	1	25

A major concern in many countries is also of waste import, particularly of hazardous wastes. Export from industrialized countries to low-income countries circumvents strict waste disposal regulations implemented in the country generating these wastes. Often this is highly organized, as informal, though illegal, transactions between an exporter and importer using false documentation (e.g. Mackenzie, 1989). Such waste export/import practices are difficult to detect, but important for situation assessment as disposal of such wastes is likely to pose a risk of groundwater contamination. It is therefore often necessary to collect information on both formal and informal (i.e. illegal) waste composition and loading.

Landfilled refuse is rich in microorganisms. Mature sites may be compared to large bioreactors in which the organic content of the waste is decomposed anaerobically. Most of the organisms that carry out these processes are harmless saprophytes. An analysis of household waste in the United Kingdom showed that over 4 per cent of the waste comprised disposable nappies (diapers) of which about one-third may be soiled with faeces. Domestic waste also contains bloodstained materials, such as sanitary pads, tampons and discarded wound dressings, and animal wastes, such as dog faeces and soiled cat litter. The potential for pathogens within this mixture of sources is extremely high. Pathogens may also be transported to landfill sites by vermin (rats) and other scavengers, in particular seagulls.

The fate of pathogens in landfill sites is not understood. Although it is generally assumed that most are rapidly inactivated by the conditions that prevail in the landfill environment, the potential of leachate and runoff from landfill sites to transport pathogens into local water resources should be addressed in situation assessment.

12.3.2 Leachate production

Most waste deposited in landfills is not inert. Degradation of many components of waste including food, paper and textiles consumes oxygen thereby changing the redox potential of the liquid present and potentially influencing mobility of other constituents. Plastics, glass and metal compounds tend to be less reactive and degrade more slowly. Under some conditions, metals may, however, become rapidly mobilized (see Chapter 4).

Percolating rainwater provides a medium in which waste, particularly organics, can undergo degradation into simpler substances through a range of biochemical reactions involving dissolution, hydrolysis, oxidation and reduction, processes controlled to a large extent within landfills and dumps by microorganisms, primarily bacteria. Table 12.1 indicates that the largest fraction of disposed waste is organic matter (e.g. food, paper), which has a well-documented degradation path. Mechanisms regulating mass transfer from wastes to leaching water, from which leachate originates, can be divided into three groups of processes:

- hydrolysis of solid waste and biological degradation;
- solubilization of soluble salts contained in the waste;
- suspension of particulate matter.

The first two groups of processes, which have the greatest influence on the composition of leachate produced, are associated with the stabilization of waste.

Initially, organic matter in the form of proteins, carbohydrates and fats, is decomposed under aerobic conditions (i.e. oxidized), through a series of hydrolysis reactions, to form carbon dioxide and water together with nitrates and sulphates via a number of intermediate products such as amino acids, fatty acids and glycerol. Such oxidation reactions are exothermic, so temperatures in the landfill become elevated. Carbon dioxide is released as a gas or is dissolved in water to form carbonic acid (H_2CO_3) which subsequently dissociates to yield the bicarbonate anion (HCO_3^-) at near neutral pH.

Aerobic decomposition of organic matter depletes the waste deposit of oxygen (O_2) as buried waste in the landfill or refuse dump becomes compacted and circulation of air is inhibited. As oxygen becomes depleted, it is replaced as the oxidizing agent by, in succession, nitrate (NO_3^-), manganese (as MnO_2), iron (as $Fe(OH)_3$) and sulphate (SO_4^{2-}). In general, the aerobic stage is short, no substantial volumes of leachate are produced, and aerobic conditions are rapidly replaced by anaerobic conditions. The main stages of anaerobic digestion are (i) acetogenic (acid) fermentation, (ii) intermediate anaerobiosis, and (iii) methanogenic fermentation, all three of which can be operating simultaneously in different parts of the landfill.

Acetogenic fermentation brings about a decrease in leachate pH, high concentrations of volatile acids and considerable concentrations of inorganic ions (e.g. Cl^-, SO_4^{2-}, Ca^{2+}, Mg^{2+}, Na^+). As the redox potential drops, sulphate is slowly reduced, generating sulphides, which may precipitate iron, manganese and heavy metals that are dissolved by the acid fermentation. Decrease in pH is due to production of volatile fatty acids (VFAs) and to high partial pressures of carbon dioxide (CO_2), whilst the increased concentrations of anions and cations results from leaching (lixiviation) of easily soluble organic material present in the waste mass. Breakdown of organic material reduces the redox potential to <330mV, which allows the next stage of the process to become initiated. Leachate from this phase is characterized by high values of biological oxygen demand (BOD) (commonly >10 000 mg/l), high BOD_5/COD ratios (commonly >0.7), acidic pH values (typically 5-6) and ammonia (NH_3) due to hydrolysis and fermentation in particular of proteins.

Intermediate anaerobiosis commences with a gradual increase in the methane (CH_4) concentration in the gas, coupled with a decrease in H_2, CO_2 and VFAs. Conversion of the VFAs leads to an increase in pH values and to alkalinity, with a consequent decrease in the solubility of calcium, iron manganese and the heavy metals, which are probably precipitated as sulphides. Ammonia is released but is not converted to nitrate in such an anaerobic environment.

Methanogenic fermentation, the final stage in the degradation of organic wastes, operates within the extremely limited pH range of 6-8. At this stage in the degradation process, the composition of leachate is characterized by almost neutral pH, and low concentrations of volatile acids and TDS, indicating that solubilization of the majority of organic components is almost complete, although waste stabilization will continue for several decades. The biogas being produced has a methane content of generally >50 per cent, whilst ammonia continues to be released by the acetogenic process. Leachate produced at this stage is characterized by relatively low BOD values, and low ratios of BOD/COD.

Degradation processes convert nitrogen into a reduced form (ammonium), and bring about mobilization of manganese and iron and also liberation of hydrogen sulphide gas. Production of methane indicates strongly reducing conditions with a redox potential in the order -400 mV. Unlike carbon dioxide, methane is poorly soluble in water.

Due to the decomposition of organic matter, leachate derived from landfills or dumps comprises primarily DOC (Table 12.2), largely in the form of fulvic acids (Christensen *et al.*, 1998). The solubility of metals in leachate is enhanced through complexation by dissolved organic matter. The solubility of organic contaminants in waste may also be slightly enhanced through the presence of high levels of organic carbon in leachate. The range of organic compounds that may be found in groundwater affected by landfill leachate is shown in Box 12.3 and Table 12.3. Hydrophobic compounds may be mobilized through leachate, as they adsorb to organic carbon in solution. For example, benzene- and naphthalene-sulphonates comprise between 1-30 per cent of the DOC in landfill-leachates analysed in Switzerland (Riediker *et al.*, 2000).

Box 12.3. Organic contaminants in groundwater affected by landfill leachate in Germany

A study investigating groundwater from 250 different municipal waste sites in Western Germany (Kerndorff *et al.*, 1992) identified a wide range of organic contaminants within 10-100 m downgradient of the deposit, some of which occurred in a large number of samples and attained concentrations well into the range of mg/l (Table 12.3). Benzene and its alkyl derivatives (four compounds) constitute the majority of the seven non-halogenated contaminants. The highest mean concentration was obtained for volatile halogenous substances, predominantly for dichloromethane (DCM) (≈ 38 mg/l), *cis*-1,2-DCE (≈ 22 mg/l), VC (≈ 1.7 mg/l), and trichloroethene (TCE) (≈ 1 mg/l). The high concentrations associated with these VOC confirm the significance of this class of substances as major emissions from waste disposal sites

Table 12.2. Key characteristics of landfill leachates from England, Germany and USA (all values in mg/l except pH) (Ehrig, 1982; Robinson *et al.*, 1982; Fetter, 1993)

Parameter	United Kingdom	Germany	USA
pH	6.2-7.4	6.1-8.0	5.4-7.2
TDS	Not analysed	Not analysed	2180-25 900
COD	66-11 600	3000-22 000	1120-50 500
BOD	<2-8000	180-13 000	100-29 200
Total organic carbon	21-4400	Not analysed	427-5890
Ammonia-nitrogen	5-730	741	26-557
Total phosphorous	<0.02-3.4	5.7	0.3-117
Chloride	70-2780	2-119	180-2650
Iron	0.1-380	15-925	2.1-1400
Manganese	0.3-26.5	0.7-24	0.03-25.9
Calcium	165-1150	80-1300	200-2100
Magnesium	12-480	250-600	120-780

Table 12.3. Organic contaminants in landfill leachate in Germany from 250 sites (adapted from Kerndorff *et al*., 1992)

Parameter	**Frequency of detection (%) (1)**	**Concentration (µg/l)** Mean	**Parameter** Maximum
TeCE	70.4	56.1	6500
TCE	55.6	1010	128 000
cis-1,2-DCE	30.1	22 100	411 000
Benzene	29.1	141	1800
1,1,1-TCA	22.8	16.5	270
m/p-Xylene	22.8	39.9	447
TCM	22.0	76.2	2,800
1,2-DCA	18.8	107	210
VC	17.7	1690	12 000
Toluene	16.5	73.2	911
DCM	14.9	38 100	499 000
CTC	14.4	1.2	23
4-Methylphenol (*p*-cresol)	13.7	42.0	283
Chlorobenzene	12.9	52.9	388
2-Methylphenol (*o*-cresol)	12.9	10.0	63
1,2-DCB	12.2	1.4	6.6
1,4-DCB	12.2	31.9	265
Naphthalene	12.1	2.2	13
Ethylbenzene	11.3	32.2	160
o-Xylene	9.5	13.8	69
2,4,6-Trichlorophenol	8.9	3.2	24
3,5-Dimethylphenol	8.1	16.2	61
Phenol	8.1	2.2	5.6
1,3-DCB	7.8	11.5	74
trans-1,2-DCE	7.5	57.1	135
Isopropylbenzene (cumol)	5.6	2.4	4.7
1,1-DCA	5.4	52.7	110
Acenaphthene	4.8	6.3	32
2,4-Dichlorophenol	4.8	3.5	17
3-Chlorophenol	4.8	12.7	23
p-Cymol[*p*-$CH_3C_6H_4CH(CH_3)_2$]	4.4	1.9	3.5
2-Ethyltoluene	4.4	0.6	1.0
2,4,5-Trichlorophenol	3.9	7.1	31
1,3,5-Trimethylbenzene	3.3	1.7	4.0
Phenanthrene	3.2	1.5	4.4
Tribromomethane	3.1	3.0	6.0

(1) Number of samples: 90 to 277

12.3.3 Leachate migration

In unsealed landfills above an aquifer, waters percolating through landfills and refuse dumps often accumulate or mound within or below the landfill (Figure 12.1). This is due to production of leachate by degradation processes operating within the waste, in addition to the rainwater percolating down through the waste. The increased hydraulic head developed promotes downward and outward flow of leachate from the landfill or dump. Downward flow from the landfill threatens underlying groundwater resources whereas outward flow can result in leachate springs yielding water of a poor, often dangerous, quality at the periphery of the waste deposit. Observation of leachate springs or poor water quality in adjacent wells/boreholes are indicators that leachate is being produced and is moving. Leachate springs represent a significant risk to public health, so their detection in situation assessment is critical in order to prevent access to such springs.

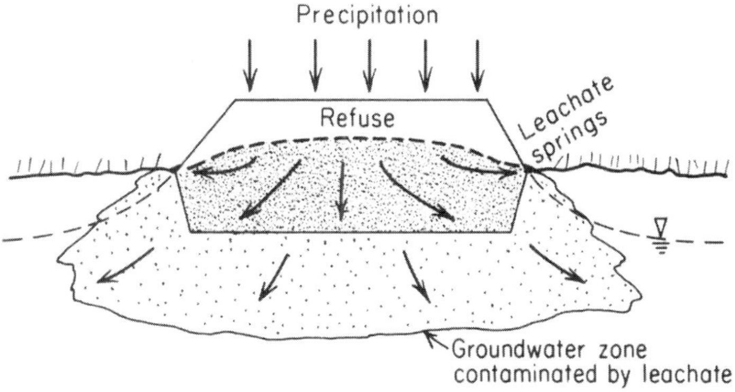

Figure 12.1. Conceptual diagram of leachate migration from a landfill (Freeze and Cherry, 1979; reprinted by permission of Pearson Education, Inc., Upper Saddle River, NJ)

One method used to reduce the generation of leachate and, hence, hydraulic heads generating flow from a closed landfill is to place a capping of low permeability material (e.g. clay or high density polyethethylene) over the waste deposit in order to reduce infiltration of rainwater. These should be recorded in situation assessment because if a landfill is capped to impede rainwater ingress, reducing leachate volumes, a more concentrated leachate will be generated. Also, microbial and biochemical reactions will be inhibited thereby prolonging the degradation process and the activity of the waste possibly for decades or even centuries. Groundwater pollution potential from older capped landfills may therefore be higher than from younger, open landfills.

Leachate migration is also affected by the manner in which waste is deposited. Compaction of waste prior to deposition reduces its permeability, whereas regular application of a topsoil cover between the loading of waste to landfills induces layering. These characteristics inevitably give rise to preferential flow paths through landfills. Johnson *et al.* (1998) found, for instance, that residence times for rainwater entering a landfill varied from a period of a few days to several years. This is reflected in the frequently temporal nature of leachate springs, which can appear in wet seasons but subsequently disappear in dry seasons to leave patches of discoloured soil (Jefferis, 1993). Inspections of potential leachate production should, therefore, focus on periods towards the end of wet seasons or following excessive rainfall events. Further, situation assessment needs to account for uncertainties in both the prediction and monitoring of leachate migration from landfills and dumps, in consequence of the complex hydrogeology of waste deposits. This is further addressed in Chapter 24 in relation to problems of planning and management.

Despite the complexity of leachate migration through landfills, fundamental aspects of subsurface contaminant transport, reviewed in Section 12.4, can practically be applied to the movement of leachate-derived contaminants from a landfill or refuse dump. These include the thickness of the unsaturated zone, the permeability and moisture content of the earth materials within the unsaturated zone, and the hydraulic conductivity and local hydraulic gradient of geological units in the saturated zone. Poorly conductive units underlying the landfill or refuse dump, e.g. clay-rich material or the presence of an installed artificial liner inhibit leachate migration. On the other hand, discontinuities such as fissures and joints in the subsurface or faults or holes in a liner, dramatically increase leachate flow. For situation assessments, access to hydrogeological information (see Chapter 8) as well as information on design and condition of potentially installed lining system (see also Chapter 24) from both beneath and downstream of landfills, is vital.

Equally important as understanding the magnitude and direction of leachate flow is recognition of the significant biochemical changes that occur, as strongly reducing leachate (redox potential <-100 mV), mixes with shallow underlying groundwater, which is mildly to strongly oxidizing (redox potential >+100 mV). These changes, illustrated in Figure 12.2 represent a reversal of the reducing reactions which take place in the landfill, and give rise to a series of redox zones in the leachate plume adjacent to the landfill in the reverse order to the sequence described in Section 12.3.2. The leachate plume thus becomes less reducing and organic carbon in the leachate is rapidly oxidized to CO_2 through contact with oxygenated groundwater.

The leachate plume undergoes continuous transition in the direction of groundwater flow (Figures 12.2 and 12.3) until conditions are reached where it is no longer anaerobic, and attains redox levels identical to background levels in the aquifer. In this transition zone, chemically reduced species such as methane and ammonia disappear, and aqueous nitrogen and sulphur are converted into their oxidized forms of nitrate and sulphate respectively. Iron is oxidized and precipitates as hydrous iron oxide, whereas in contrast, manganese, which is soluble over a wider range of electrochemical (i.e. redox) conditions, remains in solution longer (i.e. travels further with the leachate plume). Consequently, analysis for these compounds and comparison with background levels elsewhere in the aquifer can indicate the presence and extent of the plume. Significantly, a number of detailed studies of leachate plumes indicate that they rarely extend more than a few hundred metres from the landfill before all but a handful of the most persistent contaminants are completely attenuated (e.g. Christensen *et al.*, 1994: Robinson *et al.*, 1999). To determine the vertical extent of the plume often multiple depth sampling boreholes are required as indicated in Figure 12.3.

Migration of reactive constituents in leachate, such as microorganisms, organic solvents and metals, is inhibited through biochemical reactions in the plume (e.g. precipitation, volatilization), and by the interaction of these constituents with the geological materials forming the aquifer matrix (e.g. adsorption, cation exchange), as discussed in Chapters 3 and 4. These processes reduce contaminant concentrations in local groundwater by removing contaminants from solution.

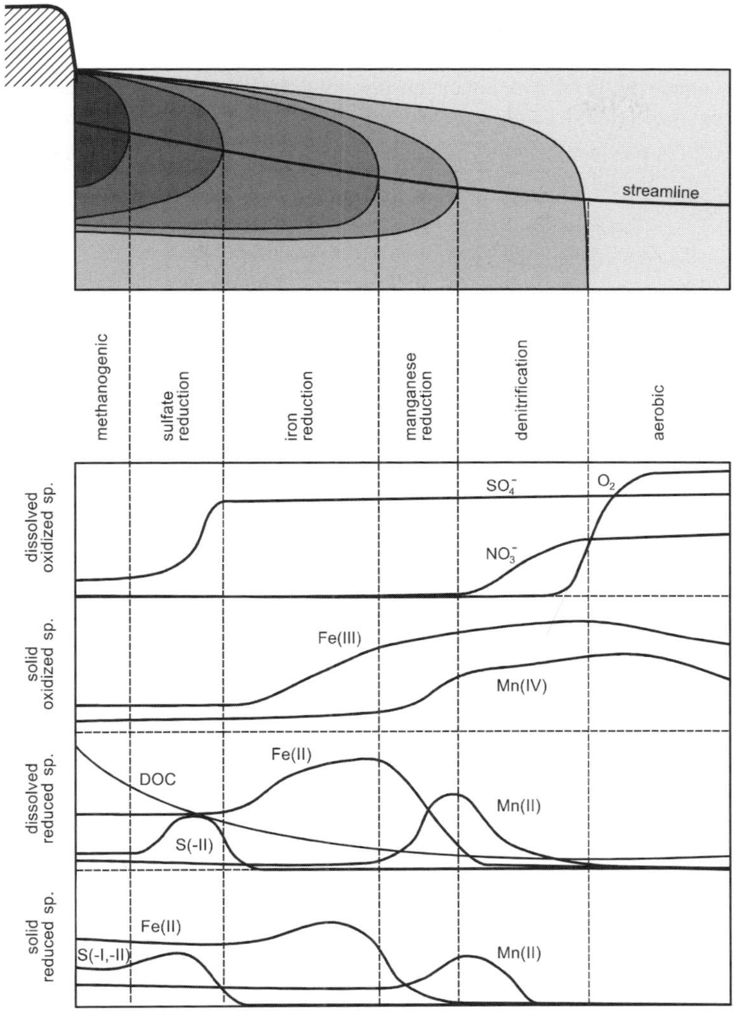

Figure 12.2. Schematic redox zonation in an originally aerobic aquifer downgradient from a landfill, and the distribution of redox-sensitive species along a streamline in the plume

Concentrations of unreactive (i.e. conservative) species in leachate can, however, only be reduced through dispersion and dilution. The extent to which dilution can reduce the concentrations of waste-derived contaminants in the leachate plume adjacent to the landfill or dump, depends upon the magnitude of both groundwater and leachate flows, together with the relative concentrations of contaminants in both the leachate and in the natural groundwaters of the aquifer upstream of the landfill (see Section 12.4).

As leachate migrates from a waste deposit in the direction of groundwater flow, the plume disperses (i.e. spreads due to differing contaminant flow paths and flow velocities), and also diffuses through the aquifer. Concentrations of both reactive and conservative contaminants decrease with distance along the groundwater flow path (Figure 12.3). It should, however, be recognized that exceptions to this general trend occur when a contaminant is transformed into a more toxic compound, as occurs in the dehalogenation of perchloroethylene/tetrachloroethene (PCE) to TCE. It should be noted that the concentration of a pollutant at any point removed from its source may vary throughout the year due to seasonal influences on recharge and release of the contaminant, or reaction times governed by variations in factors such as temperature.

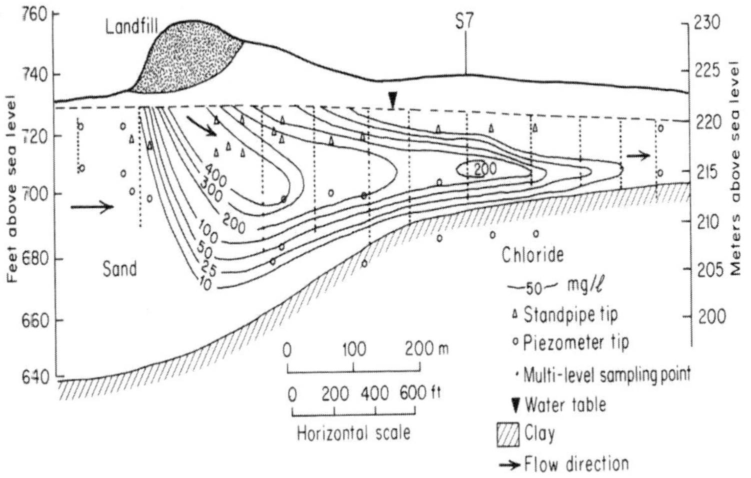

Figure 12.3. Mixing of landfill leachate with shallow groundwater in a sandy aquifer underlain by clay, as indicated by chloride concentrations (Freeze and Cherry, 1979; reprinted by permission of Pearson Education, Inc., Upper Saddle River, NJ)

12.4 ASSESSING GROUNDWATER CONTAMINATION ASSOCIATED WITH WASTE SITES

The checklist below provides guidance on how to approach assessing the likelihood of groundwater contamination through wastes and landfills found in a given drinking-water catchment. Much information for estimating pollution potential can be gleaned from amounts and types of wastes deposited, site management and site location in relation to aquifer vulnerability. As discussed in Chapters 2 and 8, this approach is not always easy, as the hydraulic conductivity and hydraulic gradient are crucially dependent on whether the aquifer has an intergranular or fissure permeability. Flow velocity can be several orders of magnitude higher in the latter. Also different contaminants may migrate at different velocities.

A number of countries use drinking-water protection zone concepts (Chapter 17) to delineate boundaries within which activities such as waste disposal are banned. Their delineation faces the same problems of understanding the hydrology of the setting. However, where protection zones exist, they are valuable for situation assessment which would begin with checking implementation (i.e. whether waste disposal is indeed being kept outside of the protection zone). Also, reviewing the information basis for their delineation will help to understand both the hydrogeological setting as well as the quality of the information base available for determining aquifer vulnerability.

Where hydrogeological understanding is poor and means to improve it are limited, a default approach to assessing pollution potential from wastes is to investigate distances between waste disposal and drinking-water abstraction, and to assess the potential hazard on the basis of the current general body of knowledge on landfill leachate plume attenuation and migration. A number of studies monitoring unlined landfills in operation before the advent of containment landfills have been ongoing over the past 20 years (e.g. Christensen *et al.*, 1994; Blight, 1995; Robinson *et al.*, 1999; Williams, 1999; Williams *et al.*, 1999; Ball and Novella, 2003; Butler *et al.*, 2003). These show that leachate plumes do not usually exceed a length of 1000 m, even in fast-flowing aquifers over periods in excess of 50 years after the initial wastes were deposited, and even within geological media with supposedly poor attenuation potential, such as sandy overburden. The processes of degradation and attenuation operating within the plume result in the front of the plume becoming stationary as degradation processes keep pace with. The migration of plume, and most pollutants, even complex organic compounds, degrade rapidly within the plume and are attenuated within a few hundred metres (Christensen *et al.*, 1995; Hancock *et al.*, 1995).

In the process of developing a GIS model for landfill site selection (Allen *et al.*, 2001), a survey of buffering distances used in various site selection criteria indicated that for individual dwellings with their own water wells in rural areas, a distance of 500 m was widely used, and except in extreme cases this would constitute a safe distance from a landfill for a water abstraction point. This distance could be reduced considerably on the upstream side of the landfill, if the direction of groundwater flow is known. Similarly, studies of leachate plumes (Christensen *et al.*, 1994) indicate that they do not tend to exceed the width of the landfill, so the plume does not fan out from the landfill in the direction of groundwater flow. Where hydrogeological information is available, such as the type of aquifer, groundwater flow direction and flow velocity, considerably smaller buffer distances of the order of 100-200 m would be adequate on the upstream and lateral sides of the landfill. On the whole, when assessing whether a landfill is safely distant from a water abstraction point, a distance of 500 m would in most cases be adequate, whilst a distance of 1000 m would be extremely conservative.

In contrast to many other human activities which cause diffuse groundwater pollution potential, landfill concentrates this to point sources. This facilitates assessing their pollution potential through screening and monitoring programmes which do not necessitate sophisticated chemical analyses of the wide range of potentially occurring pollutants, but rather select a few persistent substances, such as NH_3 and Cl, to detect and characterize leachate plume migration. Such an approach reflects the major influence that landfill leachate can exert on the abundance and concentration of individual substances present in groundwater. This is primarily valid for organic contaminants, which are almost exclusively anthropogenic, their presence in groundwater often indicating the influence of a waste site, but it is also valid for naturally occurring inorganic groundwater constituents, the content of which is increased by landfill leachate. Leachate migration can therefore be assessed by analysing the concentrations of common inorganic parameters in groundwater downgradient from a landfill in relation to their concentrations in groundwater sampled sufficiently upgradient, i.e. where it is not influenced by the landfill-derived contamination.

In order to rank the impact on groundwater of the leachate migrating from a landfill, Kerndorff *et al.* (1992) use a cf, representing the ratio of the measured concentration in the groundwater 10-100 m downgradient of the landfill to the concentration in the uncontaminated groundwater upgradient of the site. If the site is not leaking, or if the substance measured is not involved in the leakage event, the ratio should be 1.0. However, if the substance is leaking from the site, the ratio will increase to a value greater than 1.0. Thus the larger the leakage event, the larger the resultant cf. This approach identifies specific inorganic substances (those with the highest mean cfs) likely to be associated with landfill leakage events and therefore suitable for the indication of groundwater contaminations caused by landfills. In the above example, they proved to be the following: arsenic with a cf_{mean} of 122, ammonium with 65.5, cadmium with 26.9, nitrite with 25.7, boron with 21.6, chromium with cf_{mean} of 15.8, and nickel with 14.8. However, in using this approach it must be remembered that substances with high cfs are not necessarily those with the highest hazard potential nor those with the highest loads. They merely indicate the potential occurrence of groundwater contamination from a landfill with high loads of substances which may be hazardous due to their toxicity and/or persistency if they move through the aquifer towards a water supply.

12.5 CHECKLIST

NOTE ▶ *The following checklist outlines information needed for characterizing waste disposal and landfill activities in the drinking-water catchment area. It supports hazard analysis in the context of developing a Water Safety Plan (Chapter 16). It is neither complete nor designed as template for direct use but needs to be specially adapted for local conditions. The analysis of the potential of groundwater pollution from human activity requires combining the checklist below with information about socioeconomic conditions (Chapter 7), aquifer pollution vulnerability (Chapter 8), and other specific polluting activities in the catchment area (Chapters 9-11 and 13).*

ⓘ **Is waste disposed in the drinking-water catchment area?**

- ✓ Compile an inventory of sanitary landfills and legal or illegal uncontrolled waste disposal sites or dumps
- ✓ Compile an inventory of sites potentially producing special types of waste, such as health care facilities, cemeteries, scrap yards, slaughterhouses, industries (consider checklist for Chapter 11)
- ✓ Compile an inventory of sites storing, processing or treating wastes

- ✓ Compile historic data from the areas and facilities of interest
- ✓ For each inventory, identify relevant procedures, processes, responsibilities (who is in charge?) and substances/products in use
- ✓ Evaluate whether disposal sites were selected according to aquifer vulnerability and physical conditions in the catchment area (e.g. water table, soil, hydrogeology); consider checklist for Chapter 8
- ✓ …

What kind and which amounts of waste are disposed is the drinking-water catchment area?

- ✓ Estimate the amount of wastes produced and deposited in the drinking-water catchment
- ✓ For given deposits, assess the type and content of wastes (e.g. domestic, industrial, hospital) deposited
- ✓ Assess the likelihood of disposal of hazardous substances (e.g. from industry or hospitals)
- ✓ Estimate the amount and type of waste collected and deposited on controlled sites (sanitary landfills), unregulated dumps or that is randomly scattered
- ✓ Check for indication of illegal wastes imported from other countries and their nature
- ✓ …

What is the condition of the disposal sites?

- ✓ Evaluate siting, design, construction and technical condition of individual waste disposal sites in relation to aquifer vulnerability and physical conditions in the catchment area (e.g. water table, soil, hydrogeology); consider checklist for Chapter 8
- ✓ Check whether containment structures are in place and intact (e.g. lining)
- ✓ For sites with hazardous wastes, assess particularly the adequacy of protective structures in place, e.g. defence wells, drainage, containment (see Chapter 24)
- ✓ Assess the type of wastes and wastewater generated by these facilities and whether specific structures exist for separate collection of hazardous wastes or wastewater
- ✓ Identify key structural and technical strengths and weaknesses of individual disposal sites in relation to their groundwater pollution potential (see also Chapter 24)
- ✓ …

ⓘ **Are good management practices in place?**

Note: See Chapter 24 for the background information for these items

- ✓ Check whether waste management concepts are in place, e.g. for waste reduction and waste separation
- ✓ Assess whether implementation of such waste management concepts is satisfactory
- ✓ Check whether regulations are well known by administration and other staff
- ✓ Check whether broader environmental management concepts pertinent to waste disposal are understood
- ✓ Identify key strengths and weaknesses of the management practices implemented
- ✓ Assess whether containment structures for hazardous agents are intact and monitored at adequate intervals
- ✓ Check whether regular information is distributed, and whether training with respect to handling of wastes is adequate
- ✓ Check whether principles of good practice are followed by health care and research units working with highly infectious material and/or hazardous substances
- ✓ …

ⓘ **Are hazardous events likely to increase groundwater pollution potential?**

- ✓ Evaluate whether and how storm water events would enhance transport of pollutants to the aquifer
- ✓ Evaluate which spills and accidents are likely to cause groundwater pollution
- ✓ …

ⓘ **Is drinking-water abstracted in proximity to waste disposal sites?**

- ✓ Determine the direction and magnitude of the local hydraulic gradient, and whether drinking-water wells are upgradient or downgradient of the waste depository
- ✓ Assess distance between (formal and informal) waste disposal sites and drinking-water wells (see Chapter 8)
- ✓ Check adequacy of wellhead protection measures, wellhead construction and maintenance as well as sanitary seals used (see Chapter 18) to prevent ingress of contaminants from waste disposal sites
- ✓ …

Are groundwater quality data available to indicate pollution from waste disposal activities?

- ✓ Find out if leachate and/or groundwater monitoring programmes are in place around waste disposal sites
- ✓ Check whether seasonal leachate patterns are expected in relation to precipitation
- ✓ Compile data from local or regional waste disposal surveys, research projects or previous monitoring programs
- ✓ Check need and options for implementation of new or expanded monitoring programs likely to detect contamination from waste disposal facilities
- ✓ ...

What regulatory framework exists for waste disposal?

- ✓ Compile information on national, regional, local or catchment area specific legislation, regulations, recommendations or common codes of good practices on siting, construction, operation, maintenance of sites
- ✓ Check whether a regulatory framework exists for waste avoidance, waste separation, and particularly for waste disposal, and whether enforcement appears sufficient to protect groundwater
- ✓ Check whether the regulatory framework adequately addresses environmental and specifically groundwater protection
- ✓ Identify gaps and weaknesses known which may encourage specific pollution problems
- ✓ If wastes are imported, check whether this is due to stricter regulations in the country of origin, and whether the imports are legal
- ✓ ...

Documentation and visualization of information on waste disposal practices.

- ✓ Compile summarizing report and consolidate information from checklist points above
- ✓ Compile summary of types and amounts of substances expected from the specific waste disposal sites and sites potentially producing special types of waste
- ✓ Map formal and informal waste disposal sites and sites potentially producing special types of waste, preferably including suspected 'hot spots' of contamination (use GIS if possible)
- ✓ ...

12.6 REFERENCES

Ali, M., Cotton, A. and Westlake, K. (1999) *Down to Earth: Solid waste disposal for low-income countries*, WEDC, Loughborough.

Allen, A.R. (2001) Containment landfills: The myth of sustainability. *J. Eng. Geol.*, **60**, 3-19.

Allen, A., Brito, G., Caetano, P., Costa, C., Cummins, V., Donnelly, J., Fernandes, C., Koukoulas, S., O'Donnell, V., Robalo, C. and Vendas, D. (2001) *The Development of a GIS Model for the Location of Landfill Sites in Ireland and Portugal*. Final Report Interreg 2C Atlantic Region, EA-B1IRE-No.2.15, Cork.

Attahi, F. (1999) Abidjan, Côte d'Ivoire. In *Managing the Monster: Urban waste and governance in Africa*, (ed. A.G. Onibokun), pp. 11-48, IDRC, Ottawa.

Ball, J.M. and Novella, P.H. (2003) Coastal Park Landfill: leachate plume migration and attenuation. In *Proc. Sardinia 2003, Ninth International Landfill Symposium*, (eds. T.H. Christensen, R. Cossu and R. Stegmann), CISA Publisher, Cagliari.

Barrett, A. and Lawlor, J. (1995) *The Economics of Waste Management in Ireland*, Economic and Social Research Institute, Dublin.

Blight, G.E. (1995) Ground-water pollution from a landfill in a semi-arid climate. In *Proc. Sardinia 95, Fifth International Landfill Symposium* (eds. T.H. Christensen, R. Cossu, and R. Stegmann), vol. 3, pp. 593-599, CISA Publisher, Cagliari.

Butler, A.P., Brook, C., Godley, A., Lewin, K. and Young, C.P. (2003) Attenuation of landfill leachate in unsaturated sandstone. In *Proc. Sardinia 2003, Ninth International Landfill Symposium*, (eds. T.H. Christensen, R. Cossu and R. Stegmann), CISA Publisher, Cagliari.

Christensen, T.H., Kjeldsen, P., Albrechtsen H-J., Heron, G., Bjerg, P.L. and Holm, P. (1994) Attenuation of landfill leachate pollution in aquifers. *Crit. Rev. in Environ. Sci. Technol.*, **24**, 119-202.

Christensen, T.H., Nielsen, P.H. and Bjerg, P.L. (1995) Degradation of organic chemicals in a leachate pollution plume: an in-situ experiment. In: *Proc. Sardinia 95, Fifth International Landfill Symposium*, (eds. T.H. Christensen, R. Cossu, and R. Stegmann), vol. 3, pp. 621-628, CISA Publisher, Cagliari.

Christensen, J.B., Jensen, D.L., Grøn, C., Filip, Z. and Christensen, T.H. (1998) Characterisation of the dissolved organic carbon fraction in landfill leachate-polluted groundwater. *Water Res.*, **32**(1), 125-135.

Corry, J.E.L. (1978) Possible sources of ethanol ante- and post-mortem: its relationship to the biochemistry and microbiology of decomposition. *J. Appl. Bacter.*, **44**, 1-56.

Council of European Communities (1980) Directive on the protection of groundwater from certain dangerous substances. *Official J. European Communities*, L20(26), 43.

Douglas, T. (1992) Patterns of land, water and air pollution by waste. In *Managing the Human Impact on the Natural Environment*, (ed. M. Newson), pp. 150-171, John Wiley & Sons, New York.

European EA (2003) *Europe's Environment: The third assessment*. Environmental Assessment Report, No. 10, pp. 151-164, EEA, Copenhagen.

Ehrig, H.J. (1982) Quality and quantity of sanitary landfill leachate. *Waste Man. Res.*, **1**, 53-68.

Fetter, C.W. (1993) *Contaminant Hydrogeology*, Prentice-Hall, New Jersey.

Fourie, A.B. and Morris, J.W.F. (2003) The irrelevance of time as a criterion for aftercare provision. In *Proc. Sardinia 03, Ninth International Waste Management and Landfill Symposium*, (eds. T.H. Christensen, R. Cossu and R. Stegmann), CISA Publisher, Cagliari.

Freeze, R.A. and Cherry, J.A. (1979) *Groundwater*, Prentice-Hall, New Jersey.

Hancock, J.S., Phillips, J.R. and Seignior, M. (1995) In *Proc. Sardinia 95, Fifth International Landfill Symposium*, (eds. T.H. Christensen, R. Cossu and R. Stegmann), vol. 3, pp. 611-620, CISA Publisher, Cagliari.

Hjelmar, O., Johannsen, L.M., Knox, K., Ehrig, H.J., Flyvbjerg, J., Winther, P. and Christensen, T.H. (1995) Composition and management of leachate from landfills within the EU. In

Proceedings Sardinia 95, Fifth International Landfill Symposium, (eds. T.H. Christensen, R. Cossu and R. Stegmann), vol. 1, pp. 243-262, CISA Publisher, Cagliari.

Jefferis, S.A. (1993) Old landfills: perceptions and remediation of problem sites. In *Waste Disposal by Landfill*, (ed. R.W. Sarsby), pp. 93-106, Balkema, Rotterdam.

Johnson, C.A., Richner, G.A., Vitvar, T., Schittli, N. and Eberhard, M. (1998) Hydrological and geochemical factors affecting leachtae composition in municipal solid waste incinerator bottom ash Part I: The hydrology of landfill Lostorf, Switzerland. *J. Contam. Hydrol.*, **33**, 361-376.

Kerndorff, H., Schleyer, R., Milde, G. and Plumb Jr., R.H. (1992) Geochemistry of groundwater pollutants at German waste disposal sites. In *Groundwater Contamination and Analysis at Hazardous Waste Sites*, (eds. S. Lesage and R. Jackson), vol. 4, pp. 245-271, Environmental Science and Pollution Control Series, Marcel Dekker Inc., New York.

Kruempelbeck, I. and Ehlrig, H-J. (1999) Long-term behaviour of municipal solid waste landfills in Germany. In *Proc. Sardinia 99, Seventh International Waste Management and Landfill Symposium*, (eds. T.H. Christensen, R. Cossu and R. Stegmann), vol. 1, pp. 27-36, CISA Publisher, Cagliari.

Lusugga Kironde, J.M. (1999) Dar es Salaam, Tanzania. In *Managing the Monster: Urban waste and governance in Africa*, (ed. A.G. Onibokun), pp. 101-172, IDRC, Ottawa.

Mackenzie, D. (1989) If you can't treat it, ship it. *New Scientist*, **122**, 1658.

National Rivers Authority (1992) *Policy and Practice for the Protection of Groundwater*, NRA, Bristol.

Onibokun, A. and Kumuyi, A.J. (1999) Ibadan, Nigeria. In *Managing the Monster: Urban Waste and Governance in Africa*, (ed. A.G. Onibokun), pp. 49-100, IDRC, Ottawa.

OECD (1993) *Environmental Data, Compendium 1993*. OECD, Paris.

OECD (1997) *Environmental Data, Compendium 1997*. OECD, Paris

Pacheco, A., Mendes, J.M.B., Martins,T., Hassuda, S. and Kimmelmann, A.A. (1991) Cemeteries - a potential risk to groundwater. *Water Sci. Technol.*, **24**(11), 97-104.

Riediker, S., Suter, J.-F. and Giger, W. (2000) Benzene- and naphthalene-sulphonates in leachates and plumes of landfills. *Water Res.*, **34**(7), 2069-2079.

Robinson, H.C., Barber, C. and Maris, P.J. (1982) Generation and treatment of leachate from domestic wastes in landfills. *Water Pollution Control*, **54**, 465-478.

Robinson, H.C., Carey, M., Watson, G. and Gronow, J. (1999) In-situ monitoring of the unsaturated zone beneath Stangate East Landfill Site: Sixteen years of detailed data. In *Proc. Sardinia 99, Seventh International Waste Management and Landfill Symposium*, (eds. T.H. Christensen, R. Cossu, and R. Stegmann), vol. IV, pp. 117-124, CISA Publisher, Cagliari.

Röhrs, L.H., Fourie, A.B. and Blight, G.E. (2000) Is 30 years an appropriate after-care period for Municipal Landfills. In *Proc. 'WasteCon 2000', Somerset West, South Africa*, vol. 2, pp. 369-378, Institute of Waste Management, Somerset West.

Sangodoyin, A.Y. (1993) Considerations on contamination of groundwater by waste disposal systems in Nigeria. *Environ. Technol.*, **14**, 957-964.

Spongberg, A.L. and Becks, P.M. (2000) Inorganic soil contamination from cemetery leachate. *Water, Air and Soil Pollution*, **117**(1-4), 313-327.

Tchobanoglous, G., Thiesen, H. and Vigil, S. (1993) *Integrated Solid Waste Management Engineering Principles and Management Issues*, McGraw-Hill.

Teale, T.P. (1881) *Dangers To Health. A Pictorial Guide to Domestic Sanitary Defects*, Churchill, London.

Üçisik, A. S. and Rushbrook, P. (1998) The Impact of Cemeteries on the Environment and Public Health: An Introduction Briefing. World Health Organization, Copenhagen.

US EPA (1974) *Safe Drinking Water Act*, Code: 42 USC. s/s 300f et seq., Washington, DC.

Wall, D.K. and Zeiss, C. (1995) Municipal waste degradation and settlement. *J. Env. Eng.*, **121**, 214-223.

West, J. M., Pedley, S., Baker, S. J., Barrott, L., Morris, B., Storey, A., Ward, R. S. and Barrett, M.H. (1998) *A Review of the Impact of Microbiological Contaminants in Groundwater*. UK EA, R&D Technical Report, Bristol.

Western, K. (2004) Health considerations in cases of mass fatalities. In *Management of Dead Bodies in Disaster Situations*, Disaster Manuals and Guidelines Series, Nº 5, Pan American Health Organization, Washington, D.C.

Williams, G.M. (1999) Natural attenuation of contaminants in a lacustrine sand aquifer at Villa Farm, UK. In *Proc. Sardinia 99, Seventh International Landfill Symposium*, (eds. T.H. Christensen, R. Cossu and R. Stegmann), Vol IV, pp. 229-236, CISA Publisher, Cagliari.

Williams, G.M., Boland, M.P., Higgo, J.W., Ogilvy, R.D., Klinck, B.A., Noy, D.J. and Hart, P. (1999) Effect of old landfills on groundwater quality. In *Proc. Sardinia 99, Seventh International Landfill Symposium* (eds. T.H. Christensen, R. Cossu and R. Stegmann), vol. IV, pp. 125-132, CISA Publisher, Cagliari.

13

Traffic and transport: Potential hazards and information needs

A. Golwer and R. Sage

Vehicles and traffic routes are a widespread potential source of groundwater contamination. In some countries, economic development is currently swiftly leading to rapid increase of motorization and concomitant increases in pollution. Impacts on human health through contaminants in groundwater from traffic are generally substantially lower than the direct health effects through, for example, injuries and deaths in accidents, air pollution, noise and stress.

NOTE ▶ *Traffic and transport related activities and the environment in which they take place vary greatly. Health hazards arising from traffic and transport related activities and their potential to pollute groundwater therefore need to be analysed specifically for the conditions in a given setting. The information in this chapter supports hazard analysis in the context of developing a Water Safety Plan for a given water supply (Chapter 16). Options for controlling these risks are introduced in Chapter 25.*

 Edited by O. Schmoll, G. Howard, J. Chilton and I. Chorus. ISBN: 1843390795. Published by IWA Publishing, London, UK.

Traffic associated operations and facilities to recognize in situation assessments include roads, airfields, railway lines, inland waterway transportation (rivers, lakes, canals) where surface water strongly influences groundwater, as well as pipelines for crude oil and oil derivatives (Figure 13.1).

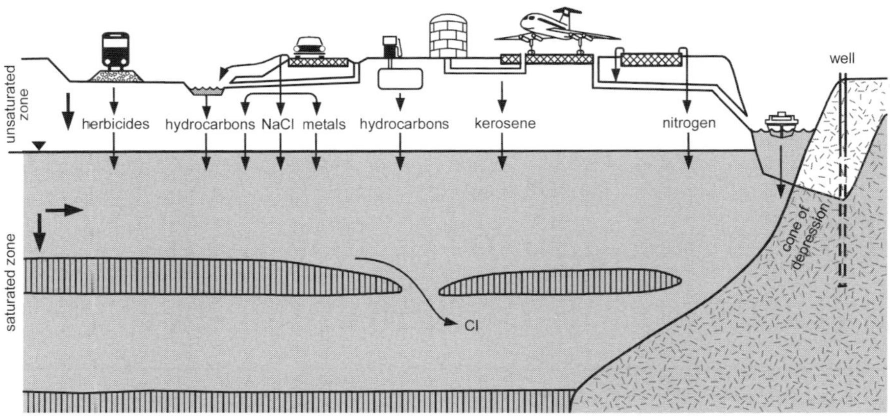

Figure 13.1. Most important groundwater contamination pathways from traffic and transport

A range of emissions of organic and inorganic substances from traffic settings may reach soils, sediments from drainage systems or surface waters via both water and air. Some of these may partly reach groundwater. The sources of groundwater pollution from the four traffic sectors may be classified according to types of traffic as well as different aspects of activity related to traffic, as outlined in Figure 13.2.

In addition to traffic routes as linear sources of emissions, transport installations and facilities, especially petrol stations, railway stations, airfields, inland harbours, car scrap yards and abandoned vehicles, may be substantial point sources of groundwater pollution. In some regions vehicle traffic, fuel storage and spills through accidents have been identified as important sources of traffic-related contaminants in groundwater.

13.1 GROUNDWATER POLLUTANTS FROM TRAFFIC

The volume of traffic obviously has a significant influence on the quantity and type of traffic-related pollution. Average daily traffic volume is an appropriate criterion for classifying roads in categories with different levels of potential pollution risk. Golwer (1991) proposes a classification of risk ranging from low (less than 2000 vehicles per day) to high (more than 15,000 vehicles per day). The traffic-related substance groups most frequently polluting groundwater are mineral oil products (including fuel additives such as methyl tertiary-butyl ether; MTBE), in many cases herbicides, and in specific situations heavy metals and de-icing agents.

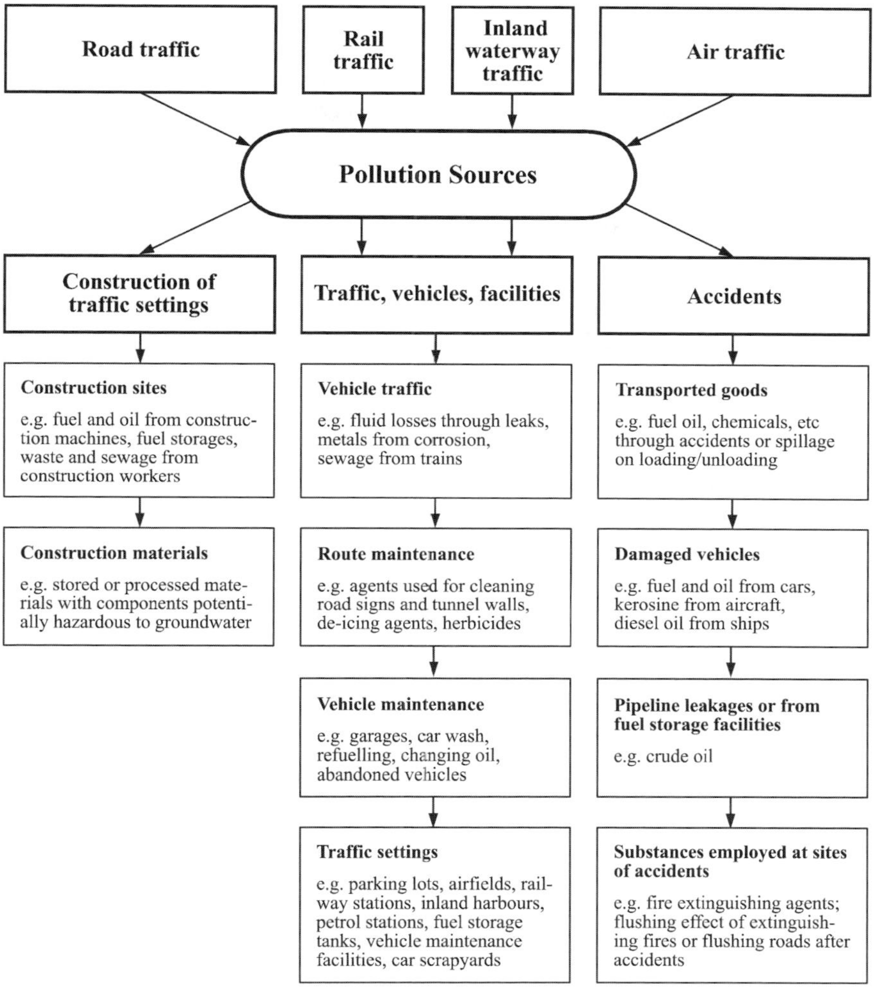

Figure 13.2. Traffic sectors and potential groundwater pollution sources

The condition and technical design of the traffic related infrastructure are critical in determining their pollution potential. Issues for situation assessment include assessing whether this infrastructure was designed to prevent or minimize pollution as well as whether or not maintenance and repair are conducted regularly. For instance, oil separators that have become full of grit will no longer work. Porous pavement materials may aid in minimizing direct runoff, but could carry pollutants directly to vulnerable groundwater sources. Other issues to consider in situation assessment are whether containment features are in use on transfer and fuelling locations, for example, how sewerage is moved from ships and trains to a point of treatment, whether there is a possibility of spillage and where such spillage would end up.

The input of dissolved mineral oil hydrocarbons – and of their decomposition products – into groundwater is dependent on the biological half-life of individual substances as well as climate and time of year. On busy roads in winter, in areas where the unsaturated zone provides little protection, mineral oil hydrocarbons have caused substantial groundwater contamination. Mineral oil hydrocarbons affect groundwater not only along roads, but particularly around petrol stations, airfields and railway stations. Accidents involving the transport of mineral oil and leaking long-distance pipelines (Morganwalp, 1994) have resulted in heavy local contamination, and have also affected groundwater quality through reduction in oxygen content (Schwille, 1976). Further, accident responses involving large volumes of water for extinguishing fires or removing contaminants may increase contaminant transport to the aquifer. Other contaminant sources such as abraded particles from tyres as well as exhaust gas and evaporation losses, generally have a low impact on groundwater though they may be re-deposited on soil surfaces and leached into the soil.

In addition to these fuel-derived contaminants, anti-knock agents have been widely detected in groundwater. These include lead and later MTBE and toluene which have replaced lead since the 1970s in the USA and later also in Europe. These are highly water-soluble and have been widely reported as pollutants related to accidental spillages and leaks from filling stations. Their health-relevance is briefly discussed in Chapter 4. Further, they serve as suitable tracers for traffic-related contamination.

For traffic safety on roads and airfields at temperatures below freezing, de-icing agents are widely employed. On roads, de-icing salts (mainly sodium chloride and lower levels of calcium chloride and magnesium chloride) have been used widely since the beginning of the 20th century. These are particularly found to be polluting in rural areas where roads are drained to soakaways. Whilst chloride itself is not a health hazard, its presence can be an indicator of other pollution from traffic. However, with greatly increased chloride concentrations, the mobility of zinc and cadmium can be increased through the formation of chloro-complexes (Bauske and Goetz, 1993).

On airports, nitrogenous de-icing agents, e.g. urea, have been employed since the mid 1970s. This has led to local high nitrate concentrations in groundwater and in some cases has required extensive remediation measures. Degradation of nitrate may produce nitrite and also ammonium. In the 1990s, non-nitrogenous agents (sodium and potassium acetates and formates) increasingly replaced nitrogenous de-icing agents. Propylene glycol and diethylene glycol are used for de-icing aircraft, which if not properly contained can also lead to pollution of both surface and groundwaters. In climates with frost, situation assessment should include airports as areas potentially contaminated not only with fuel, but also with de-icing agents.

Training areas for fire fighting on airports may give rise to special cases of pollution, especially if drainage from the area is not contained properly and not removed for treatment. The use of chemicals to aid in fire fighting and the large volumes of water involved can lead to contamination of the surrounding area. Even if the fire training areas are connected to a sewer system, leakage from this system and spray drift from the area can result in localized contamination of the soil and underlying aquifers. The associated chemicals will depend on the type of fire extinguisher agents used. Carbon tetrachloride/tetrachloromethane (CTC) and anionic synthetic detergents are frequent pollutants from such operations.

Railways, roads, parking lots and airports are sometimes kept clear of vegetation with repeated herbicide application. These have been detected in groundwater in their original form or as degradation products. The quantity of herbicide used at traffic facilities per unit area is often higher than in agriculture, and herbicides can swiftly enter surface waters and groundwater through drainage systems (Schweinsberg *et al.*, 1999). This organic substance group therefore poses a pronounced and widespread contamination hazard for groundwater, which is frequently unrecognized due to lack of targeted investigations. The substances used, e.g. atrazine (now banned in some European countries), diuron and bromacile may be alternated to avoid plant resistance to the active ingredient. In the United Kingdom, lengths of railway track crossing catchment areas for public supply boreholes, and draining to critical stretches of rivers, have been identified as potential sources of pollution.

Heavy metals specifically emitted from traffic are lead, cadmium, chromium, copper, nickel and zinc. Among these, generally even from sites with heavy traffic volume, only zinc and copper intermittently reach groundwater in amounts causing considerably increased concentrations, although a wider range may occur. In the groundwater downgradient of a basin collecting and infiltrating runoff from a very busy motorway (daily average of 125 560 vehicles in 1995) into porous soils since 1973, many substances were found in elevated concentrations as compared to upstream (Golwer, 1999). Even if concentrations are below levels hazardous to human health, their increase indicates a pollution pathway and the need for due regard to spillages and other releases.

13.2 TRAFFIC- AND TRANSPORT-RELATED ACTIVITIES POLLUTING GROUNDWATER

In addition to traffic and transport itself, construction and maintenance of transport associated facilities may cause considerable pollution of groundwater (Figure 13.2). Construction of transport lines and sites (including airports and harbours as well as roads, railways and canals) can change flowpaths and percolation patterns by relocating soil. It can lead to pollution through injuring the protective layers above the aquifer, as well as through construction activity. Refuelling and associated spillage often cause localized contamination and in sensitive areas construction of such refuelling points should be assessed for leakage and spillage. In addition to accidental spillages on the route line there may also be pollution from provisional quarters for construction workers, particularly with human excreta. As these are often only temporary facilities, they may not be subject to the same level of scrutiny or assessment as a permanent installation. Siting of such construction compounds should be subject to pollution risk assessments and adequate mitigation measures implemented.

Maintenance measures with the potential to cause pollution may include washing of tunnel surfaces which produces runoff with high contaminant concentrations, or detergents used in cleaning traffic signs. Other facilities such as refuelling depots, filling stations, car wash facilities, roadside cafes and service stations all have the potential to pollute groundwater. In vulnerable sites, situation assessment includes checking whether they are constructed properly, e.g. with spillage and runoff containment and adequate treatment.

Major pollution incidents have resulted from leakage from underground storage tanks and pipelines. These may not be detected for some time and once detected can require a significant amount of remedial action to clear up the pollution. In most cases, fuel products do not move very far from the point of pollution (at low field velocities less than 100 m) unless a preferential pathway is available. However, dissolved compounds and degradation by-products can pollute significant volumes of aquifer and be difficult to remediate. Fuel additives such as MTBE are extremely water soluble and can be detected in low concentrations over very wide areas.

13.3 PATHWAYS OF POLLUTANTS INTO GROUNDWATER

Though vehicle traffic often emits substantial amounts of airborne organic substances which deposit on the soil surface, it has been shown that this generally results in negligible impact on groundwater quality, since deposition rates from air are relatively low and the protective effect of the unsaturated zone results in sufficient reduction of concentration (Schleyer and Raffius, 2000). Nitrogen oxides from vehicle exhaust emission, however, contribute considerably to acidification of precipitation in some areas. Where soils have poor buffering capacity, this pollution can influence particularly the quality of shallow groundwater by mobilizing metals such as aluminium. Thus situation assessment may need to include vulnerability to soil acidification and mineral mobilization in areas with heavy air pollution through traffic.

Dispersion of traffic-borne pollutants from surfaces into water occurs intermittently and is strongly linked to rainfall and temperature. Obviously, at temperatures below freezing, pollutants will not be transported in water. At rainfalls of less than 0.5 mm, runoff from impermeable traffic-area surfaces generally does not occur. However, pollutants deposited on surfaces during dry periods or frost accumulate and the first rainfall or snow melt flushes these into the soil or into runoff collection systems. Most pollution from traffic routes generally occurs within 10 m of the route, and thus they can be considered as linear sources.

Where runoff from paved traffic surfaces is collected in retention basins, these may themselves be significant sources of groundwater pollution through seepage or leakage. Beneath infiltration basins percolation may occur either intermittently or continuously, usually in a vertical direction.

In the zone of percolation substance dispersion depends on distribution of precipitation and soil temperature. Because of the filtration effect, particles and particle adsorbed substances (e.g. platinum, lead, polycyclic aromatic hydrocarbons) are largely retained in the upper centimetres and decimetres of soils. Infiltration rates can be considerably higher alongside permeable traffic surfaces and on the edge or bottom of infiltration basins than in adjacent areas with soils and rocks of comparable texture, as the larger quantity of water per unit of area often leads to increased permeability through solution and removal of fine particles by the increased velocities. Dead leaf matter can be broken up by tyre action and concentrated on the sides of roads, leading to enhanced leaching of humic acids. Atmospheric deposition of pollutants accumulating on paved surfaces can also be a pathway for pollutants into the groundwater system.

Once pollutants reach groundwater, their dispersion and concentration chiefly depend on substance loading and its characteristics, as well as on the hydrogeological characteristics in the specific setting, as discussed in Chapters 2 and 4. Substance characteristics (especially water-solubility, volatility, sorption, as well as biological and photochemical degradation) are of decisive importance for retention, conversion and dilution processes during subsoil passage.

Climatic conditions fundamentally influence substance dispersion and, to some extent indirectly, substance inventory, as explained above. Climate also determines rates of degradation (which are frequently temperature-dependent and thus higher in the tropics), patterns of substance dispersion into groundwater, patterns of dilution and human activities, such as the extent of application of de-icing agents or herbicides.

13.4 CHECKLIST

NOTE ▶ *The following checklist outlines information needed for characterizing traffic and transport related activities in the drinking-water catchment area. It supports hazard analysis in the context of developing a Water Safety Plan (Chapter 16). It is neither complete nor designed as a template for direct use but needs to be specially adapted for local conditions. The analysis of the potential of groundwater pollution from human activity requires combination of the checklist below with information about socioeconomic conditions (Chapter 7), aquifer pollution vulnerability (Chapter 8), and other specific polluting activities in the catchment area (Chapters 9-12).*

ⓘ **Are transport related facilities and traffic routes present in the drinking-water catchment area?**

- ✓ Compile inventory of roads, parking lots, refuelling stations, fuel storage tanks and pipelines, car wash, car scrap yards, airfields, railway lines and stations, inland harbours, canals
- ✓ Evaluate siting of individual traffic facilities in relation to aquifer vulnerability; consider checklist for Chapter 8
- ✓ Check for presence of abandoned sites (particularly for storage of fuels and maintenance chemicals) that could still be contaminating groundwater
- ✓ …

How high is traffic intensity and its potential to pollute groundwater?

- ✓ Assess vehicle traffic information on traffic volume (e.g. daily traffic volume on roads), transport of hazardous goods, accident rates
- ✓ Assess whether accident prevention measures (e.g. crash barriers, speed limits and their implementation) are adequate in relation to traffic volume and groundwater vulnerability
- ✓ Check availability and implementation of accident response plans to restrict groundwater contamination with fuel or hazardous goods
- ✓ …

Are the transport-related facilities and traffic lines in good condition?

- ✓ Assess adequacy of design, construction and technical condition of transport-related facilities with respect to protecting groundwater from pollution, e.g. pavement, drainage system for collection of runoff, runoff treatment, containment for filling stations and car wash runoff, collection and disposal of sewage on trains and ships
- ✓ …

Is there transport related construction in the drinking-water catchment area?

- ✓ Consider impact of physical change of protective layers and permeability to shallow aquifers caused by construction
- ✓ Assess risk of aquifer pollution with fuel, lubricants and hydraulic oil from construction machines
- ✓ Assess availability of sanitation for construction workers
- ✓ Check waste removal and disposal from construction sites
- ✓ Check whether there is storage and processing of potentially hazardous construction materials
- ✓ Assess design of and protection for re-fuelling areas
- ✓ …

What maintenance practices for traffic routes might contaminate groundwater?

- ✓ Check whether application of herbicides on railway lines, de-icing agents on roads and airfields, cleaning agents for road signs or tunnel walls is practised
- ✓ Check adequacy of design, construction and maintenance of storage and cleaning facilities for these agents, e.g. for herbicides used on railway lines

- ✓ Evaluate whether good practice is used in the handling and application of such agents, whether these are regularly checked and verified, and whether their application is minimized to ensure no excess is available for pollution
- ✓ …

Are hazardous events likely to increase groundwater pollution potential?

- ✓ Evaluate whether (and how) storm water events would enhance transport of pollutants to the aquifer
- ✓ Evaluate which spills and accidents are likely to cause groundwater pollution
- ✓ …

Is drinking-water abstracted in proximity to traffic facilities?

- ✓ Assess distance between traffic facilities and drinking-water abstraction (see Chapter 8)
- ✓ Check adequacy of wellhead protection measures, wellhead construction and maintenance as well as sanitary seals used (see Chapter 18) to prevent ingress of contaminants from traffic
- ✓ …

Are groundwater quality data available to indicate pollution from traffic?

- ✓ Compile data, particularly chemical analyses, from local or regional surveys, research projects or previous monitoring programmes
- ✓ Check for implementation of new or expanded monitoring programmes likely to detect contamination from traffic
- ✓ …

What regulatory framework exists for traffic?

- ✓ Compile information on national, regional, local, or catchment area specific legislation, regulations, recommendations, or common codes of good practices on siting, construction, operation, maintenance of traffic facilities and on restrictions, ban or prohibition of substances applied on traffic settings
- ✓ Compile regulations and permits to transport hazardous substances
- ✓ Check whether the regulatory framework adequately addresses environmental and specifically groundwater protection

✓ Identify gaps and weaknesses known which may encourage specific pollution problems
✓ …

Documentation and visualization of information on traffic settings.

✓ Consolidate information from checklist points above and compile in a summarizing report
✓ Map traffic routes, preferably including average daily traffic volume, road drainage and runoff retention ponds (use GIS if possible)
✓ Map locations of airports, parking lots scrap yards, filling stations and pipe line routes, preferably including suspected 'hot spots' of contamination (use GIS if possible)
✓ …

13.5 REFERENCES

Bauske, B. and Goetz, D. (1993) Effects of de-icing-salts on heavy metal mobility. *Acta Hydrochim. Hydrobiol.*, **21**, 38-42.

Golwer, A. (1991) Belastung von Böden und Grundwasser durch Verkehrswege. *Forum Städte-Hygiene*, **43**, 266-275.

Golwer, A. (1999) Beeinflussung der Grundwasserbeschaffenheit durch Straßenverkehr in Wasserschutzgebieten. *Geol. Jb. Hessen*, **127**, 131-146.

Morganwalp, B.W. (1994) Groundwater contamination from injection wells and pipelines. In *Groundwater Contamination and Control*, (ed. U. Zoller), pp. 399-423, Marcel Dekker Inc., New York.

Schleyer, R. and Raffius, B. (2000) Grundwassergefährdung durch organische Luftschadstoffe. *DVWK-Materialien*, **1**/2000, 385.

Schweinsberg, F., Abke, W., Rieth, K., Rohmann, U. and Zullei-Seibert, N. (1999) Herbicide use on railway tracks for safety reasons in Germany. *Toxicology Letters*, **107**, 201-205.

Schwille, F. (1976) Anthropogenically Reduced Groundwaters. *Hydrol. Sci. Bull.*, **XXI** (4), 629-645.

Section III

Situation analysis

14

Assessment of groundwater pollution potential

J. Chilton, O. Schmoll and S. Appleyard

Using the knowledge of the regional groundwater conditions and the nature of the human activities and their possible impacts on groundwater quality provided by Sections I and II, this chapter describes the assessment of groundwater pollution potential, and illustrates the approach with two case studies. The intention of the chapter is to indicate the general scope and scale of what is required to assess groundwater pollution potential, rather than to provide detailed technical guidance on how it should be done. This more detailed technical material, aimed more at the practising professional actually carrying out such an assessment, can be found in, for example, Foster *et al.* (2002) and Zaporozec (2002). Chapter 15 builds on this assessment to provide guidance on establishing groundwater management priorities to reduce the impact of pollution, either by increasing the protection measures at drinking-water sources (Section IV) or by controlling pollution sources and the activities causing pollution (Section V).

Assessment of groundwater pollution potential in a given drinking-water catchment may be conducted under a wide variety of conditions and at varying spatial scales and levels of sophistication (Foster and Hirata, 1988). As a result, the assessment can produce a wide range of outcomes. The conditions may range from simple settings with almost self-evident identification of one or two key hazards (e.g. high density of poorly sealed latrines on a shallow aquifer) to highly complex urban and industrial scenarios with

 ISBN: 1843390795. Published by IWA Publishing, London, UK.

diverse human activities, in which the key sources of pollution are difficult to identify. This complexity may perhaps be coupled with small-scale variations in geology and hydrogeological conditions, rendering vulnerability assessments equally demanding. This broad variation means that assessments of pollution potential can require an equally broad range of sophistication, ranging from reconnaissance surveys of the major potential sources of groundwater pollution to detailed surveys of chemical or microbial pollutant loads and even to simple modelling of, for example, the leaching potential of pesticides used in the catchment. This implies that experienced professionals from the hydrogeology and environmental engineering disciplines will normally be needed, both to help decide on the level of sophistication required, and to undertake the assessment itself.

NOTE ▶ *This chapter indicates the general scope and scale of what is required to assess groundwater pollution potential, rather than providing detailed technical guidance on how it should be done. In the context of developing a Water Safety Plan this chapter supports risk assessment for groundwater fed drinking-water supplies.*

14.1 THE OVERALL ASSESSMENT PROCESS

Prior to undertaking the assessment of groundwater pollution potential, the first exercise is to decide upon the area to be protected around the drinking-water source. This may include the entire extent of the aquifer system in which the source occurs at one extreme, or may involve delineating the specific catchment or zone of influence of the supply source or sources in question. This will normally be the hydrogeologically-defined capture zone from which the recharge is derived, as in the Barbados case study (Section 14.5), or may just be a simple radius around the source. Guidance is given in Chapter 17 on the establishment of such zones. Once the area of investigation has been defined, then the process described below can be undertaken. However, it is important to note that groundwater catchments do not always follow surface water catchments and may cross both local and national administrative political boundaries. Thus selection of the correct and appropriate area is critical. Near national borders, it may even involve international discussions and agreements for successful implementation of groundwater protection and the associated control measures or other management responses.

In view of the complexity of factors affecting pollutant migration and the uniqueness of each field situation, it would be logical to treat each activity or source on individual merit and undertake independent field investigations to assess pollution potential (Foster and Hirata, 1988). However, because of the high cost of such investigations, simpler but consistent procedures for assessing pollution potential at modest cost are needed. The reader should not, therefore, be unduly discouraged by the complexity of hydrogeological conditions and pollutant behaviour. Using the understanding of the former gained from Chapters 2 and 8, and of the latter from Chapters 3, 4, and 9 to 13, useful assessments of groundwater pollution potential can be achieved on the basis of existing information combined with some field reconnaissance.

As shown in Figure 14.1, the potential for groundwater pollution to occur is determined by the interaction between the microbial or chemical pollutant loading which is being, or might be, applied to the subsurface environment as a result of one or more of the types of human activity described in Chapters 9-13, and the aquifer vulnerability, which depends on the intrinsic physical characteristics of the soil and strata separating the aquifer from the land surface, as described in Chapter 8.

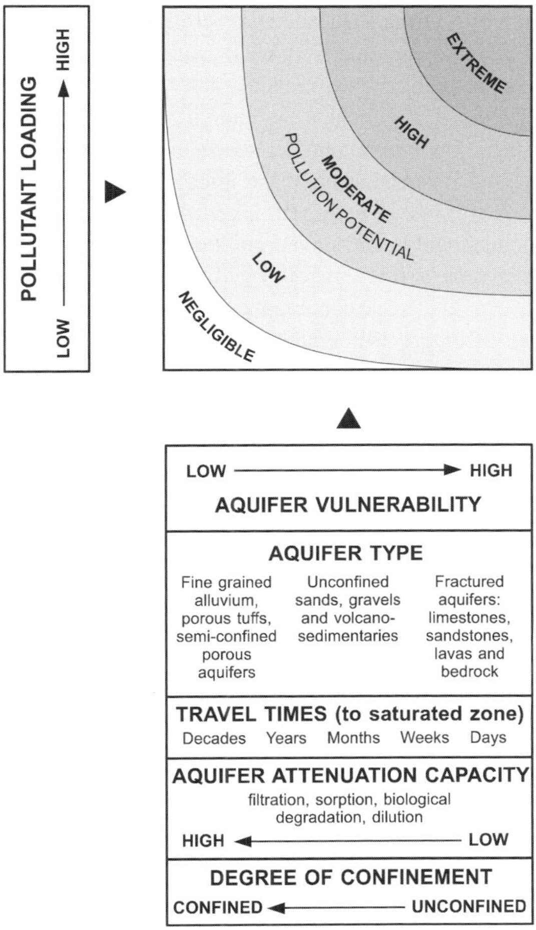

Figure 14.1. Groundwater pollution potential (adapted from Foster and Hirata, 1988)

The factors that define aquifer vulnerability (Chapter 8) are summarized here for convenience along the horizontal axis of Figure 14.1, and it is the identification and characterization of the factors that determine pollutant loading on the vertical axis that is the subject of this chapter.

The matrix in Figure 14.1 does not assign quantitative scores, but rather depicts a relative classification of pollution potential, and the components of both pollutant loading and aquifer vulnerability can have broad ranges from low to high. Thus a combination of high pollutant loading and high aquifer vulnerability provide the most extreme pollution potential in the top right corner of the figure. Adopting this approach, it is possible to envisage situations in which an aquifer is highly vulnerable, but there is little or no danger of pollution because there is no pollution load, or vice versa. Both are consistent in practice. The former might occur on an uninhabited coral limestone island, and the latter where an urban area with many small pollution sources is separated from an underlying deep aquifer by a thick sequence of impermeable clays or silts.

Whether the pollution potential derived in this way will be translated into a serious quality impact producing problems for drinking-water supplies using groundwater will depend on several factors. These include the mobility and persistence of the pollutants within the groundwater flow regime and the scope for further dilution in the saturated zone. The economic and financial scale of the impact will depend on the value of the groundwater resources affected, including the investment and operating costs in abstracting the water and delivering it to consumers, and the cost of finding alternative supplies, as well as the broader societal and environmental value of the groundwater where, for example, there are many small-scale community or private wells and boreholes.

If the relationship in Figure 14.1 could be fully quantified in probability terms, it would become a more formal indication of the likelihood that groundwater in an aquifer would become polluted at concentrations above respective guideline values. While this may be possible for some diffuse pollution sources such as agriculture and unsewered sanitation, experience suggests it is much more difficult to quantify microbial and chemical pollutant loads to groundwater for most point sources. Further, given the uncertainties about pollutant behaviour outlined below and in Sections I and II, a qualitative or semi-quantitative assessment of the potential for pollution of groundwater to occur is, for many settings, the best that situation analysis can achieve. This is, however, likely to be more than adequate as a basis for initiating consideration of actions for protecting groundwater, and for focussing more detailed investigation or monitoring on the activities or sources judged to be the most significant.

14.2 COMPONENTS OF ASSESSMENT OF POLLUTANT LOADING

The series of questions that need to be answered in an assessment of pollutant loading are shown in Figure 14.2. The six questions and the associated components of the assessment are presented one above the other, and linked sideways by arrows to the box representing pollutant loading to denote that they are not necessarily part of a sequential decision process. The information needed to answer these questions must come from a survey or inventory of likely pollutant sources, including identification, location and characterization of all sources, including where possible their historical evolution (see checklists at the end of chapters of Section II). Further discussion of data collection procedures and design and implementation of pollution inventories is provided by Foster *et al.* (2002) and Zaporozec (2002).

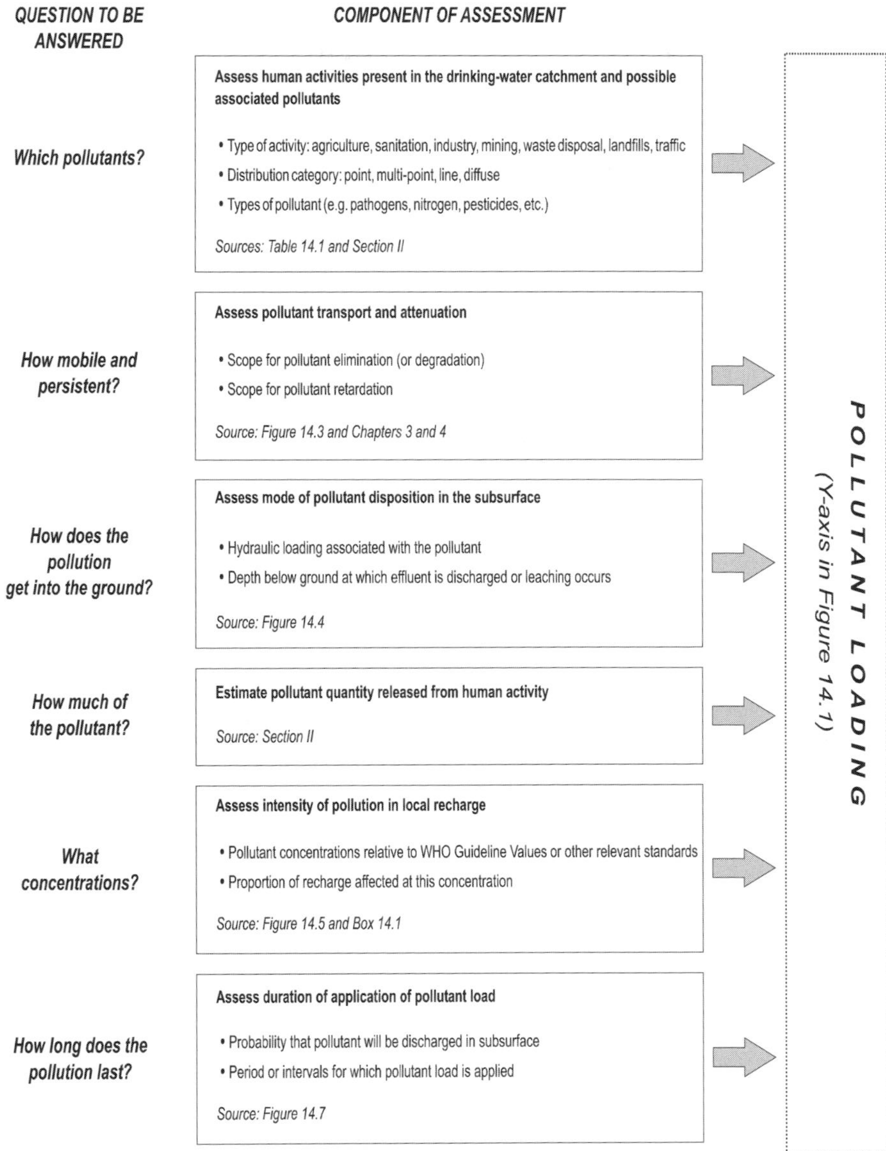

Figure 14.2 Components of assessment of pollutant loading

The information needed to answer the first question in Figure 14.2 is summarized in Table 14.1, in which the human activities in Chapters 9-13 are listed with many of the main types of pollutants and their category of distribution as point, line or diffuse sources. Table 14.1 also indicates which of the activities are accompanied by significant hydraulic loading by additional volumes of water, and for which of them the protective soil layer is by-passed in the method of usage or disposal of the potential pollutants.

Table 14.1. Summary of activities potentially generating subsurface pollutant loading (modified from Foster *et al*., 2002)

Type of human activity	**Character of pollutant loading**			
	Distribution category	Main types of pollutant	Relative hydraulic loading	By-pass of soil zone
Agriculture (Chapter 9)				
Cultivation with:				
Agrochemicals	D	*NP*		
Irrigation	D	*NPS*	+	
Sewage sludge and wastewater	D	*FNOS*	+	
Animal feedlot operation:				
Leakage from unlined effluent lagoons	P	*FN*	++	+
Land discharge of effluent	P-D	*FNS*	+	
Sanitation (Chapter 10)				
Leakage from on-site sanitation	P-D	*FN*	+	+
Land discharge of sewage	P-D	*FNOS*	+	
Leakage from sewage oxidation lagoons	P	*FNO*	++	+
Sewer leakage	P-L	*FNO*	+	+
Industry (Chapter 11)				
Leakage from effluent lagoons (process water)	P	*OMS*	++	+
Tank and pipeline leakage	P	*OM*	+	+
Accidental spillages	P	*OM*	++	
Land discharge of effluent	P-D	*OMS*	+	
Well disposal of effluent	P	*OMS*	++	+
Mining (Chapter 11)				
Mine drainage discharge	P-L	*MSA*	++	+
Leakage from sludge lagoons (process water)	P	*MSA*	++	+
Leaching from solid mine tailings	P	*MSA*		+
Oilfield brine disposal	P	*S*	+	++
Waste disposal and landfill (Chapter 12)				
Leaching from waste disposal/landfill sites	P	*NOMS*		++
Traffic (Chapter 13)				
Highway drainage soakaways	P-L	OMS	++	++
Tank leakage	P	*O*	+	+
Application of chemicals	P-L	*PS*		
Groundwater resource management (Chapter 8)				
Saline intrusion	D-L	*S*		
Recovering water levels	D	*OSA*		
Drawdown of pollutants due to abstraction	D	*OMS*		
Wellhead contamination	P	*FN*		++

Distribution category: P – point; D – diffuse; L – linear. Main types of pollutant: *F* – faecal pathogens; *N* – nutrients; *O* – organic compounds including chlorinated solvents or aromatic hydrocarbons (BTEX); *P* – pesticides; *M* – metals; *S* – salinity; *A* – acidification

Relative hydraulic loading: + to ++ (increasing importance; relative volume or impact of water entering with pollution load). By-pass of soil zone: + to ++ (with completeness of by-pass of soil and depth of penetration into unsaturated and saturated zones)

A simple approach to illustrating four of the components of Figure 14.2 is shown in Figures 14.3 to 14.5 and 14.7. These are based on diagrams originally developed by Foster and Hirata (1988), but modified to refer to the groups of human activities described in Chapters 9-13 and summarized in Table 14.1.

The four diagrams presented here are intended to be used conceptually, i.e. to provide a general and largely relative indication of which features of the selected activities contribute most to the potential for pollution of groundwater to occur. Qualitative interpretation of the four diagrams will help to indicate where efforts to improve the information base should be concentrated. Experience of assessing pollutant loading potential, including in the two case study examples, suggests that, in many situations, the complexity of human activities, industrial processes and waste disposal practices means that careful and detailed investigations using the checklists from Chapters 9-13 are required.

Pollutant mobility and persistence

Figure 14.3 helps to answer the question about mobility and persistence by locating the main classes of pollutants according to their potential for degradation and elimination or pathogen inactivation and die-off respectively, and/or retardation by the processes described in Chapters 3 and 4. The former include chemical reactions such as precipitation and complexation as well biodegradation, and the latter comprise adsorption, filtration and cation exchange. Thus mobile and persistent pollutants such as chloride and nitrate are relatively little affected by these attenuation processes in aerobic conditions in the aquifer and the overlying strata. For these pollutants, dilution will usually be the main attenuation process that operates. More readily degraded and retarded pollutants such as pesticides, bacteria and viruses can be significantly restricted from reaching aquifers if the overlying strata have adequate attenuation capacity in terms of clay and organic carbon content and microbial activity.

While Figure 14.3 is helpful conceptually, it is emphasized that this general guide has limitations. Firstly, pollutants within the simple groupings given in Figure 14.3 may behave differently, and for hydrocarbons and pesticides the tables in Chapter 4 and the references from which they were derived can provide more specific information. Secondly, while the aerobic, alkaline conditions specified in the title are the most common and widespread in groundwaters, changes in pH and Eh conditions can have important influences on mobility and persistence (Lyngkilde and Christensen, 1992; Christensen *et al.*, 1996). Eh, the redox potential, is a measure (usually in millivolts) of the intensity of oxidizing or reducing conditions within a groundwater system. Positive potentials indicate that the system is relatively oxidizing, and negative potentials represent reducing conditions (Hem, 1985). The influence on pollutant groups of reductions in pH or Eh, representing more acidic and reducing conditions respectively, is shown in Figure 14.3. Thus most metals become significantly more mobile in acidic and/or reducing conditions, and ammonium and nitrate are mutually sensitive to the oxidation status of the groundwater. Some pollutants, in particular arsenic and chromium, have several forms of natural occurrence. These can have different chemical valencies or oxidation status, and their possible behaviour in response to changes in acidity or oxidation status is too complex to be represented simply on Figure 14.3.

A final note of caution is that the figure can be considered broadly applicable to aquifers in which groundwater movement is intergranular or in a network of small fractures (Chapter 2). In the very rapid groundwater flow conditions of karstic aquifers, however, water and microbial and chemical pollutants may be moving so fast that the processes of elimination and retardation do not have time to operate.

Even allowing for these cautionary notes, the likelihood of reaching groundwater can be assessed in qualitative terms for potential pollutants identified or anticipated from the situation assessment and from Table 14.1. This likelihood increases from bottom left to top right of Figure 14.3.

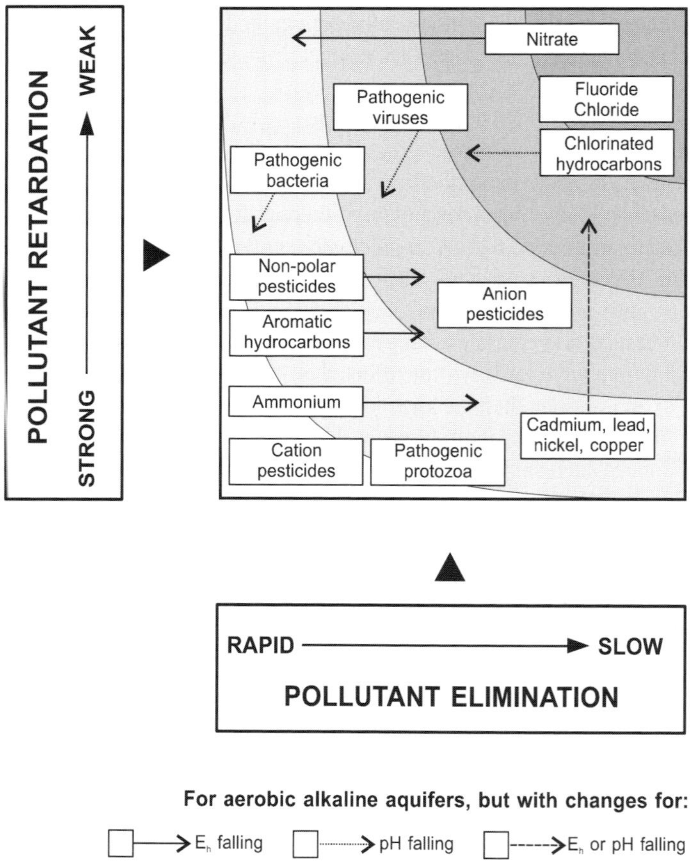

Figure 14.3. Characterization of mobility and persistence of pollutants in aerobic, alkaline conditions (adapted from Foster and Hirata, 1988)

Mode of disposition

The next characteristic of pollutant loading potential (Figure 14.2) is the mode of disposition, i.e. how the pollutant enters the subsurface. This is a combination of the hydraulic loading or surcharge associated with or imposed by the pollution source, and the depth below the ground surface at which either effluent is discharged or leaching from solid residues occurs, and is illustrated in Figure 14.4.

Thus conventional, rainfed agriculture takes place at the ground surface with a hydraulic loading only from infiltrating rainfall. Irrigated agriculture may produce significantly higher hydraulic loading, depending on the number of crops per year and type of crops, and the irrigation methods used. Many other potential sources of pollution originate below the soil – including leaking sewers, unsewered sanitation, industrial effluents and soakaway drainage from highways. Depending on the precise mode of disposition, the pathway to groundwater may be greatly shortened for several of these, as depicted in Figure 14.4. Unlined landfills can generate highly concentrated leachates which may enter the subsurface at considerable depth if the waste has been disposed of in excavations formerly used for quarries or pits. Disposal of industrial effluents into old wells or specially constructed disposal wells transports pollutants directly to the water table, often with significant local hydraulic loading (Figure 14.4). An accidental spill at the land surface may have apparently limited potential to cause groundwater pollution. However, highways, railways and airfields often drain to the subsurface by soakaways, greatly facilitating more rapid transport of pollutants to groundwater (Figure 14.4). The potential to cause pollutant loading, and likely scale of impact, increases from bottom left (small loading at the ground surface) to top right (high hydraulic loading close to or at the water table).

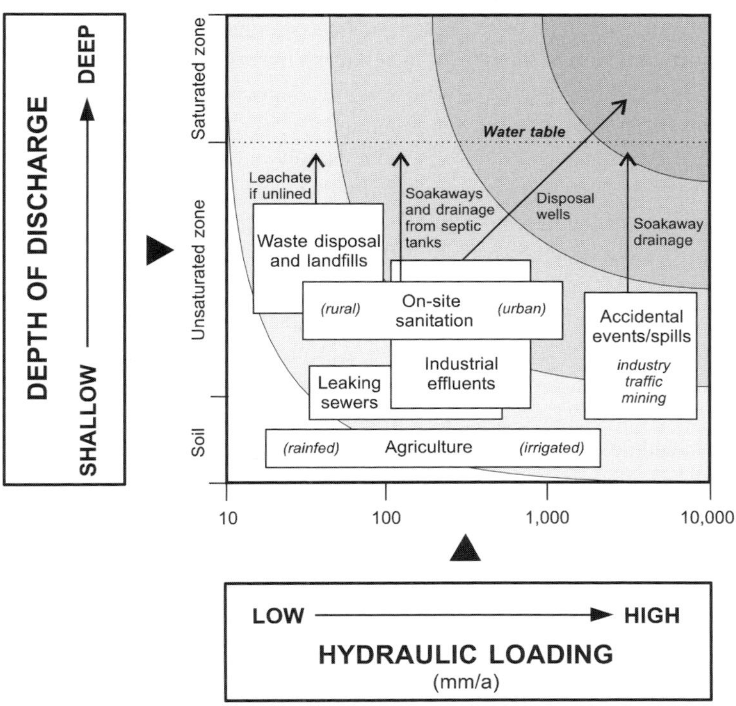

Figure 14.4. Characterization of mode of pollutant disposition (adapted from Foster and Hirata, 1988)

Pollutant quantity

Within a general assessment of potentially polluting activities, it is clearly desirable to obtain as much information as possible about each of the components shown in Figure 14.2. Thus in answer to the question about how much of the pollutant is initially released by or leached from the human activity in question, it would ideally be possible to estimate the actual quantities of pollutants at the time and place of release into the environment. For pathogens or faecal indicators these would usually be expressed in counts of organisms per 100 millilitre, and for chemicals in units such as litres or kilograms of, for example, D/dense/light non-aqueous phase liquid (D/LNAPLS) spilled in an accident or leaked from a tank or pipeline, kilograms of nitrogen fertilizer applied per ha, or cubic metres of landfill leachate and the concentrations of pollutants in the leachate. Because of the infinite variety of scope and scale of human activities that potentially generate pollutants (Table 14.1), a simple tabulation of quantities, estimation methods or published sources that could be used to answer the question 'how much pollutant?' is not feasible. While some indications of pollutant quantities for some activities such as nitrogen fertilizer applications and leaching losses (Conway and Pretty, 1991), unsewered sanitation (ARGOSS, 2001) and landfill leachates (Stuart and Klinck, 1998) can be obtained from literature sources, for the most part estimates of pollutant quantities must be attempted for the specific situation encountered.

For a few types of individual point sources, estimating the quantity of pollutant released may be a simple matter of site observation, for example, the volume of liquid leaked from a ruptured road or rail tanker, or a catastrophic failure of a fuel tank, tailings dam or similar structure of known volume. For the majority of even apparently 'simple' point sources, pipelines, broken sewers, landfills, lagoons, dams and tanks, leakage is likely to have been slow, intermittent or continuous over considerable time periods, and from storage or conveyance systems with largely unknown but probably variable discharges and pollutant concentrations. Further, the complexity and considerable diversity of many of the major potential pollutant sources and the fact that, especially in urban areas, sources may be spatially distributed over a large area, such as to cause a complex mosaic of many small pollution sources, means that the load from and consequent impact of any individual one is very difficult to isolate. In the case of the city of Leon in Mexico, for example, which is a centre of leather processing, some 530 tanneries of varying sizes are distributed throughout the urban and suburban areas (Chilton *et al.*, 1998). These use varying and largely unknown amounts of processing chemicals including chromium salts and solvents, and discharge polluting wastes. Some discharge their polluting wastewaters into the urban sewer network, and some directly to the ground. Separate collection and treatment of the chromium-rich and highly saline effluent from these dispersed, small-scale industries would be difficult and costly and would require relocation of the tanneries to a designated industrial area. For these and similar small-scale, dispersed industrial and commercial activities, mapping their locations may be feasible, if time-consuming (Zaporozec, 2002). The next step of finding out about the volumes and concentrations of effluents may, however, be much more difficult, not least because the owners and managers may be unwilling to pass on the information if their waste disposal practices are not environmentally sound, or even illegal. In the Barbados case study described at the end of this chapter, it was relatively easy to locate the few hazardous industries, but very difficult to find out about their actual disposal practices because most were probably discharging effluents directly into the underlying coral limestone aquifer.

In practice, therefore, experience shows that estimating the amounts of pollutants released is highly problematic for all but the simplest scenarios. Nevertheless, the ideal information requirements outlined above should still guide the pollutant source and load surveying process. They constitute the basis for subsequent detailed investigations of the most important pollutant sources and loads, by inspection of premises, processes and waste disposal practices and the sampling of effluents. The checklists provided in Chapters 9 to 13 address these information requirements.

Intensity of pollution

If the amount of pollutant released into the environment is known or can be estimated, what are the likely pollutant concentrations or faecal indicator or pathogen numbers at the pollutant source? This component, the intensity of pollution, is depicted in Figure 14.5, showing that pollutants from various human activities may be introduced into the environment at a wide range of concentrations, up to many orders of magnitude greater than those that would be acceptable for drinking-water or environmental standards. There is in fact a whole spectrum of occurrence, from industrial spills or traffic accidents in which a completely undiluted pollutant may be released at the surface, to the impact of agriculture and on-site sanitation, which is likely to produce pollutant concentrations in the same range as, or up to 10 or 100 times greater than guideline values. The former is a point source which directly impacts only a very small proportion of the total volume of recharge (Figure 14.5) and is potentially diluted in the larger water volume within the groundwater flow system, whereas the latter may impact a high proportion of the total recharge (Figure 14.5), depending on the land use distribution in the recharge area and the proportion of land that is cultivated and fertilized. The figure also shows that unsewered sanitation will affect an increasing proportion of the recharge as the density of installations increases from rural to suburban and urban areas.

While conceptually helpful, Figure 14.5 is again a simplification of what is actually a rather more complex situation. Firstly, most of the human activity boxes as pollution sources in the figure include several distinct pollutant groups (Table 14.1). As an example, source concentrations of agricultural pesticides in the soil at the time of application may be three or four orders of magnitude greater than guideline values, whereas nitrate would rarely be more than one order of magnitude greater. Similarly for on-site sanitation or leaky sewers, faecal indicator or pathogens at source may range up to several orders of magnitude per 100 ml, whereas initial nitrogen concentrations would be much less excessive. Initial concentrations are high, but subsequent attenuation processes are very active. Secondly, the horizontal scale extends to very high concentrations to accommodate high solubilized concentrations in groundwater adjacent to spillages of NAPL organic compounds that have relatively low WHO guideline values. For other sources, the range of concentrations may encompass the dissolved solute concentration in the very localized recharge water originating from the source. While generally indicating which activities affect only a small part of the catchment recharge but at high or very high concentrations, and which affect more of the recharge but at more modest concentrations, the figure may not be exactly comparing like with like. Potential polluting activities identified in the situation analysis should be located on Figure 14.5, and their pollution loading potential increases from bottom left to top right.

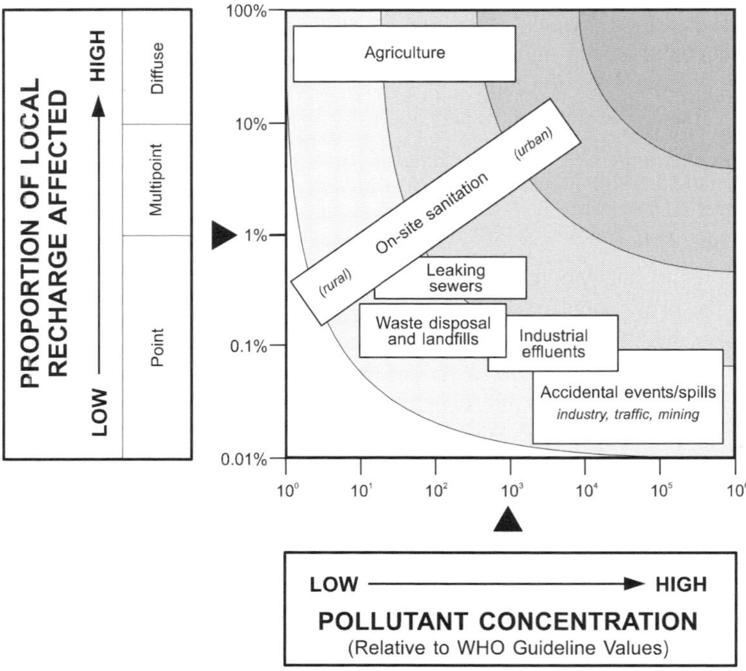

Figure 14.5. Characterization of intensity of pollution (adapted from Foster and Hirata, 1988)

For some diffuse sources of pollution, semi-quantitative estimates of the likely concentrations of persistent pollutants in local recharge may be possible, given many simplifying assumptions (Foster and Hirata, 1988). Approaches to doing this have been developed for chloride and nitrate from unsewered sanitation (Box 14.1) and for nitrate and pesticides from cultivated land (Foster *et al.*, 2002) (see Sections 14.5 and 14.6).

Box 14.1. Estimation of nitrogen loading from unsewered sanitation

Estimates of the possible concentrations of nitrate in the local groundwater recharge in areas of unsewered sanitation can be made for aerobic groundwater systems using the following equation (Foster and Hirata, 1988):

$$C = \frac{1000 \cdot a \cdot A \cdot f}{0.365 \cdot A \cdot U + 10 \cdot I} \qquad \text{(Eqn. 14.1)}$$

where: C is the concentration in mg/l of nitrate in the recharge (expressed as nitrate-nitrogen), a is the amount in kilograms of nitrogen excreted per person each year, A is the population density in persons/ha, f is the proportion of the excreted nitrogen leached to groundwater (reflects and integrates both the condition of latrines and the vulnerability of underground to nitrogen leaching), U is the non-consumptive portion of total water usage in l/person/day, i.e. the amount returned to the sanitation system, and I is the natural rate of infiltration for the area in mm/a.

Using this equation Figure 14.6, shows sensitivity of nitrate concentrations in recharge to variations in I, u and f, and indicates that many urban and suburban settings are capable of producing troublesome nitrate concentrations in the underlying groundwater.

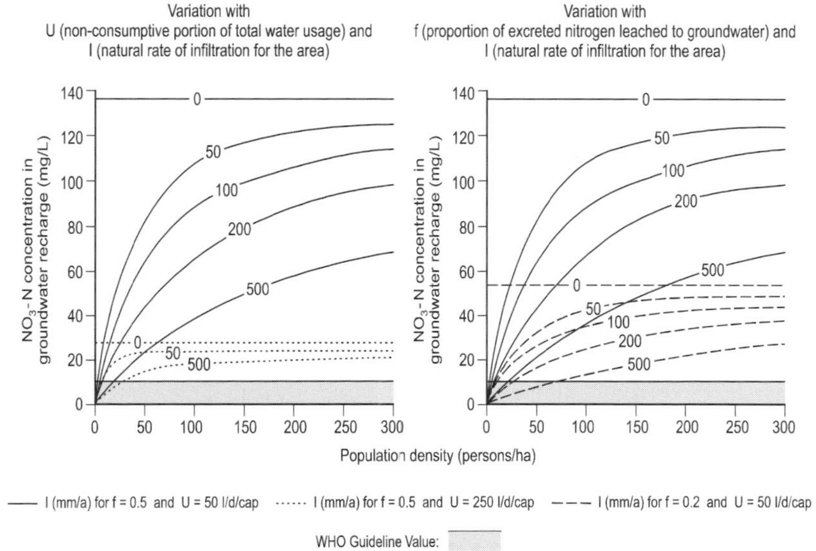

Figure 14.6. Estimation of the potential nitrate-nitrogen load in groundwater recharge in areas of unsewered sanitation (adapted from Foster and Hirata, 1988)

Overall population density and the proportion using unsewered sanitation comes from basic demographic information obtained using the checklist in Chapter 10, and the regional rate of natural infiltration (a proportion of the total rainfall) from the checklist at the end of Chapter 8. The amount of nitrogen produced per person (5 kg) in excreta each year is known from the literature (ARGOSS, 2001). Greatest uncertainty surrounds the proportion of the excreted nitrogen that will be oxidized and leached in the local groundwater recharge. A range of 0.2 to 0.6 is generally considered to be possible in shallow aerobic aquifers (Kimmel, 1984), and the actual proportion depends on the type and condition of installation, the per capita water use, the amount of volatile losses from the nitrogen compounds, the amount of nitrogen removed in cleaning and the geological setting and hydrochemical conditions. In some karstic limestone aquifers, almost all of the nitrogen deposited in sanitation systems may be oxidized and leached. The application of this approach is described in the Barbados case study (Section 14.5).

For high population densities in urban and suburban areas, local infiltration and recharge could be decreased by reductions in permeability in some areas due to surface sealing by built-up areas, and increased in others where collected urban drainage is disposed of to the subsurface, and also increased by leakage from water mains. In mixed areas of sewered and on-site sanitation, leakage of poor quality water from the sewers may further complicate the estimation.

For most point sources of pollution, the best than can usually be achieved is a relative pollution potential ranking based on the class of pollutants likely to be involved and the possible hydraulic loading, given the major uncertainties about the concentrations of pollutants in industrial effluents, mine waste and landfill leachate and their precise mode of disposition in the subsurface.

Duration of pollution

The final consideration (Figure 14.2) is how long the pollution has been going on for, or is likely to continue. Figure 14.7 provides an indication of the time over which the pollutant load is applied and the likelihood that pollution loading will occur. Thus an accident, spillage or catastrophic leak from a damaged tank may be of very short duration (Figure 14.7) and not penetrate into the subsurface if emergency action to contain the pollutant is taken quickly. If emergency action to deal with accidental pollution is not taken quickly, then solid or liquid pollutants may remain where they were released, either at or below the ground surface, and subsequently be subjected to leaching to groundwater. The slow solubilization of subsurface non-aqueous phase liquid (NAPL) sources, potentially lasting decades, even centuries, is a key concern in this regard. Delays in dealing with accidents thus tend to move such accidental pollution sources upwards and to the right in Figure 14.7.

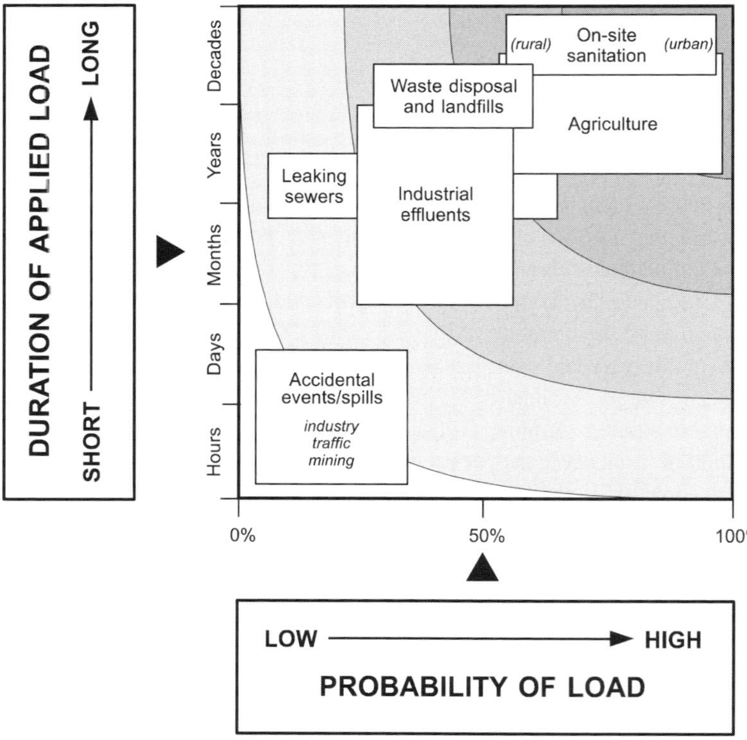

Figure 14.7. Characterization of the duration of contaminant load (adapted from Foster and Hirata, 1988)

Diffuse sources of pollution may also persist for many years or decades and, as in the case of the Perth study (Section 14.6), increase significantly with time as the city has grown and developed. These may be highly likely to cause a pollutant load (Figure 14.7), the magnitude of which will be determined by the characteristics of the other components described above. Thus as for the other components, the pollution potential increases from bottom left to top right.

14.3 OUTCOME OF ASSESSING POLLUTION POTENTIAL

The outcome of assessing the pollution loading potential would be a list of the pollutants expected to reach the aquifer in combination with a semi-quantitative assessment of their respective concentration levels. The process shown in Figure 14.2 provides the information for positioning activities and pollutants on the Y-axis of Figure 14.1. As indicated in Figure 14.1, these then interact in varying and complex ways with the different intrinsic factors used to describe aquifer vulnerability in Chapter 8, and this interaction determines the level of groundwater pollution potential.

The six components of Figure 14.2 each contribute to the overall assessment of pollution loading potential on the right hand side of the figure. The interaction between these components is, however, complex. As an example, if the water table is very close to the ground surface, then unsewered sanitation using pit latrines or septic tanks may enter directly into groundwater, with greatly enhanced likelihood of pollution. Similarly, discharge of highway drainage to soakaways may produce high hydraulic loadings and direct connection to the groundwater, with consequent high pollution potential. This complexity of interaction means that it is conceptually unwise to try to combine them into a single index of overall subsurface pollutant load, and technically difficult to do so. Even where the authors have combined them into a five class qualitative rating scheme (Zaporozec, 2002), they have depicted the intermediate steps in the process so that the dominant factors can be identified. Combined indices can have the result of producing a similar overall ranking for activities and subsurface conditions that are very different, but for which one dominant factor has been largely responsible for the outcome. This is also a concern for vulnerability depiction, and one of the reasons for the debate mentioned in Chapter 8 between those who define aquifer vulnerability in relation to a universal pollutant, and those who would prefer to define vulnerability separately for various classes of pollutant. Combined indices may also mask the components for which control measures would be most effective in reducing pollution potential, i.e. replacing a highly toxic and persistent chemical in an industrial process with a less environmentally threatening compound, or changing the mode of disposition to lessen the potential for pollution to occur.

A further complicating factor is that pollution potential will itself change with time, as human activities at the ground surface change. This is particularly important in the urban and suburban areas illustrated in the two case studies, and it is important to have at least a qualitative indication of both historical and future changes in likely pollution sources. These can affect all of the groups of human activities described in Chapters 9-13; increases in fertilizer use, development of irrigation, new pesticides, replacement of on-site sanitation with sewerage systems, developing or declining industries, changes in industrial processes, effluent treatment and disposal, closure of mines. Stricter environmental legislation may require responses that in turn also reduce potential pollutant loads. Surveys or inventories to assess the situation and provide the answers to the questions in the checklists are likely to need regular reassessment to confirm their continuing validity, and updating of information where there are major changes.

Changes in pollution loading can lead to major changes in groundwater quality. As an example, long-term increases in the application of nitrogen fertilizers to crops increase the leaching losses, leading to accumulation of nitrogen in the soil and unsaturated zone, and produce widespread upward trends in groundwater nitrate concentrations in many regions of the world (Chapter 9). Abating or reversing trends is often the objective of legislative control, and the withdrawal of the herbicide atrazine for non-agricultural weed control in the United Kingdom in 1992 has produced a reduction in concentrations in some of the public supply sources drawing groundwater polluted by this activity. However, responses of observed groundwater quality to changes in pollutant load are often delayed because of the slow movement of water and pollutants along the groundwater pathway. Further, in the case of removal or reduction of a pollution source at the ground surface, there may be significant amounts of pollutant accumulated in the unsaturated zone en route to groundwater. Pollutant concentrations in groundwater may continue to rise long after the source has been removed, and reversal of the trend may not occur for many years. This situation is seen at other groundwater supply sources affected by atrazine pollution from non-agricultural usage.

Where specific pollutants can originate from more than one major human activity, determining their origin is desirable, otherwise protection and control measures may be directed at the wrong source and hence be apparently ineffective. The most important of these are probably chloride and nitrate, which can be derived from leaking sewers, landfills, unsewered sanitation, livestock farming and agricultural fertilizers. High chloride concentrations may also indicate saline intrusion in coastal areas or the use of salt for road de-icing in cold climates. These pollutants can thus be indicators of impact from both rural and urban activities. Nitrate in particular can be problematic, as unsewered sanitation and agriculture often occur in close proximity. This situation occurs in both case studies and is common in many locations. Because of its importance for drinking-water quality, the nitrate needs to be traced back to its source so that control measures can be correctly targeted. A method that has been successfully applied is to use the distinctive isotopic signatures of nitrogen from animal and human excreta and from inorganic fertilizers to characterize the nitrate observed in the groundwater (Heaton, 1986; Aravena *et al.*, 1993; Exner and Spalding, 1994; Rivers *et al*, 1996), and hence its origin. The distinctions are not, however, always unambiguous as denitrification can also modify the isotopic signature of the nitrate in groundwater. Alternatively, trace elements associated with high groundwater nitrate concentrations, such as zinc and boron, or pharmaceuticals, may be indicative of a sewage rather than an agricultural source (Lerner and Barrett, 1996).

Where an observed or anticipated pollutant may have originated from numerous small sources as, for example, in a large industrialized city, especially one with a long and complex industrial history, then it is likely to be technically difficult and unrealistically expensive to determine the precise origins, locations and characteristics of pollution sources (Rivett *et al.*, 1990; Ford and Tellam, 1994). In such circumstances efforts are better directed at protection and control of all potentially hazardous sources rather than trying to prove the precise origins of the pollution.

14.4 USING GROUNDWATER QUALITY MONITORING TO SUPPORT THE ASSESSMENT

A qualitative categorization of pollutant loading is an essential element of assessing pollution potential and ultimately the urgency of management responses to protect public health. Evidence of pollution from any existing groundwater quality data is highly valuable to support, confirm or validate the assessment of pollution potential, and where such data are available, they should always be used. In Perth, Australia, (Section 14.6) groundwater quality monitoring data have existed for many years, and in Barbados (Section 14.5) groundwater quality monitoring was established to complement the assessment of pollution potential. The role of groundwater quality data in relation to other sources of information is discussed in more detail in Chapter 6.

In some circumstances, however, groundwater quality data become highly significant or even essential for risk assessment and management decisions, i.e. where assessment of pollution potential proves to be difficult or inconclusive. This could arise if vulnerability is uncertain due to a lack of essential components of the data specified in Chapter 8, or if pollutant loading is very difficult to assess because there are many, small, dispersed or superimposed sources whose types and amounts of pollutants are unknown. The latter situation will often be encountered in urban and periurban areas in developing countries, where industrial and commercial activities are widely dispersed and individually small, informal or unregistered and extremely difficult to assess, as was the experience in the Barbados case study summarized below.

If there are no groundwater quality data and an inventory of pollution sources proves too difficult or inconclusive, then some selective groundwater quality sampling and analysis from existing abstraction sources can help to provide a rapid assessment of pollution potential. If such a preliminary reconnaissance survey shows evidence of serious pollution, then human resources will need to be made available to start characterizing the main pollutant sources. There is in fact a close link between assessing pollution potential and monitoring, as preliminary surveys of groundwater quality and pollution sources are both important to assist parameter selection in the establishment of long-term routine groundwater quality monitoring programmes (Chapman, 1996; UNECE, 2000).

14.5 THE BARBADOS CASE STUDY

The small but densely populated Caribbean island of Barbados is almost totally dependent on groundwater for public water supply to the resident population and large numbers of visiting tourists. The groundwater resources of such island communities are often limited and of high value, the need for protection is readily apparent and degradation of the groundwater quality would have serious implications, as alternative supply options are limited and/or very costly. Although the study was undertaken from 1987-1992, it can still be used to illustrate both the assessment of pollution potential outlined in this chapter, and the approaches to and results of the assessment of information needs in Section II.

Existing groundwater protection measures and the reasons for the study

The vulnerability of the coral limestone aquifer of Barbados has long been recognized. To protect the bacteriological and chemical quality of groundwater used for public supply, the Barbados government established a policy of Development Control Zoning around existing and proposed public supply sources in 1963. Five zones were delineated (Figure 14.8) based on a simplified estimation of pollutant travel time through the aquifer. A travel time of 300 days for Zone 1 was selected to be significantly greater than the subsurface survival time of enteric bacteria, and a 600-day travel time was selected for Zone 2. Controls on potential pollution sources such as soakaway pits and septic tanks for domestic and industrial wastewater, fuel storage and industrial development are imposed within the zones (Table 14.2). In 1963, this was an important and farsighted piece of legislation, representing one of the earliest examples of a groundwater protection policy. However, there had been little or no groundwater quality monitoring since its introduction from which the effectiveness of the zoning could be evaluated, and this was a principal objective of the study.

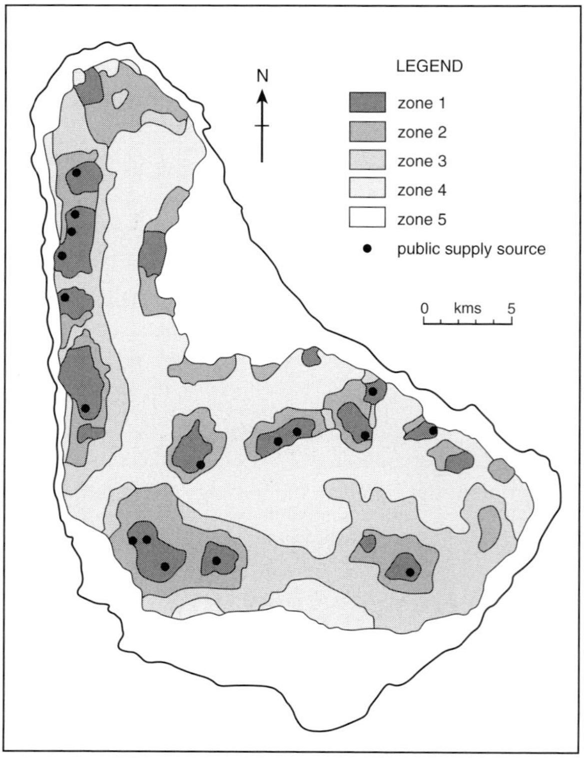

Figure 14.8. Control zones in Barbados (Chilton *et al.*, 2000)

Table 14.2. Principal features of development control zone policy (modified from Chilton *et al.*, 1991)

Zone	Definition of outer boundary	Maximum depth of soakaways	Domestic controls	Industrial controls
1	300 day travel time	None allowed	No new housing or water connections. No changes to existing wastewater disposal except when Water Authority secures improvements	No new industrial development
2	600 day travel time	6.5 m	Septic tank of approved design, discharge to soakaway pits. Separate soakaway pits for toilet effluent and other domestic wastewater. No storm run-off to sewage soakaway pits. No new petrol or fuel oil tanks	All liquid industrial wastes to be dealt with as specified by Water Authority. Maximum soakaway pit depths as for domestic waste (column 3)
3	5-6 year travel time	13 m	As above for domestic wastewater. Petrol or fuel oil tanks of approved leak proof design	
4	All high land	No limit	No restrictions on domestic wastewater disposal. Petrol or fuel oil tanks of approved leak proof design	
5	Coastline	No limit	No restrictions on domestic wastewater disposal. Siting of new fuel storage tanks subject to approval of Water Authority	

Drinking-water Supply

Barbados covers about 430 km^2 and its population is stable at about 250 000. The overall population density is 6/ha, but ranges from 30/ha in the urban southern and western coastal areas to 4/ha in the rural areas (Chilton *et al.,* 1991). At the time of the study, seventeen large abstraction wells operated by the Barbados Water Authority supplied about 115 ml/d and more than 95 per cent of the population were connected to mains water supply. Only a small proportion, mostly along the southern coastal area, had mains sewerage, although the development and extent of mains sewerage has increased since the time of the study. There were a few private wells, but many are no longer in regular use and the remainder were used for irrigation. The small number of public supply wells helped to facilitate the drinking-water catchment protection policy outlined above.

Groundwater conditions

The coral limestone forms a highly productive aquifer about 100 m thick, which contains a freshwater lens up to about 25 m thick, but thinning to 3 m close to the coast. The permeable nature of the aquifer is demonstrated by the almost total absence of surface drainage and the presence of karstic caves. As is typical for such coral limestones, soils are thin and were expected to provide little protective cover. Perusal of soils data and discussion with local agriculturalists indicated, however, that because of the frequent volcanic activity in the Caribbean region, soils are better developed, thicker and with more clay than might have been anticipated. As a consequence, the whole surface of the limestone aquifer was considered to have high, but not extreme, vulnerability to pollution (Figure 14.1). This somewhat simplified the assessment of pollution potential, as there was no requirement for defining and mapping vulnerability within the study, which therefore concentrated on the Y-axis of Figure 14.1 and the questions in Figure 14.2.

Annual rainfall varies from 1200 mm/yr at the coast to 2200 mm/yr in the interior, providing an equivalent range of some 250-650 mm/yr of recharge. The high annual recharge is roughly equal to the total groundwater storage of this relatively small aquifer. This is an uncommon feature, and implies short residence times and rapid groundwater renewal. Given the lack of surface streams, defining the groundwater catchments was not easy, and the imprecise boundaries followed the rather poorly defined surface watersheds and the buried topography of the coral limestone aquifer.

Approach to the assessment of pollution potential

The study comprised assessments of likely pollution sources in the Belle and Hampton drinking-water catchments, which together provided 90 per cent of the public water supplies of the island, and cover 55 and 67 km^2 respectively. The work was undertaken primarily by the Barbados Ministry of Health's EED, with technical support from the BGS funded by DFID and the Caribbean Programme Office of the Pan American Health Organization. The results of the assessment for Barbados are summarized in Table 14.3 by giving the qualitative and semi-quantitative responses to the questions in Figure 14.2.

In such a small and intensively developed island, pollution threats could be expected from urban, industrial and agricultural activities (Table 14.1). That these three were likely to be important enough to warrant detailed investigation was apparent from the first reconnaissance drive through the catchments, which indicated the range of urban, suburban and rural population densities, the dominance of sugarcane cultivation supplemented by horticulture and the wide variety and distribution of small-scale industries. Using Table 14.1, pathogens, nitrate from fertilizer and sanitation, agricultural pesticides and salinity were readily identified as potential pollutants requiring further assessment incorporating the components outlined in Figure 14.2.

To evaluate the likely impact of human habitations, information about population distribution, sanitation coverage and types was obtained by the Environmental Engineering Division (EED) project team from existing census data and from the records of the Public Health Inspectorate. For industry, an initial survey used the yellow pages business section of the local telephone directory to identify industrial and commercial premises in the catchments. These were each visited by the EED team, using a questionnaire to obtain information about the industrial chemicals and processes used and the methods of waste disposal. A second detailed survey of the farms and estates in both catchments also used questionnaires and follow-up site visits, to determine cultivation practices, cropping regimes, fertilizer applications and pesticide usage. Support in the design and interpretation of the agricultural survey was provided by staff of the Agriculture Department and of the Government Analytical Laboratory. For both the industrial and agricultural surveys, many additional sources of information – other government departments, universities and the National Oil Company, for example, were consulted, emphasizing that even in relatively small catchments such as these, complex land use and human activity means that the inputs of many technical disciplines are needed and a wide range of institutions are likely to have useful information. Being relatively small catchments, there were no problems related to differences between hydrological and administrative boundaries, and being a small island community, inter-institutional awareness, knowledge and cooperation were fortunately good.

Table 14.3. Assessment of pollutant loading potential for Barbados

Activity (Table 14.1)	**Pollutant (Table 14.1)**	**Component of pollutant loading**						**Aquifer vulnerability (Chapter 8)**	**Pollution potential (Figure 14.1)**	**Observed concentrations in groundwater**
		Pollutant mobility and persistence (Figure 14.3)	Mode of disposition (Figure 14.4)	Quantity of pollutant (Section II)	Intensity of pollution (Figure 14.5)	Duration of application (Figure 14.7)	Overall outcome			
Agriculture: *cultivation with agrochemicals*	Nitrate	High to extreme, mobile and persistent in aerobic limestone	Negligible to low, but with some increased loading by supplementary irrigation of vegetables	Fertilizer applications to sugarcane 130 kg N/ha/yr, but leaching losses known only from literature; horticulture variable, but probably higher	Low to moderate, estimated nitrate concentrations of 1-3 times WHO GV in recharge to 35-50% of catchment land	High to extreme, sugarcane long-established and horticulture increasing	Moderate to high, horticulture needed further checking	High	High	4-8 mg NO3-N/l, but after dilution with recharge from non-cultivated land
	Pesticides	Low to moderate based on properties of com-pounds used (Chapter 4)	Negligible to low (as above)	Herbicide applications of up to 2 kg/ha obtained from surveys; insecticide applications low and selectively targeted	Low to moderate, concentrations of 10-100 times WHO GV, depending on proportion leached	High overall, but individual compounds used for varying times	Moderate to high, for mobile, persistent, long-used, e.g. atrazine	High	Mode-rate to high, depending on com-pound	Atrazine always present at low concentrations and up to 3 µg/l

Activity (Table 14.1)	Pollutant (Table 14.1)	Component of pollutant loading						Aquifer vulnerability (Chapter 8)	Pollution potential (Figure 14.1)	Observed concentrations in groundwater
		Pollutant mobility and persistence (Figure 14.3)	Mode of disposition (Figure 14.4)	Quantity of pollutant (Section II)	Intensity of pollution (Figure 14.5)	Duration of application (Figure 14.7)	Overall outcome			
On-site sanitation	Nitrate	High to extreme	Low (rural) to moderate (urban), but use of soakaways increases potential	5 kg per person,4 persons/ha (rural) and 30 persons/ha (urban)	Negligible to low, concentrations in local recharge could be up to 5 times WHO GV	High to extreme	Low (rural) to moderate (urban)	High	Low to Moderate	4-8 NO_3-N/l in rural areas, up to 15 NO_3-N/l in densest urban areas
	Pathogens	Low to moderate	Low to moderate (as above)	Unknown	Low to moderate	High to extreme	Moderate to high	High	Moderate	Up to 50% positives *E. coli* and up to 5 x 10^2 max./100 ml
Industry: *land and well disposal of effluent*	Solvents	Low to moderate	Low, but becoming high with disposal in old wells	Small amounts used by small industries, but unable to quantify	Low to moderate, but could not estimate likely concentrations	Low to moderate, usage not long established or continuous	Low	High	Low	None found in limited sampling
Solid waste disposal	Metals	Low	Low, small shallow landfills	Mostly household waste with little industrial component	Probably low	Moderate, landfills; long-established, some now unused	Low	High	Low	Limited monitoring confirms low concentrations
Traffic: *accidental spills*	Aromatic hydrocarbons	Low to moderate	Moderate to high, roads drained to soakaways	Regular and frequent transport of unrefined oil by road from oilfield to port	Low to moderate	Low, well-established contingency to clear roads quickly	Low	High	Low	No monitoring

Potential pollution threats identified and evidence of impact

The survey results were compiled in spreadsheets and plotted in map form for both catchments, and the map of potential pollution threats for the Belle catchment is shown in Figure 14.9. Potential for groundwater pollution from all three major categories of activity were identified and pollution loadings estimated for unsewered sanitation (Box 14.1) and fertilizer usage. Thus the high density of unsewered sanitation in the urban part of the Belle catchment compared to the rural parts of both catchments presented a threat of nitrate and microbial pollution, which was confirmed by the results of the associated groundwater quality monitoring. Some industries, such as dry cleaning, paint distribution and photographic processing, were identified as using hazardous chemials and disposing of small volumes of untreated effluents directly into the coral limetone aquifer. The mode of disposition of these small amounts of industrial effluents into soakaways or old wells thus provided a notably high potential to pollute groundwater (Figure 14.4 and Table 14.3). However, they were few, widely dispersed and of very small scale, and no significant impact was detected on groundwater quality in the associated sampling programme. The most widespread and likely threats from industrial and commercial premises were presented by fuel stations and by small vehicle repair workshops.

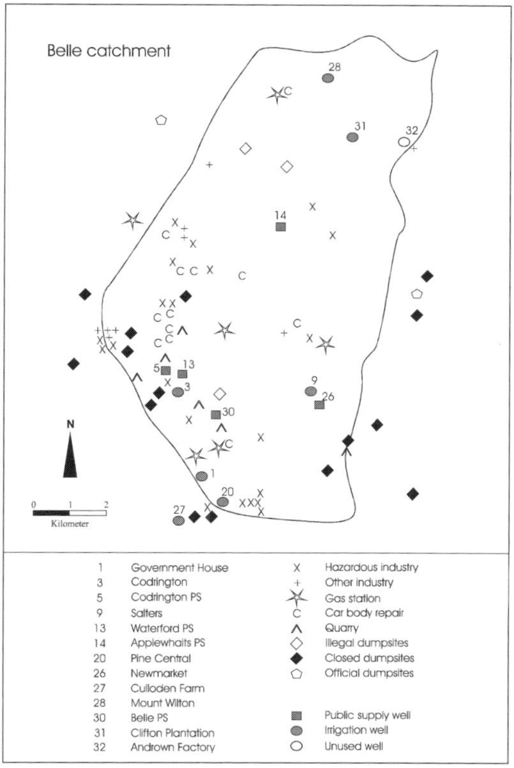

Figure 14.9. Potential pollution sources in the Belle catchment, Barbados (adapted from Chilton *et al.*, 2000)

The agricultural survey highlighted a move away from traditional cultivation of sugarcane to much more varied cropping, especially horticulture. Sugarcane is an efficient user of nitrogen fertilizer as it grows continuously, but vegetables and flowers have shorter growing seasons and are often less efficient users of soil nitrogen. They are also often grown in Barbados with supplementary irrigation, and for these two reasons higher nitrogen leaching may be expected (Figure 14.4). Groundwater nitrate concentrations of 4-8 mg NO_3-N/l were observed everywhere in the rural areas of both catchments, indicating universal but modest impact from agriculture, but there was no evidence of an overall increasing trend during the five years of the study or indeed when the results of the continuing groundwater quality monitoring programme were reviewed later by Chilton *et al.* (2000).

Estimates of nitrate concentrations in recharge reflecting the nitrogen loading from on-site sanitation were made as shown in Table 14.4, using the equation given in Box 14.1. The annual recharge is known to be higher over the hilly rural interior than at the more urbanized southern coastal belt, and per capita wastewater generation is assumed to be slightly lower in the rural areas that are, nevertheless, largely connected to the mains water system. Observed nitrate groundwater concentrations support these estimations, remaining in the range 4-8 mg NO_3-N/l in the whole of Hampton and the northern part of Belle and only rising above 10 mg NO_3-N/l in the southern, urban part of the Belle catchment with the superimposed nitrogen loading from the more dense on-site sanitation facilities (Chilton *et al.*, 1991; 2000). Sampling for faecal coliforms at the same wells supports this interpretation, with more frequent and higher counts broadly correlating with the higher nitrate concentrations.

Table 14.4. Estimated nitrate concentration in recharge affected by on-site sanitation

Components of mass balance calculation: $C = \dfrac{1000 \cdot a \cdot A \cdot f}{0.365 \cdot A \cdot U + 10 \cdot I}$	**Urban (Belle)**	**Rural (Belle and Hampton)**
a Nitrogen load (kg N/person/year)	5	5
A Population density (persons/ha)	31	4
F Proportion of nitrogen leached	0.6	0.6
U Per capita wastewater generation (l/d)	250	200
I Annual recharge (mm/yr)	300	500
C Concentration in recharge (mg/l NO_3-N)	16	2-3
Range of observed nitrate concentrations (mg/l NO_3-N)	4-10	4-8

Horticulture has a much greater variety of pests associated with it than sugarcane – and hence a wider range of herbicides and insecticides are used in their cultivation. A combination of pesticide usage data and published physicochemical properties – solubility in water, persistence as defined by soil half-lives and mobility from partition coefficients (Chapter 4) – was used to estimate susceptibility to leaching to groundwater and hence to select pesticides for monitoring. Pesticide sampling in the monitoring programme detected the almost ubiquitous presence of atrazine, but at low concentrations. This is one of the herbicides most widely used in sugarcane cultivation, but there was little evidence of the main insecticides used. This probably resulted from the wider area (of sugarcane) to which atrazine is applied, compared to the more limited areas of horticulture, and from the mode of application. Atrazine is a soil-applied herbicide, whereas most of the insecticides are foliar (applied to the plants themselves) and the former is thus more likely to be leached to groundwater.

During the assessment, additional potential pollution sources in the form of poorly maintained oil wells, illegal dumping of solid waste and highway drainage became apparent from visual inspection of the catchments, conversations with residents and organizations. In particular, transport by road tanker of crude oil from the oilfield to the port terminal presented a hazard of traffic accident, spillage and drainage directly into the coral limestone aquifer.

Outcome of the assessment of pollution potential

The assessment successfully identified the main potential pollution sources from urban domestic waste disposal and agricultural activities (Table 14.3). As the whole of the aquifer outcrop is considered to have high vulnerability, the outcome of the pollution loading assessment translates to a position on the Y-axis of Figure 14.1 and thence a ranking of groundwater pollution potential (Table 14.3). The associated monitoring programme established in the study has confirmed that elevated concentrations of nitrate and pesticides do result. Although they occasionally exceed guideline concentrations in private wells, they are lower in public supply wells. The Development Control Zone policy appeared to have been successful in protecting the island's groundwater, which had remained of good quality in spite of the dense population and rapid development. It was recommended from the results of the study that the Development Control Zone policy should not be relaxed, despite pressure from housing and industrial developers, and that the assessment should be extended to the remaining catchments along the island's west coast. It was, however, difficult to evaluate the bacteriological impact from simple monitoring because the public supply sources were equipped with in-well chlorination that prevented collection of pre-chlorination samples. More costly purpose-built sampling boreholes within the control zones would be needed for this.

The Barbados case study can also be used to illustrate key features of the situation analysis from the chapters and associated checklists in Section II (Table 14.5).

While the development control zone policy had clearly served Barbados well, the study identified the main threats, and highlighted the need for continuing vigilance to protect the island's groundwater. Management priorities identified in the study are shown in Table 14.5. While the resident population is unlikely to grow, water demand probably will, partly to meet the growing tourism industry, and this may bring land use changes, such as more golf courses and increasing local demand for horticultural products. Since the study, waterborne sewage has been extended to significant parts of the centre of Bridgetown and the southern coast. This helps protect the coastal zone by reducing pollutant loading to the groundwater discharging to the sea. Waterborne sewage may also need to be targeted at areas where housing is encroaching into the development control zones. This means that, to keep pace with changing circumstances, the situation analysis and pollution potential assessment should be updated, probably on a five-yearly basis. A further recommendation was that the assessment of pollution potential should be extended to the catchments of the wells along the west coast (Figure 14.8).

Table 14.5. Key features of the situation analysis for the Barbados catchments

Component	Key features	Specific difficulties
Collecting information (Chapter 6)	Relatively few institutional stakeholders, and good communication and cooperation between them	EED staff had many other tasks: difficult to find time for survey work
Socioeconomic setting (Chapter 7)	Dense population, relatively high economic status and relatively high environmental awareness	Increasing water demand and high per capita consumption
Hydrogeology, vulnerability and susceptibility to abstraction (Chapter 8)	Whole aquifer/catchment vulnerable: no need for vulnerability mapping Small island, thin freshwater lens, possibility of saline intrusion Large public supply wells with protection zones	Defining catchment boundaries in absence of surface waters
Agriculture (Chapter 9)	Trend from sugarcane to horticulture – increasing range of pesticides	None, good data on crops, fertilizer and pesticide use
Human excreta and sanitation (Chapter 10)	Large difference in housing density between rural and urban areas, significant loading from the latter Easy to distinguish the few sewered areas at the coast from the larger and separate unsewered areas	None, good data for population density and water usage
Industry (Chapter 11)	Small-scale and widely dispersed, mostly commercial and light industry Little usage of potentially polluting materials, but sometimes with poor effluent handling	Easy to map industrial and commercial premises and obtain information on types of pollutant, but very difficult to obtain effluent types, amounts and disposal methods
Waste disposal (Chapter 12)	Small landfills for domestic waste in old quarries	Some former landfills with unknown contents, but outside the study catchments
Traffic (Chapter 13)	Pollution potential evident, particularly during road transport of oil to the port terminal	
Existing water quality data (Chapter 14)	Almost none: groundwater quality monitoring established as part of the study	Good local analytical facilities but some staff constraints for sampling due to other EED tasks
Establishing groundwater management priorities (Chapter 15)	Recommended that controls should not be relaxed despite development pressures Continued assessment of agricultural activity as crops are changing: potential for increase of nitrate and/or pesticide pollution from horticulture, some with irrigation Better implementation of good practice for handling, treating and disposing of industrial effluents Better implementation of good practice for oil transport and traffic accident responses Extend sewage collection in the most densely populated areas Abstraction well controlled to prevent saline intrusion but maintain monitoring of salinity	

14.6 THE PERTH CASE STUDY

Perth is the only large urban centre in Western Australia and is dependent on groundwater for about 70 per cent of all water use, and about 50 per cent of its public supply. Groundwater is pumped from both a regional unconfined aquifer and deep confined aquifers. The latter are well protected and recent initiatives to urbanize their recharge area have been successfully fended off, and the land has remained in government ownership, being used only for a limited amount of forestry. To restrict future development initiatives, discussions on establishing protection zones have been initiated.

The shallow aquifer, however, has been affected by contamination. As the city has expanded, the urban area has encroached on to land in the recharge area which was previously under rural land use or completely undeveloped. The dramatic land use changes which have accompanied the rapid growth of Perth are shown in Figure 14.10. As this is a common situation elsewhere in the world where cities are expanding rapidly, the experience gained in Perth is considered very appropriate as a case study to illustrate how the risks of contamination of groundwater resources important for water supplies can be assessed using the principles outlined in this chapter.

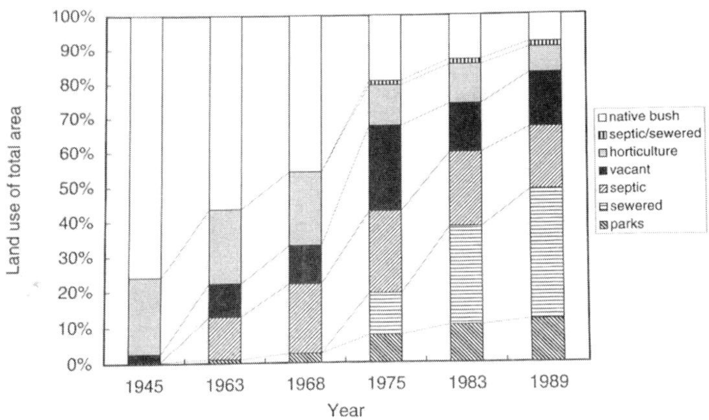

Figure 14.10. Development of land use in the Perth area (adapted from Barber *et al.*, 1996)

Socioeconomic setting

Although it has a population of only 1.3 million, Perth's metropolitan region covers an area equivalent to many large European cities (Figure 14.11). This is because Perth is very much a garden city, with most of the population living in detached houses with large gardens of lawn and exotic shrubs on 500 to 1000 m^2 blocks. This makes the overall population density (230/km^2) lower than the other major cities in Australia and much lower than equivalent cites elsewhere in the world.

Hydrogeological conditions

Fortunately, Perth overlies a very large fresh groundwater resource that forms an important component of the city's water supply and maintains ecosystems around environmentally significant lakes and wetlands. Groundwater occurs in an unconfined aquifer throughout the region, and in several confined aquifers. The groundwater in storage represents some 500 years of current annual abstraction. Boreholes of up to 1000 m depth supply water with a salinity of only 180 mg/l total dissolved solids (TDS).

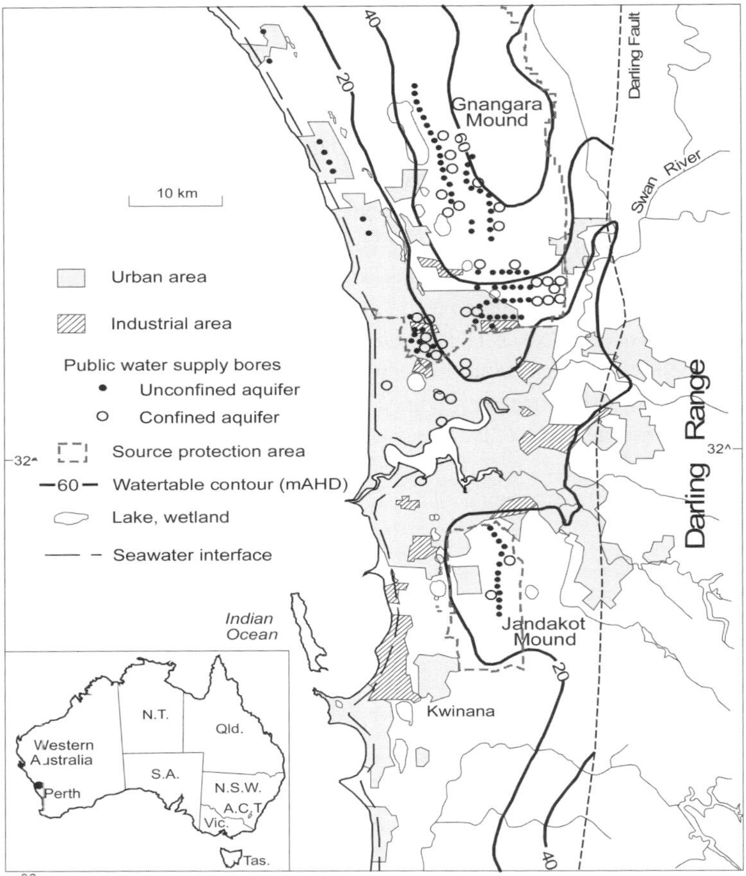

Figure 14.11. Location map of Perth showing major groundwater features

However, the shallow groundwater beneath urban areas is highly vulnerable to pollution owing to the sandy soil and the generally shallow water table, and in some areas pollution has restricted groundwater use and has had an adverse impact on wetlands. Further, the growth of the urban area has overtaken wellfields which were previously located in areas of rural land use, and has compromised water quality.

Water supply situation

The gardens in Perth require watering for about six months of the year because of the region's dry Mediterranean type climate, and are responsible for up to 80 per cent of domestic water use. The shallow water table and unconsolidated sand aquifer mean that groundwater is easily available to most private properties. As a consequence, beside the public water supply wells (mapped in Figure 14.11), there are about 135 000 privately owned small diameter boreholes or dug wells with spear points, which are used for garden irrigation in Perth. This has greatly reduced the demand from public water supply schemes. The average in-house usage of water of drinking-water quality in Perth is about 120 l/capita/day. Households that do not have access to domestic bores will typically use another 110 l/capita/day of water of drinking-water quality for watering gardens. The application rate of water from domestic bores during summer is of the order of 12 l/day/m^2/house.

Impact of urban development on groundwater quality

From Table 14.1, the principal hazards likely to cause groundwater pollution are commercial and industrial point sources within the industrial areas of Perth (Figure 14.11), and widespread but low levels of diffuse pollution from fertilizer use on gardens and from septic tank leachate. The potential point sources of pollution include a range of light industries, petrol service stations and pest control depots (Appleyard, 1995a), which have either accidentally or deliberately disposed of wastes, and about 100 former waste disposal sites (Hirschberg, 1993a; 1993b). Pollution surveys suggest there are about 2000 known or suspected sources of groundwater pollution within the whole region, and contaminant plumes from 100-1000 m or more in length have extended from some of these sites through residential areas where private boreholes are used (Benker *et al.*, 1996). Water quality surveys have detected a wide range of contaminants in shallow private boreholes near many of these sites, commonly at levels that exceed national drinking-water criteria. Although this groundwater is generally not used for drinking, other routes of exposure, such as droplet inhalation or eating irrigated produce have not been thoroughly assessed. Groundwater contamination in at least one private borehole was sufficiently severe to be toxic on prolonged skin contact and to kill plants irrigated with the water (Appleyard, 1995a).

There is also widespread leaching of nitrate from fertilizer use on gardens, and of nitrate, ammonia and bacteria from septic tanks. Estimates of pollutant loading suggest that about 1600 tonnes of nitrogen and 480 tonnes of phosphorus are applied annually to lawn areas in Perth (Sharma *et al.*, 1996). Although much of the phosphorus is bound up in soil profiles, up to 80 per cent of the applied nitrogen may leach to the water table (Sharma *et al.*, 1996). About 160 tonnes of nitrogen are discharged by groundwater to the Swan River each year, and up to 10 tonnes of nitrogen for each kilometre of coastline is discharging annually into the marine environment (Appleyard and Powell, 1998).

Nitrate concentrations in groundwater beneath Perth generally exceed 1 mg/l, and are often greater than 10 mg/l NO_3-N. The nitrate originates from these sources, and concentrations generally increase with the age of urban development (Appleyard, 1995b; Barber *et al.*, 1996). Gerritse *et al.* (1990) estimated that between 80 and 260 kg/ha of nitrogen fertilizer is applied in urban areas of Perth each year, which is comparable to many agricultural application rates. Using these rates as a basis, from their estimate of nitrate loading they suggested that concentrations in groundwater in these areas should be at least 40 mg/l NO_3-N. Observed concentrations are generally much lower, suggesting that significant denitrification must be taking place in the aquifer. The broad scatter of nitrate concentrations in groundwater in urban areas of Perth indicates that denitrification is not occurring uniformly, and/or that nitrogen inputs are not uniformly distributed.

Fertilizer use is particularly high in the horticultural areas located on the fringes of the urban region, as more than one crop is grown per year. Typically, 500-1500 kg/ha of nitrogen (mostly as poultry manure) are applied to crops each year (Lantzke, 1997), and this exceeds the capacity of plants to take up nitrogen by a factor of 4-7 (Pionke *et al.*, 1990). A combination of high intensity (Figure 14.5) and high potential (Figure 14.7) help to make horticulture an activity with high potential for polluting the underlying groundwater (Table 14.6). This reflects the situation in Barbados, where the rapid growth of horticulture was identified as a potential pollution hazard requiring further assessment (Table 14.4), and the presence of intensive horticulture should always be given careful attention in the assessment process outlined in this chapter. As a consequence in Perth excessive nitrate leaching occurs and nitrate concentrations up to 100 mg/l NO_3-N have been observed directly beneath horticultural areas.

Denitrification is favoured where aquifer redox potentials are less than about 300 mV, but there is some evidence that redox potentials increase in the older urban areas in Perth due to the sustained impact of urban recharge processes (Appleyard, 1995b). This may mean that current nitrate concentrations in groundwater beneath urban areas are not sustainable given current fertilizer usage, and that future nitrate concentrations beneath a large part of the Perth metropolitan area will exceed drinking-water guidelines.

A further source of nitrogen as well as faecal pathogen pollution of shallow groundwater beneath Perth is the use of septic tanks. Currently about 25 per cent of the urban area which was mainly developed in the 1950s and 1960s, is unsewered. There is a programme to replace septic tanks with sewer connections, but existing groundwater pollution from this source will take many years to dissipate.

Table 14.6 illustrates how the processes outlined in this chapter can be used to assess the relative hazards that particular pollutants in Perth's groundwater pose to its use as a source of drinking-water. As for Barbados, Table 14.6 provides qualitative and semi-quantitative responses to the questions in Figure 14.2. The assessment provides an indication of the potential magnitude of the loadings posed by a specific pollutant, and on how abundant it is likely to be in groundwater based on land use, aquifer vulnerability and measures of the rate at which the contaminant is discharged to groundwater.

Based on the approach outlined in this chapter, Table 14.6 indicates that the pollutants of most concern in shallow groundwater in Perth are pathogens discharged by septic tanks, high concentrations of nitrate in horticultural areas and benzene concentrations near petrol service stations. This assessment process can then be used to select management strategies to ensure that these pollutant sources do not affect the health of water consumers. This has been done in Perth by establishing drinking-water SPAs and wellhead protection zones where land use can be strictly controlled to minimize the risk of contamination. The risk of contamination from septic tanks is also being managed by progressively installing sewers in the few remaining areas of Perth which are currently unsewered, and the risks of contamination from service stations are being reduced with new codes of practice that require double lined underground storage tanks for fuels with intensive monitoring. A number of measures are also being implemented to reduce nitrate contamination by horticulture. These include training programs for farmers, changing land use in very sensitive areas and requiring farmers to manage their activities according to nutrient management plans.

Table 14.6. Assessment of pollutant loading potential for Perth (shallow aquifer only)

Activity (Table 14.1)	**Pollutant (Table 14.1)**	**Component of pollutant loading**						**Aquifer vulnerability (Chapter 8)**	**Pollution potential (Figure 14.1)**	**Observed concentrations in groundwater**
		Pollutant mobility and persistence (Figure 14.3)	Mode of disposition (Figure 14.4)	Quantity of pollutant (Section II)	Intensity of pollution (Figure 14.5)	Duration of application (Figure 14.7)	Overall outcome			
Gardening: *cultivation with agrochemicals*	Nitrate	High to extreme, mobile and persistent in shallow sandy aquifer	Negligible to low	1600 tonnes of nitrogen applied annually to lawn areas in Perth (80-260 kg/ha/a with high leaching losses)	Moderate, estimated nitrate concentrations in the range of WHO GV in groundwater and increasing	High, many months per year over decades, increases with age of urban development	Moderate to high	High, in consequence of sandy soils and shallow water table; proximity of activities to some wellfields	High	1-10 mg/l NO_3-N often less than estimated, attributed to denitrification in some areas, increase with age of local urban development
	Pesticides	Low to moderate, based on properties of compounds used (Chapter 4)	Negligible to low	Less than 50 tonnes annually at 0.1 kg/ha/a	Low, disperse except when residues poured down drains	Low, intermittent use in summer	Low		Low	Typically <1 μg/l
Horticulture	Nitrate	High to extreme, mobile and persistent in shallow sandy aquifer	Negligible to low	200-1000 kg/ha of nitrogen applied to multiple crops each year	High to extreme, application rates may exceed plant uptake 4-7-fold	High to extreme, all year	High to extreme	High, in consequence of sandy soils and shallow water table; proximity of activities to some wellfields	High to extreme	Up to 100 mg/l NO_3-N
	Pesticides	Low to moderate, based on properties of compounds used (Chapter 4)	Low	0.25-0.5 kg/ha/a	Low to moderate, most degrade in soil	High, used all year	Low to moderate, depending on proper-ties of pesticide		Low to moderate	Typically <10 μg/l

Activity (Table 14.1)	Pollutant (Table 14.1)	Component of pollutant loading						Aquifer vulnerability (Chapter 8)	Pollution potential (Figure 14.1)	Observed concentrations in groundwater
		Pollutant mobility and persistence (Figure 14.3)	Mode of disposition (Figure 14.4)	Quantity of pollutant (Section II)	Intensity of pollution (Figure 14.5)	Duration of application (Figure 14.7)	Overall outcome			
On-site sanitation: *septic tank leachate*	Nitrate	High to extreme	Low (rural) to moderate (urban), use of soakaways increases potential which decreases in urban areas with new sewers	5 kg per person, population density 2-3/ha, load 10-15 kg N/ha/a in unsewered areas (25% of total)	Negligible to low, estimated concentrations in local recharge less than WHO GV	High to extreme	Low to moderate	High, in consequence of sandy soils and shallow water table; proximity of activities to some wellfields	Low to moderate	3-25 mg/l NO_3-N in urban areas
	Pathogens	Low	As above	Generally low to moderate pathogen loads unless septic tanks poorly constructed and/or badly maintained	Low to moderate	High to extreme	Moderate		Moderate, decreasing in urban areas, progressive replacement of septic tanks with sewer connections	Faecal coliforms detected in 25% of boreholes assessed, public warnings to use water for gardens only and not for drinking
Light industry	Solvents, metals, cyanide	Low to moderate	Generally low, but locally high with disposal in wells or in soakaways	Small amounts of solvents used; historically most metal finishing wastes disposed to ground	Low	Low, intermittent usage	Low, mostly histori-cal pollution	High, in consequence of sandy soils and shallow water table; proximity of activities to some wellfields	Low	Limited sampling indicates generally less than solvents 1 μg/l, locally up to 2000 μg/l, toxic metals generally <10 μg/l.

Activity (Table 14.1)	**Pollutant (Table 14.1)**	**Component of pollutant loading**						**Aquifer vulnerability (Chapter 8)**	**Pollution potential (Figure 14.1)**	**Observed concentrations in groundwater**
		Pollutant mobility and persistence (Figure 14.3)	Mode of disposition (Figure 14.4)	Quantity of pollutant (Section II)	Intensity of pollution (Figure 14.5)	Duration of application (Figure 14.7)	Overall outcome			
Former waste disposal sites	Metals	Low	Low, small shallow landfills	Mostly household waste with little industrial component	Probably low	Moderate, landfills long-established, some now unused	Low		Low	Limited monitoring confirms low concentrations
Former pest control depots and historical regional pest control	Pesticides	Generally low but locally moderate due to disposal of wastes to ground	Very locally (within 300 m of disposal sites) moderate, but generally low	Intermittently applied; historical application of dieldrin to all house pads, fences, wooden poles, etc.	Generally low, diffuse application; high rates at very localized sites	Low, intermittent for modern pesticides, historical use of organochlorine pesticides	Low	High, in consequence of sandy soils and shallow water table; proximity of activities to some wellfields	Generally low, but isolated areas with moderate risk	Generally <1 µg/l, within 300 m of pest control depots generally <10 µg/l
Petrol service stations	Benzene, benzo(a) pyrene	Low to moderate	Moderate to high when water table is shallow due to under-ground leaks without evaporation	Continuous transport of refined hydro-carbons to petrol stations	Moderate to high, most commonly detected industrial contaminants in groundwater	Generally low, but high in zones up to 500 m in direction of flow for service stations	Moderate		Moderate, locally high	Benzene levels generally <1 µg/l, can exceed 100 µg/l up to 500 m in direction of groundwater flow from service stations

14.7 REFERENCES

Appleyard, S.J. and Powell, R.J. (1998) Preserving urban bushland and growing local plants to protect water resources. In *Proc. of 'Managing our Bushland' Conference*, pp. 72-75, Perth.

Appleyard, S.J. (1995a) Investigation of ground water contamination by fenamiphos and atrazine in a residential area: source and distribution of contamination. *Ground Water Monitoring and Remediation*, **15**, 110-113.

Appleyard, S.J. (1995b) The impact of urban development on recharge and groundwater quality in a coastal aquifer near Perth, Western Australia. *Hydrogeology J.*, **3**(2), 65-75.

Aravena, R., Evans, M.L. and Cherry, J.A. (1993) Stable isotopes of oxygen and nitrogen in source identification of nitrate from septic systems. *Ground Water*, **32**(2), 180-186.

ARGOSS, 2001. *Guidelines for assessing the risk to groundwater from on-site sanitation*. BGS Commissioned Report, CR/01/142, Keyworth.

Barber, C., Otto, C.J., Bates, L.E., and Taylor, K.J. (1996) Evaluation between land-use changes and groundwater quality in a water-supply catchment, using GIS technology: the Gwelup wellfield. *Hydrogeology J.*, **4**(1), 6-19.

Benker, E., Davis, G.B., Appleyard, S.J., Barry, D.A. and Power, T.R. (1996) Trichloroethene (TCE) contamination in an unconfined aquifer underlying a residential area of Perth, Western Australia. *Hydrogeology J.*, **4**(1), 20-29.

Chapman, D. (ed) (1996) *Water Quality Assessments: A guide to the use of biota, sediments and water in environmental monitoring*. 2nd edn, E & F Spon, London.

Chilton, P.J., Vlugman A.A. and Foster S.S.D. (1991) A groundwater pollution risk assessment for public water supply sources in Barbados. In *Tropical Hydrology and Caribbean Water Resources*, pp. 279-289, American Water Resources Association, Bethesda, Maryland.

Chilton, P.J. Stuart, M.E., Escolero, O., Marks, R.J., Gonzales, A. and Milne, C. J. (1998) Groundwater recharge and pollutant transport beneath wastewater irrigation: the case of Leon, Mexico. In *Groundwater Pollution, Aquifer Recharge and Pollution Vulnerability*, (ed. N.S. Robins), pp. 153-168, Geological Society of London Special Publication 130, London.

Chilton, P.J., Vlugman, A.A. and Headley, J. (2000) Groundwater quality monitoring tailored for pollution risk assessment in Barbados. In *International Workshop on Information for Sustainable Water Management, Monitoring Tailor Made III*, (eds. J.G. Timmerman *et al.*), pp. 297-306, Nunspeet.

Christensen, J.B. Jensen, D.L. and Christensen, T.H. (1996) Effect of dissolved organic carbon on the mobility of cadmium, nickel and zinc in leachate polluted groundwater. *Water Res.*, **30**, 3037-3039.

Conway, G.R. and Pretty, J.N. (1991) *Unwelcome Harvest: agriculture and pollution*. Earthscan Publications, London.

Exner, M.E. and Spalding, R.F. (1994) N15 identification of nonpoint sources of nitrate contamination beneath cropland in the Nebraska Panhandle: two case studies. *Appl. Geochem.*, **9**, 73-81.

Ford, M. and Tellam, J.H. (1994) Source, type and extent of inorganic contamination within the Birmingham aquifer system, UK. *J. Hydrology*, **156**, 101-135.

Foster, S.S.D. and Hirata, R. (1988) *Groundwater Pollution Risk Assessment: A methodology using available data*. WHO/PAHO-CEPIS Technical Manual, Lima.

Foster, S.S.D., Hirata, R., Gomes, D., D'Elia, M. and Paris, M. (2002) *Groundwater Quality Protection: A guide for water utilities, municipal authorities and environment agencies*. World Bank, Washington, DC.

Gerritse, R.G., Barber, C. and Adeney, J.A. (1990) *The Impact of Residential Urban Areas on Groundwater Quality, Swan Coastal Plain, Western Australia*. Commonwealth Scientific and Industrial Research Organisation, Division of Water Resources, Water Resources Series 3, Victoria.

Heaton, T.H.E. (1986) Isotopic studies of nitrogen pollution in the hydrosphere and atmosphere: a review. *Chemical Geol.*, **59**, 87-102.

Hem, J.D. (1985) *Study and Interpretation of the Chemical Characteristics of Natural Water*. Water Supply Paper 2254, 3rd edn, US Geological Survey, Washington, DC.

Hirschberg, K-J.B. (1993a) *Municipal waste disposal in Perth and its impact on groundwater quality*. Geological Survey of Western Australia, Report 34, pp. 97-109.

Hirschberg, K-J.B. (1993b) *The Location and Significance of Point Sources of Groundwater Contamination in the Perth Basin*. Geological Survey of Western Australia, Report 34, 37-46.

Kimmel, G.E. (1984) Nonpoint contamination of groundwater on Long Island, New York. *NRC Studies in Geophysics – Groundwater Contamination*, **9**, 120-126.

Lantzke, N. (1997) *Phosphorus and Nitrogen Loss From Horticulture on the Swan Coastal Plain*. Agriculture WA Miscellaneous Publication 16/97.

Lerner, D.N. and Barrett, M.H. (1996) Urban groundwater issues in the United Kingdom. *Hydrogeology J.*, **4**(1), 80-89.

Lyngkilde, J. and Christensen, T.H. (1992) Redox zones of a landfill leachate plume. *J. Contaminant Hydrology*, **10**, 273-289.

Morris, B.L., Ahmed, K.M. and Litvak, R.G. (2002) *Evolution of Developing City Groundwater Protection Policies: Stakeholder consultation case studies in Bangladesh and Kyrghystan*. IAH Congress, Mar Del Plata, Argentina.

Morris, B.L., Litvak, R.G. and Ahmed, K.M. (2001) Urban groundwater protection and management: lessons from two developing city case studies in Bangladesh and Kyrghystan. In *Proc. of NATO Advanced Science Workshop Series*, Baku Azerbaijan, June 2001.

Pionke, H.B., Shanna, M.L. and Hirschberg, K.J.B. (1990) Impact of irrigated horticulture on nitrate concentrations in groundwater. *Agri. Ecosystems and Environ.*, **32**, 119-132.

Rivers, C.N., Barrett, M.H., Hiscock, K.M., Dennis, P.F., Feast, N.A. and Lerner, D.N. (1996) Use of nitrogen isotopes to identify nitrogen contamination of the Sherwood Sandstone aquifer beneath the city of Nottingham, United Kingdom. *Hydrogeology J.*, **4**(1), 90-102.

Rivett, M.O., Lerner, D.N., Lloyd, J.W. and Clark, L. (1990) Organic contamination of the Birmingham aquifer, UK. *J. Hydrology*, **113**, 307-323.

Sharma, M.L., Herne, D.E., Byrne, J.D. and King, P.G. (1996) Nutrient discharge beneath urban lawns to a sandy coastal aquifer, Perth, Western Australia. *Hydrology Journal,* **4,** 103-117.

Stuart, M.E. and Klinck, B.A. (1998) *A Catalogue of Leachate Quality for Selected Landfills From Newly Industrialised Countries*. BGS Technical Report, WC/98/49, Keyworth.

UNECE (2000) *Guidelines for monitoring transboundary groundwaters*. RIZA, Lelystad.

Zaporozec, A. (ed) (2002) *Groundwater Contamination Inventory: A methodological guide*. IHP-VI Series on Groundwater No. 2. UNESCO, Paris.

15

Establishing groundwater management priorities

O. Schmoll, I. Chorus and S. Appleyard

Management priorities for protecting groundwater will vary widely between settings. Once the pollution potential has been assessed for a given catchment (as discussed in Chapter 14), the priorities for management will depend on the public health burden this pollution is expected to cause currently and in the future. This will determine the urgency with which preventative or rehabilitating management responses are needed.

Management responses will also vary widely, as the feasibility of technically appropriate interventions will depend on the social and economic context in each setting. Also, choices can often be made among a range of responses to a problem, and participatory approaches are likely to lead to different decisions by different communities.

Situation analysis and management decisions for a groundwater catchment need to be based on sound information and on a clear, well-documented decision-making process. Uncertainties and gaps in the information base need to be transparent. This chapter discusses criteria for determining management interventions in relation to their urgency for protecting public health, and in relation to their feasibility for a given catchment.

 Protecting Groundwater for Health: Managing the Quality of Drinking-water Sources. Edited by O. Schmoll, G. Howard, J. Chilton and I. Chorus. ISBN: 1843390795. Published by IWA Publishing, London, UK.

15.1 ENSURING THE SUITABILITY OF INFORMATION

Establishing a well-documented inventory of all of the available information for a proposed groundwater management area is usually the first step in assessing the pollution risks if groundwater is used as a source of drinking-water. Such an inventory forms the basis of decisions about how potential water quality health risks will be managed. Chapter 6 and the checklists at the ends of Chapters 7-13 provide guidance on what type of information might be included in such a catchment inventory.

Decisions about managing risks to human health usually require the following factors to be addressed to be effective and sustainable (adapted from US Congress Commission, 1997).

- *The decision making process is iterative* – in general, health-risk management decisions need to be periodically reviewed as further information becomes available. An iterative approach helps ensure that management strategies remain up to date with new scientific findings, technological developments, and with national or international best practices.
- *Decision making is participatory* – broad participation with a range of interested or affected parties improves the quality and diversity of opinions that inform the decision-making process. Participation also increases the likelihood that risk management decisions will be accepted and implemented by the relevant parties.
- *Decision making is well-informed* – risk management decisions generally need to be based on information from a variety of sources and on different types of information. Information used may include scientific data, anecdotal records, information about regulatory requirements and socioeconomic information for the region.
- *Decision making is contextual* – management decisions must be appropriate to the social and economic realities of the specific region. Simply adopting practices developed in other parts of the world where there are differences in the level of expertise or available resources to implement management decisions generally does not work. Management practices developed locally are more likely to be sustainable.
- *Decision making is holistic* – focusing management decisions on only one issue may not lead to better health outcomes for communities in the longer term. In general, protecting and managing drinking-water quality should be seen in the context of being one item in a package of measures to protect the health of communities.

Ensuring that information gathered addresses the above factors generally helps prevent many of the pitfalls that often affect decisions made about how drinking-water supplies are managed, particularly the following two:

- crucial information gaps lead to poorly informed decisions which cause resources to be wasted on ineffective measures or even lead to health problems as illustrated in Box 15.1;
- decisions on measures urgently needed to improve public health are sometimes not taken or are unduly postponed because information gaps are used as an excuse for not allocating resources to the measures.

Box 15.1 Cholera Epidemic in Peru in 1991

Andersen (1991) reports that during the 1980s a decision by local water officials to stop chlorinating water pumped from many of the wells in Lima, Peru contributed to more than 300 000 cases of cholera and over 3500 fatalities. The reason given for the decision against disinfection was the perception of a cancer threat from chlorination by-products.

Studies indicating a small statistical lifetime risk of cancer from trihalomethanes and other chlorination by-products were interpreted as demonstrating a more significant health risk than the threat from waterborne pathogens. It is possible that the scale of the epidemic could have been greatly reduced had all of the information about the respective health risks of waterborne pathogens and chlorination by-products been available and understood by the relevant decision makers, and had the relative risk been assessed.

In general, there will always be gaps in the information gathered, and management decisions will often have to be made based on some degree of uncertainty. However, if there are potentially significant risks to human health, the lack of information should not be used as an excuse to delay reasonable and cost effective measures to protect health. If concerns are substantial although the risk level is uncertain, consideration should be given to short-term measures that can be implemented to protect health. This may involve temporary provision of an alternative drinking-water source until the information gaps have been sufficiently closed and the supply demonstrated to be safe, or appropriate control measures have been implemented. In some cases both risks and potential management measures may be self-evident. These generally relate to elements of good environmental practice which can be implemented without the requirement for a detailed pollution potential assessment.

Detecting information gaps during the situation analysis is an iterative process which feeds back into improving the quality of the available information, and helps ensure the inventory is relevant to the local region and water supply. Although obtaining additional information generally requires more investment, it is important to recognize that this effort is usually fairly minor in relation to the resources that may be required to implement some engineered management measures. The additional information may also be important in determining the most cost-effective solution. It is often crucial to be able to convince those responsible for financing management decisions such as funding bodies, government, or donor agencies, to fund such investigations before beginning an activity. This may be facilitated through ongoing consultation with these bodies, and by ensuring that both the information inventory and the criteria used in assessing its value as a basis for selecting management options are well documented and supported by the water consumers.

15.2 PRIORITIZING POLLUTANTS IN GROUNDWATER WITH RESPECT TO URGENCY OF MANAGEMENT RESPONSES

In supply settings where polluted groundwater may affect drinking-water quality, adequate management responses for the protection of public health are required. The determination of their urgency involves a site specific prioritization of individual pollutants in relation to their sources (e.g. polluting activities) on the basis of the following two aspects:

- the extent of the existing groundwater pollution level (e.g. from monitoring data), or the current or predicted groundwater pollution potential of a contaminant in a given setting (as defined in Chapter 14);
- the public health burden, i.e. severity and extent of health consequences.

Management responses for contaminants with the greatest public health burden and the highest pollution level or potential should receive higher priority than those whose health impacts are mild or whose occurrence in groundwater is unlikely. This prioritization approach is conceptually depicted in Figure 15.1.

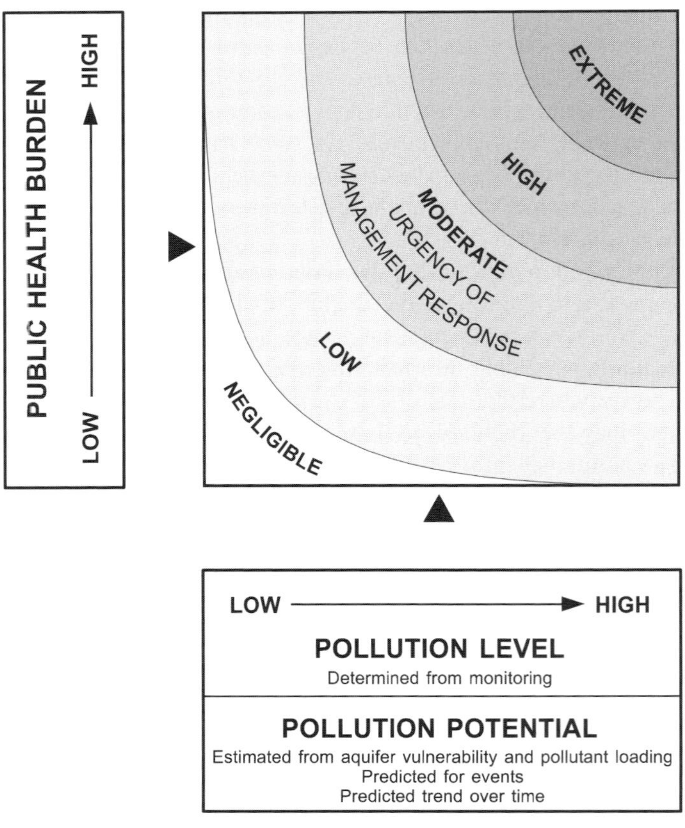

Figure 15.1. Urgency of management responses required to protect public health

The scheme in Figure 15.1 is similar to other risk ranking matrixes commonly used for relating the consequence (e.g. severity and extent) of impacts to the likelihood of an event occurring (WHO, 2004; SGWA, 1998; Deere *et al.,* 2001; MOH NZ, 2001). However, it is different with respect to the role of events causing groundwater pollution: while in general reviewing the likelihood of hazardous events is of crucial importance for risk assessment, groundwater pollution tends to be a more continuous process, e.g. from leaky sewers, poorly sited or designed latrines, waste disposal sites or agricultural activities. Although a pollution event may suddenly occur on the soil surface (e.g. through a tank truck accident or a farmer applying manure), occurrence in the aquifer is often more continuous, and the time pattern with which the pollutant may appear in the aquifer depends on hydraulic loading (i.e. rainfall patterns) as well as on aquifer vulnerability. Nitrate is an example that highlights how discrete contamination events on the surface may lead to continuous contamination of the aquifer.

The pollution potential, as defined in Chapter 14, encompasses both aquifer vulnerability and pollutant loading (in terms of pollutant load and hydraulic load) and thus already includes an assessment of the probability that polluting events on the soil surface result in groundwater contamination. Therefore, for determining the urgency of rehabilitating or preventative management responses, the scale in Figure 15.1 is not event likelihood, but rather pollution potential in terms of the probability of groundwater pollution occurring.

Tables 15.1 and 15.2 give an example of a simple prioritization matrix which applies the general concept of Figure 15.1. The classification scales in Tables 15.1 and 15.2 are based on a qualitative ranking rather than having quantitative values. This reflects the uncertainty of estimating pollution potential and public heath burden. Such a ranking scheme therefore has a relative nature: the aim is not a quantification of risk in absolute terms, but to identify the management responses that need to be most urgently addressed.

The approach shown in Figure 15.1 as well as in Tables 15.1 and 15.2 can in principle be expanded to a quantitative or semi-quantitative assessment by reducing uncertainty through improving the knowledge base. For the pollution potential, this would require analyses of the range of contaminants in the groundwater on a regular basis as well as monitoring or modelling of peak concentrations during events and changes over extended time spans to identify trends in order to calculate the dose the population would receive from drinking-water. For the public health burden, the categories for the size of the population affected and the severity of the effect would be calculated and expressed by the concept of Disability Affected Life Years (DALY) as a common public health unit which summarizes all health outcomes caused by a certain disease agent (i.e. chemical or pathogen) and provides an estimate of the burden of disease of this agent (Havelaar and Melse, 2003; WHO, 2004).

For most settings, however, the management of groundwater resources will be substantially improved by a simple approach of setting priorities based on a relative ranking of the urgency of issues as discussed above. Guidance for estimating the pollution potential is provided in some detail in Chapter 14 and in Section II of this monograph. Specifically for chemicals, Thompson *et al.* (in prep.) also provide guidance on deriving priorities for management from the use of chemicals and the conditions in the catchment. The following will briefly address some general aspects of ranking the public health burden of contaminants which frequently occur in groundwater.

Table 15.1. Simple prioritization scheme: consequence and probability scales (adapted from WHO, 2004; MOH NZ, 2001)

Scale for assessing severity and extent of impacts on public health		
Public health burden (consequence scale)	*Description*	*Position on Y-axis in Figure 15.1*
Insignificant	Insignificant	Low
Minor	Minor impact for a small population	↓
Moderate	Minor impact for a big population	
Major	Major impact (potentially lethal) for a small population	
Catastrophic	Major impact (potentially lethal) for big population	High
Scale for assessing the probability of groundwater pollution occurring		
Pollution potential (probability scale)	*Description*	*Position on X-axis in Figure 15.1*
Insignificant	Well protected aquifer and insignificant pollutant/hydraulic load	Low
Low	Low aquifer vulnerability and minor pollutant/hydraulic load	↓
Moderate	Low aquifer vulnerability and significant pollutant/hydraulic load	
High	High aquifer vulnerability and significant pollutant/hydraulic load	
Very high	High aquifer vulnerability and substantial pollutant/hydraulic load	High

Table 15.2. Simple prioritization scheme: ranking matrix for determining the urgency of management responses (adapted from WHO, 2004; MOH NZ, 2001)

Pollution potential (probability)	**Public health burden (consequence)**				
	Insignificant	Minor	Moderate	Major	Catastrophic
Very high	High	High	Extreme	Extreme	Extreme
High	Moderate	High	High	Extreme	Extreme
Moderate	Low	Moderate	High	Extreme	Extreme
Low	Low	Low	Moderate	High	Extreme
Insignificant	Low	Low	Moderate	High	High

Ranking of contaminants according to their public health burden

Ranking microbial and chemical contaminants in groundwater in terms of their public health burden depends to a large extent on site-specific factors, and it is not possible to develop an absolute ranking scale that will fit all cases. However, as a general rule the importance of contaminants in groundwater can be ranked in the following decreasing order:

- waterborne pathogens (Chapter 3);
- naturally occurring groundwater constituents such as fluoride and arsenic (Chapter 4.1);
- nitrate (Chapter 4.2);
- industrial chemicals such as chlorinated or aromatic hydrocarbons (Chapter 4.3), pesticides (Chapter 4.4) or metals (Chapter 4.5);
- pharmaceuticals and endocrine disruptors (Chapter 4.6).

This ranking is indicative only but is typical of both the observed occurrence of contaminants in groundwater and the severity of the health burden that they cause. It can be used as a preliminary ranking if no other site-specific information is available.

Waterborne pathogens pose a much greater *immediate* threat to public health than chemical contaminants and are generally considered to have the highest priority for a management response before chemical contamination issues are considered. In situations where the pollution potential for pathogens is also considered to be high, the implementation of management measures is considered to be extremely urgent (Figure 15.1). Conversely, in situations where the pollution potential is considered to be low (for instance, due to the fact that groundwater is being pumped from a well-constructed deep tubewell in a porous medium aquifer), a management response will be considered to be much less urgent, and addressing other contaminants, such as some accumulating chemical contaminants, might have a higher priority.

As discussed in Chapter 4, fluoride and arsenic n groundwater supplies may have significant effects on public health. These contaminants are generally of natural origin, and in many situations their concentration is low in groundwater. Their presence in health-relevant concentrations in a given groundwater supply depends on local geological factors and on specific hydrochemical conditions (Chapter 8) being present in the aquifer to allow these chemicals to be mobilized from sediments or bedrock into groundwater. Therefore it is essential to specifically address the possible presence of natural groundwater constituents in the catchment-specific situation analysis. This should include undertaking specific sampling and chemical analysis for these constituents, and obtaining information about the local geological conditions to determine whether there is a risk of land use or groundwater pumping increasing the concentrations of fluoride or arsenic in groundwater. This information will help determine where these chemical constituents should plot on the matrix in Figure 15.1.

In some situations, nitrate is of concern to public health because of potential health effects on bottle-fed infants (i.e. metHb, particularly in the presence of simultaneous microbial contamination; see Chapter 4.3), the large inputs of this chemical to groundwater in areas with intensive agriculture and/or on-site wastewater disposal, and its tendency to accumulate in aquifers. The potential health consequences of nitrate contamination will vary depending on the size of the population exposed (which in turn strongly depends on social factors, i.e. prevalence of breast feeding and parental knowledge) and this will affect where this contaminant will plot on the scheme in Figure 15.1.

With the exception of localized major point sources of contamination, exposure to chemical contaminants such as heavy metals, organic pollutants (e.g. chlorinated or aromatic hydrocarbons), and pesticides through drinking-water is usually a less immediate threat to public health. One aspect is that the major exposure pathways to these substances are usually air pollution and food rather than drinking-water (WHO, 2004). For protecting public health, management of these greater sources of contamination might be more urgent than addressing their occurrence in water. Although implementing management measures to deal with these contaminants is usually less urgent for most groundwater supplies, there may be circumstances in which a situation assessment may indicate the probable presence of very high concentrations in groundwater due to local contamination, raising the management priority of one or more of these chemicals.

Pharmaceuticals, endocrine disruptors and most pesticides rarely occur in groundwater in concentrations that have been shown to be hazardous to human health. In most cases the ranking following Figure 15.1 will therefore result in a rather low urgency for management response. However, in many societies there is increasing concern over traces of pharmaceuticals and pesticides found in groundwater used as drinking-water sources. Such issues may be addressed in the priority setting matrix by including other criteria, such as public perception and value judgements, in addition to public health burden. Where public concern over pharmaceuticals or pesticides makes the headlines, management measures may indeed be ranked as urgent. However, scientifically substantiated assessment of the public health impact will always need to be the most important criterion for setting priorities and for defining the urgency of management responses in order that resources are not diverted from more pressing public health problems.

In summary, the simple ranking scheme presented in Figure 15.1 and in Tables 15.1 and 15.2 is based on the relative likely health burden caused by a number of contaminants, and not their absolute magnitude, and thus simply indicates management priorities. Waterborne pathogens may cause immediate illness and their presence in water requires an urgent management response either in the catchment or at the water supplier's operational level (i.e. treatment). Chemical contaminants usually require long-term exposure to cause health effects. Consequently, management measures for chemicals are of less urgency in most circumstances and measures may be delayed until management responses for the control of pathogen contamination have been implemented.

Assessment of persistent contaminants

Unlike waterborne pathogens, which survive in groundwater for a limited period of time (Chapter 3), some chemical contaminants may persist in groundwater over a long period of time, with little or no attenuation. In situations where there is a continuing source of chemical contamination and the loading rate at which the chemical is leached into groundwater is greater than the rate at which the contaminant is removed by physical, chemical or biological processes in the aquifer, concentrations of the contaminant in groundwater may progressively accumulate with time.

The most common example of this behaviour is nitrate contamination in groundwater in regions where intensive agriculture is poorly managed. Nitrate concentrations in groundwater in these regions often continue to increase over many decades if appropriate land use management practices are not implemented. Accumulation has also been observed for various industrial chemicals (e.g. chlorinated solvents) and some pesticides (e.g. atrazine).

For such contaminants rehabilitation of groundwater quality is typically difficult and costly, and interventions may be needed before concentrations reach health-relevant levels. The process of prioritizing management responses therefore needs to address the issue of contaminant accumulation by estimating trends or using a prediction of the contaminant concentration in groundwater at some future time if no management action is taken to stop or reduce the polluting activity or practice. An outcome of such an assessment would be to indicate whether the contaminant could become an urgent management priority unless the source of contamination were to be removed or the loading rate greatly reduced. Such an assessment could be used to help set priorities for progressively changing land use within a catchment area to protect groundwater quality in the longer term or to introduce changes such as appropriate chemical handling practices in small or large industries.

Specific advice on predicting long-term nitrate concentrations in groundwater in urban and periurban environments where there is limited information can be found in Lerner (2000) and ARGOSS (2001).

15.3 SELECTION OF MANAGEMENT OPTIONS

The last step in the situation analysis for a groundwater supply is to select possible management options that are appropriate for the magnitude of the health risk posed by a specific contaminant. The range of possible measures for protecting groundwater from becoming polluted through human activities is discussed in some detail in Sections IV and V of this monograph.

Once the urgency of an intervention to protect public health has been determined, apart from the technical adequacy of the measure, selecting effective management options also needs to consider the following aspects:

- delayed response time between management interventions (i.e. removing the contaminant source) and measurable reduction of aquifer pollution;
- barriers in place in addition to groundwater protection (e.g. treatment);
- socioeconomic feasibility of management responses.

Aquifer response time

Aquifers tend to respond only slowly to changes in contaminant loading due to retention processes such as adsorption and desorption to soil particles (see Chapter 4). Moreover, due to slow flow rates and long retention times of water in many aquifers, elevated contaminant concentrations may remain present for many years. This effect has particularly been shown for contamination with chemicals such as nitrate or some pesticides (e.g. atrazine). For example, the response time between successfully implemented measures targeting the reduction of nitrogen loading from agriculture in the catchment area and measurable reduction of nitrate concentration in groundwater can range between a few years and several decades (Behrendt *et al.*, 2000).

In situations where groundwater used for drinking-water supply is contaminated and where the implementation of management measures is assessed as urgent for protecting public health, the delay in response needs to be taken into consideration. As discussed below, measures in addition to those protecting or rehabilitating the aquifer from polluting activities might be needed (e.g. water treatment, change of source) in order to provide safe drinking-water.

The delay in response is an inherent property of the groundwater system that cannot be changed by any management measure. It is therefore important to recognize that control measures for reducing chemical contamination (e.g. nitrate, pesticides) in groundwater are likely to show results in the long term only and that short-term 'success stories' are rarely achievable. When planning and implementing management actions, this issue needs to be adequately communicated both to communities using the resource for drinking-water supply and to funding agencies and politicians in order to avoid misplaced expectations and disappointment that may impair the political or financial support.

Multiple barriers

Wherever possible during the process of selecting management options, it is important not to select just one measure and rely on it for the long-term protection of public health. In particular, relying on water treatment alone to prevent health problems from contaminated groundwater may be a high-risk management strategy for some contaminants, particularly microbial contaminants. This could have significant public health consequences if the treatment system fails or mistakes are made by the operators of the system and water treatment is ineffective (O'Connor, 2002).

As depicted in Figure 15.2, risks to public health can generally be minimized when a group of complementary management measures are implemented together to ensure that there are several barriers in place between a potential source of contamination and a water consumer, particularly when the contaminant can have a major impact on health. The presence of several barriers means that the overall water supply is protected from a system failure or human error at any one point in the system because there are backup protection measures, and thus the water supply becomes a fail-safe system. This multi-barrier principle is one of the basic principles of drinking-water hygiene (WHO, 2004).

Typical barriers in groundwater supplies include:

- management practices in the catchment area to reduce contaminant inputs from human activities into groundwater (see Chapters 21-25);
- source-water protection through control of land use in protection zones (see Chapter 17);
- adequate design, construction and maintenance of water supply wells (see Chapter 18);
- treatment of pumped groundwater (LeChevalier and Au, 2004);
- ensuring adequate disinfection residual in the water distribution system;
- protection and maintenance of the distribution system (Ainsworth, 2004).

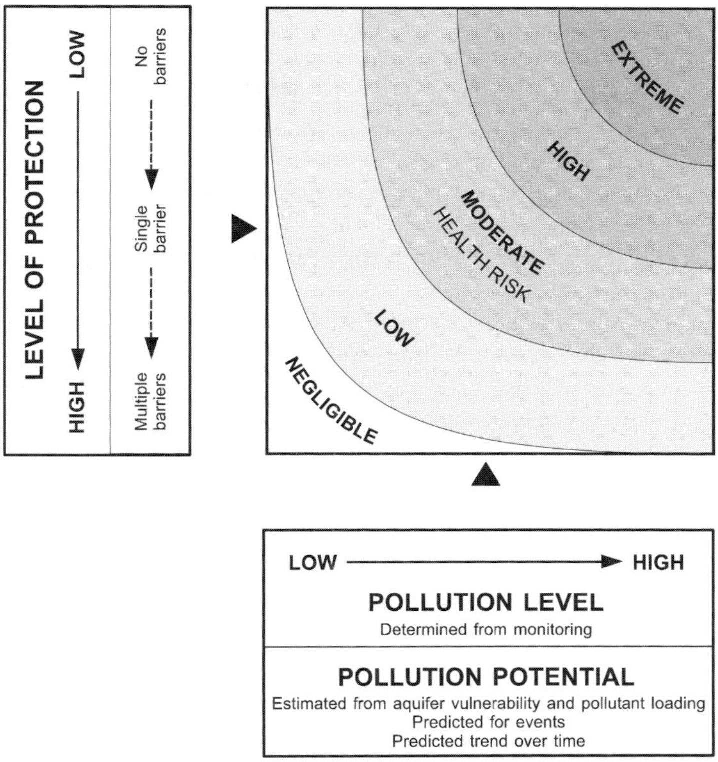

Figure 15.2. Reducing health risks by using multiple barriers to protect a water supply (adapted from Hrudey, 2001)

The extent to which these barriers are needed will depend on an assessment of the possible health outcomes of contaminants, the size of the population relying on the water supply, the resources available for management, and an assessment of the costs of implementing management practices against the possible benefits in the specific situation.

Many low-income countries lack the resources or expertise to implement a broad range of groundwater protection measures, and in some countries it can be difficult to convince key political decision makers of the importance of groundwater protection for long-term water safety. In such situations water treatment or the provision of another source of water are important immediate health protection measures.

Socioeconomic feasibility

Groundwater protection measures do not have to be expensive, and small incremental changes over a period of time can greatly improve the quality of groundwater or avoid degradation of quality. Even simple inexpensive measures such as ensuring that defecation is not carried out within 10 m of a water supply well and ensuring that surface run-off is diverted away from the well will significantly reduce public health risks from such a water supply (ARGOSS, 2001; Howard *et al.,* 2003). Other groundwater protection measures will require investments, e.g. constructing latrines, improving seals on wellheads or improving drainage of roads. Funding for these may or may not be available. Designating protection zones in the immediate vicinity of a wellhead or extended further into the aquifer's catchment may be inexpensive to the authority doing so, but – as discussed in Chapter 5 – can disrupt the livelihoods of inhabitants of the land above the aquifer or can have substantial economic consequences for them. In some settings, though interventions may appear appropriate in theory, institutional capacity is too weak to implement them.

Unless a catchment is largely uninhabited and unused by humans, the aquifer protection measures introduced in Sections IV and V of this monograph will often intervene in the way people are currently doing things, and will work only if the stakeholders in the catchment are willing to make changes. This pertains to land use in general, but also to practices, e.g. in agriculture, sanitation or handling and storing hazardous chemicals. Therefore, while for a given setting the urgency of mitigating or preventing groundwater contamination is determined from an assessment of pollution potential and public health burden (Section 15.2), selecting appropriate management measures requires a further step, i.e. assessing their feasibility in the specific setting. Feasibility depends on a variety of factors, such as cultural values, public perception, education, land tenure rights, socioeconomic status, legal requirements and institutional capacities (see Chapters 5 and 20). These factors need to be evaluated in relation to the management responses envisaged. The result of such an evaluation may be that socioeconomic measures are a crucial part of the management package, equally or even more important than the actual technical measures to protect the aquifer.

As discussed in some detail in Chapter 5, public communication and consultation is not only an important tool for assessing the potential of aquifer pollution and the feasibility of measures suggested by scientists and engineers. Beyond this, participation of the population affected by groundwater protection measures is key to developing approaches that will be supported locally and can therefore be implemented more readily. Where people's livelihoods are positively affected by a protection measure, rather than the measure being perceived as only having a negative impact, particularly in the short term, interventions are more likely to be accepted and maintained. It is important to communicate that increasing costs arise not only from the protection of the resource, but also from its deterioration. In this context, there may be conflicting interests between different stakeholders in the catchment. For example, abandoning a resource and piping in water from a more distant one in order to be able to continue a polluting activity may be in the interest of some, but others who could not afford a more expensive supply would prefer protection. The planning of groundwater protection measures will need to consider how the needs of all stakeholders can best be incorporated into the policy, and how costs can be minimized while maximizing protection.

Figure 15.3 conceptually depicts that both urgency of an intervention and the feasibility in a given setting will determine which type of management action can be taken. Management responses to technically similar problems may be very different between settings, depending on what is locally feasible.

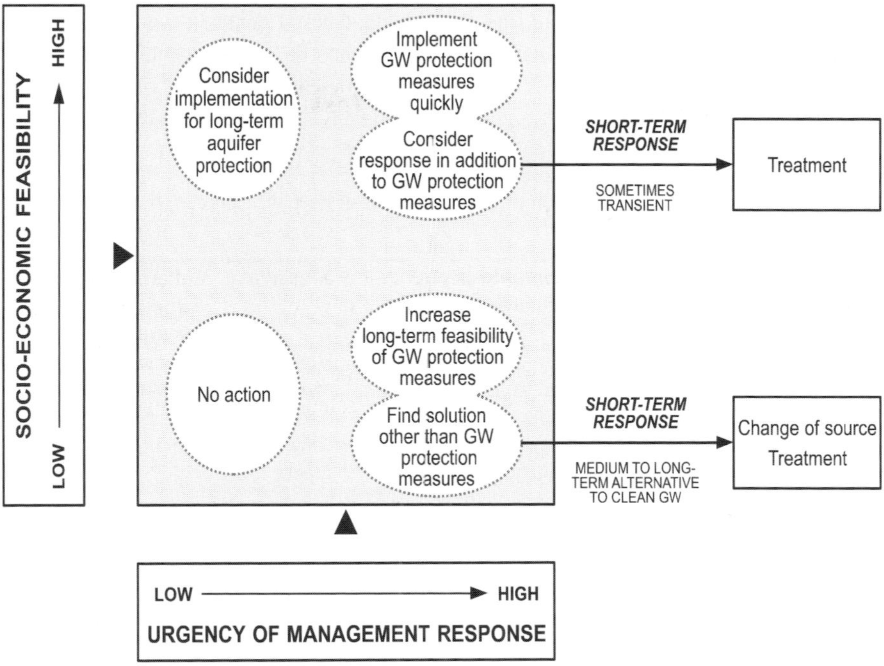

Figure 15.3. Selection of management responses in relation to their urgency and feasibility

The upper right-hand corner of Figure 15.3 shows that where urgent groundwater protection measures are assessed to be feasible, the key management action is their rapid implementation. Additional short-term measures at the point of consumption (e.g. treatment) may be necessary until the groundwater quality shows a response to the protection measure (see example settings A and C in Table 15.3). Where interventions are urgent, but it is not feasible to improve groundwater in the medium term (bottom right-hand corner of Figure 15.3), the primary response to protect public health will be management actions other than groundwater protection, as safe groundwater is not likely to become available in the short term (see example setting B in Table 15.3). However, these tend to involve either use of a more distant source or treatment, both of which are costly. In settings in which an aquifer can still be rehabilitated, an extended process of community consultation and discussion may therefore lead to additional action for groundwater protection or remediation in the longer term. As discussed in Chapter 5, feasibility may be increased by socioeconomic measures such as compensation payments for restrictions on land use, but also by e.g. improving tenure rights to make protection attractive in the longer term.

A different group of situations are those in which management responses against contamination are assessed to be less urgent (see Chapter 15.1). If feasibility is also low (e.g. for protecting an aquifer from low levels of pesticides), no action would be taken other than improving risk communication to the public if there is a concern (bottom left-hand corner of Figure 15.3). However, where measures to prevent such pollution are feasible (upper left-hand corner of Figure 15.3; see example setting E in Table 15.3), implementing such measures would be appropriate. Such a decision would follow the precautionary principle. Precautionary action is likely to be more feasible in settings in which the public perception values the resource and clean water is a widely accepted goal, and particularly where the overall public health and socioeconomic conditions enable such a priority to be set.

Table 15.3. Examples for the selection of management responses in relation to their urgency and feasibility

Groundwater pollution problem	Urgency	Groundwater protection measure	Feasibility	Alternative and/or supplementary measures
Example setting A with periodic detection of high *E. coli* counts in poorly constructed wells in a shallow vulnerable aquifer adjacent to an open defecation area	Extreme	Improve construction and maintenance of wells Construct latrines downgradient from wells	High → upper right-hand corner of Figure 15.3	Additional short-term water treatment (e.g. boiling, disinfection)
Example setting B with periodic detection of high *E. coli* counts in poorly constructed wells in a shallow vulnerable aquifer adjacent to an open defecation area	Extreme	Improve construction and maintenance of wells Construct latrines downgradient from wells	Low → bottom right-hand corner of Figure 15.3	Abandon wells and rebuild in uncontaminated area, or provide alternative water source (e.g. use of water tankers) Public consultation for increasing feasibility of aquifer protection
Example setting C with periodic detection of high *E. coli* counts from contaminated run-off in poorly constructed wells	High	Ensure setback distances to sources of pollution Improve construction and maintenance of wells	High → upper right-hand corner of Figure 15.3	Public communication about wellhead protection
Example setting D with nitrate contamination from agriculture in properly constructed tubewells	Moderate	Implement control measures for stock density as well as for application of fertilizers and manure	Low → bottom right-hand corner of Figure 15.3	Provide appropriate bottled water for bottle-fed infants; and/or encourage breast feeding
Example setting E with pesticide contamination from agriculture in properly constructed tubewells	Low	Implement training programme for farmers on good practice in choice, application and disposal of pesticides	High → upper left-hand corner of Figure 15.3	Improve risk communication to the public

Management actions for the contaminants ranked as urgent may result in simultaneous remediation of those with a lower urgency ranking, e.g. where nitrate loading from human excreta occurs together with faecal indicators, measures reducing the latter are likely to also reduce the former. Such benefits need to be included in the case for measures that are being proposed.

Generally, where socioeconomic conditions indicate that the availability of financial resources for advanced drinking-water treatment is low, maintaining groundwater quality as a cheap and safe resource not requiring treatment may be particularly important. Under these circumstances it is probable that the situation analysis would identify aquifer and/or wellhead protection as a high priority.

The important exception to the general scheme shown in Figure 15.3 is where groundwater contains toxic natural constituents such as fluoride or arsenic at concentrations that are of health concern. In these situations, groundwater protection measures will not reduce the concentrations of these constituents and water treatment or providing alternative sources of drinking-water are the only effective management options. This does not mean that groundwater protection measures are abandoned in these situations, as other possible contaminants derived from land use will still need to be managed to prevent contamination of the water supply even if it is being treated to remove the natural chemicals such as arsenic or fluoride.

15.4 DOCUMENTATION AND REPORTING

Comprehensive and easily understood documentation of the situation assessment is important as it enables both the decision-making process for a specific supply and the information on which the decisions were made to be clearly communicated to water consumers, regulatory and funding agencies and other stakeholders. In particular, documentation is important for the following reasons:

- Documentation of the sources of information that were used to make management decisions enables the quality of the information to be assessed. This can help identify whether there are any major gaps in the information used which may affect both the assessment of groundwater pollution risks and the measures selected to manage groundwater quality.
- Potential water consumers and the general public have a right to know what potential contamination risks occur in an existing or proposed groundwater supply and about how these risks will be managed. Water consumers need assurance that their water supply will not pose a public health threat, and demand a high level of accountability from water suppliers and government regulatory agencies for decisions made about drinking-water safety (CELA, 2001).
- A report of the results of the groundwater source assessment is a powerful tool to initiate discussions with a wide range of stakeholders about the need to protect the quality of groundwater in a region.
- Good documentation allows decisions about water quality management to be easily evaluated and updated as new information becomes available.

- Good documentation together with ongoing consultation with the community and key decision-makers helps secure funding and community support for implementing groundwater protection measures to ensure that drinking-water will not affect public health.

The whole process of the situation analysis should be documented formally in a report by the situation assessment team. As well as describing the results of the situation analysis and the criteria for decisions made in the assessments, the report should include the technical details such as key persons involved in conducting the situation analysis, when the information was collected, the sources of data (e.g. from site inspection, statistics, government bodies, universities, published/unpublished information), contact persons for information, and the location and format of storage of the information. A key component of this report will often be a series of maps of the catchment area showing the aquifers and groundwater conditions, vulnerability, groundwater supplies, land use and human activities which highlight the major potential sources of groundwater pollution. If possible, these should be in GIS format at a common scale so that they can be easily overlaid and updated.

15.5 REFERENCES

Ainsworth, R. (ed.) (2004) *Safe Piped Water: Managing microbial water quality in piped distribution systems*, IWA Publishing, London.

Anderson, C. (1991) Cholera epidemic traced to risk miscalculation. *Nature*, **354**, 255.

ARGOSS (2001) *Guidelines for Assessing the risk to Groundwater from On-site Sanitation*. BGS Commissioned Report CR/01/42. British Geological Survey, Wallingford.

Behrendt, H., Huber, P., Kornmilch, M., Opitz, D., Schmoll, O., Scholz, G. and Uebe, R. (2000) Nutrient emissions into river basins of Germany. *UBA-Texte*, 23/00, Berlin.

CELA (Canadian Environmental Law Association) and Walkerton Citizens (2001) *Tragedy on Tap: Why Ontario needs a safe drinking water act*. Report prepared for the Walkerton Inquiry. http://www.cela.ca/water/sdwaoverview.pdf (accessed April 29, 2005).

Deere, D.A., Stevens, M., Davison, A.D., Helm, G. and Dufor, A. (2001) Management strategies. In. *Water Quality: Guidelines, standards and health. Risk assessment and management for water-related infectious disease*. (eds. L. Fewtrell and J. Bartram), pp. 257-288, London, IWA Publishing.

Havelaar, A.H. and Melse, J.M. (2003) *Quantifying Public Health Risk in the WHO Guidelines for Drinking-water Quality: A burden of disease approach*. RIVM report 734301022/2003, Bilthoven.

Howard, G., Pedley, S., Barrett, M., Nalubega, M. and Johal, K. (2003) Risk factors contributing to microbiological contamination of shallow groundwater in Kampala, Uganda. *Water Res.*, **37**, 3421-3429.

Hrudey, S.E. (2001) Drinking water quality: a risk management approach. *Water*, **28**(1), 29-32.

LeChevalier, M. and Au, K-K. (2004) *Water Treatment and Pathogen Control: Process efficiency in achieving safe drinking-water*, IWA Publishing, London.

Lerner, D.N. (2000) *Guidelines for Estimating Urban Loads of Nitrogen to Groundwater*. Report produced for the UK Ministry of Agriculture, Fisheries and Food. http://www.shef.ac.uk/gprg/publications/pdf/urban.pdf (accessed April 29, 2005).

MOH NZ (New Zealand Ministry of Health) (2001) *How to Prepare and Develop Public Health Risk Management Plans for Drinking-water Supplies*, Wellington. http://www.moh.govt.nz/moh.nsf/0/cca65c18b2e29251cc256a7900082b9c?OpenDocument (accessed April 29, 2005).

O'Connor, D.R. (2002) *Report of the Walkerton Inquiry. Part 1: The events of May 2000 and related issues*. Report by the Ontario Ministry of the Attorney General. http://www.attorneygeneral.jus.gov.on.ca/english/about/pubs/walkerton/ (accessed April 29, 2005).

SGWA (Swiss Gas and Water Industry Association) (1998) *Branchenspezifisches Qualitätsmanagement für Versorgungsunternehmen*. SVGW, Zurich.

Thompson, T., Fawell, J., Kunikane, S., Jackson, D., Appleyard, S., Kingston, P., Callan, P. and Hashizume, H. (in prep.) *Chemical safety of drinking-water: assessing priorities for risk management*. http://www.who.int/water_sanitation_health/dwq/cmp/en/ (accessed 17 October 2005).

US Congress Commission on Risk Assessment and Risk Management (1997) *Framework for Environmental Health Risk Management* (vol 1), and *Assessment and Risk Management in Regulatory Decision-making* (vol 2). Report of the US Congress, Washington, DC.

WHO (2004) *Guidelines for Drinking-water Quality: Recommendations*, vol. 1, 3rd edn,World Health Organization, Geneva.

Section IV

Approaches to drinking-water source protection management

16

Water Safety Plans: Risk management approaches for the delivery of safe drinking-water from groundwater sources

G. Howard and O. Schmoll

The delivery of safe drinking-water requires actions to be taken throughout the water cycle from the catchment to the point of consumption. The focus of any programme designed to deliver safe drinking-water should therefore be the effective management and operation of water sources, treatment plants and distribution systems (whether piped or manual). This will demand action by water suppliers, environmental protection agencies and health bodies.

WHO (2004) outlines that the delivery of safe drinking-water is most effectively achieved a Water Safety Framework, that encompasses three elements:

1. Establishing health based targets for drinking-water based on evaluation of health concerns.
2. Developing a management system to meet these targets that is termed Water Safety Plan (WSP) and consists of

- An assessment of the water supply system to determine whether the water supply chain (from source through treatment to the point of consumption) as a whole can deliver water that meets the health based targets.
- Identification and operational monitoring of the control measures in the drinking-water supply that are of particular importance in securing drinking-water safety.
- Preparation of management plans documenting the system assessment and monitoring plans and describing actions to be taken in normal operating and incident conditions, including upgrading documentation and communication.

3. A system of independent surveillance that verifies that the above are operating properly.

The establishment of health-based water quality targets would typically be led by the health sector taking into account local health burdens. These are not discussed in detail in this text, but would include many of the considerations highlighted in Section I and would be established in relation to a level of public health risk determined as being tolerable. This may use epidemiological and quantitative risk assessment procedures, which can be applied to both chemicals and microorganisms (Haas *et al*., 1999). More detail is available on establishment of water quality targets in the third edition of the WHO *Guidelines for Drinking-water Quality* (2004) and in Havelaar and Melse (2003).

The surveillance component would typically be undertaken by a regulatory agency, which in practice may be the health, environment or local government sector. This component would incorporate many of the issues identified in Sections II, IV and V of this monograph as means of monitoring of performance. Further descriptions of approaches to be used are available in a variety of texts linked to the WHO *Guidelines for Drinking-water Quality* (WHO, 1997; 2004; Howard, 2002).

The second activity of the Water Safety Framework is termed a Water Safety Plan (WSP) (WHO, 2004; Davison *et al.,* 2005) which is typically the responsibility of the water supplier, with support from and collaboration with other sectors as discussed below. The focus of this chapter is the application of WSPs within groundwater supplies and catchments. This covers both supplies where the whole WSP will apply to the groundwater, where the point of delivery is at the abstraction point (e.g. tubewell with handpump) and those where groundwater forms only part of the overall WSP (e.g. borehole connected to a distribution system). The principles and methods used in WSPs draw on other risk management and quality assurance methods. In particular they are based on the Hazard Analysis and Critical Control Process (HACCP) approach applied in the food industry.

As described by the multiple barrier principle, source protection is the first stage in the production of safe drinking-water quality (WHO, 1993; 2004). When sources are managed effectively, subsequent treatment costs are minimized and the risks of exposures resulting from failures in treatment processes are reduced. Therefore, source and resource protection is vital for efficient risk management (WHO, 2004). Protection measures should be put in place that have been shown to be effective in improving water safety as the first stage of a plan for managing the safety of drinking-water.

16.1 END-PRODUCT TESTING AND THE NEED FOR A RISK MANAGEMENT APPROACH

The traditional approach to water quality management placed a great emphasis on the routine monitoring of water quality. The results of analysis were compared against acceptable concentrations in order to evaluate performance of the water supply and to estimate public health risks (Helmer *et al.*, 1999). The focus of attention was on end-product standards rather than ensuring that the water supply was managed properly from catchment to consumer. Although operation and maintenance of water supplies has been recognized as important in improving and maintaining water quality, the primary aim of water suppliers, regulators and public health professionals has been to ensure that the quality of water finally produced met these standards.

This reliance on end-product testing has been shown to be ineffective for microbial quality of water, as evidence has emerged of significant health impact from the consumption of water meeting national standards (Payment *et al.*, 1991). In part this is because most national standards have been set using bacterial indicators that are very different from viral and protozoan pathogens.

The quality of the source protection measures is an important component in controlling whether pathogens may be present in the final drinking-water. For instance, one study concluded that the degradation of surface water catchments was an important factor in waterborne disease transmission (Hellard *et al.*, 2001). The outbreak of *E. coli* O157:H7 and *Campylobacter jejeuni* from drinking-water in Walkerton, Ontario appears to have resulted from a combination of improper protection of the groundwater source and a failure to maintain adequate chlorination (O'Connor, 2002). The example from Walkerton particularly emphasizes the need for multiple barriers in water quality management.

End-product testing has a further weakness in that the number of samples taken is typically very small and not statistically representative of the water produced in a domestic supply. The focus on end-product testing has meant that action is only initiated in response to a failure in relation to the specified water quality standard. However, this typically means that the water has been supplied and may have been consumed before the results of the test are known and the increased risk to health identified. As a result, outbreaks occur and rates of endemic disease remain higher than when good practice in relation to water quality management is emphasized. The reliance on end-product testing is therefore not supportive of public health protection and whilst it retains a role in assessing water safety, it should not be the sole means by which risks are managed (WHO, 2004).

16.2 SCOPE OF WATER SAFETY PLANS

Water quality management elements such as documented operational procedures, monitoring process control measures and sanitary inspection have complemented end-product testing in many water supplies for a long time. Beyond these, the need for a comprehensive quality assurance approach based on sound scientific evidence and understanding the risks in a given supply system has been increasingly recognised.

Quality assurance procedures are being applied more formally to water supplies, including the use of HACCP and approaches based on the generic ISO 9000 Quality Standard.

The use of HACCP for water quality management was proposed by Havelaar (1994), following international codification of the principles for the food industry (Codex Alimentarius Commission, 1996; NACMCF, 1992). Subsequent initiatives have addressed the application of these principles to the broader control of infectious disease from water and wastewater exposures (Fewtrell and Bartram, 2001). The application of HACCP principles have also been further described in relation to specific water supplies (Barry *et al.*, 1998; Deere and Davison, 1998; Gray and Morain, 2000; Deere *et al.*, 2001; Bosshart *et al.*, 2003; Howard, 2003; Wülser and Trachsel, 2003). These experiences were used as basis for the Water Safety Plan approach in the third edition of the WHO *Guidelines for Drinking-water Quality*.

The development and implementation of a WSP would typically be the responsibility of a water supplier, although in many cases other stakeholders may have responsibilities that must be fulfilled. Such plans should address all aspects of the water supply under the direct control of the water supplier and focus on the control of water production, treatment and distribution to deliver drinking-water. In some situations, the water supplier will control the catchment and therefore will be able to identify and implement control measures within the catchment. In other situations, the water supplier may not control the catchment and therefore some aspects of control will require actions by other stakeholders. These may still be incorporated within the WSP provided that processes are set in place for communication of the findings of monitoring, and actions are identified in the case of non-compliance. In these situations, the implementation of a WSP provides a sound platform for the water supplier to take an active role in initiating and developing stakeholder involvement for the protection of drinking-water sources (Box 16.1).

WSPs can be defined for utility operated water supplies using mechanized boreholes, disinfection and piped distribution; or for simple point sources of water where water is collected by hand and transported back to the home manually (WHO, 2004). In the case of small water supplies, the WSP may be defined by an external agency and be applied either through a generic WSP for a technology type or be developed for an individual supply using very structured guidance (APSU, 2005; MOH NZ, 2001; SGWA, 2003; WHO, 2004). However, it will be expected that the activities required under the WSP will be the responsibility of the water supply operator.

Although this monograph deals with groundwater sources and their protection for public health, a key value of WSPs is that they address the full water supply chain from source to consumer. WSPs therefore demand action is taken in water sources and their catchments (whether groundwater or surface water), in treatment steps (if any are applied), subsequent distribution and household storage and treatment.

WHO (2004) identifies that the development of an effective WSP requires (Figure 16.1):

- assembling a team that understands the system and can undertake an initial assessment of the system with regard to its capability to supply water meeting the specified targets;

- identifying where contamination can occur and what measures can be put in place to prevent, reduce or eliminate contaminants (control measures);
- validation of methods employed to control hazards;
- putting in place a system for monitoring and corrective action to ensure that safe water is consistently supplied;
- periodic verification that the WSP is being implemented correctly and is achieving the performance required to meet the water quality targets.

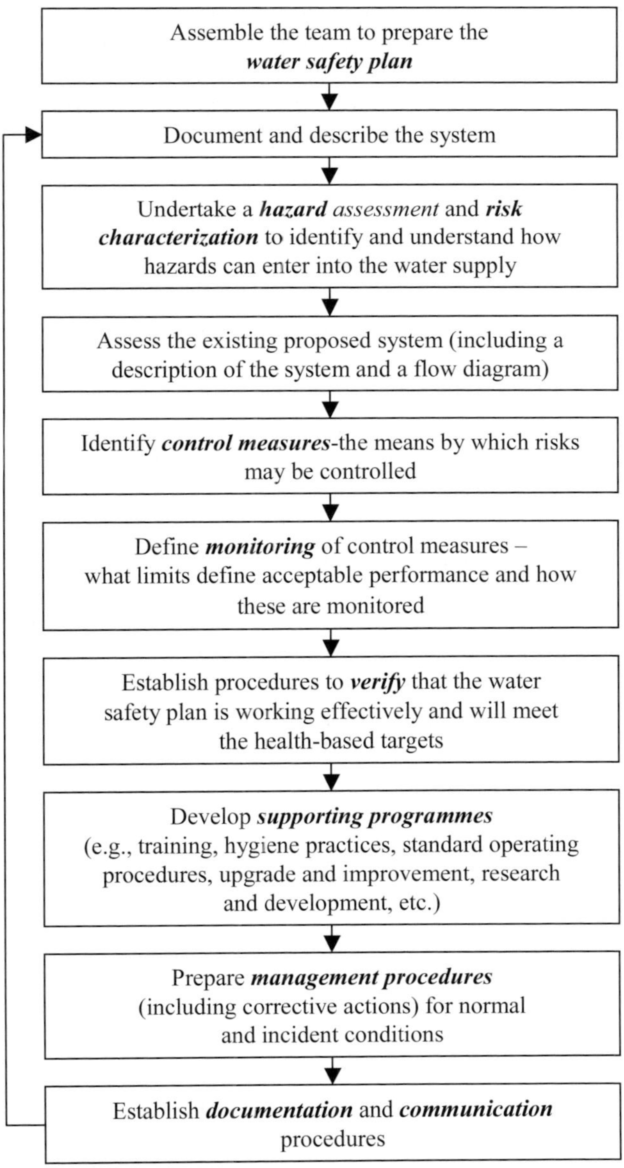

Figure 16.1. Steps in the development of Water Safety Plans (adapted from WHO, 2004)

16.3 PRELIMINARY STEPS FOR DEVELOPING WATER SAFETY PLANS

16.3.1 Assembling the team

The first stage of a WSP is to assemble a team of experts who will undertake the assessment of the water supply from catchment to consumer. This should be a multi-disciplinary team including managers, scientists (e.g. hydrogeologists, microbiologists, chemists) engineers (e.g. from operations, maintenance, design and capital investment) and technical staff involved in the day-to-day operation of the supply. The latter are essential as very often it is those members of staff who undertake work on the system every day who have the greatest knowledge about the problems that exist. A senior member of the team (usually the risk manager) should be appointed to help guide and direct the team in the study and this person should be able to either make decisions regarding investment or be able to influence others in the improvement of performance.

The development of the WSP and supporting programmes (which will typically involve actions by other stakeholders, such as environmental protection agencies) is generally most effectively implemented when the skills required are drawn from a range of stakeholders. For groundwater, this will include representatives of agencies responsible for assessing the impact of pollution and implementing controls on land-use. This may be particularly important when identifying control measures within catchments where the water supplier does not own the land. Thus the WSP team can act as catalyst for collaboration with different stakeholders and establish a sense of mutual ownership for controlling contaminants at their source.

16.3.2 Describing the water supply

The next stage in developing a WSP is to describe the water supply. In the case of groundwater supplies, this means providing information on aspects such as the depth to the water table, nature of the lithology of the aquifer and unsaturated zone from drilling logs, technologies used to abstract water, pump type and depth and the draw-down on pumping. This stage should also clearly identify whether alternative water sources exist in the community should there be need to take the source off-line at any time to effect corrective action.

The next step is to prepare a detailed flow diagram. The purpose of this stage is to provide the basis of understanding the hydrological environment and the subsequent distribution of drinking-water. The flow diagram should indicate the flow of water from the recharge to the abstraction point, the nature of the aquifer and recharge areas, flow times and vulnerability maps where available (Chapters 2 and 8 provide more detailed information on how the hydrological environment can be characterized). This stage is concerned with defining the hydrogeological conditions in order to understand what natural processes may affect the quality of water. The distribution of final water (whether piped or manual) should also be indicated on the flow diagram.

Two examples of simple flow diagrams – one for a simple setting with a shallow borehole and one for a more complex setting including treatment and piped distribution – are shown below in Figures 16.2 and 16.3.

Finally, the flow diagram is verified in the field. This will involve site inspection and for groundwater the use of conservative tracers and hydrogeological models. As groundwater flow is often complex, the process of verifying the flow diagram may be ongoing and it can be expected that the understanding of hydrogeological conditions will improve over time as more information becomes available.

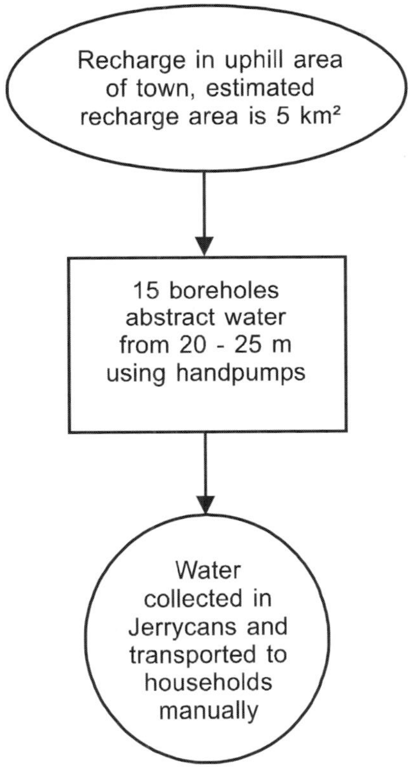

Figure 16.2. Simple flow diagram for small point water source

16.3.3 Identifying intended uses and vulnerability of the users

The intended use of the water supply should be defined to ensure that the requirements this may place on quality are incorporated into the WSP. In some cases, there will be more than one use for the water (i.e. domestic, industrial, irrigation) that may compete for allocation of resources where these are scarce. The WSP and supporting programmes should be clear in defining control in relation to drinking/domestic use. It is also important to consider whether there are particularly vulnerable groups using the water (i.e. immuno-compromised, elderly, infants) and to consider the socioeconomic conditions and vulnerability of different groups using the water. This will be linked to the development of the WSP and in particular relates to the hazard analysis and corrective actions.

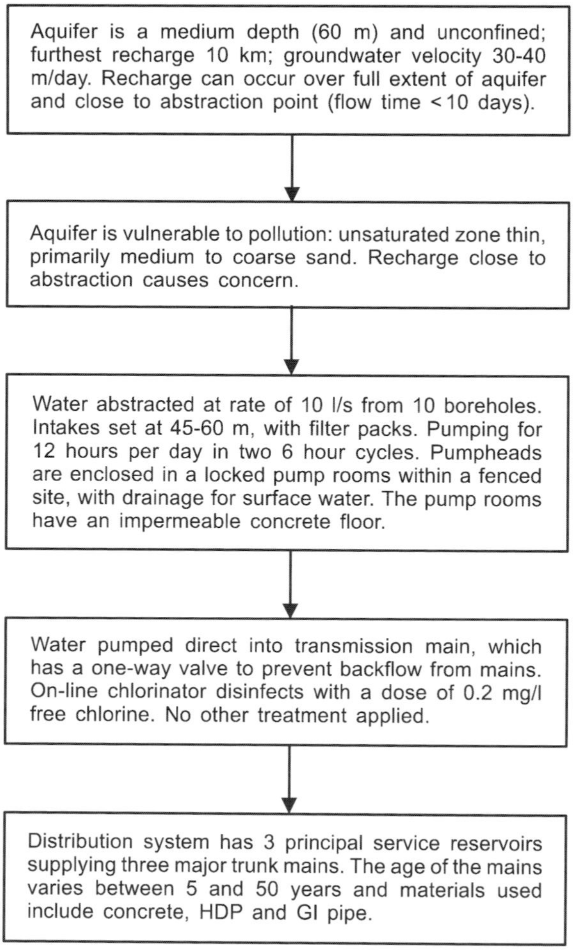

Figure 16.3. Simple flow diagram for large groundwater abstraction linked to piped distribution

16.4 HAZARD ANALYSIS

Once the system has been described, a hazard analysis should be performed. A hazard is a biological, chemical, physical or radiological agent that has the potential to cause harm to health. The simplest method of undertaking a hazard analysis is to perform a sanitary survey or catchment assessment to identify all the sources of potential hazards. Chapters 9-13 of this book provide information on the likely hazards that will be derived from different polluting activities and sources of hazards as well as indicative checklists to help assess their relevance in a given setting. The sanitary survey or catchment assessment should lead to the preparation of a map that provides details on where water sources and sources of pollutants exist within the recharge area. An example of a map in a simple setting is shown in Figure 16.4.

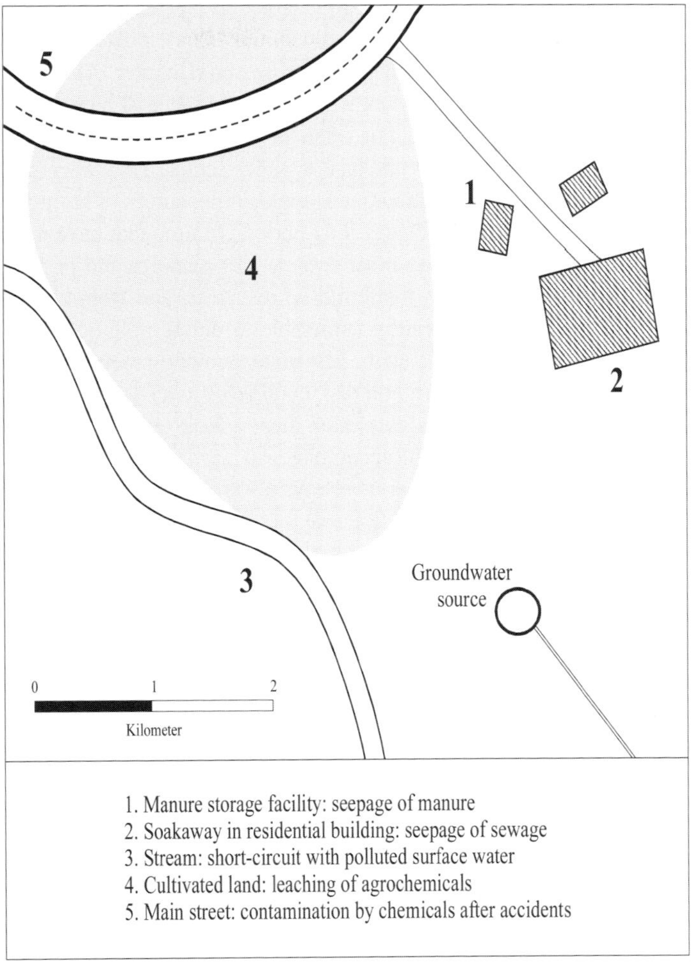

Figure 16.4. Simple map showing potential pollution sources to a small groundwater abstraction point (adapted from SGWA, 2003)

When undertaking a hazard analysis, it is often more effective to consider hazardous events rather than the specific hazards that affect the water supply. A hazardous event is an incident or situation that can lead to the presence of the hazard, and thus describes how a hazard could enter the water supply. For instance, a hazardous event could be that pathogens from human faeces enter groundwater from poorly constructed and sited septic tanks, or that hazardous chemicals leach into groundwater when spilled by accident at an industrial site. The advantage of using a hazardous event approach is that the probability of the event occurring can be considered, as the presence of a source of hazards within the drinking-water catchment area does not automatically mean that the hazards will be found at a groundwater abstraction point.

The probability with which hazards reach the aquifer depends on hydrogeological conditions as described in the concept of aquifer vulnerability in Chapter 8 and on their behaviour in the sub-surface as described in Chapters 3 and 4. By combining sanitary survey or catchment assessment data, vulnerability and pollutant behaviour the probability of occurrence of contaminants can be assessed (Chapter 14).

As discussed in Section 15.2, in groundwater the occurrence of hazards may not only be of an episodic (or event) nature but often is continuous and causes longer-term pollutant loading and attention should be paid to times scales. They may also be related to diffuse or multiple point sources rather than single point sources of hazards. Chemicals that are intentionally released to land over long periods of time that have the potential to accumulate in the aquifer, e.g. nitrate from fertilizers or manures and pesticides used in agriculture are examples of hazards that build up over time and from a diffuse area. In settings where the aquifer is moderately vulnerable to nitrate pollution, it is likely that controlled use of fertilizer in line with best management practices will not cause groundwater quality deterioration. In contrast, continuous over-fertilization in the same setting over years or even decades can cause heavy long-term nitrate pollution. For substances having the potential to accumulate in groundwater, the definition of hazardous events (e.g. over-fertilization) is always also related to the protection of future source water safety, as the present situation may not result in short term deterioration of water quality.

For natural chemical constituents affecting human health (e.g. arsenic and fluoride), the hazard analysis should include an assessment of the geological setting to evaluate whether it is likely that there will be any naturally occurring chemicals present at levels that will pose a risk to public health. The potential impact of land use on mobilization of chemicals should also be considered at this point. An initial comprehensive water quality assessment remains a further essential component of the hazard assessment for chemicals.

The hazard analysis will direct subsequent stages of Water Safety Plan development to ensure that control measures are managed, upgraded or put in place in such a way that they will effectively control identified hazards. Undertaking hazard analysis and assessing the risk posed by the hazards identified is essential: simply applying control measures without considering which hazards are the most important will mean that risks posed by some hazards are likely to remain high and resources will be wasted in controlling hazards that may be irrelevant for the specific water supply.

16.5 SYSTEM ASSESSMENT

The system assessment stage of the WSP development uses the information gained in the system description and hazard analysis in order to assess the risk of hazards occurring in drinking-water, and the potential of the system to control them. It is a first step in determining whether the water supply is able to meet the health based targets or other water quality targets defined for the water supply and if not, what investment of human, technical and financial resources would be required to improve the supply.

System assessment is largely the subject of this book, with Section II introducing the hazards, Sections IV and V the control measures and Section III discussing how to combine information for assessing the risk of hazards occurring in groundwater. The following section revisits this in terms of using these to develop a Water Safety Plan.

System assessment in the WSP-context will include further barriers such as treatment that determine whether the risk can be controlled.

For example, if a borehole abstracts water from a deep aquifer that has a significant unsaturated zone, limited human development over the aquifer and the potential to use legislation to control activities, the system is likely to be capable of meeting established targets and control measures in the catchment can be identified. In this case, the system assessment has not shown the need for any significant upgrade. By contrast, if a borehole abstracts water from a karstic aquifer where there is extensive human development over the aquifer and no disinfection, the system may not be able to meet the targets without investment at least in a treatment step at the borehole. This example illustrates when a systems assessment will identify that an upgrading of the system is required and the investment needs. The system assessment, therefore, may identify immediate investment requirements essential for meeting the targets. It is unlikely that all control measures will require substantial infrastructure improvements and therefore even in situations where improvements are needed, some readily implemented control measures can be identified, monitored and managed.

Risk assessment and prioritising hazards for control

The definition of control measures should be based on a ranking of risks associated with the occurrence of each hazard or hazardous event. Section 14.2 of this book discusses how to assess the potential for contaminants to occur in groundwater. However, in context of developing a Water Safety Plan, prioritising hazards will be more comprehensive and will involve the assessment of hazards in whole supply system, including steps after the abstraction of groundwater (such as disinfection and distribution). A more generic approach to assessing risks and prioritising hazards is therefore briefly introduced here.

Risk is the likelihood of identified hazards causing harm in exposed populations in a specified time frame, including the magnitude and consequence of that harm. Those hazardous events with the greatest severity of consequences and highest likelihood of occurrence should receive higher priority than those hazards whose impacts are mild or whose occurrence is very uncommon.

There are a variety of means by which prioritization can be undertaken, but most rely on applying expert judgement to a greater or lesser degree. The approach discussed below uses a semi-quantitative risk scoring matrix to rank different hazardous events. This approach has been applied in risk assessment, and specifically for drinking-water in Australia and Uganda (Deere *et al.*, 2001; Godfrey *et al.*, 2003); similar approaches have been used in other countries, such as New Zealand and Switzerland (SGWA, 1998; MOH NZ, 2001).

Within this approach, severity of impact is categorized as three major types of event: lethal (i.e. significant mortality affecting either a small or large population); harmful (i.e. morbidity affecting either a small or large population); little or no impact. Table 16.1 shows the definition of a set of variables for likelihood/frequency of occurrence and combined severity/extent assessment with appropriate weighting of variables. Table 16.2 indicates the final overall score of all possible combinations of the conditions. Table 16.3 gives an example of that approach for a scenario in an alluvial aquifer.

Table 16.1. Examples of definitions of hazardous event terms that can be used for risk scoring (modified from WHO, 2004)

Description	Definition	Weighting
Likelihood or frequency of occurrence		
Almost certain	Once per day	5
Likely	Once per week	4
Moderate	Once per month	3
Unlikely	Once per year	2
Rare	Once every 5 years	1
Severity of consequence or impact		
Catastrophic	Potentially lethal to large population	5
Major	Potentially lethal to small population	4
Moderate	Potentially harmful to large population	3
Minor	Potentially harmful to small population	2
Insignificant	No impact or not detectable	1

Table 16.2. Example of a simple risk ranking matrix (modified from Deere *et al.*, 2001 and WHO, 2004)

Likelihood or frequency of occurrence	Severity of consequence or impact				
	Insignificant	Minor	Moderate	Major	Catastrophic
Almost certain	5	10	15	20	25
Likely	4	8	12	16	20
Moderate	3	6	9	12	15
Unlikely	2	4	6	8	10
Rare	1	2	3	4	5

While some approaches use a scoring method as indicated by the numbers in Table 16.2 others prefer non-numerical classifications describing the risk (as indicated by the shading; see also Table 15.2 in Chapter 15). It should be stressed that when using the scoring approach it is the relative ranking based on the numerical categories rather than the numbers themselves that is important. Furthermore, in using such approaches common sense is important to prevent obvious discrepancies arising from applying the risk ranking, for instance events that occur very rarely but have catastrophic effects should also be a higher priority for control than those events that have limited impact on health, but occur very frequently.

The risk ranking approach allows the relative importance of different hazardous events to be systematically evaluated. This supports decision makers to define priorities for control within their water supply and can therefore maximize the cost-effectiveness of the WSP.

The outcome of system assessment is an identification of priorities for controlling risks, of existing control measures that have been found to be of importance for controlling risks, and of gaps in safety for which new control measures need to be identified or existing ones upgraded.

Table 16.3. Example of hazardous events identified and assessed for an alluvial aquifer

Process step	Hazardous event	Hazard type	Likelihood	Severity	Risk score
Alluvial aquifer	Water pumped during a storm event results in contaminated surface water from catchment run-off being drawn into aquifer	Microbes and chemicals (nutrients and potential pesticides from agricultural practices)	Unlikely (2)	Catastrophic (5)	10
	Cattle grazing near wellhead and rain events result in contaminated surface water entering the wellhead	Microbes and chemicals (mainly nutrients)	Moderate (3)	Catastrophic (5)	15
	Draw down of aquifer causing naturally occurring chemicals to enter water	Chemicals	Rare (1)	Major (4)	4

16.6 CONTROL MEASURES

In the context of a Water Safety Plan, control measures are tightly defined as those steps in drinking-water supply that directly affect drinking-water quality and that collectively ensure that drinking-water consistently meets health-based targets. Therefore, they are the basis on which control is assured and therefore they should always function reliably. Control measures are activities and processes applied to prevent hazard occurrence within the water supply chain or at the pollutant source, which control the risk posed by the hazards or hazardous events identified on a continuous basis. Examples of control measures for groundwater protection are provided in Chapter 17, for the immediate protection at the abstraction point in Chapter 18, for hydrological management in Chapter 19 and for specific polluting activities in Chapters 21-25.

Control measures in groundwater sources can take one of two forms:

- those that use natural attenuation processes to reduce or remove/inactivate pollutants (e.g. adsorption, filtration, predation, microbial degradation, die-off);
- those that prevent or reduce pollution of the aquifer's recharge area or ingress of pollutants into the water supply.

Measures such as groundwater protection zones (Chapter 17) and hydrological management (Chapter 19) combine both forms of control measure. In these cases, release of pollution within recharge areas is controlled (although often not entirely prohibited) within prescribed areas related to the time taken to travel to the water abstraction point, and rely on attenuation, die-off and hydrodynamic dispersion to achieve reductions in pollutant loads. When protection zones are used, there may be extensive control on specific activities such as restriction on traffic, settlements, agricultural activity and other human activity as discussed in Chapter 17.

Measures such as wellhead and sanitary completion discussed in Chapter 18, and avoidance, reduction and treatment of pollutants at their source described in Chapters 21-25, represent the second type of control measure. These are all designed to prevent contamination from either entering the supply or being released into the environment rather than relying on natural processes to remove or reduce pollutants.

The control measures should be able to influence the quality of groundwater and to be amenable to control through action, preferably by the water supplier. For instance, rainfall often exerts a profound influence on shallow groundwater and numerous studies have shown that rainfall is a principal factor in water quality deterioration (Wright, 1986; Gelinas *et al.,* 1996; Howard *et al*., 2003). Rainfall cannot be translated into a control measure because action cannot be taken directly to reduce rainfall. A series of control measures can be defined, however, that relate to the importance of rainfall in causing contamination. For example these would include providing diversion ditches to prevent inundation of the abstraction point by contaminated surface water, maintaining infrastructure integrity at the abstraction point and controlling pollutant releases through the use of protection zones. Details of required measures are outlined in Chapters 17 and 18.

WHO (2004) note that control measures included within the WSP should have the following basic characteristics:

- a monitoring system and operational limits can be defined that describe the performance of the control measure and which can be either directly or indirectly monitored;
- corrective actions can be identified as a response to deviations in control measure performance that are detected by monitoring;
- corrective action will protect water safety by ensuring that control is re-instated (this can be bringing the control measure back into compliance, enhancing the control measure or by implementing additional control measures);
- detection of deviations and implementing corrective actions can be completed sufficiently rapidly to prevent the supply of unsafe water.

Control measures should not be vague or imprecise as otherwise the management action to be informed from the assessment of control status will be difficult to define. For instance, control of agricultural pollution in a catchment may be a general control measure, but would need to be translated into a set of specific actions, such as a seasonal restriction on the application of fertilizers and manure, and restriction on feedlots within a specified distance. An example for controlling nitrate pollution in groundwater from agriculture by means of integrating the water supplier's WSP and the activities of a Regional Nitrate Committee in Switzerland is given in Box 16.1.

Individual control measures will usually be designed for a specific hazardous event(s) but it is unlikely a single control measure will provide assurance for all hazardous events and situations; rather a suite of different control measures may be necessary. Table 16.4 below provides examples of likely control measures appropriate to groundwater sources.

Many of the control measures for groundwater protection may have to be implemented through the relevant stakeholders rather than by the water supplier. The water supply agency, however, may take the lead as part of the WSPs supporting programmes (see Section 16.10) in initiating specific controls in order to protect the quality of the drinking-water source.

Box 16.1. Measures for controlling nitrate pollution from agriculture in the Lenzburg water supply

The municipality of Lenzburg, Switzerland as well as the surrounding municipalities, draw drinking-water from the same groundwater recharge area. Each municipality operates its own abstraction wells. Part of the area upstream of the abstraction zone is used for agriculture, and the rest of the upstream area is covered by forest. Due to over-fertilization by agriculture, high nitrate concentrations were measured at all abstraction wells in the recharge area. These nitrate concentrations increased from 15 mg NO3/l in the early 1960s towards the Swiss drinking-water standard of 40 mg NO3/l by the early 1980s.

The Nitrate Committee: On the initiative of the Lenzburg water supply (LWS), all municipalities that were affected by increasing nitrate concentrations in drinking-water, or that encompassed large areas with agricultural use in the recharge area, collectively formed a Regional Nitrate Committee in 1987. Each municipality appointed one district council representative plus one local farmer as members of the Committee. In addition, a professional advisor from each of the Cantonal Water Authorities and the local Agricultural Advice Centres, plus a geologist, were co-opted to provide support to the Committee. The Nitrate Committee is chaired by the LWS. All costs that result from the committee's activities are shared between all the municipalities involved according to a pre-established formula.

N-min measurements: The Committee as its first activity organized public advisory consultations for local farmers. In parallel, a nitrate regulation was elaborated that did not come into force in the municipalities involved but was set up as a target to be met by farmers on a voluntary basis. The regulation included detailed maps at the single plot scale showing the results of soil analyses and N-min measurements. N-min values represent the amount of plant-available nitrogen at the beginning of the vegetation period, and therefore provide the basis for defining an adequate amount of fertilizer to be applied. N-min measurements are repeated every year, and the results are made available to farmers, in conjunction with fertilization recommendations for individual crops, through the Agricultural Advice Centre. All samples, measurements and advisory consultations are funded through the Nitrate Committee and are therefore free of charge for the farmers.

Subsidies for intercropping: As nitrate leaching occurs mainly between seasons when fields lie fallow between two crops, the Committee recommended and supported intercropping with nitrogen binding cover crops. If farmers fulfil specific pre-conditions, such as prohibitions of tillage operations for a limited time period, minimum duration of intercropping depending on the rotation of crops, data recording by means of a field calendar, and use of recommended intercropping species, then the Committee supports those activities by paying intercropping subsidies per cultivated area unit, i.e. 400 Swiss Francs per ha in 2003. The plot-tailored fertilization together with intercropping has stopped the trend of increasing nitrate pollution of the aquifer, e.g. nitrate concentrations declined to 25-30 mg NO3/l in 2003.

Municipality liaison: The role of farmer representatives in the Nitrate Committee is of great significance because they liaise between the farming community and the Committee, and thus with the LWS. On the one hand they represent the farmers' interests to the Committee. On the other hand they have a duty to advise their farming colleagues in the municipality of the recommendations of the Committee. Furthermore, they have to check whether the statements made by farmers in their applications for intercropping subsidies are valid, and whether all pre-conditions have been fulfilled. If the liaison person approves each farmer's application, then the Nitrate Committee will pay the subsidies for intercropping.

Lenzburg's Water Safety Plan: The WSP of the LWS identifies long-term nitrate accumulation in the aquifer as a priority hazard. However, the LWS cannot exert direct control on polluting activities by farmers, as it does not own the farmland in the recharge area. Therefore the WSP explicitly refers to the activities of the Nitrate Committee, i.e. controlling nitrate pollution (through fertilization recommendations and granting subsidies for intercropping), monitoring compliance with those control measures, and imposing sanctions to farmers in case of non-compliance (corrective action). Additionally to the integration of the Committee's activities into the WSP, LWS carries out six-monthly inspections of the recharge area as means of verification that control system works effectively.

Table 16.4. Examples of control measures to protect the quality of drinking-water

Element	Control measure
Wellhead completion	Sanitary seal (prevention of direct ingress) Fencing around area Surface water diversion ditches Quality of concrete works Wastewater drainage
Land use planning	Protection zones (designated and limited uses, protective requirements) Control of human activities within drinking-water catchment Set-back distances Minimum safe distance (latrine-source) Animal access control Grass cover maintained in immediate area

16.6.1 Validation of control measures

Validation is an essential component of a WSP. Validation is an investigative activity to assess the effectiveness of individual and combinations of control measures in reducing the risk posed by hazards or hazardous events. It therefore obtains evidence on the performance of control measures and ensures that the information supporting the WSP is correct. In some cases, the performance of an individual control measure may depend in part on the performance of another previous control measure and this must be borne in mind when defining performance criteria. This reflects the multiple barrier principle that is advocated as part of effective risk management of drinking-water quality.

The efficacy of each control measure in reducing or eliminating the risk of exposure to pollutants should be measured directly against the hazard that it is designed to mitigate. This requires a research stage where the performance of the control measures, individually and in combination, is rigorously evaluated with regard to the hazards they are expected to control. This stage can be undertaken in a variety of ways and in many cases it is best to utilize all four. These are:

- evaluation of existing literature (including Chapters 3 and 4 of this book) to determine recorded survival, attenuation and dilution of pollutants in the type of groundwater to be utilized;
- rigorously designed field assessments of water quality and influences on quality;
- laboratory experiments using model groundwater;
- modelling of pollutant transport in groundwater.

In many cases, the existing literature (including this book) provides much of the information required to define control measures. However, as local conditions may vary significantly from those reported in the literature, the use of well-designed field assessments is often an effective way of gaining additional information to suit local conditions. Such assessments should be based on intensive assessment of a representative sample of water sources and evaluate data in order to define the importance of different factors in causing contamination (Howard *et al.*, 2003). Where quality is poor, making improvements in the water sources or in reducing a pollutant source can also provide a useful way to measure efficacy.

The selection of parameters undertaken in such an assessment should relate to the principal pollutants of concern. For chemicals, this process is relatively straightforward as the substances of concern can be analysed and system performance in elimination therefore be validated. For microbial quality, it is preferable for validation to be based on assessments of control measures in relation to pathogens. It may be difficult to undertake analysis of a wide range of pathogens and therefore validation may focus on a small number of representative pathogens whose control would provide confidence that all pathogens of a similar nature would also be controlled. However, it may be necessary to undertake evaluations using indicator organisms as described in the example from Uganda in Chapter 18. Wherever possible, the range of microorganisms should reflect the range of likely pathogens. These may include *E. coli* and faecal streptococci as faecal indicator bacteria, bacteriophages as index organisms for viruses and spore forming bacteria as process indicators (Ashbolt *et al.*, 2001). The primary purpose when using such organisms is to assess the impact of the control measures on the levels of the organisms and to use the resulting information on reductions as an indication of the likely impact of the control measures on pathogens. The criterion for their choice therefore is their similarity in retention by control measures as compared to groups of pathogensin order to indicate system performance in pathogen removal.

The outcome of validation is an assessment of how well the control measures in place or envisaged for introduction are likely to meet the heath based targets. This may include identifying the need for system upgrade as well as particular emphasis on monitoring and maintenance of control measures identified as being key to safety of a given supply system.

16.6.2 Establishing operational limits

For each control measure, operational limits of performance should be defined. These are quantifiable levels of performance that provide an indication of whether the control measure is functioning correctly (in compliance) or is not providing effective control (out of compliance). The operational limits for each control measure should be identified during validation, and their definition should be based on sound science but also take into account practical considerations regarding limits of detection and ease of measurement. Of utmost importance is ensuring that the operational limit is related to an action that can be taken to bring the control measure back into compliance.

Operational limits may be upper limits, lower limits or an envelope of performance measures and are typically simple process indicators that can be interpreted at the time of monitoring and where action can be taken in response to a deviation (WHO, 2004). For instance, a groundwater protection zone may be defined as a control measure and within this zone the discharge of faecal material from sanitation facilities is strictly controlled. The operational limits in this case will be the absence of sources of faecal material (e.g. from septic tanks or pit latrines) within the protection zone. If a new on-site sanitation facility is constructed within the protection zone, the operational limit is exceeded and therefore corrective action should be taken to remove the facility or ensure that the design prevents contamination from occurring. The results of monitoring in relation to this operational limit (e.g. presence or absence of on-site sanitation facilities) can be interpreted immediately on observation and a clear line of action can be defined in response to the deviation.

Control measures may also be defined related to pumping rates if it has been shown that the draw-down would substantially alter the protection zone above a certain level of pumping. In this case, the operational limit will relate to the pumping regime (possibly both in terms of allowed discharge and in terms of duration of pumping). Other operational limits that can be defined would include stock density in relation to risks of increased nitrate (Chapter 21) and simple measures of wellhead or sanitary completion related to drainage (Chapter 18).

When defining operational limits, it is important to avoid situations where exceeding the operational limit will result in immediate health risks. It is better to establish operational limits that are more conservative and still allow preventative actions to be taken. If operational limits cannot be defined, it is likely that the measure identified should be considered as being part of a supporting programme. For some control measures, further limits may be established as 'critical' limits at which exceedance represents a confidence in water safety is lost and urgent action is required. Such limits tend to be related primarily to treatment processes, for instance disinfection, rather than source protection measures.

16.7 OPERATIONAL MONITORING

Operational monitoring assesses the performance of control measures at appropriate time intervals. It is essential within WSPs to ensure that the control measures employed remain in compliance with the operational limits. An emphasis is placed on simple techniques , which describe process controls that allow rapid, and easy measurement and whose findings can be interpreted at the time of monitoring with actions identified in response to non-compliance (WHO 2004; Davison *et al.*, 2005). Examples of monitoring parameters include turbidity control to sanitary inspection (Howard *et al.,* 2001; Payment and Hunter, 2001). This requirement for operational monitoring means that the analysis of indicator organisms would not be included as monitoring parameters, although they would be used in verification (see Chapter 16.12). By contrast, for chemical hazards it may be appropriate to test for the substance of concern if the results is available within sufficient time to allow for corrective action before hazard break-though, although the analytical method may be different from that used for verification.

Operational monitoring should be able to quantify changes in performance of the control measure in relation to the operational limits and is therefore linked directly to process control or management actions prior to an increase in the risk posed by a hazard. Selection of monitoring parameters should relate to their reliability and sensitivity in estimating performance of the control measure in relation to the operational limits. The frequency of monitoring will depend on the nature of the control means and may in some circumstances be continuous and on-line, whilst in others may be relatively infrequent. In the case of groundwater sources, much of the monitoring will be based on inspection of controlled activities in the catchment area (Box 16.1) and the integrity of sanitary completion measures, rather than routine water quality testing.

Monitoring may include testing for specific water quality parameters. This may be on the source water or within the catchment area or surrounding specific pollution activities, for instance around mine tailings or landfills. It will be important to define whether monitoring should be of the pollutants of direct concern or whether sentinel chemicals or other properties of water can be used as surrogates. This may in some cases include monitoring of water levels or redox conditions if this will provide good information about increases in contamination risks.

Where water quality parameters are included in monitoring (for instance nitrate in relation to agricultural pollution or inorganic chemicals derived from landfill leachate) it is unlikely to be continuous and will be determined in relation to their adequacy for monitoring the control measure in specific settings. In each setting, monitoring may become more targeted during times of known elevated risk (e.g. seasonal influences on nitrate release). In some settings however, monitoring of such parameters may be very frequent for instance where aquifers are highly vulnerable and have rapid transit times, thus allowing very short time periods for corrective actions.

In summary, the indicators used in operational monitoring systems should be:

- *specific* – the indicator should relate to a particular control and not to a broad set of interrelated factors;
- *measurable* – it should be possible to translate the control status into some form of quantifiable assessment, even if data collection is based on semi-quantitative or qualitative approaches;

- *accurate* – the indicator must provide an accurate reflection of the control measure status in relation to the operational limits and be sensitive to changes that are of relevance and changes that may lead to exceeding the water quality targets;
- *reliable* – the indicator should give similar results each time it is measured;
- *transparent* – the process of selection of the monitoring indicator, the method and frequency of measurement and the interpretation of the results should be transparent and accepted by all stakeholders.

16.8 CORRECTIVE ACTIONS

Effective management implies definition of actions to be taken in response to variations that occur during normal operational and incident conditions. Such corrective actions should be defined for each control measure and documented in the WSP. Corrective actions are those interventions that will be undertaken in immediate response to control measures moving outside the operational limits defined. It is important that when developing the WSP such corrective actions are identified from the outset. Identification of corrective actions should not wait until a failure has occurred as this defeats the objective of risk management. However, lessons learnt from incident conditions may lead to improvement of corrective actions and thus these will not be static. Equally, corrective actions may also be refined based on experiences from other water supplies.

Corrective actions may be simple operational interventions, for instance if an inspection identifies problems with the fence or deterioration in concrete protection works around a borehole, immediate action should be taken to effect repairs. It may also involve more complex enforcement processes, for instance if stock densities are seen to increase in the catchment area, then action should be initiated to ensure that farmers reduce these (for instance through legal notices). They may involve interventions that are not possible to implement immediately or that will take some time to take effect, for example when there has been an accidental spill of chemicals that has reached the aquifer and which will require remediation through pumping and treating.

Corrective actions may include longer-term action, for instance the redefinition of groundwater protection zones as more information becomes available regarding groundwater flow and pollutant movement. In some cases, the corrective action may be limited to an increase in monitoring of a specific contaminant. For instance, if leaching from mine tailings or landfills has increased but the consequences are as yet unknown, it may be appropriate to install a monitoring network to monitor movement and behaviour in the first instance to determine whether further action is required.

Corrective actions may also include closing down a particular abstraction point until the contamination has been effectively removed or has passed through the aquifer. However, this option should only be considered when there are alternative water supplies available. If the description of the water source concludes that there is no viable alternative to the groundwater source, then it is essential that other corrective actions (e.g. treatment) can be implemented immediately to prevent public health risks.

Where a public health risk from contamination occurs despite the presence of control measures, this implies that further control measures must be defined and implemented.

This will involve investigating the cause of the contamination leading to the public health risk and from this data defining a set of new control measures to combat this risk. It should be noted that it is likely that new risks will be identified over time, and these should be assessed by periodic system validation (see Section 16.6.1). Furthermore, levels of tolerable risk and the health-based targets established may change.

16.9 VERIFICATION

Verification is a separate process to operational monitoring. It provides a final check on the overall safety of the drinking-water supply chain. Verification is not designed to be a routine frequent assessment but a periodic evaluation of the performance of the WSP as a whole. For utility supplies, verification is undertaken by the water supplier as well as independently by the surveillance agency. For community-managed water supplies, verification is likely to only be undertaken by the surveillance agency. Verification will typically involve a number of actions including audit of the implementation of the WSP and water quality analysis.

Audits of WSPs are designed to assess whether these have been appropriately designed, documented and implemented. As part of a typical audit, the records of monitoring and actions taken to ensure control is maintained are reviewed by inspectors who also inspect the infrastructure and results of monitoring to ensure that the WSP is being adhered to. Such audits will also typically assess whether communication (both within the supply organization and to regulators and users of the water) have been undertaken in a timely and appropriate manner following guidelines set out in the WSP. Audits can be equally applied to utility supplies (when internal verification will also be assessed) or community-managed supplies. In the latter case, the audit is likely to use different tools but still focus on whether the monitoring is being performed appropriately, whether control measures are functioning and whether this information is shared within the community. Water quality analysis is also likely to be included within verification programmes. Analysis of microbiological indicators is retained in such approaches, but would be undertaken less frequently than in systems relying largely on end-point testing. The range of microorganisms would be expected to increase to take account of the diversity of pathogens being controlled, yielding information of greater value in assessing performance of water quality management measures (Ashbolt *et al.*, 2001).

Careful consideration should be given to the selection of the indicator organisms used for verification. *E. coli* remains the indicator organism of choice (WHO, 2004) and in many situation thermotolerant coliforms can be used as a surrogate. However, where possible other indicator organisms should also be considered used. Other indicators include faecal streptococci, and bacteriophages. It may also be of value to undertake tracer studies in order to verify whether the land use control measures will provide adequate protection. This could also be linked to hydrogeochemical models and contaminant propagation models where these are adequately calibrated and reliable.

Chemical testing may be included both in monitoring and verification, but the techniques used may vary depending on the objective of the testing. The parameters used in verification should be evaluated at the same time as validation in order that they can be calibrated against an acceptable risk of exposure. Verification is likely to include periodic analysis of the presence and concentration of substances in groundwater. This may be done at the source, within monitoring networks established around the abstraction point or monitoring around the sources of pollution. The design of appropriate sampling networks is critical and should provide sufficient detailed information to ensure that preventative actions can be deployed in a timely manner.

16.10 SUPPORTING PROGRAMMES

In addition to process control measures put in place to assure safety, further activities are required in order to ensure that safe drinking-water can be assured, including activities which have to be undertaken by institutions and agencies other than the water supplier. These supporting programmes are as essential to the delivery of safe drinking-water as are the control measures and monitoring identified in the WSP.

There are a number of types of supporting programmes, some examples are:

- a water supplier's documented policy and commitment to provide high-quality water supplies;
- appointment of a senior member of staff as the risk or water safety manager who is responsible for ensuring the safety of drinking-water produced;
- establishment of internal allocation of roles and responsibilities for assessment and management of risks;
- established internal communication strategy within the utility to ensure information from monitoring is acted upon promptly and appropriately;
- training provided to community operators;
- design and construction codes of practice as well as codes of good hygiene practice established and enforced;
- information exchange with regulators and other stakeholders;
- a risk communication strategy to provide information to the public in times of elevated risk;
- customer complaint procedures;
- implementation of Good Laboratory Practice, including calibration of monitoring equipment;
- staff training and awareness programmes;
- securing stakeholder commitment to the protection of groundwater;
- development of training and education programmes for communities whose activities may influence source water quality;
- establishment of collaboration contracts with farmers or farmers' associations (which may include financial incentives);
- training of catchment inspectors;
- mapping of catchment characteristics (e.g. land use; vulnerability; protection zones).

The proper implementation of supporting programmes is essential for effective control of public health risks from water supplies and should be accorded adequate priority.

In situations where the water supplier does not own the land that forms the catchment, and thus has no direct control over activities in it, many control measures related to source protection may become part of the supporting programmes. Specific examples may include, but are not limited to:

- development and implementation of catchment management plans;
- controlled density of stock in pastoral areas;
- controlled application of fertilizers and pesticides in the catchment;
- controlled access for the general public to pollution-sensitive areas in the catchment;
- development of groundwater quality models.

16.11 DOCUMENTATION

The final part of the development of the WSP is to document the process, considerations and criteria leading to assessments, and to ensure that people responsible for implementing the WSP have a point of reference. Documentation is also important as part of monitoring the effective implementation of the WSP. Therefore record keeping of monitoring and actions taken is an essential component of the plan.

Tables 16.5 and 16.6 provide examples of WSPs, one for a mechanized borehole and one for a tubewell fitted with a handpump. These WSPs are generic and are designed to provide the reader with a view of the type of material that may be developed. They are not designed as finished plans for immediate implementation, but as a framework within which WSPs can be developed. More detailed descriptions of WSPs may be found both in Water Safety Plans: Managing drinking-water quality from catchment to consumer (Davison *et al.*, 2005) and the *Guidelines for Drinking-water Quality*, third edition (WHO, 2004).

Table 16.5. Model Water Safety Plan for mechanized borehole (based on Davison *et al.*, 2005)

Hazardous event	Cause	Risk	Control measure	Operational limits		Monitoring			Corrective action	Verification
				Target	*Action required if*	*What?*	*When?*	*Who?*		
Ingress of contaminated surface water directly into borehole	Poor well-head completion	Unlikely/ major	Proper wellhead completion	1 m concrete apron around wellhead Lining extends 30 cm above the apron Drainage ditches in place	Lining stops at ground level Apron damaged or cracked Ditches full, faulty or absent	Sanitary inspection	Monthly	Operator	Extend lining Repair apron Clean and repair drainage ditches	Sanitary inspection *E. coli* Faecal streptococci Bacteriophages
Ingress of contaminants due to poor construction or damage to the lining	Poorly maintained wellhead completion	Moderate/ major	Proper wellhead completion	Top 5 m of the annulus sealed Rising main in good condition	Annulus sealed for less than 3 m Colour changes Increased pumping required to raise water	Sanitary inspection Water clarity CCTV	Monthly	Operator	Insert seal around annulus Replace worn and corroded rising mains Use materials less likely to corrode (e.g. plastics)	Sanitary inspection Analysis of colour, iron and turbidity CCTV
Borehole area is inundated with contaminated surface water	Lack of diversion ditches	Unlikely/ major	Good drainage around wellhead	Diversion ditches of adequate size, in good condition and clear of rubbish	Ditch has rubbish or shows signs of wear	Sanitary inspection	Weekly	Operator	Repair and clean ditch Increase size of ditch using	Sanitary inspection
Pumping leads to increased leaching of contaminants	Pumping induces increased leaching of chemicals	Unlikely/ moderate	Pumping regime	Leaching of contaminants is within predicted range	Evidence of increased leaching of contaminants	Monitoring of key contaminants of concern Hydrochemical models	Monthly	Operator	Modify pumping regime Treatment	Hydrochemical models Monitoring contaminants of concern

Hazardous event	Cause	Risk	Control measure	Operational limits		Monitoring			Corrective action	Verification
				Target	*Action required if*	*What?*	*When?*	*Who?*		
Contaminated shallow water drawn into aquifer	Hydraulic connection exists between shallow and deeper aquifers allowing draw-down into deeper aquifer	Almost certain/ moderate	Control pumping regimes Set intake at depth	No evidence on induced leakage	Evidence of shallow water drawdown (e.g. shallow wells start to dry up)	Colour (appearance) Taste Odour Electric conductivity	Weekly	Operator	Set intake deeper (microbes) Water treatment (microbial) or blending (chemicals)	*E. coli* Faecal streptococci Bacteriophages Nitrate Tracer studies Hydrological models
Rapid recharge by rivers, streams and ponds	Hydraulic connection exists between surface water and aquifers	Unlikely/ major to catastrophic	Set intake at greater depth	Rapid recharge does not occur or cannot reach intake	Evidence of rapid recharge from surface water bodies	Surface water levels Colour Electric conductivity	Daily	Operator	Set intakes at greater depth or modify pumping regimes	*E. coli* Faecal streptococci Bacteriophages Pathogen assessments Nitrate
Pumping increases safe distances beyond current protection zone boundaries	Pumping increases cone of depression extends minimum travel time distance beyond protection zone	Unlikely/ moderate	Protection zones	Protection zones include influence of drawdown on groundwater flow	Drawdown increases distance equivalent to travel time set	Water table levels surrounding borehole when pumping	Annual	Operator	Extend ground-water protection zone to account for the change in distance	Tracer tests Hydrogeological modelling Tracer tests Analysis of key microbial and chemical contaminants controlled in protection zones

Hazardous event	Cause	Risk	Control measure	Operational limits		Monitoring			Corrective action	Verification
				Target	*Action required if*	*What?*	*When?*	*Who?*		
Backsiphonage from pipe into borehole	No back-flow preventer installed	Likely/ minor	Backflow preventer on mains	Backflow preventer installed	Lack of backflow preventer	Inspect pumping works	Installations Periodic checks	Constructor Operator	Backflow preventer installed	Audit of wellhead and pumping works
Failure in disinfection process	Disinfection process fails	Unlikely/ major to catastrophic	Effective chlorination with contact time	Ct value adequate and residual produced	Lack of residual	Monitoring chlorine dosing and residual	Daily/ hourly	Operator	Take pump off-line and repair disinfection unit	Audit of results *E. coli* Faecal streptococci Bacteriophages
Mobilisation of toxic chemicals and elution of viruses	Changes in land use and increased recharge through irrigation leads to mobilisation and elution	Rare/ minor to moderate	Land use control, in particular managing irrigation	Little artificial recharge through irrigation, pH and Eh of water stable	Significant changes in land use Increased use of irrigation	Land use pH of groundwater Redox (Eh)	Weekly	Operator	Reduce artificial recharge	*E. coli* Faecal streptococci Bacteriophages Chemicals of concern
Leaching of microbial contaminants into aquifer	Leaching of faecal material from sanitation, solid waste, drains	Moderate	Protection zones and set-back distances	Lateral separation defined on basis of travel times and hydrogeology	Latrines/sewers built or solid waste dumps within separation distance	Sanitary inspection: inspection of protection zone, electric conductivity, sewer leakage	Monthly	Operator	Remove pollutant sources Improve sanitation design Reduce sewer leakage Insert cut-off walls around sewers	Inspection *E. coli* Faecal streptococci Bacteriophages Nitrate Chloride Tracer tests

Hazardous event	Cause	Risk	Control measure	Operational limits		Monitoring			Corrective action	Verification
				Target	*Action required if*	*What?*	*When?*	*Who?*		
Groundwater contains naturally occurring chemicals	Geological setting means chemicals present at toxic levels	Moderate	Source selection	Use of groundwater with no natural chemical at harmful levels	Evidence of natural contaminants	Risk assessment of geological setting Initial assessment of water quality	Before installation	Constructor	Use alternative source Treatment	Risk assessment Water quality assessment Monitoring of chemicals of concern
Agricultural pollution: nitrate	Use of inorganic or organic fertilizers stock density	Unlikely/ minor	Protection zone	Nitrate vulnerable zones defined for aquifer prevent excessive leaching	Evidence of increasing nitrate levels	Monitor nitrate in groundwater Monitor fertilizer applications Monitor stock densities	Monthly	Supplier Environment Agency	Control of fertilizer applications Blending of drinking-water	Nitrate levels in ground-water Audit fertilizer applications Audit stick densities
Agricultural pollution: pesticides	Pesticides leached into the ground-water	Unlikely/ minor	Protection zone	Pesticide applications controlled in recharge area	Evidence of increa-sing pesticides in water Evidence of pesti-cide application at high-risk locations and times	Monitor pesticide applications	Monthly	Supplier Environment Agency	Control of pesticide applications	Pesticide levels in ground-water Audit pesticide applications
Pollution from urban areas contaminates groundwater	Poorly sealed drains cause recharge of ground-water	Moderate/ minor	Protection zones	Drainage water unable to recharge groundwater	Poorly constructed drains increase potential for recharge	Inspection	Weekly	Operator	Ensure all drains properly sealed in recharge or vulnerable areas	Audit drainage channel design, construction and maintenance

Hazardous event	Cause	Risk	Control measure	Operational limits		Monitoring			Corrective action	Verification
				Target	*Action required if*	*What?*	*When?*	*Who?*		
Leaching of chemicals from landfill sites into ground-water	Leaching of chemicals from landfills, waste dumps, industrial discharges to ground	Moderate/ minor	Protection zone	Landfills are sanitary and properly sealed Landfill presence controlled on basis of travel times and hydrogeology	Monitoring around pollutant sources indicate increasing pollution migration	Monitor for key contaminants around pollutant sources Monitor bills of lading	Weekly/daily	Waste managers Environment Agency Supplier	Move pollutant sources Improve pollution containment Monitor network around pollutant sources	Inspection Analysis of chemical composition of pollution Analysis of water quality Audit bills of lading for composition of waste
Pathogens from hospital wastes contaminate ground-water	Poor disposal of hospital wastes allows direct ingress of leaching into ground-water	Unlikely/ catastrophic	Proper hospital waste disposal	Hospital wastes with pathogenic material incinerated	Hospital waste disposal in dumps or ground containers	Monitor hospital waste disposal methods	Daily	Water supplier Health authorities	Ensure all pathogenic material incinerated or sterilized	Audit hospital waste disposal
Industrial discharges contaminated ground-water	Poorly disposed of industrial waste can inundate groundwater source or leach into aquifer	Moderate/ minor	Waste containment and treatment	Effective disposal methods prevent spills and leaching	Waste disposal methods do not provide security against inundation and leaching	Monitor containment methods at industrial sites	Monthly	Supplier Environment Agency	Ensure all industrial waste is properly contained and treated at the site	Audit industrial wastewater treatment plants

Table 16.6. 'Model' Water Safety Plan for boreholes fitted with handpumps (based on Davison *et al.*, 2005)

Hazardous event	**Cause**	**Risk**	**Control measure**	**Operational limits**		**Monitoring**			**Corrective action**	**Verification**
				Target	*Action*	*What?*	*When?*	*Who?*		
Ingress of contaminated surface water directly into borehole	Poor wellhead completion	Unlikely/ major	Proper wellhead completion measures	1 m concrete apron around wellhead Lining extends 30 cm above the apron Drainage ditches in place	Lining stops at ground level Apron damaged or cracked Ditches full, faulty or absent	Sanitary inspection	Monthly	Community operator	Extend lining Repair apron Clean and repair drainage ditches	Sanitary inspection *E. coli* Faecal streptococci Bacteriophages
Ingress of contaminants due to poor construction or damage to the lining	Poorly maintained wellhead completion	Moderate/ major	Proper wellhead completion	Top 5 m of the annulus sealed Rising main in good condition	Annulus sealed for less than 3 m Colour changes Increased pumping required to raise water	Sanitary inspection Water clarity	Annual/as need arises	Community operator	Insert seal around annulus Replace worn and corroded rising mains Use materials less likely to corrode (e.g. plastics)	Sanitary inspection Analysis of colour and iron
Borehole area is inundated with contami-nated surface water	Lack of diversion ditches	Unlikely/ major	Good drainage around wellhead	Diversion ditches of adequate size, in good condition and clear of rubbish	Ditch has rubbish or shows signs of wear	Sanitary inspection	Monthly	Community operator	Repair and clean ditch Increase size of ditch using	Sanitary inspection
Contamination introduced as handpump requires priming	Priming water contaminated	Almost certain/ minor	Use direct handpump or clean water for priming	Water for priming stored in secure container	Priming water comes from contaminated source or is stored poorly	Inspection	Weekly	Community operator	Select handpump that does not require pumping	Test priming and borehole water for *E. coli* and faecal streptococci

Hazardous event	**Cause**	**Risk**	**Control measure**	**Operational limits**		**Monitoring**			**Corrective action**	**Verification**
				Target	*Action*	*What?*	*When?*	*Who?*		
Contaminated shallow water drawn into aquifer	Hydraulic connection exists between shallow and deeper aquifers allowing draw-down into deeper aquifer	Almost certain/ minor	Pumping regimes do not induce leaching	No evidence of drawdown of shallow groundwater	Evidence of shallow water drawdown (e.g. shallow wells start to dry up)	Colour Taste Odour Inspection	Annual/as need arises	Community operator	Set intake deeper (microbes) Water treatment (microbial) blending (chemicals)	*E. coli* Faecal streptococci Bacteriophages Nitrate Tracer studies Hydrological models Electric conductivity Redox potential
Leaching of microbial contaminants into aquifer	Leaching of faecal material from sanitation, solid waste, drains	Moderate	Provide adequate set-back distances defined on travel time	No sources of faecal material within set-back distance	Latrines/sewers built or solid waste dumps within separation distance	Inspection by community	Monthly	Community operator	Move pollutant sources Improve sanitation design Reduce sewer leakage	Inspection *E. coli* Faecal streptococci Bacteriophages Nitrate Chloride Tracer studies

Hazardous event	Cause	Risk	Control measure	Operational limits		Monitoring			Corrective action	Verification
				Target	*Action*	*What?*	*When?*	*Who?*		
Groundwater contains naturally occurring chemicals	Geological setting means chemicals present at toxic levels	Moderate	Select ground-water with acceptable levels of natural chemicals	Water quality assessments indicate water quality is acceptable	Evidence of natural contaminants	Risk assessment of geological setting Water quality assessment	Before construction Periodic evaluation	Water development agency	Use alternative source Treatment of water	Risk assessment Water quality assessment Monitoring of chemicals of concern
Leaching of chemicals into groundwater	Leaching of chemicals from landfills, waste dumps, discharges to ground	Moderate/ minor	Provide adequate set-back distances defined on travel time	No sources of chemicals within set-back distance	Pollutant discharges within set-back distance	Inspection by community	Monthly	Community operator	Move pollutant sources Improve pollution containment	Inspection Analysis of chemical composition of pollution Analysis of water quality

16.12 REFERENCES

APSU (2005) Technology water safety plans for rural water supplies in Bangladesh (1st edition). www.apsu-bd.org (accessed 2005).

Ashbolt, N.J., Grabow, W.O.K. and Snozzi, M. (2001) Indicators of microbial water quality. In *Water quality: standards, guidelines and health*, (eds. L. Fewtrell and J. Bartram), pp. 289-307, WHO, IWA Publishing, London.

Barry, S.J., Atwill, E.R., Tate, K.W., Koopman, T.S., Cullor, J. and Huff, T. (1998) Developing and implementing a haccp-based programme to control cryptosporidium and other waterborne pathogens in the alameda creek watershed: case study. In *AWWA Annual Conference, Texas Water Resources* vol. B, pp. 57-69, Dallas.

Bosshart, U., Kaiser, H-P. and Stettler, R. (2003) Hazard Analysis and Critical Control Points at the Zurich Water Supply. In: *2003: Water safety*, (eds. O. Schmoll and I. Chorus), pp. 73-86, Conference Abstracts, Umweltbundesamt Texte 74/03, Berlin.

Codex Alimentarius Commission (FAO/WHO) (1996) *Report of the Twenty-Ninth Session of the Codex Committee on Food Hygiene*, Washington DC.

Davison, A., Howard, G., Stevens, M., Callan, P., Fewtrell, L., Deere, D. and Bartram, J. (2005) *Water Safety Plans: Managing drinking-water quality from catchment to consumer*. WHO, Geneva.

Deere, D., Stevens, M., Davison, A., Helm, G. and Dufour, A. (2001) Management Strategies. In *Water Quality: Guidelines, standards and health – Assessment of risk and risk management for water-related infectious disease*, (eds. J. Bartram and L. Fewtrell), pp. 257-288, WHO, IWA Publishing, London.

Deere, D. and Davison, A. (1998) Safe water – are food guidelines the answer? *Water*, **25**, 21-24.

Fewtrell, L. and Bartram, J. (eds.) (2001) *Water Quality: Guidelines, standards and health – Assessment of risk and risk management for water-related infectious disease*. WHO, IWA Publishing, London.

Gelinas, Y., Randall, H., Robidoux, L. and Schmit, J-P. (1996) Well water survey in two Districts of Conakry (Republic of Guinea) and comparison with the piped city water. *Water Resour.*, **30**(9), 2017-2026.

Godfrey, S., Niwagaba, C., Howard, G. and Tibatemwa, S. (2003) *Water Safety Plans for Utilities in Developing Countries - A case study from Kampala, Uganda*. WEDC, Loughborough University, www.lboro.ac.uk/wedc/iram (accessed April 29, 2005).

Gray, R. and Morain, M. (2000) HACCP Application to Brisbane Water. *Water*, **27**, 41-43.

Haas, C.N., Rose, J.B. and Gerba, C.P. (1999) *Quantitative Microbial Risk Assessment*. John Wiley and Sons, New York.

Havelaar, A.H. (1994) The application of HACCP to drinking-water supply. *Food Control*, **5**(3), 145-152.

Havelaar, A.H. and Melse, J.M. (2003) *Quantifying Public Health Risk in the WHO Guidelines for Drinking-Water Quality: A burden of disease approach*. RIVM report 734301022/2003, Bilthoven.

Hellard, M.E., Sinclair, M.I., Forbes, A.B. and Fairley, C.K. (2001) A randomised, blinded, controlled trial investigating the gastrointestinal health effects of drinking-water quality. *EHP*, **109**(8), 773-778.

Helmer, R., Bartram, J. and Galal-Gorchev, H. (1999) Regulation of drinking-water standards. *Water Supply*, **17**(3/4), 1-6.

Howard, G. (ed.) (2002) *Water supply surveillance - a reference manual*. WEDC, Loughborough University.

Howard, G. (2003) Water safety plans for small systems: Applying HACCP principles for cost-effective monitoring in developing countries. *Water, Sci. Technol.*, **47**(3), 215-220.

Howard, G., Bartram, J., Schuab, S., Deere, D. and Waite, M. (2001) Regulation of microbiological quality in water cycle. In: *Water Quality: Standards, guidelines and health*, (eds. L. Fewtrell and J. Bartram), pp. 377-393, IWA Publishing, London.

Howard, G., Pedley, S., Barrett, M., Nalubega, M. and Johal, K. (2003) Risk factors contributing to microbiological contamination of shallow groundwater in Kampala, Uganda. *Water Res.*, **37**(14), 3421-3429.

MOH NZ (2001) *How to Prepare and Develop Public Health Risk Management Plans for Drinking-water Supplies*. Ministry of Health, Wellington.

NACMCF (1992) *Hazard Analysis and Critical Control Point System*. National Advisory Committee on Microbiological Criteria for Foods (NACMCF), Washington, DC.

O'Connor, D.R. (2002) *Report of the Walkerton Enquiry: The events of May 2000 and related issues*. Queen's Printer for Ontario, Toronto.

Payment, P. and Hunter, P. (2001) Endemic and epidemic infectious intestinal disease and its relationship to drinking water. In *Water Quality: Standards, guidelines and health*, (eds. L. Fewtrell and J. Bartram), pp. 61-88, IWA Publishing, London.

Payment, P., Richardson, L., Siemiatycki, J., Dewar, R., Edwardes, M. and Franco, E. (1991) A Randomised trial to evaluate the risk of Gastrointestinal disease due to consumption of drinking water meeting current microbiological standards. *Am. J. Public Health*, **81**(6), 703-708.

SGWA (Swiss Gas and Water Industry Association) (1998) *Branchenspezifisches Qualitätsmanagement für Versorgungsunternehmen*. SVGW, Zurich.

SGWA (2003) *Recommendations for a simple quality assurance system for water supplies* (W 1002 e). SVGW, Zurich.

WHO (1993) *Guidelines for Drinking-Water Quality*: *Recommendations*, 2nd edn, vol. 1, WHO, Geneva.

WHO (1996) *Guidelines for Drinking-Water Quality*: *Health and other supporting criteria*, 2nd edn, vol. 2, WHO, Geneva.

WHO (1997) *Guidelines for Drinking-Water Quality: Surveillance and control of community supplies*, 2nd edn, vol. 3, WHO, Geneva.

WHO (2004) *Guidelines for Drinking-Water Quality: Recommendations*, 3rd edn, vol. 1, WHO, Geneva.

Wright, R.C. (1986) The seasonality of bacterial quality of water in a tropical developing country (Sierra Leone). *J. Hygiene*, **96**(1), 75-82.

Wülser, R. and Trachsel, H. (2003) Safeguarding the drinking water quality: using the example of the Industriellen Werke Basle. In *Water Safety*, (eds. O. Schmoll and I. Chorus), pp. 59-72, Conference Abstracts, Umweltbundesamt Texte 74/03, Berlin.

17

Groundwater protection zones

P. Chave, G. Howard, J. Schijven, S. Appleyard, F. Fladerer and W. Schimon

The protection of groundwater sources used for domestic supply requires actions at both the wellhead (as described in Chapter 18) and the wider aquifer, and they should be closely linked to form a continuum of measures. Unless the groundwater catchment area is under the control of the water supplier, implementing the full suite of measures will require actions by multiple stakeholders and intersectoral collaboration is essential for success.

Many countries have developed and implemented policies for preventing the pollution of groundwaters. These commonly involve regulatory control of activities which generate or use polluting materials, or control of the entry of potential pollutants into vulnerable surface and underground waters. However, protection zones are not applied in all countries, despite a recognition of their desirability (Bannerman, 2000). This may be due to a number of factors, including the lack of sufficiently detailed information regarding the hydrogeological environments (Taylor and Barrett, 1999; Bannerman, 2000), or existing land uses that impede enforcement of such a concept. Furthermore, poverty, uncertain tenure and limited capacity to provide compensation packages suggests that such approaches may be difficult to implement particularly in developing countries.

Protection zones are particularly effective to control pollution from diffuse sources (e.g. agriculture or traffic), while the prevention or control of point sources of pollution may be achieved through rather straight-forward approaches such as permit systems or other legal controls on the quantity, types of substances and places where discharges may take place. The prevention of groundwater pollution from diffuse sources is more problematic because the sources are less easy to identify and the impact is more difficult to control. Thus effective regulatory control of diffuse pollution often relies upon prohibition or restrictions of polluting activities in specific protected areas where impacts on groundwater sources are likely to be serious.

This chapter provides a review of the concepts of protection zones and provides examples of different ways in which these may be applied. Simple, pragmatic approaches are described as well as more complex approaches involving assessments of vulnerability of the aquifer. The smaller scale approach of well-head protection and sanitary completion in order to prevent contaminant ingress through short-circuiting is discussed in Chapter 18.

NOTE ▶ *This chapter introduces options for controlling risks by implementing protection zones. The information presented here supports defining control measures and their management in the context of developing a Water Safety Plan (Chapter 16).*

Water suppliers and authorities responsible for drinking-water quality will usually have a key role in the definition of control measures involved in the designation and delineation of protection zones, but they will rarely be the only actors responsible for implementation and monitoring. This rather requires close collaboration of the stakeholders involved.

17.1 THE CONCEPT OF A ZONE OF PROTECTION

The concept of a zone of protection for areas containing groundwater has been developed and adopted in a number of countries. Many have developed guidelines for water resource managers who wish to delineate protection areas around drinking-water abstraction points (e.g. Adams and Foster, 1992; NRA, 1992; US EPA, 1993). In general, the degree of restriction becomes less as the distance from the abstraction point increases, but it is common to include the area of the whole aquifer from which the water is derived in one of the zones, and to restrict activities in such areas in order to give general long-term protection.

Commonly, zones are delineated to achieve the following levels of protection:

- A zone immediately adjacent to the site of the well or borehole to prevent rapid ingress of contaminants or damage to the wellhead (often referred to as the wellhead protection zone).

- A zone based on the time expected to be needed for a reduction in pathogen presence to an acceptable level (often referred to as the inner protection zone).
- A zone based on the time expected to be needed for dilution and effective attenuation of slowly degrading substances to an acceptable level (often referred to as the outer protection zone). A further consideration in the delineation of this zone is sometimes also the time needed to identify and implement remedial intervention for persistent contaminants.
- A further, much larger zone sometimes covers the whole of the drinking-water catchment area of a particular abstraction where all water will eventually reach the abstraction point. This is designed to avoid long term degradation of quality.

The number of zones defined to cover these function varies between countries, usually from 2-4. By placing some form of regulatory control on activities taking place on land which overlies vulnerable aquifers, their impact on the quality (and in some cases quantity) of the abstracted water can be minimized. The concept can be applied to currently utilized groundwaters and to unused aquifers which might be needed at some time in the future. Legislation not directly related to pollution prevention, such as those related to planning, industrial production and agriculture, may be used to adjust or limit the extent to which activities that could impact upon the aquifer take place in the protection zone. In order to implement such policies, there must, of course, be adequate supporting legislation available to control these activities. As noted in Chapters 5, 7 and 20 such legislation may need to consider compensation packages to account for potential lost earnings of land users whose activities may be controlled to protect underlying groundwater.

17.2 DELINEATING PROTECTION ZONES

Groundwater protection zones have developed historically, using a variety of concepts and principles. Although some include prioritization schemes for land use, all aim at controlling polluting activities around abstraction points to reduce the potential for contaminants to reach the groundwater that is abstracted. Criteria commonly used for these include the following:

- *Distance*: the measurement of the distance from the abstraction point to the point of concern such as a discharge of effluent or the establishment of a development site.
- *Drawdown*: the extent to which pumping lowers the water table of an unconfined aquifer. This is effectively the zone of influence or cone of depression.
- *Time of travel*: the maximum time it takes for a contaminant to reach the abstraction point.
- *Assimilative capacity*: the degree to which attenuation may occur in the sub-surface to reduce the concentration of contaminants.
- *Flow boundaries*: demarcation of recharge areas or other hydrological features which control groundwater flow.

Approaches using such criteria range from relatively simple methods based on fixed distances, through more complex methods based on travel times and aquifer vulnerability, to sophisticated modelling approaches using log reduction models and contaminant kinetics (Figure 17.1). Uncertainty of the underlying assessment of contamination probability is reduced with increasing complexity.

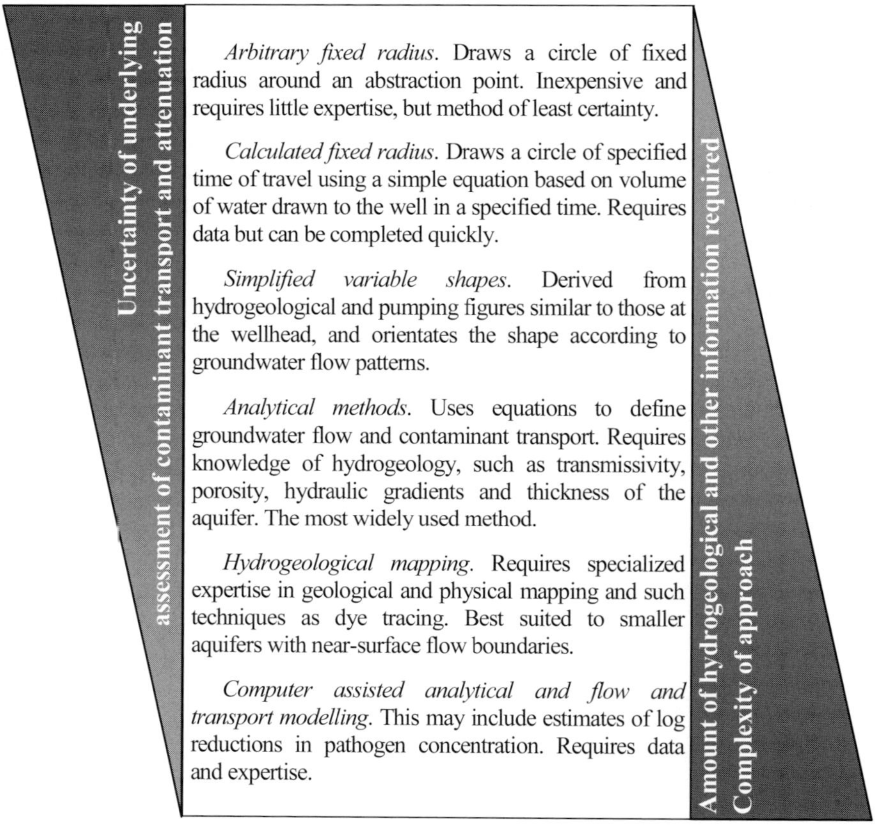

Figure 17.1. Approaches to delineating groundwater protection zones

In order to address some of the fundamental weaknesses in fixed distance approaches, more sophisticated protection zones can be defined based primarily on travel time of water through the saturated zone. For this purpose tracers are often used to acquire information about flow velocities and directions, and an overview of available tracer methods is given in Box 17.1.

Travel time approaches are more realistic in that they attempt to incorporate more empirical evidence, usually related to expected die-off of microbes or dilution of chemicals in defining the land area to be protected. Commonly time criteria are established that provide confidence that the concentration of contaminants will have been reduced to an acceptable level. Although such approaches are better able to reflect local conditions, there remain considerable uncertainties in the degree of protection afforded. In particular these approaches may not be the most cost-effective as they fail to take into account removal of contaminants through attenuation.

Box 17.1. Tracers used in defining groundwater protection zones

A key element in defining groundwater protection zones when using quantitative approaches is to identify tools that allow identification of basic hydrogeological parameters, such as flow rates and patterns, and to predict how pollutants will move through the subsurface. The latter is of particular importance as a means of quantifying the impact of attenuation and dilution.

The use of tracer tests is therefore highly recommended to acquire information about flow velocities and directions, hydraulic connections and hydrodynamic dispersion. Tracer substances can be divided in to two main groups: natural and artificial tracers. Natural tracers are already present in the study area and do not have to be added artificially to the system whereas artificial tracers have to be injected. The most common natural tracers are environmental isotopes and chemicals, organisms and physical effects such as temperature. Artificial tracers are dyes (fluorescent and non-fluorescent), salts, radioactive tracers, activable isotope tracers and particles (spores, bacteria, phages, microparticles, etc.). Table 17.1 provides a summary of selected tracers that are commonly used.

Table 17.1. Tracers commonly used in groundwater

Tracer	Examples	Advantage	Disadvantage	Comment
Natural environmental isotopes (stable/unstable)	^{2}H, ^{18}O, ^{3}H, ^{3}He, ^{4}He, ^{39}Ar, ^{85}Kr, ^{36}Cl, ^{13}C, ^{14}C, ^{34}S, ^{15}N, ^{234}U	No artificial input needed Huge spatial and temporal interpretation possible	Expensive measuring techniques due to low concentrations Complicated interpretation	Omnipresent substances (no artificial input required) Useful for calculation of mixing proportions, ages and travel times
Radioactive tracers	^{3}H, ^{51}Cr, ^{60}Co, ^{82}Br, ^{131}J, ^{24}Na	Low chemical impact on the environment Disappearance due to radioactive decay Easy and economic detection	Possible radiation during artificial input of the tracer More complicated evaluation	Have been applied as artificial tracers both in surface and groundwater with satisfying results; especially useful for sewage water with high amounts of suspended particles
Fluorescent dyes	Uranine	Economic Non-toxic Very low sorptivity High solubility in water	Sensitive to light and oxidizing substances Strong pH-dependence Difficult evaluation if Uranine is already in the hydrologic system	Very good tracer analysing groundwater-flow and flow-velocities Uranine should be restricted to groundwater in reasonable low concentrations

Tracer	Examples	Advantage	Disadvantage	Comment
Fluorescent dyes (continued)	Rhodamine B	Low sensitivity to light and pH High solubility in water	Carcinogenic High sorpitivity	Good tracer for short term tests and surface water with low contents of suspended organic and mineral particles
	Amidrhodamin G	Low sensitivity to light and pH Low sorpitivity High solubility in water Easy to measure parallel to Uranine		Good tracer for ground- and surface-water
Bacteria	*E. coli*, faecal streptococci, sorbitol fermenting bifido-bacteria	Transport behaviour models pathogenic bacteria movement	Limited persistence of sensitive indicator bacteria May have environmental rather than faecal source	Would not usually be injected directly as a tracer but monitored in relation to known hazard sites to determine impact
Bacteriophages	F-specific RNA bacteriophages, coliphages	Transport behaviour similar to viruses can be used as either index organism or process indicator	Isoelectric point and sorption dependent upon pH and need to ensure	Appropriate especially for investigating transport behaviour of viruses in order to define groundwater detection zones
Spores	*Clostridium perfringens*	Long survival times which can mimic more robust pathogens	Potential for interference by natural populations	Spores are often dyed or prepared to facilitate its transport behaviour and detection

The most sophisticated approaches to groundwater protection zone definition are based on calculated log-reductions in microbial concentrations or reductions in chemical concentrations that can be achieved through attenuation and dilution as contaminants move through the soil, unsaturated and saturated zones. These approaches require much greater knowledge of local conditions and the expected reductions that may be achieved through attenuation. They do, however, provide much more realistic estimates of the land area where control should be exerted on polluting activities, and thus may be components of quantitative risk assessments. These may involve assessment of the hazard arising from a particular activity, examination of the vulnerability of the underground water to pollution, and consideration of the possible consequences which would occur as a result of contamination.

Local conditions determine the choice of method as this depends upon the amount of expertise and data available. Technical considerations should include ease of applicability, extent of use, simplicity of data, suitability to the area's hydrogeological character and accuracy required for decision-making purposes. The choice should also be related to relevance to the protection goal, and therefore may also include approaches that employ prioritization schemes for land use. Within each of the approaches adopted, it is important to also bear in mind the importance of other factors such as other sanitation provisions, economic impact and social norms.

The following sections briefly discuss approaches to defining and characterizing protection zones that have been adopted in different countries. Depending on the level of technical expertise and objectives of the groundwater protection, they are based chiefly on distance or travel time approaches (Section 17.3), or include more hydrogeological information to assess vulnerability (Section 17.4). A recent development is to assess contaminant loading and attenuation in order to use a risk assessment for protection zone delineation (Section 17.5). A supplementary criterion used in some countries is to include an assessment of current and future land use priorities in developing groundwater protection schemes (Section 17.6).

17.3 FIXED RADIUS AND TRAVEL TIME APPROACHES

The simplest form of zoning employs fixed-distance methods where activities are excluded within a uniformly applied specified distance around abstraction points. These methods use expert judgement and experience and have been widely applied. There is limited direct scientific evidence to underpin most fixed-distance approaches, as they do not take into account local hydrogeological conditions and aquifer vulnerability or the interaction between adjacent wells and the impact that this may have on local flow conditions. This reduces the confidence in the degree of protection that is provided. These approaches are often used when there is limited information on the hydrogeology of an area and are a practical means of ensuring a measure of immediate protection.

Fixed radius approaches are used in a number of countries for defining a protection zone around the immediate vicinity of the wellhead, chiefly designed to protect the wells from pollution by short cuts. For example, in Germany this zone is set at a minimum of 10 m for wells, 20 m for springs and 30 m for wells in karst aquifers. The Swiss, Danish and Austrian protection schemes also use an innermost zone of 10 m radius. In Australia the wellhead protection zone is a concentric area comprising the operational compound surrounding for the well and is often, but not always, defined as a 50 m radius within which the most stringent controls on land use and materials apply.

Distance approaches to define protection zones targeting effective attenuation of pathogens and/or substances to acceptable levels, often underpinned by travel time concepts, are also used. This may follow the calculated fixed radius or variable shape approach (see Figure 17.1). In practice travel times are not always determined for each specific setting, and both approaches may be used together, as is the case in Ireland and Denmark (see below).

They may also be supplemented by analytical methods and hydrological modelling, if sufficient scientific expertise and data is available. The delineation of protection zones can then be based on such issues as the recorded or modelled movement of pollutants through the groundwater area. In such cases, zones may not be simple concentric circles around abstraction points, but their boundaries follow the calculated time of travel of chosen parameters. This may be important in heavily developed areas where the imposition of restrictions within a defined area may have economic repercussions.

Examples from a number of countries are summarized in Table 17.2. These examples highlight how fixed distance and travel time approaches are used in practice in different countries, and selected approaches among these are discussed in the following. In some countries, however, fixed radius and travel time approaches are supplemented by more sophisticated methods as discussed in the following sub-sections.

Table 17.2. Comparative table of examples of protection zone dimensions

Country	**Wellhead protection zone or inner zone**	**Middle zone**	**Outer zone**
		Travel time and/or radius of zone	
Australia	50 m	10 years	Whole catchment
Austria	<10 m	60 days	Whole catchment
Denmark	10 m	60 days or 300 m	10-20 years
Germany	10-30 m	50 days	Whole catchment
Ghana	10-20 m	50 days	Whole catchment
Indonesia	10-15 m	50 days	Whole catchment
Ireland	100 days or 300 m	-	Whole catchment or 1000 m
Oman	365 days	10 years	Whole catchment
Switzerland	10 m	Individually defined	Double size of middle zone
United Kingdom	50 days and 50 m minimum	400 days	Whole catchment

Ireland

In Ireland, individual public water supply sources are identified and protection zones established around them – termed Source Protection Areas (SPA). Two SPAs are delineated – an inner protection area and an outer protection area (DoELG, 1999). Both areas may be identified either on the basis of a simple zoning using an arbitrary fixed radius where scientific and geological data is in short supply, or using hydrogeological methods based on local data or modelling.

Inner protection areas are intended to protect the source from the effects of an activity that could have an immediate effect on water quality, and is defined as a 100-day time of travel from any point below the water table. 100 days is chosen by Ireland as a conservative limit to allow for the heterogeneous nature of Irish aquifers and to allow for the attenuation and die-off of bacteria and viruses which may live beyond 50 days. In some karstic areas it is not possible to identify 100-day boundaries, in which case the whole aquifer becomes a SPA. If the arbitrary fixed radius method is used, 300 m is taken as an equivalent distance. The outer protection areas covers the zone of the aquifer, the recharge of which supports the long-term abstraction of the individual source (or the complete catchment if this is the contributing area), or, using the arbitrary fixed radius method, 1000 m.

In this example, although travel time is used as the underlying concept for defining the protection zone, simple practical measures based on a broad knowledge of the groundwater system are used to define protection zones. Generally, such approaches may have particular value for small supplies where gaining access to hydrogeological expertise may be difficult or expensive.

Ghana

In crystalline rock terrains such as that found in Ghana, the protection of boreholes cannot be simply achieved by establishing protection zones. This is because heterogeneous materials developed in the weathered zone and in fractures in the bedrock provide viable flow paths for contaminants from indiscriminately located latrines, waste dumps and other pollution sources at far away places (Bannerman, 2000). The high groundwater velocities would result in groundwater protection areas covering the major parts of communities' aquifers and hence may make them impractical to achieve.

In Ghana, a pragmatic time-of-travel approach has been adopted with which to define protection area boundaries. Three protective zones are designated. Zone I covers an area of radius 10-20 m around a production well and is designed to protect it against short-circuit contamination at the well site. Zone II is situated around Zone I, and comprises the zone between the well field and a line from which the groundwater will flow at least 50 days until it reaches the production well. The choice of this travel time for Ghana was developed from experience elsewhere though it may not be applicable under all conditions. Zone III is a buffer zone between the recharge area and Zone II. If the water is produced from a spring, the zone should not be less than 20 m on the upstream (uphill side) of the water source.

United Kingdom

In the United Kingdom decisions on protection zones are taken on the basis of assessing the likely impact of a pollutant and the degree to which attenuation occurs in the geological strata influencing the source. According to the national groundwater protection policy (NRA, 1992), three distinct protection zones are recognized in the vicinity of abstraction points:

The Inner Source Protection Zone (Zone I) is located immediately adjacent to the groundwater source, and is designed to protect against the effects of activities which would have an immediate outcome on the source, in particular in relation to the release of pathogens into groundwater. It is defined as the area within which water would take 50 days to reach the abstraction point from any point below the water table, subject to a minimum of 50 m radius from the source.

The Outer Source Protection Zone (Zone II) is an area defined by a 400 day travel time to the source. It is based upon the time needed for the attenuation of slowly degrading pollutants. In England and Wales this is further modified for aquifers of high water storage capacity, such as sandstones, to allow for Zone II to cover either the area corresponding to 400 days, or the whole of the recharge area, calculated on the basis of 25 per cent of the long term abstraction rate for the source.

There is a further zone (Zone III) which covers the whole of the catchment area of the source, based on the area needed to maintain abstraction assuming that all water will eventually reach the abstraction point. In some cases, where the aquifer is confined, it is possible that the protection area is remote from the site of the source.

Denmark

Denmark has used a protection system which takes account of existing abstraction wells and utilizes two zones. The first is a 10 m fixed radius zone immediately surrounding the abstraction point to provide for technical and hygienic protection. The second zone of 60 days travel time or 300 m radius acts as an outer protection area to take account of contaminants which degrade more slowly.

Problems in dealing with pesticide contamination have also led to the consideration of a 10-20 year zone in which pesticides would be controlled. Evidence of continuing problems with groundwater quality, particularly in respect of pesticide contamination and rising nitrate levels, led the Danish Government to adopt a three zone system in 1998 to prioritize the expenditure of money and effort in controlling, particularly, point sources of pollution (Stockmarr, 1998) (discussed below in Section 17.6).

Germany

In Germany guidelines on the definition of zones are available through a code of practice (DVGW, 1995). It defines three zones. The Well Field Protection Zone (Zone I) is designed to protect individual wells and their immediate environment against any contamination and interference and has fixed dimensions of 10 m. A Narrow Protection Zone (Zone II) aims to provide protection against contamination by pathogenic bacteria and viruses and is based on a 50 day travel time. Due to the area of land required to meet the 50-day criterion, fixing a boundary is often not possible in karst terrains, mainly for economic reasons (for example where existing development would have to be removed). In such cases, Zone II may be smaller, but should in any case comprise all areas from which an increased risk to the karst aquifer may emanate.

A Wide Protection Zone (Zone III) serves to protect wells against long-range impairments, notably against contamination by non-degradable or less readily degradable chemical or radioactive substances, and usually covers the entire subsurface catchment area. If the catchment area is very large, with a boundary more than 2 km from the well, it may be sub-divided into Zone III A and Zone III B, with different levels of land use restrictions.

The Code of Practice also addresses particular cases such as the definition of protection zone boundaries for very large catchment areas or when several wells are located in the same catchment area. In general, the size of the area to be placed under protection is dependent upon the abstraction and recharge rates in the catchment area, the higher the abstraction rate the larger the protection zones to be defined. The Code of Practice also includes guidance for the definition of protection zone boundaries in the case of water production from several (geo)hydraulic systems and in the case of artificial recharge.

Australia

The Australian wellhead protection plan is a system of groundwater protection which involves four components. These comprise a set of actions to ensure that the well is properly designed and constructed (known as 'well integrity assurance') the setting up of wellhead protection zones, an appropriate monitoring system, and contamination or land use control (ANWQMS, 1995).

The wellhead protection zones are based on the definition of concentric protection zones around the wellhead. Zone I encompasses the operational compound surrounding the well, and is often, but not always, defined as a 50 m radius area within which the most stringent controls on land use and materials apply. Zone II is arbitrarily defined as the maximum distance a contaminant particle would have travelled if it took 10 years to reach the well. Zone III corresponds to the regional protection area where greater than 10 years travel time is available. This is usually the catchment area of the contributing aquifer.

Oman

In some countries, where water is in short supply and resources are very limited, protection zones are used primarily to ensure that there is adequate control over abstraction rates. This applies particularly to arid countries

For example in Oman, because of problems of water derogation, the water resources Council in 1983 decided that no wells should be constructed within 3.5 km of a motherwell of a water supply system (*falaj*). The choice of size of the protection zone was a pragmatic solution rather than being based on hydrogeological principles. Since that date the protection of groundwater has been accomplished by the adoption of National Water Development Areas – water protection zones designated for the general protection from contamination, over-extraction, intrusion by seawater and adverse development.

The schemes used a colour-coded zoning system to identify specific limitations on future developments and progressively on existing activities. Such zones were a response to already perceived potential problems and were useful in providing guidance on future developments within the water protection zones. However they had limited success in dealing with existing development due to the problems of applying retrospective controls. In response a new scheme using technically derived zones based on time-of-travel periods has been developed to accommodate this (Government of Oman, 1991).

The establishment of major government wellfields in urban areas to meet public water supply needs was followed by the recognition that these needed careful protection both as a water resource and from pollution (Government of Oman, 1991). As a further refinement of the earlier water development area zoning system described above, a revised water protection zone concept utilizing three distinct zones with relevant regulation of activities within them has been adopted. The three zones use 365 days as the time of travel to define the boundary of an innermost protection zone surrounding an abstraction point such as a well. A second tier protection area which uses a 10 year time of travel to define the boundary is established as a middle protection zone, whilst the extent of the third and outermost protection zone is delineated by the catchment boundary.

Indonesia

An integrated approach to ensure proper drinking-water quality in urban centres of Indonesia has been developed by the Indonesian-German governmental cooperation on drinking-water quality surveillance. This concept includes protection zones to protect and maintain water resources in their initial function and allotment by a natural and preventive approach. The zones are based on fixed distances for Zone I and on travel time for Zone II, using hydrogeological mapping and a flow path model where protection zones of different categories are defined. The following zones are applied:

- Zone I is defined as the area surrounding the spring/well within a radius of 10-15 m, which is fenced and where any activity that has interaction with the aquifer is prohibited.
- Zone II is the boundary that is defined by 50 days travel time, to provide protection against bacteriological contamination. In order to determine the boundaries, a hydrogeological survey is conducted for each spring and well. Besides the restrictions mentioned under Category III, all possible activities causing bacteriological contamination are prohibited.
- Zone III includes the whole catchment area based on topographical boundaries where the application of water hazardous pesticides, the infiltration of liquid waste, human settlements with unorganized discharge of the waste water within the catchment area and waste disposal are restricted. Clustering of several springs/wells in one catchment area is possible.

17.4 APPROACHES USING VULNERABILITY ASSESSMENTS

A number of countries (e.g. the United Kingdom, Australia and Ireland) have introduced vulnerability assessment of groundwaters into their protection policies (for a discussion of the concept of vulnerability see Chapter 8). Such vulnerability assessments correspond to the concept of system assessment in the context of developing a Water Safety Plan. They can refine protection categories defined by fixed distance and/or travel time approaches and allow a differentiated management response within a protection area. Such systems are also useful outside of drinking-water protection zones for long term planning of the protection of groundwater resources. Further, they provide guidance to organizations concerned with major works activities that could cause problems of groundwater contamination, such as the siting of new industrial or urban developments.

The example of Ireland highlights how vulnerability assessments have been included in protection plans. The Irish Environmental Protection Agency has proposed a protection zone identification scheme based upon the division of the entire land surface according to the vulnerability of the underlying groundwater to contamination (DoELG, 1999). In this system vulnerability depends upon the time of travel of contaminants through the strata, the relative quantity of contaminants which can reach the groundwater and the attenuation capacity of the local geology. These factors are dependant upon the subsoil characteristics, whether the contamination source is point or diffuse source and the thickness of the unsaturated zone. Assessing these factors results in classification of the vulnerability of a given area as extreme, high, moderate or low. Such ratings are based on judgement, experience and available scientific information. The resultant map shows the vulnerability of groundwater to pollution from contaminants released at 1-2 m below the surface. Where deeper discharges are made, site-specific local conditions would have to be taken into account. The characteristics of the contaminants are not considered. This vulnerability classification is not only used for drinking-water resources, but also applied to the whole land surface of the country.

For drinking-water resources, the resultant map is then overlain with the simple map of the inner and outer Source Protection Areas derived as discussed above in Section

17.3 (Figure 17.2). This results in a map showing the vulnerability of both the inner and the outer SPA. While the inner SPA will usually be too small to contain more than one or two vulnerability categories, the outer zone might encompass all four. This map is the basis for defining the level of protection to be implemented for each area (Section 17.7)

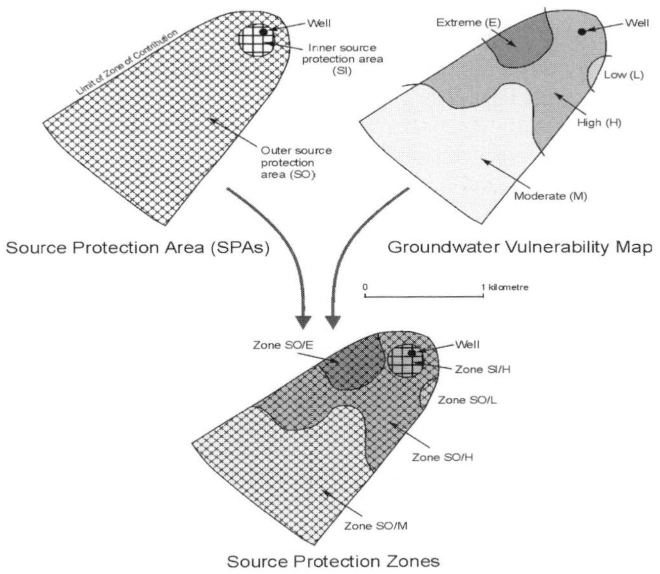

Figure 17.2. Delineation of source protection zones around a public supply well from the integration of the SPA map and the vulnerability map (DoELG, 1999)

17.5 A RISK ASSESSMENT APPROACH FOR DELINEATING PROTECTION ZONES

From 2001, a new policy for production of safe drinking-water in The Netherlands has been incorporated into legislation. This approach sets the health-based target of a maximum acceptable infection risk of one per 10^4 persons per year associated with drinking-water consumption. It then uses dose-response relationships for pathogens to determine maximum allowable pathogen concentrations in drinking-water (Regli *et al.*, 1991). In the case of viruses, it is based on the dose response relationship of rotavirus and poliovirus 3, as a worst-case. The maximum allowable concentration is 1.8×10^{-7} viruses per litre. Together with data on the occurrence of virus concentrations in surface water this implies that they need to be reduced by a factor of 5-8 $\log_{10}$ in order to produce drinking-water in which maximum allowable concentrations are not exceeded. Drinking-water companies that use surface water as a source (approximately one third of the total drinking-water production in The Netherlands) are therefore obliged to conduct a risk analysis to demonstrate adequate drinking-water treatment. Vulnerable groundwater systems may also be subject to this risk assessment. This raises the question whether current protection zones of 60 days of travel time are sufficient and actually what travel times and travel distances are needed to comply to the risk level of 10^{-4} per person per year.

Therefore, and as a first step in a vulnerability analysis of Dutch groundwater well systems to virus contamination, a hypothetical case was simulated to calculate the travel distance and time that are required for sufficient protection against virus contamination (Schijven and Hassanizadeh, 2002a; 2002b). The conditions assumed are given in Table 17.3 below. In this simulation a sewage pipe was continuously leaking virus. The virus was diluted and transported with the groundwater that was abstracted by a single well (radial flow). This hypothetical case was based on data from a field study on deep well injection (Schijven and Hassanizadeh, 2000) and a number of conservative assumptions.

Table 17.3. Conditions applied in the Dutch study for calculating required travel times and distances to adequately protect groundwater wells in unconfined shallow sandy aquifers against virus contamination (Schijven and Hassanizadeh, 2002a; 2002b)

Condition assumed for the model calculation	Evaluation of assumptions used
Shallow sandy aquifer High permeability Groundwater table 0.5-1 m below surface Depth of aquifer: 20-30 m	Sources of contamination do occur directly in the aquifer
Unconfined	Absence of protecting confining layers is typical Local differences occur in the thickness of confining layers due to irregularities and effects of erosion are regarded as a considerable source of uncertainty for protection
Temperature: 10 °C	Typical value for Dutch aquifers from 1 m below surface at groundwater table
pH 7-8	Typical values
Bacteriophage MS2 as a model virus	Represents poorly adsorbing viruses
Anoxic conditions	Do occur and result in the absence of favourable sites for attachment like ferric oxyhydroxides Low inactivation rate of MS2 (0.024 day^{-1}) has been demonstrated
Saturated conditions	Result in less attachment or inactivation compared to unsaturated conditions
Point source of contamination at water table	Worst case assumption, as horizontal transport is shortest pathway
Continuously leaking sewage pipe ($1 m^3/h$)	Realistic scenario for a steady state where low leakage rate remains unnoticed
Approximately 200 enteroviruses per litre in raw wastewater	Average value for concentrations in raw wastewater in The Netherlands
Maximum allowable virus concentration at abstraction well of 2×10^{-7} viruses per litre	Based on drinking-water consumption, virus infectivity and probability of infection of 10^{-4} per person per year
Required reduction of virus concentration 9 $\log_{10}$	Based on measured source concentration and maximum allowable concentration at the well

Under the anoxic conditions of the deep well injection study minimal removal of virus was observed, i.e. there was little attachment of virus to the grains of sand and little inactivation of virus. The same conditions were assumed to apply as well to a selection of six unconfined sandy aquifers. The absence of confining layers together with the shallowness of the aquifers and unfavourable conditions for attachment make it a reasonable assumption qualifying these groundwater well systems as relatively vulnerable.

These and other conditions applied to calculating the required travel times and distance are listed in Table 17.3. Concentrations of enteroviruses in raw domestic wastewater from the leaking pipe need to be reduced by 9 $\log_{10}$ at the point of groundwater abstraction. A steady state solution of a transport model incorporating attachment and inactivation was applied to calculate travel times and distances to achieve this.

Virus concentration was found to be reduced by 3.1-4.0 $\log_{10}$ at the abstraction well due to mixing with groundwater from all directions (radial flow). To account for an additional 5.0-5.9 $\log_{10}$ removal of virus by attachment and inactivation, residence times of about 8 to 15 times longer than the current guideline of 60 days appeared to be needed, depending on abstraction rates, aquifer thickness and sand grain size. At a higher transport velocity, removal with distance is less, but this is partly compensated by a higher dilution factor.

Although this hypothetical case was partly built on conservative assumptions, it strongly indicates that a 60-day protection zone is insufficient by far to protect against virus contamination from a nearby leaking sewage pipe. The situation may even be worse. Concentrations of noroviruses in raw wastewater (Lodder *et al.*, 1999) were found to be 10^4 to 10^6 RNA-containing particles per litre as determined by PCR, which is 10^2 to 10^4 times higher than that of enteroviruses as determined by tissue culture. However, it is uncertain what part of the RNA containing particles is actually infectious.

Compared to the removal capabilities of sandy aquifers, removal of viruses in karst, fractured bedrock and gravel aquifers may be lower. Such aquifers are identified as sensitive to faecal contamination by the US EPA's proposed Ground Water Rule (US EPA, 2000). These aquifers have in common that more permeable pathways exist that allow very high flow rates of viruses (Rossi *et al.*, 1994; Paul *et al.*, 1995; 1997). In such pathways, attachment will be very low. Due to the high transport rate (short travel times), inactivation will also be minimal. In gravel, removal of slug-injected bacteriophages T7 and H40/1 was only 2 $\log_{10}$ over a travel distance of 50 m (Rossi, 1994; Rossi *et al.*, 1994). This is about the same removal rate as for MS2 in a sandy anoxic aquifer. In fact, T7 and H40/1 were probably removed more effectively than MS2, considering the coarseness of gravel. Even considerable removal may be found in fractured rock, e.g. about 6-$\log_{10}$ removal of MS2 over a distance of 20 m in limestone (Paul *et al.*, 1997) or 1-$\log_{10}$ removal of MS2 and PRD1 over a distance of 0.5 m in a clay-rich till (Hinsby *et al.*, 1996). Nevertheless, it is obvious that preferred pathways, like fractures and breaches, will contribute greatly to the uncertainty in assessing the removal capabilities of a certain aquifer.

As these examples highlight, using a risk assessment approach for delineating protection zones requires an understanding of the elimination capacity of the unsaturated zone and the pathogen levels expected to reach the well. Often this information will not be available specifically for a given setting, and estimates can be derived from assessing pollution potential as discussed in Chapter 14.

17.6 PRIORITIZING SCHEMES FOR GROUNDWATER PROTECTION

In situations where land use pressures are high – e.g. for increasing agricultural production or where land for building is at a premium – and such land is also liable to overly the available water resource, systems of prioritization are necessary to control development of the land in such a way that the availability and quality of water supplies is not jeopardized. The benefit of prioritization approaches is that they promote cost-effective application of protection zones to take into account the need to balance economic development and resource protection. Thus they may be used as a further criterion in defining management responses, supplementing hydrogeological criteria such as travel times and vulnerability assessments. This is currently practiced in some countries, and examples are given below.

Western Australia

In Western Australia groundwater resources used for public supply are protected from pollution by being proclaimed Underground Water Pollution Control Areas and using by-laws to control activities which could potentially pollute such resources. Instead of using simply an assessment of vulnerability to pollution, the Western Australian system recognizes that water source objectives vary dependent upon the strategic importance of the source, its vulnerability and other competing land uses. The result is a three tiered priority-based system with management objectives for each priority area. Besides vulnerability, these include such issues as designated beneficial uses (for example drinking, irrigation, industrial, recreation or ecosystem protection), water quality, social, economic and ecosystem value, and current and planned land use. This assessment enables the areas on the vulnerability map to be classified in terms of the requirements for protection, and allow action levels to be set to give the required protection.

The city of Perth overlies a large fresh groundwater resource (see also Chapter 14.6 for further details). Groundwater forms an important component of the city's water supply, providing 70 per cent of water used, and also maintaining ecosystems around environmentally significant lakes and wetlands. The groundwater occurs as an unconfined aquifer throughout the region, and in several confined aquifers. The shallow groundwater in urban areas is highly susceptible to contamination owing to the sandy soil, and in some areas this has restricted groundwater use, and has had an adverse impact on wetlands. The growth of the urban area has overtaken well fields previously located in areas of rural land use, and has compromised water quality. Land use in these areas is now controlled by Priority SPAs. There are three types of protection areas:

- *Priority 1* (P1) SPAs are defined to ensure that there is no degradation of water quality used for public supply. P1 areas are declared over land where the provision of the highest quality public drinking-water is the prime beneficial land use. P1 areas include government owned land where there is no development, or use is limited to forestry or sylviculture.

- In *Priority 2* (P2) SPAs previously existing land uses are regulated to ensure that there is no increased risk of pollution to groundwater quality. P2 areas are declared over land where low intensity development (such as rural) already exists. Provision of public water supply is a high priority in these areas, but there may be some degradation of water quality.
- *Priority 3* (P3) is declared over land where water supply needs co-exist with other land uses such as residential, commercial and light industrial developments. Protection of groundwater quality in P3 areas is achieved through management guidelines rather than restrictions on land use.

In Western Australia a corridor plan is in operation. In this plan, urban development takes place in northwest, southwest, southeast and eastern corridors ensuring that the central part of the coastal plain, where the groundwater recharge areas are located, will be essentially undeveloped, thus providing a further layer of long term protection. Future expansion of the public water supply will take place by extending the well fields north and south over the groundwater mounds.

Tunisia

In a further development of the protection zone concept for groundwater resource management in Tunisia, in essence formalizing the Western Australian approach, economic and social value factors have been introduced into the assessment of the need to protect groundwaters (Findikakis *et al.*, 1998). This is a useful concept where supplies are very scarce, and where alternatives are limited, for example in arid countries. The system uses three groups of criteria which take into account the physical nature of the resource, its vulnerability to pollution or depletion by over-abstraction and the socioeconomic value of the aquifer. This latter is an important factor where aquifers are in isolated regions and where they form the main water supply source. The socioeconomic value is based on an economic indicator that identifies the relative economic importance of the supply taking into account the level of economic production dependent upon the source, and the number of people dependent upon it.

Denmark

Since 1998, Denmark defines three zones in relation to value for use, the most critical of which comprise areas of special interest for drinking-water (Stockmarr, 1998). These are defined as areas sufficiently large to supply the population in the future, taking account of other water uses. Such zones will be established in each administrative county and will eventually cover about 15 to 30 per cent of the total land area. Areas of minor interest for drinking-water are areas where groundwater is already heavily contaminated, and which represent areas of land within which such activities as landfill operation should be concentrated. These areas are generally expected to be a minor zone along the coastline where abstraction is not generally practised. The third zone will comprise most of the remaining land areas and represent land which may become important water supply areas over the next 20 to 30 years, known as areas of interest for drinking-water.

The areas are identified by reference to the classification of groundwater resources taking account of precipitation and evaporation, median river water flows, run-off, groundwater potential and catchment areas, relevant geological features, land use, and so forth, and maps will show the groundwater resource divided into the three categories.

The resultant areas of special interest for water resources are then subject to limitations on the use of land use for activities such as the location of industry or urban development.

United States of America

A draft prioritization scheme was developed by the US EPA (1986). Although this was never finalized and implemented, the approach may be of interest to readers of this monograph. The scheme combines vulnerability, quality and the resource's value to society. Three classes are identified as set out in Figure 17.3 below. Different levels of management of the overlying land are applicable to each class of groundwater under this scheme.

Classifying groundwaters under this system involves delineating a segment of the groundwater body to which the classification criteria applies. This is known as the Classification Review Area and comprises a two-mile radius from the boundaries of the activity that may affect the particular groundwater (such as the edge of a contaminated area or the proposed abstraction point). The review area is not necessarily a regulatory area at this stage.

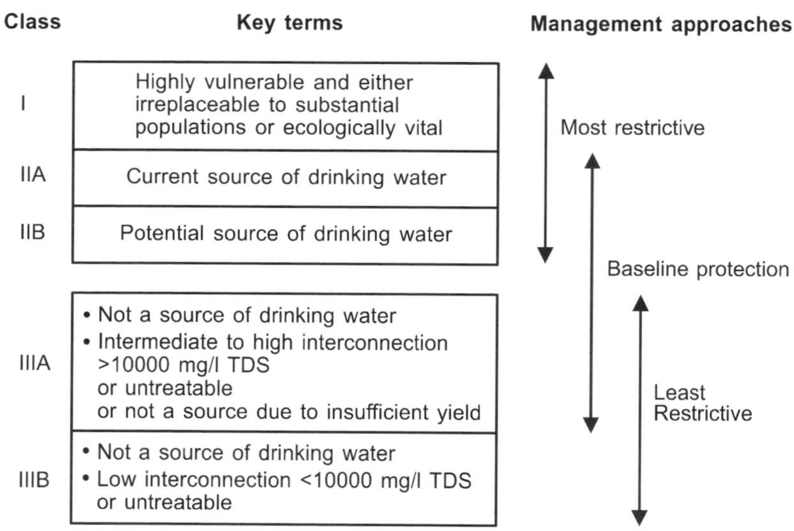

Figure 17.3. US EPA classification scheme (US EPA, 1986)

Where important sources of water are concerned, i.e. the Class I category of groundwaters, a ranking system (DRASTIC) is used to identify further the vulnerability in order to enable suitable protection procedures to be applied. The method yields a single numerical value, referred to as the DRASTIC index. The use of DRASTIC is commensurate with the idea that groundwater vulnerability should not vary according to the type of activity that is being evaluated. This system represents a common methodology which may be used on an interstate basis.

As an alternative means of assessing vulnerability, qualitative assessment may sometimes be an option, wherein the selection of vulnerability might be based on site setting, professional experience of the user, the availability of data, or previous experience. However, this option does not permit the use of referred tests and methods or other numerical criteria or decision steps.

Most of the States in the USA also have individually developed groundwater classification systems, as shown in Table 17.4.

Table 17.4. Groundwater classes based on usability and/or quality criteria used in some States of the USA (US EPA, 1985)

State	No. of classes	Criteria for classification
Connecticut	4	1. Suitable for drinking without treatment; 2. May be suitable for drinking without treatment; 3. May have to be treated; 4. May be suitable for waste disposal practices
Florida	4	1. Single source aquifers suitable for potable use; 2. Potable use TDS <10 000 mg/l; 3. Non-potable use from unconfined aquifers; 4. Non-potable use from confined aquifers
Guam	3	1. Drinking-water quality; 2. Saline; 3. Size criteria
Maryland	3	1. TDS <500 mg/l; 2. TDS 500-6000 mg/l; 3. TDS >6000 mg/l
Massachusetts	3	1. Drinking-water quality; 2. Saline; 3. Below drinking-water quality
Montana	4	1. Suitable for drinking-water; 2. Marginally suitable for drinking-water; 3. Suitable for industrial or commercial; 4. May be suitable for some uses
New Mexico	2	1. TDS <10 000 mg/l; 2. TDS >10 000 mg/l
New York	3	1. Potable use; 2. Saline water 250-1000 mg Cl/l; 3. Saline water >1000 mg Cl/l
North Carolina	5	1. Drinking-water; 2. Brackish water >20 feet below surface; 3. Fresh water <20 feet below surface; 4. Brackish water <20 feet below surface; 5. Not suitable for drinking
Vermont	2	1. Drinking-water; 2. All other groundwaters
Wyoming	7	1. Domestic; 2. Agricultural; 3. Livestock; 4. Aquatic life; 5. Industry; 6. Hydrocarbon and mineral deposits; 7. Unsuitable for any use

17.7 MANAGING LAND USE AND HUMAN ACTIVITIES IN PROTECTION ZONES

The beneficial use of protection zones relies upon the ability to restrict polluting activities in them. Commonly this is achieved through the activation of legislation which is available under the land use planning or pollution control regimes of the country. The designation of the zone triggers specific requirements, which are met by enacting relevant restrictions or introducing permitting systems. Often it is not necessary to introduce new legislation. The designation of the protection zone may require that the body which administers planning or pollution control laws takes action to ensure that they are applied rigorously and deal with the particular concerns brought about by recognition of the special characteristics of the protected area. However, this may not be trivial. Stricter application of existing legal requirements may require changes of habitually established land uses (e.g. horticulture with intensive pesticide application), and this may have substantial socioeconomic implications. Therefore, new designation of protection zones may require programmes that include compensation payments or other forms of financial support of current land users affected by the change.

Furthermore, the implementation of measures to control activities in a drinking-water catchment may be facilitated by integrating them into a Water Safety Plan (WSP), as this helps communicate their importance for achieving the quality targets. Further, developing catchment control measures in a WSP-team together with stakeholders involved in activities in the catchment improves their understanding of these issues and can thus improve their sense of ownership and responsibility for protecting the catchment.

In addition to identifying and designating the protection zones or vulnerable areas, it is important to provide guidance on activities which are either acceptable, unacceptable or need to be controlled in the various zones. Restrictions on land-use and other human activities may become control measures in a WSP, and compliance can be monitored through visual inspections in the drinking-water catchment. This is particularly feasible in some countries where such lists are extensive and very specific. In others general guidance is issued.

In the following, examples will be discussed that show different concepts of managing authorization or restrictions of land use and human activities in protection zones.

Western Australia

In the Western Australian system where activities are planned to take place within the P1, P2 and P3 priority zones (see Section 17.6), reference to specific guidance on compatible, incompatible and conditional activities must be given. Activities which are compatible may be undertaken without restriction. Those activities identified as being incompatible with the objectives of the priority classification can only be carried out after a formal EIA has been carried out. Conditional activities require appropriate site management practices and referral to the Water and Rivers Commission (which is responsible for water quality) for assessment on a case specific basis.

As examples, Table 17.5 lists some of the commercial activities which need to be assessed if they are to be permitted in groundwater protection areas in Western Australia. Similar tables exist for industrial activities, agriculture, urban development, education and research, mining and mineral processing, animal and plant processing, waste treatment and a number of other categories.

Table 17.5. Examples of commercial developments subject to control in water protection zones in Western Australia (based on WRc, 1996)

Land use	Priority 1	Priority 2	Priority 3
Aircraft servicing	Incompatible	Incompatible	Conditional
Airports or landing grounds	Incompatible	Incompatible	Conditional
Amusement centres	Incompatible	Incompatible	Compatible
Automotive businesses	Incompatible	Incompatible	Conditional
Boat servicing	Incompatible	Incompatible	Conditional
Catteries	Incompatible	Compatible	Compatible
Caravan and trailer hire	Incompatible	Incompatible	Conditional
Chemical manufacture/formulation	Incompatible	Incompatible	Conditional
Consulting rooms	Incompatible	Incompatible	Compatible
Concrete batching and cement products	Incompatible	Incompatible	Conditional
Cottage Industries	Conditional	Conditional	Compatible
Dog kennels	Incompatible	Conditional	Conditional
Drive-in/take-away food shops	Incompatible	Incompatible	Compatible
Drive–in theatres	Incompatible	Incompatible	Compatible
Dry cleaning premises	Incompatible	Incompatible	Conditional
Dye works	Incompatible	Incompatible	Conditional
Farm supply centres	Incompatible	Incompatible	Conditional
Fertilizer manufacture/bulk storage depots	Incompatible	Incompatible	Conditional
Fuel depots	Incompatible	Incompatible	Conditional
Garden centres	Incompatible	Incompatible	Compatible
Laboratories (analytical, photographic)	Incompatible	Incompatible	Conditional
Markets	Incompatible	Incompatible	Compatible
Mechanical servicing	Incompatible	Incompatible	Conditional
Metal production/finishing	Incompatible	Incompatible	Incompatible
Milk transfer depots	Incompatible	Incompatible	Conditional
Pesticide operator depots	Incompatible	Incompatible	Incompatible
Restaurants and taverns	Incompatible	Incompatible	Compatible
Service stations	Incompatible	Incompatible	Conditional
Shops and shopping centres	Incompatible	Incompatible	Compatible
Transport and municipal works depots	Incompatible	Incompatible	Conditional
Vehicle parking (commercial)	Incompatible	Incompatible	Compatible
Vehicle wrecking and machinery	Incompatible	Incompatible	Conditional
Veterinary clinics/hospitals	Incompatible	Incompatible	Conditional
Warehouses	Incompatible	Incompatible	Conditional

Germany

In Germany the Code of Practice for drinking-water protection areas includes, for the various zones, a listing of potential hazards and the resultant use prohibitions. Not all hazards listed in this catalogue will apply to the catchment area of a given well which is to be placed under protection and therefore local conditions are always considered in the vulnerability assessment. Table 17.6 provides a summary of controlled activities in the code of practice. These only constitute recommendations that need not necessarily be followed if local conditions so warrant.

Table 17.6. Examples of activities controlled in water protection zones in Germany (based on DVGW, 1995)

Zone type	Zone category	Controlled or prohibited activities
Wider protection zone	*Zone III B*	Industrial estates Pipeline systems for the conveyance of substances constituting a hazard to water Central sewage treatment plants, release of waste water to the ground Waste disposal facilities Agriculture (animal husbandry, application of fertilizers and pesticides) Air fields, Military facilities Sites for freight handling (freight railway stations, truckheads) Use of leachable substances constituting a hazard to water Mining
	Zone III A	*Hazards listed for Zone III B, plus:* Local sewerage systems Discharge of waste water into surface waters Transportation systems, unless waste water generated by these systems is piped out of Zone III A Petrol stations, motor racing Extraction of minerals and rock (near-surface resources) Penetration of strata overlying groundwater (e.g. civil engineering excavations), drilling operations Use of pesticides on road and railway areas
Outer zone	*Zone II*	*Hazards listed for Zone III A, plus:* Roads, railway lines and similar facilities for transportation Transportation of radioactive or other substances constituting a hazard to water Storage of fuel oil and diesel fuel, storage of fertilizers and pesticides Construction sites Livestock grazing Transportation of sewage or waste water Contaminated surface waters Release of storm water to the ground Swimming and camping facilities Shooting and blasting operations
Inner zone	*Zone I*	*Hazards listed for Zone II, plus:* Any type of traffic (whether vehicle or pedestrian) Use for agriculture or forestry Use of fertilizers and pesticides

United Kingdom

The situation in the United Kingdom is handled rather differently. Whilst the Environment Agency (EA) has its own direct powers under water pollution control legislation to authorize industrial activities which discharge to groundwater, it has no control over the general authorization or prohibition of other activities.

In order to give guidance to those with responsibility for such developments, a series of policy statements has been issued for the guidance of these organizations, primarily local councils which issue permissions within the context of land use planning legislation. The EA, acting as a statutory consultant at the planning consultation stage, would object to a number of activities in groundwater protection zones unless specific precautions were applied through the planning permission granted by the local authority. This gives a wide-ranging opportunity for the planning authority to insert specific protective measures into any permissions which it may grant. Because the policy statements are of a general nature, the EA is able to take account of local situations in the advice it gives to local authorities. Where other activities may be under consideration by governmental or similar responsible bodies (dealing with, for example, changes in farming practice, or pesticide formulation and use) the requirements for protective measures can be introduced at an early stage of the development. The restricted activities include polluting industries such as:

- waste management and landfill;
- activities which interfere with groundwater flow such as quarrying and gravel extraction;
- mining;
- construction of highways, railway cuttings and tunnels;
- borehole construction;
- field drainage that intercepts recharge water and any other activity that interconnects naturally separate aquifers;
- waste disposal to land;
- disturbance or redevelopment of contaminated land as a result of former industrial activities;
- the application of liquid effluents, sludges and slurries to land;
- the discharge of sewage effluent, industrial effluent, contaminated surface water into underground strata;
- other activities such as production, storage and use of chemicals, farm wastes, oil and petroleum.

Ireland

The Irish Groundwater Protection Scheme (DoELG, 1999) uses the vulnerability rating discussed in Section 17.4 as a basis for determining the level of protection (response) within the inner and outer zone. Four levels of response are defined for activities within a protection zone: acceptable (R1); acceptable in principle though subject to specified conditions (R2); not acceptable in principle though specified exceptions might be allowed (R3); and not acceptable (R4). Whereas activities within drinking-water protection zones will usually be classified as R4, an R3 rating is possible if vulnerability is low. R1 and R2 responses may be used outside of drinking-water protection zones, also depending on vulnerability.

A useful element of the Irish scheme is that it explicitly addresses uncertainty of classifications, depending on the quality of the hydrogeological and other information available. Regulatory bodies are invited to revise zone maps as information improves, and a bias towards ensuring protection may be addressed by a developer through providing new information which would enable the zoning to be altered and – if that proves adequate – the regulatory response correspondingly changed.

Indonesia

As part of the Indonesian groundwater protection approach, local regulations need to be developed and enacted which describe both protection zone boundaries and corresponding land use restrictions. The development of the local protection scheme requires an evaluation of contaminant sources within each zone as well as effective control measures for protecting the groundwater source. On the basis of a numeric scoring system the urgency of individual control or protection measures and related costs are ranked in order to prioritize individual activities. Based on this the head of district or governor issues a decree for each spring or well in which protection zone boundaries are marked and restrictions are defined in detail.

Implementation is part of the regional development plan. A multisectoral team of governmental and non-governmental experts is entrusted to plan and evaluate the progress of the establishment of the water protection zones. Community participation is a key issue in processing the protection zones. Financing comes from the local governments.

Currently the system is applied in three districts at Lombok Island and is under development in three other provinces. An example of the approach is given Table 17.7. The Indonesian Drinking Water Surveillance regulation stipulates the application of this system on a nationwide scale.

Table 17.7. Protection scheme for the spring-fed Narmada water supply (Lombok Island, Indonesia)

Protection zone	Identified contaminant sources	Generally restricted activities in protection zone	Control measures for the protection of the drinking-water source	Evaluation of control measures			Implementation priority
				Measure	*Cost*	*Total*	
Zone I: Fixed radius of 10-15 m	*Farm and rice fields*: microbial and chemical contamination from manure and chemical fertilizers and pesticides	All activities that directly impact on water quality, such as bathing and washing in ponds and streams	Repairing of the fence that protects the spring box	3	1	4	Priority I
	Fish ponds: microbial and chemical contamination due to short-circuits between ponds and groundwater	Any use of fertilizers and manures	Implementation of a training programme on best farming practices with regard to the use of fertilizers	3	3	6	Priority I
	Private drinking-water wells: direct ingress of microbial contamination due to unsanitary construction of wells or unsanitary practices	All activities that impact on water quality in ponds, i.e. solid and/or liquid waste disposal	Inspection and surveillance of farming practices	3	3	6	Priority I
	Solid waste disposal at the Sumberawan Temple: leaching of chemicals into groundwater	Any disposal of solid waste at the Sumberawan Temple site	Maintenance and surveillance of spring box	3	3	6	Priority I
			Provision of moveable solid waste bags at Sumberawan Temple	3	2	5	Priority I
Zone II: Boundaries: up-stream 315 m; downstream 40 m; side boundaries from 160-250 m	As in Zone 1 above	Use of manure and chemical fertilizers in excess application rates, i.e. in contradiction to best farming practices	Implementation of a training programme on best farming practices with regard to the usage of fertilizers and pesticides	2	3	5	Priority II
	Deforestation: devastation of the recharge area	Felling forest	Promotion of usage of environment friendly pesticides	2	2	4	Priority III
		Change of land use	Development of locally adapted best farming practices	1	3	4	Priority III
			Inspection and surveillance of farming practices	2	3	5	Priority II
			Afforestation programme	2	1	3	Priority IV

Control measure ranking: Very urgently required = 3; Urgent = 2; Less urgent = 1; Cost ranking: High cost = 1; medium cost = 2; low cost = 3

17.8 MONITORING AND VERIFICATION OF PROTECTION ZONES

Groundwater protection zones may be a key component of a WSP (see Chapter 16) for a given groundwater supply, and protection zones would typically be control measures in this context. This would subject them to operational monitoring for assessing whether or not the required restrictions on land use and controls of human activities are in place, and to verification for checking whether they are indeed effectively protecting groundwater at the point of abstraction. However, monitoring implementation and verification of water quality are equally important for supplies that are not using a WSP.

NOTE ▶ *The implementation of protection zones is effectively supported if the stakeholders involved collaboratively develop management plans that define their delineation and the activities allowed within zones, and that document monitoring procedures, which corrective actions should be taken both during normal and during incident conditions, and responsibilities, lines of communication as well as documentation procedures.*

The implementation of control measures to enforce compliance with protection zone requirements is substantially facilitated by an environmental policy framework (see Chapter 20).

Table 17.8 provides examples of control measures that may be used for protection zones, regardless as to whether or not this is done in the context of a WSP. It also includes suggestions for monitoring and verification of the example control measure given. For example, adequate protection zone delineation in order to protect the abstraction point from contamination with pathogens and/or chemicals could be validated by using tracer studies. Protection zone monitoring would focus on checking whether the required restrictions in land use and human activities are being adhered to. Groundwater quality monitoring in this context would serve to verify the efficacy of the specific protection zone concept, i.e. both its design and implementation.

NOTE ▶ *Options for monitoring suggested in Table 17.8 focus on the control measures rather than on groundwater quality.*

Comprehensive groundwater quality monitoring programmes are a supplementary aspect of monitoring with the purpose of providing verification of the overall efficacy of protection zones.

Table 17.8. Examples of control measures for groundwater protection zones and options for their monitoring and verification

Examples of control measures for protection zones	Options for their monitoring and verification
Define zone of protection for microbial quality, e.g. based on travel time and local hydrogeological conditions, vulnerability assessments or risk assessments	Conduct tracer tests (validation of delineation) Monitor land use and activities within zone to ensure compliance with use restrictions Verify protection efficacy with microbial indicators (faecal streptococci; *E.coli,* bacteriophages)
Define zone of protection for chemical quality, e.g. based on travel time and local hydrogeological conditions, vulnerability assessments or risk assessments	Conduct tracer tests (validation of delineation) Monitor land use and activities within zone to ensure compliance with use restrictions Verify protection efficacy with specifically selected potential contaminants
Define zones vulnerable to nitrate contamination	Monitor fertilizer (inorganic and organic) applications and manure applications, potentially also stock density Verify with chemical analysis
Control pumping to ensure effect of draw-down does not increase risks of leaching	Pumping tests to measure draw-down Monitor water levels around pumping wells with piezometers Audits of pumping
Prioritization of aquifers for protection zones	Priority of aquifers indicated on maps and reports Site inspection to verify compliance

17.9 REFERENCES

Adams, B. and Foster, S.S.D. (1992) Land-surface zoning for groundwater protection. *JIWEM*, **6**, 312-320.

ANWQMS (1995). *National Water Quality Management Strategies: Guidelines for groundwater protection in Australia.* Agriculture and Resources Management Council of Australia and New Zealand, Canberra.

Bannerman, R.R. (2000) Conflict of technologies for water and sanitation in developing countries. In *Water, Sanitation and Health*, (eds. I. Chorus, U. Ringelband, G. Schlag and O. Schmoll), pp. 167-170, IWA, London.

DoELG (Department of the Environment and Local Government) (1999) *Groundwater Protection Schemes*. Environmental Protection Agency and Geological Survey of Ireland, Department of Environment and Local Government, Dublin.

DVGW (1995) *Code of Practice W101 for Drinking Water Protection Areas Part 1, Protective Areas for Groundwater*. German Association of Gas and Water Experts, Germany.

Findikakis, A.N., Stephens, J.L. and Mamou, A. (1998) *An Aquifer Classification System in Support of Groundwater Management Strategy Development.* International Water Resources Engineering Conf. Memphis, Tennessee.

Government of Oman (1991) *Water Resources of the Sultanate of Oman.* Ministry of Water Resources, Oman

Hinsby, K., McKay, L.D., Jorgensen, P., Lenczewski, M. and Gerba C.P. (1996) Fracture aperture measurements and migration of solutes, viruses and immiscible creosote in a column of clay-rich till. *Ground Water*, **34**, 1065-1075.

Lodder,W.J., Vinjé, J., van der Heide, R., de Roda Husman, A.M., Leenen, E.J.T.M. and Koopmans, M.P.G. (1999) Molecular detection of Norwalk-like caliciviruses in sewage. *Appl. Environ. Microbiol.*, **65**, 5624-5627.

NRA (1992) *Policy and Practice for the Protection of Groundwater National Rivers Authority*, NRA, Bristol.

Paul, J.H., Rose, J.B. and Jiang, S.C. (1997) Evidence for groundwater and surface marine water contamination by waste disposal wells in the Florida Keys. *Water Res.*, **31**, 1448-1454.

Paul, J.H., Rose, J.B., Brown, J., Shinn, E.A., Miller, S. and Farrah S.R. (1995) Viral tracer studies indicate contamination of marine waters by sewage disposal practices in Key Largo, Florida. *Appl. Environ. Microbiol.*, **61**, 2230-2234.

Regli, S., Rose, J.B., Haas, C.N. and Gerba, C.P. (1991) Modeling the risk from Giardia and viruses in drinking water. *J. Am. Water Works Assoc.*, **213**, 76-84.

Rossi, P. (1994) Advances in biological tracer techniques for hydrology and hydrogeology using bacteriophages. Ph.D. Thesis, Faculty of Sciences, Botanical Garden Microbiology Laboratory, University of Neuchatel.

Rossi, P., De Carvalho-Dill, A., Muller, I. and Aragno, M. (1994) Comparative tracing experiments in a porous aquifer using bacteriophages and fluorescent dye on a test field located at Wilerwald (Switzerland) and simultaneously surveyed on a local scale by radio-magneto-tellury (12-240 kHz). *Environ. Geol.*, **23**, 192-200.

Schijven, J.F. and Hassanizadeh, S.M. (2002a). Virus removal by soil passage at field scale and groundwater protection of sandy aquifers. *Water Sci. Technol.*, **46**,123-129.

Schijven, J.F. and Hassanizadeh, S.M. (2002b) Errata: *Water Sci. Technol.*, **46**(3) 123-129. Virus removal by soil passage at field scale and groundwater protection of sandy aquifers. *Water Sci. Technol.*, **46**(6-7), 411-411.

Schijven, J.F. and Hassanizadeh, S.M. (2000) Removal of viruses by soil passage: overview of modeling, processes and parameters. *Crit. Rev. Environ. Sci. Technol.*, **30**, 49-127.

Stockmarr, J. (1998) Danish Hydrological landscape planning: In *How to Cope with Degrading Groundwater Quality*, (eds. T. Hilding-Rydevik and I. Johanson), pp. 57-73, Swedish Council for Planning and Coordination of Research, Stockholm.

Taylor, R. and Barrett, M. (1999) Urban groundwater development in sub-Saharan Africa. In *Proc. 25th WEDC Conference*, Addis Ababa, Ethiopia, WEDC, Loughborough University, pp. 203-207.

US EPA (1985) *Selected State and Territory Ground-water Classification Systems*, Office of Groundwater Protection, US EPA, Washington, DC, pp. 100-103.

US EPA (1986) *Guidelines for Groundwater Classification Under the EPA Groundwater Protection Strategy Office of Groundwater Protection, Final draft*, US EPA Washington, DC.

US EPA (1993) *Guidelines for Delineation of Wellhead Protection Areas*, EPA 440/5-93-001, Washington, DC.

US EPA (2000) *National Primary Drinking Water Regulations: Groundwater Rule; Proposed Rules*. Federal Register, **65**(91) May 10, 2000, pp. 30193-30274.

WRc (1996) *Land Use Compatibility in Public Drinking water Source Areas: Water quality protection note*, WRc, Perth.

18

Sanitary completion of protection works around groundwater sources

G. Howard, S. Godfrey and T. Boonyakarnkul

The proper sanitary completion of groundwater sources is of particular relevance to the microbial quality of water. It is essential to prevent the direct contamination of groundwater at the point of abstraction or resulting from rapid recharge pathways close to the source. Where contamination is allowed to directly enter the groundwater source or reach groundwater close to the point of abstraction, the travel time may be too limited to ensure adequate die-off and the processes of attenuation may not be effective in reducing the numbers of pathogens (Robertson and Edberg, 1997).

Sanitary completion is also important in preventing direct chemical contamination, but often does not provide the same degree of protection. The subsurface leaching and transport of mobile and persistent chemical contaminants means that land use controls will be required to limit risks. This is illustrated, for instance, by studies in a small town in Uganda that showed little contamination by microbial contaminants, but significant increases in nitrate derived from faecal sources (Barrett *et al.*, 2000a). Large-scale protection measures, such as designation of groundwater protection zones, are discussed in Chapter 17.

Sanitary completion refers to the protection works at the abstraction point and the immediate surrounding areas. It is sometimes also referred to as wellhead protection, although this would usually cover a wider area around the well than covered in this

 Protecting Groundwater for Health: Managing the Quality of Drinking-water Sources. Edited by O. Schmoll, G. Howard, J. Chilton and I. Chorus. ISBN: 1843390795. Published by IWA Publishing, London, UK.

chapter. In this chapter, sanitary completion includes the underground and above ground construction of the abstraction facility as well as the immediate area surrounding the abstraction point.

NOTE ▶ *This chapter introduces options for controlling risks through sanitary completion. The information presented supports defining control measures in the development of a Water Safety Plan (Chapter 16).*

18.1 SANITARY COMPLETION AND HEALTH

The direct contamination of groundwater sources resulting from poor sanitary completion has been linked to both endemic disease and outbreaks. Such contamination is present in both developed and developing countries. For instance, Olson *et al.* (2002) describe an outbreak of *E.coli* O157:H7 in Alpine, Wyoming, including cases of haemolytic uraemic syndrome, which was related to consumption of water from a poorly protected spring which sanitary surveys had identified as being at risk from contamination by surface water. Poor sanitary completion measures also appear to have played a role in the Walkerton outbreak in Canada (O'Connor, 2002). In developing countries, the use of poorly protected groundwater sources has been linked to acute diarrhoeal disease (Trivedi *et al.,* 1971; Nasinyama *et al.*, 2000). Good sanitary completion measures have been shown to be necessary to maintain the quality of water and protect public health (US EPA, 1993; Pedley and Howard, 1997; Robertson and Edberg, 1997).

The effectiveness of sanitary completion in reducing risks of pathogens is profound as it provides a barrier to direct contamination of the source (Robertson and Edberg, 1997). The degree to which risks will be reduced, however, varies between pathogen types and aquifer types and there is a need for multiple interventions to act as barriers to most pathogen types.

For many aquifers, good sanitary completion measures will control the majority of risks posed by protozoa. Sanitary completion will greatly reduce the risks from bacteria in alluvial aquifers, but significant risks will remain in fracture flow aquifers where the enforcement of protection zones and, possibly, disinfection will be required. Sanitary completion measures will in general provide much less protection against risks posed by viruses, with protection zones and disinfection being required to reduce risks.

Most sanitary completion measures do not significantly add costs onto good standard design practice. There are cost implications, however, in ensuring that effective maintenance is performed to prevent basic protection measures from deteriorating and becoming ineffective. In some cases, cost considerations may be important with regard to selecting whether improvement of sanitary completion measures or alternative interventions will be the preferred option. For instance, where an aquifer is subjected to low-level or intermittent microbial contamination, it may be more cost effective to chlorinate the water prior to distribution than to try to deepen the borehole.

18.2 THE NEEDS FOR EFFECTIVE CONTROL MEASURES IN SANITARY COMPLETION

Sanitary completion typically includes a number of essential control measures to prevent the contamination of groundwater. Failures in such control measures have been reported from a variety of situations in both developed and developing countries (Lewis and Chilton, 1984; Lloyd and Helmer, 1991; Platenberg and Zaki, 1993; Daly and Woods, 1995; Gelinas *et al.*, 1996; Howard *et al.,* 2003). In addition to the immediate protection works at the abstraction point, the appropriate sealing of abandoned wells is also noted as essential to protect functioning groundwater sources (Rojas *et al.,* 1995; Robertson and Edberg, 1997).

Failures in sanitary completion measures may result from poor construction and in particular lack of adherence to basic quality standards. For example poor jointing on casings of boreholes, incorrect selection and placement of grouting, poor selection and installation of gravel packs, poorly mixed concrete used for linings and aprons may all result in seepage of contaminated water into groundwater sources (Howsam, 1990; US EPA, 1993).

Some drilling techniques lead to increased risks because they do not allow for grouting around the casing to be used (ARGOSS, 2001). Failure to consider the pH of the groundwater may lead to corrosion and rapid deterioration of rising mains, resulting in loss of water and abandonment of the supply (Leake and Kamal, 1990). In addition, methods of water lifting can present a direct route of contamination such as through the priming of handpumps with contaminated water (MacDonald *et al.,*1999).

Failures in sanitary completion may also result from poor maintenance (Lloyd and Bartram, 1991; Lloyd and Helmer, 1991; Platenberg and Zaki, 1993; US EPA, 1993; Daly and Woods, 1995; Howard *et al.*, 2003). In many cases specific measures constructed to protect a groundwater source fail because other measures, such as fences and diversion ditches, have not been maintained. The failure to maintain ditches and fences can result in increased access to the groundwater source, increased stress and erosion on the other protection measures and increased likelihood of inundation by surface water.

Control measures as part of sanitary completion should be identified and implemented in the planning, design, construction, operation and maintenance of an abstraction facility. As the risks to groundwater sources can be described using the source-pathway-receptor model (see Table 8.8 in Chapter 8.5.2), control measures can be categorised as: controlling the source of hazards, e.g. faecal material from a pit latrine overlying an aquifer and close to an abstraction point, and controlling pathways to avoid direct or very rapid ingress of contaminated water, e.g. through cracks in the casing of boreholes, improperly sealed apron surrounding the headwall of a dug well or borehole, eroded backfilled area of a protected spring, abandoned dug wells and borrow pits.

Control measures both for sources and for pathways include indirect measures to decrease the likelihood of a hazard or pathway developing, such as a fence around the water source to prevent access of animals or humans which could be a source of hazard (through defecation) or cause a pathway (through causing damage to the source or the immediate surrounding area).

In many cases, a combination of control measures addressing hazard sources and contamination pathways is necessary. Sanitary completion provides one barrier to contamination from such sources, but should be integrated with proper pollution containment practices and other environment engineering interventions (such as improved drainage) to be effective.

18.3 CONTROL MEASURES IN SANITARY COMPLETION: PLANNING AND DESIGN

The initial design of a groundwater abstraction facility is crucial in determining how protected the source will be. Some background information and a number of basic considerations should be taken into account at this stage.

Planning site and design in relation to the hydrogeological environment

The first step in sanitary completion is to understand the nature of the hydrogeological environment – where and how many aquifers exist, what type of aquifers exist, expected yields, depth and nature of the overburden and the degree of interconnection between different aquifers (Chapter 8). It is also important to assess how the water will be abstracted – are there springs or must the groundwater be abstracted through sinking a well or borehole into the ground? This information can then be used to make basic decisions such as the type of technology to be used, the depth of abstraction and additional protection measures required.

Where aquifers are deep or multiple aquifers are found, setting the intake deeper is likely to improve the microbial quality of water. In many aquifers, in particular relatively fine-grained aquifers, there is far less vertical movement of water (and therefore pathogens) than horizontal movement. The increase in travel times for relatively small increases in depth may be many tens or hundreds of days (ARGOSS, 2001). This increases the potential for die-off of pathogens and potentially greater dispersion; although in the latter case sophisticated models may be required to predict this. It may also increase the potential for attenuation, although this cannot be relied upon.

Sinking tubewells into deeper (usually older) aquifers may also be an important way of avoiding chemical contamination in shallow groundwater, as is the case in relation to arsenic contamination in Bangladesh (Ahmed *et al.*, 2002). Where tubewells are deepened it is important that shallower layers are cased off to prevent ingress. Often the incremental cost of deepening a well is relatively low in comparison to the overall capital investment and thus yields a significant cost-benefit. Deepening tubewells requires ascertaining whether there is no or very limited hydraulic connection between contaminated shallow and uncontaminated deeper aquifers. Hydraulic connection between aquifers is relatively common in aquifers found in weathered basement rocks and may also occur in alluvial aquifer sequences with no defined aquitard or aquiclude. Where hydraulic connections exist, deepening a tubewell may limit the improvement of water quality, as induced leakage from shallow aquifers may still lead to contamination.

Planning control measures in designing abstraction may be hampered by lack of hydrogeological information. For example in fracture aquifers it may be difficult to determine the level of risk posed to a deep aquifer by a contaminated shallow aquifer. Geophysical investigation and detailed assessment may provide some, but possibly not all, the answers required during the design stage. In such cases, monitoring as part of validation of the design chosen is particularly important.

Planning site, design and operational control measures in relation to the outcome of hazard assessment

As discussed in Section II and Chapter 14 of this book, a critical step before embarking on the design of a groundwater source is to evaluate what hazards exist close to the proposed site and their potential to be attenuated or diluted. This includes determining whether particular aquifers are contaminated and therefore whether their use as a drinking-water supply is justified.

Where the situation assessment identifies existing contamination of a well or spring, or a high potential for pollution from activities and conditions too close to the abstraction facility, control measures can either be identified towards removing the cause of the hazard(s) (see also Section V), or towards changing the site or depth of the well. While removing hazards would be the preferable, in practice population density and/or severity of contamination may make relocation of wells more feasible.

Whilst an emphasis should be placed on ensuring microbial quality of water, attention should be paid to the chemical quality of different groundwaters. Assessing whether particular aquifers contain toxic levels of chemicals (e.g. arsenic) or whether the levels of chemicals will affect the acceptability of water to consumers (e.g. high iron or manganese levels) or cause unacceptable operational problems (e.g. very hard waters) is critical in the design process. The acceptability of water is a particular problem as this may lead households to reject the use of an otherwise safe source and use contaminated sources for drinking. This not only fails to meet basic health needs for low-risk drinking-water, but also represents a significant waste of resources.

In cases where the hazard only represents a risk under certain pumping conditions, the pumping regime could be defined as control measure in order to reduce the influence of the hazard. This is unlikely to be satisfactory, however, as there may be considerable uncertainty both in the abstraction model used as basis for decisions, and in operational monitoring and corrective action to ascertain that this pumping regime is always adhered to.

If the hazard cannot be removed and changes in design of the source are not possible, post-abstraction disinfection is likely to be an effective control measure. In some cases, it will be more effective to use a lower microbial quality of water and then apply treatment at household or community level and/or implement a health education programme dealing with steps available at the household level to reduce the risks. Also, a residual risk may have to be retained if contamination is relatively low, other routes of disease transmission are more significant than water and are therefore other interventions are a greater public health priority where resources are insufficient to simultaneously improve drinking-water quality.

18.3.1 Drainage and fencing

Control measures are important to protect abstraction facilities against the potential for inundation by contaminated surface water or damage by animals or overland flows caused by heavy rainfall by diverting surface water away from the headworks. For protected springs this diversion should be located above the protection works and should direct the water into a drainage ditch downstream and away from the spring. For dug wells and boreholes, diversion ditches should circle the headworks and drain the water away from the source. In designing the ditch, the topography and likely overland flows should be evaluated to ensure that the depth of the ditch is adequate to remove all stormwater.

Diversion ditches should be located some way from the groundwater source, but not so far that significant overland flow will be generated within the area between the ditch and the headworks. A general rule of thumb is a minimum of 6 m and preferably 10 m for boreholes and dug wells and up to 20 m for protected springs (Morgan, 1990).

Restricting access by both humans and animals to the headworks is also important to reduce risks of contamination and thus, where possible, water sources should be enclosed by a fence. However, this needs to be balanced against cultural norms, for instance fencing of community water sources in Bangladesh is often not practiced because this may be interpreted as restricting the use of the source.

The wellhead of boreholes serving a piped distribution system should be located within a locked building which only the operation staff of the water supplier should have access to. Where users must collect water directly at the borehole or dug well source, fencing is still required and access should be restricted to only one or two entrances. For springs, the whole backfilled area should be fenced and inaccessible as users will collect water from outlets on the spring box. Where the spring feeds a gravity piped water system, the whole spring protection works should be fenced off and access limited to the community operator. All valve and junctions boxes should have concrete lined sides and a lockable lid.

18.3.2 Design of boreholes

Boreholes or tubewells may be shallow (5-45 m) or deep (up to several hundred metres). The choice of pump (hand, mechanized or electric submersible) to withdraw the water will depend on the hydraulic (or pumping) head in the pump, with handpumps being typically constrained to depths of 45 m or less. Where confined or semi-confined aquifers are used, the water table may rise considerably higher than the depth of the well and a handpump may still be used despite the well being physically relatively deep. Where mechanized or electric submersible pumps are used, they are typically linked to a distribution system. An example of a shallow borehole is shown in Figure 18.1. Selection of appropriate design such as the use of geotextile stockings, telescopic screen or external gravel packs can improve filtration and reduce potential sanitary risk (Driscoll, 1986).

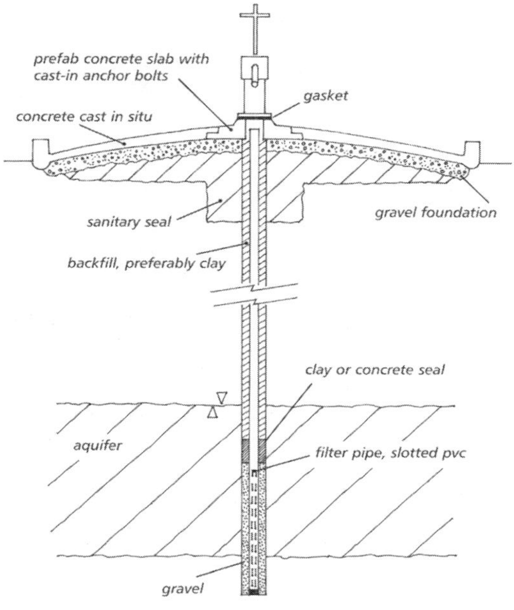

Figure 18.1. Design of a shallow borehole with handpump

For all boreholes or tubewells ensuring proper sanitary completion of the above ground infrastructure is essential to prevent direct ingress of contaminated surface water. Key components are to provide a casing over the unsaturated zone and over the upper part of the aquifer which may be expected to dewater during pumping. It is important to provide a bentonite grout seal for at least the top 1-3 m, which should be continuous with a concrete apron surrounding the top of the borehole (Driscoll, 1986). The apron must be in good condition with cracks and faults repaired rapidly.

Sanitary completion of tubewells/boreholes will be dependent on the method of drilling. For instance, MacDonald *et al.* (1999) note that the use of the sludger method commonly employed in the alluvial aquifers in Bangladesh increases susceptibility to contamination via routes close to the tubewell because it precludes sealing the annulus between the casing and drilled tubewell. However, as the formation typically collapses around the casing, the susceptibility can be reduced (Ahmed *et al.*, 2002).

Boreholes are usually fully developed prior to commissioning to ensure adequate flow using a variety of techniques. Well development is not typically designed to improve water quality, but care is needed when using some techniques (notably hydrofracturing and acidization) to avoid the creation of preferential flow paths in consolidated formations that could allow rapid transport of contaminants.

18.3.3 Design of dug wells

Most hand-dug wells are shallow (typically 20 m or less in depth) although wells as deep as 120 m have been constructed (Watt and Wood, 1977). They are often more vulnerable to contamination than boreholes, thus while some shallow dug wells have mechanized pumping, the majority (particularly those in developing countries) have water abstraction through some form of handpump, windlass or rope and bucket system. A typical design is shown in Figure 18.2.

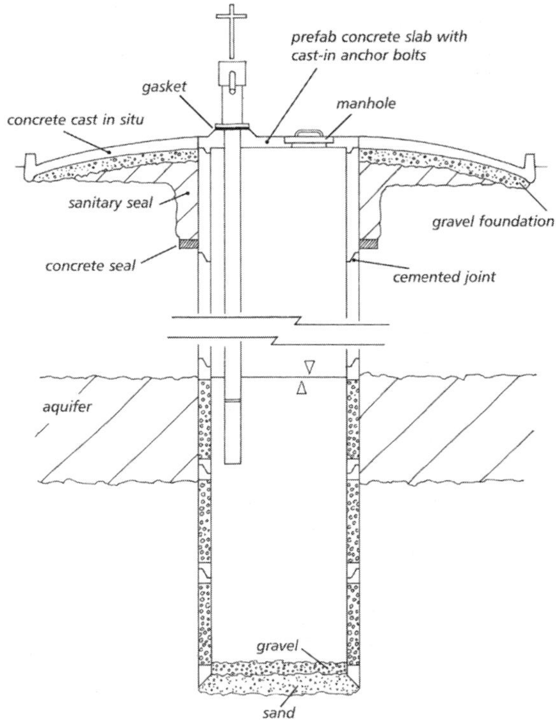

Figure 18.2. Design of a dug well with handpump

Hand-dug well designs usually have some form of lining over the unsaturated zone. In order to secure a year-round supply, caissons may be sunk below the water table to prevent drying. The design should include an apron surrounding the top of the well (usually of 1-3 m radius) with lining extended 30-50 cm above the top of the apron to provide protection against direct ingress of surface water. It is preferable that a cover is put on the well to prevent direct contamination of the water (Collins, 2000).

Studies by Lewis and Chilton (1984) note that the design, construction, operation and maintenance of the apron results in a direct reduction in levels of contamination. Dug wells can be backfilled with a sanitary seal of between 1-3 m, which increases travel time resulting in increased die off rates of pathogens. However, backfilling of wells is difficult if deepening of the well is required during drought periods. Alternative techniques such as curbing (attachment of section stabilizers) can be used to prevent movement of the shaft section of well and therefore not disturb the sanitary seal (Watt and Wood, 1977).

The means of abstraction should minimize the potential for introducing contamination from dirty containers. This may include using a handpump or other sanitary means of withdrawing water from the well such as a rope and washer pump, which have been shown to be effective in reducing levels of contamination (Gorter *et al.*, 1995). (See Section 18.5.1 for more detail about risks associated with pumps.) Where a windlass, rope and pulley system with a bucket is used, then only one bucket should enter the well and hygiene education should emphasize the need to keep the well bucket from coming into contact with the ground.

Hand dug wells often represent particular problems for sustaining good quality water, as it is difficult to ensure that very shallow water cannot enter the lining during wet periods. There are a number of different linings that may be used, including precast concrete, concrete cast in-situ and brick linings (Collins, 2000). Each of these methods gives varying degrees of sanitary protection.

Where water quality is difficult to maintain, additional improvements have been made to dug wells. These include the addition of a small sand filter set inside a box at the base of the well, a permeable base plate or ongoing chlorination of the water in the well (Lloyd and Helmer, 1991; WHO, 1997; Godfrey *et al.*, 2003). Chlorination has proven to be effective in post-emergency situations where other technology alternatives are unavailable but its effectiveness in terms of sustainability is questionable (Rowe *et al.*, 1998; Godfrey, 2003).

18.3.4 Design of protected springs

A spring is a natural groundwater source which is protected by providing a concrete headwall or spring box around the eye of the spring (where water emerges) that prevents direct contamination (WHO, 1997; Howard *et al.*, 2001; Meuli and Wehrle, 2001). There are a number of designs for protected springs, all of which utilize some form of retaining wall or spring box with an excavated area backfilled with loose material to encourage spring flow towards the outlet. A protective cover usually overlies the excavated area and the area is fenced for some distance to prevent direct access by humans and animals. One design that has been used in periurban areas is shown in Figure 18.3.

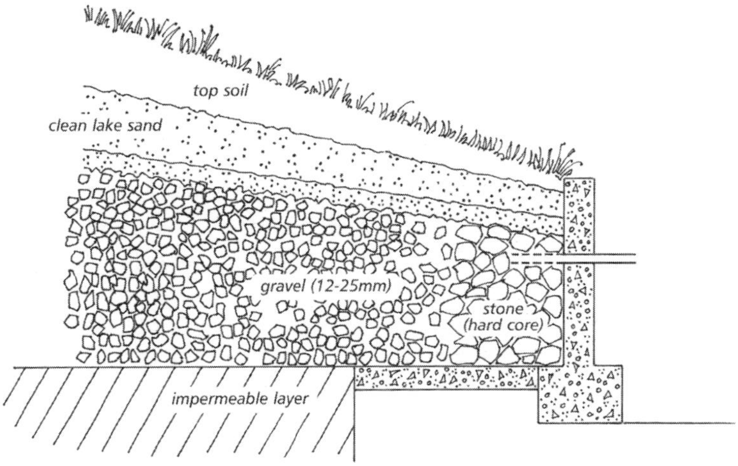

Figure 18.3. Cross-section of the backfill of a protected spring (Howard *et al.*, 2001)

Where protection is poor, contamination may occur at the point of emergence due to recharge by contaminated water in the immediate area. Thus the proper protection of the spring eye becomes vital. At most springs, the eye of the spring is excavated and the area backfilled with loose material. The filter media should be sufficiently fine to provide reasonable filtration of the groundwater entering from the spring eye and any surface water percolating through the immediate area: usually gravel although finer media may be required in more polluted areas.

It is important that this filter is overlain by an impermeable layer, commonly clay but can be a concrete cover, to reduce direct infiltration of surface water, and the whole area grassed (Howard *et al.,* 2001; Meuli and Wehrle, 2001). The filter media should be placed in the backfill area from the base of the excavation up to the expected highest level of wet season water table rise (only applicable in gravity springs).

18.3.5 Design of infiltration galleries

Infiltration galleries come in a variety of forms – they may run alongside rivers or other surface water bodies or may tap a spring line. They can be used as a part of a treatment train or may provide water directly via a shallow well or from a gravity-fed piped water supply. Infiltration galleries have been used in many countries and often have long life spans, for instance an infiltration gallery has been in operation in Lima, Peru for over 100 years and still provides high quality water with limited maintenance (Rojas *et al.*, 1995).

When using an infiltration gallery it is important to ensure that the collector pipe is laid at an adequate depth to ensure a year-round supply. The collector pipe should be surrounded by a gravel pack designed to reduce the velocity of water entering the drain to ensure that suspended sediments are removed. It is preferable that the intake holes be on the underside of the collector pipe to increase the flow path length. However, it is recognized that in most cases inlet holes will be required on the full pipe for hydraulic reasons and that the gravel pack must be laid properly. The interior of infiltration galleries will be self-cleaning if the velocity is at least 1 m per second.

18.4 CONTROL MEASURES IN SANITARY COMPLETION: CONSTRUCTION AND MATERIALS

The construction process and materials used are critical in ensuring that proper sanitary completion is achieved. Substandard work should be rejected. Poor construction quality allows faults to develop at the abstraction point. It is essential that technicians undertaking water source construction are properly trained and that guidelines for construction (for instance concrete mixes, rising main materials, etc.) are provided and followed.

The materials used can be critical to prevent water quality deterioration. Cement should be of good quality and within the recommended date of use. Sand and gravel should be clean and mixed in the proportions specified in the design. Reinforcing materials should be free of rust and dirt to ensure that a firm bond is formed with the concrete and care should be taken in selecting the gauge of reinforcing materials.

An important part of the construction process is quality control. This requires periodic checking and auditing of field practices to ensure that they are consistent with stated quality goals and objectives that the construction agency has set itself. Such quality control is necessary in all situations, whether construction is undertaken by the public or private sector. In all cases, but particularly where work is contracted to a third party, it is essential that there is evidence that the quality of construction is adequate. This may take the form of inspection and signing off a contract prior to full payment, or unannounced site visits.

18.4.1 Pumps and rising mains

For dug wells and tubewells, the selection of the rising main material is important. Galvanized iron rising mains should be avoided where water is relatively acid water because they are likely to corrode and lead to abandonment of the use of the handpump or the source. Where suction pumps are used, it is important that pumps are selected which have a non-return foot valve and do not require priming water to be added. As priming water is often taken from surface water or other stored household water, it may be contaminated (ARGOSS, 2001). Where priming water must be used, then it is important that only water collected from the well and stored in a covered container is used.

18.4.2 Cleaning of facilities prior to commissioning

For boreholes and dug-wells, good hygiene should be practiced by the team during construction. However, as some contamination will almost always remain, the wells should be thoroughly cleaned and disinfected prior to use and after maintenance tasks within the well.

For dug wells, the lining and caisson walls should be scrubbed with a chlorine solution prior to commissioning; after this washing down with chlorinated water should be sufficient. Where a handpump is installed on a dug well, the rising main should be filled with a chlorine solution and left to stand for at least one hour and preferably overnight.

Disinfection of boreholes requires filling the casing with a chlorine solution and leaving it to stand for at least one hour and preferably overnight. In both cases, the chlorinated water should be pumped to waste before use.

18.5 CONTROL MEASURES IN SANITARY COMPLETION: OPERATION AND MAINTENANCE

Whilst good design and construction will do much to ensure that wellhead protection is adequate, ensuring that it remains in good condition through ongoing preventative maintenance and repair is essential. This applies equally to springs and wells of large utilities and to small community or household supplies. The inspection routine should be defined in a management plan and include the recording of any deterioration detected and the action to be taken by whom and when.

For example, where pumps are used (whether handpump or mechanized), a stock of tools and spares should be kept by the operator so that repairs can be carried out quickly. Inherent to this is developing an effective supply chain for spares. In South Asia this has been successful as the small-scale private sector has been able to meet demand. In Africa, developing adequate supply chains has been more problematic, leading to relatively large numbers of boreholes being non-functional. In more developed countries, operators would normally have a store of the requisite tools and spares or would be able to source these quickly.

Proper training of operators of a supply is of critical importance for them to have and have the skills and knowledge to undertake at least basic preventative maintenance and perform minor repairs. More than one operator per source should be trained to ensure that maintenance and repairs can still be undertaken even if an operator moves away from the area or cannot undertake work at a particular time. For utility supplies, a number of operators may be identified who work at the supply on a rotational basis. Operators should have access to guidance and information about maintenance and repairs – e.g. specifying frequencies for replacement or worn parts and giving detailed information of repair procedures.

Where possible, the operators of water supplies should receive ongoing support from technical or professional support staff. Very often, even limited support in terms of regular visits to a supply to undertake an inspection and to meet with the operators of the supply can be very effective in sustaining good operation. This is particularly important for sustaining good quality small water supplies in both developed and developing countries and in rural and urban areas (Bartram, 1999; Holden, 1999).

In addition to basic maintenance and repairs of equipment, it is important that basic cleaning tasks are routinely undertaken. This involves cleaning and repairing diversion ditches, ensuring that wastewater ditches from springs do not become blocked and allowed to flood the source and ensuring the fence remains in good condition. Such tasks are best defined in management plans and usually are not onerous if done regularly. They can make a crucial difference in water quality control. Such activities should be supported by inspections of the site by the operator.

Experience shows that in order to sustain operation and maintenance some form of contribution from the users for the upkeep of the water source is very effective. In rural areas of low-income countries this may involve the contribution of labour. Other communities, particularly those in wealthier countries and those in urban areas of developing countries, may rely on payment by the users for the water services supplied. Most communities are willing to pay for water services providing the charges are realistic and the service meets the demands of the users. Routine payment is often preferred, as systems that operate solely on the collection of fees once a breakdown has occurred will mean that faults take longer to repair, although the latter approach has been found to work in some communities, for example in Eastern Uganda.

In both cases, community organization is often key to ensuring that maintenance procedures are supported. This may take the form of a committee that oversees the operation of the water supply. In many low-income countries, such a committee may be specific to the water source and it is preferable to ensure that the members are representative of the different interest groups in the community and in particular that women's concerns are adequately addressed. In higher-income countries such a committee may be a subcommittee from a local council or government at the local level. For instance in Chile user committees have been set up for all water supplies constructed by the regional water supply company using subsidies from the Government. These committees are supported by training programmes provided by the regional water supply companies who provide training to managers and operators of the supplies.

18.6 ASSESSMENT OF SANITARY COMPLETION AND ESTABLISHING PRIORITY RISK FACTORS

The state of sanitary completion can be assessed using inspection methodologies, as described further below. These are particularly important in the context of system assessment to determine risks and priorities for upgrading abstraction facilities as well as for defining control measures in the context of developing a Water Safety Plan (WSP). Sanitary inspections may also be used in verification via a surveillance programme using standardised approaches (Howard, 2002; WHO, 1997). Examples of such forms are commonly available, for instance in Volume 3 of the *Guidelines for Drinking-water Quality* (1997). In both cases, water quality data would also typically be collected to allow combined analysis of the effectiveness of the control measures.

Sanitary inspection methods may also be used in the routine operational monitoring of the water source as part of a WSP. Sanitary inspection approaches for routine monitoring in developed countries are likely to be the same as those used in assessment. In developing countries, other tools such as simple pictorial monitoring tools may be more effective. Routine monitoring may include some analysis of basic water quality parameters, particularly if chlorination is practiced, but this is dependent on the skill of the operators and funds for supporting such analysis.

18.6.1 Sanitary inspection

Sanitary inspection provides an easy but effective means of both assessing and monitoring sanitary completion, particularly when this employs a standardized and quantifiable approach (Lloyd and Bartram, 1991; Lloyd and Helmer, 1991; WHO, 1997). Unless a standardized approach is adopted, problems are commonly found in comparing the findings between different inspectors (WHO, 1997; Howard, 2002). This leads to inaccurate and unreliable results and limits the potential for subsequent analysis of the data. A quantified approach allows an overall risk score to be calculated in order to assess the state of supply systems and to identify priorities for action. It also permits comparisons between different source types once the data is converted into a percentage risk.

Sanitary inspections should be undertaken frequently, at least as often as samples are analysed for verifying water quality and in some cases more often. Risks are not static, they change over time as new development occurs in the area and are sometimes due to poor maintenance practices. Certain risks may also be important only seasonally, for instance the collection of surface water uphill of a groundwater source may only occur during wet periods. Therefore inspections may be required in both wet and dry seasons.

Most sanitary inspections involve a series of simple questions with Yes/No answers. As the questions are usually framed in such a way that a positive answer indicates the presence of a risk, typically a score is allocated for a positive answer and no score for a negative answer. Adding up the positive answers provides an overall sanitary risk score. An example of a sanitary inspection form is given in Box 18.1 below. Other examples are available from volume 3 of the *Guidelines for Drinking-water Quality* (WHO, 1997).

In the form in Box 18.1, questions 7, 8 and 10 refer to potential sources of faeces in the environment; questions 1, 2 and 3 refer to direct pathway factors; and, questions 4, 5, 6 and 9 refer to indirect factors. The analysis of these factors in relation to water quality provides useful information regarding which remedial and preventative actions are required for the specific water source. Data collected this way can further be aggregated and evaluated across a range of abstraction facilities of a given region in order to identify key risk factors.

Box 18.1. Example of a sanitary inspection form (based on Howard, 2002)

I. Type of Facility: PROTECTED SPRING

1. General Information: Division: Parish:
2. Code Number:
3. Date of Visit:
4. Water sample taken? Sample No.:
   Faecal Coliform/100 ml:

II. Specific Diagnostic Information for Assessment

		Risk
1.	Is the spring unprotected?	Y/N
2.	Is the masonry protecting the spring faulty?	Y/N
3.	Is the backfill area behind the retaining wall eroded?	Y/N
4.	Does spilt water flood the collection area?	Y/N
5.	Is the fence absent or faulty?	Y/N
6.	Can animals have access within 10 m of the spring?	Y/N
7.	Is there a latrine uphill and/or within 30 m of the spring?	Y/N
8.	Does surface water collect uphill of the spring?	Y/N
9.	Is the diversion ditch above the spring absent or non-functional?	Y/N
10.	Are there any other sources of pollution uphill of the spring (e.g. solid waste)?	Y/N

Total Score of Risks: /10 (Risk score 0-3=low; 3-5=medium; 6-8=high; 9-10=very high)

III. Results and Recommendations

The following important points of risk were noted (list nos. 1-10):
Comments:
Signature of Health Inspector/Assistant:

18.6.2 System assessment through sanitary inspection as a management tool

Sanitary inspections provide a useful management tool for communities, water supply agencies and surveillance bodies. The value of the sanitary inspection is that it provides a longer-term perspective on the risks of contamination, gives an overview assessment of how effective operation and maintenance has been and which system upgrade is needed. Such information can help in directing resources for improvement of the infrastructure and for improved training of water supply operators. Sanitary inspections also provide an additional means of assessing the differences in water quality from different types of water sources thus helping overall national and regional planning and policy-making (Bartram, 1999; Howard, 2002). This type of analysis is likely to be undertaken by a utility or surveillance body rather than an operator of a supply.

In a number of countries, the combined analysis of sanitary risk scores and level of contamination has proved to be an effective way of prioritizing which water supplies receive investment (Lloyd and Helmer, 1991; WHO, 2004). In many cases there is a broad relationship between the overall sanitary risk score and level of contamination (Lloyd and Bartram, 1991; Lloyd and Helmer, 1991). However, such approaches do not necessarily identify which are the most important specific factors to address as the system of sanitary inspection provides each risk factor with equal weighting, despite awareness that this is unlikely to be the case.

It is often useful to be able to determine the importance of different risk factors in order to direct investment and action on those improvements in the source that will yield the greatest improvements in water quality. Such an approach is often particularly useful in order to assess whether microbial contamination of groundwater derives from poorly sited and constructed sanitation facilities or from poor maintenance of sanitary completion measures. Leaching from on-site sanitation has been identified in some cases to be the major cause (Boonyakarnkul and Lloyd, 1994; Rahman, 1996; Massone *et al.*, 1998; Melian *et al.*, 1999). Other research from a number of countries indicates that poor sanitary completion was more important in microbial contamination than subsurface leaching from hazards such as pit latrines (Gelinas *et al.*, 1996; Cronin *et al.*, 2002; Howard *et al.*, 2003) as described further in Section 18.6.3 below. This is particularly the case in situations where there are a number of sources of human faecal matter in the environment such as refuse pits and dumps, open defecation and widespread occurrence of animal faecal matter (Barrett *et al.*, 2000b; Chidavaenzi *et al.*, 2000). Furthermore, it is often important to determine the influence of other factors such as rainfall and population density, which may affect contamination risks (Wright, 1986; Gorter *et al.*, 1995; Barrett *et al.*, 2000a; Howard, 2002).

18.6.3 Establishing the importance of different risks due to poor sanitary completion

There are a number of approaches that have been used to investigate the relationships between individual risks identified through sanitary inspection and water quality outcomes using statistical methods to analyse the data. These approaches range from the use of simple reporting of the frequency of risks in relation to specified water quality targets to the use of contingency tables and logistic regression. In order to undertake such analysis, it is important that water quality data and sanitary inspection data are available and can be paired.

In undertaking analysis of the relationship between sanitary risk factors and water quality outcomes, it is useful to compare risks in relation to water quality targets, as the failure to meet specified targets would trigger action. Cronin *et al.* (2002) present the analysis of data from two sites in Kenya and Mozambique, where the frequency of reporting of individual risks identified in inspections of sanitary completion measures were compared against samples with results above and below the median concentration of thermotolerant coliforms. This is shown in Table 18.1 below. This analysis indicated that poor sanitary completion of wells was more important in leading to contamination than subsurface leaching from sources of faecal material.

Table 18.1. Risk factors relating to higher levels of microbial contamination in dug wells in Kisumu, Kenya (based on Cronin *et al.*, 2002)

Risk factor	Percent of samples < median TTC/100 ml	Percent of samples > median TTC/100 ml	Difference
Plinth <1.5 m	83	100	+17
Well wall sealed	83	91	+8
Surface waste within 30 m	83	91	+8
Ponding on plinth	50	55	+5
Drainage channel inadequate	100	100	0
Well cover unsanitary	92	91	-1
Latrines within 10 m	55	58	-3
Open water within 20 m	64	67	-3
Ponding within 3 m	92	82	-10

Other analyses have used concentrations of indicator organisms in water to define a water quality target based on international guidelines or national standards. In this approach, for each risk factor the difference in frequency of reporting of each risk factor is compared between when the target is met and when it is exceeded with the difference providing an indication of whether there is a relationship and the strength of relationships found. Howard *et al.* (2003) describe such an analysis of water quality and sanitary risks in shallow protected springs in Kampala, Uganda shown in Table 18.2.

It is often useful to undertake further analysis of the data to assess the strength of the relationships between risk factors and water quality. In studies from Thailand, Boonyakarnkul and Lloyd (1994) developed a Sanitary Hazard Index (SHI), which related the intensity of faecal contamination associated with individual risk factors identified from sanitary inspection. These authors were able to identify which factors had the highest SHI and concluded that this should provide direction in relation to the priority accorded to reducing the presence of individual risk factors. The authors noted that there was a difference between those factors with the highest SHI and those that were most commonly reported.

Combined analysis of water quality and sanitary inspection data can also be undertaken using a range of non-parametric tests, which is common in the analysis of water resources data (Helsel and Hirsch, 1992). The use of dedicated software packages will assist in undertaking such analysis, but are not essential. Such analysis often incorporates other data such as rainfall and population density that are considered important in controlling quality.

Table 18.2. Sanitary inspection and water quality data for protected springs in Uganda

Risk factor	Percent reported when <1 cfu/100 ml	Percent report when ≥1 cfu/100 ml	Difference
Masonry defective	8	17	+9
Backfill eroded	29	67	+38
Collection area flooded	79	83	+4
Fence faulty	83	100	+17
Animal access within 10 m	79	100	+21
Latrine less than 30 m uphill	4	0	-4
Surface water collects uphill	46	100	+54
Diversion ditch faulty	79	100	+21
Other pollution uphill	46	83	+37

One example of non-parametric statistical tests is a contingency table of odds ratios. To make this analysis, variables with continuous data (e.g. water quality, rainfall and population density) must be converted into binomial categorical data. In the case of water quality targets the resulting variable will be whether the target was complied with or was exceeded (often simply expressed as either Yes or No). For rainfall data, a new variable may be whether rain was recorded within a specified time period or whether a certain depth of rainfall occurred.

An example of a contingency table is given below in Table 18.3 taken from analysis performed by Howard *et al.* (2003), which combines analysis of sanitary risks and water quality objectives for faecal streptococci and thermotolerant coliforms in protected springs in Uganda.

In the example of Table 18.3, two water quality objectives have been selected to allow the data to be analysed: the absence of faecal streptococci and less than 10 cfu/100 ml thermotolerant coliforms, the latter being a more realistic target for non-chlorinated community-managed water supplies. Odds ratios exceeding 1 show a positive relationship between the risk factor and exceeding the water quality target.

For both water quality targets the analysis demonstrates that localised pathways combined to sources of pollution and rainfall lead to contamination. Furthermore, in this setting thermotolerant coliform contamination appears to result from a more complex set of factors than faecal streptococci but is still primarily linked to poor sanitary completion.

This data can be further analysed through logistical regression (Howard *et al.*, 2003). Using the same data shown in Table 18.3, logistic regression models were developed and are shown in Table 18.4. The regression models included all co-variates where odds ratios showed relationships significant at least to the 95 per cent level. Although not significant at least to the 95 per cent level for faecal streptococci, latrine proximity within 30 m was forced into the model as this was still deemed a plausible route of contamination.

Table 18.3. Contingency table for protected springs in Uganda (adapted from Howard *et al.*, 2003)

Variable	FS >0 cfu.100ml^{-1}			TTC >10 cfu.100ml^{-1}		
	Odds ratio	p	95% CI	Odds ratio	p	95% CI
Faulty masonry	1.216	0.475	2.42	1.506	0.075	1.4
Backfill area eroded	4.135	0.000	5.8	2.762	0.000	2.73
Collection area floods	0.619	0.085	0.71	0.603	0.035	0.53
Fence absent or faulty	9.492	0.008	48.26	3.496	0.138	17.64
Animal access <10 m	3.627	0.202	25.73	1.366	0.756	9.64
Surface water uphill	2.203	0.014	2.95	3.933	0.000	4.36
Diversion ditch faulty	0.755	0.369	0.98	1.324	0.263	1.35
Other pollution uphill	3.75	0.041	12.3	5.728	0.029	26.23
Latrine <30 m uphill of spring	1.938	0.057	2.85	1.759	0.036	1.94
Latrine <50 m uphill of spring	0.838	0.531	0.98	0.738	0.198	0.17
High population density	4.49	0.000	5.43	4.708	0.000	4.75
Waste <10 m uphill of spring	1.971	0.028	2.53	2.557	0.000	2.63
Waste <20 m uphill of spring	2.437	0.001	2.78	3.085	0.000	3.03
Waste <30 m uphill of spring	1.547	0.191	2.17	1.896	0.031	2.4
Rainfall within previous 2 days	4.966	0.000	6.29	3.827	0.000	3.75

Table 18.4. Logistic regressions for protected springs in Uganda (adapted from Howard *et al.*, 2003)

Model	Model log estimate	Variables	Log estimate	Standard error	df	p-value
Faecal streptococci >0 cfu/100 ml	343.27	Constant	2.63	0.36	1	<0.001
		Eroded backfill	-0.8	0.29	1	0.006
		Faulty fence	-1.94	0.88	1	0.027
		Surface water uphill	-1.07	0.32	1	0.001
		Rainfall within 2 days	-1.34	0.27	1	<0.001
Thermotolerant coliforms >10 cfu/100 ml	338.11	Constant	2.06	0.37	1	<0.001
		Eroded backfill	-0.72	0.34	1	0.034
		Collection area flooded	0.57	0.29	1	0.047
		Surface water uphill	-0.7	0.32	1	0.031
		High population density	-1.02	0.35	1	0.003
		Rainfall within 2 days	-1.64	0.29	1	<0.001

Both regression models indicate contamination resulting from rapid recharge close to the springs and suggest that it is poor sanitary conditions at the spring itself that represent the greatest problems for the microbial quality of water. It is likely that this occurs through both direct inundation and very rapid recharge through preferential flow paths. In both cases, the principal sources appear to be waste dumps and surface water rather than latrines. This agrees with other studies that point to the importance of refuse dumps for the presence of indicator organisms (Chidavaenzi *et al.*, 2000). In a study of wells in rural Mozambique, Godfrey *et al.,* (2005) found that there was a pulse response of microbial contamination to rainfall events. Soil and engineering studies indicated that localised pathways were likely to be the primary cause of contamination rather than contamination due to aquifer pathways (Godfrey *et al.*, 2005).

The findings of Howard *et al.,* (2003) and Godfrey *et al.*, (2005) are in agreement with other studies into the causes of microbial contamination of shallow groundwater supplies, which have tended to emphasize direct ingress rather than subsurface leaching of contaminants in causing contamination (Rojas *et al.*, 1995; Gelinas *et al.*, 1996). These findings emphasise the importance of sanitary completion of groundwater sources.

The influence of sanitary completion on controlling quality may vary with different technologies and areas. For instance, studies in Thailand by Boonyakarnkul and Lloyd (1994) concluded that on-site sanitation factors led to the greatest Sanitary Hazard Index and were therefore priority risks to resolve. In Uganda, the major control on quality in tubewells appeared to be the proximity and location of on-site sanitation rather than wellhead completion (Howard *et al.*, 2003). By contrast, studies in Bangladesh reported that wellhead completion was more important than subsurface leaching from on-site sanitation (MacDonald *et al.,* 1999; Ahmed *et al.,* 2002).

The results of these studies support the validation of control measures, an essential step within a WSP (see Chapter 16). The performance of a WSP may be assessed by repeating the above analysis after upgrading sanitary completion to address faults. identified.

18.7 CONTROL MEASURES FOR SANITARY COMPLETION OF GROUNDWATER SOURCES

The design, construction, operation and maintenance requirements for groundwater sources can be translated into a series of control measures or points at the wellhead or spring protection works. Key control measures for different types of groundwater source are shown in Table 18.5 below. Planning measures to control the presence of hazards in the catchment area or immediate vicinity of a well or spring are discussed in more detail in Chapters 18-25.

NOTE ▶ *In water supplies developing a Water Safety Plan (Chapter 16), system assessment would identify which control measures exist, their effectiveness and which need to be upgraded or newly introduced. Management plans would document why specific control measures were chosen, how their performance is monitored and which corrective action should be taken both during normal operations and during incident conditions when monitoring indicates loss of control.*

Table 18.5. Examples of control measures for sanitary completion and options for their monitoring and verification

Process step	Examples of control measures for sanitary completion	Options for their monitoring and verification
PLANNING	Plan site and depth of abstraction to avoid presence of hazards and pathways for their ingress into the water source, e.g. prevent presence of faecal material within set-back distance Plan pumping regime to avoid leaching of contaminants into the aquifer by providing sufficient distance from sources of contaminants	Review (applications for) permits for construction of new abstraction facilities or for reconstruction and upgrade of existing ones
DESIGN AND CONSTRUCTION	Ensure good drainage around wellhead or spring, e.g. • with ditches to divert runoff away from the wellhead or backfill area of a spring • for wells with an apron to direct spills away from the wellhead • for springs with good drainage of wastewater away from the spring area Design wellhead or spring area protection to prevent direct contamination, e.g. with • Fencing to exclude animals from wellhead or spring backfill area • apron extending around the wellhead at least 1 – 1.5 m from casing • for boreholes ensure that join between apron and casing or lining is sound • for dug wells ensure wellhead is raised by at least 0.3 m and covered by slab • for springs ensure backfill area behind spring box or retaining wall is protected, e.g. with grass cover Ensure sanitary completion of lining, e.g. • with lining extending at least 30 cm above the apron • with seal sufficiently extended below ground level: at least 1.5 m for boreholes with handpump and 5 m for mechanised boreholes • for boreholes with rising main in good condition • for dug wells by proper construction and use of mortar seal on lining, ensure lining stays in good condition (no weep holes during rainfall !) Ensure adequate choice and good condition of structures, e.g. • for boreholes that pumps are firmly attached to the wellhead • for dug wells install handpump or other sanitary means of abstraction	Sanitary inspection of design and condition
OPERATION AND MAINTENANCE	For boreholes and wells, ensure pumping regime does not exceed amounts allowed for during planning	Meter or estimate amount of water abstracted
	For dug wells ensure hygienic use of handpump or other means of withdrawing water	Regular inspection of condition and of use. Periodic analysis of microbial indicators.
	Ensure regular maintenance and cleaning of well or spring environment, e.g. removal of debris blocking diversion ditches or those removing wastewater from the vicinity of springs; repair of fences; repair of structures such as aprons, covering flaps, handpumps	Review inspection reports for compliance to management plans. Periodic analysis of microbial indicators.

Table 18.5 focuses on control measures for the design and construction of wells and springs which are specific to sanitary completion. For the operation of abstraction facilities, maintenance and repairs are crucial control measures for keeping contaminants out, and management plans to define the scope and timescales of such activities are important to support that they are regularly carried out.

Regardless of whether or not any of these control measures are part of a WSP, their monitoring and verification is crucial to ensure that they are in place and are effective. Table 18.5 therefore includes options for surveillance and monitoring of the control measure examples given. As most of the control measures for sanitary completion involve issues of design and maintenance, many of them are most effectively monitored by regular inspections and through reviewing inspection and maintenance reports. The periodic analysis of microbiological indicator organisms is also crucial to the verification and validation of protection measures. In this context, management plans are an important tool to ascertain that inspection and maintenance activities are regularly carried out. This aspect of monitoring focuses on checking whether the controls are operating as intended, rather than on contaminant concentrations in groundwater.

NOTE ▶ *Options for monitoring suggested in Table 18.5 focus on the control measures rather than on groundwater quality.*

Comprehensive groundwater quality monitoring programmes are a supplementary aspect of verification of the efficacy of sanitary completion.

18.8 REFERENCES

Ahmed, K.M., Khander, Z.Z., Lawrence, A.R., MacDonald, D.M.J. and Islam, M.S. (2002) An investigation of the impact of on-site sanitation on the quality of groundwater supplies in two peri-urban areas of Dhaka, Bangladesh. In *ARGOSS Assessing risk to groundwater from on-site sanitation: scientific review and case studies*, BGS Commissioned Report CR/02/079N, pp. 37-67, BGS, Keyworth.

ARGOSS (2001) *Guidelines for assessing the risk to groundwater from on-site sanitation*, BGS Technical Report, BGS, Wallingford.

Barrett, M., Howard, G., Pedley, S., Taylor, R. and Nalubega, M. (2000a) A comparison of the extent and impacts of sewage contamination on urban groundwater in developed and developing countries. In *Water, Sanitation and Health*, (eds. I. Chorus, U. Ringelband, G. Schlag and O. Schmoll) pp. 179-186, IWA, London.

Barrett, M.H., Johal, K., Howard, G., Pedley, S. and Nalubega, M. (2000b) Sources of faecal contamination in shallow groundwater in Kampala. In *Groundwater: Past achievements and future challenges*, (eds. O. Sililo, S. Appleyard and M. Barrett), pp. 691-696, Balkema, Rotterdam.

Bartram, J. (1999) Effective monitoring of small drinking water supplies. In *Providing Safe Drinking Water in Small Systems*, (eds. J.A. Cotruvo, G.F. Craun and N. Hearne), pp. 353-365, Lewis, Washington, DC.

Boonyakarnkul, T. and Lloyd, B.J. (1994) *The Impact of Rehabilitation on Tubewell Water Quality Using Combined Sanitary Inspection and Bacteriological Analysis*. Paper presented at the Second Asian Regional Workshop on Groundwater Contamination, Adelaide.

Chidavaenzi, M., Bradley, M., Jere, M. and Nhandara, C. (2000) Pit latrine effluent infiltration into groundwater: the Epworth case study. *Water, Sanitation and Health*, (eds. I. Chorus, U. Ringelband, G. Schlag and O. Schmoll) pp. 171-178, IWA, London.

Collins, S. (2000) *Hand-dug Shallow Wells*. SKAT Publications, St. Gallen.

Cronin, A., Breslin, N., Taylor, R.G. and Pedley, S. (2002) Assessing the risks to groundwater quality from on-site sanitation and poor sanitary well completion, in 'Eocsan- closing the loop'. In *Proc. Second International Conference of Ecological Sanitation*, pp. 431-436, Lubeck, April 2003.

Daly, E.P. and Woods, L. (1995) Groundwater quality in the Nore River Basin in the southeast of Ireland. In *Groundwater Quality*, (eds. H. Nash and G.J.H. McCall), pp. 131-144, Chapman and Hall, London.

Driscoll, F.G. (1986) *Groundwater and Wells*, Johnson Screens, St Paul, Minnesota.

Gelinas, Y., Randall, H., Robidoux, L. and Schmit, J-P. (1996) Well water survey in two Districts of Conakry (republic of Guinea) and comparison with the piped city water. *Water Res.*, **30**(9), 2017-2026.

Godfrey, S. (2003) Appropriate chlorination techniques for wells in Angola. *Waterlines*, **21**(4), 6-9.

Godfrey, S., McCaffrey, L., Obika, E.U. and Becks, M. (2003) The effectiveness of point source chlorination in improving water quality in IDP camps in Angola. *JCWEM*, **17**(3), 149-152.

Godfrey, S., Timo, F. and Smith, M. (2005 Relationship between rainfall and microbiological contamination of shallow groundwater in Northern Mozambique. *Water SA*, **31**(4), 609-614.

Gorter, A.C., Alberts, J.H., Gago, J.F. and Sandiford, P. (1995) A randomised trial of the impact of rope-pumps on water quality. *J. Trop. Med. and Hygiene*, **98**, 247-255.

Helsel, D.R. and Hirsch, R.M. (1992) *Statistical Methods in Water Resources*, Elsevier Science, Amsterdam.

Hix, G.L. (1993) Using mud rotary to drill monitoring and remediation wells. *Water Well J.*, **47**(5), 80-85.

Holden, R.M.L. (1999) Circuit rider training program in first nation communities. In *Providing Safe Drinking Water in Small Systems*, (eds. J.A. Cotruvo, G.F. Craun and N. Hearne), pp. 513-521, Lewis, Washington, DC.

Howard, G. (ed.) (2002) *Urban Water Supply Surveillance - A reference manual*. WEDC, Loughborough University.

Howard, G., Mutabazi, R. and Nalubega, M. (2001) Rehabilitation of protected springs in Kampala, Uganda. In Water, Sanitation and Hygiene: challenges of the millennium, *Proc. of the 26th WEDC Conference*, Dhaka, Bangladesh, pp. 20-23.

Howard, G., Pedley, S., Barrett, M., Nalubega, M. and Johal, K. (2003) Risk factors contributing to microbiological contamination of shallow groundwater in Kampala, Uganda. *Water Res.*, **37**(14), 3421-3429.

Howsam, P. (1990) Well performance deterioration: an introduction to causes. In *Water Wells: Monitoring, maintenance and rehabilitation*, (ed. P. Howsam), pp. 19-24, E&F Spon, London.

Leake, C.C. and Kamal, T. (1990) Well deterioration in Sindh, Pakistan. In *Water Wells: Monitoring, maintenance and rehabilitation*, (ed. P. Howsam), pp. 59-71, E&F Spon, London.

Lewis, W.J. and Chilton, P.J. (1984) Performance of sanitary completion measures of wells and boreholes used for rural water supplies in Malawi. In *Challenges in African Hydrology and Water Resources*, p. 235, IAHS, Harare.

Lloyd, B. and Bartram, J. (1991) Surveillance solutions to microbiological problems in water quality control in developing countries. *Water, Sci. Technol.*, **24**(2), 61-75.

Lloyd, B. and Helmer, R. (1991) *Surveillance of Drinking Water Quality in Rural Areas*, Longman, London.

Macdonald, D., Ahmed, K.M., Islam, M.S., Lawrence, A. and Khandker, Z.Z. (1999) Pit latrines - a source of contamination in peri-urban Dhaka? *Waterlines*, **17**(4), 6-8.

Massone, H.E., Martinez, D.E., Cionchi, J.L. and Bocanegra, E. (1998) Suburban areas in developing countries and their relationship to groundwater pollution: a case study of Mar del Plata, Argentina. *Environ. Man.*, **22**(2), 245-254.

Melian, R., Myrlian, N., Gouriev, A., Morau, C. and Radstake, F. (1999) Groundwater quality and rural drinking-water supplies in the Republic of Moldova. *Hydrogeology J.*, **7**, 188-196.

Meuli, C. and Wehrle, K. (2001) *Spring Catchment*, SKAT, St Gallen.

Michael, A.M. and Khepar, S.D. (1989) *Water Well and Pump Engineering*. TATA Press, New Delhi.

Morgan, P. (1990) *Rural water supplies and sanitation – A text from Zimbabwe's Blair Research Laboratory*, Macmillan Publishers, London.

Nasinyama, G.W., McEwen, S.A., Wilson, J.B., Walter-Toews, D., Gyles, C.L. and Opuda-Asibo, J. (2000) Risk factors for acute diarrhoea among inhabitants of Kampala District, Uganda. *S. African Med. J.*, **90**(9), 891-898.

O'Connor, R. (2002) *Report of the Walkerton Enquiry, Part One: The events of May 2000 and related issues (a summary)*. Ontario Ministry of the Attorney General, Ontario.

Olsen, S.J., Miller, G., Breuer, T., Kennedy, M., Higgins, C., Walford, J., McKee, G., Fox, K., Bibb, W. and Mead, P. (2002) A Waterborne Outbreak of *Escherichia coli* O157:H7 Infections and Hemolytic Uremic Syndrome: Implications for Rural Water Systems. *Emerging Infectious Diseases*, **8**(4), 370-375.

Pedley, S. and Howard, G. (1997) The Public Health Implication of Groundwater Microbiology. *Q. J. Eng. Geology*, **30**(2), 179-188.

Platenberg, R.J.P. and Zaki, M. (1993) Patterns of water quality in rural areas of Assyut Governate, Egypt. *Water Sci. Technol.*, **27**(9), 55-66.

Rahman, A. (1996) Groundwater as source of contamination for water supply in rapidly growing megacities in Asia: case of Karachi, Pakistan. *Water Sci. Technol.*, **34**(7-8), 285-292.

Robertson, J.B. and Edberg, S.C. (1997) Natural protection of spring and well drinking water against surface microbial contamination. *Critical Reviews in Microbiology*, **23**(2), 143-178.

Rojas, R., Howard, G. and Bartram, J. (1995) Groundwater Quality and Water Supply in Lima. In *Groundwater Quality*, (eds. H. Nash and G.J.H. McCall), AGID Report No. 16, pp. 159-167, Chapman and Hall, London.

Rowe, A., Angulo, F., Roberts L. and Tauxe, R. (1998) Chlorinating well water with liquid bleach was not an effective water disinfection strategy in Guinea Bissau. *Internat. J. Environ. Health Res.*, **8**, 339-340.

Trividi, B.K., Gandhi, H.S. and Shukla, M.K. (1971) Bacteriological water quality and incidence of waterborne diseases in a rural population. *Indian J. Med. Sci.*, **25**, 795-801.

US EPA (1993) *Wellhead Protection: A guide for small communities*, US EPA, Washington, DC.

Watt, S.B. and Wood, W.E. (1977) *Hand-Dug Wells and Their Construction*, Intermediate Technology Publications, London.

WHO (1997) *Guidelines For Drinking-water Quality: Surveillance and control of community water supplies*, 2nd edn, vol. 3, WHO, Geneva.

WHO (2004) *Guidelines For Drinking-water Quality: Recommendations* 3rd edn, vol. 1, WHO, Geneva.

Wright, R.C. (1986) The seasonality of bacterial quality of water in a tropical developing country (Sierra Leone). *J. Hygiene*, **96**(1), 75-82.

19

Hydrological management

J. Chilton and W. Alley

The management of groundwater resources is a broad and complex subject. Basic elements of hydrological management include water conservation measures to keep abstraction at a sustainable level or economic analysis of over-exploitation impacts. The lack of water suitable for domestic uses can have serious public health consequences, and the health sector may need to ensure that domestic water requirements are taken account of in overall water resources management. This may include, for example, participating in the negotiation of groundwater allocations with competing users, such as the agricultural sector.

The overall sustainable management of water resources, i.e. the quantity of water available for use, is largely outside the scope of this monograph. However, there is often a relationship between groundwater quantity and quality, and there are some situations and circumstances in which poor management of groundwater resources can have consequences for groundwater quality that, as pointed out in Chapter 8, can be severe and sometimes difficult and costly to reverse. This chapter briefly discusses management approaches to dealing with the types of groundwater resources degradation outlined in Chapter 8.

Effective management of groundwater resources requires integration of the most important hydrogeological and socioeconomic elements that determine the interactions between land and water use and groundwater systems. The functions required for management need to be identified, and if they are not already enabled through existing institutions, these may need to be strengthened or new institutional arrangements developed to allow appropriate combinations of legal, social and financial instruments to be used to manage groundwater resources. A more detailed discussion of the institutional issues related to groundwater resource management can be found in Chapters 5 and 7 as well as in Feitelson and Haddad (1998), Salman (1999) and Foster *et al.* (2000).

NOTE ▶ *This chapter introduces options for controlling risks potentially caused by abstraction through hydrological management. The information presented here supports defining control measures and their management in the context of developing a Water Safety Plan (Chapter 16).*

Water suppliers and authorities responsible for drinking-water quality often have a key role in hydrological management, but in many settings, other stakeholders also abstract groundwater. In such settings, the definition of control measures for hydrological management will require close collaboration of the stakeholders involved, including public authorities.

19.1 MANAGING ABSTRACTION TO PREVENT SALINE INTRUSION

A major quality issue caused by poor groundwater management is saline intrusion (see Chapter 8.6.2). There are numerous examples of saline intrusion where heavy groundwater abstraction from productive coastal limestone or alluvial aquifers for urban, industrial or agricultural usage has produced serious intrusion of saline water into these aquifers, often stretching far inland, and one example is given in Box 19.1.

The widespread prevalence and costly economic consequences of severe saline intrusion have led to the development of sophisticated approaches to its investigation, particularly by means of geophysics and numerical modelling, and to its management and control. The most obvious and technically easiest approach to managing saline intrusion is to restrict pumping to allow natural recharge from the hinterland to help maintain or re-establish natural conditions. If urban, commercial and agricultural activity in the coastal zone has become dependent on groundwater then this is economically and practically difficult to do. Reducing groundwater abstraction may require a combination of regulatory and fiscal measures (Salman, 1999), including licensing and charging for abstraction, raising energy prices or reducing subsidies, licensing and controlling new borehole construction and managing crop prices, imports and exports. A useful introduction to these options is provided by Foster *et al.* (2000).

Box 19.1. Saline intrusion in the Greater Jakarta area

The population of Greater Jakarta has risen from 1.5 million in 1950 to an estimated 11-12 million in 2004. This rate of increase is typical of cities in the region and, in this situation, water supply provision is a continuous challenge with respect to both quantity and quality. In 1985 the total water demand of the city was 450 million m^3 and was met by surface water from rivers and reservoirs and from groundwater. Some 200 million m^3 were drawn by innumerable shallow wells and 50 million m^3 from deep wells. The latter is estimated to have risen to 70 million m^3 by 2000.

Hydrogeological setting: The base of the aquifer system in the Greater Jakarta area is formed by consolidated Miocene sediments, which outcrop at the southern boundary of the basin (Figure 19.1). The material filling the basin consists of marine Pliocene and Quaternary fan and delta sediments that are up to 300 m thick. Thin sandy layers, only 1-5 m thick, interbedded within the predominantly silty and clayey sediments form the productive parts of the aquifer system (Djaeni *et al.*, 1986). On the basis of the vertical variation in hydraulic conductivity in these sediments, the overall system can be divided into a shallow unconfined aquifer zone down to 40 m, a middle zone from 40 m to 140 m and a deep, strongly confined aquifer zone below 140 m.

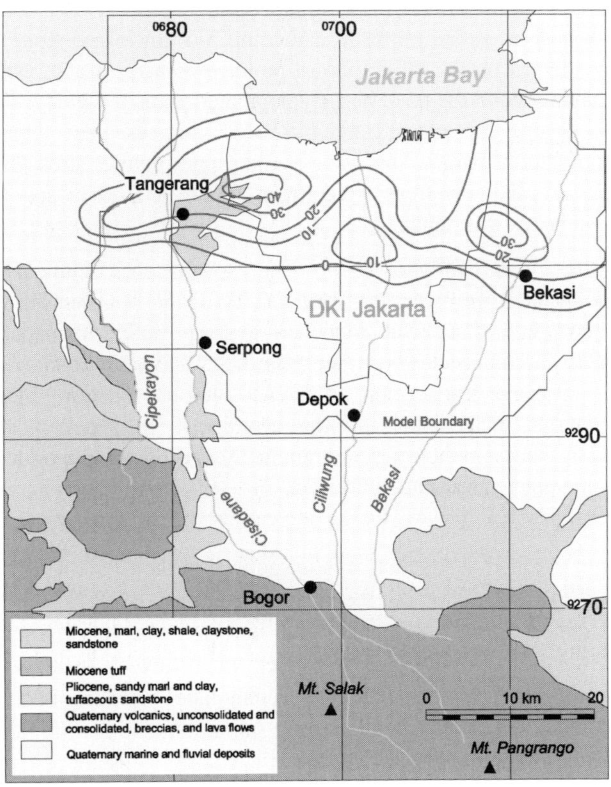

Figure 19.1. Geology of the Jakarta-Bogor area

Impacts of abstraction on groundwater levels: In 1900 the groundwater pressure head in the deep aquifer was at 5-15 m above sea level in the northern and central districts of Jakarta, and the few existing wells were generally overflowing. From 1900 to 1970, piezometric heads dropped at 0.1-0.2 m/a, increasing to more than 1 m/a in some areas of the city by the 1980s, and had dropped to 10-40 m below sea level by 1997 (Figure 19.1).

Groundwater in the deep aquifer moves from the recharge area in the south which has an average annual rainfall of 2900 mm, to the discharge area of the coastal plain with an average rainfall of about 1700 mm/a. This rainfall is more than enough to replenish the shallow aquifer in normal rainy seasons, although in dry years the combination of increased abstraction and lack of recharge causes some water table decline in the shallow aquifer. However, the lateral inflow of about 15 million m^3/a does not balance the abstraction referred to above, and the continuous drop in piezometric head indicates that downward leakage cannot balance the abstraction either because the vertical hydraulic conductivity of the confining layers is only 0.001 m/d. The area of groundwater depletion continues expanding laterally and vertically, and land subsidence is beginning to occur as a result (CCOP, 1999).

Impacts on groundwater quality: The groundwater chloride content ranges from 200 mg/l in the south to 600 mg/l further north (CCOP, 1999). The shallow aquifer suffers from saline intrusion extending 3-6 km inland from the coast, and is also affected by the infiltration of polluted recharge water, especially in the industrial areas and the most densely populated parts of the city. The deeper aquifer is less severely affected and in the coastal zone there appears to be quality stratification, with saline water in the upper aquifer (less than 100 m below ground), relatively fresh water from 100 m to 200 m and increasing salinity again below this depth.

Managing the impacts of abstraction: The seawater encroachment would be more pronounced if the hydraulic conductivity of the sediments was greater. The aquifer system has been modelled to help define management and monitoring requirements. Further increases in groundwater abstraction would have major negative impacts on an already serious situation. Management recommendations include moving the centres of groundwater abstraction southwards further from the coast, and gradually closing down production wells in the area endangered by saline intrusion.

A variant of controlled abstraction was introduced in the Chalk aquifer on the south coast of England in 1957, following the very dry summers of 1949 and 1956 in which water levels declined dramatically and saline intrusion threatened, especially around Brighton (Headworth and Fox, 1986). In winter, when there is recharge and the most active groundwater flow is towards the sea, pumping takes place from boreholes close to the coast to intercept this flow without causing saline intrusion. In summer, at times of lower groundwater levels and less active flow, the coastal wells are rested to reduce the danger of saline intrusion and pumping takes place further inland, exploiting the storage in the aquifer. As a result, average groundwater levels close to the coast have recovered by 4-5 m. Although there are significant additional costs for pumping and distribution, which are taken account of in the abstraction scheduling, the overall policy has worked well because service reservoirs in the distribution system have been closely controlled to optimize pumping, the new supply sources added have allowed the abstraction to be better distributed and the general lack of private supply boreholes helps to permit comprehensive management by the water utility (Jones and Robins, 1999).

As an alternative to reducing abstraction, methods of increasing local recharge close to the coast have been developed. Novel methods include the use of seepage barriers by placing recharge basins or lines of recharge wells between the coast and the areas of groundwater abstraction. These help to restore groundwater levels by creating an artificial recharge mound, which increases heads and pushes the saline interface down and back towards the coast.

Another method involves the use of abstraction barriers, by pumping from a line of defence wells to form a cone of depression in the water table between the coast and the valuable freshwater abstraction sources further inland, to intercept the saline water.

All of these control measures require regular monitoring to check that they are in place and functioning. They also require validation to check their effectiveness through groundwater quality monitoring. As this can usually be done simply by measuring electrical conductivity, monitoring is relatively inexpensive, either from regular field measurements of discharged groundwater or by observing the movement of the saline interface in suitably located observation boreholes. To support adequate monitoring being carried out, monitoring regimes are best explicitly defined in management plans.

19.2 MANAGING ABSTRACTION TO CONTROL INDUCED POLLUTION

The widespread occurrence in thick aquifer sequences of downward leakage of shallow polluted groundwater to deeper horizons providing public supply is also mentioned in Chapter 8.6.2. This leakage may be induced or accelerated by heavy and prolonged pumping from depth, such that polluted groundwater penetrates to depth more quickly than anticipated. In the case of Santa Cruz illustrated in Chapter 8.6.2, however, using chloride and nitrate as indicators of the polluted water, the pollution front does not appear to have penetrated below about 90 m depth (Morris *et al.*, 2003) despite the heavy pumping from the deep public supply boreholes. This is probably because the continuing abstraction from the shallow aquifer for private water supplies intercepts, abstracts and recycles a proportion of the shallow, polluted water. Though in this case the situation developed coincidentally, this is what good management of this aquifer system would require and, provided that the shallow groundwater is not used for potable supply, both the shallow and deep groundwater abstraction are able to continue.

In such situations, controlling abstraction will be only one option in a range of possible approaches to managing groundwater quality. In the Santa Cruz case, other management activities that would improve the situation would be directed at reducing the pollutant loading. Measures would include, for example, extending mains sewerage to the most vulnerable and permeable strata underlying the urban area to reduce the loading from unsewered sanitation, locating polluting industries away from these areas and/or improving their practice of handling pollutants and disposing of effluents, as well as using the planning process to encourage new city expansion over less vulnerable, clay-covered areas. Hydrological management may include encouraging the use of the shallow water for non-potable uses, lining wastewater or drainage canals or installing mains sewerage, and possibly the development of new wellfields beyond the urban boundaries to meet increasing demand.

Pollution of the upper part of an aquifer sequence threatening deeper potable supply boreholes can also occur in areas of intensive agriculture. In the United Kingdom, the impact of rising nitrate concentrations in groundwater has been felt first in the uppermost parts of the aquifers. When the problem was first observed, engineering solutions such as deepening the abstraction boreholes or increasing the length of casing to shut out shallow groundwater were advocated and tried. Because the impact of pollution on groundwater quality is such a long-term process, such engineering measures are at best temporary solutions, and reduced abstraction and blending with lower-nitrate water has become more common.

19.3 MANAGEMENT OF ARTIFICIAL RECHARGE AND WASTEWATER USE

The main purposes of artificial recharge of groundwater are to store surface water and reclaimed municipal wastewater or storm water for future use and to reduce, stop or reverse declines of groundwater levels, in order to prevent eventual exhaustion of the groundwater resource in the long term (Asano and Cotruvo, 2004). An aquifer can be used to store surface water in times of excess, which is recovered in times of shortage, particularly where the supply of water varies greatly over the year. Evaporation losses from underground storage are much less than from surface storage reservoirs, and environmental impact is likely to be lower. Artificial recharge can also be used to prevent saltwater intrusion in coastal aquifers and for oxygenation of an aquifer to change the quality of groundwater.

Recharge is also incidentally achieved, for example in irrigation and land treatment. Rivers and canals are often used to carry uncontrolled and untreated domestic and industrial effluents in and close to cities, and recharge from them, via unplanned percolation or infiltration, has a major impact on the underlying groundwater quality (Gooddy *et al.*, 1997; Morris *et al.*, 2003; Asano and Cotruvo, 2004). Recharge with wastewater – both designed and unintentional – may be an important element of the management of water quantity in an aquifer.

A broad range of technologies and methods is used to recharge the water into the ground. Artificial recharge is becoming increasingly important, and is a large and growing subject with an extensive literature (NRC, 1994; Asano, 1998; Dillon, 2002; Aertgeerts and Angelakis, 2003; Asano and Cotruvo, 2004).

19.3.1 Source water for recharge

Artificial recharge requires a sufficient and reliable source of water, and its choice is related to the anticipated usage of the additional groundwater resources. Source waters used in practice include rivers, collected rainwater, water previously used for cooling or heating purposes in production, water previously used for other purposes in production, or untreated or treated wastewater. In a review of 45 case studies of aquifer recharge, Pavelic and Dillon (1997) found that 71 per cent used natural source waters from rivers, lakes or groundwater, 20 per cent used treated sewage effluent and 9 per cent used urban storm water run-off.

Where there is a large variation through the year in the amount of available water for recharge, large storage volumes to cope with peak discharges may be required before recharge occurs. As wastewater is often predictably available irrespective of the seasons and climatic variations, it has great attraction as a source (NRC, 1994).

Surface water, storm water and particularly wastewater used for aquifer recharge usually contain a wide range of pathogens and chemical contaminants. Infiltration through the soil and unsaturated zone can greatly improve the quality (Bouwer and Rice, 1984; Idelovitch and Michail, 1984), and where artificial recharge is undertaken with poor quality water, the recharge process itself usually provides an element of water treatment. On the other hand, contaminants introduced through recharge may lead to long-term degradation of aquifer quality. The use of aquifer recharge therefore requires a careful assessment of the risk of degrading the groundwater quality, particularly if it is a source for drinking (see also Chapter 14). The nature and concentrations of pathogens and chemical contaminants (Chapters 3 and 4) present in the water used for recharge, aquifer vulnerability of the setting (Chapter 8), as well as recharge methods employed, determine the need and level of pre-treatment and the residence time necessary in the unsaturated zone for removal to be sufficiently effective so that the water can be used as drinking-water without a health risk. With some types of recharge techniques or installations, there may also be a health hazard at the surface, for example farmers irrigating with wastewater, animals and children playing at infiltration lagoons that are not fenced.

Particularly in those instances where disposal of wastewater is the primary objective and aquifer recharge is only an incidental or unintentional side-effect, little or no consideration may have been given to the possibility that infiltration of water and pollutants can have a major impact on the underlying groundwater. Improving the management of these for the objective of sustainable use of an aquifer as drinking-water resource will also require an assessment of current contamination as well as of aquifer vulnerability (see Chapters 8 and 14), and introducing or upgrading wastewater treatment prior to infiltration may prove necessary (see Section 19.3.2).

For wastewater, Table 19.1 shows a range of recharge techniques, proposes a level of treatment necessary to avoid aquifer degradation, and shows that, in many cases, the actual level of treatment falls short of that required to prepare the wastewater.

Table 19.1. Summary of wastewater treatment, use and disposal practices which lead to intentional or unintentional groundwater recharge (modified from Foster *et al.*, 1994)

Recharge technique/process	**Primary objective**	**Level of treatment**		**Ground-water recharge**
		Recommended	Frequently encountered	
Stabilization/oxidation ponds	Treat, dispose, sometimes treat/use	P, S	P	inc, acc
Infiltration well/lagoon/pit/trench		P, S	P, S	des, inc
Land-drain infiltration		P, S	S	des, inc
Land-spreading and evaporation		P, S	R, P	inc
Agricultural/amenity irrigation	Use	P, S, T	R, P	inc, acc
Riverbed seepage	None	-	R, P, S	acc
Bank filtration	Treat/use	P, S	R, P, S	des
Deep injection wells	Dispose	P, S, T	S, T	des

R = raw, P = primary, S = secondary, T = tertiary, des = designed, inc = incidental, acc = accidental

19.3.2 Recharge techniques

For recharge to be sustainable and safe, systems need to be suitable for the water types available and the underlying hydrogeological conditions, and the choice of site and system are important control measures (see Chapter 19.6). Techniques for applying water for infiltration include well injection, sprinkling onto the land surface and infiltration basins

Well injection does not retain contaminants through soil filtration. This technique is often used where groundwater is deep or where the topography or the existing land use makes surface application impractical or too expensive; or when direct injection is particularly effective in creating freshwater barriers in coastal aquifers against intrusion of saltwater (Asano and Cotruvo, 2004). Traditionally it consists of at least two wells, one for injection of the water and one for recovery of the recharged water. Recently, well injection that uses the same well for injection and recovery of recharged water, a technology known as aquifer storage and recovery, has become more commonplace. For conducting deep well injection safely, detailed hydrogeological information (e.g. extent of the aquifer, water quality in the aquifer, regional groundwater flow) are necessary and need to be carefully related to the quality of the water to be injected, particularly if the aquifer is used for drinking-water supply.

Sprinkling recharge water onto the land surface and allowing it to infiltrate downward into the unsaturated zone may effectively remove contaminants by the treatment effect of soils. This is a simple method suitable for water with a low suspended solid load, which only requires sprinkling equipment, a pump and pipes and hoses to deliver the water. The sprinkling system can be located at a fixed point, and water is sprinkled from that point by rotating the sprinkler around it. Another method is by moving the sprinkling point on a wagon/truck. By using a wagon or truck the water can be distributed more evenly at the surface than when sprinkling is limited to single points.

In basin infiltration the water is recharged from a basin created at the surface. Water is pumped into the basin from which it infiltrates slowly into the unsaturated zone. This is also a simple system where the only requirements are a pump and pipes/tubes to deliver the water, and an excavator to create the basin. One advantage of using a basin instead of the sprinkling system is that as long as the basin is filled with water the infiltration will continue at the maximum infiltration rate, provided the basin is maintained to avoid clogging.

19.3.3 Management of recharge to reduce pollution risk

A further requirement for artificial recharge to be effective is an adequate volume of permanent or seasonal groundwater storage to be able to accept the recharging water. The unsaturated zone requires a certain thickness so that slow infiltration allows sufficient residence time to achieve sufficient removal of the possible pollutants. The likelihood of pollution occurring depends on the scale and mode of infiltration, the quality of the water and the hydrogeological conditions, and it is necessary to understand the relationship between these to establish approaches to management that reduce the risk of groundwater pollution (see also Chapter 14).

Selection of scale and mode of infiltration

The selection of methods for recharge depends on many factors, including land availability, soil type, hydrogeological conditions, available finances and level of technology and the need for subsequent recovery of the water (Asano, 1998). The nature and hydraulic conductivity of the soil or subsoil and the required loading exercise major control over the method used. Wright and Rovey (1979) determined that for wastewater loading rates greater than 20 mm/d, basin or lagoon infiltration is applicable on sandy soils. Below this loading, over-irrigation and overland flow methods should be used, and can be applied on more silty or clayey soils. The former involves the application of excess irrigation loads and the latter implies allowing the applied water to flow over the irrigated land. The best soils for infiltration have hydraulic conductivities in the range 0.1 to 2 m/d. Below this range, very fine-grained soils will limit the rate of percolation and above this, coarse-grained soils permit rapid infiltration but the residence times may not be optimum for pollutant removal. The relationship between infiltration capacity of the soils and infiltration method (Foster *et al.*, 1994) and the availability of suitable land for the construction of facilities is thus a key factor in the planning process.

Quality of the infiltrating water

The quality of the water will also influence the effectiveness of recharge, and highly treated wastewater effluents or source water will infiltrate at the highest rates. As already mentioned above, pre-treatment may be necessary to remove particles (e.g. bacteria, algae or inorganic particles) or compounds that cause clogging of the injection wells, sprinkler systems or the upper part of the unsaturated zone. The purpose of such pre-treatment is to prolong the time a plant can function properly. Another pre-treatment process is the addition of oxygen if the water contains very high concentrations of organic material which have to be biodegraded, and the water does not contain sufficient oxygen for the removal process.

In some countries (e.g. the USA) aquifer recharge systems are legally required to use only treated wastewater effluents, even if contaminants from primary effluents would potentially be equally attenuated by infiltration and soil aquifer treatment (Lance *et al.*, 1980; Carlson, 1982).

Processes affecting water quality

During infiltration of recharge water, quality can undergo changes from the following:

- microbial constituents can be attenuated by filtration and sedimentation, and can grow or decay;
- organic compounds in the source water can be adsorbed onto, or ion exchange with, the soil particles, volatilize to the air phase, be subjected to biodegradation, and be chemically transformed;
- inorganic compounds can also be adsorbed to, or ion exchange with, the soil, volatilize to the air phase, be transformed by redox processes, precipitated or dissolved, and participate in biodegradation of the organic compounds;
- solids in the source water can be attenuated by filtration and sedimentation.

Chemical transformations are not generally important processes in these situations because the rates are usually very low compared to the residence time in an artificial recharge system. In contrast, growth and decay of bacteria always occur, and some of the most important pollutant attenuation processes in recharge (biodegradation of organic compounds) are enhanced by the presence of suitable bacteria. Provision of suitable organic substrate for bacterial activity may be one reason why the quality of primary effluent is as effectively improved as that of secondary effluent, and why there is a dilemma as to whether disinfection by chlorination before recharge is desirable, as it may destroy soil microbial populations that take part in the elimination processes. Some of the processes interact with each other: for example, carbon dioxide produced by biodegradation is an acid and consequently changes the pH of the system, which in turn affects microbial activity.

The use of soil aquifer treatment has been shown to greatly reduce bacteria and viruses, suspended solids, organic carbon, total nitrogen and phosphorus (Bouwer and Rice, 1984; Idelovitch and Michail, 1984; Jimenez and Chávez, 2004). Even when untreated wastewater is used primarily for irrigation, but with incidental recharge (Table 19.1), infiltration through the soil and unsaturated zone removes many or even most of the pollutants. In the case of the very extensive irrigation with collected but untreated wastewater in Mexico (Jimenez and Chávez, 2004), metals, organic carbon, nutrients and pathogens are at least partially removed in the distribution system, the soil and the aquifer. In detailed field investigations at Leon, salinity, the most unaffected and conservative of pollutants in the wastewater, was observed to have penetrated deep into the aquifer and affected the quality at 300-400 m deep public supply boreholes within the wastewater irrigated area (Chilton *et al.*, 1998).

Agricultural or amenity irrigation with wastewater makes maximum use of the attenuation properties of the soil. However, in many of the types of facilities listed in Table 19.1, the soil is disturbed or removed, and this will reduce the capacity for pollutant attenuation by the processes outlined above and described in more detail in

Chapter 4. This is why there is a clear relationship between type of facility used and level of treatment required.

Operational considerations in artificial recharge and wastewater use

Operational management of artificial recharge should be based on sound scientific understanding and follow a general multiple barrier approach. These barriers would include ensuring that the design level of treatment is adequate for the particular setting, is secure, verification that the expected improvement in quality in the subsurface does in fact occur, and that post-abstraction treatment in the case of potable supply use is also properly maintained. Among the most important aspects of the management of artificial recharge facilities is to ensure that the anticipated attenuation processes do indeed occur, and that nothing happens to lessen their effectiveness. In fact, management usually has three main objectives, which have to be balanced to optimize the operation of the scheme:

- maintain the efficiency of infiltration close to the rates envisaged in the design of the system;
- maximize quality improvement by the physical, chemical and biological processes outlined above;
- ensure avoidance of aquifer contamination.

These objectives can often be in conflict, as rapid infiltration may mean insufficient residence time for these renovating processes to operate. Thus if clogging of recharge basins is a problem which is managed by periodic removal of the clogged layer, then care has to be taken as the organic mat formed on the floor of the basin is a key component in the attenuation processes. Algal and bacterial growth in basins or in the water source used can also rapidly reduce infiltration rates, and may also encourage the precipitation of calcium carbonate or iron salts, which can seal lagoons or injection wells. Usually, this is limited to a zone a few centimetres thick. For injection wells, this is dealt with by conventional well rehabilitation methods and chlorination, and the precipitates may be removed from the well screens by introducing acids, agitating the mixture within the well and pumping out the residue.

For basins, clogging can be addressed by periodic drying out, or by removing the uppermost soil layer. In an empty basin or in a sprinkling plant the clogged soil can be taken away and discarded or the infiltration capacity can be regained by harrowing or ploughing. If the basin is full of water the clogged soil can be excavated by use of specially designed machinery or equipment. This has the advantage that the clogged soil can be taken away without disturbing the operation of the plant. Where specific organic compounds are removed from the recharge water by sorption to the soil of the basin rather than by biodegradation, attention is needed to potential saturation of the soil: if the soil's capacity for removing such contaminants is exceeded and cannot be regained, it can no longer be used for artificial recharge, and the topmost layer may need to be removed.

Management of the periodic flooding and drying of basins is a key factor in the success of wastewater recharge projects, as this controls the alternation of oxidizing and reducing conditions which are so essential to the biochemical processes involved in attenuating nitrogen from the recharge water (Bouwer and Rice, 1984). Schemes should be large enough and have sufficient separate basins to allow this to occur. In practice, the inflow volume often increases beyond the design capacity as growing urban areas generate more wastewater, and this compromises correct operation. The wetting and drying cycles themselves have to be carefully managed as, especially in very hot and arid climates, prolonged drying of the soil may damage the bacterial populations and decrease the efficiency of bacterial attenuation processes until they have a chance to re-establish themselves.

As these operational considerations are of critical importance for the safety of aquifer recharge schemes, the specific control measures identified for a given setting should be documented in a management plan together with the operational monitoring system that is to be used to ascertain that they continuously operate as specified (see also Chapter 19.6).

19.4 BANK INFILTRATION

Riverbank infiltration is a long-established method of obtaining drinking-water supplies by using the attenuation capacity of the subsurface close to a riverbed as a natural filter through which river water is drawn. In Europe, over 100 years of experience of bank filtration has accumulated since the first such installations at Düsseldorf on the Rhine and Nijmegen on the Waal in the 1870s. The review by Grischek *et al.* (2002) indicates the scale and importance of bank filtration. In the Rhine basin, some 20 million people receive their drinking-water supplies from bank filtration, and such facilities provide 45 per cent of drinking-water in Hungary, 50 per cent in Slovakia, 16 per cent in Germany and 5per cent in the Netherlands. Berlin is 75 per cent dependent on bank filtration of lake water, and the cities of Düsseldorf and Budapest are totally dependent on bank filtration. A useful summary of the features of selected bank filtration systems is given by Grischek *et al.* (2002).Bank filtration schemes usually consist of a gallery or line of wells or boreholes located parallel to and a short distance from a surface water body. Pumping lowers the groundwater level adjacent to the surface water body, inducing river water to move through the aquifer to the wells or boreholes. Current understanding of the processes of pollutant attenuation processes in bank filtration schemes is largely based on empirical knowledge. There are no common guidelines on the conditions necessary for the optimization and adequate protection of bank filtration schemes. As a rough approximation, minimum travel time approaches have often been used, i.e. that the wells or boreholes should be located at a sufficient distance from the river to allow 50 days flow time from riverbed to abstraction point (Grischek *et al.*, 2002), or 30-60 days as suggested by Huisman and Olsthoorn (1983). This has traditionally been considered sufficient for pathogen removal. The distance from the river bed to the abstraction wells is not directly reflected in the travel time, which also depends on the aquifer thickness, abstraction rates and the nature of the riverbed material. In practice, travel times range from a few days to more than ten years, and separation distances from a few tens of metres to several hundred metres (Grischek *et al.*, 2002). More recent risk assessment approaches differentiate between groups of pathogens and chemicals, and they measure or model the likely retention of contaminants in the underground of the respective setting in relation to the given conditions (see also

Chapters 3 and 4). As with other measures to reduce contaminant concentrations to safe levels it is important to validate the individual system's performance against the quality targets given.

The factors which determine the success of bank filtration schemes are a reliable source of water in the river, acceptable river water quality, as well as sufficient permeability of the river bed deposits and the adjacent alluvial formations. Provided that the permeability of the stream bed and aquifer are high and the aquifer is of reasonable thickness, then large amounts of water can be abstracted from the wells without serious adverse affects on the groundwater levels on the inland side of the wells (Huisman and Olsthoorn, 1983).

It is possible to distinguish river water from groundwater at many such installations by the use of isotopic or chemical tracers. Investigators have used chloride, boron, ^{18}O and, more recently, organic and pharmaceutical compounds as tracers of the treated sewage component of major rivers. The proportion of water drawn from the river varies from 10 per cent to nearly 100 per cent, and is often more than 50 per cent. At least some of the water in a bank filtration well is inevitably drawn from the aquifer independently of the bank filtrate. It may then be necessary to protect the groundwater from pollution by imposing suitable source protection zones around the wells or line of wells. If the aquifer material is sufficiently permeable to permit large volumes of water to move through it from the river, it is probably also permeable enough to be vulnerable from pollution at the ground surface. It may be necessary, therefore, to impose controls on river water quality and, as in the case of the river bank filtration wells which provide part of Budapest's municipal supply, groundwater protection measures which impose some controls on urban and agricultural activities within their capture zones.

River waters often carry considerable amounts of suspended matter, which can be left on the river bed and cause clogging. To help prevent rapid clogging, the rate at which the surface water enters the aquifer should be kept low, which is why multiple abstraction points are normally used. Sometimes the banks and bed of the river are scraped or dredged clean during periods of low flow to remove accumulated silt, clay and organic matter, in much the same way as for artificial recharge basins mentioned above. As with infiltration basins for artificial recharge, this has to be done with care, as the filter skin formed on the bed of the river plays a key role in restricting pollutant transport, and its careless total removal might allow breakthrough of previously attenuated pollutants.

Bank filtration installations close to large rivers usually need to be well protected from the danger of flooding to prevent short-circuiting or direct ingress of polluted floodwaters. In many cases, this is achieved by bringing the well casing or caisson some distance above the permanent ground level, so that access is raised above likely flood levels.

19.5 VALIDATION OF ARTIFICIAL RECHARGE AND BANK INFILTRATION SCHEMES

Approaches to recharging aquifers with potentially contaminated surface water – and particularly those schemes involving wastewater use – require validation of the management approach to ensure that the water quality targets are met. Monitoring programmes for validation need to be tailored to the specific problems identified in a situation assessment for the given setting. For bank filtration, validation of the hazard analysis requires repeated sampling and analyses of surface water quality and of groundwater between the river and the abstraction boreholes or wells to investigate which pollutants may enter the groundwater flow pathway and how well they are attenuated before the water reaches the abstraction wells. For designed wastewater recharge schemes, validation would usually include analyses of the quality of the incoming wastewater, stages of the treatment process and the final water before use. The monitoring of the quality of recharge water and groundwater at established recharge schemes such as those in Arizona, USA (Bouwer and Rice, 1984) and the Dan region in Israel (Idelovitch and Michail, 1984) have provided much of the evidence for the long-term ability of soil-aquifer treatment to remove a range of pollutants, and these data are also valuable for validation of the design of the systems.

Depending on contaminants expected from the water used for recharge and/or infiltration, parameters for validation may include:

- microorganisms (e.g. faecal indicators or individual pathogens such as bacteria, viruses or protozoa);
- specific organic compounds expected to be present in water used for recharge;
- organic matter (TOC, BOD, etc.) especially if the recharge water is sewage;
- regular drinking-water parameters (pH, conductivity, nitrate, nitrite, ammonium, sulphate, iron, etc.);
- if present in high density in the surface water used, algae and cyanobacteria (cell counts);
- if cyanobacteria are present in high densities, the potential for breakthrough of their toxins should be assessed (e.g. through periodic screening following heavy blooms);
- suspended solids;
- in climates with seasonal change, water temperature as an indicator of the water's travel time.

19.6 MONITORING AND VERIFICATION OF CONTROL MEASURES FOR HYDROLOGICAL MANAGEMENT

Table 19.2 summarizes selected examples of the measures proposed above for ensuring sustainable groundwater supply without compromising its quality. These begin with planning the choice of sites for groundwater abstraction and determining permissible amounts in relation to natural recharge. Where artificial recharge or bank filtration schemes are intended, they should be planned and operated in relation to aquifer vulnerability. Checking that design and construction are conducted adequately – following the plans – is important because many of the structures for abstraction or artificial recharge are installed underground, where they will rarely be accessible for inspection later. Further, operational controls are critical to ensure that amounts abstracted are in compliance with plans and permits. Likewise, operational controls for artificial recharge or bank filtration address water volumes infiltrated as well as regimes for removing clogging layers.

Regardless of whether or not any of these control measures are part of a WSP, their monitoring and verification is crucial to ensure that they are in place and are effective. Table 19.2 therefore includes options for surveillance and monitoring of the control measure examples given. Most of these focus on checking whether the controls are operating as intended, rather than on contaminant concentrations in groundwater.

NOTE ▶ *The implementation of control measures in hydrological management such as the examples suggested in Table 19.2 is effectively supported if the stakeholders involved collaboratively develop management plans that define the control measures and how compliance is monitored, which corrective action should be taken both during normal and during incident conditions, responsibilities, lines of communication as well as documentation procedures.*

The implementation of control measures to protect drinking-water aquifers from quality impairments caused by abstraction is substantially facilitated by an environmental policy framework (see Chapter 20).

In addition to monitoring of the functioning of control measures, overall monitoring programmes are important to verify comprehensively that groundwater abstraction is not mobilizing contaminants or inducing saline intrusion. For artificial recharge or bank filtration schemes, groundwater monitoring is particularly important to verify that these are not introducing pollutants into aquifers used for drinking-water abstraction, i.e. that the management concept is adequate and safe. This would typically include faecal indicators and potentially pathogens of particular concern in the water infiltrated. Where surface water used for infiltration is contaminated by chemicals from industry, household use or small-scale enterprises, occasional validation of the efficacy of their removal would be part of a monitoring programme to ensure overall groundwater safety.

NOTE ▶ *Options for monitoring suggested in Table 19.2 focus on the control measures rather than on groundwater quality. Analysis of selected parameters in groundwater which indicate drawdown of the water table or migration of contaminants is suggested where this is the most effective operational control.*

Comprehensive groundwater quality monitoring programmes are a supplementary aspect of monitoring with the purpose of providing verification of the efficacy of overall drinking-water catchment management.

Table 19.2. Examples of control measures for hydrological management and options for their monitoring and verification

Process step	Control measures for hydrological management	Options for their monitoring and verification
PLANNING	Plan abstraction and manage demand in relation to rates of recharge to ensure sustainable use (incl. in relation to amounts used by other stakeholders, e.g. for irrigation)	Review (applications for) permits for construction of new wells Monitor and record volumes abstracted Conduct tests with tracers and/or indicator organisms to validate adequate choice of site
	For artificial recharge or bank filtration: determine adequate design and choice of site in relation to the quality of river water used and pollutant attenuation in the subsoil	Monitor quality of surface water infiltrated Validate contaminant removal efficiency
	For some settings with seasonal recharge: establish seasonally variable abstraction regimes adapted to recharge patterns	Monitor groundwater levels Monitor discharge, conductivity and/or movement of saline interface
	Develop water conservation measures to help limit abstraction	Monitor water usage of different users (e.g. domestic, incl. mains leakage, irrigation and industry)
DESIGN AND CONSTRUCTION	For bank filtration: ensure adequate protection of wellheads to avoid contamination through flooding	Wellhead inspection as described in Chapter 17
	To avoid saline intrusion: establish hydraulic barriers, e.g. through defence wells or pumping regimes	Monitor operation of defence wells and pumping regimes Monitor movement of interface of saline/non-saline groundwater (e.g. through conductivity recording in observation wells)
OPERATION	Control abstraction in relation to recharge	Record volumes abstracted Monitor groundwater levels Maintain licensing system and abstraction records
	For artificial recharge and bank filtration: quantity management to maintain quality (i.e. functioning attenuation processes), e.g. through preventing hydraulic overloading and/or adapting amounts to seasonal patterns	Record quantities delivered to recharge facility and quantities abstracted Monitor infiltration rates
	For artificial recharge and bank filtration: ensure performance through adequate maintenance of facilities to avoid clogging of pores, as well as break-through of contaminants	Monitor quality of recharge water Record frequency and depth of removal of clogging layer and monitor infiltration rates particularly after removal
	For control of downward moving polluted groundwater: intercept through abstraction of shallow groundwater	Record volumes of shallow groundwater abstracted Early warning monitoring of an easily recorded parameter which indicates shallow groundwater reaching deeper aquifer horizons

19.7 REFERENCES

Aertgeerts, R. and Angelakis, A. (Eds.) (2003) *State of the Art Report Health Risks in Aquifer Recharge Using Reclaimed Water*, WHO, Geneva.

Asano, T. (1998) *Water Quality Management Library: Wastewater reclamation and reuse*, vol. 10, Technomic Publishing, Lancaster, Pennsylvania.

Asano, T. and Cotruvo, J.A. (2004) Groundwater recharge with reclaimed municipal wastewater: health and regulatory considerations. *Water Res.*, **38**, 1941-1951.

Bouwer, H. and Rice, R.C. (1984) Renovation of wastewater at the 23rd Avenue rapid-infiltration project. *J. Water Pollution Control Federation*, **56**, 76-83.

Carlson, R. (1982) Rapid infiltration of primary and secondary effluents. *J. Water Pollution Control Federation*, **54**(3), 270-280.

CCOP (1999) *Coastal Environmental Geology of the Jakarta Reclamation Project and Adjacent Areas*. Case Study Report No 2, Jakarta Bay Project, CCOP, Bangkok.

Chilton, P.J. Stuart, M.E., Escolero, O., Marks, R.J., Gonzales, A. and Milne, C.J. (1998) Groundwater recharge and pollutant transport beneath wastewater irrigation: the case of Leon, Mexico. In *Groundwater Pollution, Aquifer Recharge and Pollution Vulnerability*, (ed. N.S. Robins), pp. 153-168, Geological Society of London Special Publication 130.

Dillon, P.R. (ed.) (2002) *Management of Aquifer Recharge for Sustainability*. Swets and Zeitlinger, Lisse.

Djaeni, A. Hobler, M., Schmidt, G., Soekarni, P. and Soefner, B. (1986) Hydrogeological investigations in the Greater Jakarta area. In *Proc. Salt Water Intrusion Meeting*, Delft, pp. 165-176.

Feitelson, E. and Haddad, M. (1998) *Identification of Joint Management Structures for Shared Aquifers*, World Bank Technical Paper 415, Washington, DC.

Foster, S.S.D., Gale, I.N. and Hespanhol, I. (1994) *Impacts of Wastewater Use and Disposal on Groundwater*, BGS Technical Report, WD/94/55.

Foster, S.S.D., Chilton, P.J., Moench, M., Cardy, F. and Schiffler, M. (2000) *Groundwater in Rural Development: Facing the challenges of supply and resource sustainability*. World Bank Technical Paper No. 463, World Bank, Washington, DC.

Gooddy, D.C., Lawrence, A.R., Morris, B.L. and Chilton, P.J. (1997) Chemical transformations of groundwater beneath unsewered cities. In *Groundwater in the Urban Environment*, (ed. P.J. Chilton), vol. 2, pp. 405-410, Balkema, Rotterdam.

Grischek, T., Schoenheinz, D., Worch, E. and Hiscock, K. (2002) Bank filtration in Europe; and overview of aquifer conditions and hydraulic controls. In *Management of Aquifer Recharge for Sustainability*, (ed. P.R. Dillon), pp. 485-488, Swets and Zeitlinger, Lisse.

Headworth, H.G. and Fox, G.B. (1986) The South Downs chalk aquifer: its development and management. *J. Inst. Water Eng. Sci.*, **40**, 345-361.

Huisman, L. and Olsthoorn, T.N. (1983) *Artificial Groundwater Recharge*, Pitman, Boston.

Idelovitch, E. and Michail, M. (1984) Soil-aquifer treatment - a new approach to an old method of wastewater reuse. *J. Water Pollution Control Federation*, **56**, 936-943.

Jimenez, B. and Chávez, A. (2004) Quality assessment of an aquifer recharged with wastewater for its potential use as drinking source: 'El Mezquital Valley' case. *Water Sci. Technol.*, **50**(2), 269-276.

Jones, H.K. and Robins, N.S. (eds) (1999) *The Chalk Aquifer of the South Downs*, Hydrogeological Report Series of the BGS, Wallingford, Oxon.

Lance, J.C., Rice, R. and Gilbert, R. (1980) Renovation of wastewater by soil columns flooded with primary effluent. *J. Water Pollution Control Federation*, **52**(2), 381-388.

Morris, B.L. Lawrence, A.R., Chilton, P.J., Adams, B., Calow, R. and Klinck, B.A. (2003) *Groundwater and its Susceptibility to Degradation: A global assessment of the problem and options for management*. Early Warning Report Series, RS 03-3, UNEP, Nairobi.

NRC (1994) *Groundwater Recharge Using Water of Impaired Quality*. National Academy Press, Washington, DC.

Pavelic, P. and Dillon, P.J. (1997) *Review of International Experience in Injecting Natural and Recycled Water into Aquifers for Storage and Reuse*. Centre for Groundwater Studies, Report No. 74, Adelaide.

Salman, S.M.A. (1999) *Groundwater: Legal and Policy Perspectives*. World Bank Technical Paper No. 456, Washington, DC.

Wright, K.R. and Rovey, C.K. (1979) Land application of waste - state of the art. *Ground Water*, **17**, 47-61.

Section V

Approaches to pollution source management

20

Policy and legal systems to protect groundwater

P. Chave, G. Howard, P. Bakir, S. Appleyard and B. Hoque

This chapter deals with the policy and legal environment within which groundwater protection and management should operate. Effective policies for groundwater protection must take account of the institutional and cultural environment in the country, the interrelationship of quantity and quality of groundwater, financial viability of any proposed measures for protection and acceptability of the measures to society. Policies that are considered reasonable in some countries may not be acceptable in others. Therefore, where proposed policy builds on experience elsewhere, it is important that local values of the society are taken into account. For this reason, it is essential that effective policy development includes the public, government agencies and other stakeholders potentially affected at the earliest possible stage.

The overall process of developing and implementing policies and strategic management for groundwater protection may follow the route shown in Figure 20.1.

While step 3 of this process is discussed in some detail in Chapters 5 and 7 and criteria for developing specific protection concepts are discussed in Chapter 17, this chapter reviews the overall framework of governmental policy and institutions that facilitates their implementation. Such a framework provides an important context in

 Protecting Groundwater for Health: Managing the Quality of Drinking-water Sources. Edited by O. Schmoll, G. Howard, J. Chilton and I. Chorus. ISBN: 1843390795. Published by IWA Publishing, London, UK.

which fragmental local initiatives and actions can be amalgamated in to a comprehensive national or regional policy.

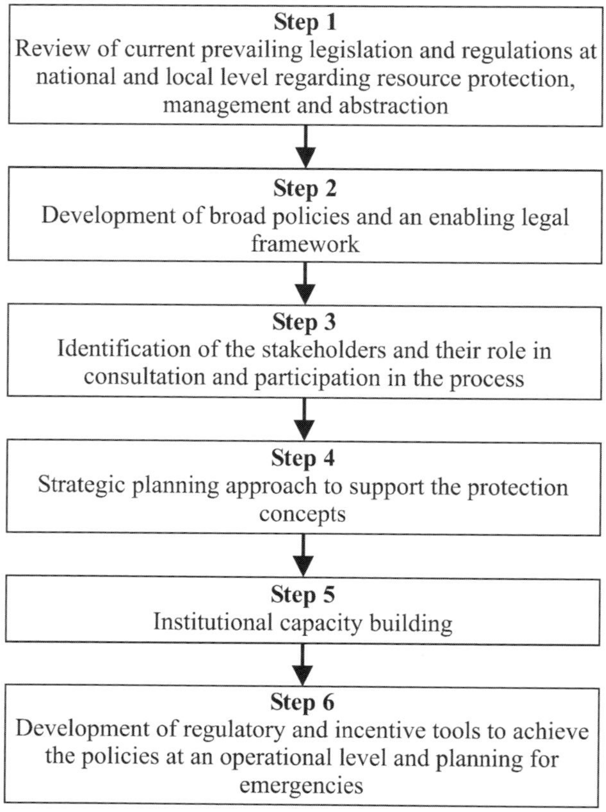

Figure 20.1. Flow chart for developing groundwater protection policies

20.1 GROUNDWATER PROTECTION POLICIES

Policies need to be applied in a properly understood and constituted framework so that their application is clear and their effectiveness is assured. OECD (1989) developed the DPSIR causality framework (shown in Figure 20.2) to enable the basic issues in policy development to be identified and possible impacts of proposed solutions to be tested. The DPSIR framework involves five principal steps. The Driving forces describe the human activities, such as the intensification of farming and chemical industry production or development of land for housing, which may lead to significant threats to the groundwater quality or quantity (as described in Section II). The Pressures describe the stresses that the developments place on a particular aquifer in terms of its possible uses. The State of the aquifer is described in terms of its quality and hydraulic condition and the Impact shows the outcome of loss of the source, for example the need to find alternative drinking-water sources if an aquifer becomes unusable. Responses describe the policies that have been or are being developed to deal with the problem. By using the DPSIR cycle it is possible to decide which of several alternative policies might be the optimum solution. Although not directly included within the DPSIR framework, processes to review effectiveness of responses are also critical (see Chapters 15 and 16).

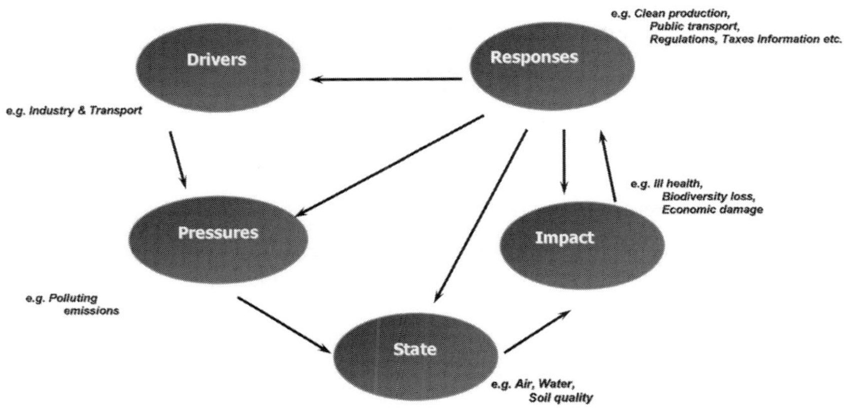

Figure 20.2. DPSIR framework (EEA, 1998)

UNECE (2000) has described how this approach is incorporated in the policy development as being a set of seven key steps:

1. Identify the functions or uses of the groundwater.
2. Identify the issues and problems (particularly health related problems).
3. Establish a function-issues table to see whether the issues are in conflict with the functions of the groundwater systems.
4. Establish management objectives using the function-issue table for priority setting based upon urgency and technical/financial means (see also Chapter 15).
5. Ensure that suitable information is collected on place and time dependent factors (on the groundwater body, the stages of the management programme, etc.), as discussed in Chapter 6.
6. Use the DPSIR concept to examine the detailed relationships and causalities.
7. Make a checklist with criteria that have to be met linked to the measurable factors identified above.

The European Union's Framework Directive for Water as one example of policy for groundwater protection – in this case not primarily targeting its use as drinking-water source – is described in Box 20.1.

Box 20.1. Example of policy for groundwater protection – European Union Framework Directive for Water

The European Union (EU) water policy has been based on six basic principles:

- a high level of protection;
- application of the precautionary principle;
- the prevention of pollution;
- the rectification of pollution at source;
- adoption of the polluter pays principle;
- the integration of environmental protection into other policies such as agriculture, transport and energy.

In the context of the application of these principles to the protection of groundwater, in 1992 the Council of Ministers (Resolution 92/C 59/02, OJ C59/2 6.3.92) recognized the dangers of falling groundwater levels and long-term problems associated with the pollution of certain aquifers, e.g. for providing drinking-water. As a result the Commission revised the Groundwater Directive by incorporating it into a general freshwater management policy.

Consequently, the EU has adopted a new approach to water policy. The Water Framework Directive (EU, 2000) expands the scope of water protection to all waters, surface waters and groundwater and requires the achievement of 'good status' for all waters by a certain deadline. The regime will require the overall management of water based on river basins, with a combined approach of using emission limit values for the control of discharges and water quality standards applicable to the natural waters. The Directive also concerns water pricing and ensuring that the citizen is more involved in decision taking.

The main features of this policy, which Member States of the EU will be obliged to transpose into their domestic legislation, include the recognition that underground water, as part of the whole water cycle, plays a part in maintaining a sustainable ecosystem and drinking-water supply and that water quantity and quality are inextricably linked. The policy will take account of the natural flow within the hydrological and hydrogeological cycles when determining action which is aimed at improving or maintaining the water's good status.

The polluter pays principle will be incorporated through the use of appropriate economic instruments with which to control water usage and pollution levels, and the whole system must be managed on a river basin basis – incorporating the requirements of groundwater and surface water as an integrated whole. The policy demands forward planning, through the development and publishing of river basin management plans with a significant degree of public involvement in such processes.

20.1.1 Institutional issues for policy development

An effective first step for developing a groundwater protection policy is to establish a policy task force that draws together the key institutions with an interest in the use and management of groundwater resources. This taskforce would include e.g. Environment, Health, Agriculture, Industry, Local Government and (where there are cross-border aquifers) Foreign Affairs. Such an intersectoral task force requires representatives from senior levels of Government who are able to develop and define policy, thus it may be composed of the most senior civil servants who report directly to the Ministerial level.

A lead agency would be identified to coordinate the definition of a policy and strategy for groundwater protection and management. This agency should be mandated by the inter-ministerial body, would typically fall under a Ministry of Environment or Water, and may take the form of a Commission.

Once the institutional environment has been reviewed, rationalization can be considered as a means to develop a flexible and effective approach. This rationalization will involve identifying which organization should take lead responsibility for groundwater protection and consideration of how this should be structured, and relationships between local and national components of the organization.

Rationalization may result in the removal of responsibilities or power from some organization, which is often a difficult process. It is essential that before major changes are implemented, there is proper consideration of how this will be undertaken, what the implications will be for staff and for the organization as a whole and how this will be managed in the most effective manner.

The relationships between the organization responsible for groundwater protection and other organizations that either use or potentially pollute groundwater will need to be defined. This will include not only the Government environment, but also NGOs and the general public. The means of reporting will have to be transparent and accountable and multiple reporting mechanisms may well be required.

This institution will need to be supported by appropriate legislation and provided with adequate powers to develop and enforce regulations and laws. Such powers may include aspects such as issuing permits and abstraction licences, control of land use, defining protecting areas and establishing minimum construction requirements. The groundwater protection body must have adequate numbers of staff and resources in order to be able to monitor groundwater resources, collect essential data and ensure that they can undertake inspections and serve enforcement notices. The performance of the protection agency is highly dependent on the staff it contains and without proper training and support, it will be difficult to recruit and retain motivated staff of high calibre.

The importance of resolving institutional arrangements from the outset of developing groundwater protection should not be underestimated. Effective institutional frameworks and clear and accountable systems of responsibilities greatly facilitate achieving the objectives of groundwater protection.

20.1.2 Capacity-building to support institutional delivery

To support an intersectoral approach, interdisciplinary training is often required to ensure that staff has the necessary competence and skills to resolve groundwater issues and to work with other disciplines and sectors. Capacity-building ensures that staff is fully aware of current policies for groundwater, understand the issues related to groundwater management and have the expertise to provide advice and implement protection measures. It is also critical that they understand what other disciplines and sectors need to contribute to groundwater management. This is particularly important where water suppliers do not own groundwater catchments and activities are required by a number of sectors to ensure groundwater that is acceptable for use in drinking-water supplies (see also Chapter 16). Box 20.2 highlights an example from India where the training of staff was recognized as a critical aspect of improving institutional capacity to protect groundwater.

Box 20.2. Capacity-building in India (based on OECD, 1989)

Between 1990 and 1997, a series of training events on groundwater management was run for senior engineers and scientists aimed at improving the sustainability of drinking-water supplies. Each year, up to 18 candidates were drawn from a variety of backgrounds and locations across India. For formal training they were first sent to the United Kingdom, followed by a period of fieldwork in a selected catchment within India to observe local conditions and apply the lessons learnt to identify polluting activities as well as potential solutions. The formal training considered all aspects of groundwater development from drilling techniques to groundwater modelling and included source protection and pollution prevention modules. As the course material was developed, trainers were coached in the context and presentation of this course material to allow transfer of the training element to India. The intensive course helped promote interchange between the different disciplines and requirements and expanded understanding of the interdependence of the different work streams.

The fieldwork was undertaken over a period of five to eight weeks each year within a different State to allow a number of geological environments and levels of development to be covered. In all areas, the catchments were under stress from competing demands for limited resources or at risk from actual or threatening pollution. In the short time available the aim was for the delegates to gather sufficient information to allow an assessment of the catchment, describe the geological and hydrogeological environment, identify issues and promote solutions to aid management of the system.

The most important solution identified by the candidates was to treat the catchment as an integrated unit for both planning and management purposes. Bringing the different users and departments together was seen as a major step to sustainable development and the appropriate allocation of scarce water resources. Education of local people in both aspects of hygiene and availability of water featured highly in improving public health. It was also recognized that further training and motivation of Government staff was required, and that improvements in agricultural practices could help to conserve water and make its application more effective.

Overall groundwater availability was a major topic. Understanding its location and movement was key to successful development. The need for appropriate monitoring and sampling programmes was recognized. The provision of new, and maintenance of existing, water harvesting and retention structures to increase recharge and groundwater storage came high on the list of priorities. Often, existing structures were located in inappropriate positions or were installed for another purpose. With more integrated planning these could have been built to serve a number of purposes more effectively. Many could be modified at relatively minor costs to achieve these ends. Identification of optimum drilling locations using more modern techniques was encouraged, as in many villages some hand pumps ran dry early in the dry season due to inappropriate locations.

Recognition of the value of water and its declaration as a National Asset were also seen as important issues. This helps avoid wastage and misuse, as well as promoting better utilization of existing resources. Aspects of the ownership of water and the infrastructure to capture and distribute it were considered to be important, as were pollution prevention matters. These ranged from simple wellhead protection to management of discharges from industry. At the end of the fieldwork period, the candidates prepared and presented their findings to an invited audience of local (as well as National) government representatives, village heads, NGOs and others contacted as part of the study. They then returned to their previous posts, often to be promoted, and have used their training to encourage better communication and awareness in their local areas.

20.1.3 International groundwaters

The UN Convention on the Protection and use of Transboundary Watercourses and International Lakes, signed in Helsinki in 1992 (UN, 1992) recognizes the difficulties of protecting water bodies, including groundwaters, which cross international borders. The Convention requires all signatory countries to:

- prevent, control and reduce pollution of waters which may have a transboundary impact;
- to ensure that these waters are used with the aim of ecologically sound and rational water management;
- to use such waters in a reasonable and equitable way;
- to ensure that conservation of ecosystems is achieved.

The Convention requires the adoption of prevention, control and reduction programmes for water pollution, and the establishment of monitoring systems. Bilateral and multilateral cooperation is essential to the successful protection of such waters and riparian countries are expected to enter into agreements over such issues as joint monitoring programmes and conduct joint programmes for the prevention, control and reduction of transboundary impacts. Warning and alarm systems are required to inform countries of any critical situations that have a cross-border impact.

A Protocol on Water and Health to this Convention was agreed by an interministerial conference in London in 1999. This Protocol links the issues of human health, water resources and sustainable development and targets *'the promotion, at all appropriate levels, of human health and well-being within a framework of sustainable development, including the protection of water ecosystems and through preventing, controlling and reducing water-related diseases*'. It emphasises the need to create legal, administrative and economic frameworks to reach these targets, including the development of water management plans, and explicitly

- includes groundwater among other water environments;
- includes WHO Guidelines as translated into national and international legislation;
- addresses the protection of water used as source for drinking-water and the development of effective water management systems including controlling pollution;
- includes promoting understanding of public health aspects by those responsible for water management and vice versa promoting the understanding of basic principles of water management, supply and sanitation among those responsible for public health;
- requires the development of effective networks to monitor and assess water-related services (WHO and UNECE, 2001).

The Protocol also explicitly includes a number of principles, including the precautionary principle, particularly towards preventing outbreaks and incidents of water-related diseases, using water resources in such a way that the needs of future generations are not compromised, the polluter pays principle and access to information.

UNECE has issued further guidance concerning groundwater management (UNECE, 2000). The development and implementation of cross-border policies for groundwater depends upon institutional aspects, which include the arrangements and responsibilities for cooperation. The convention requires that socioeconomic conditions in riparian countries should be taken into account in deciding upon institutional arrangements.

As discussed in Section II, to support management of transboundary groundwaters, it is often useful that action plans, which include quantified targets and arrangements for mutual assistance, are drawn up by riparian states taking into account items such as:

- land and groundwater uses, including the possibility that restrictions or bans on certain activities may be imposed;
- zoning criteria, including the concept of protection zones;
- economic activities, paying attention to their impact on groundwater;
- pollution and abstraction of groundwater, to include the necessity of monitoring and the sustainability of abstractions.

To deal with transboundary groundwaters, the Convention requires the establishment of Joint Bodies to take on the task of monitoring and assessment of the effectiveness of the agreed measures as shown in the Box 20.3.

Box 20.3. Tasks of Joint Bodies (based on UNECE, 2000)

- Collect, compile and evaluate data to identify pollution sources;
- develop joint monitoring programmes;
- draw up inventories and exchange information;
- establish emission limits for waste water and evaluate effectiveness of controls;
- elaborate joint water quality objectives and criteria for preventing, controlling and reducing cross boundary impacts;
- develop action programmes for pollution reduction from point and diffuse sources;
- establish early warning systems;
- serve as a forum for exchange of information;
- promote cooperation and exchange of information on BAT;
- participate in EIAs;
- coordinate activities of others.

The Merske Brook Case Study: An example of international cooperation

An example of the problems relating to international groundwater bodies and steps taken to overcome these is found in the Merske catchment area of which 3185 ha are located in Belgium and 2792 ha in the Netherlands. The Merske Brook is fed by deep groundwater and infiltration from an agricultural area. The regional groundwater flow is from Belgium into the Netherlands, but there are separate local systems where groundwater from Belgium flows into the Netherlands over longer timescales. The catchment can be subdivided into an area of deep groundwater exfiltration in the brook valley, a strongly dehydrating area on which agriculture is practiced and a high infiltration area used for forestry and agriculture. Deteriorating water quality and lowering of water tables represent the problems. In order to help in resolving these problems, cross-border catchment area committees have been established with the aim of bringing about a cross-border water policy. Known as 'Markcomite' the Merske Committee was established in 1994 and has set up a number of research projects to aid water improvements. However, the problems of setting up and running such cooperation should not be underestimated. In this example it was found that clear differences existed between the two countries in their internal water management regimes which needed to be resolved. Belgium manages its groundwater at national level, whereas the Netherlands utilizes three levels – national, provincial and local water board levels. In terms of water management the Netherlands had various measurement networks already installed, supplying data to all its administrative levels, whereas in Belgium there were fewer measurement systems. However, the Dutch data were not generally mutually compatible. The availability of data was not consistent, some being digital others not. Tackling differences in institutional structures and data organization is frequently among the first tasks of transboundary water committees and major harmonization of surveillance programmes may be an important step of their work.

20.2 LEGISLATIVE FRAMEWORK FOR GROUNDWATER PROTECTION

The availability of appropriate instruments to enable the implementation of groundwater protection policies is essential. The important issue of groundwater ownership must be dealt with and revisions to water rights may be an initial stage in the implementation of policies.

20.2.1 Environmental legislation

The risk of groundwater pollution from commercial or industrial activities or urban development can be reduced through incorporating groundwater protection strategies in environmental legislation (Patrick *et al.*, 1987). Equally, individual development proposals can be assessed either through a formal EIA process and/or by systems of licensing. Both EIAs and licensing can help ensure that potentially polluting activities are located in areas where the risk of groundwater pollution is minimal. Licensing can also ensure that activities conform to best management practices (BMPs). BMPs generally define a set of standard operating procedures and design standards for a particular land use to ensure that the risk of pollution from accidental spillage or over-application and misuse of chemicals is minimized. Environmental licensing may also prescribe ongoing groundwater monitoring and may require industries to undertake groundwater remediation. Such measures need to be accompanied by enforcement with meaningful penalties, which has proven successful in many countries.

Environmental legislation can also help manage pollution from past land uses. Many jurisdictions have provisions for surveying and managing contaminated sites. Information on contaminated sites is maintained in publicly accessible databases, for example various databases have been established by the US EPA. Owners of contaminated sites in the USA (irrespective of whether they were the polluter or not) are usually required to clean up contaminated soil and groundwater before land is sold and redeveloped.

Some jurisdictions have closely aligned environmental and planning legislation in two important ways. Some environmental legislation allows entire planning schemes to be subject to EIA, and there are formal links between different pieces of legislation to allow this to happen. It is also possible in some jurisdictions to create regional environmental protection policies that can support groundwater quality and thus drinking-water quality protection. These policies will generally set out beneficial uses to be protected and protection objectives, one of which may be protecting groundwater as drinking-water resource. Such planning schemes may establish water quality standards and set prescriptive controls on land uses.

20.2.2 Legislative reform

In many countries, much of the existing planning and environmental protection legislation was drawn up before groundwater protection was a significant issue. In some jurisdictions management of groundwater pollution has been compromised because there were many government agencies with overlapping responsibilities whose decisions about particular development proposals were poorly coordinated. The management of groundwater quality issues has been improved by either reforming the way that government agencies work together using existing legislation, or by reforming legislation and restructuring government agencies. The New Zealand Resource Management Act (1991) is a good example of a piece of legislation where planning, environmental and natural resource management issues are all coordinated through a single piece of legislation.

Legislative reform alone may not improve groundwater quality protection unless there is the political will to effectively implement it (Foster *et al.*, 1992). For political action to occur, for example attaining strong support for allocating more finance to groundwater protection measures, will require that the public perceives the benefits. This poses special problems for groundwater. Frequently, the benefits of groundwater protection measures are not obvious until well into the future, whereas any controls may affect some people immediately. Communities may be tempted to postpone protection measures until the degradation of groundwater quality used for drinking-water supply becomes so severe that widespread concern amongst the general public or specific interest groups prompts them into action. However, postponing groundwater protection measures often leads to more costly and intractable problems in the long run. Educating the general community about the importance of groundwater protection is one of the most effective ways of influencing the political process to implement protection measures.

20.2.3 The law relating to groundwater ownership and abstraction rights

A fundamental legal issue, which invariably needs attention, is the question of ownership of underground water. This is because the introduction of protection measures by the State usually involves the enactment of controls that apply to groundwater or to activities on the ground above the aquifer. These controls may infringe existing rights and customs and it may be necessary to modify or withdraw the existing rights of individuals in order to enact the necessary changes. Ownership varies from country to country. In many countries common law recognizes a link between water and land ownership (Howarth, 1992), however, in others ownership of water resources lies with the State.

Where there is a link between land ownership and water, rights typically relate to the right to abstract and use water, including drinking-water abstraction, rather than rights of ownership of the water itself. In order to allow full control through a planned protection regime it is often necessary to alter the legal status of such water rights. Where plans are being developed to use water resources to their full extent, for example by the conjunctive use of surface and groundwater, existing restrictions on the general availability of groundwater caused by the imposition of individual water rights may be unacceptable. Such rights may have to be extinguished so that access to all the available water in a territory is guaranteed.

Abstraction rates may have a fundamental influence on water quality. The control of abstraction, often seen as a matter of the protection of quantity, is also an important issue relating to quality (see Chapters 8 and 19). Control of abstraction requires a sound legal basis and good enforcement. The different legal systems in different countries are very varied, but for all approaches it is essential that the law is clear as to whom regulations apply.

As a general worldwide trend, there is a realization that individual rights need to be subordinate to the protection of quantity and quality of groundwater resources, and 'rights' are being cancelled in favour of 'permissions' to undertake activities such as the abstraction and use of groundwater.

In the United Kingdom for example, where there are no established rights to riparian ownership of underground water, but there are long-standing common law rights for its abstraction and use, Section 24 of the Water Resources Act 1991 requires abstractors to obtain a licence from the EA to drill a borehole and abstract underground water. Once granted, this gives an indefinite right to withdraw water. The use of the water may be specified as for drinking, irrigation, and so forth. However, under new proposals to ensure the long-term sustainability of aquifers, these rights will be modified by central government and time limits will be placed on the permit. The government originally proposed a limit of 15 years when new authorizations are granted but, following extensive consultation, the time limit may be varied to reflect an individual catchment situation depending upon the local availability of water.

Time limits on licences to abstract are imposed in some other countries. In South Africa, individual ownership has been extinguished by decree under the Water Act (1998), which takes the view that underground water is a common resource. Whilst individual ownership rights to water are withdrawn, rights to abstract and use water are granted through a licensing procedure. Licences are issued on a five-year cycle and for a maximum time of 40 years, determined by the use to which the water is put. A reserved quantity of 25 litres per person per day is retained before other uses are authorized to ensure that people have access to sufficient water.

In Arizona, a dry state of the USA, Active Management Areas are established where groundwater is under threat. The use of groundwater is generally subjected only to rules on reasonable usage, but where it is in short supply, water management rules are applied and new abstractions of groundwater are subjected to a permitting procedure.

20.3 CONSULTATION AND PARTICIPATION

The general public can only participate in decision-making on environmental health issues if it has access to information. Often what is needed first in programmes to protect groundwater is to ensure that all staff has the understanding and the tools to establish a dialogue and atmosphere of trust with stakeholders. A Canadian expert on participation has stated:

> *'The level and quality of participation by the public will be no better than that of the staff in the proponent's organization... The development of a relevant public participation policy is often part of the pre-work needed before launching a pro-active program with the organization's external publics.'* (O'Connor, 1993).

The general manager of the Montgomery, Alabama, Waterworks, emphasizes the point of ensuring all staff know how to listen:

> *'A key aspect of our program involved sitting back, getting out of the driver's seat and becoming a one-vote stakeholder during the decision process, even though our agency was providing funding for the program... All of this inspired openness, trust and full buy-in among the 25 different stakeholder groups involved in the Catoma Creek Watershed.'* (Water and Wastewater International, 1999).

Understanding the needs of the community, whether it be an entire nation or a small village, is a prerequisite and part of the process of establishing a dialogue. Professionals in public and private agencies developing policies must understand what the public wants from them and be clear about what they want from the public. From that knowledge, they can then begin to develop mutual trust, common goals and plans. This requires that the breadth of views are represented. Continuous information gathering, and reconfirmation with the stakeholders of conclusions on knowledge, attitudes and practice is important for ensuring accuracy of assessments, monitoring of progress and change, and establishing trust among stakeholders. Transparency of information dissemination is extremely important and requires the development of communication plans. In Bangladesh for example, communities had limited information with which to make decisions, and it was recognized that improving their access to information would strengthen their ability to fight for their rights (see Box 20.4).

Box 20.4. Bangladesh – community initiative in regulating industrial pollution

A detailed study of various publicly owned fertilizer and pulp plants showed that *'even very poor people in Bangladesh can negotiate pollution reduction and compensation when the damage is evident and they have economic alternatives.'* The survey revealed *'a pattern of informal regulation which has remarkably similar characteristics. Fish kills, paddy crop damage and poisoned drinking-water provide a straightforward, but limited, basis for damage estimation by downstream communities. Plant staff members live and work near these communities, and are therefore potentially subject to social pressures ranging from harassment through ostracism to outright violence.'* However, community pressure was only effective in areas where community members had alternative employment. In addition, *'the affected communities are hampered by poor information. In some cases they cannot identify the offending polluter; they have little basis for assessing pollutant risk; and they generally know little about the cleanup options faced by polluting firms.'* In these cases, the communities need outside help from an NGO, government agency or concerned business in order to obtain more accurate information to support their claims (Huq and Wheeler, 1993).

Similarly, a fundamental tenet of integrated water cycle management (including all waters) that is being developed in NSW, Australia is the inclusion of all identified stakeholders at the outset of the process. Agency staff develop and disseminate information initially with the local water utility and then with the government agencies and local interest groups. This process has ensured ownership of the process by all involved as it is possible for everyone to understand how they impact on the water cycle. Information is then used to refine the process and the adopted control measures that are decided on by the local water utility and the community it serves.

Community participation is an ongoing process of information gathering, dialogue and negotiation. In the United Kingdom for example, where water and wastewater services have been privatized, this was accompanied by creating the Ofwat National Customer Council and Service Committees. Their mandate is to ensure that companies continue to supply good quality drinking-water, look after the environment, keep average prices low, improve customer service, and particularly to ensure this dialogue between customers and providers (OFWAT, 1998).

In developing countries, community participation in the provision of water supplies and sanitation has been shown to be effective in rural areas and is increasingly noted as successful in urban areas (World Bank, 1993; IRC, 1995; Satterthwaite, 1997; WHO and UNICEF, 2000). Community participation in water resource management has been less widely applied but is increasingly noted as successful, particularly within local communities. Experience is far more limited with processes of dialogue with communities about national and international water resource management despite the urgent need in many parts of the world to ensure that this occurs.

Cultural values of groundwater

In some countries, groundwater has specific cultural meanings for part or all of the population. It is important that these values are included in public participation programmes on groundwater quality issues involving and/or affecting indigenous communities. Some of these concepts are illustrated below with the example of participation programmes involving aboriginal communities in Western Australia.

Under Western Australian Aboriginal Heritage legislation, it is an offence to disturb sites of special cultural significance to aboriginal communities, which generally include land near springs, groundwater-dependent wetlands and waterways. It is a requirement for all major development proposals to ensure that adequate consultation has taken place with relevant aboriginal communities, including local custodians who are able to 'speak for the land'. Consultation has to take place in a culturally sensitive manner and is usually mediated by anthropologists.

In developments that utilize large amounts of groundwater and can cause groundwater contamination such as large irrigation projects, it is important that aboriginal communities are involved at an early stage in the planning stage to ensure the protection of cultural values (Macintyre and Dobson, 1998; Yu, 1999). Allocation plans for groundwater resources can then be developed ensuring that sufficient water is allocated for the maintenance of cultural and environmental values before divertible resources are determined. Cultural values (often the maintenance of specific water levels in wetlands) have to be determined by consultation with relevant communities, and sufficient research and site-specific investigations have to be undertaken to convince these communities that groundwater quality will not be affected by the development (Yu, 1999).

Consultation is also needed when assessing and remediating existing groundwater problems. One example is the metropolitan region of Perth, the largest urban centre in Western Australia with a population of 1.3 million. There are a large number of heritage sites in this area, particularly near rivers and wetlands. Consultation is required to ensure that the location of monitoring boreholes to assess groundwater contamination and that remediation using in-ground structures will not disturb sites of cultural importance.

20.4 LAND USE PLANNING AND MANAGEMENT

There is a long recognized relationship between land use and pollution of groundwater, although this may take decades to be noticed. Once pollution of an aquifer has occurred, it is extremely difficult to clean up and it is rarely possible to return an aquifer to a pristine condition. For this reason, the best practice is prevention through the regulation of land use in areas that overlie groundwater flow systems.

This section discusses both the advantages and the limitations of land use management for protecting groundwater resources and provides an introduction to the mechanisms and approaches commonly used to control land use. Land use management to protect groundwater quality usually involves a combination of approaches and the particular mix used will vary considerably.

20.4.1 Regulatory approaches to controlling land use in sensitive areas

Land uses and economic activities in sensitive areas, particularly in drinking-water catchments, need to be subject to some form of government regulatory control, and require approvals to proceed. Land use can be managed through a variety of tools including national or regional planning regulations, environmental legislation and local government by-laws.

Although planning legislation varies considerably from country to country, it is commonly organized in a hierarchical manner. Broad policies and principles are set at the national level (or at an international level, where there is international grouping, such as the European Union). Local regulations are established at a state or regional government level, and local government is responsible for town planning and regulating local zoning and the subdivision of land. Most planning controls to protect groundwater quality are implemented by local governments, but groundwater protection issues can be incorporated into national planning policies and regional planning regulations, as they are in many states in the USA. The Statement of Planning Policy for the Jandakot region in Perth, Western Australia shown in Box 20.5, provides an example of a regional policy that recognizes the importance of groundwater protection.

Controls on land zoning and subdivision imposed by local governments can be very effective tools for protecting groundwater. Zoning consists in dividing a locality into areas where the allowed land uses are specified and can be used both to define the kind of land uses permitted and to regulate the permitted uses. Zoning can be used, therefore, to direct future development towards defined objectives (groundwater protection usually being only one of many reasons for controlling land use). Typical zoning requirements for groundwater protection generally limit permitted uses, for example to low density residential development with limited use of septic systems, or leaving land as public open space. Such controls require continued monitoring to ensure that the requirements are maintained through time. The protection zone concept for drinking-water catchments is discussed in more detail in Chapter 17.

Box 20.5. The Jandakot regional groundwater protection regulations (based on Boyd *et al.,* 1999)

The protection of groundwater quality is recognized as an important element in land use management in the Perth Metropolitan Region Scheme, the planning framework for the city of Perth. The Jandakot mound is one of the recharge areas for the region and includes land currently reserved for the protection of groundwater quality. Here, the purpose of the regional policy is to:ensure: that development in the area is compatible with the long-term use of groundwater for public water supply and ecosystem maintenance; ensure that land uses with potential detrimental effects on groundwater resources are brought under planning control; provide guidance on planning requirements for development proposals; guide local governments in amending their town planning schemes; acquaint affected landholders with the proposed changes in planning controls.

Regional plans like that for Perth can help protect groundwater by ensuring the appropriate location and density of specific types of development, and by ensuring that waste disposal sites are located appropriately. Regional plans can also help guard against inconsistent decisions being made at a local government level when individual development proposals are viewed in isolation.

Controls on land zoning and subdivision imposed by local governments can be very effective tools for protecting groundwater. Zoning consists of dividing a locality into areas where the allowed land uses are specified and can be used both to define the kind of land uses permitted and to regulate the permitted uses. Zoning can be used, therefore, to direct future development towards defined objectives (groundwater protection usually being only one of many reasons for controlling land use). Typical zoning requirements for groundwater protection generally limit permitted uses, for example to low density residential development with limited use of septic systems, or leaving land as public open space. Such controls require continued monitoring to ensure that the requirements are maintained through time. The protection zone concept for drinking-water catchments is discussed in more detail in Chapter 17.

Where they are successfully applied both zoning and subdivision regulations are most useful for controlling future development. However, they have little effect for groundwater protection purposes in areas with previous development that led to pollution. Some countries do, however, use such methods to control further degradation of protected areas. For example, Chilean environmental legislation includes the concept of saturated zones where no further development is permitted when one or more environmental standards have already been surpassed (Government of Chile, 1994). In Perth, Australia, there are programmes to replace septic tanks with sewer connections, but existing groundwater contamination will take many years to dissipate. There are also media campaigns promoting the wise use of water and fertilizer, and of the benefits of using local native plants in gardens, which do not need fertilizer (Appleyard and Powell, 1999). A number of community action groups are also becoming interested in the issue.

20.4.2 Other land use measures for pollution control

Governments are often reluctant to impose new controls and regulations on existing activities to or impose retrospective legislation because of the potential for economic disruption. Other forms of intervention are often required to protect groundwater quality, usually taking the form of financial incentives or penalties.

In particularly sensitive areas, governments may decide to purchase activities considered to be potential sources of pollution, or offer significant financial incentives for industries to relocate. Practices that cause groundwater pollution may be given incentives for change. For example, tax incentives to use a specific type of fertilizer or pesticide that is less susceptible to leaching or degrades in the soil more quickly, can be an effective tool. Imposing penalties for any pollution above a certain standard also has a role to play.

Market based approaches seek to relate the cost of contamination to the cause so that the price mechanism can be used to restrict the amount of contamination that reaches groundwater. Stringent application of the polluter pays principle should also lead to polluting activities paying for the monitoring and reporting of contamination levels in groundwater. Other market mechanisms such as insurance bonds may provide added incentives for avoiding the pollution of groundwater in the longer term.

20.5 TOOLS FOR POLLUTION CONTROL

There are a number of specific tools and incentives that may be employed to maximize the impact on groundwater protection policies and regulations. This may include the setting of specific end-of-pipe control, establishing integrated pollution control measures, use of prohibitions and the use of codes of practice. Some of the tools are discussed more specifically in Chapters 21-25. Water quality objectives may also be determined to provide a mechanism to reduce pollution of groundwater.

Within all these approaches, the use of incentives is often as effective as the use of prohibitions or controls. By providing evidence of benefits derived from reduced pollution, many industries will be interested in changing practices. However, the use of regulations remains important, but will only be as effective as the degree to which these are enforced. Regulations and standards, without the back-up of inspection and enforcement regimes are largely worthless. Similarly, regulations that are applied discriminately may send the wrong message. For instance, towns and sewage services are seen as an easy target for regulators when in fact it is often the agricultural sector that is the major polluter. In some countries, the agricultural sector may be politically powerful and may seem to be immune from regulatory enforcement in part because of difficulties in enforcing regulations relating to diffuse sources. Tools for pollution control thus need to be equitable and uniformly enforced.

Legislation developed for general environmental protection and pollution control can be employed to deal with activities which affect the quality of groundwater used as drinking-water source. For example, laws which are used to regulate the quality of discharges to watercourses so that the quality of water is not impaired can be extended to groundwater. Permitting procedures for the discharge of materials from factories and wastewater treatment plants are commonly used to control point source pollution and these procedures can be applied specifically to protect groundwater. Control measures for diffuse sources of pollution are more difficult to implement but the use of land use planning procedures, codes of practice, and general attention to the pollution risks from activities may help to reduce the risks of groundwater pollution. Pollution from road construction, quarrying, landfill and oil installations, for example, can be tackled by agreeing to necessary precautionary measures with the developers and operators of the systems. Agricultural policies can also be developed that control the release of pollutants, through for instance control of fertilizer and pesticide applications, as noted in Chapter 21.

20.5.1 End of pipe controls

Legal remedies to groundwater pollution from point sources generally operate in one of two ways – control of the discharge or control of the process from which the discharge originates. In situations where it is possible to identify a discrete discharge from a pipe or other such structure a 'permitting' regime may be established to control discharges of polluting materials into watercourses or into the ground. In other cases, point source pollution (for instance from pit latrines) may be controlled in sensitive areas by establishing specific design and construction criteria.

It may be considered necessary for an industrial or commercial organization to dispose of its liquid effluents into or over the ground above an aquifer. In such cases the permitting system may give adequate control over the quantity and content of the effluent so that the groundwater is protected. Commonly used as end-of-pipe controls, such permits are issued under legislation by the pollution control authorities following an application for permission to discharge. The law usually requires an application to be made by the discharger, stating the likely rate and constituents of the discharge, and the authorizing body must take steps to consult interested parties and examine the likely effect on the water before the permit is granted.

In the United Kingdom for example, Section 85 of the Water Resources Act 1991 makes it an offence to allow the entry of poisonous, noxious or polluting material into controlled waters (that is, most naturally occurring surface and underground waters), but a discharge may be made following the issue of a permit under Section 88 of the Act. A permit is granted after receipt of an application, local and national advertising of the proposal, consultation with affected persons and statutory bodies and after consideration of the likely effects of the discharge on the relevant water. A permit so granted may place strict limits on the constituents of any discharge and the manner in which it is discharged to prevent any deterioration in the receiving water.

If the effluent contains particularly toxic or dangerous substances the issue of a permit may have to be refused, as the risk of contaminating the groundwater to such an extent that it could not be used for drinking purposes would be too high. This situation has been dealt with in the European Union through the adoption of a specific directive, the

Groundwater Directive (80/68/EEC) (EU, 1980), which prohibits the direct discharge of dangerous substances into these waters, requires permits for indirect discharges (i.e. discharges which percolate through the unsaturated zones), and which also limits the discharge of other substances which may lead to pollution. This directive is reflected in the local legislation in each of the Member States through a process of legal transposition.

20.5.2 Integrated pollution control approach for industry

Although the use of permitting is an effective way of controlling known effluent discharges, groundwater is easily contaminated by other routes. There are many examples where spillage of materials or poorly designed storage facilities at industrial sites (Chapter 11) have caused pollution and the careful control of discharge points has not given adequate protection from activities on the site. Such problems extend to the pollution caused by the disposal of solid wastes. To overcome such lack of control of potentially polluting activities an alternative approach has been developed (see also Chapter 23). In this concept, rather than limiting legislative control to the permitting of individual discharges, the overall process itself is the subject of a permitting regime. The approach uses the principle that all possible environmental impacts of all activities on the site should be considered, taking account of the effects of the installation and its discharges to air, land and water. Prior authorization must be obtained before the installation is allowed to operate, and where a permit is granted and there are discharges to the environment, the principle of using the best available techniques is applied to prevent or minimize the extent of the discharges. This system has now been adopted in the European Union, for example, through the Integrated Pollution Prevention and Control Directive (EU, 1996).

Experience of using this manner of control has been gained in a number of countries. For example, in the United Kingdom the Environmental Protection Act 1990 introduced Integrated Pollution Control to a specified range of the most polluting industries; in France the Law on Classified Installations of 19 July 1976 defines discharge thresholds for air and water and the technical conditions which must be fulfilled in order to be granted permission to undertake the activity; and in Sweden the Environmental Protection Act of 1969 covers emissions to air and water and noise emissions and is based on integrated pollution prevention principles. At present this regime is used for larger installations because of the amount of work involved in assessing the pollution potential and in enforcing the conditions of the resulting rather complex permit, but the principle is capable of adoption for any size or type of plant, and could be used to limit the construction of undesirable installations where groundwater is particularly at risk.

Such an approach is particularly valuable for groundwater protection because the procedure of assessing the impact of the process on the environment enables the identification of risks of pollution from diffuse inputs as a result of spillages from storage facilities or operations within the plant and also allows an assessment of the impact on groundwater of such activities as ground disturbance during construction. The approach requires attention to be given to the possible risks from closure of the site and any clean-up measures required at this stage. This approach has central principles in common with the WSP approach (see Chapter 16), such as system assessment for identifying risks and process control. It is therefore a good basis for developing a WSP to include aquifer protection.

20.5.3 Prohibitions

In some cases it is not possible to provide adequate protection of groundwater by means of a permitting regime. This is because either the activity is regarded as too unpredictable to enable enforceable conditions to be added to the permit, or because the activities are intrinsically too dangerous to public health to permit them to be carried out in an area used for water supply. In such cases the use of a prohibition notice may be required. The legal basis of this must be clear, however, and the law needs to be clear on precisely what is prohibited, and there must be suitable penalties to encourage people to obey the prohibitions together with adequate enforcement.

Some countries use prohibitions as a precursor to the issue of a permit, so that the legal position is that the activity is prohibited unless a permit is in force, or unless specific standards are met. For example, the Nigerian National Environmental Protection Act of 1991 prohibits the release of hazardous or toxic substances into the air, water or land unless limits set by the national Agency are met.

In the United Kingdom prohibition notices may be issued at any time by virtue of the Environment Act 1995 in respect of activities that are considered likely to cause pollution. Such notices are short-term prohibitions requiring the person on whom they are served to take action to deal with a problem.

20.5.4 Prevention of diffuse pollution of groundwater through Codes of Practice

In some cases, the issue of specific legal direction to avoid pollution is not possible. This is particularly the case for non-point source pollution. For example, it would be very difficult to control agricultural activities (Chapter 21), any of which might cause contamination of groundwater, by the issue of laws covering all the possible activities involved. In such cases the issue of codes of good practice offer an alternative. However, approaches may be developed to consider the point of drainage of a sub-catchment into other catchments, in which case the point of drainage can be seen as the point source. Under this approach, targets may be set for sub-catchments and the activities within that sub-catchment regulated or required to operate within BMPs accordingly.

In the United Kingdom a Code of Good Agricultural Practice has been issued (MAFF, 1991), and this has been given statutory status which means that if a farmer causes pollution and is taken to court, the question of whether or not he has obeyed the Code may be used as a material fact when deciding upon his penalty.

Codes of Practice can be devised and successfully operate in a wide variety of situations which could affect groundwater quality and quantity, including such diverse areas as road building, mineral excavation, fuel storage and use and many more. The codes should identify best practice in the context of the prevention of groundwater pollution. This is an important commercial reality for water suppliers since guidelines,

BMPs and codes of practice generally represent the current accepted body of knowledge of their profession, making their application essential to demonstrate all reasonable precautions and due diligence (Davison *et al.,* 1999).

20.5.5 Prevention of diffuse pollution of groundwater through regulations

Although many activities are difficult to control through a legal permitting regime, it is possible to identify some situations in which it is possible to be more precise about the methods used to control them and to issue legally binding regulations or decrees. These usually apply to particularly discrete issues. For example the problem of slurry storage and slurry use on farms is a major problem to groundwater quality but the problems and their solution can be readily identified. They relate to the design and use of storage facilities, and how and when farm slurry can safely be spread on land in such a way that water pollution is avoided. The Control of Pollution (Silage, Slurry and Agricultural Fuel Oil Regulations) 1991 of the United Kingdom is an example of a legally binding regulation that sets out detailed guidance on storage and usage requirements. Such a regulation would also have to be obeyed if the farmer was abiding by the Code of Good Agricultural Practice but is a legally binding obligation in its own right. Similar issues for which governments implement such regulations include specific requirements for the storage and transport of hazardous chemicals or for car-washing facilities (see Chapters 23 and 25).

20.5.6 Water quality objectives

It is also possible to establish statutory water quality objectives for the water body, with the compliance of such objectives a legal obligation. Water quality objectives define a quality of water that must be met continuously for the whole water body taking into account the range of needs (suitability for water supply, irrigation, industry and ecological needs). Permits for waste discharge and disposal are then issued on the basis of whether the discharge will cause deterioration in water quality so that the objectives are no longer met. The advantage of setting such objectives is that it allows an overall framework for water quality management across the water body and places individual controls within a broader framework. Thus the nature of individual discharge permits within water quality objectives requires that not only the impact on the immediate area is considered, but also wider impacts on the water body as a whole.

In the groundwater context applying such objectives may be more problematic, as the effect of individual discharges on the mass of groundwater in an aquifer is very difficult to predict and the proportion of the total amount of a contaminant found in an aquifer that can be reasonably allocated to individual polluters is often difficult. The use of statutory water quality objectives is therefore little used for groundwater, although the new Water Framework directive of the EU proposes to establish objectives related to groundwater status, including quality and quantity.

20.5.7 Controls on product specifications

For some pollutants, for instance pesticides, controls on production and use may be effective in reducing the risk of groundwater contamination.

This approach is being taken in a number of cases. In Canada, for example the Pest Control Products Act 2002 regulates the distribution of those pesticides which are likely to influence groundwater quality in a non-specific way, and the Pest Management Regulatory System controls the production of pesticides. Such a system, although remote from the point of influence, is believed to provide a holistic preventative measure against groundwater pollution from pesticides. A rather similar indirect control on pesticides has been introduced in the EU through the so-called Uniform Principles Directive (91/414/EEC) in which a uniform authorization process is used to approve the active ingredients in pesticide formulations before they are placed on the market.

20.6 ENFORCEMENT

The value of regulations is dependent on the degree to which they are enforced. An essential prerequisite for groundwater protection regulation, therefore, is that the organization responsible for protecting groundwater has a clear legal mandate, including the powers to take action against organizations or individuals who breach the regulations. This applies whether this organization operates at a national or local level and whether groundwater is dealt with on a catchment, region or whole country basis.

Once the legal basis on which groundwaters are to be protected has been established and the policy has been implemented, there must be means of ensuring that those affected continue to comply with the provisions. The regulatory organization needs to be granted powers to inspect and to take action against non-compliance. The organization may be the same as that which grants permits or controls pollution. Alternatively it can be a separate enforcement agency or, as in some countries (e.g. Italy), it could be a branch of the civil police. The organization will require legal powers to:

- enter property and land
- inspect and collect data
- prosecute or levy fines.

A key issue to be resolved is the determination of what constitutes a breach of compliance of the legal requirements. The enforcement agency has a duty to determine whether this has occurred. A decision must be taken as to when such a breach warrants enforcement action, and the agency requires legal powers to proceed. Legal powers may consist of verbal or written warnings, formal notices, administrative acts and fines, invoking criminal sanctions. In some cases civil law may be used against a polluter. The enforcement regime should contain a practical mechanism for ensuring that compliance is improved. Enforcement is itself part of a cycle as shown in Figure 20.3.

Permits and licences have to be drawn up in such a way that their conditions are achievable. Compliance control of discharges and abstraction must be enacted by suitable protocols, inspection visits, sampling, and other means of verification. An important aspect of enforcement is promotion of the concept that licence holders should take responsibility for meeting their conditions. The need to use enforcement action through the courts or by other means should be a last resort. The possible need to review and change legislation as a result of experience is also an integral part of this cycle.

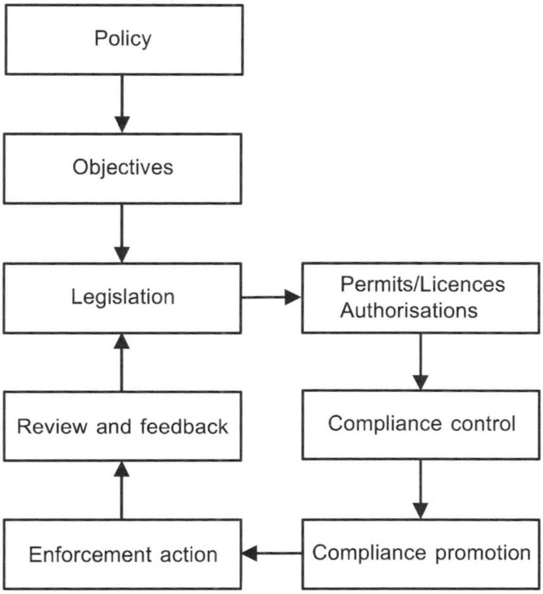

Figure 20.3. Regulatory cycle (adapted from Glaser, 1996)

20.7 MANAGEMENT PLANS FOR DISASTERS AND INCIDENTS

Disasters and major incidents affect all countries and the development of management and preparedness plans for such situations is important in protecting public health. Extreme conditions often place both the use and the management of groundwater under considerable stress, as the quality and quantity of water from groundwater sources may be affected. They may also become more important in the supply of water during disasters as other sources become heavily contaminated and therefore good management of groundwater where it is used for drinking purposes is essential. It is important to develop intersectoral plans to protect groundwater during these events and to ensure that different institutions have clearly defined roles and understand their responsibilities in responding to disasters.

The management challenges in extreme conditions will vary depending on the nature and extent (both spatial and temporal) of the condition, local factors, available resources and available information about appropriate technologies. Without the development of management plans that can be quickly and effectively implemented, extreme events may cause great suffering to the affected populations and may compromise the viability of groundwater resources in the longer term.

Disasters can be natural, including geological (earthquakes, volcanic eruptions, landslides, tsunamis) and meteorological (tropical cyclones, floods, droughts), or man-made (social disruption, war, industrial accidents, ecological mismanagement). Globally, tropical cyclones, floods and earthquakes are shown to be the most frequent types of disasters with earthquakes and tropical cyclones the most deadly regarding human life.

The severity of a disaster depends on the magnitude of the event and the vulnerability of the population and infrastructure. The people affected by a disaster are often faced by increasing health risks and are often more vulnerable to water related diseases. For instance, post-disaster diarrhoea epidemics frequently occur due to lack of access to safe and adequate volume of domestic water. The causes of this reduced access are varied and include damage to shallow groundwater sources and pollution of shallow aquifers. The consequences of poor preparedness may be very significant for public health as shown in Box 20.6.

Good disaster preparedness and mitigation plans have important implications for groundwater protection and health. Ensuring that appropriate measures are put in place to reduce the likelihood of large-scale disruption or pollution of supply and developing rapid response plans – for instance emergency chlorination programmes – will reduce the impact of the disaster on water supply and public health.

Controlling widespread contamination of the shallow aquifer may not be easy. However, remediation is sometimes not as difficult as assumed. For instance in Bangladesh it was found that the quality of water from flooded hand pumps usually became better after it is pumped over several hours.

Disaster preparedness

Effective disaster preparedness involves a range of stakeholders, including Government departments responsible for health, social services, water and local Government as well as communities, the private sector and NGOs. It is usually most effective when a lead agency is identified that takes responsibility for coordinating and planning the emergency response and for ensuring that different agencies are aware of the support available and the different roles they are expected to play.

Emergency actions are usually looked upon as being short-term measures. However, proper disaster management has immediate to long-term implications for appropriate protection, utilization and sustainable development of the groundwater resources. A properly developed disaster preparedness programme is an essential first step in this management.

The prorgramme should begin with a survey and mapping of all water facilities. It may be of value to utilize GIS as a means of storing and presenting data in order to define vulnerable areas and priority interventions. This will allow proper planning for the disaster response to be undertaken and should indicate special needs for groundwater protection and identification of points where resources must be available.

Based on field surveys and assessed needs, the activities and procurement of the materials possible within the available resources should be managed in consultation with the stakeholders. Proper utilization of the disaster information and warning centres should be part of the groundwater protection preparedness initiatives. Attempts should be made to integrate the preparedness and response to disaster into the national planning and policy framework. Developing a strategy for community awareness-raising is essential to support disaster preparedness.

Box 20.6. Flooding and groundwater contamination in Bangladesh after cyclones

During the cyclones in 1991 and flood in 1998 in Bangladesh, many tubewells were damaged and remained under water over several days and this led to deterioration in the physiochemical and microbial quality of the water. This probably occurred due both to direct ingress at the tubewells themselves as a result of inundation and a much wider gross contamination of the shallow aquifer. As the majority of the Bangladeshi population drinks water from tubewells, the damage and contamination of the tubewells represented a major public health crisis (Siddique *et al.,* 1991). The numbers of people using the non-flooded tubewells increased significantly leading to long queues and long distances to tubewells. The reduced access to safe/usable tubewells affected the availability of water and consequently hindered personal and kitchen hygiene practices. As a result, diarrhoea epidemics were observed.

One of the problems that Bangladesh faced was poor preparation for the effects of such a disaster, despite the regular occurrence of such events. The limited knowledge of groundwater management meant that little protection was provided to prevent damage to the infrastructure and remediation of widespread contamination of the shallow aquifer. Water was transported from other areas to the affected areas even though the water from non-flooded handpumps was safe. This created panic and unnecessary water shortage among the local people as they abandoned the local groundwater for the transported water. Moreover, the bacteriological quality of transported water was worse than the local handpump water and therefore represented a higher risk to public health.

20.8 REFERENCES

Appleyard, S.J. and Powell, R.J. (1999) Preserving urban bushland and growing local plants to protect water resources. In *Proc. of the Urban Bushland Conference, Managing our Bushland,* pp. 72-75, Perth, October 1999.

Boyd, D., Tomlinson, A. and Carman-Brown, A. (1999) Innovative approaches to groundwater protection on the Jandakot Mound. In *Proc. of an Internat. Conf. on Diffuse Pollution, Solutions and Innovations*, pp. 466-476, Perth, May 1999.

Davison, A.D., Davis, S. and Deere, D.A. (1999) *Quality Assurance and Due Diligence for Water: Can HACCP deliver?* Presented at Cleaner Production in the Food and Beverage Industry. AWWA Conference, Hobart 1-3 September, 1999.

EEA (1998) *Europe's Environment: The Second Assessment.* European Environment Agency. Copenhagen.

EU (1980) *Council Directive 80/68/EEC on the protection of groundwater against pollution caused by certain dangerous substances*, OJ L20 17.12.79, Brussels.

EU (1996) *Council Directive 96/61/EEC concerning integrated pollution prevention and control,* OJ L257 10.10.96, Brussels.

EU (2000) *Directive 2000/60 /EC of the European Parliament and of the Council of 23 October 2000 establishing a framework for Community action in the field of water policy*, OJ L 327 /1 22.12.2000, Brussels.

Foster, S., Adams, B., Morales, M. and Tenjo, S. (1992) *Estrategias par la protección de aguas subterráneas*, Centro Panamericano de Ingeniería Sanitaria y Ciencias del Ambiente, Lima

Glaser, R. (1996) Dilemmas in compliance. *Industry and Environment*, **19**, 16-21.

Government of Chile (1994) *Ley Sobre Bases Generales del medio Ambiente*, Diario Official, Santiago, 9 March.

Howarth, W. (1992) *Wisdom's Law of Watercourses*, 5th edn, Shaw & Sons, Kent.

Huq, M. and Wheeler, D. (1993) *Pollution Reduction Without Formal Regulation.* Environmental Department Divisional Working Paper No. 1993-1939, Pollution and Environmental Economics Division, World Bank, Washington, DC.

IRC (1995) *Water and Sanitation for All: A world priority*, vol. 2, Achievements and Challenges, Ministry of Housing, Spatial Planning and the Environment, The Hague.

Macintyre, K. and Dobson, B. (1998) Factoring aboriginal environmental values in major planning projects. In *Proc. of Australian National Environmental Law Association*, Perth, 1998.

MAFF (Ministry of Agriculture, Fisheries and Food) (1991) *Code of Good Agricultural Practice for the Protection of Water*. Ministry of Agriculture Forestry and Fisheries, London.

O'Connor, D.M. (1993) *Ten Lessons Learned in 22 Years of Public Participation*. Presented at the Second Annual Meeting of the International Association for Public Participation Practitioners held at Kananaskis, August 29-September 1, 1993.

OECD (1989) *Water Resources Management, Integrated Policies*, Organization for Economic Cooperation and Development, Paris.

OFWAT (1998) *Public involvement and consultation.* www.ofwat.gov.uk (accessed April 29, 2005).

Patrick, R., Ford. E. and Quarles, J. (1987) *Groundwater Contamination in the United States*, Philadelphia, University of Philadelphia Press.

Satterthwaite, D. (1997) Urban Poverty: reconsidering its scale and nature. *IDS Bulletin*, **28**(2), 9-23.

Siddique, A.K., Baqui, A.H., Eusof, A. and Zaman, K. (1991) Floods in Bangladesh: pattern of illness and causes of death. *J. Diarrhoeal Disease Res.*, **9**, 310-314.

UN (1992) *Convention on the Protection and Use of Transboundary Watercourses and International Lakes*, United Nations Economic Commission for Europe, Geneva.

UNECE (2000) *Guidelines on Monitoring and Assessment of Transboundary Groundwaters.* RIZA, Inst. For Inland Water Management and Wastewater Treatment, United Nations Economic Commission for Europe Task Force for Monitoring and Assessment Project, Lelystad.

Water and Wastewater International (1999) Long-term watershed restoration plan to reduce pollution. *Water and Wastewater Internat.*, **14**(2),10.

WHO and UNECE (2001) *The Protocol on Water and Health: What it is, why it matters.* World Health Organization Regional Office for Europe, Copenhagen.

WHO and UNICEF (2000) *Global Water Supply and Sanitation Assessment 2000 Report.* World Health Organization/UNICEF, Geneva/New York.

World Bank (1993) *Water Resources Management*, World Bank policy paper, World Bank, Washington, DC.

Yu, S. (1999) *Ngapa Kunangkul - Living Water: A report on the aboriginal cultural values of groundwater in La Grange sub-basin.* Unpublished report prepared by the Centre for Anthropological Research, University of Western Australia.

21

Agriculture: Control and protection

S. Appleyard

Agricultural activities can contaminate vulnerable aquifers with a range of pathogens and hazardous substances. The scale of their groundwater pollution potential is different from other human activities, as animal manure, agrochemicals, sewage sludge or wastewater are intentionally applied to land, often to large areas. Contaminants relevant to human health include a range of pathogens carried by farm animals but also infectious to humans, nitrate and a wide range of pesticides. Land clearing and irrigation practices also impact aquifer vulnerability and hydraulic loading and thus affect pollution potential.

There are a wide range of measures that can be implemented to reduce the impacts of agricultural production on groundwater quality. These include structural measures such as the construction of treatment facilities for wastewater from intensive animal feeding operations or adequately sized, sealed and bounded sites for pesticide mixing and cleaning of equipment; and operational measures such as applying the correct amount of fertilizer at times of the year when plant uptake occurs, or matching irrigation to crop needs. Control measures in planning address the type of agricultural land use in relation to aquifer vulnerability and use. Examples include restricting or limiting stock density and type of crop. In general agricultural management practices are aimed at reducing the pollution of groundwater by either minimizing the availability of pollutants (source reduction), by retarding the transport of water, pathogens and nutrients through the soil profile, or by chemically or biologically transforming chemical pollutants (e.g. pesticides) into less toxic materials within the soil profile.

 Protecting Groundwater for Health: Managing the Quality of Drinking-water Sources. Edited by O. Schmoll, G. Howard, J. Chilton and I. Chorus. ISBN: 1843390795. Published by IWA Publishing, London, UK.

These management practices are often referred to as good management practices. However, the term "good" is often a highly subjective and site-specific label: what is adequate in one area may not work in a different location due to differences in physical conditions, or to cultural factors which may restrict the adoption of a particular measure by local farmers.

Management practices generally cannot solve water quality problems in isolation, but are used in combinations to build management practice systems (US EPA, 2000). For example, soil testing is a good practice for nutrient management but, to be fully effective, it also requires estimates of realistic yield, good water management, appropriate planting techniques, proper nutrient selection, rates and placement. A set of practices does not constitute an effective management system unless the practices are selected and designed to function together to achieve specific water quality objectives reliably and efficiently. Their documentation in a management plan is important to define routines of monitoring whether practices are being adhered to and are functioning as intended.

In general, changes in agricultural practices will only occur if there is some incentive for new techniques to be adopted. This usually means that the measures are affordable for farmers, and that they can either see cost savings in implementing the measures, or that financial incentives are offered by government agencies, water suppliers or consumers for implementing the measures. The provision of a system of branding goods as environmentally responsible products may also be of benefit to the growing market for organic and green produce.

The implementation of good management practices and control measures to protect drinking-water catchments from contamination through agricultural activities is substantially facilitated by an agricultural policy targeting sustainable use of resources. Many measures for this broader environmental target will encompass the protection of groundwater used for drinking-water. They may include but are not restricted to:

- training and education programmes to increase local awareness of agricultural impacts on groundwater and drinking-water quality;
- establishment of catchment management groups with involvement of local community, relevant government agencies and local politicians;
- conversion programmes of arable land to unfertilized grassland.

Vice versa, the development of control measures for agriculture in drinking-water protection zones in some countries has pioneered the development of approaches to environmentally sound agricultural practices.

This chapter presents information on management practices that have proven to be effective in controlling groundwater pollution, and looks at how they can be implemented both at the farm scale and at the catchment or watershed scale. It is not intended to be a comprehensive account of all agricultural pollution control measures, but sufficient information is provided for local authorities to develop practices suited to their local conditions, usually in collaboration between the sectors responsible for public health, water management, agriculture and environment.

NOTE ▶ *In developing a Water Safety Plan (Chapter 16), system assessment would review the efficacy of control measures and management plans for protecting groundwater in the drinking-water catchment from agriculture. Chapter 9 provides the background information about the potential impact of agriculture on groundwater and provides guidance on the information needed to analyse these hazards.*

This chapter introduces options for controlling risks from agriculture. As the responsibility for agriculture usually falls outside that of drinking-water suppliers, close collaboration of the stakeholders involved, including the authorities responsible for agriculture, is important to implement, upgrade and monitor these control measures. This may be initiated by the drinking-water sector, e.g. in the context of developing a Water Safety Plan or of designating protection zones (see Chapter 17).

21.1 PATHOGEN MANAGEMENT ON AGRICULTURAL LAND

As discussed in Chapter 9, a number of pathogens occurring in animal manure may also cause illness in humans. These include bacteria (e.g. *E. coli* O157:H7, *Leptospira, Yersinia enterocolitica, Campylobacter* spp., *Listeria moncytogenes*, *Salmonella* spp., *Clostridium perfringens*), viruses (e.g. hepatitis E) and protozoa (e.g. *Cryptosporidium parvum, Giardia lamblia*). Pathogens may further be introduced through wastewater irrigation or use of sewage sludge on agricultural land. Although filtration through the subsoil may attenuate them more readily than nitrate or agrochemicals, where they do break through into drinking-water aquifers, pathogenic microorganisms are likely to present the most immediate and serious threat to public health. Consequently in such settings, implementing management measures to deal with this issue often has a high priority in relation to contamination from agrochemicals. Many of the management measures that are implemented for controlling groundwater contamination by pathogens, however, will also prevent contamination by chemicals derived from agricultural land use.

In general, there are four specific points in a farm management system where targeted control measures will help prevent the transport and proliferation of microorganisms that may be carried by stock but can cause disease in humans. These points are:

- *Avoiding the import of pathogens into farms to prevent a specific disease from becoming established and proliferating in a farm setting.* The main sources of pathogen import include new stock, the purchase of contaminated feeds, importing contaminated drinking-water, infected farm workers, contaminated soil and manure carried between farms on farm machinery or tools, and the introduction of diseases by pests and wildlife.

- *Interrupting the cycle of pathogen amplification and proliferation on a farm.* Pathogens can be circulated on a farm through poor storage and handling practices for feeds, drinking-water, and animal wastes.
- *Safe waste management.* Poor waste management practices can contribute to animal re-infection and can greatly increase the risk of groundwater becoming contaminated with pathogens. The risk of waterborne disease is particularly high in areas where fresh human excrement is applied as a fertilizer and for soil amendment.
- *Preventing pathogen export or transport from a farm.* Without careful management, pathogens can be exported from a farm to initiate a cycle of infection and drinking-water contamination in other areas.

Management measures that are effective in minimizing pathogen proliferation at one or more of these points are outlined in Table 21.1, and are described in more detail below. A more detailed overview on zoonotic waterborne pathogens in animal reservoirs and their control on farm level is given by Gannon (2004). In general, management measures to reduce pathogen levels on farms should not be implemented in isolation, but rather an integrated pathogen management system should be established where linked management measures that target one or more of these control points work together to break the infection cycle that can indirectly lead to the contamination of drinking-water.

Table 21.1. Control measures for addressing pathogen proliferation and transport on a farm (adapted from Rosen, 2000)

Measure	**Management area**			
	Import control	Proliferation of pathogens on farm	Waste management	Export control
Composting wastes		X	XXX	X
Constructed wetland			X	XXX
Filter strips				XXX
Riparian buffers				XXX
Sediment traps				XXX
Waste management system (including storage, treatment ponds, waste reuse)		XXX	XXX	
Irrigation water management				XXX
Prevent stock access to wellheads and streams by fences		XXX	XXX	XXX
Regulatory control of stock feed sales	XXX			
Quality assurance system for preparing stock feeds	XXX			
Washing farm machinery	XXX			XXX

xxx = direct control of pathogens; x = indirect control of pathogens

Although the above control measures will greatly reduce the numbers of animal-borne pathogens on a farm, it is unlikely that specific pathogens will be totally eliminated from an agricultural environment. Consequently, additional management measures are generally required to protect groundwater supplies from microbial contamination in agricultural areas to prevent the spread of waterborne diseases. Measures include structural features and land use practices that are applied at all scales from the immediate vicinity of wells and springs used for water supply, to the entire groundwater recharge area. In regions where implementation is difficult (e.g. due to lack of financial resources), it is recommended that control measures in drinking-water catchment areas are implemented as a matter of priority before general catchment-wide measures are implemented.

A number of control measures are given below, in order of increasing distance from water supply wells. :

- *Wellhead construction*: microbial contamination of water in wells can be minimized by ensuring that dug wells are surrounded by an impermeable apron (concrete or other material) constructed above ground level to prevent the ingress of surface runoff. Contamination of tubewells can be minimized by ensuring that casing is constructed above groundwater level and has no cracks that will allow water to enter. The annular space surrounding the casing should be sealed with cement to prevent water ingress (for details see Chapter 18).
- *Wellhead inspections and maintenance:* regular, systematic inspection of the structural condition of wells and of activities in the immediate vicinity of water supply wells will help reduce the risk of pathogen contamination caused by construction problems (for details see Chapter 18).
- *Drainage management:* drains and bunds can help ensure that contaminated surface runoff is diverted away from water supply wells and springs (for details see Chapter 18).
- *Backflow prevention:* the use of check valves or other devices on pipe connections to wells can help prevent potentially contaminated water siphoning back into a well.
- *Stock access and waste storage:* stock should be excluded from areas where wells have been constructed by the use of fences. Manure or other waste materials should not be stored or applied within the stock exclusion zone.
- *Water table access:* karstic features (such as dolines and caves) or abandoned mine shafts where there is a direct connection between the land surface and the water table can be fenced-off and drainage diverted to prevent contaminated runoff flowing directly to the water table without being filtered in the soil profile.
- *Storage of manure and other waste materials:* storage areas should have sufficient capacity to accommodate livestock manure during the rainy season or winter months when land application is not possible. The risk of groundwater contamination can be minimized by ensuring these materials are stored on a bounded impermeable surface (preferably covered with a roof). Liquid manure can be stored in covered storage tanks or in appropriately sized and lined wastewater treatment ponds. Covered storage facilities reduce nutrient losses (e.g. ammonia) (EA, 2001).
- *Animal waste treatment:* there are several treatment techniques that can be employed by farmers to ensure that pathogen levels are reduced before animal manures and sewage sludges are applied to agricultural land. They include composting, air drying, lagoon storage, aerobic digestion and lime stabilization.

Composting is one of the most effective techniques (using within-vessel, static aerated pile or windrow composting methods). If the composting process is well managed, pathogen levels can be reduced by more than 4 logs (Sobsey *et al.*, 2003; Gannon *et al.*, 2004). To ensure that pathogens are killed, temperatures within the compost pile should be between 45 °C and 55 °C (with short periods exceeding 45 °C) (EA, 2001). If the compost pile is not covered during the composting process, up to 50 per cent of the carbon and 20-30 per cent of the nitrogen may be lost from the material, reducing the effectiveness of the composted waste as a fertilizer (Goss *et al.*, 2001).

- Another effective option for treating animal wastes to remove pathogens is to process them in an anaerobic digester to generate biogas that can be used for household use. Once again, the temperature inside the biogas generator should exceed 55 °C to ensure pathogens are eliminated from the wastes. Depending on design and operating conditions, pathogen reduction typically is more than 4 logs (Sobsey *et al.*, 2003; Gannon *et al.*, 2004). Aerobic and anaerobic biological treatment processes (including composting and anaerobic digestion) that operate at mesophilic conditions (i.e. below 35-45 °C) are unlikely to reduce pathogen levels by more than 1-2 logs (Sobsey *et al.*, 2003).
- Lime treatment is another option for treating animal wastes. Pathogen levels in animal wastes can be reduced by a factor of 1000 to 10 000 if sufficient lime is added to raise the pH of the material to 12 for at least a 2-hour period (Sobsey *et al.*, 2003). Long-term storage of manure for several months can also reduce pathogen levels in wastes. Desiccation or air drying to very low moisture levels (<1 per cent) will typically result in more than 4 log reductions. At moisture levels of 5 per cent however, pathogen reduction will typically be less than a factor of 10 only (Sobsey *et al.*, 2003; Gannon *et al.*, 2004). Protozoa such as *Cryptosporidium* and *Giardia* may persist in cyst form.
- *Manure application* – ensuring that manure is not applied immediately upslope of water supply wells or when there is a high risk of rainfall will reduce the risk of groundwater contamination by microbes and nitrate. Incorporation of manure into soil rather than simply applying the material on the soil surface will also greatly reduce the risk of microbes being transported in surface runoff and being washed into poorly constructed wells or through bedrock fractures or karstic features in areas where the soil cover is thin or absent.
- *Pasture maintenance* – ensuring that pasture is maintained in good condition by controlling stock density and through appropriate rotation periods will reduce the mobility of microbes in surface runoff and water percolating into the soil during rainfall events. This may include fences to protect particularly vulnerable features from stock and excreta.

Depending on the size and type of agricultural enterprise, management plans may be useful to define these control measures, their operational monitoring, critical limits and corrective actions (e.g. in animal waste treatment, duration, temperature or pH effectively inactivating pathogens), maintenance of facilities, responsibilities and documentation.

21.2 NUTRIENT MANAGEMENT ON AGRICULTURAL LAND

In the context of groundwater protection for health, nitrate is the only relevant nutrient, and the following discussion will therefore focus on control measures for managing nitrate application. However, nutrient management will often also include phosphorus in order to protect surface waters from eutrophication and its consequences for water quality. Effective nutrient management restricts the movement of nitrogen and phosphorus compounds through soil profiles by minimizing the amount of nutrients that can be leached below crop root zones (source control). This is usually achieved by developing a nutrient budget for the crop, applying nutrients at the correct time using appropriate application methods, applying only sufficient nutrients to produce the crop, and considering specific environmental risks that may be posed by a specific site (e.g. the presence of karst features, etc.). The focus of nutrient management is to increase the efficiency by which applied nutrients are used by crops and thereby reducing the risk of leaching. In many cases, the implementation of nutrient management measures results in less fertilizer or manure being used on crops, reducing overall crop production costs and thus often creating a benefit for the farmer.

The main principles that should apply in nutrient management on crops to protect groundwater quality are:

- determine realistic yields for crops under local soil and climatic conditions (preferably accounting for soil variations on a field-by-field basis) – i.e. not trying to force crops with excessive fertilizer application;
- account for nutrients available to crops from all sources before applying additional fertilizer or manure (i.e. ensuring that only sufficient nutrients are available for crop growth);
- synchronize nutrient applications (particularly nitrogen) with crop needs: nitrogen is most needed during active crop growth, and nitrogen applied at other times is easily leached from soils.

Management practices that address these principles are addressed below.

Preparation of nutrient management plans

Nutrient management plans are often required by local authorities to ensure farmers are using nutrients in an efficient way at the farm scale that prevents groundwater pollution. A plan should contain information about the nature and distribution of soils on the farm, their nutrient status, nutrient leachability and soil erodability, and a description of the proposed and past agricultural practices on the farm, and the proposed measures to prevent groundwater pollution and the erosion of soils from the property. If crops are irrigated, then sufficient rainfall, evaporation and irrigation information should be provided to determine the soil-water balance and assess the risk of nutrient leaching.

Nutrients need to be applied with fertilizers and manures in adequate amounts for a particular crop and close to the time when they are needed for crop growth to minimize the potential for leaching of excess soluble nutrients. Determining an appropriate fertilizing regime for nitrogen, as part of the nutrient management plan, requires application of only the amount of nitrogen that is needed by the crop minus any available nitrogen pools in the soil or from crop residues: The optimum nitrogen fertilizing rate for field crops, applied as chemical fertilizer or manure, can be roughly estimated as follows (modified from Feldwisch and Schultheiß, 1998):

Nitrogen uptake by harvested crop in kg N/dt (see Table 21.2 for examples)

- Realistic estimate of expected yields of a particular crop variety under local soil and climate conditions (dt/ha)

+ Surcharge for non-harvestable residues, i.e. roots, stubble, leaf fall: depending on the crop, this rate ranges between 20-50 kg N/ha

– Mineral nitrogen pool in the soil at the beginning of the growing season: the nutrient status of soil is best determined from soil analyses, or is estimated from previous experience or long-term records

– Nitrogen delivery from preceding or residual crops remaining in fields (e.g. straw, leaves, herbage) (see Table 21.3 for examples)

– Nitrogen delivery from intercrops/catch-crops: depending on the crop, this rate ranges between 0-40 kg N/ha

– Nitrogen delivery from soils, i.e. from mineralization processes of organic nitrogen during the vegetation period (see Table 21.4 for examples).

In general, less fertilizer should be applied to lighter, sandy soils than clay-rich soils. As a general guide, nitrogen applications in excess of 140 kg/ha on sandy soils and 200 kg/ha on loamy soils in any 12 months commonly lead to nitrate concentrations in groundwater exceeding drinking-water guidelines. An effective approach to avoid excess nitrate leaching into groundwater is to apply manures only at rates necessary to meet crop phosphorus needs, with any additional nitrogen requirements being met with the application of chemical fertilizers or the use of legumes in crop rotations. The amount of fertilizer required to provide 100 kg of nitrogen to soils for a variety of types of fertilizers commonly used in agriculture is summarized in Table 21.5. The use of Global Positioning Systems on agricultural equipment can help optimize application rates to suit variations in soil type across a field.

In addition to determining the appropriate rate, the following measures should be taken when developing a fertilizing regime for a particular crop:

- selecting the appropriate type of fertilizer for a particular crop type and suited for the expected growth rate;
- selecting the appropriate method of applying fertilizer for a specific crop type which ensures maximum nutrient uptake (for example, row fertilizing, top fertilizing, foliar fertilizing);
- ensuring that only crops adapted for local conditions are grown;
- ensuring that the nutrient content of manure and crop residues are properly evaluated to allow a comprehensive nutrient budget to be determined.

In groundwater recharge areas, it is often necessary to reduce fertilizer application rates to protect groundwater quality, and the total nutrient demand of the crop should be considered.

Table 21.2. Nutrient uptake for a variety of crops (modified from Feldwisch and Schultheiß, 1998)

Crop	Harvested crop *corn (I)/straw (II) beet (I)/leaves (II) tuber(I)/herbage (II)*	N-uptake[3] (kg/dt)			P-uptake[3] (kg/dt)			K-uptake[3] (kg/dt)		
		I	II[1]	total	I	II[1]	total	I	II[1]	total
Feeding wheat	1.0:1.0	1.8	0.50	2.3	0.35	0.13	0.48	0.50	1.2	1.7
Wheat	1.0:1.0	2.1	0.50	2.6	0.35	0.13	0.48	0.50	1.2	1.7
Wheat for blending	1.0:1.0	2.4	0.50	2.9	0.35	0.13	0.48	0.50	1.2	1.7
Rye	1.0:1.0	1.5	0.50	2.0	0.35	0.13	0.48	0.50	1.7	2.2
Triticale	1.0:1.0	1.8	0.50	2.3	0.35	0.13	0.48	0.50	1.4	1.9
Feeding barley	1.0:1.0	1.7	0.50	2.2	0.35	0.13	0.48	0.50	1.4	1.9
Brewer's barley	1.0:1.0	1.4	0.50	1.9	0.35	0.13	0.48	0.50	1.4	1.9
Oats	1.0:1.0	1.5	0.50	2.0	0.35	0.13	0.48	0.50	2.5	3.0
Spelt	1.0:1.0	1.6	0.50	2.1	0.35	0.13	0.48	0.66	1.4	2.1
Maize	1.0:1.0	1.5	1.0	2.5	0.35	0.13	0.48	0.42	1.7	2.0
Winter rape	1.0:2.0	3.3	1.4	4.7	0.79	0.35	1.1	0.83	4.2	5.0
Sunflower	1.0:3.5	2.8	1.5	4.3	0.70	0.40	1.1	2.0	6.3	8.3
Linseed and flax	1.0:1.8	3.5	0.80	4.3	0.53	0.13	0.66	0.8	1.3	2.1
Field pea[2]	1.0:1.4	3.6	1.5	5.1	0.48	0.13	0.61	1.2	2.2	3.4
Field bean[2]	1.0:1.4	4.1	1.5	5.6	0.53	0.13	0.66	1.2	2.2	3.4
Soya bean[2]	1.0:1.4	5.8	3.7	9.5	0.70	0.57	1.3	1.4	3.3	4.7
Potato (early)	1.0:0.3	0.45	0.10	0.55	0.07	0.01	0.08	0.50	0.17	0.67
Potato (other)	1.0:0.3	0.35	0.10	0.45	0.06	0.01	0.07	0.50	0.17	0.67
Sugar beet	1.0:0.8	0.18	0.28	0.46	0.04	0.04	0.08	0.21	0.42	0.63
Fodder beet	1.0:0.3	0.14	0.11	0.25	0.03	0.01	0.04	0.37	0.12	0.49
Silage maize	-	-	-	1.4	-	-	0.59	-	-	1.7

[1] Related to unit harvested crops. [2] No nitrogen fertilization necessary for legume crops. [3] The amount of nitrogen, phosporus and potassium removed from the soil by growing crops.

Table 21.3. Nitrogen delivery by crop residues (based on Feldwisch and Schultheiß, 1998)

Preceding crop/ residual crop	N-delivery (kg N/ha)
Cereals, flax, sunflower, maize for silage	0
Potato, grain maize, annual grass or perennial ryegrass, fallow land (in rotation) without legumes	0-10
Rye, wild mustard species, mustard	10-20
Beet leaves, annual grass or perennial ryegrass (>1 year)	20-30
Grain-legumes (leguminous crops), clover, ley-farming, Lucerne (alfalfa), fallow land (in rotation) with leguminous crops	30-40
Field vegetables, land planting (>1 year), temporary grassland	40-50

Table 21.4. Estimates for mineralization rates for organic nitrogen for a variety of soil types (based on Feldwisch *et al.*, 1998)

Location	Mineralization rate (kg N/ha/a)
Humic soils associated with a shallow water table	>100 (in first 4 years) 20-50 (>4-20 years)
Sandy cover soils on moorland	20-50
Colluvium	20-50
Ploughing up of grassland	>100 (in first 4 years) 20-50 (>4-20 years)

Table 21.5. Nitrogen content of commonly used fertilizers (DEP and WRc, 2000; US EPA, 2000)

Fertilizer type	Fertilizer required to provide 100 kg of nitrogen (kg)
Inorganic	
Ammonium nitrate	294
Ammonium nitrate and urea	312
Ammonium sulphate	476
Urea	217
Aqua ammonia	500
Anhydrous ammonia	122
Ammoniated superphosphate	2000
Monoammonium phosphate	769
Diammonium phosphate	556
Urea and ammonium phosphate	357
Organic	
Cattle manure	2000-5000
Horse manure	1250-5000
Poultry manure	667-2000

Use of soil surveys

Soil surveys are often carried out to identify variations in soils across the farm, determine likely variations in crop productivity, and identify environmentally sensitive sites. This allows management plans to be developed that take into account local variations. Aerial photographs and existing soil maps are often used to undertake the survey. If the farm is located in a catchment where agriculture is known to have caused nutrient contamination, a nutrient management plan is usually required to help prevent further deterioration of water quality. This will again allow for more responsive management plans to be developed that reflect local variations and conditions.

Appropriate timing of fertilizer application

Fertilizers or manures should be applied during growing seasons when plant uptake is at a maximum. They should not be applied at times when heavy rainfall or melting snow can leach nutrients below plant root zones, making the nutrients unavailable for crops.

Fertilizer or manure applications can be matched to crop uptake rates by splitting the nutrient budget over several applications, use of slow release fertilizers, fertigation (i.e. including small amounts of nutrients in irrigation water) or by applying denitrifying inhibiters to reduce nitrogen loss in soils. Manures should be applied uniformly in accordance with crop needs, and surface applications to no-till cropland should be avoided.

Maintenance of buffers or protection zones around sensitive areas

Protection of drinking-water supplies and sensitive environmental features in agricultural regions may require appropriate buffers to be established to allow nutrient and other contaminant levels to be attenuated. Land in such buffer or protection zones should not be a source of nutrient pollution. Possible land uses include reserves of native vegetation, tree lots or parks used for passive recreation (see also Chapter 17).Their protection and maintenance may be designated in a management plan in order to avoid degradation and loss of function.

Use of crop sequences to minimize nitrogen leaching

Nitrogen leaching to groundwater can be minimized by maintaining a permanent crop cover on fields comparable with permanent grassland. This requires detailed planning of crop sequences. Crops that are especially effective in removing and therefore preserving excessive nitrogen from soils (catch-crops) can be incorporated into the crop sequence. Effective catch-crops include rape, mustard, sunflowers and different grasses that can all bind between 75 and 160 kg N/ha. Optimum nitrogen removal rates can be achieved if catch-crops are sown soon after main crops are harvested and by maintaining a dense vegetation cover (Feldwisch and Schultheiß, 1998).

21.3 MANAGEMENT OF WASTEWATER AND HUMAN EXCRETA USED ON LAND AND IN AQUACULTURE

Although human excreta has been used to enhance soil fertility for several thousand years wastewater has only been widely used as a source of water and nutrients in agriculture over the last few decades. Wastewater from reticulated sewerage or domestic systems is being increasingly used in many countries due to the increasing scarcity of water resources and the high cost of chemical fertilizers. When well managed, the use of wastewater in agriculture will have a minimal impact on groundwater quality. Important control measures for the use of wastewater are set out below.

Nutrient management

Wastewater is a significant source of nutrients, and the set-back distances to sensitive features, rate and timing of application of effluent should be in accordance with management measures for fertilizers and manures set out in Section 21.1. Although wastewater may not contain a sufficient amount of all the nutrients essential for crop growth, care must be taken to ensure that the use of fertilizer supplements does not exceed crop requirements and lead to leaching of chemical contaminants into groundwater. Depending on its source, the composition of wastewater may vary considerably, and frequent monitoring is required to ensure that its nitrogen content in particular continues to be matched to crop uptake rates.

Application of wastewater and excreta

Wherever possible, wastewater should not be applied to crops by flood irrigation as this irrigation method increases the risk of chemical contaminants and pathogens being leached through the soil profile into groundwater. Preferred irrigation methods to protect groundwater quality are:

- controlled irrigation through furrows
- irrigation by sprinklers
- localized irrigation through drippers

Untreated wastewater and raw excrement should not be applied to crops because of the risk to health of agricultural workers from potential contact with pathogens on crops, and guidelines for the quality of wastewater should always be followed (WHO, 1989; Havelaar *et al.*, 2001). Pathogen levels in wastewater can be greatly reduced by holding effluent in stabilization ponds that allow a retention time of 12-18 days, or by chlorination after secondary treatment and filtration (Horan, 1991). Pathogen levels in excreta can also be reduced by storage, or through composting with other organic material (Franceys *et al.*, 1992). Details on the requirements for sewage sludge quality are discussed further in Chapter 22. Care should be taken when using treated wastewater to consider aquifer vulnerability, e.g. where soils are thin or locally absent, where there are open wells, sinkholes or other voids that provide a direct conduit to the water table, or where there is a shallow water table because of the risk of contaminating groundwater by pathogens. Determining an acceptable depth to groundwater should take into account hydraulic load and soil conditions as well as depth of the water table (see Chapter 14).

Controlling the source of wastewater

Sewage effluent from catchments with a large component of industrial waste may contain high concentrations of arsenic, heavy metals, pesticides, solvents or hydrocarbons that have the potential to cause groundwater contamination if the wastewater is used for irrigation. Wherever possible, industrial wastewater should not be used on land with crops for human consumption, where soils are thin or locally absent, where there are open wells, sinkholes or other voids that provide a direct conduit to the water table, or where the water table is shallow because of the increased risk of contaminating groundwater. Most regulations for sludge disposal include requirements to assess local conditions before site approval and application of sludge to land (US EPA, 1995; Pedley and Howard, 1997). Controlling the source and quality of sewage effluent, and restricting its use on certain crop types or in certain areas can be more readily accomplished when:

- there are appropriate waste disposal laws and the society is law abiding;
- a public body controls the management of wastes;
- an irrigation project has strong central management;
- there is adequate demand for the crops allowed under crop restriction and where they fetch a reasonable price;

- there is little market pressure in favour of crops not permitted for irrigation by wastewater of a particular quality.

Setback distances for fish ponds

Only treated wastewater should be used in aquaculture and guidelines have been established for these requirements based on overall exposure (WHO, 1989; Havelaar *et al.*, 2001).

Where nonetheless such practices are reality, set back distances can be defined to protect groundwater wells constructed near fish ponds filled with untreated wastewater and fertilized with raw excrement. Consequently, fish ponds should be treated as sources of microbiological contamination that are comparable with sanitation systems. Groundwater supplies can be protected from pathogens from these sources by ensuring that the distances between fish ponds and water supply wells (setback distances) are set using the principles discussed in Chapters 17 and Chapter 22. The latter also provides details of appropriate treatment of wastewater and sludge prior to reuse in aquaculture.

In general, setback distances for fish ponds will be greater than for sanitation systems as the hydraulic loads from leaking ponds is likely to be substantial. This may lead to a localized up-coning of groundwater mound and substantially increasing the local hydraulic gradient and groundwater flow rates. Groundwater is also likely to flow laterally from fish ponds, so wells thought to be upgradient of the ponds may also be susceptible to microbial contamination where there is a significant cone of depression caused by pumping regimes. The risk of pathogen contamination of groundwater can be reduced by slowing the leakage rate of water from ponds through the use of low permeability liners in the ponds (compacted clay or synthetic materials). In densely populated areas establishing adequate setback distances may not be achievable and fish ponds continue to receive wastewater. In such settings, potential health risks should either be managed by treating water pumped from wells near fish ponds, or providing the local population with alternative water sources.

21.4 NUTRIENT AND PATHOGEN MANAGEMENT ON GRAZING LAND

The effects of livestock grazing on groundwater quality can be very variable, and are dependent on site specific conditions such as climate, vegetation density, grazing density and the duration of grazing. The risk of nutrient enrichment of groundwater quality is generally low in semi-arid or arid areas where livestock and animal wastes are uniformly distributed at low densities in the landscape. The risks of groundwater contamination may also be low in humid areas if grazing land is managed correctly. For any grazing measure to work, it must be tailored to fit the needs of the local vegetation and terrain, type of livestock and the culture of the local farming community. However, the following control measures are generally effective in reducing nutrient and pathogen contamination from grazing.

Maintaining vegetation cover on rangeland

Maintaining good quality pasture on rangeland is essential to help hold nutrients in soils, to prevent the erosion of soil and to prevent livestock congregating in small areas of good pasture thereby increasing the risk of localized groundwater contamination problems. This requires maintaining appropriate stocking densities to preserve a uniform cover of native pasture and trees (particularly in semi-arid or arid areas) or planting perennial pasture species adapted for local conditions. In areas with seasonal high temperatures, the preservation of native trees or the establishment of shade trees is important to provide sufficient shade cover to stop livestock congregating in small areas. Maintaining a well distributed water supply for livestock also helps prevent overgrazing in small areas as well as the formation of concentrated urine patches, which on grazed pasture are often the most significant source of nitrate contamination of groundwater.

Use of fencing to exclude livestock from sensitive areas

Livestock should be excluded from riparian vegetation around watercourses, particularly in semi-arid areas where periodic flows in watercourses are often the main source of groundwater recharge. Riparian vegetation plays an important role in minimizing the movement of animal wastes and eroded soil in overland flow into watercourses. Other sensitive areas like sinkholes in karstic areas should be fenced to prevent animal access.

Limit stocking rates in groundwater recharge areas

In drinking-water catchments where the protection of groundwater quality is a high priority, animal stocking rates should be restricted to minimize the risk of groundwater contamination by nutrients and pathogens. A stocking rate of less than two horses or cows per ha has been found to be an effective protection measure in particularly sensitive areas with very sandy soils (WRc, 1998). In Germany, recommended stocking rates are 1.3 to 1.4 large animal units per ha (where an animal unit is equivalent to an animal of about 500 kg). Optimizing livestock diets also help to minimize the quantity and nitrogen content of manures.

Use of grazing management plans

The development of farm-specific grazing management plans is an effective control measure for protecting groundwater quality from this land use while maintaining or increasing the economic viability of the grazing operation. The steps to developing an effective grazing management plan are to (US EPA, 2000):

- undertake an inventory of existing resources and pasture condition;
- determine management goals and objectives;
- map out grazing management units;
- develop and implement a grazing schedule;
- develop and implement a monitoring and evaluation strategy.

21.5 MANAGEMENT OF ANIMAL FEEDING OPERATIONS AND DAIRIES

The often severe water quality problems associated with animal feeding operations and dairies are due to the high concentrations of animals and their accumulated wastes, and the large amount of wastewater that can be generated from stormwater runoff and the washdown of the facilities. The risk of groundwater pollution occurring can be minimized by the proper siting of the facilities, by waste and water management practices and by reducing the amount of nutrients excreted in wastes through a well managed feeding regime.

Siting animal feeding operations and dairies

One of the major considerations in preventing groundwater pollution from an animal feeding operation or dairy is the location of the facility. For new facilities and expansions to existing facilities, consideration should be given to siting the facility away from surface waters, areas with a high leaching potential, sinkholes or other environmentally sensitive areas, and in areas where sufficient land is available to apply wastes to soils at rates which will not affect groundwater resources.

The US EPA (2000) indicates that there are eight critical factors to be considered when siting and operating a feedlot. These are:

1. *Divert clean water*: siting or management practices should ensure that clean runoff does not touch stock holding pens or manure storage areas.
2. *Prevent seepage*: buildings and storage facilities should be designed and maintained to prevent contaminated water seeping into groundwater.
3. *Provide adequate storage*: liquid manure storage systems should be designed to safely store the quantity of wastewater produced in the feedlot plus additional runoff from intense storms (often designed for a 25 year, 24 hour storm). Dry manure should be stored under cover wherever possible to prevent the generation of additional contaminated runoff.
4. *Apply manure in accordance with a nutrient management plan*: it is important that manure use is seen as part of an overall nutrient management strategy for a farm (see above).
5. *Manage land where manure spreading is taking place*: land being used for manure disposal should be well managed to prevent erosion that can cause nutrients to be moved offsite. Management measures include proper stock control and the use of vegetated buffers and filter strips.
6. *Record keeping*: records should be kept of the amount of manure generated and the disposal method used. This will help assess the effectiveness of a nutrient management plan, and allow it to be revised if necessary.
7. *Mortality management*: dead animals should be managed in a way that minimizes impacts on groundwater. The British Columbia government in Canada recommends that animal disposal in pits is not within 120 m of wells used for water supply or within 30 m of a surface water feature. The guidelines (BC MAF, 2000) recommend that the base of disposal pits is at least 1.2 m above the seasonally highest water table, and that disposal pits have at least 1 m of earth cover.

8. *Consider the full range of environmental constraints*: when expanding an existing facility or siting a new feedlot, consideration should be given to the distance of the facility from surface water features, from areas where groundwater is highly vulnerable to contamination, and from sinkholes or other features that allow surface runoff direct access to the water table.

Solid waste management practices

Manure should be removed from animal pens at frequent intervals. It is recommended that it is stockpiled in a roofed facility on an impervious floor to prevent leaching by rainfall. Manure can be stored for an extended period until it is used on the farm or is moved offsite for use elsewhere. Maintaining low moisture content in the manure will minimize odour problems and the generation of leachate. If sufficient space is available, manure can be aerobically composted in turned piles or rows to improve its performance as a soil amending agent and to reduce pathogen levels.

As discussed in Section 21.2 above, manure must be applied to soils at rates which will minimize groundwater contamination by nitrate, and the appropriate application rate should be determined by carrying out tests on local soils.

Water and liquid waste management practices

Stormwater runoff from roofs and paved areas should be channelled away from feedlots or dairy pens using bunds and drains to ensure it does not become contaminated with animal wastes. It is recommended that stormwater is collected in a lined settling pond and can be disposed of by irrigation under suitable weather conditions.

Runoff from areas contaminated with animal wastes should be channelled to a series of ponds for storage and treatment to reduce the BOD and nutrient content before disposal by irrigation or other means. At least two fully lined ponds are required to adequately treat the wastewater. The first pond allows solid organic material to settle out and be degraded under anaerobic conditions by microorganisms. Water from this pond is decanted in a shallower pond which is aerated by wind action, and which allows the penetration of sunlight to help reduce levels of microorganisms before water is disposed of by irrigation under suitable weather conditions.

The storage and treatment ponds should have sufficient capacity to store water over winter or the wettest period of the year, and should be able to store water from intense rainfall events. Regulations for the design and capacity of storage ponds vary from jurisdiction to jurisdiction. A general guide given by BC MAF (2000) recommends that the volume of storage of ponds can be estimated by the following formulas:

Runoff from paved feedlots:

$$V = A \text{ x } (0.48\ P_m + 0.65\ P_s) \qquad \text{(Eqn. 21.1)}$$

Runoff from unpaved feedlots:

$$V = A \text{ x } (0.22\ P_m + 0.45\ P_s) \qquad \text{(Eqn. 21.2)}$$

Runoff from manure storage areas:

$$V = A \text{ x } (0.25\ P_m + 0.65\ P_s) \qquad \text{(Eqn. 21.3)}$$

where V is the volume of storage, A the area contributing to runoff, P_m the sum of six-monthly winter precipitation (rainfall plus an equivalent water depth of snowfall), and P_s the 24-hour precipitation from a storm expected once in 25 years.

Wastewater treated as outlined above will in most cases not be of a suitable quality to discharge to waterways, but may be disposed of by irrigation under suitable conditions. As discussed above, wastewater should not be applied to land unless the soil nutrient status has been determined, and its use is consistent with the nutrient management plan developed for the farm. Sufficient land area should be available for a 10 to 14 day rest period between applications on a given part of the farm, the objective being to alternate between anaerobic and aerobic conditions in the upper part of the soil (shorter periods may be possible under dry summer conditions). Crops or pasture should be maintained to take up as much nitrogen and phosphorus as possible to minimize the risk of groundwater pollution occurring.

21.6 PESTICIDE MANAGEMENT

As some pesticides in current use are toxic at low concentrations and can cause groundwater contamination if used incorrectly, these chemicals need to be stored, used and disposed of with great care to minimize groundwater contamination problems. In general, pesticides should be applied to soils at the lowest possible rate which controls the pest problem to reduce the potential for leaching. Control measures that reduce the risk of groundwater contamination by pesticides include the following:

Undertake an inventory of current and historical pest problems

Adequate pest control requires a good understanding of what insect pests and weeds are a problem in each field so that pesticide spraying can be well targeted at appropriate application rates. Agricultural extension officers, universities or specific consultants can help identify the distribution of weeds and insect pests on individual farms. The costs entailed in compiling pest inventories are usually rapidly recovered by reductions in the amount of pesticides applied to crops.

Assess the potential for the leaching and runoff of pesticides from farms

Pesticides should not be applied near features that allow direct access to the water table such as karst features, abandoned wells or drainage wells. Adequate buffers should be maintained from water supply wells, surface water bodies and other sensitive environmental features. Pesticide application rates may need to be reduced in areas with light, sandy soils and a shallow water table to reduce the risk of groundwater contamination. Application rates may also need to be reduced on heavy soils with steep slopes to prevent surface runoff that could infiltrate into soils further downslope. Cropping practices such as no-till methods can greatly reduce the risk of runoff from steep slopes on heavy soils.

Increase organic matter in soils

Increasing the organic content of soils by the application of composts or manures can greatly increase the binding capacity of soils and reduce the risk of pesticide leaching.

Proper construction of pesticide storing and mixing areas

The risk of groundwater contamination taking place is often greatest at sites that are used for storing and mixing pesticides before application to fields. The construction of dedicated storage and mixing areas can greatly reduce the risk of contamination occurring through spills. These should have an impermeable (concrete) floor and be surrounded by curbing to contain any spills, and the area should be large enough to confine at least 125 per cent of the displaced volume of liquids housed in the storage area. Runoff from the mixing area should be stored in tanks or sealed ponds to allow water to be removed for treatment or to be applied to fields.

Pesticides in small containers should be stored inside a shed on wooden palettes or shelves to keep the containers off the floor to minimize corrosion of containers. Large bulk storage tanks containing pesticides should be elevated so that leaks are easily seen. Regular monitoring and maintenance of the integrity of structures to contain pesticides are important and would be included in a management plan.

Special care needs to be taken with the handling and transfer of pesticides to the distribution containers and subsequent washing down of vehicles and equipment. It has been estimated that over 50 per cent of observed contamination can occur from these practices if executed poorly.

Use of cropping measures and biological controls to reduce pesticide use

There are a variety of cropping measures that can be used to reduce the use of pesticides on crops. There are also an increasing number of new crop varieties which only require low applications of pesticides to achieve a high resistance to fungal diseases and to damage by insects and nematodes.

Crop rotations can interrupt pest build-up by eliminating the host plant or by changing the physical conditions that allow the use of smaller amounts of pesticides. An example of this is corn-soybean rotation in which broadleaf weeds are more easily controlled in the corn crop and grass weeds are more easily controlled in soybean crop. Some plant species have allelopathic properties when used in rotations; that is, they can reduce pest populations in subsequent crops. For example, a rye cover crop may reduce weed populations in subsequent crops.

The use of trap crops can also greatly reduce the use of pesticides. These are plant species which are more attractive to particular pest species that the main crop, so pesticide applications can specifically target these plants. Trap crops can either be planted adjacent to main commercial crops, or planted at an earlier time to ensure pest control before the main crop is harvested.

Increasing the biological diversity on farms by retaining remnant vegetation or maintaining hedgerows can also indirectly reduce the need for pesticides by increasing the number of predator species. In general, agricultural monocultures create simple environments where pests have little or no competition from other species or predators. Having a broad array of plant species on farms diversifies the habitat and helps reduce pest populations. Insect predator species may be introduced into fields and the use of specific pesticides to target only pest species can enhance the effectiveness of predators.

Use of low-toxicity or biological pesticides
Highly toxic or environmentally persistent pesticides can often be replaced with less toxic or persistent equivalents. The use of less harmful pesticides can be increased by management measures such as the use of financial incentives, specific training Biological pesticides include microorganisms which are pathogens to specific insect pests, and the use of pheromones to lure or trap insects. They are very specific to a target species, are effective at low concentrations and generally pose little or no threat to human health or other species. Pheromones can also be used to disrupt the reproduction of insect pests or attract predators and parasites.

21.7 IRRIGATION WATER MANAGEMENT AND DRAINAGE

Most of the groundwater quality problems associated with irrigated agriculture are due to the inefficient use of water which increases leaching of water and soluble salts through the soil profile. Poorly managed crop irrigation and drainage schemes, including poor maintenance of irrigation structures, can be a major cause of health and environmental problems.

Irrigation scheduling practices
Proper scheduling of the application of water is a key element in the management of irrigated agriculture. Scheduling should be based on knowing the daily water use of the crop, the water-holding capacity of the soil, the lower limit of soil moisture for each crop and soil, and measuring the amount of water applied to crops. Natural precipitation should also be considered and adjustments made in irrigation schedules. Practices that help manage irrigation scheduling are:

- metering of water flow rates;
- using soil and crop water use data to determine the timing of water applications;
- reducing hydraulic load by using efficient irrigation systems.

Practices for the efficient application of irrigation water
Irrigation water should be applied in a manner that ensures an even distribution of water and efficient use by crops and minimizes runoff or deep percolation. The method of irrigation varies considerably with the type of crop grown, topography, soils and local cultural factors. However, there are several practices that are particularly effective for applying and controlling the distribution of irrigation water. These include:

- *drip or trickle irrigation systems*: water is applied at low pressure to crops with minimal evaporative losses;
- *sprinkler systems*: there are a large variety of sprinkler systems for applying water under pressure to crops;
- *water control structures*: the use of furrows, contour levees or contour ditches can help control the spread of water applied in an irrigation area;
- *irrigation field ditches*: permanent structures used to convey water from the source of supply to a field or fields in a farm distribution system;

- *land levelling*: reshaping the surface of the land to planned grades to ensure that water does not pool in certain areas causing excessive infiltration or salinization, or else runoff at high velocities causing erosion.

Efficient use of runoff water
Runoff from precipitation or excess irrigation can be captured, stored and used as part of an irrigation system to improve the overall efficiency of the system and minimize the potential for leachate infiltration.

Drainage water management
Drainage water from an irrigation system should be managed to reduce the potential for leachate to contaminate groundwater and reduce erosion. A well planned and maintained drainage system should be an integral part of the design of an irrigation system. Practices that can be incorporated into a drainage system include the use of:

- *filter strips*: an area of vegetation for removing particulate matter from runoff and wastewater;
- *surface drainage field ditches*: a graded ditch for collecting excess water in a field;
- *sub-surface drains*: a conduit such as a perforated pipe installed beneath the ground surface to collect and convey excess irrigation water.

Drainage is also commonly used in areas with a naturally shallow water table to make land suitable for agriculture. Where soils naturally contain high concentrations of sulphide minerals, wide, shallow drains should be constructed rather than conventional deep drains. This minimizes the disturbance of sulphides and reduces the risk of forming acid sulphate conditions.

21.8 MONITORING AND VERIFICATION OF MEASURES CONTROLLING AGRICULTURAL ACTIVITIES

Table 21.6 summarizes selected examples of the measures proposed above to control groundwater contamination from agricultural activities. These include planning, physical structures to prevent leachate and operational controls to ascertain implementation of management plans. In some settings, some of these control measures may be suitable for integration into the WSP (see Chapter 16) of a drinking-water supply and become subject to operational monitoring in the context of such a plan.

Regardless of whether or not any of these control measures are part of a WSP, their monitoring and verification is crucial to ensure that they are in place and effective. Table 21.6 therefore includes options for surveillance and monitoring of the control measure examples given. Most of these monitoring options focus on checking whether controls are operating as intended, rather than on contaminant concentrations in groundwater:

- For control measures in the context of planning, surveillance will review how well land use is taking aquifer vulnerability into account, e.g. with respect to siting of animal feedlots and treatment of their runoff, or whether management plans for nutrient and pesticide application exist and are adequate. Monitoring will include site inspection to assess whether plans are being implemented.

- For control measures addressing design and construction, monitoring is largely through site inspection to ascertain their adequacy and integrity, e.g. whether a feed-lot is properly drained, whether the storage tank for liquid manure is covered, whether irrigation systems are constructed as indicated in the management plan or permit, and whether maintenance of structures is adequate.
- For control measures addressing day-to-day routine operations, monitoring focuses on assessing whether the specific restrictions, limitations or management plans imposed to protect the drinking-water aquifer are being followed, e.g. by inspecting farm records on agrochemical use, counting heads of stock, or sampling the nutrient content of soils.

NOTE ▶ *The implementation of control measures such as those suggested in Table 21.6 is effectively supported if the stakeholders involved collaboratively develop management plans that define the control measures and how their performance is monitored, which corrective action should be taken both during normal operations and during incident conditions, responsibilities, lines of communication as well as documentation procedures.*

The implementation of control measures protecting drinking-water aquifers from agriculture is substantially facilitated by an environmental policy framework (see Chapter 20).

In addition to the operational monitoring of the functioning of control measures, overall groundwater monitoring is important to verify comprehensively that agrochemicals and manure are not contaminating aquifers used for drinking-water abstraction, i.e. that the management concept for the catchment is adequate and safe. With respect to fertilizers and manure this would typically include nitrate and potentially, for the latter, also indicators for pathogen occurrence or even pathogens of particular concern in the respective setting. With respect to pesticides, chemical analyses for monitoring may be facilitated by information on the range of substances typically applied in the region.

NOTE ▶ *Options for monitoring suggested in Table 21.6 focus on control measures rather than on groundwater quality. Analysis of selected parameters in groundwater which indicate leaks of containments for agrochemicals or manure is suggested where this is the most effective operational control.*

Comprehensive groundwater quality monitoring programmes are a supplementary aspect of monitoring with the purpose of providing verification of the efficacy of the overall drinking-water catchment management.

Table 21.6. Examples of control measures for agriculture and options for their monitoring and verification

Process step	Examples of control measures for agriculture	Options for their monitoring and verification
PLANNING	Define criteria for exclusion or restriction of agricultural activities (e.g. stock density, type of crop) in vulnerable drinking-water catchments (e.g. implement protection zones)	Monitor land use within vulnerable areas/protection zones and ensure that restrictions are implemented (site inspection)
	Require permits for the location, design and operation of feedlots in vulnerable drinking-water catchments	Review plans and applications for permits for agricultural activities in relation to vulnerability of drinking-water aquifer
	Require nutrient and pesticide management plans with specific limitations on amounts and timing of agrochemical, manure and sludge application	Audit nutrient and pesticide management plans
	Restrict wastewater irrigation near boreholes, in vulnerable groundwater recharge areas or in drinking-water protection zones (permit process)	Monitor selected parameters in groundwater which indicate contamination with agrochemicals and/or pathogens
DESIGN, CONSTRUCTION AND MAINTENANCE	Construct and maintain safe containments for agrochemicals and adequately sized, impermeable and bonded sites for pesticide mixing and cleaning of equipment Install and maintain safe storage tanks for liquid manure	Inspect structures and review management plans Monitor selected groundwater parameters (agrochemicals, indicator organisms) which would indicate leakage
	Prevent chemical use and animal access near features that allow direct access to groundwater (e.g. sinkholes, abandoned mineshafts, wetlands with groundwater throughflow)	Ensure features are fenced off with appropriate set-back distances through statutory controls and inspections
	Apply drip irrigation to avoid water table drawdown and overdraught	Inspect irrigation system Monitor groundwater table
OPERATION	Control implementation of restrictions on agricultural activity in vulnerable drinking-water catchments	Inspect farm records of agrochemical application Count heads of stock
	Control implementation of pathogen, nutrient and pesticide management plans (i.e. choice, amounts and timing of application)	Inspect timing and amounts of manure application; review management plan and documentation Analyse residual nitrogen or phosphorus in the soil at beginning of growing season to determine fertilization needs
	Grow winter cover crops to consume excess soil nitrogen	Conduct visual site inspection
	Match irrigation to crop needs	Inspect and monitor drainage Inspect farm records on water use Audit irrigation plans

21.9 REFERENCES

BC MAF (2000) *Environmental Guidelines for Beef Producers*, Appendix D, Environmental guidelines, http://www.agf.gov.bc.ca/resmgmt/fppa/pubs/environ/ beef/beeftoc.htm (accessed April 29, 2005).

DEP (Department of Environmental Protection of Western Australia) and WRc (2000) *Environmental Guidelines for the Establishment and Maintenance of Turf and Grassed Areas*. Publication of the Government of Western Australia, Perth.

EA (2001) *Best Farming Practices: Profiting from a good environment.* R&D Publication 23, Bristol.

Feldwisch, N. and Schultheiß, U. (1998) Verfahren zur Verminderung der Stoffausträge aus der Pflanzenproduktion: Allgemeine ackerbauliche Aspekte. In *Handbuch zum Gewässerschutz in der Landwirtschaft*, (eds. H-G. Frede and S. Dabbert), pp. 58-100, Ecomed Verlagsgesellschaft, Landsberg.

Feldwisch, N., Frede, H-G. and Hecker, F. (1998) Verfahren zur Abschätzung der Erosions- und Auswaschungsgefahr: Auswaschungsgefahr von Nitrat. In *Handbuch zum Gewässerschutz in der Landwirtschaft*, (eds. H-G. Frede and S. Dabbert), pp. 50-57, Ecomed Verlagsgesellschaft, Landsberg.

Franceys, R., Pickford, J. and Reed, R. (1992) *A Guide to the Development of On-site Sanitation.* WHO, Geneva.

Goss, M.J., Rollins, K.S., McEwan, K., Shaw, J.R. and Lammers-Helps, H. (2001) *The Management of Manure in Ontario with Respect to Water Quality.* Report produced by the University of Guelph for the Walkerton Inquiry. www.walkertoninquiry. com/part2info/commissuepapers (accessed April 29, 2005).

Havelaar, A., Blumenthal, U., Strauss, M., Kay, D. and Bartram, J. (2001) Guidelines: the current position, In *Water Quality: Guidelines, standards and health*, (eds. L. Fewtrell and J. Bartram), pp. 17-42, IWA, London.

Horan, N.J. (1991) *Biological Wastewater Treatment Systems*. Wiley, Chichester.

Pedley, S. and Howard, G. (1997) The pubic health implications of microbiological contamination of groundwater. *Q. J. Eng. Geol.*, **30**,179-188.

Rosen, B.H. (2000) *Waterborne Pathogens in Agricultural Watersheds*. U.S. Department of Agriculture Watershed Science Institute Report. http://wvlc.uwaterloo.ca/ biology 447/modules/module8/SludgeDisposal/Pathogens_in_Agricultural_Watersheds.pdf (accessed April 29, 2005).

US EPA (1995) *Land Application of Sewage Sludge and Domestic Septage: Process design manual*. US EPA, Washington.

US EPA (2000) *National Management Measures to Control Nonpoint Source Pollution From Agriculture*. Draft report produced by the Office of Water. http://www.epa. gov/owow/nps/agmm/index.html (accessed April 29, 2005).

WHO (1989) *Guidelines for the Safe Use of Wastewater, Excreta and Greywater in Agriculture, Volume 2*. World Health Organization, Geneva.

WRc (1998) *Guidelines for Stabling and Agistment of Horses in Underground Water Pollution Control Areas*. Government of Western Australia water quality guideline. http://www.wrc.wa. gov.au/protect/policy/WQPN.htm (accessed April 29, 2005).

22

Human excreta and sanitation: Control and protection

G. Howard, B. Reed, D. McChesney and R. Taylor

The safe disposal of human excreta is essential for public health protection. The unsafe disposal of excreta is a principal cause of the transmission of pathogens within the environment and improvements in excreta management provide significant reductions in diarrhoeal disease (Esrey *et al.*, 1991; Esrey, 1996; Hutton and Haller, 2004). Access to improved sanitation lags behind access to water supply throughout much of the world and in particular within developing countries. It is estimated that over twice the number of people lack access to improved sanitation than lack access to an improved water supply (WHO and UNICEF, 2004).

Excreta disposal technologies may represent a risk to groundwater and inappropriate design, siting and maintenance of sanitation facilities can contaminate groundwater and thus lead to public health risks from drinking-water. Chapter 10 provides an overview of these risks and how these may be assessed. However, the health risks from the absence of improved excreta disposal are likely to exceed those posed by contamination of groundwater from sanitation alone, and this must be borne in mind when planning improvements in sanitation and groundwater protection. Furthermore, the lack of excreta disposal may be a direct cause of contamination of groundwater sources and improvements in sanitation

 Protecting Groundwater for Health: Managing the Quality of Drinking-water Sources. Edited by O. Schmoll, G. Howard, J. Chilton and I. Chorus. ISBN: 1843390795. Published by IWA Publishing, London, UK.

may also deliver improvements in microbial quality in groundwater (Howard *et al.*, 2003).

This chapter provides an overview of some of the options for managing groundwater pollution risks derived from sanitation, both for on-site and off-site methods. These include planning, design and construction of facilities, as well as monitoring and managing their safe operation. Important aspects discussed in the context of planning are siting decisions and infrastructure changes that will help reduce risks, and balancing the sometimes competing needs for better sanitation and groundwater protection. The end of the chapter summarizes some major control measures that can be used to provide protection of groundwater through effective management of sanitation.

NOTE ▶ *In developing a Water Safety Plan (Chapter 16), system assessment would review the efficacy of control measures and management plans for protecting groundwater in the drinking-water catchment from human excreta and sanitation. Chapter 10 provides the background information about the potential impact of wastes on groundwater and provides guidance on the information needed to analyse these hazards.*

This chapter introduces options for controlling risks from human excreta and sanitation. In some settings, water utilities or communities hold the responsibility both for drinking-water supply and for sanitation and in these, measures to control risks from human excreta may readily become part of a Water Safety Plan. Where the responsibility for sanitation falls outside that of drinking-water supplier, close collaboration of the stakeholders involved is important to implement, upgrade and monitor these control measures. This may be initiated by water suppliers and supported by the health authority which is usually responsible for the surveillance of both the drinking-water supply and sanitation facilities.

22.1 BALANCING INVESTMENT DECISIONS

Both drinking-water source and sanitation improvements are important to protect public health but, as noted in Chapter 10, excreta disposal systems may lead to groundwater contamination and therefore potentially lead to a risk to public health derived from use of this water for drinking. This has led to a debate among water and sanitation professionals regarding the need to balance risks derived from the contamination of drinking-water from sanitation systems with the health risks posed by an absence of effective excreta disposal.

All decisions about technology and interventions will have cost implications for the users and decisions may be required to balance competing risks. Very good technical solutions that users cannot afford will not deliver significant improvements in health. The evidence suggests that the nature of the competing risks and priorities for investment are different between different communities and may change over time.

The issues are often different in rural and urban areas. In urban areas, although contamination of groundwater may be greater, alternative water sources are often available within easy reach or more distant sources can be developed and water delivered to urban populations in a cost-effective manner. In rural areas, although contamination risks may be lower from reduced discharge or leaching of contaminants, alternative sources may be limited or their development to provide water to rural households may be very expensive. Therefore control of risks from sanitation design, construction and maintenance may be more critical. These issues apply equally in developed and developing countries.

In communities in developing countries where improved sanitation facilities do not exist, setting stringent requirements for sanitation facility design and construction to meet criteria to prevent groundwater pollution may be counter-productive, unless the risk is very significant. Such criteria may make sanitation improvements too expensive for many households resulting in continued lack of sanitation and ongoing disease transmission (Mara, 1996). Where the construction of sanitation will represent a very significant risk of groundwater contamination, decisions will be required as to whether changes in water supply or sanitation are more cost-effective (Franceys *et al.,* 1992).

Although microbial contamination of groundwater from excreta disposal should be minimized, some groundwater pollution may have to be accepted in order to reduce a greater health risk from a lack of excreta disposal (Cairncross and Feacham, 1993; ARGOSS, 2001). Where a groundwater source is linked to a piped water supply, public health risks derived from microbial contamination of groundwater from on-site sanitation may be controlled by treatment of the water prior to distribution. Where water is collected by hand, it may be more cost-effective to provide an alternative (piped) water supply bringing in uncontaminated water than to change sanitation designs (Franceys *et al*., 1992). If this approach is followed then action may be required to close the previously used groundwater source, as evidence from several developing countries indicates that use of untreated groundwater sources is common even when better quality alternatives are available (Ahmed and Hossain, 1997; Howard *et al.*, 2002).

In more developed countries, water source substitution may not be more cost-effective than changing the sanitation facility and the public health demands for groundwater protection may be greater than those associated with sanitation. As a result, investments in upgrading sewer systems, installing sewers into previously unsewered areas or introducing on-site sanitation systems that have a lower impact may all be preferred solutions. Experience suggests that decisions should be based on each individual case and blanket solutions are rarely applicable.

The contamination of groundwater from chemicals derived from human excreta and sewage may be more difficult to control and the treatment of drinking-water less certain to remove or reduce concentrations to an acceptable level. Thus source protection will become more important. For instance, the removal of nitrate through treatment is unlikely to be feasible in most situations and removal of pesticides and organic chemicals may be difficult and expensive. Where there is extensive nitrate contamination, blending with other low-nitrate waters may be required to reduce nitrate levels in final waters, although this is only feasible for piped water supplies where alternative sources of sufficient quantity exist. For other chemicals, blending or use of granular activated carbon may be needed to reduce concentrations to acceptable levels.

In such cases there may be greater need for source protection, particularly in those countries where disease burdens from microbial hazards from drinking-water are low. Thus in developed countries, it may be cost-effective to increase groundwater protection as the potential threats of chemical contamination of groundwater sources are relatively significant.

In developing countries, the microbial health risks posed by inadequate sanitation often so greatly outweigh the risks to health associated with possible groundwater contamination by chemicals that rising levels of contamination may be acceptable if this allows improved sanitation. However, in making these decisions, it is important to take into account long-term usage of water resources. For instance, because nitrate is conservative in many groundwaters controlling nitrate pollution is important for securing the long-term viability of the aquifer. Controlling nitrate risks for this reason is best justified where groundwater represents the long-term option as a source for domestic supply.

22.2 SELECTING THE RIGHT SANITATION TECHNOLOGY

The first stage in the management of sanitation in order to protect against groundwater contamination is to select the right technology for the local environment. This requires that data is collected on the local hydrogeological conditions (Chapter 8) and the role of groundwater in water supplies within a region or the country. Sanitation technology selection will not be based solely on concerns regarding risks of groundwater contamination. Community preferences, usual methods of anal cleansing, available resources and costs of technology options should all be considered in this process (Franceys *et al.*, 1992; Cotton and Saywell, 1998). In many cases these concerns will take priority over groundwater quality concerns, but it is important that attention is also paid to groundwater pollution risks in the decision-making process.

The links between water supply service and sanitation options cannot be ignored. As water supply service levels increase, so will water consumption, thus increasing the volume of wastewater that must be disposed of. Using sanitation technologies that are not designed to take large volumes of wastewater will not be appropriate and water-based systems or sanitation systems that separate effluent and solid material should be considered. Equally, the use of flush toilets and other water-based systems will not be appropriate where the water supply service is only a public tap, dug well or other form of communal supply. Intermediate levels of service (single on-plot tap) may be suitable for some forms of modified sewerage, but will not allow the use of conventional sewerage.

It should be noted that sewerage systems do not necessarily confer additional benefits to health or groundwater protection over those offered by on-site sanitation. As noted in Chapter 10, sewers often leak and there is significant evidence of their role in causing pollution of groundwater. Decisions on whether on-site or off-site sanitation systems will be used will also depend on cost-benefit analyses of the options, including costs for sewer maintenance, which are beyond the scope of this text. In rural areas of many developed countries, on-site sanitation options remain the most viable solution, e.g. the use of septic tanks is common in the USA (Lerner, 1996). Provided such facilities are properly sited, designed, constructed and maintained they provide a level of service equivalent to a connection to a sewer and represent only a limited risk to groundwater. Some forms of ecological sanitation are also designed for use in developed countries where water consumption is high. Therefore on-site sanitation solutions should not be viewed as limited only to developing countries.

Maintenance of sanitation facilities is an important issue for any technology chosen, and the implementation of routines for inspection and maintenance may be supported by their documentation in management plans of the stakeholders responsible for these facilities.

22.3 MEASURES FOR CONTROLLING RISKS FROM ON-SITE SANITATION

In many developing countries access to water supply and sanitation remains low and there is an urgent need to provide both improved water supply and a safe means of excreta disposal (WHO and UNICEF, 2000). In many rural and peri-urban communities (including poor marginalized communities within urban centres) it is likely that improved access to sanitation facilities will be in the form of on-site sanitation (Mara, 1996; Cotton and Saywell, 1998; ARGOSS, 2001).

In this section some key issues relating to the risks posed to groundwater quality from on-site sanitation and the potential means by which this may be reduced through design (including siting), construction and maintenance are reviewed. It is not intended to provide a detailed description of how to design and construct such facilities, but rather the specific measures that can be used to protect groundwater. For details on design, construction and operation criteria, readers are referred in particular Franceys *et al.* (1992) and in relation to urban areas, Mara (1996).

22.3.1 Siting of on-site sanitation facilities

A key strategy for the control of risks from sanitation systems and human excreta is to ensure that they are sited so that the risk posed to sources of groundwater used for drinking-water supplies is minimized. As a significant source of pathogens, the control of sanitation systems in relation to groundwater sources is a key consideration when defining groundwater protection zones for microbial quality (as discussed in Chapter 18). Sanitation systems should not, by preference, be located within the zone of protection for microbial quality and definitely not in close vicinity to a wellhead or spring (Chapter 17). However, in more densely populated areas it may not be feasible or cost-effective to remove the sources of pollution and therefore engineering improvements to the sanitation facilities may be required.

Site selection of sanitation facilities is an important control measure to protect groundwater from human excreta. In many developing countries, recommendations are often made regarding siting of latrines with respect to groundwater sources. These are often developed separately from (and usually before) groundwater protection zones.

Set-back distance recommendations range from simplistic to more sophisticated approaches based on hydrogeological conditions. Pickford (1995) notes recommendations from India for pit latrines to be located some six m downhill of the nearest water source. Such recommendations should always be treated with some caution. Although the hydraulic gradient of shallow groundwater typically follows the ground surface, it should be borne in mind that where the well is equipped with an electric submersible or other form of pump, there will be a substantial draw-down. Therefore contaminants can be drawn into the well from areas downhill and physical location may not always provide adequate protection.

In many countries, single-distance criteria are used. A distance of 15 m is a commonly used criterion, based on suggestions by Wagner and Lanoix (1958). The weaknesses in using these approaches were highlighted by Lewis *et al.* (1982) who noted that this distance may be overly conservative in some hydrogeological environments (thus limiting health gains from sanitation) and insufficient in other environments with rapid flow rates. Lewis *et al.* (1982) suggested that set-back distances should be established based on local hydrogeological conditions (such as water table depth, nature of unsaturated zone) and the hydraulic load from the latrine.

In South Africa, for example, DWAF (1997) developed a framework for selecting separation distances using contaminant risk assessment based on:

- whether the site for sanitation development overlies a major aquifer;
- the proposed use of groundwater;
- the depth to the water table;
- the type of aquifer;
- presence of existing latrines within 50 m and upgradient;
- evidence of contamination.

Other work in South Africa has recommended an approach that takes into account estimation of pollution risk based on travel time for microbes, mass balance for nitrate and using a probabilistic approach for contamination exceeding specified targets (van Ryneveld and Fourie, 1997).

One problem noted with the definition of set-back distances is that these do not always take into account that different types of technology are likely to have different levels of pollution potential. Pit latrines commonly are significantly deeper than septic tanks and tend to rely on infiltration of leachate through the surrounding soil. Pour-flush latrines have a much higher hydraulic load than dry latrines and as a result have a greater pollution potential. Septic tanks typically receive relatively large volumes of wastewater and therefore if not constructed properly, may lead to a significant hydraulic load and increased pollution potential. This is reduced through ensuring that the tank is watertight and that effluent is discharged into drain-fields and soakaways at a much shallower level.

In addition to set-back distances, hydrogeological data are often used to identify areas where groundwater-fed sources are particularly susceptible to contamination. A widely used method, for example, is known as DRASTIC (Aller *et al.*, 1987; US EPA, 1992) which employs several hydrogeological factors in order to develop an index of the vulnerability of groundwater to contamination (for more detail see Chapter 8.1.4). The method is, however, data intensive and does not include factors that relate specifically to the risk to groundwater posed by sewage and sewage-derived microorganisms. DRASTIC also does not provide any site-specific guidance. The index may, however, provide a worthwhile framework for assigning set-back distances based on different levels of risk. A framework similar to DRASTIC, designed specifically to assess the vulnerability of groundwater to contamination by sewage-derived *Cryptosporidium parvum*, has been developed in the United Kingdom by Boak and Packman (2001).

Recent research in developing countries has attempted to develop guidance without requiring detailed hydrogeological information in order to determine set-back distances for pit latrines (ARGOSS, 2001). Risk assessments are defined for three scenarios (localized microbial contamination; widespread microbial contamination; and widespread nitrate contamination). For separation distances related to microbial quality, decisions are based on a time of travel estimation that includes hydraulic and pollutant loading as well as the attenuation potential and survival of microbes. Set-back distances are estimated from a set of standard tables and figures that have been calibrated with data from field studies and analysis of published works. This approach uses a three-tier approach to risk as shown in Table 22.1 below. Figure 22.1 provides an example of the flow-chart for decision-making; Box 22.1 provides a case study of applying this methodology.

More qualitative approaches to defining separation distances using available data can be used based on statistical analysis of water quality and sanitary inspection data (ARGOSS, 2001; Howard *et al.*, 2003). These have been shown to be robust in supporting water and sanitation planning and offer an effective way to determine siting requirements based on local conditions when there is limited hydrogeological data available.

Table 22.1. Levels of pathogen risk in relation to travel time (adapted from ARGOSS, 2001)

Level of risk	Comments
Significant risk	Travel time under 25 days (breakthrough of both viral and bacterial pathogens in significant numbers possible)
Low risk	Travel time above 25 days (primarily related to the potential for viral breakthrough) but under 50 days
Very low risk	Travel time above 50 days (unlikely to have significant breakthrough of any pathogens, although low risk of viral breakthrough remains)

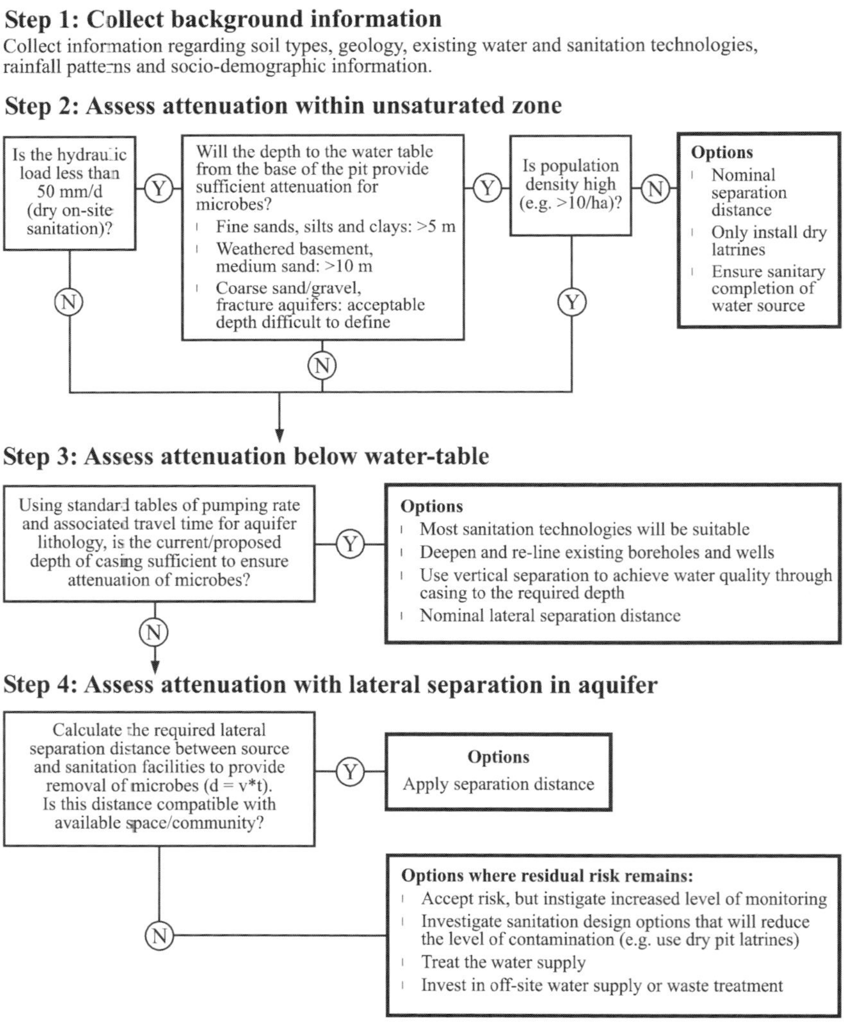

Figure 22.1. Flow chart for assessing the risk of microbiological contamination of groundwater supplies via aquifer pathways where on-site sanitation is being installed (modified from ARGOSS, 2001)

Box 22.1. Calculating the separation distance between latrines and protected springs in Wakiso District, Uganda

A study to determine acceptable separation distances between latrines and water points was undertaken in Nabweru sub-county in the District of Wakiso, Uganda, a rural area with a population density ranging from 9-12 people per ha. The population relies on non-piped sources of water (boreholes, dug wells and springs) and pit latrines for sanitation. The area is hilly with swamps in low-lying areas and most springs are found on the lower slopes.

The soil is a fine sandy silt, with clayey soils found in the swampy areas. The area receives about 1600 mm of rainfall a year in two principal wet seasons (March to June; August/September to November). The depth to water table across the area ranges from 30 m in higher areas (where boreholes and dug wells are found) to 5-8 m close to the protected springs. Pit latrine depths typically range from 6-10 m in depth.

The assessment used the ARGOSS methodology and took into account the different technology types and the geological setting. It was concluded that as dry pit latrines were used, the hydraulic load would be less than 50 mm/day and therefore there was potential for unsaturated zone attenuation. As the depth to water table in higher levels was 20 m below the base of the deepest pits, it was concluded that unsaturated zone attenuation would limit all microbial risks in these areas and a nominal horizontal separation between latrines and water sources would be adequate.

In lower-lying areas, the depth to the water table beneath the base of the pits was only 1-2 m and therefore it was concluded that the desired horizontal separation distance should be calculated.

The horizontal separation distance d was calculated using:

$$d = v*t$$

where v is velocity and t is time. The velocity was calculated by using:

$$v = K_i/\varphi$$

where K is the hydraulic conductivity (permeability), i is the hydraulic gradient, and φ is porosity.

For this area, the permeability and porosity were estimated using the lithology as a guide. A permeability of 5 m/d was selected and a typical porosity for this type of formation would be 0.1. The hydraulic gradient was assumed to be 1/100. The velocity was therefore: (5*0.01)/0.1 = 0.5m/d.

The team used the low-risk travel time (exceeds 25 days) and therefore the required horizontal distance was: 0.5*25 = 12.5 m. Subsequent analysis of the water combined with sanitary inspection confirmed that these distances were effective.

Numerical tools to support siting decisions

Numerical approaches have been developed to determine set-back distances using contaminant transport and groundwater flow models (Bear *et al.*, 1992). Commercial software including numerical codes such as MODFLOW (McDonald and Harbaugh, 1988) and FLOWPATH (Franz and Guiguer, 1990) have received widespread application under a variety of hydrogeological conditions (Cleary and Cleary, 1991; Bair and Roadcap, 1992; Taylor and Howard, 1995). A distinct advantage of numerical models is that the impact of varying different factors such as slope (topography) and pumping (abstraction) rate on time of travel can be evaluated thereby enabling estimation of generic set-back distances.

Apart from the necessary expertise, a key detraction to the use of numerical models is that they require a significant range and input of data including many parameters such as hydraulic conductivity and porosity for which there are often significant uncertainties. Furthermore, most models are run under steady-state conditions that assume uniform hydrogeological conditions even for highly dynamic parameters such as recharge and discharge (pumping). Estimation of set-back distances typically presumes that sewage-derived pathogens move at the same rate as groundwater flow and survive for similar lengths of time in groundwater (e.g. 50 days) regardless of water temperature and chemistry. Approaches based solely on travel time do not take into account that distance may also be important in determining microbial removal. Such gross assumptions are, however, common to all methods of estimating set-back distances, which are largely justified on the basis that this will provide protection against gross contamination and may be applied even where data is limited. As noted in Chapter 18, however, the use of single travel times may not be fully protective and leads to a significant residual risk to public health (Schijven *et al.,* 2002a; 2002b). Developing more complex models taking into account all the factors influencing microbial removal may increase the certainty with which set-back distances are defined at a local level, but will require significant data on local hydrogeological conditions. In some situations the limitations of available data may make collection of additional information valuable or even essential in establishing guidelines for siting of on-site sanitation and water sources. There are a number of ways of acquiring this information, including tracer tests, soil percolation tests and targeted water quality studies.

Tracer studies often provide invaluable information regarding specific characteristics of a particular site to aid siting conditions. They may also be used to develop an understanding of aquifer types and can provide useful data regarding regional groundwater flow rates and the likelihood of preferential pathways. When the principal concern is to understand groundwater flow, conservative, non-hazardous chemical tracers can be used (for instance chloride or non-hazardous dyes). Chemical tracers may be of little value when determining whether local conditions will lead to breakthrough of pathogens. If this is of interest, microbial tracers (e.g. bacteriophages; see also Box 17.1 in Chapter 17) will provide a better indication of whether there is likely to be significant attenuation within the environment being studied and thus provide greater information for the siting of on-site sanitation. This may be important when determining whether a significant risk exists from latrines located close to a water source.

Drangert (2000) refers to tracer studies in Kenya that showed where latrine pits reached the saturated zones bacteria travelled 20-30 m within a week and in areas of pronounced topography bacteria could travel distances of up to 100 m within a day. This suggests that in this case, either the water source or sanitation technology should have been changed. By contrast, in tracer studies in Uganda, breakthrough by conservative chemical tracers occurred within 24 hours, whilst no phage tracers were detected over the 120 hour sampling period, suggesting that there was greater protection against microbial contaminants than would be suggested from groundwater flow rates (Taylor *et al.*, 2004). Soil percolation tests may also provide some information regarding how rapidly effluent may infiltrate into the soil and this may be used to aid planning of siting and sizing of drainfields and soakaways. These tests will, however, provide very limited information for pit latrines, unless the test is performed at the likely depth of the latrine.

Where there is limited hydrogeological data, the use of well-designed water quality studies can provide information regarding appropriate set-back distances. This may be done by selecting a representative sample of water supplies and undertaking an assessment of water quality over a period of time, combined with detailed assessments of latrine proximity. The assessment should ensure that all seasonal differences will be covered and consideration may need to be given for more frequent sampling during the onset of the rains when water quality often shows greatest deterioration (Howard *et al.*, 2002). The best way to analyse this data is to compare latrine distance against water quality targets to determine at which distance the latrines start to exert an influence.

22.3.2 Engineering design to control pollution in high-risk areas

Where pit latrines must be used and risks of aquifer contamination are high, control measures addressing design and construction of the sanitation facilities are important to reduce risk. High-risk areas will include those areas where the water table is high or where there are very rapid groundwater flow rates (for instance fracture aquifers, gravel and aquifers with preferential flow paths).

Where there is limited space between the base of the pit and the water table, the use of sand envelopes around the base and sides of the pit are often recommended as this will help encourage an active biological community to reduce breakthrough of pathogens (Franceys *et al.*, 1992). These recommendations are based on original field and laboratory experimentation by Coldwell and Parr (1937) and later by Ziebell *et al.* (1975). The former found that a 0.25 m envelope of sand provided an effective barrier to thermotolerant coliform movement. However, although this provides confidence in control of bacterial contamination, confidence in control of viral pathogens is more limited. Ziebell *et al.* (1975) found that development of biological communities within sand envelopes took up to 100 days, suggesting an initial period of elevated risk during the first use of the latrine.

Whether the use of sand envelopes is economically feasible is questionable, for instance Pickford (1995), commenting on recommendations for a 0.5 m sand envelope in India, notes that this would lead to a four-fold increase in the volume of the excavation of a 1 m diameter pit. In such circumstances using an alternative source of water, treating the water or using an alternative sanitation design may be more cost-effective.

Where water tables are high, the use of dry latrines may be more appropriate than wet latrines, as the latter have significantly greater hydraulic loads which may exceed attenuation potential even when enhanced by the addition of sand envelopes. If wet latrines will be used because of cultural practices or preferences, then siting will be critical and consideration should be given to the use of alternative water supplies or treating existing supplies.

Vault latrines

The risks from on-site sanitation in high water table areas can also be reduced by using vault latrines (also known as cesspits). A vault latrine has a watertight lining on the pit that does not allow significant infiltration into the surrounding soil and which requires periodic emptying (Franceys *et al.*, 1992). An example of a vault latrine is shown in Figure 22.2 below.

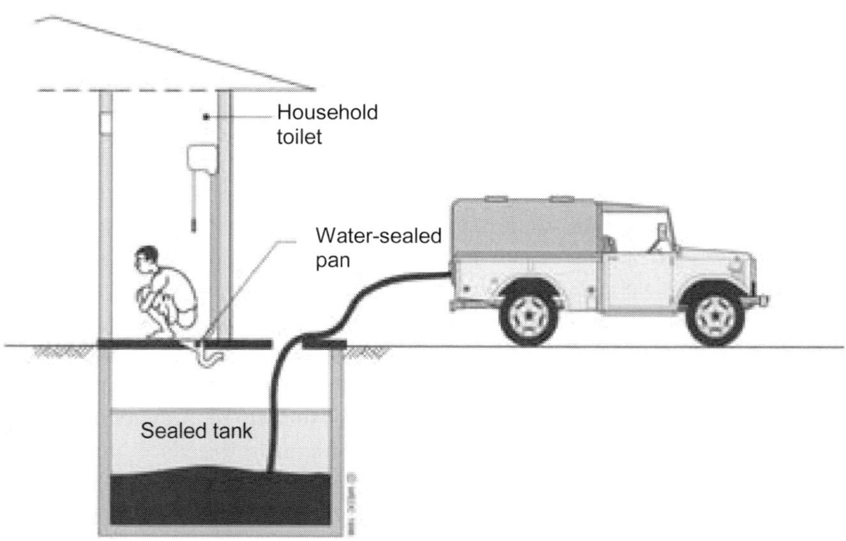

Figure 22.2. Vault latrine

Most designs retain excreta for a short period of time. Because of the volume of liquid collected in the tank, emptying is typically required every two to three weeks. Experience reported from Japan and Korea suggest that vault and vacuum systems are cost-effective and wastes can be managed properly (Pickford, 1995; Ahmed and Rahman, 2000). The use of vault latrines remains limited in many developing countries, primarily because of the high operating costs associated with emptying. Some authors suggest that they are not appropriate as construction and operation are often poor resulting in significant health risks to users and the wider community (Franceys *et al.*, 1992). However, designs continue to be developed and tested successfully (Dadie-Amoah and Komba, 2000). Proper construction can be supported by establishing and enforcing construction standards to ensure that vault latrines do not still represent a risk to groundwater. Proper operation, emptying and maintenance at adequate intervals can be supported by a management plan that describes these activities and the responsibilities for them.

A variety of low and high-cost emptiers are available and have been tested. For instance, a programme in the Kibera slum in Nairobi has shown that a small pedestrian-operated tanker (the vacutug) provided an effective service (Wegelin-Schuringa and Coffey, 2000). Similar equipment (the MAPET) was used in Dar-es-Salaam, Tanzania (Howard and Bartram, 1993; Wegelin-Schuringa and Coffey, 2000) and other equipment has been used for pit latrines in Zimbabwe (Jere *et al.*, 1995).

The final disposal of sludge should be properly managed. In Kibera, disposal of the sludge is into sewers or direct to sewage treatment works. However, enforcement is often important as previous experience in Dar-es-Salaam showed that there were concerns about burying of the sludge on-site or close to homes, with consequent risks of soil and water contamination and vector breeding problems (Howard and Bartram, 1993).

The lining of the vault pits is often susceptible to damage during the emptying process and it is important to ensure that the power of the suction pump used to empty the pit contents will not cause damage to the linings. Inspections of lining integrity are an important control measure but are difficult to enforce in many settings. They require trained inspectors with appropriate safety procedures, protective clothing and masks.

In some areas of very high water tables, pit latrines are raised to about ground level (Mara, 1996). These are typically expensive to construct and require regular maintenance and, as noted by Franceys *et al.* (1992), care should be taken to avoid seepage of effluent at or above ground level. In some situations mound latrines are constructed with a hole and bung inserted into the pit to allow rapid emptying without the need for suction. This is acceptable if this is designed and operated correctly and has fittings to allow a pipe to be connected to a tanker. In many cases the hole and bung are poorly designed or emptying is done by households simply by opening the bung during in periods of rainfall with the pit contents discharged to the nearest storm water drain. This has been noted to occur, for instance, in low-lying areas of Kampala, Uganda. Such practices represent a significant risk to health, may cause groundwater contamination and should be avoided.

Ecological sanitation

There is increasing interest among some workers in the sanitation and groundwater sector in the use of ecologically sustainable sanitation that promotes the re-cycling of nutrients for use as fertilizers in agriculture or gardening and to avoid water pollution with nitrate and phosphorus. Ecological sanitation as it is commonly known can cover a wide range of technical options from very high to comparatively low cost. The designs can either involve separation of urine or composting and in all cases the reuse of wastes is promoted as a means of ensuring nutrients are recycled in the environment (Drangert, 2000; Winblad, 2000; Esrey, 2001).

The use of on-site methods of eco-san are likely to beneficial in relation to preventing groundwater contamination because they have a low hydraulic load. Although there are many ardent advocates of ecological sanitation, this is not a technology with zero groundwater pollution potential nor does it provide overall greater benefits than alternative options. The final disposal of wastes may still lead to problems if it is not managed properly; and the use of organic fertilizers does not mean, for instance, an immediate removal of problems with nitrate contamination. The disposal of wastes may be more problematic in urban areas where there may be more limited local demand for use in agriculture, leading to poor disposal methods and public health risks.

There is also concern regarding the safety of handling the matter and ensuring that there has been inactivation of pathogens. Of particular importance is not whether recommended practices limit such risks, but actual practice by the users in relation to their exposure to raw or poorly treated faecal material.

22.4 MEASURES FOR CONTROLLING RISKS FROM SEPTIC TANKS AND AQUAPRIVIES

A key element in both septic tanks and aquaprivies is that the tanks should be watertight to prevent subsurface leaching of contaminants into groundwater. Groundwater contamination tends to result from infiltration of effluent through drainfields, trenches and pits, the density of septic tanks in an area or poor siting in relation to vulnerable groundwater sources (Payne and Butler, 1993).

The control of the location and density of septic tanks involves ensuring that planning and development legislation takes due account of the groundwater protection needs. Payne and Butler (1993) provide a series of simple flow charts to aid planners in the United Kingdom decide whether applications for septic tank construction should be approved.

As septic tanks and aquaprivies must be emptied periodically, control of the final sludge is essential. Uncontrolled, illegal emptying or dumping of sludge into the environment may represent a significant risk to groundwater. One form of control is through the licensing of de-sludging companies and enforcing codes of practice. Management plans defining these operations, responsibilities and documentation of emptying may support good practice and surveillance.

Legislation should also include penalties for households that illegally empty their tank, although the monitoring of this may present some difficulties.

Effluents are most often disposed of through a soakaway or drainfield containing infiltration trenches (Franceys *et al.*, 1992). Soakaways require less space but infiltration trenches have several advantages, such as increased area for infiltration (through the sides of the trench), the potential for shallower construction (increasing the travel path to any underlying aquifer) and the ability to distribute flows along the trench, providing alternative infiltration routes, unlike small soakaways which may eventually become clogged. Soakaways may be either underground chambers with openings in the base and sides to allow infiltration to the soil or an underground pit filled with coarse granular material, often lined with a geotextile to prevent ingress of silt. Where underground chambers are used, surrounding the structure with sand may provide additional potential for the removal of pathogens.

Infiltration trenches are normally filled with coarse granular material with a perforated pipe in the base of the trench to distribute the effluent along it, avoiding overloading the soil at any one particular point. Designs for drain trenches should take into account infiltration rates, which should preferably be calculated on-site, but the values in Table 22.2 can be used as a general guide.

Table 22.2. Infiltration capacity of different soil types (based on Franceys *et al.*, 1992)

Type of soil	Infiltration capacity – settled sewage (l per m^2 per day)
Coarse or medium sand	50
Fine sand, loamy sand	33
Sandy loam, loam	25
Porous silty clay and porous clay loam	20
Compact silty loam, compact silty clay loam and non-expansive clay	10
Expansive clay	<10

The performance of the drainage trenches depends on the efficiency of the tank and soil conditions and some workers suggest separating sullage from toilet wastes and treating this separately (Franceys *et al.*, 1992). Alternatively the performance of the septic tank or aqua privy can be enhanced by extending the treatment process through increasing the size of the tank and introducing baffles to promote sedimentation.

22.5 MEASURES FOR PREVENTION AND CONTROL OF SEWER LEAKAGE

The easiest, cheapest, most straightforward way to manage pollution from sewers is to not let it happen in the first place. Misstear *et al.* (1996) note that leaks from sewers arise from a combination of:

- cracked and fractured pipes
- opened or displaced pipe joints
- root intrusion
- pipe deformation
- sewer collapse
- reverse gradients
- siltation
- blockages
- poorly constructed connections
- abandoned laterals left unsealed.

These may be due to ground loss, ground movement, material deterioration or poor system management. Control measures to protect groundwater from sewer leakage therefore include planning, design and construction as well as maintenance and operational controls.

The design of the sewerage system can reduce risks of leakage. Good workmanship, careful inspection of the pipes while they are being laid, the use of good quality pipe and bedding materials and careful compaction all improve the life and physical condition of the pipe. Testing the pipes after they have been laid can identify poor construction practices. Using shorter 'rocker' pipes in areas where there might be settlement allows the pipes to move rather than crack.

Pipes and pipe bridges on or above the ground need to be protected against physical damage, e.g. from accidental collisions. Burying the pipe gives a degree of protection and support. A compromise has to be made between giving the pipe sufficient cover to protect it from surface loading and burying it deeply which makes the sewer more difficult to monitor and replace, and which reduces the travel time of any leaks to the aquifer. In areas where a shallow buried pipe may be subject to damage because of heavy loads passing over the sewer route, the bedding can be replaced with concrete (either just below the pipe or completely encasing it). Routing the pipes in areas where they will not be subject to heavy loads will also increase their robustness.

If the surrounding ground is relatively impermeable, the bedding material in the pipe trench itself can become a drainage route for any leakage, especially if the pipes are laid on a slope. Placing a clay collar around the pipe can prevent the flow of wastewater along the pipe route, as shown in Figure 22.3.

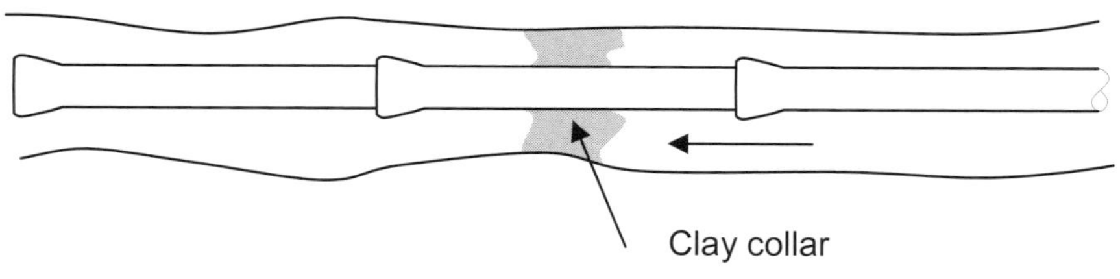

Figure 22.3. Clay collar on sewer

22.5.1 Sewer management

Ensuring that sewers do not become surcharged (i.e. flow full) reduces the possibility of sewage leaking out of the pipe. This can be achieved by ensuring the pipe is large enough for future flows, having separate pipes for surface water drainage, periodically cleaning pipes, monitoring pipes for blockages and removing them promptly, and public education, to prevent unsuitable objects being introduced into the sewer.

Ensuring that connections are made to the pipe either at manholes or using specific lateral connection pipes is preferable to using pipe saddles to connect to existing pipes, as these are more prone to leakage. Where connections are no longer required, these need to be plugged to prevent leakage.

Action can also be taken on what goes down the sewer. Charging for industrial waste can encourage users to reduce the amount of wastewater they produce or cause them to pre-treat the effluent before it enters the sewer. Some substances can be banned from being disposed of into sewers. This action will depend on a working regulatory system.

Hydrogen sulphide formation can represent a significant problem for sewers, leading to conversion to sulphuric acid which can cause corrosion of sewer pipes. Hydrogen sulphide can also lead to odour problems. Metcalf and Eddy (1991) note that hydrogen sulphide attack can be controlled by:

- controlling organic and sulphur inputs at source
- aerating the sewage
- adding chemicals
- periodic cleaning
- ventilation
- good design.

Routines for sewer management, such as monitoring for leaks and blockages, cleaning and maintenance, responsibilities and documentation are best laid down in management plans.

22.5.2 Controlling exfiltration

If exfiltration is suspected, temporary or permanent sumps can be dug into the pipe bedding to collect any leakage and monitor changes in flows. Trying to monitor flows at the end of the pipe would necessitate careful planning, as the complex and changing relationship between sewer inflows, infiltration, exfiltration and overflows can make firm conclusions difficult to arrive at. Large rises in flows during wet weather can indicate infiltration to a foul sewer or misconnections from storm water sewers, as can high night-time flows that do not correspond to water consumption.

Visual surface inspections of the sewer route may reveal major leaks (especially in a dry season) whilst a CCTV survey can give a more detailed analysis. New developments are bringing in lasers and ultrasound to improve inspection, especially below the water flowing in a live sewer (Makar, 1999). Such inspection techniques however are expensive and only provide a snapshot of the state of the sewer. CCTV and other such methods are not suitable for the small diameter pipes used in condominal sewerage systems. Flow measurements can be made at successive points along a sewer line. Assuming no inflows, an assessment can be made of leakage from the sewer. This may vary with seasons.

For public sewers possible strategies are the adoption of a double skin construction technique (although this may not prove cost effective) and the adoption of higher testing pressures. It is also important to include private sewers in a regular schedule of sewer testing especially in areas of high groundwater vulnerability. Misstear *et al.* (1996) recommend installation of shallow monitoring boreholes in environmentally vulnerable areas perceived to be at risk, as well as improved monitoring of the water supply boreholes themselves, measuring microbial indicators, nitrogen species, boron and phosphate.

22.5.3 Control of sewer leakage

If sewers are found to be leaking, but the pipes are still structurally sound, there are alternatives to replacing large amounts of the system. These include relining the pipes – either through spraying a cement coating onto the inside of the pipe or by inserting a flexible plastic liner into the pipe that can then be fixed to the walls. Localized leaks (from cracks or failed joints) can be grouted remotely to seal the hole. Where pipes are no longer structurally sound, the existing pipes can be burst and replaced with a new plastic pipe without having to dig a new trench.

22.5.4 Open drains

Sewers are not just limited to pipes below ground. Open channels may also be used. These may be unlined, pitched with stone or lined with concrete. Where the lining is not water tight, the channel can act as an infiltration trench. This may be acceptable for surface run-off, but is not recommended for conveyance of foul sewage. Surface run-off can flood the channel, causing pollution, which is a greater risk to health than the indirect route of infiltration to groundwater. Where open channels are used for foul sewage and no alternative is possible, they should be routed away from populated areas and have raised sides to limit the ingress of rainwater.

22.5.5 Surface water management

Whilst rainwater is perceived as being clean it can rapidly become polluted and transport contaminants, for example washing poorly disposed faeces into the water cycle. Flooding can cause sewer surcharge and inundate pit latrines. Therefore attention needs to be paid to the management and disposal of stormwater, as soakaways and ponds may provide a path for contaminants to the aquifer. CIRIA (2000) provides an indication of methods used to manage storm water run-off depending on the land use of the catchment and the status of the groundwater.

22.6 CONTROL MEASURES FOR SEWAGE TREATMENT

Where sewerage systems exist they need to be connected to some form of sewage treatment and disposal which protects public health from exposure both directly and through contaminating groundwater. Options range from conventional plants over alternative systems such as waste stabilization ponds and wetland systems. The use of excreta and wastewater may be a desirable option, but may also represent both a direct risk to health and a risk of groundwater contamination (WHO, 1989; Foster *et al.*, 1994; Mara, 1997). Planning and designing wastewater and excreta use should therefore relate to aquifer vulnerability. One control measure is banning or restricting wastewater irrigation within drinking-water protection zones, unless the quality of the treated wastewater is sufficient as to represent little or no risk. Controls in application and means of irrigation and use of fish ponds are discussed in Chapter 21.

22.6.1 Conventional treatment works

Sewage treatment plant outfalls chiefly pollute surface waters. Risks to groundwater will derive from the final disposal of the sludge and its location with regard to groundwater, as well as from hydraulic connections between groundwater and surface waters affected by sewage. Leaking sewage pipes can also contaminate groundwater sources.

Sludge disposal to land or landfill is widely practised and in many countries is subject to specific regulations governing acceptable practice. These will usually govern the conditions under which different types of sludge (untreated, partially treated, treated) can be disposed of and set allowable concentrations of contaminants in sludge quality guidelines and water quality objectives (WRc, 1992; US EPA, 1995a; 1995b). The risk to groundwater from land application of sludge is generally related to aquifer vulnerability and is considered to be particularly relevant to applications close to a groundwater abstraction point (Wolstenholme *et al.*, 1992) (see also Chapter 21).

Various tools have been developed to aid decision-making based on regulations. For instance in South Africa, an expert system for use in sewage sludge has been developed called Sludge Land Application Decision Support which aids decision-makers in establishing sludge disposal guidelines (Anonymous, 1997). This requires significant amounts of data on contaminant loads as well as soil and groundwater conditions. Similar guidance has been developed by the US EPA (1995a; 1995b) which provide detailed discussion of the site conditions with respect to sludge type and quality, soil and hydrogeology.

The development of site-specific plans is essential for effective management of sludge. In general, sludge disposal to land should not occur within an area where groundwater is (or needs to be) protected against microbial contaminants, and sludge disposal to land in areas protected against nitrate contamination should also be restricted.

For effluents, quality control will be achieved through application of appropriate discharge consents and water quality objectives for receiving waters, which will typically be set taking into account uses of the water (including environmental/ecological demands). This should also take into account the nature of surface water-groundwater relationships and in particular to ensure that in situations where surface water recharges groundwater, the potential for groundwater pollution is addressed.

Sewage discharges into surface waters that recharge groundwater should be controlled and discharge consents applied that will be consistent with the protection of groundwater quality. This often results in the extension of groundwater protection zones for significant distances along rivers upstream of abstraction points.

22.6.2 Waste stabilization ponds and reedbeds

Waste stabilization ponds are widely used in developing countries, but are also used in industrialized countries where there is sufficient land available (Horan, 1990; Mara, 1997). Waste stabilization pond systems are usually composed of a series of ponds with sludge treatment carried out in the first pond (which may be anaerobic or facultative), subsequent treatment through secondary facultative ponds (taking settled sewage) and a series of maturation ponds. Well-operated waste stabilization ponds produce very high quality effluent that should represent limited risk to health when discharged and often represent significantly lower operating costs than alternatives (Mara, 1997).

The risks to groundwater from waste stabilization ponds arise primarily from the risks of leaching of wastewater from the base of the ponds leading to microbial or chemical (particularly nitrate) contamination of underlying groundwater. The base of waste stabilization ponds are normally compacted to provide an aquitard layer to ensure the pond will fill with water and to prevent leakage into the subsurface. A specific lining is often also installed; conventional approaches use a puddled clay layer of 5-10 cm thickness, although polyethylene and vinyl sheeting have been used in smaller ponds (WHO-EMRO, 1987).

Although some experts believe the risk to groundwater is be restricted solely to those situations when ponds are located close to abstraction points, infiltration of up to 20 mm per day from ponds has been noted to occur in Latin America and the Caribbean (Foster *et al.*, 1993). In Lima, Peru, penetration of indicator bacteria beneath waste stabilization ponds of over 15 m has been noted, although the majority of microorganisms were removed in the top 3 m of the unsaturated zone (Geake *et al.*, 1986). This indicates the need for proper site investigations and ensuring that ponds are not located over significant aquifers used for domestic water supply.

Reedbeds are also used to treat effluent prior to discharge and can provide significant improvements in wastewater quality. The major problem for groundwater will be leaching of contaminants into the sub-surface, and studies on this issue are limited. Contamination may be significant when the wetland is influent to groundwater, but limited when the groundwater is influent to the wetland. The risk to groundwater can be controlled by locating reedbeds with sufficient distance to abstraction points.

22.7 MONITORING AND VERIFICATION OF MEASURES CONTROLLING SANITATION SYSTEMS

Measures for controlling human excreta disposal and sewerage in drinking-water catchments proposed above range from planning the choice of site and sanitation technology in relation to aquifer vulnerability and socioeconomic criteria to specific design and construction criteria for sanitation facilities and operational controls. Selected examples are summarized below in Table 22.3. Where drinking-water supply and sanitation are a joint responsibility of a utility or community, such measures may readily be integrated into a WSP.

NOTE ▶ *The implementation of control measures such as those suggested in Table 22.3 is effectively supported if the stakeholders involved collaboratively develop management plans that define the control measures and how their performance is monitored, which corrective action should be taken both during normal operations and during incident conditions, responsibilities, lines of communication as well as documentation procedures.*

The implementation of control measures protecting drinking-water aquifers from industry, mining and military activities is substantially facilitated by an environmental policy framework (see Chapter 20).

Supporting programmes specifically for protecting groundwater from human excreta may include but are not restricted to the development of guidance on sanitation technology selection, ensuring that construction quality standards are developed and enforced, and community education regarding items and substances disposed of in sewers and latrines.

Monitoring and verification of the control measures implemented is crucial to ensure that they are in place and are effective. Table 22.3 therefore includes options for surveillance and monitoring of the control measure examples given. Most of these focus on checking whether the controls are operating as intended, rather than on contaminant concentrations in groundwater. In the context of planning, surveillance will address whether plans exist and how they took criteria for siting and set-back distances in relationship to aquifer vulnerability into account. Verification will address whether these criteria are being adhered to and plans are being implemented, e.g. by site inspection to check the location of sanitary facilities and the depths of pits and sewers during their construction.

Once the planning stage has determined safe site selection and sanitation facilities, monitoring their design and construction is important to ascertain their integrity, e.g. whether latrine pits are fitted with liners or sewer pipes with clay collars. For the day-to-day routine operation of control measures, monitoring focuses on assessing whether they are functioning as they should, e.g. whether sewers or pits are leaking, whether latrines are in a condition to be used and whether sewage sludge is disposed of as approved by the permit (see Table 23.1). This can in part be achieved through inspection, and defining inspection routines in management plans is useful to support that they are regularly performed. Monitoring for leaks often requires analysing a selected parameter in groundwater near the respective sanitation facility (e.g. latrine pit or sewer) which will most effectively indicate leaks. This may be achieved with indicator organisms (e.g. faecal streptococci, *E. coli*, bacteriophages) or substances typically present in the sewage. Where sewerage is contaminated by a range of chemicals from household use and connected enterprises, monitoring overall groundwater safety would occasionally address these contaminants. This also applies to sludge contaminants of specific concern where it is applied on land.

Management plans will define monitoring systems, corrective action to be taken if leaks are detected as well as regular maintenance operations. To be effective, they include responsibilities and routines for documentation of these activities, their findings and any corrective action taken.

NOTE ▶ *Options for monitoring suggested in Table 22.3 focus on the control measures rather than on groundwater quality. Analysis of selected parameters in groundwater which indicate leaks of on-site or sewer systems is suggested where this is the most effective operational control.*

Comprehensive groundwater quality monitoring programmes are a supplementary aspect of monitoring with the purpose of providing verification of the efficacy of the overall drinking-water catchment management.

Table 22.3. Examples of control measures for sanitation systems and options for their monitoring and verification

Process step	Examples of control measures for sanitation systems	Options for their monitoring and verification
PLANNING	Require set-back distances for sanitation facilities in relation to travel time to aquifer, as adequate in local hydrogeological conditions	Review (applications for) permits for construction of new on-site sanitation systems or sewers Inspect protection zones to ensure that set-back distances are implemented
	Locate sewers outside drinking-water protection zones	Inspect sewer laying and pit construction to verify that safe distances are implemented
	Ensure sufficient distance (at least 2 m) between base of latrine pit, soakaway or infiltration trench and highest water table	Conduct tests with tracers and/or indicator organisms to verify adequate siting
	Require permits for sludge disposal or reuse options based on an assessment of aquifer vulnerability and contaminants	Review (applications for) permits for sludge disposal Conduct tests with tracers and/or indicator organisms to verify safety
DESIGN AND CONSTRUCTION	Construct and maintain vault latrine pits impermeable Fit sewers with linings to reduce breakage Fit waste stabilization ponds with linings	Inspect during construction Carry out tracer tests Monitor selected groundwater parameters (indicator organisms, substances typically occurring in the sewage) which would indicate leakage
OPERATION AND MAINTENANCE	Maintain on-site sanitation facilities in good condition and encourage use	Inspect regularly
	Maintain condition of clay collars on sewers	Inspect regularly
	Prevent sewer leakage	Run sewer leak detection programmes Review records of sewer leak detection and repairs
	Implement adequate final disposal of sludge as designated	Inspect disposal practices

22.8 REFERENCES

Ahmed, F. and Hossain, M.D. (1997) The status of water supply and sanitation access in urban slums and fringes of Bangladesh. *J. Water Supply Res. Technol. – Aqua*, **46**(1), 14-19.

Ahmed, M.F. and Rahman, M.M. (2000) *Water Supply and Sanitation: Rural and low-income urban communities*, IRC, Netherlands.

Aller, L., Bennet, T., Lehr, J.H., Petty, R.J. and Hackett, G. (1987) *DRASTIC: A standardised system for evaluating ground water pollution potential using hydrogeologic settings*. US EPA Report No. EPA-600/2-87-035, Dublin.

Anonymous (1997) *Permissible Utilisation and Disposal of Sewage Sludge*, Water Research Commission, Pretoria, South Africa.

ARGOSS (2001) *Guidelines for Assessing the Risk to Groundwater From On-site Sanitation*, BGS Commissioned Report CR/01/142, BGS, Wallingford.

Bair, E.S. and Roadcap, G.S. (1992) Comparison of methods used to delineate capture zones of wells: 1. leaky-confined fractured- carbonate aquifer. *Ground Water*, **30**(2), 199-211.

Bear, J., Beljin, M.S. and Ross, R.R. (1992) *Fundamentals of Ground-Water Modelling*, US EPA, Oklahoma.

Boak, R.A. and Packman, M.J. (2001) A methodology for the assessment of risk of *Cryptosporidium* contamination of groundwater. *Q. J Eng. Geol. Hydrogeology*, **34**, 187-194.

Cairncross, S. and Feachem, R. (1993) *Environmental Health Engineering in the Tropics: An introductory text*, 2nd edn, John Wiley and Sons, Chichester.

CIRIA (2000) *Sustainable Urban Drainage Systems – design manual for England and Wales*. CIRIA Report C522, CIRIA, London.

Cleary, T.C.B.F. and Cleary, R.W. (1991) Delineation of wellhead protection areas: theory and practice. *Water Sci. Tech.*, **24**(11), 239-250.

Coldwell, E.L. and Parr, L.W. (1937) Groundwater pollution and the bored hole latrine. *J. Infectious Diseases*, **61**, 148-183.

Cotton, A. and Saywell, D. (1998) *On-plot Sanitation in Low-income Urban Communities: Guidelines for selection*, WEDC, Loughborough University.

Dadie-Amoah, K. and Komba, S. (2000) Volta watertight VIP latrine. In *Water, Sanitation and Health*, (eds. I. Chorus, U. Ringelband, G. Schlag and O. Schmoll), pp. 361-365, IWA, London.

Drangert, J-O. (2000) Re-use - the ultimate sink? Urine-diverting toilets to protect groundwater quality and fertilise urban agriculture. In *Water, Sanitation and Health*, (eds. I. Chorus, U. Ringelband, G. Schlag and O. Schmoll), pp. 275-280, IWA, London.

DWAF (1997) *A Protocol to Manage the Potential of Groundwater Contamination From On-site Sanitation.* National Sanitation Co-ordination Office, Department of Water Affairs and Forestry, Pretoria.

Esrey, S. (1996) Water, waste and well-being: a multi-country study. *Am. J. Epidemiology*, **143**(6), 608-623.

Esrey, S.A. (2001) Towards a recycling society: ecological sanitation - closing the loop to food security. *Water Sci Technol.*, **43**(4),177-87.

Esrey, S.A., Potash, J.B., Roberts, L. and Shiff, C, (1991) Effects of improved water supply and sanitation on ascariasis, diarrhoea, dracunculiasis, hookworm infection, schistosomiasis, and trachoma. *Bulletin of the WHO*, **69**(5), 609-621.

Foster, S.S.D., Adam, B., Morales, M. and Tenjo, S. (1993) *Groundwater Protection Strategies: A guide to implementation*, PAHO-CEPIS, Lima.

Foster, S.S.D., Gale, I.N. and Hespanhol, I. (1994) *Impacts of Wastewater Use and Disposal on Groundwater*, BGS Technical Report WD/94/55, BGS, Wallingford.

Franceys, R., Pickford, J. and Reed, R. (1992) *A Guide to the Development of On-site Sanitation*. WHO, Geneva.

Franz, T. and Guiguer, N. (1990) *FLOWPATH, Version 4, Steady-state two-dimensional horizontal aquifer simulation model*. Waterloo Hydrogeologic (Waterloo).

Geake, A.K., Foster, S.S.D., Nakamatsu, M., Valenzuela, C.F. and Valverde, M.L. (1986) *Groundwater Recharge and Pollution Mechanisms in Urban Aquifers of Arid Regions*, BGS Hydrology Research Report 86/11, BGS, Wallingford.

Horan, N.J. (1990) *Biological Wastewater Treatment Systems*, Wiley and Sons Ltd, Chichester.

Howard, G. and Bartram, J. (1993) *Infrastructure and Training Needs for Sustainable Urban Sanitation*, Robens Institute and Commonwealth Secretariat, Guildford.

Howard, G., Nalubega, M., Barrett, M., Pedley, S., Kulabako, R., Haruna, R., Johal, K. and Taylor, R. (2002) Impact of on-site sanitation on groundwater supplies in Kampala and Iganga, Uganda. In *Assessing Risk to Groundwater From On-site Sanitation: Scientific review and case studies*. BGS Commissioned Report CR/02/079N. BGS, Wallingford, pp. 69-98.

Howard, G. Pedley, S., Barrett, M., Nalubega, M. and Johal, K. (2003) Risk factors contributing to microbiological contamination of shallow groundwater in Kampala, Uganda. *Water Res*., **37**(14), 3421-3429.

Hutton, G. and Haller, L. (2004) Evalution of the costs and benefits of water and sanitation improvements at the global level. World Health Organisation, Geneva, 87 pp.

Jere, M.J., Dzotizei, A.T. and Munjoma, M. (1995) Pit latrine emptying using motorised equipment. In Sustainability of Water and Sanitation Systems. In *Proc. 21st WEDC Conference, Kampala, Uganda*, (ed. J. Pickford), pp. 220-222, WEDC, Loughborough University.

Lerner, D.N. (1996) Urban groundwater an asset for the sustainable city. *European Water Control*, **6**, 43-51.

Lewis, J.W., Foster, S.S.D. and Drasar, B.S. (1982) *The Risk of Groundwater Pollution by On-Site Sanitation in Developing Countries*. International Reference Centre for Wastes Disposal (IRCWD), Duebendorf.

Makar, J.M. (1999) Diagnostic techniques for sewer systems. *J. Infrastructure Systems*, **5**(2), 69-78.

Mara, D. (1996) *Low-cost urban sanitation*. Wiley and Sons Ltd, Chichester.

Mara, D. (1997) *Design Manual for Waste Stabilisation Ponds in India*. Lagoon Technology International, Leeds.

McDonald, M.G. and Harbaugh, A.W. (1988) *A Modular Three-Dimensional Finite-Difference Ground-Water Flow Model*. U.S. Geological Survey Techniques of Water Resources Investigations TWRI 6-A1, 575 p.

Metcalf and Eddy, Inc. (1991) *Wastewater Engineering: Treatment, disposal, and reuse*, 3rd edn, (revised by G. Tchobanoglous and F.L. Burton), McGraw-Hill, New York.

Misstear, B., White, M., Bishop, P. and Anderson, G. (1996) *Reliability of Sewers in Environmentally Vulnerable Areas*, CIRIA Project Report 44, CIRIA, London.

Payne, J.A. and Butler, D. (1993) *Septic Tanks and Small Sewage Treatment Works: A guide to current practice and common problems*. CIRIA Technical Note 146, CIRIA, London.

Pickford, J. (1995) *Low-cost Sanitation: A review of practical experience*. IT Publications, London.

Schijven, J.F., Berger, P. and Miettinen, I. (2002a) Removal of pathogens, surrogates, indicators and toxins using bank filtration. In *A Monograph on Riverbank Filtration*. (ed C. Ray), University of Hawaii, Kluwer Academic.

Schijven, J.F., Hassanizadeh, S.M. and de Bruin, H.A.M. (2002b) Two-site kinetic modeling of bacteriophages transport through columns of saturated dune sand. *J. Contaminant Hydrology*, **57**, 259-279.

Taylor, R.G. and Howard, K.W.F. (1995) Averting shallow well contamination in Uganda. In Sustainability of water and sanitation systems, In *Proc. 21st WEDC Conference*, Kampala, pp. 62-65.

Taylor, R.G., Cronin, A.A., Pedley, S., Atkinson, T.C. and Barker, J.A. (2004) The implications of groundwater velocity variations on microbial transport and wellhead protection: review of field evidence. *FEMS Microbiology Ecology*, **49**: 17-26.

US EPA (1992) *Handbook: Ground water and wellhead protection*. US EPA Report No. EPA/625/R-94/001, Washington, DC.

US EPA (1995a) *Process Design Manual: Surface disposal of sewage sludge and domestic septage*. US EPA, Washington, DC.

US EPA (1995b) *Process Design Manual, Land application of sewage sludge and domestic septage*. US EPA, Washington, DC.

van Ryneveld, M.B. and Fourie, A.B. (1997) A strategy for evaluating the environmental impact of on-site sanitation systems. *Water SA*, **23**(4), 279-291.

Wagner, E.G. and Lanoix, J.N. (1958) *Excreta Disposal for Rural Areas and Small Communities*. WHO, Geneva.

Wegelin-Schuringa, M. and Coffey, M. (2000) A small pit-emptying machine: an appropriate solution for a Nairobi slum. In *Water, Sanitation and Health*, (eds. I. Chorus, U. Ringelband, G. Schlag and O. Schmoll), pp. 347-352, IWA, London..

WHO (1989) *Guidelines for the Safe Use of Wastewater, Excreta and Greywater in Agriculture, Volume 2*, World Health Organization, Geneva.

WHO and UNICEF (2004) *Meeting the MDG drinking water and sanitation target: a mid-term assessment of progress*. Geneva and New York.

WHO-EMRO (1987) *Wastewater Stabilization Ponds: Principles of planning and practice*. WHO-EMRO Technical Publication No. 10, WHO-EMRO, Alexandria.

Winblad, U. (2000) Sanitation without water In *Water, Sanitation and Health*, (eds. I. Chorus, U. Ringelband, G. Schlag and O. Schmoll), pp. 269-274, IWA, London..

Wolstenholme, R., Dutch, J., Moffat, A.J., Bayes, C.D. and Taylor, C.M.A. (1992) *A Manual for Good Practice for the Use of Sewage Sludge in Forestry*. Forestry Commission Bulletin, 107, HMSO, London.

WRc, (1992) *Sewage Sludge: Current disposal practice and future developments in selected countries*. Foundation for Water Research, Marlow, Bucks.

Ziebell, W.A., Anderson, J.L., Bouma, J. and McCoy, E.L. (1975) *Faecal Bacteria: Removal from sewage by soils*. Winter meeting ASAE, December 1975, Chicago, Illinois.

23

Industry, mining and military sites: Control and protection

C. Teaf, B. Merkel, H.-M. Mulisch, M. Kuperberg and E. Wcislo

A range of hazardous substances may be released to the environment from industrial sites, depending on specific industrial processes (see Table 11.2). Among these, the mobile compounds reach groundwater (see Chapter 4). Less mobile compounds may also contaminate groundwater where process wastewaters are discharged through soakage pits. The most common contaminants to reach groundwater in significant quantities from industrial sites are the chlorinated solvents such as trichloroethene (TCE) and perchloroethylene/tetrachloroethene (PCE) but, in specific circumstances, concentrations of many others such as chromium and petroleum constituents may be elevated. Mining can give rise to a range of inorganic contaminants and acid waters, in particular, can result in the accelerated leaching of metals into groundwater. Stored, disposed and deteriorating explosives have been found in some groundwaters below military sites. In Germany and in the USA, perchlorate used in rocket fuel has given rise to major problems. However, the most common contaminant for both military and industrial sites is probably oil from machinery and vehicles, particularly in the case of military sites.

 Protecting Groundwater for Health: Managing the Quality of Drinking-water Sources. Edited by O. Schmoll, G. Howard, J. Chilton and I. Chorus. ISBN: 1843390795. Published by IWA Publishing, London, UK.

In contrast to groundwater contamination from agriculture and off-site sanitation, larger industrial operations tend to be localized point sources of pollution. This is not the case for small-scale enterprises, particularly where these are not connected to centralized sewerage. Nevertheless, the control and protection measures proposed in this chapter for industry can in principle be applied to small-scale enterprises as well.

Military bases often resemble both industrial facilities and small cities regarding the use, storage, and disposal of a variety of chemicals, heavy metals and waste materials. Many planning and operational control measures to prevent the contamination of groundwater by chemicals used in routine military operation are the same as those for industrial sites, and they are therefore discussed together in this chapter.

A variety of effective control measures can be implemented to minimize the likelihood and the magnitude of groundwater impacts from industrial, mining and military activities in groundwater recharge zones. These measures fall into broad categories of: (i) planning, including principal site selection; (ii) engineering approaches which can be implemented in the phase of planning and designing of facilities; and (iii) operational/procedural controls which can be administrated for both new and existing facilities. Some control measures may have both engineering implications (process design) and administrative elements (modification of employee practices), e.g. efforts to substitute with less hazardous process chemicals or development of a corporate recycling plan to reduce waste volumes. Operational monitoring of control measures is important to ensure the ongoing safe storage, handling and disposal of process chemicals, maintenance supplies and waste materials (see Table 23.1). Good practice to support this includes training of personnel in proper safety and handling of these materials under routine conditions as well as in the case of spills or leaks.

All of these measures are typically directed at preventing or limiting the quantity and significance of releases. They include monitoring for early detection of releases and improvement of available containment or remedial capabilities in the event that accidental or intentional contaminant releases occur. In terms of resource allocation, there is a clear benefit to avoidance of releases or accidents. Plans and procedures for avoidance of releases are usually less costly in terms of time and money than remedial measures (i.e. the cleanup of contaminated media such as soils and groundwater) once contamination has spread over a broader area, perhaps even throughout a watershed or aquifer.

Implementing control measures for industry, mining and military sites in drinking-water catchments can be triggered by water suppliers and/or the public authority responsible for drinking-water safety, e.g. in the context of designating protection zones (see Chapter 17), or in the context of developing a WSP (see Chapter 16).

NOTE ▶ *In developing a Water Safety Plan (Chapter 16), system assessment would review the efficacy of control measures and management plans for protecting groundwater in the drinking-water catchment from contamination by industrial, mining and military activities. Chapter 11 provides the background information about the potential impact of these activities on groundwater and provides guidance on the information needed to analyse these hazards.*

This chapter introduces options for controlling risks from these activities. As the responsibility for them usually falls outside that of drinking-water suppliers, close collaboration of the stakeholders involved, including the authorities responsible for the surveillance of industry, mining and military activities, is important to implement, upgrade and monitor these control measures. This may be initiated by the drinking-water sector, e.g. in the context of developing a Water Safety Plan or of designating protection zones (see Chapter 17).

23.1 INDUSTRIAL AND MILITARY SITES

As discussed in Chapter 11, the main concern at industrial facilities as well as at smaller enterprises typically is the improper containment and handling, management or disposal of chemicals, which can lead to soil, surface water and groundwater contamination. This may be the result of active contamination routes, such as intentional dumping or inappropriate disposal activities, or may occur via passive contamination routes such as leaking tanks or broken transfer pipes. Both for industrial and for military sites, groundwater protection usually involves improvement of design and construction of facilities, modification of current practices, as well as remediation of past contamination.

23.1.1 Strategies for pollution prevention and environmental management

The protection of groundwater from industrial and military contaminants is facilitated if this can be managed within general environmental controls, i.e. in environmental management systems such the international ISO 14001 standard or pursuant to EU Regulation 761/2001. One example of a comprehensive strategy to address site-specific control measures at industrial facilities is embodied in the 1991 Integrated Pollution Prevention and Control approach of the Organisation for Economic Co-Operation and Development in Europe (Recommendation on Integrated Pollution Prevention and Control; C/90/164/Final/ 1991). This approach recommends means by which to anticipate and manage chemical handling and process-related activities that may potentially contaminate the environment, including groundwater. Recommendations include those of an engineering nature, as well as an administrative or institutional nature. The approach calls for:

- identification of existing contamination;
- design of mechanisms to detect potential future releases;
- development of plans to minimize the impacts of such releases.

Various cradle-to-grave or catchment-to-consumer management strategies similar to the Organisation for Economic Co-Operation and Development approach for chemicals, especially directed at protection of groundwater resources, have been implemented in a number of countries. Such management systems help to achieve the objective of establishing environmentally safe and groundwater-conserving practices in dealing with industrial chemicals through policy and administrative measures as well as through an appropriate management of material flows based on life cycle analysis. Active military installations, for example, are well suited for such management systems due to the controlled nature of site personnel and activities. Environmental management systems can, in the long term, replace some monitoring and control tasks, resulting in cost saving.

Audits and evaluations of products are one way in which manufacturers, distributors and users of chemicals can contribute to production and use of substances with less pollution potential (HERA, 2002). Environmental and regulatory compliance audits have been common practice in industrial and commercial settings for a decade or more under international programs of environmental management and consumer product safety such as ISO 14000 (Fredericks and McCallum, 1995; ISO, 2001). In addition, responsible and detailed labelling of consumer products is a method for linking information from the manufacturers with consumers and users to optimize environmentally sound disposal practices (US EPA, 2002b).

Similarly, the encouragement or institution of procedures for re-using waste materials can be an economically and technically sound means to reduce waste volumes and limit potential contamination by process wastes. The use of internal process modifications, or business contacts with an external waste exchange programme which converts one plant's waste output into another plant's resources are proven methods for achieving these goals. Industrial waste treatment and recovery strategies to convert or to process wastes into profitable materials have been effective in worldwide applications since at least the 1970s. These strategies are most effective where large volume wastes of specific types (e.g. spent solvents with low residual contaminants) are available from a plant, and low cost transportation is available to a plant with distillation or purification facilities.

Waste exchange, defined as the use of discarded, surplus or off-specification materials for beneficial purposes, represents one potential component of waste management options. While some materials are easily amenable to such exchanges (e.g. solvents for reclamation, metals dusts for refining), other waste sources require innovative approaches to identify users. Examples such as the oil refinery in Poland discussed in Box 23.1 illustrate the double benefit associated with the waste exchange concept. This growing trend in waste management benefits both parties and is typically facilitated by a non-profit intermediary.

Box 23.1. Waste exchange at a petroleum refinery site in Czechowice, Poland

The oil refinery case example described in Chapter 11 (Box 11.1) illustrates the potential for waste exchange as an avoidance strategy: Final disposition of the acidic petroleum sludges currently stored in the refinery's waste lagoons is an issue of concern for the refinery as it seeks to modernize its operations. With advice and guidance from an American waste exchange, the refinery sought to find a potential user for these sludges. A nearby cement manufacturer was identified as having the capabilities to co-fire the sludge in their cement curing/drying kilns, strictly for its energy value. Negotiations between the two parties led to a series of test burns using varying amounts of refinery sludge. The process was proven to be feasible and negotiations began for full-scale implementation. Successful consummation of this arrangement provided a low- or no-cost source of supplemental fuel to the cement manufacturer while providing a disposal mechanism and, potentially, income to the oil refinery. Existing waste materials will be removed from the urban area in which the refinery is located, reducing the potential for groundwater contamination, and the cement manufacturer will reduce their use of external fuel.

A further important strategy to avoid contaminating groundwater is transition to production processes that substantially reduce or totally replace the use of hazardous chemicals and/or the use of water. Such developments have been successful in many branches of production and include effluent-free steel industry, mercury free chlor-alkali electrolysis, AOX-free propendioxide production, or wastewaterless flue gas washers and cooling systems. Instalment of such production technologies often proves cost-effective for the enterprise within fairly short time spans, particularly in settings where water prices are an issue, or where the enforcement of pollution restriction legislation renders polluting practices costly.

Avoidance strategies are also important for a wide variety of substances used in industry and with the potential to adversely affect groundwater which are also are present in common household products (e.g. alcohols, petroleum hydrocarbons, chlorinated solvents, soaps/surfactants, ammonia, phthalates, paints, batteries, pesticides, adhesives). While individual quantities per household may seem small in comparison to those generated by industrial facilities, a large number of households disposing such products in landfills and/or septic systems may represent an equal or greater potential hazard (EC, 2002; Health Canada, 2002; US EPA, 2002a). In some areas, incentives encourage the production, marketing and use of less toxic and less environmentally hazardous alternatives. However, the costs and time necessary to effect changes in established behaviours can be large (EC, 2002).

23.1.2 Choice of site

A fundamental element of any strategy for prevention or avoidance of adverse impacts in groundwater recharge areas is appropriate choice of the site for a facility, including the option of relocating existing facilities. Consideration of groundwater vulnerability is invaluable in assessing the suitability of locations for new operations, and may be used in conjunction with site development plans and engineering precautions to design a facility with minimum potential aquifer impacts. An important measure in planning and choice of site is to require permits for construction and operation which specify activities, production processes and management plans. Legislation and local land use controls or zoning requirements can be effective tools to guide industrial development for achievement of minimum impact in drinking-water catchment areas. Limitations on siting in flood prone or low areas, areas of karstic terrain, close proximity to water bodies or within current or potential future drinking-water protection zones recognize that accidental releases or ongoing industrial operations in these areas rapidly affect groundwater.

Where facilities already exist in vulnerable drinking-water catchments and relocation is not an option, control measures to prevent releases of hazardous substances become particularly important, and specific controls may be required in permits for their operation to limit their hazard potential. Such requirements may include use of environmentally improved technology and products, more intensive monitoring systems, emergency response plans and prohibiting the use of specifically identified hazardous substances in their processes.

Issues of site-specific relevance include: surface topography and features; soil type and local variability; aquifer vulnerability (see Chapter 8); proximity to rivers and other water bodies; chemical type, physical form and quantity of materials handled; degree to which plant construction will require major changes to existing conditions and thus impact on aquifer vulnerability (e.g. extensive excavation, backfilling or soil relocation, pipeline installation, well construction, paving or building cover for substantial areas).

23.1.3 Design and construction for prevention of spills and leakage

A wide range of engineering measures can be applied in the design and construction of a facility as an effective defence to prevent or avoid releases of hazardous substances. These include approaches such as impermeable surfaces and secondary containment structures around tanks, double-walled pipes, alarm devices indicating overfilling of tanks or other vessels, and knock-down barriers to protect tanks or pipes from damage by vehicles (see also Box 23.2 for examples). They also include using natural (e.g. clay) or synthetic (e.g. geotextile) liners to prevent percolation from ponds or storage areas, and capturing in-plant process residues or surface run-off in properly designed and constructed holding areas prior to treatment. Often, structures to retain spilled and/or leaking fluids are important particularly for unloading stations where hazardous fluids are transferred from railway or truck tanks to on-site containers (see Figure 23.4 in Box 23.2). Prevention and mitigation of releases can also be accomplished by neutralizing, encapsulating, stabilizing or solidifying materials (e.g. process wastes, soils, sludges) to prevent or control mobility.

Canopies over stored materials, coupled with capping options designed to isolate materials or to protect them from precipitation and to prevent leaching, can also be site-specific source control measures (Figure 23.3 in Box 23.2). The appropriate degree of complexity for a capping option is related to factors including size and configuration of the capped area, toxicity and potential mobility of the materials to be addressed, duration of required isolation, and whether the surface of the capped area is to be used for secondary activities. As indicated in Table 23.1 at the end of this chapter, capping typically will be accompanied by an operational monitoring requirement to ensure that the cap (and/or companion liner system) continues to be effective at isolating the materials. Available capping options vary widely in cost, durability and effectiveness for particular applications. Such options may best be viewed as temporary, albeit long-term, solutions for which subsequent, permanent solutions are desirable.

Levels of sophistication – and thus of costs – can vary for design and construction control measures. Often, fairly simple low-cost measures effectively provide substantial protection against soil and groundwater contamination, and are valuable first steps upon which incremental improvements can build later. In the study shown in Box 23.2, short-term, medium-term and long-term measures were proposed for many of the problems identified. For example, while the long-term measure for protecting tanks against overflow of hazardous chemicals through overfilling would be to fit them with approved devices, overflow can already be quite effectively prevented by installing a simple indicator of filling level and a routine for its regular monitoring. An immediate measure would be to ensure that special care is taken when filling the tank by requiring two staff members to fill the tank together. Likewise, while double bottoms for tanks may be installed in the long run, intensified internal checks and determination of the wall thickness of the tank may improve safety in the meantime.

For all containment structures, regular maintenance and monitoring of their integrity is critical for keeping them functional. Management plans for a facility should include these activities and responsibilities for their regular performance and documentation.

23.1.4 Operational controls

For protecting groundwater from industrial contamination, controlling operations that may lead to spills and leaching is often equally important as safe containment. Operational controls address procedures for handling, using, transferring and storing substances such as properly unloading trucks or railway tankers, using safety couplings and valves, using mobile drip trays, avoiding overfilling containers and providing materials to absorb hazardous chemicals in case of spills. An important aspect is preventing joint storage of substances that may undergo chemical reactions with each other and taking properties such as auto-ignition, combustibility or corrosiveness into account. Also, labelling of tanks, containers and facilities with hazardous chemicals is necessary to allow appropriate emergency responses. A further important operational control is the implementation of emergency response plans which are regularly rehearsed by the staff of the facility.

Operational controls are best developed with operational staff and fixed in writing as standard operating procedures in a facility's management plans. Implementation is supported by checklists and forms to sign after conducting specific routines. Adequate training and qualification of staff, including the aspects of groundwater protection, as well as clear assignment of tasks and responsibilities, are prerequisites to making them work. Often the target of avoiding spills for the sake of groundwater protection is closely linked to the target of avoiding exposure for the sake of occupational health and safety, and both may be addressed within the same control measure.

Box 23.2. Technology transfer for plant-related water protection in Moldavia, Rumania and Ukraine (based on FEA, 2002)

Within the framework of the Environmental Action Program for Central and Eastern Europe which was agreed by the Ministers for Environment of the UNECE, a Technical Assistance Programme launched by the German Ministry of Environment developed a methodology for assessing water pollution hazards by industries with high water pollution potential. This used the recommendations of the International Commissions for the Protection of the Rhine (ICPR) as well as of the Elbe (ICPE) as a basis. From this assessment, short, medium, and long-term measures were identified with which the ICPR and ICPE recommendations can be met. The majority of these measures are equally relevant to the protection of groundwater and surface water. Measures relating to design and structure of facilities include the following:

Short-term measures:

- Repair and seal cracks and damage in existing sealed surfaces
- Perform internal examinations of tanks and containers
- Fill tanks and containers under the supervision of two operating persons
- Examine and prepare a concept for joint storage of hazardous substances (with the potential to react)
- Use mobile collecting basins and detachable connections for plants with transhipment (tank wagon – plant connections)

Medium-term measures:

- Provide a stop valve for open-air collecting basins connected to the wastewater system
- Demonstrate that wastewater pipelines are not leaking (Figure 23.2)
- Renovate sealed surfaces in plants for transhipment and/or storage

Long-term measures:

- Install overfill safety systems for storage containers
- Provide collecting basins for retaining water-polluting substances and fire-fighting water
- Create sealed surfaces and retaining volume for railway tank-car stations (Figure 23.4)
- Establish wastewater treatment facilities that meet quality requirements

Operational control measures include requiring the plant operator to:

- define in-plant responsibilities for taking and checking safety measures which include functional safety, impermeability of containment structures, functionning of safety equipment, documentation (in writing) of regular checks undertaken
- provide detailed reports on accidents and incidents, including causes, consequences and future preventive measures
- report releases of hazardous substances to competent authority
- define equipment for plant monitoring and related instructions for action, including prevention of accidents, water hazard potential, potential for substance release, precautionary measures and protection requirements

- use internal monitoring wherever there is a need to prevent releases of substances hazardous to water, to allow detection on time to implement contingency measures

Checklists were developed for setting up internal alarm and hazard control plans defining actions and responsibilities for types of incidents (e.g. leakage, overfilling of vessels, failures of receptacles, containers, pipelines, fires and fire-fighting water, accidents during transport of hazardous goods) as well as for different plants. This includes exercises to train accident responses at regular intervals.

Figure 23.1. Leakages at a production plant

Action proposals: Technical structures to minimize foaming; venting on a buffer tank for the retention of the foam.

Figure 23.2. Single wall pipe subways through retention room; no knock-down protection

Action proposals: Pipe installation above the retention room wall; constructing knock-down protection (big stones); regular pressure tests; street crossing over ground; double wall pipes installation.

Figure 23.3. Storage of solids

Action proposals: Creation of a reasonable canopy; moving the pipe in the bow area; renovation of the existing sealing area.

Figure 23.4. Unloading station for hazardous fluids from railroad tank cars

Action proposals: Conduct unloading with two people; build adequately sealed retention space.

23.1.5 Decommissioning of contaminated sites

When industrial and military sites are abandoned, hazardous chemicals that may leak into groundwater may unintentionally be left behind. An important control measure in drinking-water catchments therefore is proper decommissioning – potentially involving clean-up – of such sites. Issues of decontamination and remediation of sites formerly used for industrial or military purposes are often complex due to the difficulties of identifying those responsible for the pollution in order to implement the polluter pays principle. This is particularly difficult in the context of abandoned sites. Teaf (1995) and Herndon *et al.* (1995) have described the former military facilities in central and eastern Europe as a large scale example of this and reported that the technical and financial responsibility for mitigation became the burden of the host country. Similar problems occur on abandoned industrial sites. An important control measure to prevent this type of situation is to include the responsibility for decommissioning and potentially necessary remediation in plans and permits for establishing such operations.

23.1.6 Clean-up and remediation of contamination

Once a decision has been made to clean up a given site, an initial site characterization must be performed to determine the type and extent of contamination, it may be possible to use available data for preliminary decision-making. For example, after-care measures in the form of exploratory investigations, containment techniques and remedial actions (Teaf, 1995) were carried out in particular in the early 1990s in Germany for military-contaminated sites located in the vicinity of drinking-water abstraction. This included the toxicological assessment of individual constituents and groups of military chemicals, as well as assessment of their migration behaviour and biochemical, chemical and hydrolytic degradability in subsoil (e.g. to evaluate their potential to leach into groundwater).

The characterization process prior to mitigation of industrial and military sites must consider a cardinal rule: that which is not sought is never found. Although the highest concentrations of contaminant generally will be focused at the source area, the characterization and clean up efforts also must identify and evaluate the extent and continued migration of contaminant plumes in soils, groundwater or surface water. This is critical because degradation often occurs in the areas of lower concentration associated with plume fringe, which may be far from the source.

A variety of technologies exist for the remediation of soil, surface water and groundwater at industrial facilities (e.g. thermal and chemical treatments, biological remediation technologies, soil washing and filtration; see Soesilo and Wilson, 1997; Nyer, 1998; Hyman and Dupont, 2001). Depending on the type of contamination and the threat to drinking-water aquifers, natural attenuation may also be an option (see also Chapter 24). When selecting a remedial technology, the decision will be influenced by potential effectiveness, reliability, implementability, cost and time constraints. Each technology has intrinsic advantages and disadvantages that can be optimized by carefully matching site-specific conditions with a remedial technology or suite of technologies. For example, many organic contaminants (e.g. petroleum hydrocarbons) are readily degraded by microbial communities under appropriate environmental conditions (see Chapter 4). Bioremediation seeks to optimize those conditions through a variety of in situ or constructed on-site mechanisms. Biological technologies such as these take advantage of and facilitate natural processes and, as such, are often favoured and are potentially less expensive, in comparison with more technologically complex approaches. The increased time frames associated with some biological remediation technologies may be more easily accommodated at sites controlled by government entities (e.g. military installations) than at those associated with commercial or industrial enterprises. Contaminants such as petroleum products or chlorinated solvents are amenable to such efforts.

Once a release to soils, waterbody sediments or other elements of a groundwater recharge area has occurred, there are many established and new methods to prevent or limit contaminant migration in soils and to control or reverse plume expansion in groundwater. These methods include physical controls (e.g. sheet piling, trenches/slurry walls/grouting, recovery wells, air sparging), physical separation (to reduce reactions) and chemical methods for contaminant control (e.g. oxidation/aeration, reduction, permeable reactive barrier, dual phase extraction), as well as in situ or ex situ degradation by physical or biological processes. Recent advances in phytoremediation, for example, have resulted in deployments of certain tree species known as phreatophytes (e.g. poplar, willow) to intercept contaminant groundwater plumes (Quinn *et al.*, 2001). Such biological control also may enhance degradation of some organic contaminants. Maintenance and operation costs of such a system are lower than for typical engineered systems (e.g. pump and treat) over the relative lives of the systems. Depending on local and regional hydraulic effects exerted by water bodies, surface water control may be an important element of a comprehensive strategy to prevent industrial impacts in recharge areas.

The most straightforward mechanism for addressing contaminated soil, generally above the saturated zone, involves excavation and off-site disposal. However, the quantity and character of soils, as well as the associated removal, transportation and disposal costs, may limit the utility of this option. In addition, the transport of contaminated materials to another location may not relieve the original landowner of legal liability.

23.2 MINING

As with industrial activities, control measures for mining activities involve prevention as well as remediation and monitoring whether process controls are being implemented. Due to the large scale of many mining activities and milling sites, retrospective mitigation of their environmental impact is often substantially more difficult than prevention. Further, groundwater protection strategies are needed for both the active mining period and the post-mining period, and have to include the mine itself as well as mine waste, milling facilities and atmospheric emissions. Control measures may be equally necessary for small mining sites, particularly where they are numerous and potentially lead to considerable contamination of groundwater (see Chapter 11).

As for industry, choice of site is the first and often most important measure to protect groundwater. Many countries require an environmental assessment study for new mining activities exceeding a certain size (number of employees, amount of ore excavated). Ideally, intersectional collaboration in this planning phase should involve public health authorities and water suppliers to help recognize the potential impact on groundwater resources. Numerical modelling of groundwater flow, hydraulic situation before, during, and after mining activities and the impact of mining on groundwater quality is a state of the art technique and often successfully performed. Groundwater modelling is also an important tool to determine appropriate locations for monitoring wells to be drilled in the region of interest for mining, in order to record groundwater flow and quality parameters. Moreover, an Environmental Impact Assessment (EIA, Chapter 20) should be performed taking into account the vulnerability of the groundwater, the type of ore mined and processed, and other environmental threats in the region. This will lead to a more sustainable mining activity by introducing appropriate treatment and processing techniques. The EIA should cover the entire time frame, i.e. the exploration of an ore body, the mining activity, the remediation measures taken and the post-mining land use.

23.2.1 Deep mines

Constructing and operating a deep mine usually requires groundwater withdrawal. A necessary control measure to prevent water pollution in some cases is water treatment if the water contains toxic elements above a critical level. Monitoring would address on a regular basis whether treatment is in place and properly operating.

A further measure for preventing contamination is limiting the use of hazardous chemicals in ore processing and, where use is inevitable, application and handling with special care. Control measures may involve limiting, budgeting and recording the amounts of such chemicals used. Areas where heaps and tailing ponds will be constructed have to be investigated carefully including geological and hydrogeological aspects; in many cases liners (e.g. geotextile; see also Chapter 24) are useful as additional protection against contaminant leakages.

Before closing a deep mine, potential contaminants (e.g. fuel, oil, machinery) should be removed. In numerous cases where this was not done, considerable amounts of contaminants and waste in the mine have led to groundwater contamination.

Refilling of tunnels and shafts with waste rock or fly ash is a common technique to avoid land subsidence. However, it may also help in establishing lower permeability in the flooded mine and act as reactive material. The chemical nature of such fill materials should also be considered. These materials may be a potential source of contaminants (e.g. metals) in addition to mined materials. On the other hand, they may also be selected to bind contaminants: calcite may buffer low pH values, while iron (Fe^0) acts as a reducing agent, and fly ash or brown coal seem to be effective in sorption. However, little is known about long term behaviour of reactive material in underground mines. Thus the choice of adequate refilling materials is an important groundwater protection measure but long-term surveillance will often be necessary to ensure that contaminants are not released in concentrations above critical levels. Controls to ensure that adequate measures are taken for closure may include the requirement of approval of plans for such measures by government authorities or a catchment protection body.

During controlled flooding of a mine, contaminated groundwater is pumped and treated until the contamination level has decreased to acceptable concentrations. In many cases, this may require an extended period of time, and alternative passive treatment techniques might be preferable. In some cases, hydraulic isolation of the mine area might solve the problem, but this can be expensive as well. Tracer experiments are common tools to investigate the hydraulic flow pattern in a deep mine. Constructed wetlands can be used as effective and inexpensive measures to treat surface water after the first flush has reached an acceptable value of contaminants (Younger, 2000). As long as the contaminated groundwater flows at shallow depths, reactive walls (i.e. subsurface permeable barriers built with reactive materials to degrade or immobilize water-borne contaminants) may be considered as a low cost measure (Blowes *et al.*, 2000). Reactive walls or permeable reactive barriers are passive treatment systems: a ditch is excavated in an aquifer downstream of the contaminant source and refilled with permeable and reactive material (e.g. mixture of sand with iron). Since iron in its elemental form is a very strongly reducing agent, metal ions (e.g. uranium, chromium) will be transferred in their reduced redox state and in consequence precipitate. Thus groundwater leaving the permeable reactive barriers is purified by certain metals and organic contaminants efficiently and at low costs. All approaches to treating water from mines will require adequate surveillance of treatment efficacy which would be defined in a management plan.

23.2.2 Open pit mines

Since open pit mining usually destroys aquifer structure, this type of mining often has the most severe impact on groundwater on a regional scale. Legislation and governmental controls on surface mining in relation to groundwater use have been implemented successfully. To control sulphides, waste rock should be covered as soon as possible (see below). Carbonate as alkalinity buffer may be added as additional measure to compensate the pH value due to pyrite oxidation. Calcium phosphate also has been used to control acid generation (Evangelou, 1995).

The design of open pit mining activities must also account for the final shape of a mine lake. Rapid recovery of groundwater to the final level in such a lake is often targeted to minimize erosion and stability problems with the embankment. As discussed in Chapter 11, acid mine drainage may flow from the oxidized zones of aquifers and heaps towards the pit lake resulting in extremely low pH-values in the lake water. If surface water is available to fill the lake, water quality will be no problem in the very beginning as this is usually well buffered. However, hydraulic equilibrium between groundwater and the lake will establish itself with time and water quality may decline when the groundwater in contact with rock is contaminated due to the solution of secondary minerals and/or waste deposits. Therefore management action to protect groundwater from post-mining lakes and vice versa requires both consideration of these processes already in the planning phase for the activity and surveillance for the post-mining phase until new hydrological equilibrium between ground- and surface water, as well as chemical equilibrium between solids and water, have been reached. Acid mine lakes can be treated by means of liming with dolomite quicklime. The pH will rise to about six and high sulphate concentrations will decrease by the formation of gypsum. Time intervals of monitoring should relate to rates of change and may decrease as processes slow down.

23.2.3 Acid mine leachate

As pointed out in Chapter 11, acid mine leachate is one of the most severe potential groundwater impacts from this human activity. Approaches to controlling this involve keeping the oxygen supply to sulphide minerals as low as possible to avoid reactions producing sulphuric acid. This requires careful investigation of the distribution of sulphides in the mine area and its vicinity. Also, minimizing the dewatering cone of depression will reduce leachate. Refilling shafts and adits with material of fine grain size reduces the permeability in these artificial cavities and helps establish more natural groundwater levels during the post-mine period. This material also may act as reactive material, lowering the outflow of contaminants from the mine site (see Section 23.2.1). Where these measures are chosen – alone or in combination – monitoring their proper operation is needed on a regular basis to ensure their implementation. Depending on the setting, monitoring could include regular checks on pH and sulphate concentrations in order to control whether sulphide oxidation is still ongoing; on the depression cone; and on the amounts as well as type of refilling material actually used.

23.2.4 Heaps, piles, mills and tailings

Major sources of pollution from mining often are heaps, piles and tailing ponds. Waste rock and residues from ore milling and ore processing ('tailings') at new or operational mining facilities need to be handled with the same care as municipal or industrial wastes. Control measures to mitigate their impact therefore include many state-of-the-art techniques used for waste deposits, such as drainage and treatment of drainage water to meet the targeted water quality criteria, or placement of spoil heaps and tailings over areas of impermeable sediments such as clay or bedrock that will not allow leachate to reach groundwater. Alternatively a clay lining or a geo-textile fabric can be used to line the site intended for disposal of spoil and tailings, or a foundation pad can be constructed which is impermeable or has reduced permeability. In both cases care must be taken to contain or treat leachate that runs off from the site. Corresponding control measures address the function of such containments and whether they are intact. Control measures for sustainable mining may also include the addition of buffering minerals to heaps, e.g. a certain amount of lime stone or fly ash according to the amount of sulphide in the waste rock. This buffers the formation of acid mine drainage in situ.

Control measures for such approaches involve periodic assessment of whether seals are tight, and monitoring systems for groundwater quality up- and downstream will assist in verifying whether the approach taken is sufficient.

In many settings, earlier construction of heaps, piles and tailings without consideration of their impact on groundwater quality has led to problems now requiring remediation. For example, remediation of the uranium mining and milling sites which were in operation from 1946-1990 in the eastern part of Germany is costing the German Government about US$ 6.5 billion. Treatment of contaminated groundwater as well as surface water from deep mines during ongoing operation may be accomplished by means of classical treatment techniques, though this may be prolonged and costly. Thus alternative treatment techniques such as reactive walls, carbonate drains, and constructed wetlands are increasingly being used. Constructed wetlands have proved to be a promising tool for natural attenuation of mine-related contaminants (Hedin *et al.*, 1994; Younger, 2001).

Physical shaping and capping of heaps and tailings is necessary to avoid erosion, dust transport and reduction of the amount of infiltration. If radioactive ores or waste rock with radioactive components occur on-site, this must be taken into account in designing covers or caps that can act as a radon barrier as well (Merkel *et al.*, 2002). Tailings may be covered with wet or dry caps, the latter being most common. Again, control measures should ascertain that caps are in place and functioning.

Rehabilitation of old heaps and tailings requires a careful investigation of boundary conditions and impact on groundwater. This will show to what extent reshaping and capping of these heaps and tailings may be necessary to achieve slope stability, erosion protection and surface or groundwater protection. Passive water treatment techniques may be applicable for long-term protection of groundwater resources.

23.2.5 In situ leaching

Mining by in situ leaching (ISL) presents special concerns to groundwater quality since hazardous chemicals are used for the in situ extraction of ore by leaching (see Chapter 11). Approval of ISL mining by regulators should therefore require management plans which define control measures with operational monitoring systems as well as maintenance of all installations to ensure that groundwater clean up can be performed at the specific site. Monitoring is critical to ensure that no process chemicals leave the ISL mining site during operation. When ISL mining is terminated, the site should be cleaned until pre-mining or otherwise acceptable conditions have been established.

23.3 MONITORING AND VERIFICATION OF MEASURES CONTROLLING INDUSTRY, MINING AND MILITARY SITES

The control measures for industry, military sites and mining in drinking-water catchments proposed above range from planning tools in the context of broader environmental policy to specific technical measures such as structures, containments and operational controls. Selected examples are summarized in Table 23.1.

NOTE ▶ *The implementation of control measures such as those suggested in Table 23.1 is effectively supported if the stakeholders involved collaboratively develop management plans that define the control measures and how their performance is monitored, which corrective action should be taken both during normal operations and during incident conditions, responsibilities, lines of communication as well as documentation procedures.*

The implementation of control measures protecting drinking-water aquifers from industry, mining and military activities is substantially facilitated by an environmental policy framework (see Chapter 20).

Monitoring of the measures implemented is crucial to ensure that they are in place and effective. Table 23.1 therefore includes options for monitoring and verification of the control measure examples given. Most of these focus on checking whether the controls are functioning as intended, rather than on contaminant concentrations in groundwater. For planning, reviewing will address whether plans exist, are appropriate and are being implemented, particularly in the context of issuing permits for new or extended operations. Periodic auditing of plans is an effective tool for such surveillance. Likewise, reviewing of emergency response plans would assess whether they are appropriate and whether they are occasionally being used for appropriate facility training exercises.

Similarly, for control measures in design and construction, the first step is to assess whether or not they are adequate for achieving the protection target, and whether or not they are in place as indicated in the construction plan. For the day-to-day routine operation of controls, monitoring focuses on assessing whether they are functioning as they should, e.g. whether containments are sealed, mine drainage is being treated or waste management plans are being implemented.

Monitoring of controls for day-to-day operations is particularly important as these tend to slip if not taken seriously. Examples given in Table 23.1 include maintenance routines, specifications on amounts and types of chemicals to be used, safety rules for handling, transferring and storing hazardous chemicals and routines for pumping hazardous leachate from mines. Such rules will be specified in management plans and standard operating procedures. Their implementation can be monitored by checking records, e.g. of maintenance measures taken or amounts of chemicals used in process steps, as well as by occasional inspection of process steps, such as unloading tankers with hazardous chemicals or integrity of storage structures, and by interviewing technical staff on how these steps are normally performed.

NOTE ▶ *Options for monitoring suggested in Table 23.1 rarely include regular groundwater quality monitoring. Where control measures such as structures are poorly accessible, however, monitoring of selected indicator parameters in groundwater is suggested.*

Comprehensive groundwater quality monitoring programmes are a supplementary aspect of monitoring with the purpose of providing verification of the efficacy of the overall drinking-water catchment management.

Where spills and releases are suspected or where the risk that this may happen is elevated, monitoring to provide for early detection is important. Careful evaluation of both the hydrogeology and the facility operations will allow prediction of likely locations and flow patterns of initial releases. Monitoring for key parameters that would readily indicate a leak at these locations can provide early warnings. This may include groundwater sampling and analysis of selected indicator parameters that would readily reflect leakage and potential contamination. Contaminant analyses will also be an important control measure after decommissioning of industrial and military sites and in particular after clean-up and remediation of contamination. Generally, resources expended in monitoring result in reduced remedial costs (and potential enforcement) in the event of a release. In the context of monitoring for overall verification of the catchment management concept, it is often effective to include contaminants anticipated or known to occur from industry, mining and military activities in the catchment, particularly at the sites of these activities, but also at groundwater intakes.

Table 23.1. Examples of control measures for industry, mining or military sites and options for their monitoring and verification

Process step	Examples of control measures for industry, mining and military sites	Options for their monitoring and verification
PLANNING	Require permits for the location, design and operation of industries, manufacturing enterprises, mining and military sites (e.g. EIA)	Review (application for) permit with respect to adequacy of siting, planning and design as well as public consultation
	Require plans for post-operational safety of site as part of the permit for such operations which are likely to need post-closure management (e.g. mining or military training sites)	Require long-term financial commitments and post-operational management plans (e.g. for lakes resulting from open pit mining) for issuing permit
	Require environmental or chemical management plans, including waste management plans when issuing a permit (including e.g. probations or limitations of specific processes or chemicals; treatment for mines using in-situ leaching)	Review existence and adequacy of management plans; audit if possible
	Require emergency response plans for enterprises which operate with hazardous substances	Review or audit emergency response plans
	If drinking-water protection zones are designated, enforce keeping hazardous enterprises out	Conduct periodic site inspections
DESIGN AND CONSTRUCTION	Install and maintain temporary and/or permanent containment structures (tanks, caps, vaults) for storage and handling of hazardous chemicals, explosives, mine heaps, tailings and ponds	Review adequacy of design and compliance with plans and regulations Inspect sites and enterprises for compliance with plans, and structural integrity and function
	Remove or remediate contaminated soil	Analyse residual soil and groundwater samples
	Refill mine tunnels and shafts; remove/stabilize potential contaminants; remove contaminants (e.g. fuel oil), machinery before refilling	Conduct follow-up site inspection and monitoring
	Rehabilitate old heaps and tailings; treat leachate	
OPERATION AND MAINTENANCE	Control/restrict amounts and types of chemicals used in production processes and mining operations	Review records/reports of chemical use, storage of wastes and maintenance of systems Analyse in situ leachate for chemical concentrations
	Control storage, handling and disposal of high risk chemicals and wastes	Inspect compliance to codes of practice, standard operating procedures and/or chemical management plans
	Maintain containment structures for storage and handling of hazardous chemicals and explosives	Check whether maintenance plans have been signed off; occasionally inspect maintenance Monitor downstream groundwater for parameter indicating leakage
	Minimize acid leachate from mines by controlling dewatering cone of depression	Monitor water levels, pH, or sulphide
	Treat contaminated groundwater from (active or closed) mining operations until contaminant concentrations reach acceptable levels	Monitor operational parameters for treatment system chosen (e.g. condition of artificial wetland and water flow) Analyse selected contaminants in treated water
	Conduct post-operational management of sites potentially leaking hazardous substances	Inspect monitoring and maintenance by operators and evaluation of reports required by permit Monitor downstream groundwater for parameter indicating contaminant migration

23.4 REFERENCES

Blowes, D.W., Ptacek, C.J., Benner, S.G., McRae, C.W.T. and Puls, R.W. (2000) Treatment of dissolved metals and nutrients using permeable reactive barriers. *J. Contaminant Hydrology*, **45**, 123-137.

EC (2002) *Study on Household hazardous Waste with a Main Emphasis on Hazardous Household Chemicals*. European Commission WRc Reference CO-5089-2. July, 2002, Medmenham.

Evangelou, V.P. (1995) *Pyrite Oxidation and its Control*. CRC Press Boca Raton, New York, London, Tokyo.

FEA (2002) *Technology Transfer for Plant-related Water Protection in Moldavia, Romania and Ukraine*. Federal Environmental Agency of Germany, Berlin, http://www.umweltbundesamt.de/anlagen/Gewaesserschutz/english.htm (accessed April 29, 2005).

Fredericks, I. and McCallum, D. (1995) *International Standards for Environmental Management Systems: ISO 14000*. Canadian Environmental Protection, August 1995.

Health Canada (2002) *Environmental Health*. http://www.hc-sc.gc.ca/english/protection/environment.html (accessed April 29, 2005).

Hedin, R.S., Nairn, R.W. and Kleinmann, R.L.P. (1994) *Passive Treatment of Coal Mine Drainage*. Information Circular IC 9389, US Department of the Interior (Mines), Pittsburgh.

HERA (2002) *Human and Environmental Risk Assessment on Ingredients of Household Cleaning Products*. http://www.heraproject.com/files/hera_leaflet_xp_DEF.pdf (accessed April 29, 2005).

Herndon, R.C., Richter, P.I., Moerlins, J.E., Kuperberg, J.M. and Biczo, I.L. (eds.) (1995) *Clean-up of Former Soviet Military Installations*. Springer-Verlag, Berlin.

Hyman, M. and Dupont, R.R. (2001) *Groundwater and Soil Remediation: Process Design and Cost Estimating of Proven Technologies*. American Society of Civil Engineers Press, Reston, Virginia.

ISO (2001) *Draft of Joint ISO 9000 and ISO 14000 Audit Standard Published*. International Organization of Standardization, Geneva.

Merkel, B., Planer-Friedrich, B. and Wolkersdorfer, C. (2002) *Uranium in the aquatic environment*, Springer-Verlag, Berlin.

Nyer, E.K. (1998) *Groundwater and Soil Remediation: Practical methods and strategies*. Lewis Publishers, Boca Raton, Florida.

Quinn, J.J., Negri, M.C., Hinchman, R.R., Moos, L.P., Wozniak, J.B. and Gatliff, E.G. (2001) Predicting the effect of deep-rooted hybrid poplars on the groundwater flow system at a large-scale phytoremediation site. *Int. J. Phytoremediation*, **3**(1), 41-60.

Soesilo, J.A. and Wilson, S.R. (1997) *Site Remediation: Planning and management*. Lewis Publishers, Boca Raton, Florida.

Teaf, C.M. (1995) Human health and environmental risks associated with contaminated military sites. In *Clean-up of Former Soviet Military Installations*, (eds. R.C. Herndon, P.I. Richter, J.E. Moerlins, J.M. Kuperberg and I.L. Biczo), pp. 31-44, Springer-Verlag, Berlin.

US EPA (2002a) *Municipal Solid Waste and Household Hazardous Waste*. http://www.epa.gov/epaoswer/non-hw/muncpl/hhw.htm (accessed April 29, 2005).

US EPA (2002b) *Environmental Labelling Issues, Policies and Practices Worldwide*. http://www.epa.gov/opptintr/epp pubs/ envlab/chap1.pdf (accessed April 29, 2005).

Younger, P.L. (2000) The adoption and adaptation of passive treatment technologies for mine waters in the United Kingdom. *Mine Water Environ.*, **19**, 84-97.

Younger, P.L. (2001) Mine water pollution in Scotland: nature, extent and preventative strategies. *Sci. Tot. Environ.*, **265**, 309-326.

24

Waste disposal and landfill: Control and protection

A. Allen and R. Taylor

Waste disposal by landfill has led to the pollution of groundwater resources under a wide range of conditions around the globe (e.g. Sangodoyin, 1993; Ahel *et al.*, 1998; Christensen *et al.*, 1998; Afzal *et al.*, 2000). In the USA, Lee and Jones (1991) assert that approximately 75 per cent of the estimated 75 000 sanitary landfills pollute adjacent groundwater with leachate. Leachate derived from waste deposits (landfills, refuse dumps) includes a wide range of contaminants, depending on the types of wastes deposited (see Chapter 12). There is consequently a strong need, supported by legislation in many regions, to protect groundwater from the effects of waste disposal. This chapter provides an overview of current approaches towards this aim and explains their scientific rationale.

24.1 WASTE CONTROL

Control of the type and amount of waste placed in landfills is a basic measure to protect groundwater. In many countries legislation regulates the type of wastes deposited at MSW landfills: waste that is considered hazardous due to its ignitability, corrosivity, reactivity, toxicity and carcinogenicity (Sharma and Lewis, 1994) is not accepted at

 Protecting Groundwater for Health: Managing the Quality of Drinking-water Sources. Edited by O. Schmoll, G. Howard, J. Chilton and I. Chorus. ISBN: 1843390795. Published by IWA Publishing, London, UK.

MSW landfills, but is separated and removed for specialized disposal. Industrial wastes classified as hazardous include solvents and metal-rich materials. At the household level, hazardous wastes include refrigerators, paint, a range of cleaning products, batteries and such automotive products as lubricants (Tchobanoglous *et al.*, 1993).

NOTE ▶ *In developing a Water Safety Plan (Chapter 16), system assessment would review the efficacy of control measures and management plans for protecting groundwater in the drinking-water catchment from waste disposal and landfill. Chapter 12 provides the background information about the potential impact of wastes on groundwater and provides guidance on the information needed to analyse these hazards.*

This chapter introduces options for controlling risks from wastes. As the responsibility for waste disposal usually falls outside that of drinking-water suppliers, close collaboration of the stakeholders involved, including the authorities responsible for landfill, is important to implement, upgrade and monitor these control measures. This may be initiated by the drinking-water sector, e.g. in the context of developing a Water Safety Plan or of designating protection zones (see Chapter 17)..

Legislation to gradually divert most organic waste from landfills has been introduced by several countries. It is designed to reduce overall waste quantities disposed by landfill and thus also the quantity of leachate produced in landfills, since biodegradation of organic wastes is the dominant source of leachate production. Separate treatment of diverted organic waste by aerobic or anaerobic digestion will lead to the production of considerable quantities of compost. To protect the underlying groundwater, the control of run-off from larger (i.e. commercial outdoor non-reactor) compost piles is necessary.

Separation or sorting of waste for reuse or recycling (e.g. paper, bottles, cans) is another key measure in controlling and reducing waste going to landfill. Waste separation and sorting, which should preferably take place at source, provide an opportunity to reuse or recycle waste materials and to compost organics. This not only reduces the amount of refuse disposed by landfill but also leachate production. In many countries, semi-automated waste separation and sorting takes place after collection at centralized materials recovery facilities. Runoff from such facilities needs to be controlled to avoid groundwater contamination. However, recycling and reuse initiatives generally require government support as recycled items are commonly more expensive to produce than the items they replace. Government support of recycling can be financial through creation of markets or through government purchasing policy.

Open refuse dumps, where waste disposal is unsorted and unregulated, are characteristic of many countries of the developing world and represent an increased risk to groundwater quality. In such settings, waste separation and sorting for reuse and recycling is often conducted on an informal basis either at points of collection or at the dump itself. Although some of these operations are officially sanctioned, most are unofficial. They represent an important function as an informal recycling system contributing to reduction of waste to landfill. Direct exposure to waste materials may, however, present a health hazard.

A waste reduction strategy successfully used in a number of countries has been the imposition of a deposit fee refunded upon return of, for example, bottles, cans, tins, paper and other items. Taxes specifically on packaging materials have been implemented in Germany, for example, to reduce unnecessary waste volume. Household composting can be promoted for further waste reduction by subsidized sale of home composting bins.

Waste reduction strategies may also include landfill taxes. These provide an economic incentive and generate revenue. The downside to consider in planning policy is that they may also promote illegal (uncontrolled) disposal practices. For instance, imposition of a charge for acceptance of construction and demolition waste at one landfill in Ireland led to a rapid drop in the quantity of such waste delivered to the landfill.

Waste incineration can be an effective strategy to substantially reduce the amount of waste that is landfilled. Adequately engineered systems can effectively control air pollution and toxic residues, including those from flue gas cleaning systems, can be encapsulated for safe disposal. A major drawback of this strategy is its high costs, which tend to distract from investments into other options such as recycling.

Industrial wastes are often more hazardous than municipal solid waste (MSW). Unregulated disposal, often on site, was commonplace in the past and remains a problem in some regions. Major contamination of groundwater resources can be avoided through specifically designed and managed disposal of wastes from such industries as smelting, electroplating and tanning industries. A case study showing planning steps taken to remedy unregulated disposal of industrial wastes is described in Box 24.1.

Box 24.1. Planning priorities for the remediation of contaminated groundwater in the metallurgical centre of Ust-Kamenogorsk, Kazakhstan

Metallurgical plants produce fluid and solid waste that is hazardous due to high (toxic) metal concentrations. Disposal of this waste without pretreatment and in locations with limited capacity to attenuate these metal contaminants can lead to groundwater pollution. Ust-Kamenogorsk is a city in northeastern Kazakhstan with a population of 290 000 that was a centre for metallurgy and heavy industry in the former USSR. As a result of uncontrolled dumping of industrial wastes over more than 50 years, 300 000 tons of toxic arsenic mud, 6000 tons of PCB-contaminated mud and 19 million tons of slag, clinker and sludge from metallurgical processes containing mobile metal compounds were stored on permeable ground. Another eight million tons of fly ash from coal-fired heat and power plants, containing fluoride and boron have been deposited within the city area. The alluvial aquifer underlying the city is the sole water resource for the city water supply but was heavily polluted by arsenic, boron, fluoride, cadmium, copper, lead, manganese, selenium and zinc. High chloride, nitrate and sulphate contents are encountered in the contamination plumes. Contamination plumes in the groundwater clearly relate to dumps of metallurgical sludges and to seepage of process water from leaking pipe work and cracked factory floors (von Hoyer and Muff, 2001).

Comprehensive remediation of contaminated lands in Ust-Kamenogorsk is not currently feasible due not only to limited resources with which to execute the clean-up but also because of the magnitude of the problem that features a large number of pollution centres, complex nature of the contaminants and the large area of the polluted land. In the interim, an approach was necessary that seeks to improve the environment and the living conditions in the city but recognizes the need for continued industrial activity that provides jobs and tax revenue. The approach also had to be integrated into the legal framework and taxation system of the public sector, waste management and city planning, recognizing explicitly that groundwater pollution by hazardous waste is related to the production methods applied by the industry. Several social and economic sectors need to be involved in this task: the central government, the regional administration and the industrial sector. The following actions were recommended:

Political actions:

- Separation of responsibilities for hazardous waste. Government: hazardous waste inherited from the USSR era; industry: hazardous waste generated after independence
- Adaptation of an environmental fee system, with the aim to financially support the development of cleaner production technology from the fees collected from pollution
- Allocation of State Environmental Fund for the construction of environmentally safe disposal sites for hazardous waste

Technical measures (level of priority):

- Safe central drinking-water supply to all city areas from unpolluted wellfields; closure of polluted household wells (high)
- City development planning: protection of operative and potential city wellfields, concentration of industrial activities in existing industrial areas and relocation of residential areas to unpolluted areas (high)
- Capping of abandoned hazardous waste deposits with multibarrier cover in order to reduce leakage of contaminants, recultivation (high)
- Retrofitting of the zinc and lead plant in order to reduce leakage of process water (high)
- Optimization of the hydraulic containment for contaminated groundwater in the zinc and lead plant area (high)
- Solidification of PCB mud, deposition and capping with multibarrier cover, recultivation (high)
- Modernization of water quality laboratory in the East Kazakhstan Ecology Administration (high)
- Establishment of a surface and groundwater data bank and development of a groundwater flow and contaminant transport model in order to monitor the impact of remediation measures and industrial activities and to protect the city water supply (high)
- Capping of closed fly ash dumps with multibarrier cover, recultivation (low)
- Upgrading of existing groundwater monitoring network (low)

Legislation on handling, containment and disposal of hazardous wastes is in place in many countries and chiefly addresses industrial and medical waste. Pretreatment (physical, chemical, thermal or biological) of some of these is an option to reduce their hazardous impact. Separate approaches may be needed for different types of hazardous wastes, depending on the health hazards they impose, and their source. Box 24.2 uses the example of wastes from health care facilities to highlight management approaches for such specific waste sources.

Box 24.2. Managing wastes from health-care facilities

Many wastes from health-care facilities (e.g. hospitals) require pretreatment before these waste streams can be unified with other similar waste and disposed of as household waste. For example, microbiologically contaminated waste should be disinfected or sterilized before it is handled in any other way, and wastes containing cytotoxic compounds should be incinerated. Radioactive waste and associated waste waters require separate collection, as well as storage until their radioactivity have declined. For the disposal or discharge of wastes containing pharmaceutically active substances, diagnostic agents and active disinfecting substances into the environment risk, assessment is – for most compounds – similar to that for other chemicals (in particular pesticides and other biocides). Some substances require special management strategies, e.g. antibiotics and disinfectants due to their ability to foster resistance, and cytotoxic compounds of which some show mutagenic, carcinogenic and fetotoxic properties (Eitel *et al.*, 2000).

One option to reduce the environmental impact of hospital wastes on a local scale is to implement an environmental management system, e.g. according to the ISO 14000 standard which defines targets for use of specific substances and their emission. This can be combined with the introduction of health and safety management procedures (Kümmerer *et al.*, 2001). Often the process of improving the safety of medical staff can be addressed together with the reduction of environmental impacts including impacts on groundwater used as drinking-water resource.

Avoidance strategies have proven successful in health care facilities. For example, because of their low biodegradability, Freiburg University Hospital, Germany has largely eliminated products containing benzalkonium chloride or other quaternary ammonium compounds, and alcohols or aldehydes are used instead. This has considerably reduced quaternary ammonium compound concentrations in the hospital's effluent. Another important element of avoidance strategies is information for users about the potential impact of drugs, diagnostic agents and disinfectants on water quality, if they are not properly disposed of or returned to the pharmacy (e.g. safety data sheets, package inserts, specialist information for pharmacists/dispensing chemists).

For further information on safe waste management from health-care facilities see Prüss *et al.*, 1999; WHO, 2000; WHO, 2004a; WHO, 2004b.

24.2 SITING AND PLANNING OF LANDFILLS

As discussed in Chapter 12.1, little attention was historically paid to the siting of landfills. Rock quarries and open gravel pits were often exploited as they avoided the effort and expense of excavation. Landfills, nevertheless, tend to be located close to urban areas where significant volumes of municipal and industrial wastes are produced. Whether the intention is to store waste in containment landfills or employ a strategy of NA (see Sections 24.3.1 and 24.3.2), it is safest to position landfills in areas removed from groundwater drinking-water supply sources and on sites where the underlying geology is able to attenuate to some degree the leachate that is generated from the stored wastes (see also aquifer vulnerability in Chapter 8). Clay- and organic-rich materials (overburden or mudstone bedrock) are suitable as they both retard groundwater flow and interact with reactive contaminants in leachate. Unfortunately, it is commonly difficult to choose such ideal locations strictly on the basis of hydrogeological considerations as socioeconomic considerations, including the 'not in my back yard' syndrome, tend to dominate the process of selection of landfill sites. Onibokun (1999) notes that landfills and dumps in Africa commonly occur in poorer areas, where residents are often less able to prevent landfill in their own backyard, are unlikely to benefit from waste collection services, and invariably depend upon local, often poorly protected sources of water.

In many countries, a tool for commissioning a site for landfill is Environmental Impact Assessment (EIA; see also Chapter 20). These include licensing applications, public hearings, appeals and sometimes court cases. The site selection process can be assisted by a GIS approach that is integrated with rigorous geotechnical site investigations (Allen *et al.*, 2001). Areas underlain by aquifers used for drinking-water supply and their catchments would normally be identified and rejected for waste disposal during an initial exclusion step, whilst areas with favourable geological conditions would gain high positive weightings in the ensuing assessment of the residual areas remaining after the initial assessment step. A second stage to narrow down potential sites, involving considerations such as cost/distance analyses and visual impact assessments, is then followed by the geotechnical investigation, which can also fulfil the requirements of the EIA. A major requirement of an EIA at any proposed landfill site will be a detailed hydrogeological investigation to assess aquifer vulnerability and potential impacts to groundwater. Determination of the attenuation capacities of the subsurface materials underlying the site, which should also be undertaken, can enable assessment of the capacity of the site to attenuate leachate migrating from the landfill.

In many countries, applications for licenses are also required for existing landfills and dumps. Where these are unsuitably located relative to drinking-water resources or to aquifer recharge areas, and where the underlying geology gives inadequate groundwater protection, a license may be revoked or refused and the landfill forced to close. Licenses, whether granted to existing landfills or to new landfills, usually come with stipulations in the form of requirements and restrictions. Stipulations for new landfills usually include design and operational requirements based on site conditions established in an EIA, whereas for existing landfills, upgrading of groundwater protection measures and operational procedures may be demanded. Stipulations for all landfills will also include types of waste acceptable for disposal and types of waste not permitted. Unacceptable waste may include hazardous wastes, for which specialized disposal or incineration may be required. Other requirements will generally also include monitoring procedures and frequency, and possibly also maintenance provisions at closure.

A policy of minimum travel time between landfills and groundwater-fed drinking-water sources has been adopted in a number of countries, often within the concept of drinking-water protection zones, as discussed in Chapter 17. According to this policy, waste disposal facilities are prohibited in areas within a certain (e.g. 50-day) travel time (by groundwater flow) of a groundwater-fed drinking-water source. A key drawback to the travel-time criterion is that it requires an indication of the mean groundwater flow velocity in order to estimate an appropriate separation distance. In practice, this is difficult to determine due to the pronounced heterogeneity and complexity of many groundwater systems. More refined approaches base landfill site selection decisions on a more detailed understanding of groundwater conditions, and the ability of the natural environment to contain or alternatively attenuate contaminants in waste leachate. Groundwater flow conditions can be assessed from local borehole records (i.e. hydraulic gradient, aquifer thickness) and the results of pumping tests in the underlying aquifer (i.e. estimates of hydraulic conductivity). The geology of a potential landfill site and, hence, its natural capacity to restrict subsurface flow and attenuate contaminants, can be elucidated from drilling logs (if available) and geological maps of the area.

Further planning aspects critical to potential groundwater contamination are the size of a waste disposal facility and the rate at which refuse is deposited. These determine its operational lifetime. Also, planning and siting needs to include transfer stations and material recovery facilities to ensure that these are not polluting groundwater.

24.3 DESIGN STRATEGIES FOR LANDFILLS

Different approaches exist regarding strategies of protecting groundwater resources from the leachate generated by waste disposal. Two different design strategies for landfills are: *containment* and treatment of leachate on site versus *attenuation* through degradation, dilution and dispersion of leachate.

24.3.1 Containment strategy

Containment requires that all liquid and gaseous emissions produced within the landfill are contained and collected for treatment. The central aim of containment is, therefore, to minimize production of leachate by restricting access of rainwater to the waste, and to prevent its migration from the landfill of leachate produced. This is accomplished by enclosing the waste in artificial lining systems consisting of a landfill liner and cap. As a consequence, leachate drainage systems, containment ponds and leachate treatment facilities are essential additional components of modern containment landfills. Experience has shown that artificial membranes will eventually leak so modern designs usually include composite two-, three- and four-layer multibarrier clay/membrane liner systems (Figure 24.1). These multibarrier systems consist of sheets of artificial membrane, most commonly high-density polyethylene, interlayered with natural or bentonite-enriched clay layers (Seymour, 1992; Cossu, 1995). In the European Union, for example, landfill regulations make it mandatory to entomb waste using engineered lining systems except at sites with low in situ hydraulic conductivity (less than 10^{-9} m/s) (Allen, 2001).

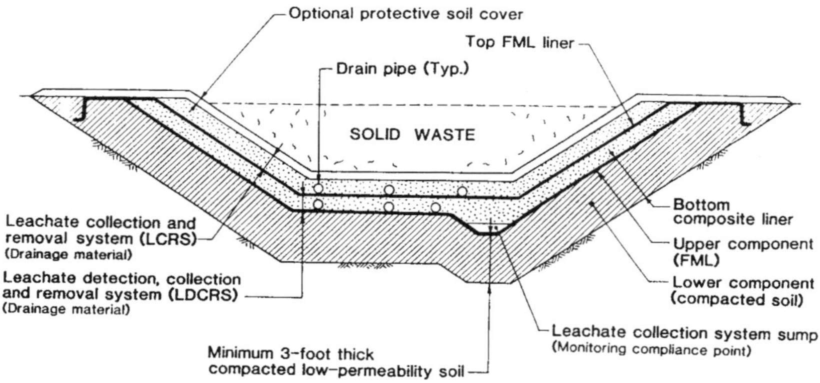

Figure 24.1. A sanitary landfill featuring an engineered lining and leachate collection system (Rowe *et al.*, 1997)

Leachate containment with engineered lining systems requires a suitable geological sub-base as a secondary barrier to groundwater when containment structures become permeable. Leakage can result from stress cracking of the membrane, cracking under cold conditions, damage, particularly from stones in the protection layer and from heavy dumping equipment, or failure of the membranes near welded seams (Rollin *et al.*, 1991; Surmann *et al.*, 1995). Synthetic liners are also susceptible to failure if installation is not subject to strict quality controls and favourable weather conditions (Averesch, 1995). Indeed, it is unlikely that any manufactured synthetic membrane is completely free of defects even prior to installation (Christensen *et al.,* 1994a). Lastly, some contaminants are able to diffuse through installed liners including intact geotextiles (Potter and Yong, 1993; Rowe, 1994).

Natural leachate containment is assumed under geological conditions where in situ hydraulic conductivity is less than 10^{-9} m/s. In practice, this is not always easy to ascertain. This is indicated by bulk hydraulic conductivities, derived from pumping or tracer tests over a larger volume, which are often several orders of magnitude greater due to preferential flow along discontinuities such as fissures (Gerber and Howard, 1996). Thus apparently low permeability strata such as glacial till and shale do not necessarily assure containment of leachate. When planning landfill based on natural leachate con-ainment, it is therefore important to validate the bulk hydraulic conductivity of the site.

An important aspect of containment is the aftercare timespan. Rates of waste degradation are a function of moisture content. Under wet conditions (i.e. uncapped landfills in climates where precipitation exceeds evaporation), an aftercare timespan of 30 years was initially assumed adequate to allow for degradation of waste to an inert state (Bookter and Ham, 1982). However, more recent work indicates that even under very wet conditions, complete degradation of waste may take at least 40-60 years (Wall and

Zeiss, 1995), and that some components (e.g. NH_4-N concentrations) may not have fallen to compliance thresholds of wastewater regulations for at least 100 years subsequent to landfill closure (Kruempelbeck and Ehlrig, 1999). On the other hand, for dry uncapped landfills in climate regimes where annual evaporative rate exceeds precipitation, giving rise to a moisture deficiency, degradation of waste may be considerably slower and estimates for dry landfills in water deficient areas suggest timescales in excess of 400 years (Röhrs *et al.*, 2000). The Gaza case study example given in Box 24.3 shows that depending on the composition of the wastes, leachate production may suffice for effective degradation.

Box 24.3. Groundwater protection in Gaza through upgrading of disposal standards

Collection and disposal of solid waste in the central part of the Gaza Strip are the responsibility of the Solid Waste Management Council – an autonomous public body governed by a board comprising the mayors of the eleven towns and villages that it serves. Services include collection and disposal of about 220 tons of MSW per day, generated by some 350 000 people. The sanitary landfill operated by this Council was designed and constructed in co-operation with the German Agency for Technical Cooperation.

At the onset of the project in 1994 a number of uncontrolled open dumpsites existed in the area under consideration. Adopting a strategy of containment, the first step in improving this situation was to assess soil and groundwater conditions at several locations. Based on this information and other factors, one of the existing dumpsites was chosen as a site for a central landfill.

In Gaza annual precipitation is between 200 mm and 450 mm; the wet season is October and April; no rainfall during rest of year; high annual evaporation between 1200 mm and 1400 mm. i.e. annual evaporative rate exceeds precipitation, representing moisture deficient conditions for most months of the year, so most experts were of the opinion that insignificant quantities of leachate would be produced.

Nevertheless, since groundwater is the main potable water source in the Gaza Strip, it was considered desirable to avoid any risk of further groundwater contamination. It was therefore decided to line the landfill site as detailed below: two asphalt liners with a bitumen mastic layer between the liners; coarse aggregates and drainage pipes to convey leachate to a storage pond; pumps and a sprinkling system for recirculation of leachate.

Decisions regarding capping, final cover and post-closure care were delayed until the quantity of leachate produced had been established. The lessons learned include:

Leachate quantities produced are high, contrary to expectation: Measurements indicate that the average leachate flow during the winter 1999/2000 to be 27.4 m^3 per day, and only slightly less during the dry summer season (25.4 m^3 per day), during which no leachate was expected at all. One possible explanation is that the composition of waste in Gaza differs from that for Western Europe. It contains more biodegradable organics with a high moisture content and more inert material, but less paper and other light fractions to absorb water. Because of its high initial moisture content this organic-rich waste biodegrades readily, promoted by the high temperatures prevailing in Gaza, producing large quantities of solubilized liquids even under moisture-deficient climate conditions, thus creating substantial quantities of leachate. Leachate samples, obtained and analysed, indicate that COD concentrations are in the range of 40 000 mg/l and BOD is about 11 000 mg/l, quite similar to values for young landfill sites in western Europe. These results indicate how incorrect conclusions can be arrived at if based on experience from quite different climatic and socioeconomic settings. Specifically, it proved to be important to have provided a lining in order to protect groundwater drinking-water resources.

The density of landfilled waste at the disposal site is exceptionally high: Based on before-and-after topographical surveys and calculations relating the volume filled to the total weight disposed of at the site, the density amounts to about 1.9 tonnes/m^3. Again this value met with disbelief from experts because this is almost twice the value at landfill sites in western Europe. Benefits of these high disposal densities are considerable:

- the life span of the site is almost doubled;
- capital costs for disposal on a per tonne basis are reduced by almost half.

This indicates that densities at disposal sites in the developing world may be considerably higher than in industrialized countries, again possibly due to the composition of the wastes, and the rapid biodegradation brought about by its high moisture content.

New approaches to final cover: The generation of leachate at the Gaza landfill site is likely to decrease substantially once the site reaches its capacity limit. The reason for this expectation is that the bulk of the leachate generated is due to the high moisture content of the waste, which promotes rapid degradation under the high temperature conditions prevailing at Gaza. When no more fresh waste is added, the moisture content will drop and degradation rates will decrease. Hence, once the site has been filled, the main source of leachate – fresh waste that has just been placed – no longer exists. This suggests that the application of an impermeable cap may not be required. It was therefore decided that the main consideration in the selection of cover material should be that it is a suitable substrate for plant growth. Organic and inert material from the landfill itself proved to be suitable for this purpose.

Appropriate solution for post-closure care: Considering that net evaporation is about 1000 mm per year, the surface area of the existing leachate storage pond is sufficient to allow evaporation of the quantity of leachate expected during the post closure period. It was therefore decided that, after closure of the site, the existing leachate storage pond be converted to an evaporation pond. This solution is inexpensive, reliable and almost maintenance-free.

Costs: As the (virtual) disposal density is about 1.9 tonnes/m3, capital costs are equivalent to about US$ 2.5 per tonne of MSW delivered to the site. The average household size in Gaza is 6.9 persons, disposal costs per household amount to about US$ 0.60 per month. This shows that relatively high standards of protection of groundwater resources can be achieved at moderate cost.

Isolating the waste from water (i.e. dry entombment) significantly reduces rates of degradation of the waste, thereby prolonging the activity of the waste and inhibiting, possibly by decades or centuries, its stabilization to an inert state (Allen, 2001). This extends the time span during which synthetic materials in artificial liners are subjected to the corrosive effects of leachate and the elevated temperatures generated by the exothermic processes operating with landfills. Recent research indicates that degradation of geomembranes occurs through oxidation processes over time (Hsuan and Koerner 1995; Koerner and Daniel, 1997).

Further considerations on the containment landfill strategy are its inherent costs and sustainability. The capital costs of installing an engineered lining and leachate collection system (Figure 24.1) will vary depending upon the size and lifespan of the landfill but are likely to prove prohibitive for many communities in low-income countries. Operational costs for one small landfill in Ireland, situated on a thick natural clay overburden deposit with a K value of $<10^{-9}$ m/s, and initially employing unlined cells, increased ten-fold when installation of a liner for future cells was required as part of its licensing stipulations. Leachate collection and treatment add significantly to the operating costs of containment landfills. There are, furthermore, the costs of other essential components of containment landfills including leak detection and landfill gas collection systems. Because of the large investment required for such technologies, landfills become economical on a large scale, promoting development of regional superdumps.

24.3.2 Attenuation strategy

The attenuation strategy allows leachate to migrate outwards from the landfill and takes advantage of the natural subsurface processes of biodegradation, filtration, sorption and ion exchange to attenuate the contaminants in leachate. The attenuation strategy is based on the dilute and disperse principle of leachate management proposed by Gray *et al.* (1974). A significant study at the time drew from a large body of field and laboratory investigations to highlight the efficacy of natural processes in attenuating leachate concentrations and indicated that, in appropriate situations, such an approach is effective in reducing the risk of pollution to water resources. This method of leachate management relied on natural low permeability and attenuation characteristics of geological barriers in the subsurface, primarily clay-rich overburden and, to a lesser extent, consolidated mudrocks, to prevent groundwater pollution by landfill leachate. The dilute and disperse principle of leachate control was superseded in the early 1980s by the containment strategy after having been discredited due to failures which occurred where the strategy was employed without adequate consideration of prevailing hydrogeological conditions.

The critical difference between the modern attenuation approach and the former dilute and disperse approach is that attenuation is an active management strategy, requiring the presence of a natural in situ or imported attenuation barrier to attenuate the leachate, whereas dilute and disperse relied on passive subsurface dilution and dispersion processes without the presence of a specific attenuation layer. Although the concepts are similar, it is now recognized that dispersion and dilution alone may not sufficiently attenuate leachate to adequately protect groundwater. More recent studies (e.g. Warith and Yong, 1991; Batchelder *et al.*, 1998; Yong *et al.*, 1999) support the conclusion that clay-rich overburden and mudrocks have the capacity to attenuate leachate. The effectiveness of the strategy is further confirmed by the fact that even within geological units of relatively high permeability and supposedly poor attenuation potential, such as sandstone, and sandy overburden, attenuation processes operate very effectively, and most pollutants are moderated within a few hundred metres (Christensen *et al.*, 1994b; Williams, 1999; Ball and Novella, 2003; Butler *et al.*, 2003).

Natural geological barriers, may be defined as low permeability clay-rich geological units (hydraulic conductivity $<10^{-5}$ m/s), which can perform the function of an attenuating layer, enabling leachate to percolate slowly downwards, simultaneously undergoing attenuation by biodegradation, sorption, filtration and ion exchange processes with the clays in the unit (Allen, 2002). Extremely low permeability geological units (hydraulic conductivity $<10^{-9}$ m/s) cannot fulfil an attenuation function as they perform in a similar manner to artificial or natural lining systems providing almost complete containment of all emissions. Similarly, geological units with higher permeability (hydraulic conductivity $>10^{-5}$ m/s) do not provide sufficient confinement to leachate and are thus unsuitable for attenuation. The optimum permeability for attenuation is in the order of 10^{-6} to 10^{-8} m/s.

It is also recognized that the rate of degradation of waste materials can be enhanced by maximizing the flow of rainwater into the landfill leading to dilution of the leachate produced from the waste. Degradation of waste follows a well-documented path, with production of both leachate and biogas which vary in composition as degradation progresses. Waste with a high proportion of organics will produce significant quantities of leachate even under moisture-deficient conditions (see case study in Box 24.3), due to solubilization of organics by microbiological and biochemical processes. Rates of degradation are promoted by a steady flow of water through the waste, which also results in production of greater quantities of leachate but of a more dilute, less toxic nature.

The fundamental assumption of the attenuation strategy is that the underlying geology is able to moderate contaminant concentrations derived from landfill leachate to acceptable levels prior to groundwater discharge in a stream or water source (e.g. well or spring). However, not all geological units are able to fulfil this function, so a site selection protocol that includes an assessment of proximity to drinking-water wells, as described in Chapter 24.2, is an essential prerequisite to the adoption of an attenuation strategy. The attenuation mechanism may still be operated in unfavourable situations if natural clay or peat material is imported and installed as a liner to improve hydraulic conductivities and attenuation potential and if leachate migration is controlled. Nevertheless in certain types of terrain, such as karstified limestone, where groundwater flow occurs primarily along secondary fissures (i.e. non-intergranular flow) and attenuation of contaminants is limited, rapid and severe pollution of groundwater can result (Edworthy, 1989) and the attenuation strategy should be avoided.

Even in favourable geological situations, leachate migration should be controlled and monitored. Control measures include leachate collection and recirculation systems in order to prevent shock loading of the receiving environment. Location of monitoring wells needs to be based on detailed hydrogeological investigations or incipient groundwater pollution may be missed due, in part, to the limited predictability of groundwater flow from waste deposits (Chapter 12.2.3). Drainage, storage and recirculation of leachate prevent build up of leachate head as a guard against shock loading of the attenuation medium. The dilute nature of the leachate allows inexpensive treatment options by reedbeds or peat beds where leachate production is excessive.

A key attraction of an attenuation strategy is avoidance of the excessive costs of containment landfills that are untenable for many countries. It also avoids the long-term costs for maintenance and aftercare monitoring, which may be required for containment landfills for decades or even centuries after the site has ceased operating, as long as the waste remains active (Mather, 1995). Apart from a drainage system and containment ponds to control the leachate head in order to prevent shock loading of the attenuating medium and a monitoring programme, attenuation landfills have little attendant costs. The key constraint of the attenuation strategy, however, is the uncertain but genuine risk of groundwater pollution by leachate if attenuation proves less effective than assumed when selecting the site or if the site is not adequately managed (e.g. with respect to drainage measures).

24.3.3 Choice of strategy

Choice of strategy is likely to differ for upgrading existing landfills, remediating historic ones threatening a groundwater resource and for planning new ones. In choosing a suitable strategy for landfilling, with protection of groundwater for use as drinking-water a high priority, the following should be borne in mind:

- Every case is unique both with respect to natural hydrogeological conditions, land use and socioeconomic requirements, and the option chosen should be the most appropriate for the specific situation.
- The choice of site needs to be based on a detailed site selection process and needs to undergo a rigorous geotechnical investigation programme including hydrogeological surveys and delineation of the attenuation potential of the underlying geology (see also Chapters 8 and 14). Regardless of whether the chosen landfill management strategy is containment or attenuation, the underlying geology should have the potential to act as a groundwater protection barrier.
- The merits of containment with probable delays in stabilization of the waste to an inert state for many decades must be balanced against an attenuation strategy that seeks to degrade and stabilize waste in the shortest time possible.
- In balancing the economics of the containment strategy against that of attenuation landfills it is essential to include the costs of all ancillary elements required for each management option, and maintenance and monitoring costs after closure.

In many countries legislation now requires containment as well as collection and treatment of all leachate produced in the landfill. The rationale for this is protecting groundwater not only for drinking-water abstraction but also for environmental objectives, including the protection of soil and groundwater ecosystems in proximity to a landfill. The NA strategy, however, explicitly accepts environmental impact within some distance downstream of the landfill, in which no drinking-water abstraction would occur.

Economic constraints often limit the feasibility of technical options, particularly in developing countries. The preface of Botswana's Guidelines for the Disposal of Waste by Landfill highlights approaches towards incremental improvements (see Box 24.4).

Box 24.4 Extract from the Preface to the Guidelines for the Disposal of Waste by Landfill, Republic of Botswana, 1997

"The principal method of waste disposal in Botswana is by land burial. The uncontrolled burial of waste however can lead to serious groundwater pollution problems. For a country almost totally reliant on its already scarce groundwater resources, it is important that the standards of waste disposal by landfill are sufficiently improved to minimize the risk of pollution to water resources, and furthermore to public health and the degradation of natural resources. Water is a public commodity, and it is not ours to pollute as we wish. Due to the regional characteristics of water, any actions by users or polluters of a water source will affect other 'innocent' people downstream of that source. Legislation is being drafted which will require all landfill sites (as well as all waste facilities, transporters and even generators) to be licensed. In this respect it is essential that the licensing authority has guidelines on which to base their licensing decisions and the specific licensing conditions which they are to impose on each individual facility.

"The underlying philosophy and guiding principles used in drawing up the guidelines are that they should be:

- regionally compatible – to avoid the situation where Botswana could become a dumping ground for the southern African region, merely because it has lower environmental standards than neighbouring countries;
- specific to Botswana – to incorporate the specific social, cultural, economic and political criteria within Botswana;
- affordable without compromising on risk – to provide optimum protection of water resources.

"Because of its sparse population, the predominance of small villages in the country, and being a largely arid country, a degree of flexibility is needed in specifying requirements. A system of graded standards has therefore been introduced where the requirements could be adjusted up or down according to the risk imposed. Graded standards, an innovation developed and used extensively by the authors of the South African minimum requirements for landfills, are applied to different categories of landfill site (categorized according to its risk of pollution) as defined by the type and quantity of wastes to be landfilled.

"In this way the standards for landfilling of waste can be improved without incurring excessive development and operation costs, and without subjecting the community to an unacceptable risk.

"These guidelines are practical and specific to Botswana yet regionally compatible, and should be widely used by waste management practitioners. They should be seen as a dynamic set of requirements which will change with time to reflect the latest in relevant landfill technology as the results of world-wide landfill research dictates".

The challenge is to choose approaches that are viable in the longer term even if they do not immediately meet optimal technical standards. Graded standards for landfill are an innovation towards this aim. In South Africa, for example, minimum requirements for landfills vary according to different categories of landfill sites that are defined by the type and the quantity of wastes that are disposed by landfill. The Gaza case study (Box 24.3) demonstrates that locally adapted solutions may be very effective. It also shows that careful planning, taking into account the uncertainties when extrapolating experience gained in different types of settings, can lead to viable options at rather low costs.

Maintenance of an inventory of all waste disposal sites including those no longer in operation is critical since the risks posed by the landfill to the quality of local groundwater remain for decades. Where poorly sited, designed or constructed landfills or informal dumps are identified as hazard, an approach to remediation is to discontinue their use, cover them and, where necessary, monitor downstream groundwater quality.

Strategies for landfill sites (including historic sites and informal dumps) that are polluting or threatening aquifers used as drinking-water source include engineered barriers to leachate migration such as cut-off walls or trenches. A further option is to install defence wells to abstract leachate with high pollutant concentrations before the leachate plume reaches the drinking-water well. In some cases, particularly where there is evidence that hazardous wastes are leaching towards a drinking-water abstraction point, digging away and relocating the waste to a more adequately designed and managed landfill may be necessary. However, due to their high costs, these measures may not be economically feasible in all situations and relocating drinking-water abstraction wells may also be an alternative option.

24.4 OPERATION AND MAINTENANCE OF LANDFILLS

Regardless of the choice of strategy, controlling the operation of landfills is important to prevent groundwater contamination. Depending on the type of landfill and the type of waste, specific operational requirements may be set out in the license and operational controls need to enforce these. It may be effective to define these requirements in a management plan jointly developed by the water supplier(s) together with the other stakeholders and the surveillance authorities involved.

A key operational control is to monitor, record and document the composition and amount of waste delivered to the landfill (e.g. through inspecting and weighing waste trucks entering the landfill) and to turn deliverers away if the waste does not meet specifications. Documentation should include the origin and composition of the waste, and potential hazard classification. Random sampling is important for enforcing compliance with license requirements, and its approximate frequency would be defined in the management plan. Furthermore, documentation of where specific types or batches of wastes within the landfill are deposited may help trace the origin of particularly problematic leachate plumes detected by monitoring and thus enable targeted remediation.

Operational activities include landfill development. In containment landfills with a cellular structure, preparation and lining of future cells will necessarily occur. Whilst an active cell is operational, capping of recently active cells that have received their quota of waste will also be conducted. It is helpful to specify the lining and capping requirements and all other engineering systems to be installed in the license issued for the landfill, and to further detail them in the management plan. Licenses will also specify the height to which waste can be accumulated as well as the final profile of the capped landfill. On the basis of projected tonnages of waste to be deposited, the date of decommissioning of the landfill can be estimated. Daily operation of landfills may include the application of cover at the end of each working day for hygienic reasons in order to reduce wind blow and accessibility for flies, birds and vermin.

Drainage of leachate is important to reduce the hydraulic head of leachate, as this promotes leakage through artificial liners and can also lead to increased leachate migration rates through a clay liner. Accumulated leachate needs to be drained to a lined leachate pond from where it may be either tankered to a leachate or wastewater treatment facility or recirculated through the landfill to promote degradation of the waste. Depending on the type of waste deposited and ensuing leachate composition, treatment approaches range from wastewater lagoons to highly sophisticated methods such as reverse osmosis.

Regular maintenance of technical installations, e.g. leachate drainage systems, caps and barriers, is important to ensure that such technical controls are functioning. Operational monitoring is critical both for technical systems and for the implementation of management plans (see examples in Table 24.1), and they should therefore include a description of the operational monitoring to be conducted for each control measure.

Periodic monitoring of groundwater quality using a system of wells located both upstream and downstream of a landfill site is normally a license requirement for highly regulated landfills. It is particularly important for landfills based on the attenuation strategy. Monitoring will focus on indicator and bulk parameters such as conductivity and organic carbon, but may also include the hazardous components such as adsorbable organic halogen compounds and metals. Exceeding of threshold values may trigger actions such as closing and capping cells in a landfill, improving containment of new cells or establishing defence wells as outlined above.

24.5 PUBLIC PARTICIPATION AND EDUCATION

Public involvement, based on the communication of the relationship between waste disposal, groundwater and health, is important in two ways: health concerns often drive opposition to establishing waste treatment facilities or landfills, and the lack of understanding of health and groundwater concerns, particularly for hazardous wastes, is often a cause for careless, informal waste dumping.

Public participation in the selection of sites for landfill and waste treatment facilities can help to overcome the not in my back yard syndrome. Comprehensive dissemination of information about the plans, including the type and amount of waste envisaged as well as the intended measures to protect drinking-water resources and the environment, is the basis for public participation. This best commences at the outset of the selection process when the public is informed as to local/regional landfill site requirements in the context of the overall waste management strategy for the region. Public meetings would then follow at various stages in the selection procedure, with details of the selection process explained, concerns allayed and questions answered. In some countries, public participation in the management of landfills is encouraged. Consultative committees composed of local community representatives and the landfill managers serve to improve public perception of landfills and promote trust between the local community and landfill operators. They may also allay suspicion and fears over health and environmental issues as well as act as a conduit for information and concerns to be passed among stakeholders.

A widespread public understanding of groundwater protection issues is the prerequisite for avoidance strategies that require waste separation at the household level. Examples for this are campaigns to promote proper disposal and recycling of hazardous wastes such as batteries, motor oil, paints and solvents, or returning unused pharmaceuticals to the pharmacy. The development of an awareness of the potential of hazardous substances to contaminate groundwater-fed drinking-water supply may be particularly important for communities with much small-scale enterprise where they rely on shallow groundwater (see also Chapter 23). Public participation based on an understanding of contamination pathways from wastes to wells may also be important in making relocation decisions, i.e. to either move a waste dump or nearby wells if the distance between them has proven unsafe.

In various countries public participation is promoted by media advertising and by education campaigns, particularly of children at the earliest levels of schooling. Education is the key to changing public attitudes towards waste issues.

24.6 MONITORING AND VERIFICATION OF MEASURES CONTROLLING WASTE DISPOSAL AND LANDFILL

The protection and control measures for waste disposal in drinking-water catchments proposed above range from planning tools in the context of broader environmental policy to specific technical measures such as containing a landfill, managing leachate or constructing defence wells and trenches where leachate is threatening an aquifer. Selected examples are summarized in Table 24.1. Among these, planning and choice of site are particularly critical for waste disposal. This includes fundamental decisions on the disposal strategy with implications for operational controls. In some settings, some of these control measures may be suitable for integration into the WSP of a drinking-water supply (Chapter 16) and become subject to operational monitoring in the context of such a plan.

Monitoring of the measures implemented is crucial to ensure that they are in place and effective. Table 24.1 therefore includes options for surveillance and monitoring of the protection and control measure examples given. Most of these focus on checking whether the controls are operating as intended rather than on contaminant concentrations in groundwater. For planning, surveillance will begin with reviewing permits, the applications for which should not only demonstrate how aquifer vulnerability has been taken into account, but also how the landfill will be designed. If the strategy is controlled attenuation, review of the information base for assessing and predicting leachate migration will be particularly important. Where containment is intended, review of the application will address the choice of liner technology and leachate management. As waste deposits have long-term implications for groundwater, the review of applications for permits will also consider future land use and development planning in relation to groundwater demands. As landfill is a highly emotive issue in many cultures, reviewing applications for permits may take this into account by addressing whether and how planning has been based on public consultation and participation.

NOTE ▶ *The implementation of control measures such as those suggested in Table 24.1 is effectively supported if the stakeholders involved in waste disposal collaboratively develop management plans that define the control measures and how their performance is monitored, which corrective action should be taken both during normal operations and during incident conditions, responsibilities, lines of communication as well as documentation procedures.*

The implementation of control measures protecting drinking-water aquifers from waste disposal and landfill is substantially facilitated by an environmental policy framework (see Chapter 20).

For protection and control measures addressing design and construction, the first step in surveillance is to assess whether or not they are adequate for achieving the protection target, and whether or not they are in place as indicated in the construction plan. For landfills based on attenuation, construction controls will be few but controlling design may be important (e.g. for ensuring that protective layers of overburden are maintained or introduced as planned). For contained landfills, monitoring the quality of the liner and particularly its installation will be important. Documentation of the design and structure of the landfill, as well as of the criteria upon which planning decisions were based, is particularly important as it provides a basis for future situation assessment.

For the day-to-day routine operation of landfills, monitoring focuses on whether the amounts and type of waste deposited are in compliance with the permit (see Table 24.1). Such monitoring can include random sampling of waste delivered by trucks and checking accompanying documents. Further important operational controls include but are not restricted to leachate drainage and recycling or treatment as well as operation of defence wells or trenches where these are needed.

NOTE ▶ *Options for monitoring suggested in Table 24.1 include monitoring downstream groundwater for selected indicators of leachate migration.*

Comprehensive groundwater quality monitoring programmes are a supplementary aspect of monitoring with the purpose of providing verification of the efficacy of the overall drinking-water catchment management.

In addition to surveillance and monitoring of the functioning of control measures, groundwater monitoring serves to verify the whole drinking-water catchment management concept comprehensively. For waste disposal, groundwater monitoring is also a measure to control whether natural attenuation is performing as anticipated, whether leachate migration is within the area expected, and whether containments are leaking. It typically focuses on indicator and bulk parameters and in some situations will address specific substances of concern.

Table 24.1. Examples of control measures for waste disposal and landfill and options for their monitoring and verification

Process step	Examples of control measures for waste disposal and landfill	Options for their monitoring and verification
PLANNING	Require permit for siting based on a hydrological assessment (i.e. aquifer vulnerability, attenuation potential) including type and amount of waste and disposal strategy	Review (application for) permit with respect to adequacy of siting, strategy chosen and design as well as public participation
	Ban inadequate disposal of hazardous wastes in drinking-water catchments	Inspect existence of illegal disposal sites Monitor waste composition on permitted dumps
	Require waste management plans in drinking-water catchments (e.g. separate collection and disposal of hazardous wastes at specifically contained and managed sites or incineration systems)	Review existence and adequacy of waste management plans
	If drinking-water protection zones are designated, enforce keeping waste disposal out	Conduct periodic site inspection
DESIGN AND CONSTRUCTION	Where necessary, construct landfill with a basic liner to prevent rapid leachate migration but which also allows for maximum circulation and dilution of leachate	Review adequacy of design and compliance with plans and regulations Inspect construction site, particularly installation of liners On-site inspection
	Construct drainage for leachate and facilities for either recycling or treating it	
	Cover or cap landfills when closed	
	Improve attenuation potential by addition of imported clay or peat	
	Where leachate migration threatens or pollutes the aquifer, construct barriers such as trenches, cut-off walls or defence wells	
OPERATION AND MAINTENANCE	Control and document types of wastes deposited	Inspect records of site and of trucks dumping waste Review substance budgets of producers/users of hazardous materials (including infectious material) Perform random sampling and analyses of waste composition
	Maintain closed landfills	Inspect function and integrity of structures
	Maintain barriers such as trenches, cut-off walls or defence wells where leachate migration threatens or pollutes the aquifer	Monitor downstream groundwater for indicator of landfill leaching
	Collect and recycle leachate in the landfill to improve decomposition	Monitor pump performance and downstream groundwater quality

24.7 REFERENCES

Afzal, S., Ahmad, I., Younas, M., Zahid, M.D., Khan, M.H.A., Ijaz, A. and Ali., K. (2000) Study of water quality of Hudiara drain, India-Pakistan. *Environ. Internat.*, **26**, 87-96.

Ahel, M., Mikac, N., Cosnovic, B., Prohic, E. and Soukup, V. (1998) The impact of contamination from a municipal solid waste landfill (Zagreb, Croatia) on underlying soil. *Water Sci. Tech.*, **37**(8), 203-210.

Allen, A. (2001) Containment landfills: the myth of sustainability. *Eng. Geol.*, **60**, 3-19.

Allen, A. (2002) Attenuation: A Cost Effective Landfill Strategy for Developing Countries. In *Engineering Geology for Developing Countries. Proc. 9th Congress of the International Association of Engineering Geology and the Environment*, (eds. J.L. van Rooy and C.A. Jermy), pp. 156-167, South African Institute of Engineering Geology and the Environment, Durban, South Africa,.

Allen, A., Brito, G., Caetano, P., Costa, C., Cummins, V., Donnelly, J., Fernandes, C., Koukoulas, S., O'Donnell, V., Robalo, C. and Vendas, D. (2001) *The Development of a GIS Model for the Location of Landfill Sites in Ireland and Portugal.* Final Report Interreg 2C Atlantic Region, EA-B1IRE-No.2.15, Cork.

Averesch, U. (1995) Specific problems in the construction of composite landfill liner systems. In *Proc. Sardinia 95, Fifth International Landfill Symposium*, (eds. T.H. Christensen, R. Cossu and R. Stegmann), vol. 2, pp. 115-130, CISA Publisher, vol 2, Calgiari.

Ball, J.M. and Novella, P.H. (2003) Coastal Park landfill: leachate plume migration and attenuation. In *Proc Sardinia 2003, Ninth International Landfill Symposium*, (eds. T.H. Christensen, R. Cossu and R. Stegmann), CISA Publisher, Cagliari.

Batchelder, M., Mather, J.D. and Joseph, J.B. (1998) The stability of the Oxford Clay as a mineral liner for landfill. *JCIWEM*, **12**, 92-97

Bookter, T.J. and Ham, R.K. (1982) Stabilisation of solid waste in landfills. *J. Env. Eng.*, **108**, 1089-1100.

Butler, A.P., Brook, C., Godley, A., Lewin, K. and Young, C.P. (2003) Attenuation of landfill leachate in unsaturated sandstone. In *Proc. Sardinia 2003, Ninth International Landfill Symposium*, (eds. T.H. Christensen, R. Cossu and R. Stegmann), CISA Publisher, Cagliari.

Christensen, J.B., Jensen, D.L., Grøn, C., Filip, Z. and Christensen, T.H. (1998) Characterisation of the dissolved organic carbon fraction in landfill leachate-polluted groundwater. *Water Res.*, **32**(1), 125-135.

Christensen, T.H., Cossu, R. and Stegmann, R. (1994a) Principles of landfill barrier systems. In *Landfilling on Waste: Barriers*, (eds. T.H. Christensen, R. Cossu and R. Stegmann), pp. 3-10, E. and F.N.Spon, London.

Christensen, T.H., Kjeldsen, P., Albrechtsen H-J., Heron, G., Bjerg, P.L. and Holm, P. (1994b) Attenuation of landfill leachate pollution in aquifers. *Critical Reviews in Environ. Sci. & Tech.*, **24**, 119-202.

Cossu, R. (1995) The multi-barrier landfill and related engineering problems. In *Proc. Sardinia 95, Fifth International Landfill Symposium*, (eds. T.H. Christensen, R. Cossu and R. Stegmann), vol. 2, pp. 3-26, CISA Publisher, Cagliari.

Edworthy, K.J. (1989) Waste disposal and groundwater management. *J. Instit. Water Environ. Man.*, **3**, 109-114.

Eitel, A., Scherrer, M. and Kümmerer K. (2000) *Handling Cyotostatic Drugs: A practical guide.* 2nd edn, Bristol Myers-Squibb, New Brunswick, München.

Gerber, R.E. and Howard, K.W.F. (1996) Evidence for recent groundwater flow through Late Wisconsinan ill near Toronto, Canada. *Canadian Geotechnical J.*, **33**, 538-555.

Gray, D.A., Mather, J.D. and Harrison, I.B. (1974) Review of groundwater pollution from waste disposal sites in England and Wales, with provisional guidelines for future site selection. *Quart. J. Eng. Geol.*, **7**, 181-196.

Hsuan, Y.G. and Koerner, R.M. (1995) *Long term durability of HDPE Geomembranes Part 1 – Depletion of antioxidants*. Geosynthetics Research Institute GRI Report #16, Folsom, Pennsylvania.

Koerner, R.M. and Daniel, D.E. (1997) *Final Covers For Solid Waste Landfills And Abandoned Dumps*. Thomas Telford Ltd, London.

Kruempelbeck, I. and Ehlrig, H-J. (1999) Long-term behaviour of municipal solid waste landfills in Germany. In *Proc Sardinia 99, Seventh International Waste Management and Landfill Symposium*, (eds. T.H. Christensen R. Cossu and R. Stegmann), vol. 1, pp. 27-36, CISA Publisher, Cagliari.

Kümmerer, K., Scherrer, M., Hubner, P., Metz, L. and Daschner, F.D. (Eds.) (2001) *Environmental Management for Hospitals, Pharmacies and Practical Surgeries*. Kenzingen.

Lee, G.F. and Jones, R.A. (1991) Landfills and groundwater quality. *Ground Water*, **29**(4), 482-486.

Mather, J.D. (1995) Preventing groundwater pollution from landfilled waste – is engineered containment an acceptable solution? In *Groundwater Quality*, (eds. H. Nash and G.J.H. McCall), pp. 191-195, Chapman & Hall, London.

Onibokun, A.G. (1999) Synthesis and Recommendations. In *Managing the Monster: Urban waste and governance in Africa*, (eds A.G. Onibokun), pp. 227-252, IDRC, Ottawa.

Potter, H.A.B. and Yong, R.N. (1993) Waste disposal by landfill in Britain: problems, solutions and the way forward. In *Waste Disposal by Landfill*, (ed. R.W. Sarsby), pp. 167-173, Balkema, Rotterdam.

Pruess, A, Giroult, E., Rushbrook, P. (eds.) (1999) Safe management of wastes from health-care activities. World Health Organization, Geneva.

Röhrs, L.H., Fourie, A.B. and Blight, G.E. (2000) Is 30 years an appropriate after-care period for Municipal Landfills. In *Proc. 'WasteCon 2000', Somerset West, South Africa, Institute of Waste Management*, vol. 2, pp. 369-378.

Rollin, A., Mlynarek, J., Lafleur, J. and Zanescu, A. (1991) An investigation of a seven years old HDPE geomembrane used in a landfill. In: *Proc Sardinia 91, Third International Landfill Symposium*, (eds. T.H. Christensen, R. Cossu and R. Stegmann), pp. 667-678, CISA Publisher, Cagliari.

Rowe, R.K. (1994) Design options for hydraulic control of leachate diffusion. In *Landfilling on Waste: Barriers*, (eds. T.H. Christensen, R. Cossu and R. Stegmann), pp. 101-113, E. and F.N.Spon, London.

Rowe, R.K., Quigley, R.M. and Booker, J.R. (1997) *Clayey Barrier Systems For Waste Disposal Facilities*. Chapman and Hall, London.

Sangodoyin, A.Y. (1993) Considerations on contamination of groundwater by waste disposal systems in Nigeria. *Environ. Technol.*, **14**, 957-964.

Seymour, K.J. (1992) Landfill lining for leachate containment. *J. Instit. Water Environ. Man.*, **6**, 389-396.

Sharma, H.D. and Lewis, S.P. (1994) *Waste Containment Systems, Waste Stabilisation, and Landfills*. J. Wiley and Sons, Chichester.

Surmann, R., Pierson, P. and Cottour, F. (1995) Geomembrane liner performance and long term durability. In *Proc. Sardinia 91, Third International Landfill Symposium*, (eds. T.H. Christensen, R. Cossu and R. Stegmann), vol. 2, pp. 405-414, CISA Publisher, Cagliari.

Tchobanoglous, G., Thiesen, H. and Vigil, S. (1993) *Integrated Solid Waste Management Engineering Principles and Management Issues*. McGraw-Hill, New York.

von Hoyer, M. and Muff, R. (2001) *Proposals For The Remediation of Contaminated Groundwater and Hazardous Industrial Waste Sites in Ust-Kamenogorsk Area, Kazakhstan*. BGR Report Tech. Coop. Proj. No. 99.0795.7, Hannover 07/2001.

Wall, D.K. and Zeiss, C. (1995) Municipal waste degradation and settlement. *J. Env. Eng.*, **121**, 214-223.

Warith, M.A. and Yong, R.N. (1991) Landfill leachate attenuation by clay soil. *Hazardous Waste and Hazardous Materials*, **8**, 127-141.

WHO (2000) Wastes from Health-care Activities. Fact Sheet No. 253, World Health Organization, Geneva.

WHO (2004a) Health-care Waste Management. Policy paper, World Health Organization, Geneva.

WHO (2004b) Health-care Waste Management. Fact Sheet No. 281, World Health Organization, Geneva.

Williams, G.M. (1999) Natural Attenuation of Contaminants in a Lacustrine Sand Aquifer at Villa Farm, UK. In *Proc. Sardinia 99, Seventh International Landfill Symposium*, (eds. T.H. Christensen, R. Cossu and R. Stegmann), vol. 4, pp. 229-236, CISA Publisher, Cagliari.

Yong, R.N., Tan, B.K., Bentley, S.P. and Thomas, H.R. (1999) Competency assessment of two clay soils from South Wales for landfill liner contaminant attenuation. *Quart. J. Eng. Geol.*, **32**, 261-270.

25

Traffic and transport: Control and protection

A. Golwer

The most frequently occurring groundwater contaminants from traffic and transport are de-icing agents, particularly salt, fuel, including fuel additives, and some persistent herbicides e.g. atrazine. The issues are, therefore, less related directly to health than to drinking-water acceptability, except in specific local circumstances where a spill due to an accident can lead to a substance draining into an area vulnerable to groundwater contamination.

A number of approaches and control measures can be used to minimize pollution of aquifers with hazardous substances originating from traffic and transport related activities. These include proper planning of new transport links and routes, protective structures and containments, control of construction works, technical improvement of vehicles, impact assessment of substances used in transportation systems (such as de-icing agents or fuel additives), improved management of maintenance activities, regulation of the transport of hazardous goods through drinking-water catchments, rapid response to accidents involving spills of hazardous substances and treatment of traffic surface run-off prior to discharge. Monitoring programmes are important for determining the success of such prevention measures.

The existence of strategies and policies for protecting the environment from traffic and transport-related emissions facilitates the development and implementation of specific control measures to protect a drinking-water catchment. While their development and implementation may be initiated by the water supply or the public authority responsible for its safety, establishing effective control measures to protect groundwater usually requires intersectoral collaboration. This includes changes in public awareness and transport policies as well as training of people employed in the traffic sector. Successful implementation of strategies for the protection of groundwater resources may require a combination of education, fiscal, regulatory and supply-orientated measures (see also Chapter 20). Economic incentives can also contribute towards improving traffic behaviour or reducing the use of environmentally harmful types of traffic.

As with other potentially polluting human activities, giving priority to the prevention of groundwater contamination avoids the need for subsequent measures to reduce or remediate groundwater pollution, which is usually much more difficult and expensive. Where necessary precautionary measures are not immediately economically feasible, incremental improvement towards long-term targets should be envisaged, particularly through taking aquifer vulnerability into account when planning new transport systems or expanding existing ones.

As discussed in Chapter 13, situation assessment will collect information on the existing traffic and transport related infrastructure together with data on its proximity to groundwater systems and designated groundwater protection zones in order to assess pollution potential of aquifers. While many protection zone concepts seek to avoid traffic systems in the inner protection zone (i.e. close to abstraction points), they may be tolerated in the outer area of protection zones, though under the prerequisite of locally appropriate, largely constructional measures of protection.

NOTE ▶ *In developing a Water Safety Plan (Chapter 16), system assessment would review the efficacy of control measures and management plans for protecting groundwater in the drinking-water catchment from traffic and transport. Chapter 13 provides the background information about the potential impact of traffic and transport and provides guidance on the information needed to analyse these hazards.*

This chapter introduces options for controlling risks from traffic and transport. As the responsibility for these activities usually falls outside that of drinking-water suppliers, close collaboration of the stakeholders involved, including the authorities responsible traffic and transport, is important to implement, upgrade and monitor these control measures. This may be initiated by the drinking-water sector, e.g. in the context of developing a Water Safety Plan or of designating protection zones (see Chapter 17).

25.1 PLANNING AND REGULATIONS

Planning is particularly important in vulnerable groundwater catchments in order to limit or restrict construction of traffic facilities or to provide robust defences against pollutants entering the groundwater systems. For existing traffic lines, this may result in upgrading protective structures as well as in operational changes such as directing transports of hazardous goods to other routes (e.g. outside of the drinking-water catchment or protection zone). In planning the construction of new or extended traffic routes and facilities, the appropriate traffic system and route should be evaluated with respect to groundwater protection requirements. In some cases the construction of new traffic routes might be dispensed with through the improved utilization of existing routes. Aquifer protection may be an important criterion for decisions on the allocation of investments to railway versus road transport.

Requiring permits for the construction of traffic infrastructure may be an effective planning tool to assess plans for their impact on drinking-water and thus for the protection of human health. In this context, an Environmental Impact Assessment (EIA; see Chapter 20) of the proposed facilities and any alternatives provide a valuable basis for decisions. These include an assessment of the vulnerability of the aquifer to substances potentially emitted from the traffic lines intended and transport-related accumulation of hazardous substances with the potential to contaminate groundwater. Designing traffic infrastructure to include adequate drainage and disposal of drainage as well as less polluting maintenance procedures is easiest and least costly if included already in the planning stage. Management plans which restrict traffic or investment into protective structures may be more easily enforced if the drinking-water catchment or its most vulnerable areas are designated protection zones (see Chapter 17).

Frequently, not only transport activities themselves, but also the temporary construction areas associated with transport infrastructure pose a significant (though short term) risk themselves (see Chapter 13) and thus should be included in such impact assessments. Where soil and rock are to be removed, care should be taken to ensure this will not remove valuable protective layers to aquifer systems, or that dams created through construction unexpectedly affect flow paths to the groundwater. Management plans for the construction activity may be effective to define sufficiently protective procedures, and surveillance, e.g. through inspection of construction sites, is often critical for their implementation.

Regulatory requirements such as EIAs facilitate the selection of alternative routes and construction methods. Guidelines can be supplied for different construction activities as well as for raising awareness for pollution prevention. In many countries, numerous technical regulations at both a regional and a national level already exist which govern the construction and drainage of traffic routes and also – at least partly – take account of groundwater protection. An example directly addressing groundwater protection are the Guidelines for road construction measures and for existing roads in drinking-water protection areas in Germany (FGSV, 2002). Regulatory approaches have induced behavioural changes in some countries, e.g. safe disposal of motor oil and restriction of car washing to contained sites at service stations.

A further important regulatory and planning approach is the development of accident response plans, particularly for transport of hazardous goods, but also for fuel spills caused by accidents, as rapid clean-up can prevent or substantially reduce groundwater contamination. To be effective, such response plans need to be tailored for the respective setting. Further, it is important to train the response with the parties that need to react quickly in the case of a spill or accident.

A number of international regulations target reducing the environmental impact of transport, and their improved coordination and harmonization facilitates implementation of many measures. For example, the construction and operation of airports is regulated internationally (ICAO, 2000).

25.2 RUNOFF CONTROL

Whether or not collection and treatment of run-off from roads and other surfaces potentially contaminated by transport and traffic is necessary, depends both on the amount of traffic and on aquifer vulnerability and use. In rural areas, scattered run-off from low density traffic routes can percolate over a wide area, and the impact on groundwater quality is frequently regarded as harmless or tolerable if, outside of water protection areas, percolation occurs through a vegetation-covered area with an unsaturated zone at least 1 m thick.

In contrast, roads in built-up areas, as well as aircraft manoeuvring areas and airport aprons, are frequently connected to a drainage system. From less polluted roads this may be directed into a separate storm sewer system, not connected to the foul sewer system, and discharged into a receiving water body. Treatment is often not necessary or may be limited to retention basins (Figure 25.1) which settle some of the particulate load. These are best constructed at least 2 m deep in order to allow evenly distributed through-flow to avoid resuspension through turbulence. Where necessary because of aquifer vulnerability, they need to be impermeable to the underground. Retention basins may also be equipped with a vertical barrier to skim off low-density liquids such as fuels that float as upper layer on the run-off which need to be removed immediately after the pollution event. Removal of solids settled to the bottom, inspection and maintenance should occur at regular intervals and after events such as storms, extended periods of drought or frost and after accidents leading to loading of fuel and oil to the basin. Management plans are useful to define these maintenance activities, including the time intervals and responsibilities for their performance.

For collected run-off from very busy roads (average daily traffic volume >15 000 vehicles) treatment may be a necessary control measure before percolation to groundwater or discharge into surface waters, through mechanical separation as shown in Figure 25.1, through further steps such as percolation through artificial wetlands, or even by retention in larger basins and eventual discharge to a sewage treatment plant. The scale of treatment required for road run-off is determined by the hazard posed. Indeed, as the pollutant load and water volume can vary considerably over time, designing treatment to cope with the variable loading and volumes may be both difficult and expensive.

In some areas, road drainage systems may be connected directly into a sewage system, and the run-off can then be treated together with domestic and industrial waste water. While this is desirable in principle, combined collection of road drainage with sewage also poses problems. Elements of traffic-specific substances – for example, heavy metals – may enrich the sewage sludge and thus compromise sludge reuse and even cause problems for its disposal. Furthermore, during heavy rainfall events, combined sewerage systems can deliver huge volumes of run-off over short periods. Where storage volume is insufficient, this may swamp the treatment process, overflow into recipient water-bodies without treatment, or even place raw sewerage on streets and other areas. Therefore, planning for construction or upgrading of systems directing run-off to sewage treatment plants must include these considerations when calculating the dimensions necessary, particularly for retention basins.

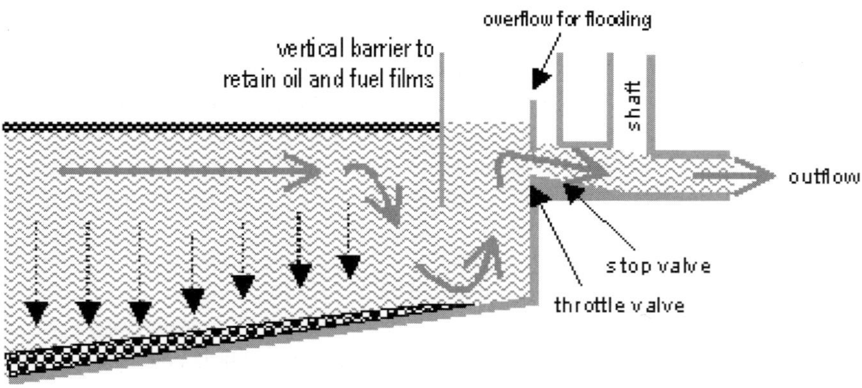

Figure 25.1. Basic scheme of a retention basin for particle sedimentation with vertical barrier to retain oil and fuel films

25.3 DESIGN AND MAINTENANCE OF PROTECTIVE STRUCTURES

The appropriate design for pollution prevention structures is best selected following a pollution risk assessment that considers both the type of traffic and the vulnerability of any aquifers in the vicinity. Legislation requiring protective structures in specified settings exists in many countries, often in combination with good practice codes or engineering guidelines (see e.g. DEFRA (2002) for the United Kingdom or FGSV (2002) for Germany). Typical pollution prevention designs incorporate the use of double-skinned tanks for fuel storage, bunding (i.e. containing) of above-ground storage facilities, adequate monitoring facilities in above and below-ground storage facilities (e.g. observation boreholes in tankpits, leak detection mechanisms in double-skinned tanks, pressure sensors in delivery infrastructure), overfill prevention systems, tanker stand areas with drainage capturing spills during delivery.

These engineered pollution prevention systems are effective only in combination with procedures for site operations, preferably described in a management plan and subjected to regular surveillance or audit. Regardless of the presence of pollution prevention structures, early detection of contaminant release is critical in protecting groundwater. This requires adequate training of operational staff to ensure that monitoring procedures are followed correctly, including regular (at least weekly) monitoring of volumes of stored fuels and other wetstock (at the most simple level, regularly conducting an audit of fuel delivered versus fuel supplied). It also requires developing staff awareness of, and interest in, the nature of plumbing and tank corrosion.

Where drainage and pollution preventing structures exist, it is essential that they be maintained in an optimum condition. For example, oil traps must be kept free of grit and other particles, otherwise they will overflow and fail in their protective capacity. Containments to prevent groundwater contamination (e.g. for fuel tanks) need to be inspected regularly to ensure integrity. Management plans would include regular maintenance programmes and operational monitoring to ensure such structures and treatment facilities are kept functional. Inspections of conditions downstream of outfalls are also recommended as part of a maintenance programme to ensure any adverse impacts are noted and dealt with on a timely basis. Improving the maintenance of roads, rail-track systems and airport operational areas forms part of effective control measures. The positive effects of road cleaning and major factors influencing cleaning efficiency were demonstrated by early investigation (Sartor and Boyd, 1972; Shaheen, 1975).

25.4 MINIMIZING USAGE OF HARMFUL CHEMICALS

Minimizing the amounts of chemicals used in maintenance of transport routes is an obvious method of reducing pollution potential. Management plans should be developed and applied to address applications of such chemicals. Subjecting the plans to regular audit helps ensure that they are implemented correctly. Restrictions on the use of particular chemicals in catchment areas will also aid in reducing pollution. For example, in the United Kingdom, following serious pollution of run-off from some stretches of railway lines with atrazine, the use of a different weed control method was adopted for designated sections of track in drinking-water catchment areas, resulting in a decrease in the levels of pesticide detected. The use of leaf and soil herbicides with quickly-degradable active substances, potentially within the framework of integrated vegetation control, represents an improvement in protective measures as compared with the current, largely preventative application of herbicides. A further example is the replacement of nitrogenous by non-nitrogenous de-icing agents on airfields.

Developments to reduce the health hazards from transport-related substances in groundwater include use of alternative chemicals less likely to pollute groundwater, environmentally more compatible fuels, mechanical (instead of chemical) snow, ice and weed clearance; these may all contribute to a coordinated pollution minimization policy. Groundwater pollution from air fields can be reduced by safer refuelling of aircrafts as well as switching fertilization of air-field lawns with highly soluble nitrogen compounds to controlled-release fertilizers, or by using more readily biodegradable pest control agents used against e.g. field voles (a burrowing rodent which needs to be controlled to reduce damage to air strips).

Regulating the nature, type and availability of maintenance materials can also aid in reducing pollution. Checklists of suitable alternatives can be provided to operators and local authorities to inform them of the benefits of their use over traditional substances known to cause pollution. Legislation banning the production, import or use of heavily polluting materials can be effective in avoiding the pollution they would otherwise cause.

25.5 ACCIDENTAL SPILLAGE AND DISPOSAL

A major pollution risk associated with traffic is spillage caused by accidents. Particularly in drinking-water catchments, the risk of accidents on roads can be lowered through technical measures (such as crash barriers, concrete skidding-walls, ramparts) and traffic-regulation measures (speed limits, overtaking bans, prohibition or restriction of vehicles with loads hazardous to water) (FGSV, 2002). National legislation banning the use and transport of specific hazardous substances or banning their transport on roads, particularly near vulnerable aquifers or in protection zones, can be an effective measure to prevent accidental pollution. Where this is not possible, issuing permits is an important control measure. They should take full account of the nature of the potential pollutant and detail emergency procedures. Where permits for the transport of hazardous goods are granted, emergency response plans to deal with accidents are important. These need to take into account that for some spills, the clean up chemicals used and subsequent washing of surfaces can introduce additional pollutants and aid in the spread of these.

Leaks from fuel storage tanks and pipelines are frequent sources of pollution. Such infrastructure should be subject to regular inspections and testing programmes. Even simple physical structures such as bunded fuel tanks (i.e. placing them inside a structure that can contain the tank's volume should it leak) can provide a significant reduction in pollution risk.

Groundwater pollution with hazardous substances from filling stations, fuel or waste transfer depots can be controlled by adequate design minimizing the risk of spillage and, if spillage should occur, accident response plans should be in place to allow rapid recovery of the spilt materials. If the facility is in the catchment of a public supply source, the responsible water authority should be informed and immediate remediation and control measures carried out, taking account of local conditions and the characteristics of the hazardous substance. These may include temporary closure of drinking-water abstraction to minimize drawing the pollutant into the aquifer, the construction of scavenger wells to remove the pollutant, or the diversion of the pollution through infiltration measures, thus producing a hydraulic barrier.

25.6 MONITORING AND VERIFICATION OF MEASURES CONTROLLING TRAFFIC AND TRANSPORT

The approaches to controlling traffic and transport in drinking-water catchments proposed above range from planning tools in the context of broader environmental traffic policy to specific technical measures such as structures, containments or the restriction of chemicals used in maintenance of traffic facilities. They also include process controls to check if transport facilities are operated properly in order to avoid contamination of drinking-water catchments. The most important measures are summarized in Table 25.1. In some settings, some of these measures may be suitable for integration into the WSP (see Chapter 16) of a drinking-water supply and become subject to operational monitoring in the context of such a plan.

NOTE ▶ *The implementation of control measures such as those suggested in Table 25.1 is effectively supported if the stakeholders involved collaboratively develop management plans that define the control measures and how their performance is monitored, which corrective action should be taken both during normal operations and during incident conditions, responsibilities, lines of communication as well as documentation procedures.*

The implementation of control measures protecting drinking-water aquifers from traffic and transport is substantially facilitated by an environmental policy framework (see Chapter 20).

In all settings, monitoring of the measures implemented is crucial to ensure that they are in place and effective. Table 25.1 includes options for surveillance and monitoring of the protection and control measure examples given. Most of these focus on checking whether the controls are operating as intended, rather than on contaminant concentrations in groundwater. For planning, surveillance will address whether plans exist, are appropriate and are being implemented, particularly in the context of issuing permits for traffic and transport infrastructure. Auditing of plans is an effective tool for such surveillance. Similarly, for measures addressing design and construction, the first verification step is to assess whether or not they are adequate for achieving the protection target, and whether or not they are in place as indicated in the construction plan. For the routine operation of controls, monitoring focuses on assessing whether they are functioning correctly, e.g. whether containments are leaking or whether restrictions on transport of hazardous goods through a catchment are being enforced (see Table 25.1).

NOTE ▶ *Options for monitoring suggested in Table 25.1 rarely include regular groundwater quality monitoring. Where containments and protective structures are poorly accessible for inspection of their integrity, however, monitoring of selected indicator parameters in groundwater may be needed to detect leakage.*

Comprehensive groundwater quality monitoring programmes are a supplementary aspect of monitoring with the purpose of providing verification of the efficacy of the overall drinking-water catchment management.

Some protection measures are difficult to monitor directly, e.g. integrity of a subterranean fuel road drainage pipes, and may most effectively be monitored by some parameter analysed in groundwater that would most sensitively indicate leakage (e.g. chloride, conductivity or in some settings, simply changes in water-table). Intensified monitoring of specific contaminants in groundwater may serve as a control measure after transport accidents involving hazardous goods or fuel spillage. Also, in drinking-water catchments with a potential for pollution by traffic and transport, overall verification of the catchment management concept would include monitoring of specific transport-related contaminants anticipated or known to occur.

Table 25.1. Examples of control measures for traffic and transport and options for their monitoring and verification

Process step	Examples of control measures for traffic and transport	Options for their monitoring and verification
PLANNING	Planning of new or expansion of existing traffic lines and facilities in relation to vulnerability of drinking-water catchments including e.g. siting, choice of materials and mode of construction, run-off collection, restriction of substances used in maintenance	Review plans with respect to the vulnerability and protection of drinking-water catchments
	Accident response plans in drinking-water catchments for releases of fuel and/or hazardous substances including lines of communication, immediate and subsequent measures	Approval, possibly audit, of accident response plans by public authority responsible
DESIGN AND CONSTRUCTION	Collect and adequately dispose wastes and wastewater during construction	Review adequacy of design and compliance with plans and regulations
	Install protective structures that minimize groundwater pollution through routine traffic and accidents, e.g. run-off collection, impermeable surface barriers, bunding of fuel tanks, crash barriers, retention and settling ponds, oil separators treatment facilities for run-off	Inspect sites regularly (including construction sites) and test functioning of facilities Assess integrity of containments, tanks, pipelines and tankers through visual inspection and leak monitoring systems
	Install specific protective structures of refuelling and vehicle maintenance stations (e.g. containment, drainage, oil separators)	
	Install terminal reception facilities for sewerage collection (e.g. from trains, busses, ships, planes)	
	Document construction details relevant for targeted response to spills, e.g. location of drainage pipes, sites for infiltration, location and construction of pipe joints	Check record drawings and documentation of construction details

Process step	Examples of control measures for traffic and transport	Options for their monitoring and verification
OPERATION AND MAINTENANCE	Maintain protective structures that minimize groundwater pollution from traffic, e.g. keep run-off drainage clear of obstacles, remove sludge from retention/settling ponds; repair sealed surfaces when damaged	Inspect integrity of structures and test functioning at regular intervals Where critical, monitor downstream groundwater for parameters indicating leakage
	Collect and adequately dispose wastewater from vehicles, terminal reception facilities, toilets; maintain sanitary facilities	Inspection of records for maintenance activities
	Maintain tanks and pipelines for fuel (e.g. kerosene, diesel, gasoline)	Regular inspection of integrity of containments (leak monitoring systems) Regular monitoring of fuel amounts delivered, stored and supplied; action plan to follow up discrepancies indicating losses
	Control amounts and types of chemicals used for maintenance of traffic lines (e.g. de-icing agents, herbicides)	Inspect records of chemical consumption, devices for use, storage of chemicals
	Devise and conduct regular staff training programmes in auditing and monitoring procedures such as to ensure early detection of leaks	Audit the number of staff trained and the frequency of that training Conduct regular checks of the efficacy of the training by testing staff response to a range of simulated scenarios Review staff performance during both simulated and real situations and modify the training if necessary
	Develop response plan for anomalies found during routine audits and monitoring	Conduct regular reviews of the plan with staff Evaluate staff response to real and simulated situations and revise the response plan if necessary
	Control traffic through protected drinking-water catchments to implement restrictions on the transport of hazardous goods as well as speed limits and bans on overtaking	Inspect records for traffic controls

25.7 REFERENCES

DEFRA (2002) *Groundwater Protection Code: Petrol stations and other fuel dispensing facilities involving underground storage tanks*. DOE, Food and Rural Affairs, London.

FGSV (2002) *Richtlinien für bautechnische Maßnahmen an Straßen in Wasserschutzgebieten (RiStWag)*. Forschungsgesellschaft für Straßen- und Verkehrswesen, Köln.

ICAO (2000) *Airport Planning Manual: Land use and environmental control*, Part 2, Doc. 9184. International Civil Aviation Organization, Montreal.

Sartor, I.D. and Boyd, G.B. (1972) *Water Pollution Aspect of Street Surface Contaminants*. US EPA Report No. R 2-72-081, Washington, DC.

Shaheen, D.G. (1975) *Contributions of Urban Roadway Usage to Water Pollution*. EPA, Report 600/2-75-004, Washington, DC.

Index

Page numbers in italics refer to Figures, those in **bold** indicate **Tables**.

 Protecting Groundwater for Health: Managing the Quality of Drinking-water Sources. Edited by O. Schmoll, G. Howard, J. Chilton and I. Chorus. ISBN: 1843390795. Published by IWA Publishing, London, UK.